The Universal Book of Astronomy

The Universal Book of Astronomy

From the Andromeda Galaxy to the Zone of Avoidance

David Darling

John Wiley & Sons, Inc.

This book is printed on acid-free paper. ♾

Published by John Wiley & Sons, Inc., Hoboken, New Jersey
Published simultaneously in Canada

For general information about our other products and services, please contact our Customer Care Department within the United States at (800) 762-2974, outside the United States at (317) 572-3993 or fax (317) 572-4002.

Wiley also publishes its books in a variety of electronic formats. Some content that appears in print may not be available in electronic books. For more information about Wiley products, visit our web site at www.wiley.com.

Library of Congress Cataloging-in-Publication Data:

Darling, David J.
The universal book of astronomy from the Andromeda Galaxy to the zone of avoidance / David Darling.
p. cm.
Includes bibliographical references and index.
ISBN 0-471-26569-1
1. Astronomy–Dictionaries. I. Title.
QB14. D37 2003
520'.3–dc21

2003013941

Printed in the United States of America

10 9 8 7 6 5 4 3 2 1

Contents

Introduction

The universe has sprung many remarkable surprises on us in the past few years. Scores of planets have been detected around other stars. Vast amounts of ice and evidence for recent liquid water have been found on Mars. Brown dwarfs, medium-sized black holes, giant Centaurs, ultra-luminous clusters, microquasars, and magnetars have been added to an already extraordinary cosmic menagerie. Dark energy has joined dark matter as a likely chief ingredient of the universe, which, against all expectations, seems to be gathering pace in its headlong outward rush. We have seen the surface of the Sun, the center of our galaxy, and the environs of dying stars in unprecedented, awe-inspiring detail. We have begun to make out the great filaments of galaxy clusters and the gaping voids that are the largest of cosmic structures.

We are lucky to find ourselves in a world of such wonders at a time when they are being revealed by powerful new instruments–the enhanced Hubble Space Telescope, the Very Large Telescope, the Chandra X-ray Observatory, and others. Yet the breakthroughs are coming so thick and fast that it is hard, even for professional astronomers let alone the interested layperson, to keep track of them. This book is intended as a comprehensive, user-friendly guide to the cosmos of the early twenty-first century. By including the latest results and conjectures alongside well-established theory and facts, and by extensive cross-referencing, it offers a way for everyone, from the nonspecialist to the seasoned researcher, to gain a better grasp of the modern universe. And grasp it we should because we are all part of this astonishing creation–our very bodies fashioned of atoms that were made inside giant stars long ago. Bring only your imagination and the courage needed to gaze up and out across 15 billion light-years!

How to Use This Book

Entries range from short definitions to lengthy articles on topics of major importance or unusual interest, and are extensively cross-referenced. They are arranged alphabetically according to the first word of the entry name. So, for example, "BL Herculis star" precedes "black dwarf." Terms that appear in **bold** type have their own entries. Metric units are used throughout, unless it is more appropriate, for historical reasons, to do otherwise. See the "Units" section that follows for conversion factors. Star charts are included at the back of the book to show the position of bright stars and other celestial objects. Readers are invited to visit the author's web site at www.daviddarling.info for breaking news in astronomy and related subjects.

Exponential Notation

In the interest of brevity, exponential notation is used in this book to represent large and small numbers. For example, 300,000,000 is written as 3×10^8, the power of 10 indicating how many places the decimal point has been moved to the left from the original number (or, more simply, the number of zeroes). Small numbers have negative exponents, indicating how many places the point has been shifted to the left. For example, 0.000049 is written as 4.9×10^{-5}.

Units

Distance

1 kilometer (km) = 0.62 miles

1 meter (m) = 3.28 feet (ft) = 39.37 inches (in.)

1 centimeter (cm) = 0.39 in.

1 m = 1,000 millimeters (mm) = 10^6 micrometers (microns, or μm)

1 angstrom (Å) = 0.1 nanometer (nm) = 10^{-10} m

1 astronomical unit (AU) = 1.50×10^8 km (the mean Earth-Sun distance)

1 light-year = 63,240 AU = 9.46×10^{12} km

1 parsec (pc) = 3.2616 light-years = 206,260 AU = 3.0857×10^{13} km

Angle

1 degree (°) = 60 arcminutes (′) = 3,600 arcseconds (″)

1 arcsecond = 1,000 milliarcsec = 10^6 microarcsec

Speed

1 km/s = 2,240 mph

Mass

1 kilogram (kg) = 2.21 pounds (lb)

1 metric ton = 1,000 kg = 2,205 lb = 0.98 long ton

$M_{sun} = 1.989 \times 10^{30}$ kg = 1,047 $M_{Jupiter}$ = 332,958 M_{Earth}

Note: In this book, *tons* refer to metric tons.

Force

1 newton (N) = 0.22 pounds-force (lbf) = 0.102 kilograms-force (kgf)

Pressure

1 bar = 0.987 atmosphere = 101,300 pascals = 14.5 lb/square inch = 100,000 N/m^2

Energy

1 joule (J) = 9.48×10^{-4} British thermal unit (Btu)

1 electron-volt (eV) = 1.60×10^{-19} J

1 GeV = 10^3 MeV = 10^6 keV = 10^9 eV

Note: Electron-volts are convenient units for measuring the energies of particles and electromagnetic radiation. In the case of electromagnetic radiation, it is customary to measure longer-wavelength types in terms of their wavelength (in units of cm, μm, Å, etc.) and shorter-wavelength types, especially X rays and gamma-rays, in terms of their energy (in units of keV, MeV, etc.). The wavelength associated with electromagnetic waves of energy 1 keV is 0.124 nm.

Temperature

C = 5/9 (F – 32)

F = 9/5 C + 32

0K = –273.16°C

1K increase = 1°C increase

A

A star

A star of **spectral type** A, white in color, with a spectrum dominated by the **Balmer series** of hydrogen. Lines of heavy elements, such as iron, are noticeable at the cooler end of the range. **Main sequence** A stars have surface temperatures of 7,500 to 9,900 K, luminosities of 7 to 80 L_{sun}, and masses of 1.5 to 3 M_{sun}; familiar examples include **Sirius, Vega,** and **Altair.** A-type **supergiants,** such as **Deneb,** may be as hot as 11,000 K and have masses up to 20 M_{sun} and luminosities of more than 35,000 L_{sun}. Among A-type **peculiar stars** are **Ae stars, Am stars,** and **Ap stars.** Also, two of the main kinds of pulsating variables, **RR Lyrae stars** and **Delta Scuti stars,** have surface temperatures in the A-star range.

Aaronson, Marc (1950–1987)

A talented young American astronomer who was killed in a tragic accident in the dome of the 4-m **Mayall Telescope** on Kitt Peak. Due to a malfunction of the emergency stop, Aaronson became trapped in the catwalk door of the rotating telescope dome when the outer stepladder closed it. Minor planet (3277) Aaronson is named in his honor.

Abell cluster

A **rich cluster of galaxies** identified in the *Abell Catalog of Rich Clusters of Galaxies,* which was the first comprehensive catalog of clusters of galaxies and, in its extended form, is still the largest. The *Abell Catalog* was first published in 1958 by the American astronomer George Abell (1927–1983), a graduate student at the time, and contained 2,712 clusters of galaxies north of declination –27° identified by Abell from his analysis of the ***Palomar Observatory Sky Survey*** photographs. To be included in the *Catalog* a galaxy cluster had to have at least 50 members in the **magnitude** range m_3 to $m_3 + 2$, where m_3 is the magnitude of the third brightest cluster member, within a radius of about 1.5 megaparsecs (just under 5 million light-years). Abell classified such clusters as regular and irregular, and ranked them on 6-point scales according to both their richness (population density) and distance (1 to 6, for closest to farthest). The *Catalog* proved of seminal importance to extragalactic astronomy, leading to a better understanding of the nature and properties of clusters of galaxies, and the identification of **superclusters** and **large-scale structure.** Following the latest (1989) revisions and extensions to include rich clusters in the southern hemisphere, *Abell's Catalog* now consists of the *Northern Abell Catalog* (clusters A1 through A2712), the *Southern Abell Catalog* (A2713 through A4076), and the *Supplementary Southern Abell Catalog* (A4077 through A5250). Among the best-known Abell clusters are the **Perseus Cluster** (A426) and the **Coma Cluster** (A1656).

aberration, optical

A flaw in the imaging properties of a lens, mirror, or optical system. There are six main types: **astigmatism, chromatic aberration** (a problem restricted to lenses), coma (see **coma, optical**), **distortion, field curvature,** and **spherical aberration.**

aberration of starlight

The difference between the observed position of a star and its true direction; this is a combined result of the

Abell cluster Abell 496 is a rich cluster of galaxies that lies about 500 million light-years away and has a mass (including dark matter) about 1,000 times greater than that of the Milky Way Galaxy. This photo shows a region some 4.5 million light-years across. *European Southern Observatory*

observer's motion across the path of the incoming starlight and the finite speed of light. There are three components: *annual aberration* (up to 20.47″) caused by Earth's revolution around the Sun, *diurnal aberration* (up to 0.3″) caused by Earth's axial rotation, and the very small *secular aberration* caused by the motion of the solar system through space.

ablation

Loss of material from the surface of a **meteoroid** as it is heated by friction on passing through an atmosphere.

absolute magnitude

(1) The **apparent magnitude** of a star or other celestial object if located at a distance of 10 parsecs (32.616 light-years), assuming there is no absorbing material between the object and the observer. Absolute magnitude is often derived from the **visual magnitude** but may be specified for other wavelengths (see **photometry**) or for all wavelengths, when it is called the absolute **bolometric magnitude**. (2) In the case of comets and asteroids, absolute magnitude is defined differently as the apparent magnitude the object would have if it were one astronomical unit from both the Sun and Earth and at a **phase angle** of 0° (i.e., fully illuminated by the Sun).

absorption

The drop in intensity of light, or of any other form of **electromagnetic radiation**, as it passes through a medium, such as Earth's atmosphere, the outer layers of a star, or dust in interstellar space. It is due to the transfer of energy by photons (particles of light) to atoms, ions, or molecules. Absorption at a specific wavelength results in **absorption lines**; otherwise it is known as *continuous absorption*. Absorption in Earth's atmosphere is one of the causes of **atmospheric extinction**; between the stars it occurs as **interstellar extinction**.

absorption line

A dark line in a **continuous spectrum** caused by **absorption** of light at a specific wavelength. In a star's spectrum, such lines are due to absorption of light by cooler gases in the outer layers of the star's atmosphere and they reveal the kind of atoms, ions, or molecules present; the **Fraunhofer lines** in the solar spectrum are the best known example. Absorption by molecules results in a **band spectrum**.

absorption spectrum

A spectrum of **absorption lines** or bands, produced when light from a hot source, itself producing a continuous spectrum, passes through a cooler gas.

acapulcoite

A primitive (little altered since its formation) **achondrite** belonging to a small group named after the only witnessed **fall** of this type, the Acapulco meteorite that fell in Mexico in 1976. Acapulcoites are made mostly of fine-grained **olivine**, **orthopyroxene**, **plagioclase feldspar**, nickel-iron metal, and the iron sulfide troilite, and are transitional between primordial chondritic matter and more differentiated rocks. Their mineral composition is between that of E and H **chondrites**, but they have an oxygen isotope pattern that sets them apart from all other known chondrite groups. Importantly, some acapulcoites contain a few relict **chondrules**, and one specimen, from Tissemoumine, Morocco, named NWA 725, shows an abundance of distinct chondrules. These distinct chondrules confirm that acapulcoites are very primitive and that they form a missing link between chondrites and achondrites. They are thought to have come from the same parent body as the closely related **lodranites**, which show signs of slightly more melting.

Acasta Formation

The oldest known outcrop of rock on Earth. It occurs about 350 km north of Yellowknife in Canada's Northwest Territories and consists of gneiss (a type of coarse-grained metamorphic rock) that dates back just under 4 billion years, to around the beginning of the period of **late heavy bombardment**.

acceleration due to gravity (g)

The acceleration with which a body falls freely in a gravitational field; also known as the *acceleration of free fall*. At Earth's surface, g varies with latitude around a mean value of 9.807 m/s^2 in part because our planet is not a perfect sphere.

accretion

The process by which particles stick together to form larger bodies; for example, dust in the **solar nebula** accreted to form **chondrules**, and **planetesimals** accreted to form planets.

accretion disk

A spinning ring of gas and dust that may form around any of a variety of stars or of other massive objects. Accretion disks have been observed, or theorized to exist, in association with **protostars**, **T Tauri stars** and other very young stars; **Vega**-type stars; and many **interacting binary** systems, including **U Geminorum stars**, **novae**, symbiotic stars, and W Serpentis stars, in which the secondary component is losing matter to the primary. Accretion disks are also believed to occur commonly around **black holes**, including stellar black holes in **close binary** systems and supermassive black holes at the centers of some galaxies. In the latter case, accretion disks are thought to place a crucial role in the **active galactic nucleus** phenomenon.

Achernar (Alpha Eridani, α Eri)

The ninth brightest star in the sky but one that is not well known to northern observers as it can only be seen at latitudes below 32°N. Its Arabic name means "river's end" and refers to its location at the southernmost point of **Eridanus**. Achernar's high spin velocity of at least 225 km/s has led to it becoming a **Be star**. Observations by the Very Large Telescope, published in 2003, reveal that it is extraordinarily flattened, with an equatorial radius 50% larger than its polar radius. Achernar also shows small, regular light variations of a type that make it a **Lambda Eridani star**.

Visual magnitude:	0.45
Absolute magnitude:	–2.77
Spectral type:	B3V
Luminosity:	over 3,000 L_{sun}
Distance:	144 light-years

Achilles (minor planet 588)

The first of the **Trojan** asteroids to be found, in 1906 by Max **Wolf**. Its discovery, at Jupiter's fourth **Lagrangian point** (L_4), preceding the planet in its orbit, confirmed Joseph-Louis Lagrange's theory that two-body systems have points where a third object of negligible mass can reside in stable equilibrium. Diameter 116 km; spectral class C; semimajor axis 5.175 AU; perihelion 4.40 AU; aphelion 5.95 AU; inclination 10.3°; period 11.77 years.

achondrite

The rarer of the two main types of **stony meteorite**, accounting for about 9% of all meteorite falls. Achondrites are made of rock that has crystallized from a molten state. They contain mostly one or more of the minerals **plagioclase feldspar**, **pyroxene**, and **olivine**, and generally, but not always, lack the small rounded inclusions known as **chondrules** that are typical of **chondrites**. Most achondrites are chemically similar to **basalts** and are thought to be the product of melting on large asteroids, moons, and planets. Soon after these bodies formed, they were heated from within and partially melted. Although this process is still active on Earth, it ended about 4.4 billion years ago on asteroids, 2.9 billion years ago on the Moon, and perhaps 1 billion years ago on Mars. Heating the primordial mixture of stony minerals, metals, and sulfides (of which chondrites are made) produced liquids, the densest of which sank to become planetary or asteroidal cores. Lighter stony minerals rose and solidified to become basaltic rocks, fragments of which were subsequently broken off by impacts and hurled into space. More than 200 of these *evolved achondrites* have been found, covering a wide range of compositions and origins. The so-called **HED group** includes the **howardites**, **eucrites**, and **diogenites**, which appear to share the same parent body, believed to be the asteroid **Vesta**. Other evolved achondrites that seem to have come from partially differentiated asteroids other than Vesta have been mostly assigned to two distinct groups known as the **angrites** and the **aubrites**. Although the majority of achondrites are of asteroidal origin, some are known to have come from the highland regions of the Moon's farside (see **lunar meteorites**) and from Mars (see **Mars meteorites**). NWA011, a meteorite found in the Sahara in 1999, is suspected of having originated on Mercury. As well as the evolved achondrites, there is an entire group of *primitive achondrites* whose members all seem to have derived from small chondritic parent bodies that only partially melted and differentiated through accretion processes or from impact events, and then rapidly cooled. Primitive achondrites vary widely in composition and fall into the following main subgroups: **acapulcoites**, **lodranites**, **brachinites**, **winonaites**, and **ureilites.**

achromatic lens

A lens with two or three closely spaced, often cemented, elements designed to produce images largely free from false color. It has at least one element of **flint glass** and another of **crown glass**–the **dispersion** of the latter compensating for the **chromatic aberration** of the former. While achromatic lenses bring two primary colors (red and blue) into focus at the same point, they leave some uncorrected chromatic aberration at wavelengths in between. This variation is negligible for most purposes, and a single focal-length value is applied to the entire visible spectrum.

A-class asteroid

A rare **asteroid** class whose members are extremely red and have a fairly high **albedo** of 0.13 to 0.35. Strong absorption in the near-infrared is taken to indicate the presence of **olivine**. Examples include the asteroids (246) Asporina (diameter 70 km) and (446) Aeternotas (diameter 52 km).

Acrux (Alpha Crucis, α Cru)

The twelfth brightest star in the sky, the southernmost first magnitude star, and the brightest and southernmost star in **Crux**. Acrux is a multiple system. A moderate telescope shows two similar **B stars** separated by 4″: α^1, a B subgiant (visual magnitude 1.33, luminosity 25,000 L_{sun}, surface temperature 28,000 K), and α^2, a B dwarf (visual magnitude 1.73, luminosity 16,000 L_{sun}, temperature 26,000 K). α^2 is a 13-M_{sun} single star but α^1 is a spectroscopic binary whose 14- and 10-M_{sun} components are separated by about 1 AU and complete an orbit every 76 days. α^1 and α^2, with a minimum separation of 430 AU, take at least 1,500 years to circle around each other. Another B subgiant lies 90″ away from the triplet but, despite its similar velocity through space, is probably a more distant star that happens to lie along the same line of sight.

Visual magnitude:	0.77
Absolute magnitude:	–4.19
Spectral type:	B0.5IV + B1V
Distance:	321 light-years

ACT Catalog

A catalog, compiled by the **U.S. Naval Observatory** (USNO), to provide accurate **proper motions** for the majority of stars in the ***Tycho Catalogue***. Since the Tycho Catalogue is based on data collected by the **Hipparcos** satellite over a relatively short period, it gives only rough values of proper motions. The USNO solved this problem by combining Tycho positions with those of the ***Astrographic Catalogue*** that date from around 1900. The results of this long baseline for computations are proper motions for 988,758 stars, covering the entire sky, that are about an order of magnitude more accurate than those published in the *Tycho Catalogue*.

active galactic nucleus

See article, page 9.

active galaxy

Any galaxy with an **active galactic nucleus**.

active optics

A computer-controlled system that compensates for the distortion of a telescope's mirrors caused by gravity, and thus allows a thinner mirror and a lighter support structure to be used. It works by monitoring the image of a guide star and sending appropriate signals to actuators behind the mirror, which in turn control moveable supports to correct the mirror's shape and alignment.

active region

A localized volume of the Sun's outer atmosphere where powerful magnetic fields, emerging from subsurface layers, give rise to various short-lived features. These features may include **sunspots** and **faculae** in the **photosphere**; **plages**, **fibrils**, and **filaments** in the **chromosphere**; and **coronal condensations** in the **corona**. **Solar flares** are also associated with active regions.

active Sun

The condition of the **Sun** characterized by unusually large numbers and size of spots, flares, and prominences. It is especially associated with maxima of the **solar cycle**.

Adams, John Couch (1819–1892)

An English mathematician and astronomer, born in Lidcot, Cornwall, the son of a tenant farmer, who predicted the existence of **Neptune**. While a student at Cambridge he wrote this note (found only after his death) dated July 3, 1841:

> Formed a design at the beginning of this week of investigating, as soon as possible after taking my degree, the irregularities in the motion of Uranus, which are as yet unaccounted for, in order to find whether they may be attributed to the action of an undiscovered planet beyond it; and, if possible, thence to determine the elements of its orbit approximately, which would lead probably to its discovery.

Having graduated brilliantly (with more than double the marks of his nearest competitor), he focused his attention on the problem of the trans-Uranian planet. In October 1845, he gave his predicted position for the new world to George **Airy**, the Astronomer Royal. But Airy procrastinated for nine months until he heard of a similar claim by Urbain **Leverrier**. He then instigated a search, but the race was lost: Neptune was found in 1846 by Johann **Galle** using Leverrier's figures. Adams remained silent throughout the bitter ensuing debate that established his precedence in the discovery. Although eventually he was offered a knighthood and the post of Astronomer Royal after Airy, Adams turned them down and remained at Cambridge as director of the observatory.

Adams, Walter Sydney (1876–1956)

An American astronomer who discovered the first **white dwarf**. The son of American missionaries, Adams was born in Syria. He followed his Dartmouth professor, Edwin Frost, to **Yerkes Observatory**, and accompanied his Yerkes director, George **Hale**, to **Mount Wilson Observatory**, where Adams served as director from 1923 to 1946. His spectroscopic studies, with Arnold Kohlschütter, of the Sun and stars led to the discovery of spectroscopic **parallax** as a stellar yardstick. This method involves measuring the relative intensities of **spectral lines** of both giant and main sequence stars to determine **absolute magnitudes** and hence distances. Adams worked

active galactic nucleus (AGN)

A **galactic nucleus** that gives off much more energy than can be explained purely in terms of its star content. AGN are found at the heart of active galaxies, including **quasars**, **Seyfert galaxies**, **blazars**, and **radio galaxies**. In addition to their great energy output, they can be highly variable. Some quasars vary in brightness over a few weeks or months, while some blazars show changes in X-ray output over as little as three hours. These fluctuations place strict limits on the maximum size of the energy source, because an object can't vary in brightness faster than the time it takes light to travel from one side of its energy-producing region to the other. The rapid flickering of AGN means that they draw their energy from a small volume, in some cases less than one light-day across. Furthermore, observations of the orbital motion of stars and other material around AGN show that a large mass, ranging up to several billion solar masses, is concentrated within its "engine room." This leads to the almost unavoidable conclusion that the central engine is a supermassive **black hole**. Since a black hole, by definition, emits nothing, the radiation from an AGN is believed to come from material heated to several million degrees in an **accretion disk** before tumbling into the black hole or, in some cases, being shot away in twin **jets** along the central engine's spin axis.

active galactic nucleus An X-ray view, by the Chandra X-ray Observatory, of the core of the active galaxy Centaurus A. It shows a spectacular jet and many pointlike sources, the majority of which are X-ray binaries. *NASA/Smithsonian Astrophysical Observatory*

Astronomers have identified about a dozen types of AGN, each with a different overall pattern of emission. Some of these types differ only in appearance; for example, if a thick dust cloud lies between us and the center, it can block radiation at frequencies between visible light and X rays, so that we see only the radio, infrared, and very high-energy radiation, plus the emission lines from outlying clouds. In at least some cases, the dust in the nucleus forms a thick ring around the center. Then the appearance of the AGN depends crucially on whether it is seen side-on, so that the center is hidden, or face-on, so that the center is visible.

AGN fall into two major divisions: radio-loud and radio-quiet. Within each is a broad range of luminosity, from AGN so weak that they're barely detectable against the light from the central stars of their host galaxy, to quasars that are more than 100 times brighter than all the stars in their hosts put together. Radio-loud AGN always produce jets in which the material may be flowing at close to the speed of light. The power of the jets (the kinetic energy flowing along them per second) roughly matches or exceeds the AGN's luminosity (total energy radiated per second). Radio-loud AGN are generally found in elliptical galaxies and almost never found in spirals. Radio-quiet AGN may also have detectable jets, but these are thousands of times weaker than the AGN's total radiant energy. They usually occur in spiral galaxies, although some of the most luminous radio-quiet AGN have elliptical hosts. The obvious differences between spirals and ellipticals occur on scales many times larger than that of the AGN, and it remains uncertain how AGN "know" which type of galaxy they're in.

In the 1980s, it became clear that nuclear activity can extend to lower levels and can appear in galaxies that superficially look normal. Many galaxies were found to show emission from their nuclei that couldn't be explained in detail by young stars, and these were dubbed **LINERs** (low-ionization nuclear emission-line regions). Debate continues as to how many kinds of objects fall into this category, but X-ray and ultraviolet observations show that some of these are indeed lower-power versions of Seyfert nuclei, complete with X-ray source and variability. The phenomenon of nuclear activity now appears to range in luminosity over six orders of magnitude and to show up in a significant fraction of bright galaxies, including our own.

with Hale on the discovery of magnetic fields in **sunspots** and used photography to measure the differential rotation of the Sun. With Theodore Dunham Jr., he codiscovered carbon dioxide in the atmosphere of Venus, and the molecules CN and CH in interstellar clouds. Adams identified **Sirius** B as the first known white dwarf, and his measurement of its gravitational redshift in 1924 helped strengthen the case for the **general theory of relativity**.

adaptive optics

An optical system that enables rapid fluctuations in a telescope's image quality, caused by atmospheric turbulence, to be corrected in fractions of a second. These fluctuations are measured by a *wavefront sensor* that uses a reference star to measure the distortions that are taking place. The reference star is typically a reasonably bright star close in the sky to the object under study but may also be the object itself or a reflected laser beam that serves as an artificial reference star. The measured distortions are then removed with a *phase corrector*–typically a very thin mirror in the light path of the telescope that can be rapidly deformed by actuators to the equivalent shape of the wavefront, which must be subtracted to produce a sharp image. Unlike **active optics**, adaptive optics provides real-time response and can result in a level of **resolution** close to the theoretical limit allowed by diffraction. It is used in conjunction with many new telescopes including the **Keck Observatory** telescopes and the **Large Binocular Telescope**.

Adhara (Epsilon Canis Majoris, ε CMa)

The second brightest star in **Canis Major**; it lies below **Sirius** at the lower right of a triangle of bright stars known to the Arabs as "the Virgins," from which its name (alternatively spelled "Adara" or "Undara") derives. It is one of the hottest of the bright stars: so luminous that if placed at the distance of Sirius (just over 8 light-years), it would shine 15 times brighter than Venus. Astronomers have used its spectrum extensively to study the nature of local interstellar matter.

Visual magnitude:	1.50
Absolute magnitude:	–4.11
Spectral type:	B2II
Luminosity:	15,000 L_{sun}
Surface temperature:	20,000 K
Distance:	425 light-years

adiabatic process
A process in which no heat enters or leaves a system. This is the case, for example, when an interstellar gas cloud expands or contracts. Adiabatic changes are usually accompanied by changes in temperature. Compare with **isothermal process**.

Adrastea
The second moon of **Jupiter**, also known as Jupiter XV; it was discovered in 1979 by the American astronomers David Jewitt and G. Edward Danielson from Voyager images. Adrastea measures 23 × 20 × 15 km and orbits 57,300 km above the Jovian cloud-tops, close to the outer edge of Jupiter's ring system. It appears to act as a **shepherd moon** and, together with **Metis**, as the source of the material for the rings.

advance of perihelion
The slow rotation of the **major axis** of a planet's orbit in the same direction as the revolution of the planet itself, due mainly to gravitational interactions with other planets. A small additional advance of Mercury's perihelion, by 43″ per century, was eventually explained as an effect of the **general theory of relativity**. In the case of **close binary** stars, the advance of pericenter may additionally be caused by mass transfer and the stars' distorted (elliptical) shapes. Advance of perihelion (or pericenter) is also known as *apsidal motion.*

Ae star
An **A star** with strong **emission lines**, usually of hydrogen, superimposed on an otherwise normal spectrum and caused by a circumstellar shell of heated material. Some Ae stars have only recently formed and may be surrounded by visible nebulosity, in which case they are known as **Herbig Ae/Be stars**.

afterglow
(1) A broad glowing arc, sometimes seen high in the western sky at twilight, caused by fine particles of dust scattering light in the upper atmosphere. (2) Lingering radiation, also called *postluminescence,* that remains after an event like a **gamma-ray burst** (which has an X-ray afterglow) or the Big Bang (the afterglow of which is the cosmic microwave background radiation).

Ahnighito meteorite
The largest part of the **Cape York meteorite** and the largest meteorite on display in any museum (The American Museum of Natural History, New York).

AI Velorum star (AI Vel star)
A type of **pulsating variable** that is very similar to a **Delta Scuti star** but has a greater amplitude (up to about 1.2 magnitude) and a period of 1 to 5 hours. The prototype, located in **Vela** 2.8° north-northeast of Gamma Velorum, is the brightest of its class, with a magnitude range of 6.4 to 7.1. AI Velorum stars were formerly known as *dwarf Cepheids.*

airglow
Diffuse radiation, extending from the near infrared to the far ultraviolet, that is continuously emitted by a planetary atmosphere; also known as *nightglow.* It is caused by the collision of charged particles and X rays from space, mainly from the Sun, with atoms and molecules high in the atmosphere. Earth's airglow varies with the time of night or day, latitude, and season, goes from a minimum at the zenith to a maximum about 10° above the horizon, and originates mainly from discrete atomic and molecular transitions that give rise to a mostly emission-line and emission-band spectrum. Emission from oxygen at 5577 Å (green) predominates at night, while yellow sodium and red oxygen emissions are prominent at twilight. Daytime airglow, although drowned by sunlight, is actually 1,000 times as intense as at night.

airmass
The path length that light from a celestial object takes through Earth's atmosphere relative to the length at the **zenith**. Airmass is 1 at the zenith and roughly 2 at an altitude of 60°. It can be calculated to a good approximation from the formula

$$A = 1.0 / [\cos(Z) + 0.50572 \times (96.07995 - Z)^{-1.6364}]$$

where Z is the zenith angle (the vertical angle of an object from the zenith).

Airy, George Biddell (1801–1892)
An English astronomer, born in Alnwick, Northumberland, who graduated at the head of his class from Cambridge in 1823 shortly after devising a way to correct **astigmatism**–a condition from which he personally suffered. In 1826 Airy was appointed Lucasian Professor of Mathematics (Newton's old position) at Cambridge, and, two years later, Plumian Professor of Astronomy. As the seventh **Astronomer Royal** (1835–1881) he turned the **Royal Greenwich Observatory** into a model of efficiency and a leading center for positional astronomy; the **transit** telescope he installed defines the location of 0° longitude on Earth. However, Airy's arrogance and disinterest in basic research held up the confirmation of an eighth planet (Neptune) based on predictions by John **Adams** and also left Greenwich a late-starter in the fields of spectroscopy and astrophysics. His precision, to the point of pedantry, extended to his labeling empty boxes "empty."

Airy disk
The central spot in the **diffraction pattern** of the image of a star at the focus of a telescope, named after George **Airy**. It is surrounded by several fine diffraction rings like the rings around the bull's-eye of a target. The size of the Airy disk, equal (in radians) to about 1.22λ times the **focal ratio**, where λ is the wavelength of light, is the same for all telescopes of a given size and is less in instruments of larger aperture. (It is one of the quirks of astronomy that bigger telescopes produce smaller images of stars.) In practice, however, atmospheric turbulence, unless compensated for, results in a false disk that is larger than the Airy disk.

Aitken, Robert Grant (1864–1951)
An American astronomer, born in Jackson, California, who worked at the **Lick Observatory** from 1895 to 1935, serving the final four years as its director. Specializing in observations of **double stars**, he discovered 3,100 new pairs including, in 1923, the faint companion of **Mira**. He published an important catalog of double stars (see ***Aitken Double Star Catalogue***), and also measured positions of comets and planetary satellites and computed their orbits. He was editor of the *Publications of the Astronomical Society of the Pacific* for many years, wrote an influential book on binary stars, and lectured and wrote widely for the public.

Aitken Double Star Catalogue (ADS)
The name by which Robert **Aitken**'s massive *New General Catalogue of Double Stars within 120 Degrees of the North Pole* is generally known. Published in two volumes by the Carnegie Institution in 1932, it superceded Sherburne **Burnham**'s 1906 catalog, contained measurements of 17,180 double stars north of declination –30°, and allowed orbit determinations that greatly improved knowledge of stellar masses.

Al Niyat (Sigma and Tau Scorpii, σ Sco and τ Sco)
Two stars of similar brightness in **Scorpius** with the same Arabic name, meaning "artery." They mark the vessels that lead away from the "heart" of the Scorpion, **Antares.** σ Sco is a multiple system (visual magnitude 2.90, absolute magnitude –3.87, distance 735 light-years) dominated by a brilliant **spectroscopic binary** consisting of an O9 (main sequence) dwarf and a B2 giant orbiting each other every 33 days at about the distance of Venus from the Sun. Another B star, two magnitudes fainter, lies 0.4″ away, while farther out is a ninth magnitude B9 companion. σ is believed to be part of the Upper Scorpius portion of the **Scorpius-Centaurus Association**, although it appears to lie almost 300 light-years from the center of this group. It is also involved with a large mass of interstellar gas that it ionizes and causes to glow. Unusual among bright naked-eye stars, σ's light is reddened and dimmed by over a magnitude by interstellar dust.

The lower of the two artery stars, τ Sco, is a B0 dwarf (visual magnitude 2.78, absolute magnitude 2.82, distance 430 light-years). It, too, is a member of the Upper Scorpius Association and was the first star, other than the Sun, to have its spectrum analyzed in detail (by Albrecht Unsöld in 1939). Its measured rotation speed is very low for a B star–less than 5 km/s, compared with a norm for this type of several hundred km/s. Astronomers suspect it may actually be a fast rotator whose spin axis happens to be pointed almost directly at us. It also shows unusual emission lines in its infrared spectrum, leading some researchers to suggest that it may be a **Be star** seen pole-on.

Albategnius (Al-Battani, Muhammad ibn Jabir) (c. 850–929)
An Arab prince, born in Batan, Mesopotamia, who was the leading astronomer and mathematician of his time. He drew up improved tables of the Sun and the Moon, measured the eccentricity of Earth's orbit and the inclination of Earth's equator to its orbital plane, and derived an accurate length for the year, which was used in the Gregorian reform of the **Julian calendar**. His observations at Rakku, made over a 40-year period, were summarized in *Movements of the Stars* (first published in Europe in 1537) and enabled **Hevelius** to discover the secular variation in the Moon's motion. In mathematics, Al-Battani introduced the use of signs. See also **Arabian astronomy**.

albedo
A measure of the reflecting power of a nonluminous object, such as a planet, moon, or asteroid. Albedo (from the Latin *albus* meaning "white") is expressed as the fraction of light and/or other radiation falling on an object that is reflected or scattered back into space; its value ranges from 0, for a perfectly black surface, to 1, for a totally reflective surface. Several different types of albedo are defined. The two main categories are *normal albedo* and *Bond albedo*. *Normal albedo*, also known as *normal reflectance*, is a measure of a surface's relative brightness when illuminated and observed vertically. Within this category, *visual albedo* refers to radiation only in the visible part of the spectrum. *Geometric albedo*, also known as *physical albedo*, is the ratio between the brightness of an object as seen from the direction of the Sun, and the brightness of a hypothetical white, diffusely reflecting sphere of the same size and at the same distance. The normal albedo of a moon or asteroid, which can be calculated if the object's apparent brightness, size, and distance are known, is an important indicator of surface composition. *Bond albedo* (named for the American

astronomer George **Bond**), also known as *spherical albedo,* is the fraction of the total incident solar radiation–the radiation at all wavelengths–that is reflected or scattered by an object in all directions; this is an important measure of a planetary body's energy balance. (See table, "Examples of Visual Albedo.")

Examples of Visual Albedo

Object/Substance	Visual Albedo	Notes
Charcoal (carbon)	0.04	Darkest common substance
Mercury	0.11	Rocky surface; no atmosphere
Earth	0.37	Land/ocean/ice surface; partial cloud cover
Venus	0.65	Complete white cloud cover
Pristine snow	almost 1.0	Brightest common substance

albedo feature

A marking on the surface of a celestial object that is significantly brighter or darker than its surroundings. As in the case of Syrtis Major on **Mars**, it need not necessarily correspond with an actual geological or topographical feature.

albedo Striking albedo variations between the dark maria and bright highland regions of the nearside of the Moon are evident in this composite picture built up from thousands of images taken in the near infrared by the Clementine probe. *NASA*

Albireo (Beta Cygni, β Cyg)

The second brightest star in **Cygnus** and widely regarded as one of the most attractive double stars in the sky. Its name is of uncertain provenance, having first appeared in a 1515 translated edition of Ptolemy's *Almagest* as *ab ireo*–far from the original Arabic name *Al Minhar al Dajajah,* meaning "hen's beak." Albireo's stellar duet, separated by 35″, makes a striking gold-blue contrast, easily seen at low telescopic power. β^1 is an orange giant **K star** (visual magnitude 3.1, surface temperature 4,100 K, luminosity 100 L_{sun}, radius 20 R_{sun}), while its partner, β^2, is a main-sequence B star (visual magnitude 5.1) that is slightly variable, rapidly rotating, losing matter, and surrounded by a gas disk of its own making. The two rotate around each other at a distance of about 4,400 AU and with a period of about 7,300 years. β^1 is itself a binary system consisting of a B dwarf in a tight orbit around an aging giant.

Visual magnitude:	2.90
Absolute magnitude:	–2.32
Spectral type:	(K3II + B8V) + B9V
Distance:	385 light-years

Alcyone (Eta Tauri, η Tau)

The brightest star in the **Pleiades** and the only one to have a **Bayer designation**. Like its bright neighbors, Alcyone is a **B star** but is somewhat evolved and considered a giant. Its high rotational speed–over 200 km/s (more than 100 times faster than the Sun)–has caused gas to spin from its equator into a surrounding light-emitting disk. This makes Alcyone a **Be star** but with a disk thicker than usual. A smaller companion lies just a few AU away.

Visual magnitude:	2.85
Absolute magnitude:	–2.41
Spectral type:	B7III
Luminosity:	1,400 L_{sun}
Distance:	368 light-years

Aldebaran (Alpha Tauri, α Tau)

The brightest star in **Taurus**, the thirteenth brightest star in the sky, the most luminous star within 100 light-years of the Sun, and the nearest **red giant**. Its Arabic name, meaning "the follower," refers to its apparent pursuit of the **Pleiades** across the sky. Aldebaran lies in front of the **Hyades** cluster but is not physically associated with it, being only half as far away. As part of a zodiacal constellation, it is close to the Sun's path, the Sun passing to the north of it around June 1; it is also regularly occulted by the Moon. Aldebaran is a low-level **irregular variable** that fluctuates by about 0.2 magnitude. Having evolved off the main sequence, it has expanded to a radius of about 40 R_{sun}–big enough to have a measurable angular diameter of 0.021″ (the apparent size of a nickel seen 50 km away). If put in place of the Sun, Aldebaran would stretch halfway to Mercury and span 20° of our sky.

Visual magnitude:	0.87
Absolute magnitude:	–0.64
Spectral type:	K5III
Surface temperature:	4,000 K
Luminosity:	350 L_{sun}
Distance:	65 light-years

Alderamin (Alpha Cephei, α Cep)

The brightest star in **Cepheus** and a subgiant **A star**. Its name comes from the Arabic *al dhira al yamin,* meaning "right arm," which apparently refers to the arm of the king whom the constellation represents. Alderamin lies close to the precessional path of the north celestial pole (see **precession** of equinoxes), so that it periodically comes within 3° of being an exact **pole star**–a status it last held in about 18,000 B.C. and will hold again about 5,500 years from now. Its unusually high rotational speed of 246 km/s at the equator (about 125 times higher than the Sun's), prevents the separation of chemical elements that is common to stars of this class. The rapid spin may also be tied to Alderamin's activity. The Sun is magnetically active because its outer third is subject to large convective currents. Such convective zones aren't generally expected in A-type stars; yet Alderamin emits roughly the same amount of X-ray radiation as does the Sun.

Visual magnitude:	2.45
Absolute magnitude:	1.57
Spectral type:	A7IV
Surface temperature:	7,600 K
Luminosity:	18 L_{sun}
Radius (est.):	2.5 R_{sun}
Mass (est.):	1.9 M_{sun}
Distance:	49 light-years

Alexandra family

A small family of main-belt **asteroids** lying at a mean distance of 2.6 to 2.7 AU from the Sun, with orbital inclinations of 11 to 12° and a variety of compositions, including representatives from classes C, G, and T. The prototype, (54) Alexandra, of spectral class C and diameter 165 km, was discovered in 1858 by Hermann Goldschmidt; its orbital details are: semimajor axis 2.71 AU, perihelion 2.18 AU, aphelion 3.25 AU, eccentricity 0.199, inclination 11.8°.

Alfirk (Beta Cephei, β Cep)

The second brightest star in **Cepheus** and the prototype **Beta Cephei star**. Its Arabic name (also spelled "Alphirk") may refer to "the two stars" (the other being Alderamin) or may come from a phrase meaning "flock of sheep." A giant **B star**, Alfirk's chief period is 4.57 hours, during which it varies from magnitude 3.16 to 3.27 and back. Like all Beta Cephei stars, however, Alfirk pulsates with multiple periods, smaller changes taking place with a variety of other periods between 4 and 5 hours, in addition to 6- and 12-day rotational modulations. It is also a **Be star** that sheds matter and has a magnetic field about 100 times stronger than Earth's. Two smaller, fainter A stars accompany it: the inner, about 45 AU away with an orbital period of some 90 years; the outer, easily seen in a small telescope, at least 2,400 AU away with a period of at least 30,000 years.

Visual magnitude:	3.23
Absolute magnitude:	–3.08
Spectral type:	B2III
Surface temperature:	26,700 K
Luminosity:	14,600 L_{sun}
Mass:	12 M_{sun}
Distance:	595 light-years

Alfvén, Hannes Olof Gösta (1908–1995)

A Swedish physicist known for his pioneering theoretical research in magnetohydrodynamics for which he shared the 1970 Nobel Prize for physics with the Frenchman Louis Néel. In the 1930s, Alfvén suggested that **sunspots** are the result of the Sun's magnetic field becoming temporarily "frozen" into the solar **plasma**. In 1942 he proposed that waves, now known as **Alfvén waves**, can travel through a plasma under conditions similar to those in the solar atmosphere.

Alfvén wave

A transverse (side-to-side vibrating) wave, similar to a sound wave, that travels along magnetic field lines in a **plasma**. Its speed of propagation, v, known as the *Alfvén speed* is given by:

$$v = \sqrt{(\mu H^2 / 4\pi\rho)}$$

where H is the magnetic field strength, ρ the plasma density, and μ the magnetic permeability.

Algieba (Gamma Leonis, γ Leo)

A magnificent binary system in **Leo** with orange-red and yellow components visible through a small telescope under good seeing conditions; its Arabic name means "the forehead." The brighter component (magnitude 2.6) is a giant **K star** with a surface temperature of 4,400 K and a luminosity of 180 L_{sun}; its partner is a magnitude 3.8 giant **G star** with a temperature of 4,900 K and a luminosity of 50 L_{sun}. The angular separation of just over 4″ means that the two stars are at least 170 AU apart–four times the Pluto-Sun distance–and have an orbital period of over 500 years.

Visual magnitude:	2.01
Absolute magnitude:	–0.92
Spectral type:	K0III + G7III
Distance:	126 light-years

Algol (Beta Persei, β Per)

The second brightest star in **Perseus** and the prototype **eclipsing binary** of the type known as **Algol stars**. Its original Arabic name *Ra's al ghul* means "the demon's head." (A "ghoul" also appears in the Arabian Nights saga.) In Greek mythology, Algol represents Medusa's head with which Perseus turned Cetus to stone, and the star is traditionally considered unlucky. It consists of a bright **B star** and a dimmer but much larger giant **K star** that orbit each other every 2.87 days (69 hours). Because the orbital plane of the two stars lies along our line of sight, the K star periodically passes in front of its smaller partner, obscuring the latter's light and causing a dimming of the combined system from magnitude 2.1 to 3.5. The minimum lasts about 5 hours before the B star comes out of eclipse and Algol's brightness returns to its peak value. A very slight dip in combined brightness happens when the B star blocks a tiny portion of the light from its companion. The first measurement of Algol's period was made by John **Goodricke** in 1783; however, the star's strange behavior was almost certainly known to Arab astronomers, especially in view of the star's bad reputation. Algol is also famed for the so-called *Algol paradox,* which is that the more evolved of the stars in the system–the K-class giant–has the *lower mass.* This seems to break the rule in **stellar evolution**ary theory that the more massive a star, the faster it will evolve. The only explanation is that, in the past, the dim companion has transferred mass to the B star. The two stars are so close–less than 0.1 AU apart–that as the giant, and initially more massive, star swelled up, it began to lose material to its more compact neighbor, which thereby ended up the heavier of the two. Algol has a third star 1.5 times the mass and radius of the Sun that orbits the close pair in 1.86 years.

Visual magnitude:	2.09 max
Absolute magnitude:	–0.18
Spectral type:	B8V + G5IV
Distance:	93 light-years

Algol star

A type of **eclipsing binary**, named after the prototype, **Algol**, that has periods of constant or near-constant brightness between minima, indicating that the two stars form a **close binary** of the *detached* or *semidetached* kind. Thousands of examples are known, with periods ranging from about 5 hours to 30 years and brightness variations of up to several magnitudes. In most cases where mass transfer takes place, it is by direct accretion rather than by an **accretion disk**. However, an extreme group of eclipsing, mass-transferring binaries, known as W Serpentis stars or "hyperactive Algols," does have accretion disks and may be in a pre-Algol-type stage. Algol stars are among the most important kind of star systems in terms of the information they provide on stellar masses and sizes.

Alhazen (Abu Ali al Hassan ibn al Haitham) (c. 965–c. 1040)

An Arab mathematician and physicist who wrote the first important book on optics since the time of **Ptolemy**, in which he rejected the older notion that light was emitted by the eye in favor of the view accepted today: light from external sources enters and is focused by the eye. His *Treasury of Optics* (first published in Latin in 1572) discusses lenses, plane and curved mirrors, and colors. Prior to this work he made a near-disastrous expedition to southern Egypt, sponsored by the Caliph al-Hakim, to study possible ways of controlling the Nile. Realizing that the river could not be easily tamed and that heads would (literally) roll when the bad news was relayed, Alhazen feigned madness upon his return and kept up the pretence until the Caliph died in 1021. See also **Arabian astronomy**.

Alhena (Gamma Geminorum, γ Gem)

A subgiant **A star** and the third brightest member of **Gemini**. Its Arabic name (alternatively given as Almeisan) refers to a brand on a horse or a camel. Alhena is a **spectroscopic binary** with a period of 12.6 years and is the brightest star ever observed to be occulted by an asteroid. In 1991, (381) Myrrha passed in front of Alhena enabling not only the asteroid's diameter (140 km) to be determined but also the fact that the dimmer companion star is a Sun-like G dwarf almost 200 times fainter than Alhena proper. Accumulated observations have shown that the companion, of about 1 solar mass, orbits the 2.8-M_{sun} primary at an average separation of 8.5 AU–approximately the size of Saturn's orbit–but ranges from as close as Earth is to the Sun to about the Uranus-Sun distance.

Visual magnitude:	1.93
Absolute magnitude:	–0.61
Spectral type:	A0IV
Surface temperature:	9,200 K
Luminosity:	160 L_{sun}
Distance:	105 light-years

Alioth (Epsilon Ursae Majoris, e UMa)

The brightest star in **Ursa Major** and the third star from the end of the **Big Dipper**'s handle; its proper name comes from a corruption of the Arabic meaning "black horse." Alioth is the brightest **Ap star** and is also an **Alpha^2 Canum Venaticorum star** with a period of 5.1 days. Its oxygen abundance is 100,000 times greater near the magnetic equator than near the magnetic poles (which are offset from the rotational equator and poles); the amount of chromium varies similarly. Although visually the brightest Ap star, Alioth has one of the weakest magnetic fields of this class–15 times weaker than that observed for **Cor Caroli**.

Visual magnitude:	1.76
Absolute magnitude:	–0.22
Spectral type:	A0V
Surface temperature:	9,400 K
Luminosity:	108 L_{sun}
Mass:	about 3 M_{sun}
Distance:	81 light-years

Alkaid (Eta Ursae Majoris, η UMa)

The second brightest star in **Ursa Major** and the end star in the handle of the **Big Dipper**. Its name (alternatively given as "Benetnasch") is a contraction of the phrase *ka'id banat al Na'ash* meaning "chief mourner of the daughters of Al Na'ash" (the latter represented by stars in the Dipper's handle) that stand by a funeral bier made of the Dipper's bowl. Alkaid is just below the temperature limit at which stars generate strong X rays because of shock waves in their winds.

Visual magnitude:	1.85
Absolute magnitude:	–0.60
Spectral type:	B3V
Luminosity:	700 L_{sun}
Mass (est.):	6 M_{sun}
Distance:	101 light-years

Allende meteorite

A **carbonaceous chondrite** that fell near the village of Pueblito de Allende in the Mexican state of Chihuahua on February 8, 1969, scattering several tons of material over an area of 48 km × 7 km. Specimens of the meteorite were found to contain a fine-grained carbon-rich matrix studded with **chondrules**, both matrix and chondrules consisting mainly of **olivine**. Close examination of the chondrules revealed tiny black markings, up to 10 trillion per cm^2, that were absent from the matrix and taken as evidence of radiation damage. Similar structures have been found in lunar basalts but not in their terrestrial equivalent, which would have been screened from cosmic radiation by Earth's atmosphere and geomagnetic field. Irradiation of the chondrules, it seems, happened after they had solidified but before the cold accretion of matter that took place during the early stages of formation of the solar system, when the parent meteorite came together. The Allende meteorite also contains fine-grained, microscopic diamonds with strange isotopic signatures that point to an extrasolar origin; these interstellar grains are older than the solar system and probably the product of a nearby **supernova**.

all-sky camera

A camera with a very wide **field of view** that can be used to observe the sky from horizon to horizon. Such instruments are used, for example, to study aurorae and for meteor and fireball patrols. They produce a circular image with the zenith at the center and the horizon at the circumference.

Almaak (Gamma Andromedae, γ And)

One of the most beautiful double stars in the sky. Its Arabic name (also written as Almach, Almak, Alamak, and Alamaak) is unrelated to the legend of **Andromeda** but refers instead to a badgerlike animal of the Middle East. Even a small telescope shows a superb pair separated by 10″: the brighter component is golden yellow, its partner is blue. The second magnitude primary, γ^1, is a bright giant **K star** with a surface temperature of about 4,500 K, a luminosity of about 2,000 L_{sun}, and a radius of 80 R_{sun}–big enough to swallow the orbit of Venus. Its fainter blue companion, γ^2, is also double, although the dual nature is more difficult to detect. The two fifth-magnitude stars orbit each other with a period of about 60 years and, despite being near their greatest separation are still only 0.5″ apart, and thus almost impossible to see separately. The brighter of these two is itself a **spectroscopic binary** with a period of 2.7 days. The three components of γ^2 are all hot main sequence stars with temperatures around 10,000 K.

Visual magnitude:	2.10
Absolute magnitude:	–3.09
Spectral type:	B8V
Distance:	355 light-years

Almagest

The Arabic title of **Ptolemy of Alexandria**'s *Syntaxis,* the writings in which he combined his own astronomical research with those of others. Although much of the work is inaccurate, even in premise, it remained the standard reference source in Europe until Nicolaus **Copernicus** published his results in the sixteenth century.

almanac

A book of tables, usually covering a period of one calendar year, that lists the future positions of the Moon, planets, and other prominent celestial objects, together with other useful astronomical data. Important examples include *The **Astronomical Almanac*** and *The Nautical Almanac.*

Alnair (Alpha Gruis, α Gru)

The brightest star in **Grus** and the thirtieth brightest star in the sky. Its Arabic name (also written "Al Na'ir") means "the bright one," and comes from a longer phrase for "the bright one in the fish's tail," since the Arabs considered the stars of Grus to be the tail of **Piscis Austrinus.** Alnair can't be seen from latitudes higher than 42° N.

Visual magnitude:	1.73
Absolute magnitude:	–0.74
Spectral type:	B7IV
Surface temperature:	13,500 K
Luminosity:	380 L_{sun}
Distance:	101 light-years

Alnath (Beta Tauri, β Tau)

A giant **B star** and the second brightest star in **Taurus.** Its Arabic name (also written as Elnath, El Nath, and Nath) means the "butting one"–appropriate to its position at the tip of the Bull's northern horn. Alnath also doubles as the gamma star of neighboring **Auriga**, although it is now never referred to as Gamma Aurigae. It is distinctive, too, in lying just 3° to the west of the *galactic anticenter,* the point in the sky that lies directly opposite the center of the Milky Way Galaxy. Alnath appears to be a **mercury-manganese star**, with a manganese abundance 25 times greater than the Sun's but with calcium and magnesium abundances that are only one-eighth solar. Another prominent example of such a star, is, by coincidence, the only other bright star shared by two constellations–**Alpheratz** (Alpha Andromedae or Delta Pegasi). Alnath has left the main sequence and will evolve, over the next million years or so, to become a cooler, orange giant.

Visual magnitude:	1.65
Absolute magnitude:	–1.37
Spectral type:	B7III
Surface temperature:	13,600 K
Luminosity:	700 L_{sun}
Radius:	5 to 6 R_{sun}
Mass (est.):	4.5 M_{sun}
Distance:	131 light-years

Alnilam (Epsilon Orionis, ε Ori)

The central and brightest of the three stars in **Orion's Belt** and the fourth brightest in the whole of **Orion.**

Visual magnitude:	1.69
Absolute magnitude:	–6.39
Spectral type:	B0Ia
Surface temperature:	25,000 K
Luminosity:	375,000 L_{sun}
Mass:	40 M_{sun}
Distance:	1,340 light-years

Alnilam is a blue-white **supergiant** that has long served as a standard star against which to compare others. Its simple spectrum also provides a useful background against which to study the intervening **interstellar medium.** It lies close to the Orion Molecular Cloud and illuminates part of it as the *reflection nebula* NGC 1990. Like most supergiants, Alnilam is rapidly shedding mass: an intense stellar wind blows from its surface at speeds up to 2,000 km/s and carries away about two-millionths of a solar mass per year. Although only about 4 million years old, Alnilam is already fusing heavy elements in its core and is doomed to explode, in the next million years or so, as a supernova.

Alnitak (Zeta Orionis, ζ Ori)

The left star in **Orion's Belt**, the fifth brightest star in **Orion**, and (in terms of apparent magnitude) the brightest **O star** in the sky. Its Arabic name comes from a phrase meaning "the belt of al Jauza." A B-type companion lies about 3″ away, the pair orbiting each other with a period of several thousand years. Like all O stars, Alnitak is a source of X rays associated with a powerful **stellar wind**; the X rays are produced when clumps of gas in the wind crash violently into one another. Alnitak is probably only about 6 million years old and will eventually become a red supergiant before exploding as a supernova. In its vicinity are several dusty interstellar clouds, including, to the south, the **Horsehead Nebula**.

Visual magnitude:	1.74
Absolute magnitude:	–5.26
Spectral type:	O9.5Ib
Surface temperature:	31,000 K
Luminosity:	100,000 L_{sun}
Mass:	20 M_{sun}
Distance:	817 light-years

Alpha Centauri (α Cen)

With the possible exception of **Proxima Centauri**, which may or may not be part of it, the nearest star system to the Sun. α Cen (also known as Rigil Kentaurus, "the Centaur's foot" in Arabic) is the third brightest star in the night sky, after **Sirius** and **Canopus**, but lies so far south that it is visible only from latitudes below 25° N. Its two bright components, A and B, consist of a yellow **G star**, similar to the Sun but about 200 million years older, and an orange **K star** of the same age as its brighter companion, moving around each other every 79.9 years in a highly elongated orbit (eccentricity 0.519) with a mean separation of 23.7 AU–a bit more than the distance between Uranus and the Sun. Their binary nature was first observed by Nicolas **Lacaille** in 1752. Both lie 4.36 light-years (41 trillion km) away. No planets have been found in the system to date. (See table, "Alpha Centauri A and B.")

Alpha Centauri The nearest star system, Alpha Centauri, as seen at infrared wavelengths by the MSX (Midcourse Space Experiment) satellite. *MSX/IPAC/NASA*

Alpha Cygni star (α Cyg star)

A type of supergiant **pulsating variable** star, of spectral type A or B, that undergoes **non-radial pulsations** with a period of 5 to 10 days and a visual amplitude of 0.1 magnitude or less, named for the prototype, Alpha Cygni (**Deneb**). The stellar oscillations may be complex, consisting of multiple pulsation frequencies as well as the fundamental one, so that the brightness changes appear irregular. The changes can only be small, however; otherwise, the star would become unstable and blow apart.

alpha particle

A **helium** nucleus (two protons and two neutrons bound together). Alpha particles are emitted by certain radioactive nuclei and also play an important role in nuclear **fusion** processes within stars.

Alpha² Canum Venaticorum star (α² CVn star)

A type of **rotating variable**, also known as a *spectrum variable*, that consists of a main-sequence star of spectral type B8p to A7p with a strong magnetic field. α² CVn stars have spectra with abnormally strong lines of silicon, strontium, chromium, and rare earths, the intensities of which vary with rotation. They also show magnetic field and brightness changes with periods of 0.5 to 160 days and visual amplitudes of 0.01 to 0.1 magnitude. It is thought the powerful magnetic field of an α² CVn star gives rise to different compositions and brightness at different parts of the stellar surface, in the same way that the Sun's (much weaker) field produces sunspots. The prototype is the companion of **Cor Caroli**.

Alphard (Alpha Hydrae, α Hya)

The brightest star in **Hydra**, and all the more prominent for being in a fairly barren region of the sky to the southwest of **Regulus**. Its Arabic name means "the solitary

Alpha Centauri A and B

	Class	Temp.	Magnitude Visual	Magnitude Absolute	Luminosity (Sun = 1)	Mass (Sun = 1)	Diameter (Sun = 1)
α Cen A	G2V	5,790 K	–0.01	4.34	1.52	1.10	1.23
α Cen B	K0V	5,260 K	1.35	5.70	0.50	0.91	0.87

one." Alphard is a mild **barium star** with a companion that, before it became a **white dwarf**, contaminated its partner with the by-products of nuclear fusion that had been dredged to its surface.

Visual magnitude:	1.99
Absolute magnitude:	–1.69
Spectral type:	K3III
Surface temperature:	4,000 K
Luminosity:	400 L_{sun}
Distance:	177 light-years

Alphekka (Alpha Coronae Borealis, α CrB)

Also known as Gemma, the brightest star in **Corona Borealis**. Its Arabic name (alternatively spelled "Alphecca"), meaning "break," refers to the broken circlet of stars that makes up the Northern Crown. Alphekka is an **Algol star**: the bright naked-eye star has a much fainter, Sun-like (G5) companion that produces a barely discernable eclipse of the primary every 17.4 days. Alphekka is also a part of the **Ursa Major Moving Cluster**.

Visual magnitude:	2.22
Absolute magnitude:	0.42
Spectral type:	A0V + G5V
Surface temperature:	9,500 K
Luminosity:	60 L_{sun}
Distance:	75 light-years

Alpher, Ralph Asher (1921–)

An American physicist best known for his theoretical work on the origin and early evolution of the universe. In 1948, together with Hans **Bethe** and George **Gamow**, he suggested how the abundances of chemical elements could be explained as a result of thermonuclear processes immediately after the **Big Bang**. This work became known as the "alpha, beta, gamma" theory. As further developed in collaborations with Robert Herman (1914–1997) and others, this concept of cosmic **nucleosynthesis** became an integral part of the standard Big Bang model. It also led to a prediction of the **cosmic microwave background.**

Alpheratz (Alpha Andromedae, α And)

The brightest star in **Andromeda** and the northeastern star of the **Square of Pegasus**. Its proper name is generally taken to mean "the horse's shoulder" or "the horse's navel," indicating that the star originally belonged more to Pegasus (in which it served as Delta Peg), although now it is formally within the boundaries of Andromeda and marks the head of the royal daughter; its alternative name is Sirrah. Alpheratz is the brightest **mercury-manganese star** and is also a **spectroscopic binary** with a period of 96.7 days; the dimmer companion seems to be about one-tenth as bright as the primary.

Visual magnitude:	2.07
Absolute magnitude:	–0.30
Spectral type:	B9IVp
Surface temperature:	13,000 K
Luminosity:	200 L_{sun}
Distance:	97 light-years

Alphonsine Tables

A set of tables giving the positions of the Sun, the Moon, and the planets, published in 1252 under the patronage of Alfonso X (1223?–1284), King of Léon and Castile, who had assembled, at Toledo, many of the world's leading astronomers. The tables were calculated by a team of astronomers using the principles set out by Ptolemy in the ***Almagest*** but incorporating more recent observations. They remained the standard in Europe for nearly 400 years, until superceded by the work of Johannes **Kepler**.

Al-Sufi, Abd al-Rahman (A.D. 903–986)

A Persian nobleman and astronomer, also known by the Latinized name Azophi, whose *Book of the Fixed Stars* (c. 964) includes a catalog of 1,018 stars, giving their approximate positions, magnitudes, and colors. It contains Arabic star names that, in corrupted form, are still in use today, and the earliest known reference to the **Andromeda Galaxy**. See also **Arabian astronomy**.

Altair (Alpha Aquilae, α Aql)

The brightest star in **Aquila**, the twelfth brightest star in the sky, and the southern point of the **Summer Triangle**; its name comes from the Arabic *al nas al tair,* meaning "the flying eagle." Altair's high proper motion will cause it to shift by as much as a degree over the next 5,000 years. It is also a fast rotator with an equatorial spin speed of 242 km/s, compared with the Sun's 2 km/s. As a result, despite having a radius greater than that of the Sun, Altair manages to complete a rotation every nine hours and is probably one of the most flattened stars known.

Visual magnitude:	0.76
Absolute magnitude:	2.20
Spectral type:	A7V
Luminosity:	11 L_{sun}
Radius:	1.7 R_{sun}
Distance:	16.8 light-years

altazimuth mounting

A telescope **mounting** that allows up and down pivoting (changes in altitude, or elevation, from 0 to 90°) together with horizontal rotation (changes in **azimuth** by ± 180°). To track an object across the sky with such an arrangement demands that the telescope move simultaneously around two axes–a complication that, in the past, led to an overwhelming preference for **equatorial mountings**. However, computer control has made altazimuthal operation so much easier that altazimuthal mountings, which are simple to construct, have been adopted almost universally for large modern instruments. The vertical axis is a fork that holds the two ends of the horizontal axis, and the horizontal base can be spun around a vertical axis at its center. Because there are no difficult angles to sustain, it takes a less massive, less expensive mount to support a given size of telescope. The restricted range of motion needed to cover the whole sky leads to another substantial saving in the size of the dome needed. Drawbacks, in addition to dual-axis tracking, are that the tracking rate varies with position in the sky and that the field of view rotates, which, although not a problem for point sources such as stars, must be addressed for any extended source by either rotating the instrument or by de-rotating the image.

altitude

The angular distance of a celestial body above or below the observer's horizon, measured vertically. Altitude is 0° at the horizon and 90° at the zenith. It is specified together with **azimuth** to give the **horizontal coordinates** of an object.

Aludra (Eta Canis Majoris, η CMa)

A supergiant **B star** and the fifth brightest star in **Canis Major**; its name comes from the Arabic *al-'Udhrah,* meaning "the Virgins," which refers to the stars of the Great Dog's lower parts. Some of the Virgins, including Aludra and **Wezen**, are in a loose **OB association** known as Collinder 121. Formed only about 12 million years ago, Aludra is already in old age, blowing a strong **stellar wind** from its surface at a speed of some 500 km/s. The material it has lost–about one-third of a solar mass–has gathered into a cloud that is detectable by its infrared emission. A seventh magnitude white star about 2′ from Aludra, easily visible in a small telescope, is not a physical companion but lies a mere 600 light-years from the Sun.

Visual magnitude:	2.45
Absolute magnitude:	–7.51
Spectral type:	B5Ia
Surface temperature:	13,500 K
Luminosity:	100,000 L_{sun}
Distance:	3,200 light-years

aluminizing

The process of coating a mirror with aluminum. The aluminum is vaporized in a vacuum, causing a film of metal, about 100 nanometers thick, to be deposited on the glass. Layers of other materials, such as silicon dioxide, may subsequently be laid down to give protection. Aluminum offers consistently high reflectance throughout the visible, near-infrared, and near-ultraviolet regions of the spectrum. Before the 1930s, when the vacuum aluminizing technique was developed, the only method available for coating mirrors was *silvering* in which the silver was deposited chemically by a process developed by the German chemist Justus von Liebig (1803–1873). The initial reflectivity of silver (89 to 93%) is slightly higher than that of aluminum throughout most of the visible spectrum, but this advantage is short-lived because silver tarnishes quickly, whereas the reflectivity of an aluminum coating drops by only a few percent per year. Silver, too, is much more expensive. See also **speculum**.

AM Canum Venaticorum star (AM CVn star)

A rare type of **cataclysmic binary** believed to consist of two **white dwarfs** in an extremely tight orbit that form an accreting, semidetached system (see **close binary**). AM CVn stars are helium-rich with no detectable trace of hydrogen, have very short orbital periods–that of the prototype is about 18 minutes–and show permanent **super-humps**.

AM Herculis star (AM Her star)

A type of **magnetic cataclysmic variable** that consists of a closely orbiting dwarf **M star** or **K star** and a super-strong magnetic **white dwarf** primary, in which the magnetic field of the primary not only prevents the formation of an **accretion disk** but also synchronizes the primary's rotation with its orbital period. Such systems are the most extremely polarized objects known (hence their alternative name *polars*), exhibiting both strong linear and, more significantly, circular **polarization**. The powerful (10 to 100 megagauss) magnetic field of the white dwarf captures infalling material from the **red dwarf** before it can form an accretion disk and forces it instead into an *accretion stream* or funnel, one part of which heads for the star's north magnetic pole and the other for the south pole. As the field lines, like those around a bar magnet, converge, they channel the streams of matter onto tiny polar accretion spots. The streams slam into the white dwarf at some 3,000 km/s and have their kinetic energy converted primarily into X rays. The white dwarf's magnetic field also locks the spin of the two stars so that they always present the same face to each other (although in about one-tenth of AM Her stars, the white dwarf's axial and orbital rotations are off by about 1%). Furthermore, the magnetic field of the white dwarf often tilts over, so that one magnetic pole points toward the direction from which the stream comes (this being the lowest-energy configuration). As a result, material flows preferentially onto that pole. Eclipses in AM Her systems provide a graphic illustration of this stream geometry. Light curves reveal that the tiny, and thus rapidly eclipsed, accretion spot at the stream-facing pole emits about half the total radiated energy of the system, most of the rest coming from the extended stream, which enters and leaves the eclipse more gradually. The optical variations in an AM Her star, described as "flickering," may range over 4 to 5 magnitudes. The prototype, AM Her, which has an orbital period of 3.1 hours, was discovered by Max **Wolf** in 1923, then listed in the *General Catalogue of Variable Stars* as an irregular variable with a range from twelfth to fourteenth magnitude. This listing remained unchanged until 1976, when the true nature of AM Her began to emerge. It was found to be the optical counterpart of an X-ray source, 3U 1809 + 50, discovered by the Uhuru satellite. Some of the optical features of AM Her's light curve are explainable in terms of the red dwarf secondary. First, the red dwarf is distorted into an egg-shape by the attraction of its companion, toward which the long axis of the egg points. When the secondary is seen broadside, it appears slightly brighter than when end on; hence, as the entire system rotates, there are two long, weak brightness maxima and two long, shallow minima per period. Second, there are sometimes brightness fluctuations due to heating of the red dwarf's surface by X rays from the primary. This "hot spot" is periodically lost from view on the far side of the rotating secondary. The short-term flickerings are due to the turbulent nature of the mass transfer in the system.

Am star

Also known as a *metallic-line star*, a type of **A star** whose spectrum has strong and often variable **absorption lines** of some metals (hence the "m"), such as zinc, strontium, zirconium, and barium, more typical of an **F star**, and deficiencies of others, such as calcium and/or scandium. These abundance anomalies are due to some elements being pushed to the surface because they are better light absorbers, while other elements sink to lower levels under gravity–an effect that requires slow stellar rotation. Normal A stars spin quickly, but most Am stars are known to be members of **close binary** systems in which the two stars slow each other down by tidal action. Familiar examples include **Sirius** and Acubens. See also **Ap star** and **mercury-manganese star**.

Amalthea

The third closest and the fifth largest moon of **Jupiter**, also known as Jupiter V; Amalthea orbits at a mean distance of 181,300 km from the center of the planet. It is extremely irregular in shape, with dimensions of about 270 × 165 × 150 km, and is heavily scarred by craters, some of which are very large in relation to its overall size. Pan, the biggest crater, is 100 km across and at least 8 km deep. Another crater, Gaea, spans 80 km and is probably twice as deep as Pan. Amalthea has two known mountains, Mons Lyctas and Mons Ida, with local relief up to 20 km. The surface is dark and reddish, apparently due to a dusting of sulfur from **Io**'s volcanoes. Bright patches of green on the major slopes of Amalthea are of an unknown nature. Measurements of the moon's mass by the Galileo spacecraft, during its final flyby in 2002, revealed a surprisingly low density, close to that of water. Since Amalthea is unlikely to be composed of ice, it is most likely a loose "rubble pile" with many empty spaces. Discovered in 1892 by Edward **Barnard**, and sometimes referred to as *Barnard's satellite*, Amalthea was the last moon in the solar system to be found by direct visual observation.

Ambartsumian, Viktor Amazaspovich (1908–1996)

A Georgian-born Armenian astrophysicist who worked at the Pulkovo Observatory; the **Byurakan Astrophysical Observatory**, which he founded and directed; Erevan University; and the Armenian Academy of Sciences, where he served as president from 1947 to 1993. While at Pulkovo he also taught at the University of Leningrad and wrote the first Russian textbook on theoretical astrophysics. Most of his research was devoted to invariance principles applied to the theory of radiative transfer, inverse problems of astrophysics, and the empirical

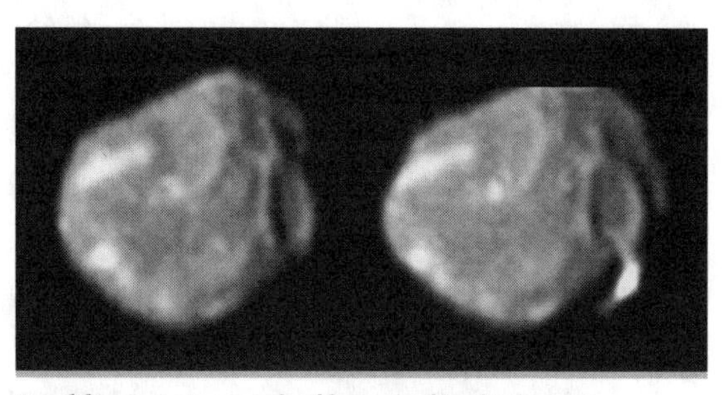

Amalthea A stereo pair of images of Jupiter's moon Amalthea taken by the Galileo spacecraft in 1999 and showing features as small as 3.8 km across. The large impact crater, near the right edge of Amalthea's disk, is about 40 km across. The bright linear streak to the left in each image may be material thrown out of the adjacent crater or, alternatively, may be a long narrow ridge. *NASA/JPL*

approach to the problems of the origin and evolution of stars and galaxies. He was first to suggest that **T Tauri stars** are very young and to propose that nearby **stellar associations** are expanding. He also showed that evolutionary processes such as mass loss are occurring in galaxies; worked on interstellar matter, radio galaxies, and active galactic nuclei; served as president of the International Astronomical Union; and hosted conferences on the search for extraterrestrial civilizations.

Ambartsumian's Knot

A dwarf galaxy in **Ursa Major** located at the end of what appears to be a bridge of matter extending from the galaxy NGC 3561 (R.A. 11h 11m, Dec. +28° 42′); the latter is interacting with the nearby elliptical Arp 105.

ambipolar diffusion

An important process in the initial stage of star formation whereby clumps of a **molecular cloud** uncouple from the interstellar magnetic field, which would otherwise resist the further gravitational collapse of the clumps (see **stellar evolution**). Magnetic fields thread all interstellar space and act upon ions (mostly protons) and electrons. These charged particles, in turn, collide with neutral atoms and thereby, in most cases, couple them to the field. However, in a molecular cloud, where the fractional ionization is very low (one part per million or less), neutral particles only rarely encounter charged particles, and so are not well coupled. Thus, the interstellar magnetic field can drag the ions through the neutral gas without acting as a significant brake on the cloud's collapse.

American Astronomical Society (AAS)

The chief organization in the United States for professional astronomers, founded in 1899 with headquarters in Washington, D.C. It publishes the *Astronomical Journal*, *Astrophysical Journal*, and a quarterly *Bulletin* (BAAS).

amino acids

Organic chemicals that can link together in chains (polypeptides) and form proteins. Dozens of different amino acids have been discovered in **carbonaceous chondrite** meteorites and unconfirmed evidence has been found for the simplest amino acid, glycine (NH_2CH_2COOH), within the **Sagittarius B2** molecular cloud near the center of our galaxy. These findings encourage the view that some of the building blocks of life as we know it are widely spread throughout the universe. See also **life in the universe**.

ammonia (NH_3)

A compound that occurs commonly in the atmospheres of **gas giants**, frozen in the nuclei of **comets**, and, as widely dispersed molecules, in **molecular clouds**.

Amor group

Also known as *Earth-grazing asteroids*, one of three groups of **near-Earth asteroids**; their orbits have perihelia between 1.017 AU (Earth's aphelion distance) and 1.3 AU (Mars's perihelion distance). The prototype of the group is the 1-km-wide (1221) Amor, discovered in 1932 by Eugène **Delaporte**; the largest members are (1036) Ganymed, with a diameter of 40 km, and (433) **Eros**. Amor asteroids vary widely in composition, which indicates they have come from different parts of the main **asteroid belt** and have been perturbed into their present orbits by either Jupiter or Mars. Close approaches to Earth and Mars can further alter their paths so that they temporarily become *Earth-crossing* **Apollo group** asteroids. About one in ten Amors will cross Earth's orbit in the course of a few hundred to a few thousand years.

amplitude

(1) The maximum departure of a cyclical quantity from its mean value. (2) The difference between the maximum and minimum magnitudes of a variable star, i.e., the total range of its brightness.

analemma

A stretched figure-eight curve obtained if solar **declination** is plotted against the **equation of time** throughout a whole year. Often found on **sundials** and globes, analemmas were originally used to calculate standard clock **time** from the apparent time as indicated by the position of the Sun. The Mars analogue to the terrestrial analemma has a teardrop shape.

Ananke

One of the outer moons of **Jupiter**, also known as Jupiter XII; it was discovered in 1951 by Seth **Nicholson**. Ananke is 28 km in diameter and orbits in a **retrograde** direction at a mean distance of 21.3 million km from the planet.

anastigmat(ic)

An optical system, with at least three elements, that is completely corrected for **spherical aberration**, coma (see **coma, optical**), and **astigmatism**.

Anaxagoras of Clazomenae (c. 500–c. 428 B.C.)

An Ionian philosopher who, at the age of 20, moved to Athens and effectively established it as the new center of Greek philosophy. For three decades he helped shape the thoughts of a number of illustrious pupils, including Pericles the statesman, Euripides the playwright, and possibly even Socrates. Anaxagoras's explanations of the Moon's light, eclipses, earthquakes, meteors, rainbows, sound, and wind seem surprisingly modern. He thought that the Moon had "a surface in some places lofty, in others hollow," the Sun was a brightly glowing rock "bigger than the Peloponnese," and the stars were other suns lying at such a distance that they appeared to give out no heat. When he was about 33, a meteorite big enough to fill a wagon landed in broad daylight near the town of Aegospotami. Anaxagoras caused a sensation by claiming it had come from the Sun. Eventually, he was charged with impiety and sentenced to death for suggesting that in place of the traditional pantheon of Greek gods there was just a single eternal intelligence, or "Nous." Fortunately, Pericles, the most respected man in Athens, put in a good word for him, and the sentence was commuted to exile. See also **Greek astronomy**.

Anaximander of Miletus (611–546 B.C.)

A Greek philosopher, student of **Thales of Miletus**, and possibly the first person to speculate on the existence of other worlds. Anaximander held that the fundamental essence of all things is not a particular substance, like water, but *apeiron,* or the infinite. It seemed reasonable to him, given this boundless creative source extending in all directions, that there might be an indefinite number of worlds existing throughout time, worlds that "are born and perish within an eternal or ageless infinity." Anaximander also pioneered the notion that Earth is not flat, suggesting it was cylindrical and that it floated free, unsupported, at the exact center of the universe, with people living on one of the flat ends. As for the Sun, he said it was as large as Earth–an audacious theory at the time. See also **Greek astronomy**.

Anaximenes of Miletus (c. 585–c. 525 B.C.)

A Greek philosopher and student of **Anaximader of Miletus**, who was the first to draw a clear distinction between planets and stars. He believed the Sun to be hot because of its quick motion around Earth, that the stars were too remote to send us detectable heat, and that the stars were fastened to a crystalline sphere. See also **Greek astronomy**.

Andromeda (abbr. And; gen. Andromedae)

Daughter of Cassiopeia; a large, prominent northern **constellation,** to the south of Cassiopeia, between Perseus and Pegasus. In mythology, Andromeda was chained to a rock to be sacrificed to Cetus (the whale) as a punishment to her mother, Cassiopeia, who had boasted that Andromeda was more beautiful than the daughters of Neptune. Andromeda's three brightest stars lie along an arc that stretches to the northeast of the square of **Pegasus**, of which **Alpheratz** forms a corner. Alpheratz and **Mirach**, the two brightest stars, have identical magnitudes, which is a coincidence found in no other constellation. Near to third-ranked **Almaak** is the outstanding feature of Andromeda–our nearest neighboring major galaxy, the **Andromeda Galaxy**. (See tables, "Andromeda: Stars Brighter than Magnitude 4.0," and "Andromeda: Other Objects of Interest." See also star chart 1.)

Andromeda Galaxy (M31, NGC 224)

The nearest large galaxy to our own, a class Sb spiral (see **galaxy classification**) and a major member of the **Local Group**. It is the most remote object normally visible to the naked eye, although keen observers can sometimes see the **Trapezium** Galaxy. Recent estimates suggest that Andromeda is bigger but less massive than the **Milky Way Galaxy.** It is accompanied by at least 10 satellite galaxies, including **M32** and **M110** (both visible through binoculars), NGC 185 (discovered by William **Herschel**), NGC 147 (discovered by Louis **d'Arrest**),

Andromeda: Stars Brighter than Magnitude 4.0

Star	Magnitude		Spectral	Distance	Position	
	Visual	Absolute	Type	(light-yr)	R.A. (h m s)	Dec. (° ′ ″)
α **Alpheratz** (Sirrah)	2.07	–0.30	B8IVpMnHg	97	00 08 23	+29 05 26
β **Mirach**	2.07	–1.87	M0IIa	199	01 09 44	+35 37 14
γ **Almaak**	2.10	–3.09	K3IIb+A0V	355	02 03 54	+42 19 47
δ	3.27	0.81	K3III	101	00 39 20	+30 51 40
51	3.59	–0.05	K3III	174	01 38 00	+48 37 42
o	3.6v	1.33	B6IIIpe+A2p	141	23 01 55	+42 19 34
λ	3.82v	1.75	G8III-Iv	84	23 37 34	+46 27 30
μ	3.86	0.75	A5V	136	00 56 45	+38 29 58

Andromeda: Other Objects of Interest

Name	Notes
Stars	
R Andromedae	A Mira variable. Magnitude range 5.8 to 14.9; mean period 409 days; R.A. 2h 24.0m, Dec. +38° 35′.
S Andromedae	See separate entry.
Z And	The prototype **Z Andromedae star**.
LL And	See **brown dwarf**.
Galaxies	
Andromeda Galaxy	M31 (NGC 224). See separate entry.
M32 (NGC 221)	A satellite of the Andromeda Galaxy. See separate entry.
M110 (NGC 205)	The second brightest satellite of the Andromeda Galaxy. See separate entry.
NGC 891	One of the best examples of an edge-on spiral. R.A. 2h 19m, Dec. +42° 7′.
Open cluster	
NGC 752	A fine object, 5° south of Almaak, best viewed with binoculars as its 100 or so member stars are scattered over a large area. R.A. 1h 57.8m, Dec. +37° 41′.
Planetary nebula	
Blue Snowball Nebula	NGC 7662. See separate entry.

the dwarf spheroidal systems And I, And II, And III, and And V, the dwarf irregular And IV (although this may be a cluster), And VI (known as the **Pegasus Dwarf**), and And VII (known as the **Cassiopeia Dwarf**). The Andromeda Galaxy is also home to more than 300 **globular clusters**, the brightest of which, G1, is the most luminous globular in the Local Group (visual magnitude 13.7); it outshines the brightest globular in the Milky Way, **Omega Centauri**, and can be glimpsed by large amateur telescopes under favorable conditions. The Hubble Space Telescope has revealed the appearance of a double nucleus in Andromeda. It may be that there really are two bright nuclei, perhaps because a smaller galaxy was cannibalized in the past, or that there is a single core partly obscured by dust.

The earliest known reference to our large galactic neighbor is by **Al-Sufi** who called it "the little cloud" in his *Book of Fixed Stars* (A.D. 964). For many years it was known as the Great Andromeda Nebula and believed to be one of the closest nebulae. William **Herschel** thought its distance would "not exceed 2,000 times the distance of Sirius" (i.e., 17,000 light-years); nevertheless, he viewed it as the nearest "island universe" like our Milky Way. In 1912, Vesto **Slipher** found the **radial velocity** of Andromeda to be the highest cosmic velocity ever measured up to that time–about 300 km/s in approach (later refined to 266 km/s)–pointing to an extragalactic nature. In 1923, Edwin **Hubble** found the first **Cepheid variable** in the Andromeda Galaxy and thus established the intergalactic distance of and the true nature of M31 as a galaxy. Because he was not aware that there are two different kinds of Cepheids, however, his distance was off by a factor of more than two–an error not discovered until 1953 following the completion of the Hale Telescope.

Visual magnitude:	3.4
Diameter (of disk):	160,000 light-years
Mass:	300 to 400 billion M_{sun}
Distance:	2.9 million light-years

anemic spiral galaxy

A galaxy with characteristics that are intermediate between those of a normal spiral and a **lenticular galaxy**. Anemic spirals have smooth, weak spiral arms and a reduced rate of star formation linked to their deficiency in neutral hydrogen gas. They tend to occur in **rich clusters of galaxies**, suggesting that they may have been stripped of interstellar material through multiple encounters with other galaxies. This raises the question of whether anemic spirals mark both a morphological *and* an evolutionary transition between normal spirals and lenticulars.

Angelina (minor planet 64)

The second largest known **E-class asteroid**, after Nysa, and one with an unusually high **albedo**; it was discovered

Andromeda Galaxy A 3 × 3 degree view of the Andromeda Galaxy with its dwarf companions M32 (upper right) and M110 (lower center). *Isaac Newton Group of Telescopes*

by E. Wilhelm **Tempel** in 1861. Whereas most E-class asteroids dominate the **Hungaria group** region of the main asteroid belt, Angelina is one of the few that orbit much farther out. Diameter 60 km, albedo 0.34, semimajor axis 2.684 AU, eccentricity 0.125, inclination 1.31°.

Anglo-Australian Observatory

An observatory that operates the Anglo-Australian Telescope and **United Kingdom Schmidt Telescope** on Siding Spring Mountain, near Coonabarabran, New South Wales, Australia. Its headquarters and laboratory are on the same campus as the **Australia Telescope National Facility** in the Sydney suburb of Epping.

angrite

An evolved **achondrite** composed mainly of **augite** together with small quantities of **olivine** and **troilite**. Angrites, named for the Angra dos Reis meteorite, which fell in Rio de Janeiro, Brazil, in early 1869, are **basaltic** rocks, often containing porous areas and many round vesicles (small cavities) with diameters up to 2.5 cm. These vesicles have been interpreted as remnants of gas-bubbles that formed before the rock crystallized. However, recent research suggests that the vesicles were originally solid spheres that became *exsolved* (separated from a crystal phase) in subsequent stages of rock formation. Both theories are consistent with a magmatic origin of the angrites, making them the most ancient igneous rocks known, with crystallization ages of around 4.55 billion years. They are thought to have formed on one of the earliest differentiated **asteroids**. By comparing the reflectance spectra of the angrites to that of several main belt asteroids, two potential parent bodies have been identified–(289) Nenetta and (3819) Robinson.

Ångström, Anders Jonas (1814–1874)

A pioneering Swedish spectroscopist at the University of Uppsala who deduced, in 1855, that a hot gas emits light at the same wavelengths at which it absorbs light when cooler–a fact demonstrated experimentally by Gustav **Kirchoff** four years later. In 1861, Ångström began a study of the solar spectrum, proving the existence of hydrogen in the Sun and mapping about 1,000 lines seen earlier by Josef **Fraunhofer**. He recorded his measurements in units of 10^{-10} m–a unit subsequently named in his honor. Ångström was also the first to examine the spectra of **aurorae**.

angular diameter

The angle that the actual diameter of an object makes in the sky; also known as *angular size* or *apparent diameter.* The Moon, with an actual diameter of 3,476 km, has an angular diameter of 29′ 21″ to 33′ 30″, depending on its distance from Earth. If both angular diameter and distance are known, *linear diameter* can be easily calculated.

angular momentum

The momentum that an object or system of objects has because of its rotation. Since angular momentum is a conserved quantity in physics, the angular momentum of an orbiting body must stay the same at all points in the orbit. Orbital angular momentum is given by multiplying together a body's mass (m), its orbital **angular velocity** (v), and the distance (r) from the body around which it is moving. Since m is constant, v increases as r decreases (and vice versa) in an elliptical orbit. This is why planets in the solar system, for example, travel faster at perihelion than they do at aphelion. Angular momentum is also conserved for an object spinning on its own axis (like a pirouetting ice skater), which explains why stars spin more slowly as they expand and faster as they contract (although any mass loss will carry away some angular momentum).

angular separation

The angle, measured from the observer, between two celestial objects, such as the components of a binary star system.

angular velocity

The rate of rotation of an object, either about its axis or in its orbit about another body. It may be measured in revolutions or degrees per unit time.

anisotropy

A variation in a property with direction. The **cosmic microwave background,** for example, shows a slight anisotropy in temperature of a few microkelvins from one tiny patch of sky to another.

annihilation radiation

Radiation produced when a particle and its **antiparticle** collide and destroy each another. In the case of collisions between electrons and positrons (antielectrons), **gamma rays** with a characteristic energy of 511 keV are produced.

annual variation

The changes in the **right ascension** and **declination** of a star due to annual **precession** and **proper motion.**

anomalous X-ray pulsar (AXP)

A type of **X-ray pulsar** with a very long axial rotation period (for a pulsar), of 6 to 12 seconds, combined with a very powerful X-ray emission that cannot be explained by such a low spin rate. Two main theories exist to explain the anomalously powerful X rays. In the first model, bits of gas blown off in the **supernova** explosion that created the pulsar fall back onto the remnant star, whose magnetic field is assumed to be no stronger than an ordinary pulsar's. As the gas collects on the surface, it becomes hot and emits X rays. In the second model, which is now strongly supported by observation, AXPs are assumed to be **magnetars**–neutron stars with ultra-strong magnetic fields.

anomaly

An angle that gives the position of an object in an elliptical orbit at any given time. The *true anomaly* is the angle between the **periapsis** of an orbit and the object's current orbital position, measured from the body being orbited and in the direction of orbital motion. The *mean anomaly* is what the true anomaly would be if the object orbited in a perfect circle at constant speed. The mean anomaly is 0° at periapsis and 180° at apoapsis, just as for the true anomaly, but at other points along the orbit the values of the mean and true anomalies differ. The mean anomaly at a given time is often used as an **orbital element**. The *eccentric anomaly* is an angle related to both the true anomaly and the mean anomaly that is encountered when solving **Kepler's equation**.

anorthosite

A coarse-grained **igneous** rock made largely of **plagioclase feldspar** (95%), with small amounts of **pyroxene** (4%), **olivine**, and iron oxides. Anorthosite makes up about 60% of Earth's crust. It was found in all the rocks returned from the Moon, including the oldest (dating back 4.4 to 4.5 billion years), and is believed to make up a significant fraction of the lunar crust.

ansa (pl. ansae)

(1) A portion of a ring that appears farthest from the disk of a ringed planet in an image. (2) One of the extremities of a **lenticular galaxy**. The word comes from the Latin for "handle," since the earliest views of Saturn's rings suggested that the planet had two handles extending on either side.

Ant Nebula (Menzel 3)

A remarkable bipolar or "butterfly" **planetary nebula** in **Norma**, whose outflow speeds, of up to 3.5 million km/hr, surpass those of any other known object of its type. Although similar in appearance to another bipolar planetary, the **Butterfly Nebula** (M2-9), the Ant has an outflow pattern that more closely resembles that of the bizarre, unstable star **Eta Carinae**. Two main theories contend to explain the intriguing double-lobe symmetry of the ejected nebula. One is that the central star of the Ant has a close companion whose strong gravitational force shapes the gas flowing out from the primary. If this is the case, the companion must be so close, at about the distance of Earth from the Sun, that it may actually orbit inside its partner's bloated atmosphere. The other theory is that as the dying star spins, its powerful magnetic field is twisted into complex shapes like strands of spaghetti. The outflowing gas is then forced to trace out these intricate patterns, making them visible in the process.

Antares (Alpha Scorpii, α Sco)

The brightest star in **Scorpius** and the sixteenth brightest star in the sky, a red **supergiant** and **irregular variable**. Its name means "rival of Mars" (Ares and Mars being, respectively, the Greek and Roman gods of war), a reference to its similar appearance to the Red Planet. In fact, since Antares lies within a zodiacal constellation, which contains the apparent path of the Sun and the planets, it is commonly mistaken for Mars. With a diameter of 800 million km, it would, if put in place of the Sun, reach about halfway to Jupiter. A fierce **stellar wind** blowing from the surface of Antares has resulted in a circumstellar gas cloud, which is illuminated by the light from a hot B-type companion star that, at fifth magnitude, hides within the supergiant's bright glare (separation 3″, period 900 years).

Visual magnitude:	0.9 to 1.1
Absolute magnitude:	–5.28
Spectral type:	M1Ib + B4V
Surface temperature:	3,400 K (primary)
Luminosity:	40,000 L_{sun}
Distance:	604 light-years

Antennae (NGC 4038 and NGC 4039)

The nearest and youngest example of a pair of colliding galaxies; the Antennae lie about 63 million light-years away in **Corvus**. Each galaxy's gravitational pull has drawn out a curved tail of stars from the other. These structures, shown in the ground-based photo, span a total of some 360,000 light-years and have been produced by a collision between the galaxies that began about 100 million years ago and is still in progress. In the more detailed image of the heart of the Antennae, captured by the Hubble Space Telescope, bright knots made up of over 1,000 young star clusters bursting to life are visible. These are infant **globular clusters**, newly born out of collisions between giant hydrogen clouds in the two galaxies. At the other end of the stellar evolutionary spectrum, X-ray images of the central regions of the galaxies, taken by the **Chandra X-ray Observatory**, have revealed dozens of bright point-like sources that are probably neutron stars or black holes tearing gas off nearby stars. These collapsed objects are the remains of large stars that formed earlier in the burst of star formation triggered by the collision and that have already died.

anthropic principle

The idea that the existence of life and, in particular, our presence as intelligent observers, constrains the nature of the universe. It was first articulated in 1961 by Robert **Dicke**, who argued that the presence of advanced carbon-based life required the universe to be roughly the age it is found to be–about 15 billion years. If it was much younger, there wouldn't have been time for sufficient interstellar levels of carbon to build up by **nucleosynthesis**. If it was much older, the "golden" age of main sequence stars and stable planetary systems would have drawn to a close. In 1973, the concept was given its now-familiar name and raised to prominence through the efforts of Cambridge physicist Brandon Carter, encouraged by a pioneer of quantum mechanics, John Wheeler. Carter distinguished between two degrees of the hypothesis: the *weak anthropic principle,* which was essentially a generalization of Dicke's idea, and the *strong anthropic principle,* which went much further and claimed that "the universe . . . must be such as to admit the creation of observers within it at some stage." It was not long before Wheeler found a crucial role for those observers in his *participatory anthropic principle.* Somehow, suggested Wheeler, we play a part in actualizing the world in which we find ourselves. Our observations today, at a quantum level, constrain the universe so that it had to have evolved in such a way that would eventually give rise to us. Needless to say, these remain controversial claims.

antimatter

Matter composed of **antiparticles**. Although it has been speculated that antimatter galaxies might exist whose total mass would help produce a matter-antimatter balance in the universe, this idea lacks observational support. Standard cosmological theory accounts for the apparent preponderance of ordinary matter in the cosmos in terms of a symmetry-breaking shortly after the **Big Bang**.

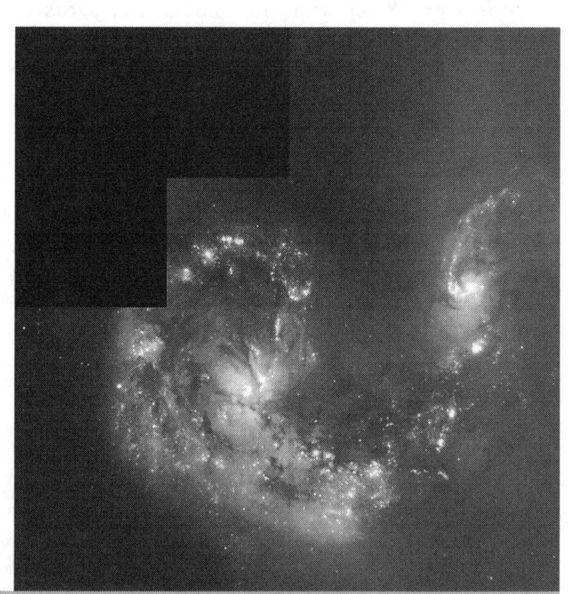

Antennae The photo on the left provides a sweeping view of the Antennae, while the photo on the right is a close-up by the Hubble Space Telescope of the activity at the center of this colliding pair of galaxies. The cores of the twin galaxies are the orange blobs, left and right of center; these are crisscrossed by filaments of dark dust. A wide band of chaotic dust stretches between the galactic cores. The sweeping spiral-like patterns, traced by bright blue star clusters, are the result of a firestorm of star birth that was triggered by the collision. *NASA/ESA/Brad Whitmore (STScI)*

Antiope (minor planet 90)

The first **asteroid** shown to be a binary system; its two components, each about 85 km across, are separated by about 170 km and complete an orbit around each other every 16.5 hours. Antiope was discovered in 1866 by the German astronomer R. Luther and is a member of the **Themis family**. Spectral class C, albedo 0.05. Orbit: semimajor axis 3.155 AU, perihelion 2.665 AU, aphelion 3.645 AU, eccentricity 0.155, inclination 2.2°, period 5.60 years.

antiparticle

A counterpart of an ordinary subatomic particle, which has the same mass and spin but the opposite charge. Certain other properties are also reversed, including the magnetic moment. An encounter between a particle and its corresponding antiparticle–for example, an electron and a positron–results in their mutual annihilation.

antitail

Part of the dust tail of a **comet** that seems to point, often like a spike, toward the Sun. This rare phenomenon is an illusion caused by the viewing geometry and typically occurs when Earth crosses the plane of a comet's orbit when the comet is relatively close to the Sun. Under these circumstances, the cometary dust, which lies in a thin sheet and lags behind the comet, may be seen edge-on as an antitail.

Antlia (abbr. Ant; gen. Antliae)

The Air Pump; a small, obscure southern **constellation**. Originally called *Antlia Pneumatica* and thought to represent the pump invented by English physicist Robert Boyle, it lies slightly north of Vela. Its brightest star, Alpha Ant, is a giant **K star** (visual magnitude 4.28, absolute magnitude –0.98, spectral type K4III, distance 366 light-years). Among its interesting deep-sky objects are the planetary nebula NGC 3132 (R.A. 10h 4.9m, Dec. –40° 11′) which is similar in appearance to the **Ring Nebula** but a magnitude brighter, and the **Antlia Dwarf Galaxy**. (See star chart 4.)

Antlia Dwarf Galaxy

A small **dwarf spheroidal galaxy**, of low surface brightness, that lies about 4 million light-years away and is an outlying member of the **Local Group**. It was only discovered in 1997. The Antlia Dwarf seems to be dominated by an old stellar population, similar to what has been found for most other dwarf spheroidal galaxies in the Local Group. However, unlike any of these, Antlia seems to contain a cloud of atomic hydrogen gas at its center with a mass of about 800,000 M_{sun}. It remains to be understood why a galaxy that is relatively gas rich has no young or intermediate-age stars.

Antlia Dwarf Galaxy The Very Large Telescope captured this detailed view of the Antlia Dwarf, along with many background galaxies, in 1999. *European Southern Observatory*

Antoniadi, Eugène Michael (1870–1944)

A Greek-French astronomer, born in Turkey, who became one of the leading planetary observers of his time. His specialty was **Mars** and in his magnum opus, *La Planète Mars* (1930), he presented a state-of-the-art summary of Martian topography that helped set the scene for the modern investigation of the planet. Early in his career, he worked at Camille **Flammarion**'s observatory at Juvisy and claimed to have seen Martian canals. As late as 1903, he maintained their "incontestable reality," although he believed them to be natural. So great had his reputation as an observer become that by the 1909 **opposition** of Mars he was given access to the largest reflector in Europe, the 83-cm refractor at the Meudon Observatory. With this, he was able to resolve the canals into streaks or borders of darker regions and concluded, "The geometrical canal network is an optical illusion." Quickly, observers in the United States, including George **Hale**, confirmed Antoniadi's opinion, and worldwide belief in the canal hypothesis went into rapid decline. Antoniadi was also a regular observer of **Mercury** and **Venus**, and his *La Planète Mercure et la rotation des satellites* (1934) was the only work on the subject for two decades. In it, he published the most detailed pre–Space Age map of Mercury based on the assumption, first made by Giovanni **Schiaparelli**, that the planet always kept the same face toward the Sun–an assumption now known to be false. He also produced detailed drawings showing spots on

Saturn and spokes in its rings, and the South Tropical Disturbance on **Jupiter**.

Antoniadi scale

A five-point scale, devised by Eugène **Antoniadi** and used by amateur astronomers to indicate the quality of the seeing: I–perfect seeing, without a quiver; II–slight undulations, with moments of calm lasting several seconds; III–moderate seeing, with larger tremors; IV–poor seeing, with constant troublesome undulations; V–very bad seeing, scarcely allowing the making of a rough sketch.

AO Cassiopeiae (AO Cas)

One of the most massive and luminous binary systems known, consisting of two **O stars** of 32 M_{sun} and 23 R_{sun}, and 30 M_{sun} and 15 R_{sun}. The two stars orbit so closely, with a period of only 3.5 days, that their surfaces are almost touching and highly distorted, and there is a continuous exchange of matter. AO Cas, also known as *Pearce's Star,* lies about 7,000 light-years away.

Ap star

An **A star** whose spectrum has unusually strong lines of some ionized metals and rare earths, pointing to a vast overabundance (10^3 to 10^6 solar values) of these elements in the star's surface layers. More generally, the term Ap star, or *peculiar A star,* has come to encompass a range of chemically anomalous stars roughly between spectral types B5 (see **Bp star**) and F5 (see **Fp star**). The elements in overabundance vary from one Ap star to another and may include manganese, mercury, silicon, chromium, strontium, europium, and others. Ap stars typically have surface temperatures of 8,000 to 15,000 K, strong magnetic fields, and low rotational rates–properties that help explain their observed chemical anomalies. The separation of elements is enabled by the slow spin and the relatively high temperature, and hence lack of **convection**. Separation happens because each ion has its own photoabsorption characteristics. If a certain element absorbs photons (light particles) more easily, it will tend to be pushed to the surface and become overabundant. Otherwise it will sink under the force of gravity and appear depleted in the star's spectrum. The strength of the magnetic field also plays a part in determining which elements are overabundant as shown by the fact that **manganese stars**–similar to Ap stars but without a strong magnetic field–have anomalies of the same order of magnitude but often not for the same elements. Variations in the spectrum of many Ap stars, associated with magnetic variations, can be understood in terms of the **oblique rotator** model.

Apache Point Observatory (APO)

An observatory located at an altitude of 2,780 m in the Sacramento Mountains, near Sunspot, New Mexico. It is privately owned and operated by the Astrophysical Research Consortium, whose members include several American universities. APO houses a 3.5-m telescope, the 2.5-m Sloan Digital Sky Survey telescope, and two smaller telescopes.

apastron

The point of greatest separation between two stars that are in orbit around each other; the converse of *periastron.*

aperture

The effective diameter of a telescope's **objective** lens or **primary mirror**. The larger the aperture, the greater the ability to collect light and detect faint distant objects. A rough guide for estimating the magnification that an aperture can handle well is 2× per mm, or 50× per inch.

aperture synthesis

A method of combining signals from a collection of individual antennae or telescopes to provide an image with a **resolution** equivalent to that of a single telescope with a size roughly equal to the maximum distance between the individual antennae.

aphelion

The point in a solar orbit that is farthest from the Sun.

apoapsis

The point in an orbit at which two objects are farthest apart. Special names, including **apastron**, **aphelion**, and **apogee**, are given to this orbital point for familiar systems.

apochromat(ic)

A lens or optical system virtually free of **chromatic aberration**, which for practical purposes means that light of at least three different wavelengths is brought to focus at the same point. The best apochromatic lenses use fluorite crystal and may correct three wavelengths with only two optical elements. However, because fluorite is expensive to manufacture, and, because of its brittleness, is difficult to grind, polish, and mount, high quality apochromatic **refracting telescopes** are costly. **Reflecting telescopes**, on the other hand, are apochromatic in performance without the extra expense.

apogee

The point in a geocentric orbit at which an object is farthest from Earth.

apohele

An **asteroid** whose orbit lies entirely inside that of Earth; the name–an unofficial one proposed by David Tholen of the University of Hawaii–is Hawaiian for "orbit" and pronounced "ah-poe-hay-lay." Although membership of this group (like that of the **vulcanoids**) remains hypothetical, it is reasonable to suppose that, on occasions, a **near-Earth asteroid** ends up as an apohele through successive perturbations by Earth and Venus. One possible candidate, discovered by Tholen and Robert Whiteley in February 1998 but only observed on two consecutive days and subsequently lost due to failure of the observing instrument, is the 40-m-wide asteroid cataloged as 1998 DK36.

Apollo (minor planet 1862)

The prototype asteroid of the **Apollo group**. It was discovered by the German astronomer Karl Reinmuth (1892–1979) in 1932, when it approached Earth to within 10.5 million km (0.07 AU), but was then lost until 1973. Apollo comes as close to Earth as 4.2 million km (0.028 AU) and makes near passes of Venus and Mars, whose orbits it crosses at perihelion and aphelion, respectively. Diameter 1.4 km, spectral class Q, rotational period 3.063 hours. Orbit: semimajor axis 1.486 AU, perihelion 0.65 AU, aphelion 2.30 AU, eccentricity 0.57, inclination 6.4°, period 622 days.

Apollo group

Also known as *Earth-crossing asteroids,* one of three groups of **near-Earth asteroids**; they have semimajor axes greater than 1.0 AU and perihelia less than 1.017 AU (Earth's aphelion distance). Some Apollo objects can approach the Sun at a distance closer than Mercury can, the record-holder being 1995 CR with a perihelion distance of 0.12 AU. Other notable members of the group include (1862) **Apollo** (the prototype), (1866) Sisyphus (the largest, with a diameter of about 8 km), (3200) **Phaethon,** (1685) **Toro**, and (4179) **Toutatis**.

Apollonius of Perge (c. 262–c. 190 B.C.)

A Greek mathematician, born in Turkey, who, in his famous book *Conics,* wrote about the parabola, ellipse, and hyperbola–the conic sections that describe the shapes of various types of orbit. Apollonius also helped found Greek mathematical astronomy. **Ptolemy** says in his *Syntaxis* that Apollonius introduced the theory of **epicycles** to explain the apparent motion of the planets across the sky. Although this isn't strictly true, since the theory of epicycles was mooted earlier, Apollonius did make important contributions, including a study of the points where a planet appears stationary. He also developed the *hemicyclium,* a sundial with hour lines drawn on the surface of a conic section to give greater accuracy. See also **Greek astronomy**.

apparent magnitude

A measure of the *observed* brightness of a celestial object; it depends on an object's actual (intrinsic) brightness, its distance from the observer, and, in the case of objects outside the solar system, the amount of absorption by intervening matter. The brighter an object appears, the smaller the numerical value of its apparent magnitude. A star that is one magnitude brighter than another (e.g., +1 versus +2) looks 2.5 times brighter. Among the brightest objects in the sky are the Sun (apparent magnitude –26.7), the full Moon (–12.6), Venus (at brightest, –4.7), Sirius (–1.44), Canopus (–0.62), Alpha Centauri (–0.27), and Arcturus (–0.05). On a clear, dark night, the unaided eye can see stars as faint as apparent magnitude +6. Unless otherwise qualified, the term is normally taken to mean apparent *visual* magnitude. See also **magnitude**.

apparition

The period during which an object in the solar system, such as a planet or comet, is visible from Earth. The term is not normally used in connection with the Sun, the Moon, or the stars.

apsis (pl. apsides)

Either of the points in the orbit of a celestial object that is farthest from or closest to the body being orbited. The closest point is the *periapsis,* the further point the *apoapsis.* The line joining these points is called the *line of apsides,* and is another name for the **major axis** of the orbit. Rotation of the line of apsides in the plane of the orbit is known as *apsidal motion* or **advance of perihelion** (in the case of a planet orbiting the Sun) or *advance of pericenter* (in the case of a binary star).

Apus (abbr. Aps; gen. Apodis)

The Bird of Paradise, originally called *Avis Indica;* a small, faint **constellation** lying near the south celestial pole, immediately below the **Southern Triangle**. (See table,

Apus: Stars Brighter than Magnitude 4.0

	Magnitude		Spectral	Distance	Position	
Star	Visual	Absolute	Type	(light-yr)	R.A. (h m s)	Dec. (° ′ ″)
α	3.83	−1.68	K5III	411	14 47 52	−79 02 41
γ	3.86	0.41	K0IV	160	16 33 27	−78 53 49

"Apus: Stars Brighter than Magnitude 4.0," and star charts 4 and 5.)

Aquarius (abbr. Aqr; gen. Aquarii)

The Water-bearer; a large but quite faint southern **constellation** and the twelfth sign of the **zodiac**. It is surrounded by Pegasus, Equuleus, and Delphinus to the north, Aquila to the east, Pisces Austrinus and Sculptor to the south, and Cetus to the west. The stars Sadachbia (Gamma), Eta Aqr, Zeta Aqr, and Pi Aqr form a small Y-shaped asterism, called the Water Jar or Urn, thought to mark the Water-bearer's cup. (See tables, "Aquarius: Stars Brighter than Magnitude 4.0," and "Aquarius: Other Objects of Interest.")

Aquarius Dwarf

A **dwarf irregular galaxy**, or, possibly, a dwarf spheroidal galaxy, that is a member of the **Local Group**. It lies about 2.6 million light-years from the Milky Way Galaxy in **Aquarius**.

Aquila (abbr. Aql; gen. Aquilae)

The Eagle (of Zeus); a distinctive **constellation** on the celestial equator that spreads across many rich star fields. Due to a lane of obscuring dust known as the **Great Rift**, the Milky Way splits in two through this constellation and also through **Ophiuchus**. Aquila has been a fruitful hunting ground for **nova**-seekers, producing four since 1899, including two of the brightest on record. One of these, in

Aquarius: Stars Brighter than Magnitude 4.0

Star	Magnitude		Spectral	Distance	Position	
	Visual	Absolute	Type	(light-yr)	R.A. (h m s)	Dec. (° ′ ″)
β Sadalsuud	2.90	–3.47	G0Ib	612	21 31 33	–05 34 16
α Sadalmelik	2.95	–3.88	G2Ib	759	22 05 47	–00 19 11
δ Skat	3.27	–0.18	A3V	159	22 54 39	–15 49 15
ζ	3.65	1.14	F6IV+F3V	103	22 28 50	–00 01 12
c^2	3.68	–0.60	K1III	234	23 09 27	–21 10 21
λ	3.73	–1.67	M2.5IIIaFe	392	22 52 37	–07 34 47
ε Albali	3.78	–0.46	A1V	230	20 47 40	–09 29 45
γ Sadachbia	3.86	0.43	A0V	158	22 21 39	–01 23 14
b^1	3.96	0.48	K0III	162	23 22 58	–20 06 02

Aquarius: Other Objects of Interest

Name	Notes
Stars	
R Aqr	See **symbiotic variable.**
V1344 Aqr	See **X-ray jet variable.**
Planetary nebulae	
Helix Nebula	NGC 7293. See separate entry.
Saturn Nebula	NGC 7009. See separate entry.
Globular clusters	
M2 (NGC 7089)	One of the most impressive of its type; it appears in binoculars as a misty patch with a central concentration, and with a 15-cm telescope is partially resolved into stars. Magnitude 6.5, diameter 12.9′, distance 37,500 light-years; R.A. 21h31m, Dec. –1° 3′.
M72 (NGC 6981)	A faint object in small telescopes. Magnitude 9.8; diameter 2′; distance 55,400 light-years; R.A. 20h 52m, Dec. –12° 39′.
Galaxies	
Aquarius Dwarf	See separate entry.
Atoms for Peace Galaxy	NGC 7252. See separate entry.

Aquila: Stars Brighter than Magnitude 4.0

Star	Magnitude		Spectral	Distance	Position	
	Visual	Absolute	Type	(light-yr)	R.A. (h m s)	Dec. (° ′ ″)
α **Altair**	0.76	2.20	A7V	16.8	19 50 47	+08 52 06
γ Tarazed	2.72	–3.03	K3II	460	19 46 15	+10 36 48
ζ Dheneb	2.99	0.95	A0Vn	83	19 05 24	+13 51 48
θ	3.24	–1.49	B9.5III	287	20 11 18	–00 49 17
δ	3.36	2.42	F3IV	50	19 25 30	+03 06 53
λ Althalimain	3.43	0.51	B9Vn	125	19 06 15	–04 52 57
β Alshain	3.71	3.02	G8IV	45	19 55 19	+06 24 24
η **Eta Aquilae**	3.87	–3.91v	F6Ibv	1,170	19 52 28	+01 00 20

Aquila: Other Objects of Interest

Name	Notes
Star	
GRS 1915+105	A star system containing the most massive known stellar **black hole**.
Planetary nebulae	
NGC 6803	A small bright ring. Magnitude 11.4; diameter 6″; R.A. 19h 29m, Dec. +9° 58′.
NGC 6891	A bright disk and faint ring. Magnitude 12; diameter 1.2′; R.A. 20h 13m, Dec. +12° 35′.
Open cluster	
NGC 6709	A loose cluster arranged in a diamond formation with a bright star at each apex; Magnitude 6.7; diameter 13′; R.A. 18h 49m, Dec. +10° 17′.

A.D. 389, became as bright as Venus; the other, in 1918, outshone **Altair**. (See tables, "Aquila: Stars Brighter than Magnitude 4.0," and "Aquila: Other Objects of Interest." See also star chart 14.)

Ara (abbr. Ara; gen. Arae)

The Altar; a small southern **constellation** that is rich in star clusters. (See tables, "Ara: Stars Brighter than Magnitude 4.0," and "Ara: Other Objects of Interest." See also star chart 14.)

Arabian astronomy

Following **Ptolemy**, **Greek astronomy** rapidly declined and ended with the Arabian conquest of Alexandria in A.D. 641. Although the magnificent library and museum were destroyed, the Arabs encouraged learning and for the next 800 years developed an important astronomical tradition of their own. Observatories were established at a number of cities including Damascus, Cairo, Baghdad, and Meragha. One of the greatest stimuli to Arabian astronomy was the need to calculate and maintain the Islamic **calendar**, which demanded new mathematical methods and more precise timekeeping. Among the greatest Arabic astronomers were Al-Farghani (?–c. 861), **Albategnius** (Al-Battani, Muhammad ibn Jabir) (c. 850–929); **Al-Sufi**, Abd al-Rahman (903–986); Abu'l-Wafa', Mohammed Al-Buzjani (940–998); Al-Quhi, Abu Sahl Wayjan ibn Rustam (c. 940–c. 1000); **Alhazen** (Abu Ali al Hassan ibn al Haitham) (c. 965–c. 1040), **Arzachel**, Al-Zarqali, Abu Ishaq Ibrahim ibn Yahya (1028–1087); Abraham bar Hiyya Ha-nasi (c. 1065–c. 1136), and Alpetragius (?–c. 1204).

Arago, (Dominique) François (Jean) (1786–1853)

A French physicist and astronomer, director of the Paris Observatory (1843–1853), and secretary of the French Academy of Sciences, who through his lectures and writings did much to popularize astronomy. Working with Augustin Fresnel (1788–1827), he discovered that two beams of light polarized in perpendicular directions do not interfere, leading to the transverse theory of light waves. In 1811 he invented the polariscope and later developed a **polarimeter**, which he used in several astronomical studies. During the solar eclipse of 1842, he examined

Ara: Stars Brighter than Magnitude 4.0

Star	Magnitude		Spectral	Distance	Position	
	Visual	Absolute	Type	(light-yr)	R.A. (h m s)	Dec. (° ′ ″)
β	2.84	–3.50	K3Ib-IIa	603	17 25 18	–55 31 47
α Choo	2.84	–1.52	B2Vne	242	17 31 50	–49 52 34
ζ	3.12	–3.11	K3III	574	16 58 37	–55 59 24
γ	3.31	–4.40	B1Ib	1,140	17 25 24	–56 22 39
δ	3.60	–0.20	B8Vn	187	17 31 06	–60 41 01
θ	3.65	–3.81	B2Ib	1,010	18 06 38	–50 05 30
η	3.77	–1.14	K5III	313	16 49 47	–59 02 29

Ara: Other Objects of Interest

Name	Notes
Open clusters	
NGC 6193	A good object for binoculars, containing about 30 stars. Magnitude 5.2; diameter 15′; R.A. 16h 38m, Dec. –48° 40′.
IC 4651	Richer but fainter than NGC 6193 and requires a small telescope. Magnitude 7.1; diameter 12′; distance 3,600 light-years; R.A. 17h 21m, Dec. –49° 54′.
Globular cluster	
NGC 6397	The brightest of three easy globulars in Ara; it can be glimpsed with the naked eye and looks like a misty patch in small instruments. At a distance of about 7,500 light-years, it may be the second closest globular to Earth after **M4**. Magnitude 5.6; diameter 25.7′; R.A. 17h 37m, Dec. –53° 39′.

polarized light from the chromosphere and corona, and determined that the Sun's limb is gaseous. Arago suggested that his student Urbain **Leverrier** investigate irregularities in the orbit of Uranus and, after Neptune was discovered, took part in arguments regarding naming the planet and, with John **Adams**, regarding priority.

Arago point

(1) One of the three commonly detectable points, called *neutral points,* along the vertical circle through the Sun at which the degree of **polarization** of diffuse sky radiation goes to zero. The Arago point (also known as the *Arago spot*) typically lies about 20° above the antisolar point (where an imaginary ray connecting the Sun and the observer meets the sky), but is found at higher altitudes in turbid air. The latter property makes the Arago distance a useful measure of atmospheric turbidity. The other two neutral points are called the *Babinet point* (15 to 20° directly above the Sun, hence difficult to detect because of solar glare) and the *Brewster point* (15 to 20° directly below the Sun). (2) In optics, a bright spot that appears in the center of the shadow of a circular disk illuminated by a point source, caused by **diffraction**.

archeoastronomy

The study of nonwritten evidence for astronomy practices and knowledge in past civilizations and societies. Artifacts of interest to the archeoastronomer include megalithic remains in northwestern Europe, especially stone circles such as Stonehenge, and rock art, depicting events such as supernovae, by Native Americans and Australian Aborigines.

Arches Cluster

A cluster of about 150 hot, young stars concentrated within a radius of about 1 light-year, making it the most compact cluster of stars in our Galaxy. It lies some 25,000 light-years away near the galactic center and is named after a series of arch-shaped filaments, detected at radio wavelengths, that lie in its vicinity. X-ray observations by the Chandra X-ray Observatory have shown an envelope of 60-million-degree gas around the cluster that is thought to be heated by intense stellar winds from the member

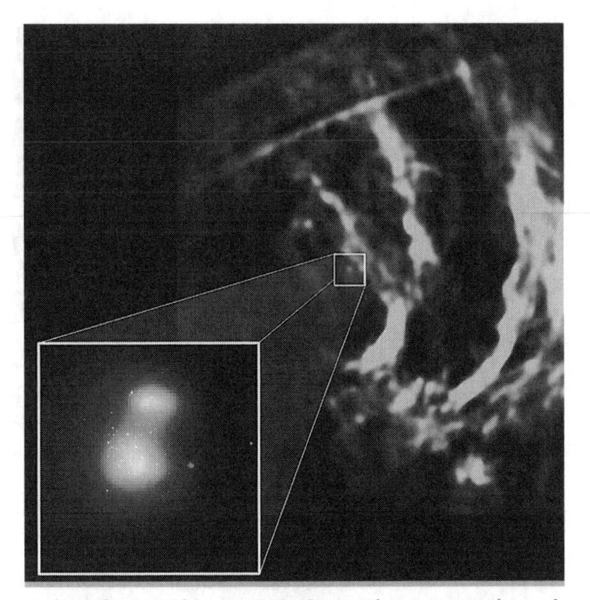

Arches Cluster This composite image shows an envelope of 60-million-degree gas around the Arches Cluster. Observations from the Chandra X-ray Observatory, shown as the diffuse emission in the inset box, overlays a Hubble Space Telescope infrared image of the same region, in which some of the individual stars in the cluster can be seen as pointlike sources. Both the X-ray and infrared observations are shown in context of the spectacular filamentary structures that appear in radio wavelengths and that give the cluster its name. The radio data were obtained using the Very Large Array. *NASA/STScI/NRAO*

stars colliding with the surrounding medium. Studies of the Arches Cluster can be used to learn more about the environments of starburst galaxies in which this process occurs on a much larger scale.

Arcturus (Alpha Boötis, α Boö)

The brightest star in **Boötes** and in the entire northern hemisphere, and the fourth brightest in the sky; its Greek name means the "bear watcher," in reference to the star's apparent pursuit of **Ursa Major** around the pole. An orange subgiant **K star**, Arcturus has a velocity somewhat higher than that of other bright stars, and comes from an older population of the Galaxy. Consistent with this, it is rather metal-poor with only about one-fifth as much iron relative to hydrogen as is found in the Sun. It is probably at the stage of fusing helium to carbon in its core. Arcturus became famous when its light was gathered by a photocell and used to switch on the 1933 World's Fair in Chicago–this light having started its journey at about the time of the previous Chicago fair in 1893. Arcturus was one of the first stars to have its **proper motion** measured.

Visual magnitude:	–0.05
Absolute magnitude:	–0.30
Spectral type:	K2IIIp
Surface temperature:	4,290 K
Luminosity:	180 L_{sun}
Radius:	23 R_{sun}
Mass:	1.5 M_{sun}
Distance:	37 light-years

Arecibo Observatory

The site of the world's largest single-dish **radio telescope**, opened in 1963. The observatory is located 19 km south of Arecibo, Puerto Rico, and is operated by the National Astronomy and Ionosphere Center, whose headquarters are at Cornell University. The surface of the telescope's 305-m fixed spheroidal dish is made from nearly 40,000 perforated 1- by 2-m aluminum panels. These are supported by a network of steel cables strung across a natural karst sinkhole. Suspended 150 m above the reflector is a 900-ton platform that houses the receiving equipment. Although the telescope is not steerable, sources within 20° of the zenith can be tracked by moving the feed antenna, which was upgraded in 1996. The Arecibo telescope is used for radio astronomy, SETI, and atmospheric studies.

Arend-Roland, Comet (C/1956 R1)

A **long-period comet** discovered on November 8, 1956, by Belgian astronomers Silvain Arend (1902–1992) and Georges Roland (1922–1991). It reached perihelion (0.32 AU) on April 8, 1957, passed within 0.57 AU of Earth on April 21, and reached a maximum brightness of magnitude –1. Comet Arend-Roland developed a tail up to 30° long in photographs and a prominent spikelike **antitail**. Its orbit was hyperbolic (eccentricity 1.0002) with an inclination of 119.9°.

areo-

Pertaining to Mars (Ares). *Areobiology* is the study of possible life on Mars. *Areocentric* describes a coordinate system centered at Mars. *Areographic* refers to positions on Mars measured in latitude from the planet's equator and in longitude from a reference meridian. *Areography* is the study of Martian surface features.

Argelander, Friedrich Wilhelm August (1799–1875)

A German astronomer who established the study of **variable stars** as an independent branch of astronomy and is renowned for his great catalog, the ***Bonner***

Dürchmusterung, listing all stars in the northern hemisphere above ninth magnitude. He later initiated the first ***Astronomische Gesellschaft Katalog***. Argelander also devised a way of estimating stellar magnitudes, initiated the subdivision of magnitudes into tenths, and began the practice of assigning capital letters to variable stars.

Argo Navis

An enormous southern **constellation**, representing Argos, the ship of Jason and the Argonauts, listed by **Ptolemy** among his original 48. In the eighteenth century, it was divided by Nicolaus **Lacaille** into **Carina** (the keel), **Puppis** (the stern), and **Vela** (the sails). There was no reassignment of ***Bayer designations*** when this took place, so, for example, Puppis lacks stars labeled Alpha, Beta, and Gamma.

argument of perihelion (ω)

The angle between the **ascending node** and the **perihelion** of an orbit around the Sun, measured in the orbital plane and in the direction of orbital motion. The argument of perihelion specifies the direction of the **major axis** of an orbit around the Sun and is one of the **orbital elements**.

Ariel

The twelfth nearest, the fourth largest, and the brightest moon of **Uranus**, also known as Uranus I; it was discovered by William **Lassell** in 1851. Ariel orbits at a mean distance of 190,930 km from the planet and has a diameter of 1,158 km. Its surface is a mixture of heavily cratered terrain, smooth plains, and systems of interconnected valleys, called *chasmata*, that are hundreds of kilometers long and more than 10 km deep. Some ridges in the middle of the valleys are probably due to upwellings of ice, while the partial submergence of some craters indicates past flooding. Although Ariel may have been internally hot and active long ago, perhaps as a result of tidal action, it is cold and geologically dead today.

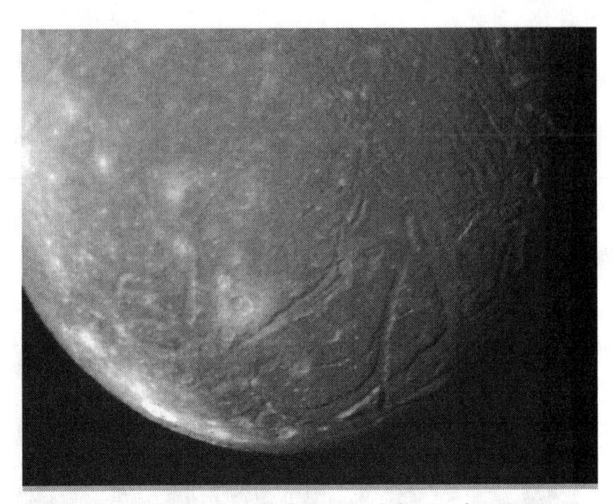

Ariel The southern hemisphere of Ariel, seen by Voyager 2 from a distance of 170,000 km on Janury 24, 1986. Most of the visible surface consists of relatively intensely cratered terrain crisscrossed by fault scarps and fault-bounded valleys (graben). Some of the largest valleys, which can be seen near the terminator (at right), are partly filled with younger deposits that are less heavily cratered. Bright spots near the limb and toward the left are chiefly the rims of small craters. *NASA/JPL*

Aries (abbr. Ari; gen. Arietis)

The Ram, whose golden fleece inspired Jason's epic voyage; a northern **constellation** and the first one of the **zodiac**, it lies between Pisces and Taurus, and appears as a small flat triangle below Andromeda, climbing the eastern sky in the northern autumn. The Sun passes through Aries from late April to mid-May. In former times the vernal **equinox** (the point where the Sun crosses the celestial equator when moving from south to north) lay within Aries and is still known as the *First Point of Aries*, even though, because of **precession**, it now lies in Pisces. (See table, "Aries: Stars Brighter than Magnitude 4.0," and star charts 1 and 6.)

Aries: Stars Brighter than Magnitude 4.0

	Magnitude		Spectral	Distance	Position	
Star	Visual	Absolute	Type	(light-yr)	R.A. (h m s)	Dec. (° ′ ″)
α **Hamal**	2.01	0.48	K2IIIabCa	66	02 07 10	+23 27 45
β Sheratan	2.64	1.33	A5V	60	01 54 38	+20 48 29
c Nair al Butain	3.61	0.16	B8Vn	159	02 49 59	+27 15 38
γ Mesarthim	3.88	–0.11		204		
γ^2	4.75		A1pSi		01 53 32	+19 17 45
γ^1	4.83		B9V		01 53 32	+19 17 37

Aristarchus of Samos (c. 310–c. 230 B.C.)
A Greek mathematician and astronomer who was the first to propose a heliocentric theory, with Earth revolving around the Sun. To explain the lack of observed stellar **parallax**, he argued that the stars must lie very far away. However, his new scheme for the solar system found little support at a time when **Aristotle**'s geocentric model was in favor and the idea of a moving Earth was frowned upon. Aristarchus also applied modern geometric methods to measure the size of celestial bodies. From a lunar eclipse, he concluded that the Moon's radius is half that of Earth (actually it is 0.28 times Earth's radius). He measured the Moon's angular diameter to be 2° (actually 0.5°) and calculated the Earth-Moon distance to be 114.6 Earth radii (actually 60.4). By noticing that the Sun and the Moon have equal angular diameters during a solar eclipse, he estimated that the Sun's distance from Earth was 19.1 times the Moon's distance from Earth (actually 390 times). See also **Greek astronomy**.

Aristotle of Stagira (384–322 B.C.)
A major Greek philosopher, student of Plato, and founder of the Lyceum in Athens. In his *De Caelo,* Aristotle deduced that Earth was spherical (from the shadow it cast during a lunar eclipse) and accepted the Earth-centered cosmic model of **Eudoxus of Cnidus** and Callipus with its concentric heavenly spheres carrying the Sun, the Moon, the planets, and the stars. Aristotle increased the number of spheres to 49 and devised a mechanical scheme to explain the observed movement of all the celestial bodies. This involved the idea of "reacting spheres" in which the outermost sphere, bearing the fixed stars, controlled the motion of the others and was itself controlled supernaturally. His worldview, modified by **Ptolemy**, who replaced the spheres with **epicycles**, was not seriously challenged for almost 2,000 years. See also **Greek astronomy**.

armillary sphere
An ancient Greek, Arabic, and medieval device used to show and observe the movements of astronomical bodies. The **celestial sphere** is modeled by a skeletal framework of rings (*armillaries,* from the Latin *armilla,* meaning "ring"), each representing an important circle, such as the **ecliptic** or celestial equator. Some of the rings may be moveable to enable the sky's appearance to be shown at different times and latitudes. Armillary spheres are also found in the form of equatorial **sundials**, in which the ring representing the celestial equator is graduated with hours.

Arp, Halton Christian (1927–)
An American astronomer noted for challenging the theory that the large **redshifts** of **quasars** and other **active galaxies** are an indication of great distance. His *Atlas of Peculiar Galaxies,* published in 1966, contains photographs of 338 peculiar and interacting galaxies taken with the **Hale Telescope**, and was instrumental in raising the issue of **discordant redshifts**. Arp received a B.S. from Harvard (1949) and a Ph.D. from the California Institute of Technology (1953).

Arrhenius, Svante August (1859–1927)
A Swedish physical chemist and Nobel Prize winner (1903), famous for his work on electrolytes, who was the first to present a detailed scientific hypothesis of *panspermia* (see **life in the universe**). In this, he argued that life arrived on Earth in the form of microscopic spores that had been propelled across interstellar space by the pressure of starlight. He published a seminal paper on the subject in 1903, followed by a popular book, *Worlds in the Making* (1908, first published in Swedish in 1906). In *The Destinies of Stars* (1918), he presented a carboniferous swamp version of **Venus** that remained in vogue for many years.

Aryabhata the Elder (A.D. 476–c. 550)
A Hindu astronomer and mathematician whose surviving masterpiece, *Aryabhatiya,* written in verse, contains one of the earliest treatments of algebra and some remarkably advanced astronomical knowledge. It argues that the planets move in elliptical paths, the apparent rotation of the heavens is due to Earth's axial rotation, and the Moon and planets shine by reflected sunlight. Aryabhata also correctly explained the causes of solar and lunar eclipses–improving upon the then-popular Indian belief that eclipses were the work of a demon called Rahu–and gave a value for the length of the year that is just a few hours longer than the one accepted today. See also **Indian astronomy**.

Arzachel (Al-Zarqali, Abu Ishaq Ibrahim ibn Yahya) (1028–1087)
The foremost Spanish Arab astronomer of his time. He carried out a series of observations at Toledo and compiled them as his famous *Toledan Tables,* correcting geographical data from **Ptolemy** and Al-Khwarizmi. The *Tables* were translated into Latin in the twelfth century. Arzachel was the first to prove conclusively the motion of the aphelion relative to the stars, measuring it as 12.04″ per year–remarkably close to the modern value of 11.8″. He also invented a flat **astrolabe**, known as a *safihah,* details of which were published in Latin, Hebrew, and several European languages. Copernicus, in his *De Revolutionibus Orbium Clestium* expresses his indebtedness to Arzachel and **Albategnius** and quotes their work several times. See also **Arabian astronomy**.

Asbolus (minor planet 8405)
A **Centaur**, about 80 km in diameter, upon which a young craterlike feature was detected using the Hubble Space Telescope. The extreme brightness of this feature compared with the rest of the surface suggests that it was formed by an impact within the past 10 million years. Orbit: perihelion 6.84 AU, aphelion 28.98 AU, eccentricity 0.618, inclination 17.6°, period 75.8 years.

ascending node
The point in an orbit where a body traveling from south to north crosses a reference plane, such as the plane of the **ecliptic** (in the case of a solar system object) or the celestial **equator**. The opposite point in the orbit, where the body moves from north to south across the reference plane, is the *descending node.* The *longitude of the ascending node* is an **orbital element**.

Asclepius (minor planet 4581)
A small asteroid, a few hundred meters across, of the **Apollo group** that can approach Earth's orbit to within 600,000 km; it was discovered in 1989 by the American astronomers Henry Holt (1929–) and Norman Thomas (1930–). Orbit: semimajor axis 1.023 AU, perihelion 0.66 AU, aphelion 1.39 AU, inclination 4.9°, period 1.03 years.

ashen light
A faint glow sometimes visible on the unlit side of **Venus** when it is in the crescent phase. Its cause is uncertain but may be similar to that of terrestrial **airglow**.

aspect
The apparent position of any of the planets or the Moon relative to the Sun, as seen from Earth. The four main aspects are **conjunction**, **greatest elongation**, **opposition**, and **quadrature**.

asterism
A distinctive pattern of stars, such as the **Big Dipper** or the **Summer Triangle**, that forms part of one or more **constellations**.

asteroid
See article, pages 39–42.

asteroid belt
A region of space, extending from 2.15 to 3.3 AU from the Sun–between the orbits of Mars (1.5 AU) and Jupiter (5.2 AU)–where the great majority of **asteroids** are found. It marks the transition from the inner to the outer solar system. Within the belt, the distribution of asteroids is nonuniform, with concentrations in asteroid groups and families, and also relatively empty zones known as **Kirkwood gaps**.

asthenosphere
The uppermost layer of the **mantle** of a rocky planet, located below the **lithosphere**. On Earth, this zone of soft, easily deformed rock lies at depths of 100 to 700 km.

astigmatism
A lens or mirror defect in which the size and shape of an image vary for different points of focus. Light passing through different parts of an astigmatic lens, for example, is focused at different distances beyond the lens, so that the image of a point can appear variously as a short horizontal or vertical line or an ellipse. The best focus is a small circle known as the *circle of least confusion.* See **aberration, optical**.

astrobiology
The study of **life in the universe**; also known as *exobiology* or *bioastronomy.*

astrobleme
A highly eroded meteorite crater on Earth, of which about 150 are known. The largest are the 200-km-wide Sudbury Crater in Ontario, Canada, and the 180-km-wide **Chicxulub Basin** in Mexico. While some astroblemes are almost 2 billion years old, about 60% were formed within the past 200 million years. See also **Earth impact craters**.

astrochemistry
The study of the chemical interactions between the gas and **dust** interspersed between the stars. Astrochemistry began with the observation of neutral hydrogen's **21-centimeter line**, which revealed an abundance of hydrogen between the stars. Since that time, more than 130 types of **interstellar molecules** have been detected. Scientists are now expanding their search to more complex, organic (carbon-rich) molecules that may hold the key to how life began. **Giant molecular clouds** (GMCs) contain enormous numbers of molecules; however, throughout most of their volume, pressures, densities, and temperatures are extremely low–a tiny fraction of those found on Earth. Molecular evolution in space involves chemical reactions that take place in the icy coatings of dust grains. Forged in the cores of stars, then returned to the interstellar medium by way of stellar winds, planetary nebulae, and supernovae, elements such as carbon, oxygen, hydrogen, and nitrogen combine to form hydrogen cyanide, water, and ammonia. Evidence is mounting that these molecules could, in turn, combine to produce simple **amino acids**, one of the main chemical building blocks of life. For decades, astronomers have debated whether the molecules of life were formed in the depths of space or evolved from

asteroid

A rocky or metallic object, smaller than a **planet** but bigger than a **meteoroid**, that orbits the Sun or another star; also known as a *minor planet.* More than 20,000 have been given official designations, most of them in the **asteroid belt** between the orbits of Mars and Jupiter (see table, "The First 20 Asteroids Discovered"). Asteroids outside this belt include the **Trojans**, which share Jupiter's orbit, and the **near-Earth asteroids**. A number of different asteroid groups have been distinguished on the basis of their similar orbital characteristics. Other subplanetary objects, considered distinct from asteroids, are **Centaurs**, **Kuiper Belt objects**, and **comets**, although classification is ambiguous in some cases. Some of the smaller moons in the solar system appear to be captured asteroids, including the two moons of Mars and a number of the outer moons of the four gas giants.

Asteroids range in size from a few meters to over 900 km across, and vary greatly in composition (see table, "The 10 Largest Asteroids"). Although none is visible to the naked eye, many can be seen at times with binoculars or small telescopes, including the four largest: (1) **Ceres**, (2) **Pallas**, (4) **Vesta**, and (10) **Hygiea**. The numbers, which are part of the asteroids' designations, give the order of discovery. Thirty known asteroids exceed 200 km in diameter and the census of asteroids larger than 100 km in diameter is believed to be virtually complete. In the 10 to 100 km range, probably about half await discovery. However, of the estimated 1 million asteroids bigger than 1 km across, only a tiny percentage is known. The total mass of all the asteroids, most of which is concentrated in the main belt, is about one-twentieth that of the Moon and about three times that of Ceres. Some asteroids, such as Ceres, Pallas, and Vesta, are nearly spherical; others, like (15) Eunomia (see **Eunomia family**), (107) Camilla, and (511) Davida, are quite elongated; still others, such as (4769) **Castalia**, (216) **Kleopatra**, and (4179) **Toutais**, have bizarre shapes. Several asteroids, including (243) **Ida**, (45) **Eugenia**, and (762) Pulcova, are known to have small moons of their own. The discovery of these moons is important because it enables an accurate determination of the parent asteroid's mass and average density. The density then gives a clue to the asteroid's makeup–either in terms of composition or of structure. There are also **binary asteroids**, such as (90) **Antiope** and, possibly, (1620) **Geographos**, in which two components of roughly equal size orbit each other at very close range. Several asteroids have been studied by passing space probes, including Ida and (951) **Gaspra** (by **Galileo**), and (253) **Mathilde** and (433) **Eros** (by **NEAR-Shoemaker**).

Most asteroids move in orbits that are somewhat more inclined and eccentric than those of the major planets (with the exception of Pluto)–the orbit of an average main-belt asteroid being inclined at about 10° to the plane of the ecliptic with an eccentricity of about 0.15. But some asteroids, such as (3200) **Phaethon** and (944) **Hidalgo**, have highly inclined and/or elliptical paths, suggesting they may be defunct cometary nuclei. Rotational periods of asteroids range from 2.3 hours to 48 days, but in more than 80% of cases are 4 to 20 hours. **Albedos** vary from just under 0.02 to over 0.5, with the majority of asteroids tending toward the lower (dark) end of this range. Low-albedo asteroids are generally found in the outer half of the asteroid belt, while higher-albedo objects tend to occupy the inner half. This fact stems from compositional differences, which in turn are related to how far from the Sun asteroids of different types formed.

Asteroids are thought to be the remnants of stillborn planets. According to this idea, the newborn Jupiter gravitationally scattered nearby large planetesimals–accreting lumps of matter in the embryonic stage of planet formation–some of which may have been as massive as Earth is today. Some of these big planetesimals strongly perturbed the orbits of the planetesimals in the region of the asteroid belt, raising their mutual velocities to the average 5 km/s seen today. As a result, what had been mild accretionary collisions in the future belt region became catastrophic disruptions. Only objects larger than about 500 km in diameter could have survived 5 km/s collisions with objects of comparable size. Ever since, the asteroids have been collisionally evolving so that, with the exception of the largest, most present-day asteroids are either remnants or fragments of past impacts.

While breaking down larger asteroids into smaller ones, collisions expose deeper layers of asteroidal material. If asteroids were compositionally homogeneous, this would have no noticeable result. Some of them, however, became *differentiated;* in other words, after they formed from primitive material in the solar nebula, they were heated (by radioactive decay or other means) to the point where their interiors melted and geochemical processes occurred. In some cases, temperatures became high enough for iron to form. Being denser than other materials, the iron sank to the center, forming an iron core and forcing **basalt**ic lavas to the surface. At least one asteroid with a basaltic surface, Vesta, survives to this day. Other differentiated asteroids were disrupted by collisions that stripped away their crusts and mantles and exposed their iron cores. Still others may have had only their crusts partially stripped away, which exposed surfaces such as those visible today on the A-, E-, and R-class asteroids.

Collisions gave rise to the **Hirayama families** and at least some of the planet-crossing asteroids. Tiny fragments from the latter enter Earth's atmosphere to become sporadic meteors, while larger pieces survive passage through the atmosphere to end up as **meteorites**. The very largest produce craters such as the **Barringer Crater**, and one may have been responsible for the extinction of the dinosaurs. Luckily, collisions of this sort are rare. According to current estimates, a few asteroids of 1-km diameter collide with Earth every 1 million years.

The 10 Largest Asteroids

No.	Name	Diameter (km)
1	**Ceres**	933
2	**Pallas**	608
4	**Vesta**	538
10	**Hygeia**	450
31	Euphrosyne	370
704	**Interamnia**	350
511	**Davida**	323
65	Cybele	309
52	**Europa**	289
3	**Juno**	288

The First 20 Asteroids Discovered

No.	Name	Year	q	a	P	e	i	D	A
1	**Ceres**	1801	2.55	2.99	4.610	0.079	10.60	933	0.054
2	**Pallas**	1802	2.11	3.42	4.607	0.237	34.85	608	0.074
3	**Juno**	1804	1.98	3.35	4.358	0.257	13.00	288	0.151
4	**Vesta**	1807	2.15	2.57	3.630	0.089	7.14	538	0.229
5	Astraea	1845	2.10	3.06	4.139	0.187	5.34	117	0.140
6	Hebe	1847	1.93	2.92	3.778	0.203	14.77	195	0.164
7	**Iris**	1847	1.84	2.94	3.686	0.230	5.50	209	0.154
8	**Flora**	1847	1.86	2.55	3.267	0.156	5.89	151	0.144
9	**Metis**	1848	2.09	2.68	3.684	0.123	5.58	151	0.139
10	**Hygeia**	1849	2.84	3.46	5.593	0.100	3.81	450	0.041
11	Parthenope	1850	2.20	2.70	3.840	0.102	4.62	150	0.126
12	Victoria	1850	1.82	2.85	3.568	0.218	8.37	126	0.114
13	Egeria	1850	2.36	2.80	4.135	0.085	16.53	224	0.041
14	Irene	1851	2.16	3.01	4.163	0.164	9.13	158	0.162
15	**Eunomia**	1851	2.15	3.14	4.300	0.188	11.73	272	0.155
16	Psyche	1851	2.53	3.32	5.000	0.135	3.09	250	0.093
17	Thetis	1852	2.13	3.52	3.880	0.138	5.59	109	0.103
18	Melpomene	1852	1.80	2.80	3.480	0.218	10.14	150	0.144
19	Fortuna	1852	2.06	2.83	3.816	0.158	1.56	215	0.032
20	Massalia	1852	2.06	2.76	3.737	0.145	0.70	131	0.164

Key: q = perihelion (AU), a = aphelion (AU), P = orbital period (y), e = eccentricity, i = inclination (°), D = diameter (km), A = albedo

Asteroid Names

Following its discovery, an asteroid is given a preliminary designation that consists of the year of discovery, an upper case letter to indicate the half-month in that year (A = Jan 1–15, B = Jan 16–31, . . . , Y = Dec 16–31, the letter "I" being omitted), and a second upper case letter in sequence. When this sequence of 25 letters has been completed, it is repeated and followed by a sequential number, for example, 2003 FC5. A permanent designation, consisting of a number and a name, is given to asteroids whose orbits have been accurately determined. The number represents the order of discovery, starting from 1 Ceres. The name is generally proposed by the discoverer and submitted to the **International Astronomical Union** for approval.

Asteroid Class

Any of a number of categories that an asteroid can be placed in based on its reflectance spectrum and albedo, which are indicators of surface composition. The distribution of the various classes throughout the asteroid belt is highly structured, suggesting that many asteroids formed at or near their present distances from the Sun and are representative of the composition of the solar nebula (not including hydrogen and helium) at these locations. Class S (silicaceous) asteroids are more prevalent in the inner part of the main asteroid belt, giving way to Class C (carbonaceous) in the middle and outer parts of the belt. Together, these two types account for about 90% of the asteroid population. Class M (metal) objects are especially concentrated in the middle of the belt, while dark

Asteroid Groups*

		Orbital Characteristics (AU)		
Region	**Group/Family**	**Semimajor Axis**	**Perihelion**	**Aphelion**
Near-Earth				
Aphelion greater than Earth's perihelion	Aten	<1.00		>0.983
Perihelion less than Earth's aphelion	Apollo	>1.00		<1.015
Never crosses Earth's orbit	Amor	>1.00	1.017–1.30	
Before main belt				
Crosses Mars's orbit	Mars-crosser		1.30–1.666	
L5 Lagrange point of Mars	Mars Trojan	1.524		
Resonance 2:9 with Jupiter	Hungaria	1.81–1.99		
Main belt				
Between resonances 1:4 and 2:7	Flora	2.12–2.25		
Between resonances 2:7 and 1:3	Phocaea	2.25–2.50		
Between resonances 2:7 and 1:3	Nysa-Polana	2.41–2.50		
Between resonances 1:3 and 2:5	various	2.50–2.82		
Between resonances 2:5 and 3:7	Koronis	2.82–2.95		
Between resonances 3:7 and 4:9	Eos	2.95–3.00		
Between resonances 4:9 and 1:2	Themis	3.00–3.27		
Resonance 4:7 with Jupiter	Cybele	3.31–3.75		
Resonance 2:3 with Jupiter	Hilda	3.83–4.00		
After main belt				
Resonance 3:4 with Jupiter	Thule	4.28		
L5 Lagrange point of Jupiter	East Trojan	5.06–5.31		
L4 Lagrange point of Jupiter	West Trojan	5.08–5.28		
After Jupiter				
Crosses Jupiter's orbit	Jupiter-crosser	>5.2	<5.2	

*Does not include Centaurs, Kuiper Belt objects, or Oort objects.

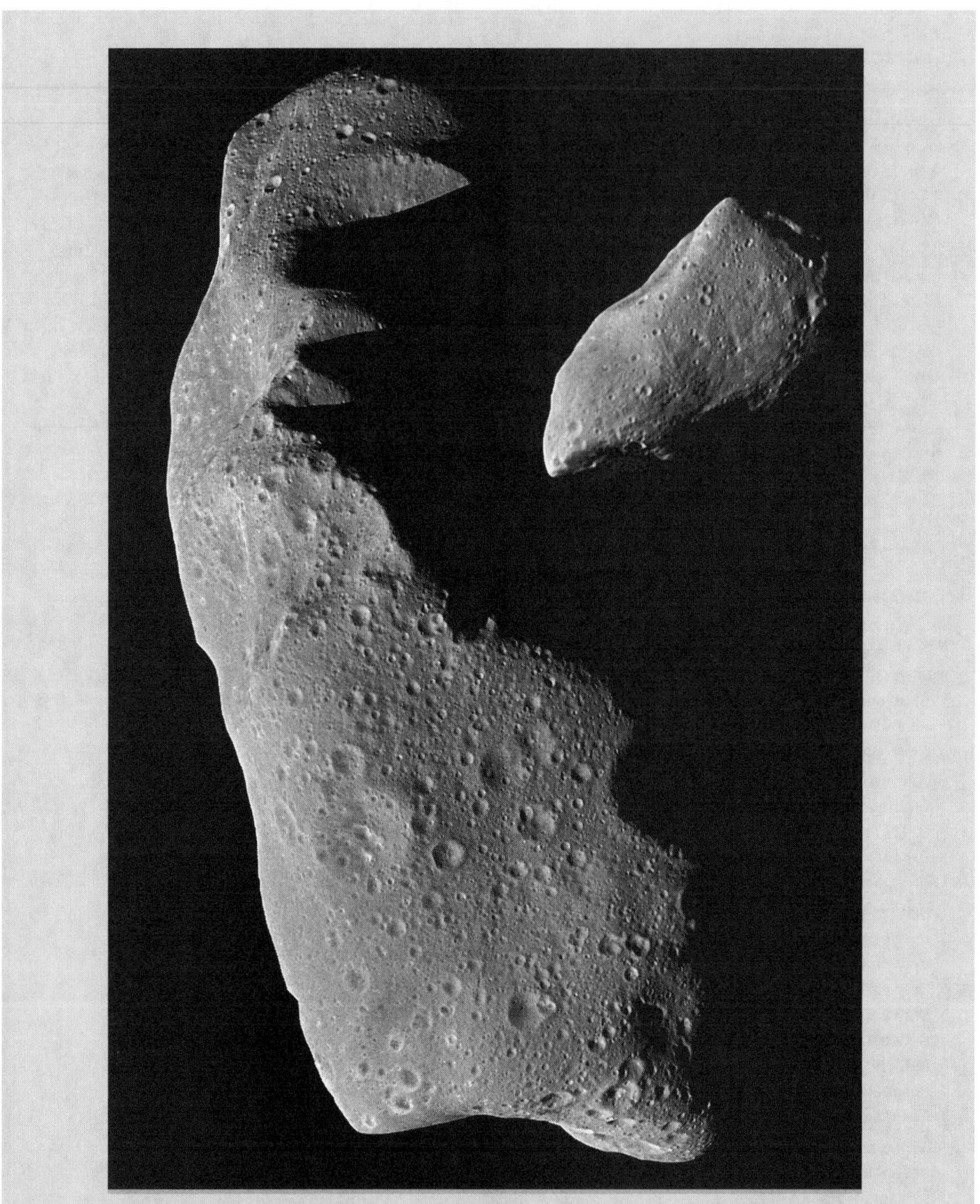

asteroid The asteroids Ida (left) and Gaspra (right) shown to the same scale. These images were taken by the Galileo spacecraft while en route to Jupiter. Gaspra was imaged on October 29, 1991, at a range of 5,300 km; Ida was snapped on August 28, 1993, from a range of just over 3,000 km. Gaspra is about 17 km across, while Ida measures 9.3 × 12.7 × 29.9 km. *NASA/U.S. Geological Survey*

asteroid The asteroid Eros seen from a distance of about 100 km by NEAR-Shoemaker on September 5, 2000. The knobs sticking out of the surface near the top of the image surround a boulder-strewn area and are probably remnants of ancient impact craters. The faint grooves that run diagonally across the surface may have formed during a collision between the asteroid and a smaller body. *NASA/Johns Hopkins University*

reddish asteroids of classes P and D become more common at the outer edge of the belt and beyond.

The most widely used taxonomic scheme, proposed by American astronomer David Tholen and Italian astronomer Maria Barucci in 1989, identifies 14 different classes–A, B, C, D, E, F, G, M, P, Q, R, S, T, and V (see **A-class asteroid**, etc., for separate entries on each of these types). However, there is disagreement, on the one hand, about whether all of the Tholen-Barucci classes are distinct and, on the other, about whether additional categories are needed. To add to the confusion, because most **meteorites** are known to be fragments of asteroids, there is an ongoing effort to match meteorite classification with asteroid classification. This is difficult because whereas meteorites can be analyzed in the lab to determine their exact chemical composition and petrologic type, most of the information about asteroids comes from remote spectroscopy.

Asteroid Group

A collection of asteroids that move in generally similar orbits. Within a group there may be one or more *families* of asteroids whose orbital characteristics are so much alike that they almost certainly came from the breakup of a single parent body. Most groups and families (generally named after the first-discovered member) are found in or near the main asteroid belt. They include the **Hungaria group**, **Flora family**, **Nysa-Polana family**, **Phocaea group**, **Koronis family**, **Eos family**, **Themis family**, **Cybele group**, and **Hilda group**. The three largest families (Eos, Koronis, and Themis) have been determined to be compositionally homogeneous. If the asteroids belonging to them are considered to be fragments of a single parent body, then these parent bodies probably had diameters of 100 to 300 km. The smaller families have not been as well studied because they have fewer and smaller members; however, it is known that some of the smaller families are compositionally inhomogeneous and that, at least in some cases, what are observed are pieces of a geochemically differentiated parent body. Beyond the main belt are the Jupiter **Trojans**, while inside the belt are the **Mars-crossers** and **Mars Trojans**. Even closer to the Sun are three groups of **near-Earth asteroids**: the **Amor group**, **Aten group**, and the **Apollo group**. (See table, "Asteroid Groups," page 41.)

scratch on the surface of the young Earth and of other planets. While this issue remains hotly debated, evidence is mounting that at least some of life's precursor molecules are formed between the stars and are then delivered to the surface of new worlds aboard comets, asteroids, and meteorites. In hopes of unmasking more evidence for this scenario, astronomers are searching for amino acids in the cold, molecular gas found in some regions of our own Milky Way Galaxy. One such region is **Sagittarius B2**.

Astrographic Catalogue

Part of the international *Carte du Ciel* program designed to photograph and measure the positions of all stars brighter than magnitude 11.0. (The brightest stars are missing due to their images being grossly overexposed and, therefore, not measured.) In total, over 4.6 million stars were observed, many as faint as thirteenth magnitude. This project was started in the late 1800s, and most observations were made between 1895 and 1920. Data from the *Astrographic Catalogue* was recently used by the U.S. Naval Observatory in producing the ***ACT Catalog.***

astrolabe

An Arabic and medieval European sighting instrument on an **altazimuth mounting**, used to determine the elevation above the horizon of celestial objects. It comprises two or more flat, metal, calibrated disks, attached so that both or all can rotate independently. For early navigators and astronomers it served as star chart, compass, clock, and calendar. It was eventually replaced by the sextant as a navigational device. The *Danjon astrolabe* is a type of portable solstitial armillary, modified for stellar observation. Suspended by a small hook or eye, the instrument consisted initially of a single ring that hung in a vertical plane. Pivoted at the center of the ring was a rod equal in length to the ring diameter, carrying sights at either end. When aligned on a star or planet, an angular scale inscribed on the armillary ring showed the object's altitude.

astrometric binary

A **binary star** in which the presence of an unseen or unresolved companion is revealed by small side-to-side irregularities in the movement of the visible component.

astrometry

The branch of **astronomy** that deals with measuring the positions of celestial objects, especially stars. Astrometrists measure **parallaxes** and **proper motions**, which allow astronomers to determine stellar distances and velocities.

Astronomer Royal

An honorary title held by a prominent British astronomer, created in 1675 by King Charles II when the **Royal Greenwich Observatory** was founded. (See table, "The Astronomers Royal.") Before 1971, the Astronomer Royal was also director of the Royal Observatory, but after that year these became separate appointments. The title of Astronomer Royal for Scotland was created in 1834 and held by the director of the Royal Observatory, Edinburgh, until 1995 when the appointments were separated.

Astronomical Almanac

An annual joint publication of the **U.S. Naval Observatory** (USNO) and (since the closure of the **Royal Greenwich Observatory**) the U.K. Nautical Almanac Office, which contains information on the positions of the Sun, the Moon, and the planets and their satellites, phases of the Moon, eclipses, sunset and sunrise, bright stars, and locations of observatories. It began in 1980 through the merger of *The American Ephemeris and Nautical Almanac,* produced by the USNO between 1855 and 1980, and *The Astronomical Ephemeris,* which originated in Britain in 1766 as *The Nautical Almanac and Astronomical Ephemeris.*

astronomical clock

A highly accurate clock that keeps either **sidereal time** or **Universal Time**. Formerly, such instruments were checked by timing the passage of clock stars across the meridian using meridian or transit telescopes and a book of clock errors. Today, atomic clocks are used instead.

The Astronomers Royal

	Lived	Served
John **Flamsteed**	1646–1719	1675–1719
Edmond **Halley**	1656–1742	1720–1742
James **Bradley**	1693–1762	1742–1762
Nathaniel Bliss	1700–1764	1762–1764
Nevil Maskelyne	1732–1811	1765–1811
James Pond	1767–1836	1811–1835
George Biddell **Airy**	1801–1892	1835–1881
William Henry Mahoney Christie	1845–1922	1881–1910
Frank Watson **Dyson**	1868–1939	1910–1933
Harold **Spencer Jones**	1890–1960	1933–1955
Richard van der Riet Woolley	1906–1986	1956–1971
Martin **Ryle**	1918–1984	1972–1982
Francis Graham-Smith	1923–	1982–1990
Arnold W. Wolfendale	1927–	1990–1995
Martin Rees	1942–	1995–

astronomical constants
Precisely measured fundamental quantities, such as the solar **parallax**, the constant of **aberration**, and the **obliquity** of the ecliptic.

astronomical extinction
See **atmospheric extinction** and **interstellar extinction**.

Astronomical Society of the Pacific (ASP)
An international organization for professional and amateur astronomers, founded in 1889 and based in San Francisco. It publishes the amateur-level *Mercury* magazine and the research journal *Publications of the Astronomical Society of the Pacific.*

Astronomische Gesellschaft Katalog (AGK)
A standard catalog of star positions. AGK1 was initiated by Friedrich **Argelander** in 1867, prepared using data from meridian circles at 17 observatories around the world, and published between 1890 and 1954. It listed the positions of some 200,000 stars down to ninth magnitude between declinations 80° and –23°. A second version, AGK2, that used photographic data from Bonn and Hamburg, was started in the 1920s and published between 1951 and 1958. The latest version, AGK3, published in 1975, lists 183,145 stars north of declination –2° and includes both positions and proper motions. All AGK positions are calculated relative to reference stars, 21,499 of which were used for AGK3 and listed in AGK3R.

astronomy
The science of space beyond Earth's surface and lower atmosphere. In addition to observational astronomy, it includes such branches as **astrometry**, the measurement of the positions and movements of celestial bodies; **astrophysics**, the physics of stars, galaxies, and other objects in the universe; **celestial mechanics**, which specializes in orbit calculation, including the influence of gravity, mass, acceleration, and inertia; and **cosmology**, which investigates the origins and evolution of the universe.

astrophotography
The marriage of photography and astronomy. Long exposures on hypersensitized films (see **hypersensitization**) show much that cannot be seen during real-time observation. Images of faint nebulae and galaxies can continue to improve during exposures ranging from several minutes to many hours, revealing detail that can't be seen without tremendous light amplification. Such photography requires excellent motorized mounts to correct for Earth's rotation, perfect alignment, and periodic correction during the exposures. In professional astronomy, photography has been superceded by digital imaging using **CCD** detectors and other electronic equipment.

astrophysics
A branch of **astronomy** that deals mainly with the physics of stars, stellar systems, and interstellar material. The alliance of physics and astronomy, which began with the advent of **spectroscopy**, made it possible to investigate *what* celestial objects are and not just *where* they are.

astroseismology
The study and measurement of acoustic vibrations in stars in order to learn more about stellar structure and evolution. Astroseismology, an outgrowth of **helioseismology**, had to await the development of instruments capable of detecting radial (up and down) movements in a star's surface of a few meters per second over distances measured in light-years. Such instruments are now available and are also used to search for **extrasolar planets**. The first star, other than the Sun, in which evidence of sound waves was found is the near-solar twin, **Alpha Centauri** A, the nearest star visible to the naked eye, which pulsates with a 7-minute cycle, similar to that observed in the Sun. Astronomers have also picked up seismological activity in more distant and nonsolarlike stars, including Xi Hydrae, a giant star with a radius of about 10 R_{sun} and a luminosity of about 60 L_{sun}, lying at a distance of 130 light-years. Xi Hya, it turns out, behaves like a subultra-bass instrument, oscillating with several periods of around 3 hours.

asymmetric drift
(1) The tendency of a **population** of stars to have a mean rotation velocity around the **galactic center** that lags behind that of the local standard of rest. (2) The negative of the mean *V* velocity (one of the components of **space velocity**) of a stellar population. In general, it increases with random motion within a population, so that the older the population, the more negative the *V* velocity. For example, a young thin disk population has an asymmetric drift of 0 km/s, whereas a halo population has an asymmetric drift of about 200 km/s.

asymptotic giant branch (AGB)
A region of the **Hertzsprung-Russell diagram** that lies above and parallel to the **red giant** region. It is occupied by evolved stars of intermediate to low mass (less than an initial mass of 8 M_{sun}) that have a dormant, helium-filled core surrounded by a **helium-burning** shell, on top of which lies a **hydrogen-burning** shell. Initially, the hydrogen-burning shell produces most of the star's energy output. However, the hydrogen shell eventually dumps enough helium "ash" onto the helium shell that the latter

undergoes an explosive event called a *thermal pulse.* Although this pulse is barely noticed at the surface of the star, it serves to increase the mass of the carbon/oxygen core, so that the size and luminosity of the star gradually increases with time. As the star climbs the AGB, a **stellar wind** develops in the star's envelope that blows the outer layers into space at a rate of 10^{-8} to 10^{-4} M_{sun} per year. Within this wind, **dust** particles (crucial to the development of interstellar clouds and, eventually, protoplanetary systems) are formed from carbon material dredged up from the core by **convection** currents. Also through this mass loss, AGB stars avoid ending as **supernovae**. When the envelope of the star is nearly gone, a time of enhanced loss with a rapid velocity produces a **planetary nebula** and eventually leaves behind a **white dwarf** of 0.6 to 0.7 M_{sun}.

Atacama Large Millimeter Array (ALMA)

When completed, by 2010, this will be the largest and most sensitive telescope in the world at millimeter and submillimeter wavelengths. A joint United States–European project, ALMA will consist of at least sixty-four 12-m dish antennas at an altitude of 5,000 m near Cerro Chajnantor in northern Chile. The array will be configurable with baselines up to 10 km, be capable of imaging in all atmospheric windows between 10 mm and 350 microns, and have 10 times the resolution of the **Very Large Array** (VLA) or the **Hubble Space Telescope**. The U.S. side of the project is run by the **National Radio Astronomy Observatory**, while the European side is a collaboration between the **European Southern Observatory**, the French Centre National de la Recherche Scientifique, the German Max Planck-Gesellschaft, the Netherlands Foundation for Research in Astronomy and Nederlandse Onderzoekschool Voor Astronomie, and the United Kingdom Particle Physics and Astronomy Research Council. In June 2002, work began to assemble the first of the ALMA prototype antennae at the ALMA Test Facility located on the VLA site in New Mexico. See also **submillimeter wave astronomy**.

ataxite

A rare variety of **iron meteorite** (designated type D) made almost entirely of **taenite**, a solid solution of iron and 27 to 65% nickel; the Greek name means "without structure" and refers to the lack of a visible **Widmanstätten pattern** (spindles of **kamacite** visible only microscopically). Strangely enough, the largest meteorite known, the **Hoba West Meteorite**, belongs to this rare structural class. Compare with **octahedrite** and **hexahedite**.

Aten group

One of the three groups of **near-Earth asteroids**; Atens have semimajor axes less than 1.0 AU and aphelia greater than 0.983 AU, so that they orbit mostly inside Earth's orbit. The prototype for the group, (2062) Aten, discovered in 1976 by the American astronomer Eleanor Helin, has a diameter of less than 1 km, and is of class S.

Atlas

(1) The second moon of **Saturn**, also known as Saturn XV; it lies near the outer edge of Saturn's A-ring at a mean distance of 137,670 km from the planet's center and is probably a **shepherd moon**. It measures approximately 40 × 20 km. Atlas was discovered in 1980 by Richard Terrile from Voyager 1 photos. (2) The star 27 Tauri, the second brightest star in the **Pleiades**, with a visual magnitude of 3.63.

atmosphere

The gaseous layer that surrounds an astronomical body, such as a planet or a star. The ability of a planet to retain a substantial atmosphere depends on the strength of the planet's gravitational field and its temperature. Mercury, for example, has no atmosphere because its gravity is quite feeble and it is very hot. Pluto, on the other hand, develops a thin atmosphere near perihelion because, although its gravity is weak, its temperature is low enough to prevent gas molecules, released from the surface, from escaping quickly into space. Of the moons of the Solar System, only **Titan** has a dense atmosphere. See also **stellar atmosphere**.

atmospheric extinction

The reduction in the intensity of light from a celestial body due to absorption and scattering by Earth's atmosphere. It increases from the zenith to the horizon and affects short wavelengths more than long wavelengths, so that objects near the horizon appear redder than they do at the zenith. The brightness of a star in the zenith will be reduced by only about 0.3 magnitudes, whereas the extinction at 20° altitude is about 0.9 magnitudes and at 10° altitude about 1.6 magnitudes. See also **interstellar absorption**.

atmospheric refraction

The shift in apparent direction of a celestial object caused by the bending of light rays as they pass through Earth's atmosphere. It increases the observed altitude of an object and is greatest (just over half a degree) for objects on the horizon.

atmospheric window

A range of electromagnetic wavelengths to which Earth's atmosphere is largely or partially transparent. All spectral regions are affected to some extent by absorption in the atmosphere but there are two nearly transparent ranges, the *optical window* and the *radio window,* and several narrow, partial *infrared windows.*

The *optical window* allows visible light through, from red, as far as the A band (see **Fraunhofer lines**) of molecular oxygen (O_2) at 7600 Å, to violet and a little beyond, as far as the ozone (O_3) cut-off at 2950 Å. Shorter wavelengths are blocked by atoms and molecules of oxygen, nitrogen, and other gases, and by **geocoronal** hydrogen and helium.

The *radio window* spans a wavelength range from about 1 mm to about 30 m. Lower wavelengths are reflected by the **ionosphere**, while shorter wavelengths suffer increasing amounts of molecular absorption.

Several narrow *infrared windows* exist at micrometer (μm) wavelengths, the photometric designations of which are J (1.25 μm), H (1.6 μm), K (2.2 μm), L (3.6 μm), M (5.0 μm), N (10.2 μm), and Q (21 μm). There are also small but usable windows at 350 and 460 μm. Because water vapor is one of the main absorbers of infrared, observatories for studying infrared must be situated in particularly dry or mountainous regions where the effect of water vapor is reduced and/or the atmosphere is thinner.

atom

The fundamental unit of a chemical element. An atom consists of a nucleus, containing protons and neutrons, and electrons, which occupy shells that surround the nucleus and are centered on it.

atomic number

The number of protons in the nucleus of an atom of a chemical element.

Atoms for Peace Galaxy (NGC 7252)

A peculiar **elliptical galaxy**, about 300 million light-years away in **Aquarius**, which is thought to be the product of a merger between two **disk galaxies** that took place about 1 billion years ago. Its nickname comes from its prominent looplike structure, made of stars, that resembles a schematic diagram of an electron orbiting an atomic nucleus. (In December 1953, President Eisenhower made his "Atoms for Peace" speech to foster peaceful applications of nuclear energy.) The central region of NGC 7252 is home to more than 500 **ultra-luminous clusters**, which appear as bluish knots of light, mostly 50 million to 500 million years old, and each typically about 60 light-years across. These unusual, young-looking globular clusters are the most obvious fruits of the suspected **galaxy merger**, originating from the infall of gas and subsequent burst of star formation that the merger triggered. Their discovery provides an important key to understanding how all **globular clusters** form in ellipticals and also how giant ellipticals originate. The globular clusters found in NGC 7252 are thought to resemble the progenitors of similar clusters that orbit the Milky Way Galaxy. Since globular clusters normally contain ancient red giants, they provide a fossil record of the formation and evolution of galaxies. Globular clusters contain about 1 million stars each, arranged in a tight, spherical swarm and are generally found to be about 12 billion years old. However, the ultra-luminous clusters found in NGC 7252 contain hot bluish stars. Because these blue stars are short-lived, the clusters in NGC 7252 are estimated to be mostly 50 million to 500 million years old. During the past decade, the hypothesis that spiral galaxies can collide and merge to form elliptical galaxies has become increasingly popular. One argument against this theory is that elliptical galaxies have more globular clusters than would be expected if disk galaxies were simply combined, since disk galaxies have relatively few clusters. However, this problem goes away if, when disk galaxies collide, new globular clusters can be manufactured. A small disklike structure at the heart of NGC 7252 also supports the notion that giant ellipticals grow out of spiral mergers. The pinwheel-shaped disk, deep inside Atoms for Peace, bears an uncanny resemblance to a face-on spiral galaxy

Atoms for Peace Galaxy The core of the Atoms for Peace Galaxy (NGC 7252), captured by the Hubble Space Telescope in 1992. It shows a striking mini-spiral disk of gas and stars and about 40 exceptionally bright, young globular star clusters. The strong spiral structure is 10,000 light-years across and the entire picture corresponds to a region about 46,000 light-years across. The clusters are some 60 light-years in diameter—the same size as globular clusters that orbit our own Galaxy. *NASA/ESA/Brad Whitmore (STScI) et al.*

(see photograph), yet it is only 10,000 light-years across–about one-twentieth the size of the total galaxy. That this structure is a remnant of the collision between two galaxies is also supported by the fact that the mini-spiral is rotating in a direction opposite to the rest of the galaxy. Within a few billion years, the gas in NGC 7252 will be exhausted and the galaxy will look like a normal elliptical galaxy with a small inner disk.

Atria (Alpha Trianguli Australis, α TrA)

A giant **K star** and the brightest star in **Triangulum Australe**. There are indications of strong flare activity on its surface, and a small and faint binary companion has recently been detected.

Visual magnitude:	1.91
Absolute magnitude:	–3.62
Spectral type:	K2IIb or K2IIIa
Luminosity:	2,000 L_{sun}
Distance:	416 light-years

aubrite

A variety of evolved **achondrite**, also known as an *enstatite achondrite,* that is calcium-poor and composed mainly of the silicate mineral orthopyroxene **enstatite** ($MgSiO_3$). Aubrites are named for the small Aubres meteorite that fell near Nyons, France, in 1836. Due to their typical light-colored **fusion crusts**, their white interiors, and their fragile compositions, most aubrites are witnessed falls or finds from the blue-ice fields of Antarctica. As well as large white crystals of enstatite, they contain small, varying amounts of **olivine**, nickel-iron metal, **troilite**, and a variety of exotic accessory minerals, pointing to a magmatic origin under highly reducing conditions. Most aubrites are heavily **brecciated**, indicating a violent history for their parent body. Comparisons of the aubrite spectra to the spectra of asteroids have revealed striking similarities between the aubrites and the main belt asteroid (44) Nysa (see **Nysa-Polana family**) and other E-class objects. A small near-Earth asteroid, (3103) Eger, which is the only known E-class NEA, is suspected of being the actual parent body of the aubrites.

augite

A **pyroxene**, a complex aluminous silicate of calcium, iron, and magnesium, crystallizing in the monoclinic system, and occurring in many igneous rocks, particularly those of basaltic composition.

Auriga (abbr. Aur; gen. Aurigae)

The Charioteer (possibly Erechthonius, son of Vulcan); among the brightest northern **constellations**, it lies midway between Perseus and Ursa Major in a region crossed by the Milky Way, and is the site of the *galactic anticenter* (the point in the sky diametrically opposite the center of the Galaxy). Gamma Aur, or El Nath, is shared with Taurus and is now universally referred to as Beta Tau. **Epsilon Aurigae** and **Zeta Aurigae** are remarkable eclipsing binaries. Together with Eta Aur, they make up the Kids, an asterism that lies just to the south and slightly ahead of **Capella** (the "she-goat"). The most northerly of the trio, Epsilon, is a late addition; in antiquity Zeta and Eta were known respectively as "the western kid" and "the eastern kid." (See tables, "Auriga: Stars Brighter than Magnitude 4.0," and "Auriga: Other Objects of Interest." See also star chart 2.)

aurora

A glow in a planet's **ionosphere** caused by interaction between the planet's magnetic field and charged particles from the Sun; *aurora* is Latin for "dawn." Aurorae have been observed on Earth, Jupiter, Saturn, Mercury, and Uranus, and are expected to occur in some form on all worlds that have magnetospheres. They are distinguished from **airglow** by their confinement to magnetic polar and subpolar regions, and their sporadic occurrence. Different auroral colors stem from emission by different atmospheric gases. On Earth, the northern and southern hemisphere versions are known as *aurora borealis* and *aurora australis,* respectively. Aurorae change in brightness, shape, color, dynamics, and location in response to changes in the state of the magnetosphere. This variability shows up most dramatically during an auroral **substorm**.

Australia Telescope National Facility (ATNF)

A set of eight radio telescopes that can be operated individually or as a group. Six of the telescopes make up the Australia Telescope Compact Array (ATCA), located at the **Paul Wild Observatory** near Narrabri. Each of these antennae has a reflecting surface with a diameter of 22 m. A further 22-m antenna, known as the **Mopra Telescope**, is located near Coonabarabran. The Australia Telescope also includes the 64-m radio telescope at the **Parkes Observatory**, which has been in use since 1961.

averted vision

Looking at a faint object "out of the corner of the eye"–with peripheral vision–in order to detect it more easily.

Auriga: Stars Brighter than Magnitude 4.0

Star	Magnitude Visual	Magnitude Absolute	Spectral Type	Distance (light-yr)	Position R.A. (h m s)	Position Dec. (° ′ ″)
α **Capella**	0.08	−0.48	G5IIIe+G0III	42	05 16 41	+45 59 53
β **Menkalinan**	1.90	−0.11	A2IV	82	05 59 32	+44 56 51
θ	2.65	−0.98	A0IIIpSi+G2V	173	05 59 43	+37 12 45
ι Hassaleh	2.69	−3.29	K3II	512	04 57 00	+33 09 58
ε **Epsilon Aurigae**	3.03v	−5.95	F0Iae	2,040	05 01 58	+43 49 24
η	3.18	−0.96	B3V	219	05 06 31	+41 14 04
δ	3.72	0.55	K0III	141	05 59 32	+54 17 05
ζ **Zeta Aurigae**	3.69v	−3.23	K4II+B8V	789	05 02 29	+41 04 33

Auriga: Other Objects of Interest

Object	Notes
Stars	
AE Aur	See **runaway star.**
RT Aur	A **Cepheid variable**; magnitude range 5.4 to 6.6, period 3.7 days.
T Aur	Nova 1882; reached magnitude 4.1, now at 15.8. Discovered by Scottish amateurastronomer, T. D. Anderson.
Diffuse nebulae	
Flaming Star Nebula	IC 405. See separate entry.
IC 410	Resembles the **Rosette Nebula**, with a small embedded open cluster, NGC 1893,and a dark patch of obscuring dust at its center. R.A. 5h 22m, Dec. +33° 27′.
Open clusters	
M36 (NGC 1960)	About 60 stars. Magnitude 6.3; diameter 12′; R.A. 5h 32m, Dec. +34° 7′.
M37 (NGC 2099)	About 500 stars, some 150 of which shine at magnitude 12.5 or brighter. Magnitude 6.2; diameter 20′; R.A. 5h 49m, Dec. +32°32′.
M38 (NGC 1912)	A cruciform-shaped cluster, located near both the Flaming Star and IC 410. Magnitude 7.4; diameter 20′; R.A. 5h 25m, Dec. +35° 48′.

This method takes advantage of the fact that the most optically sensitive part of the retina (the layer of cells at the back of the eye) is off-center. Two kinds of light-detecting cells make up the retina: cones and rods. The cones support high-light color vision, while the rods are for low-light vision. In the human eye there are about 5 million cones and 100 million rods. The cones occupy a small spot (the fovea) centrally located on the retina, whereas the rods surround the cones and cover a much greater area (the macula). An observer can multiply light sensitivity many times by slightly averting the head so that the projected image from an eyepiece falls onto the retinal region of high light-sensitivity. See also **dark adaptation.**

Avior (Epsilon Carinae, ε Car)

The third brightest star in **Carina**, a **spectroscopic binary** consisting of a main sequence **B star** (temperature 23,000 K, mass 16 M_{sun}, radius 6 R_{sun}) and a giant **K star** (temperature 4,100 K, mass 4.6 M_{sun}, radius 20 R_{sun}), separated by 0.46". This corresponds to a distance of 90 AU–about 2.25 times the mean distance of Pluto from the Sun.

Visual magnitude:	1.86
Absolute magnitude:	–4.58
Spectral type:	K3III + B2V
Luminosity:	6,000 L_{sun}
Distance:	632 light-years

axial period

The time taken for a body to spin completely around its **axis** relative to some fixed point. In Earth's case, this is one **day**.

axis

An imaginary straight line through a celestial body, around which it rotates.

azimuth

The angular distance to the foot of the vertical circle through a celestial body, measured from north around the observer's horizon. Azimuth is 0° for an object due north, 90° due east, 180° due south, and 270° due west. It is specified together with **altitude** or, occasionally, **zenith** distance, to give the **horizontal coordinates** of a body.

B

B star

A large, luminous, blue-white star of **spectral type** B with a surface temperature of about 11,000 to 30,000 K. The spectrum is characterized by **absorption lines** of neutral or singly ionized helium, the **Balmer lines** of atomic hydrogen, especially at the cooler end of the range, and lines of singly ionized oxygen and other gases. Main sequence B stars, such as **Spica** and **Regulus**, have masses of 3 to 20 M_{sun} and luminosities of 100 to 50,000 L_{sun}. Often they are found together with **O stars** in **OB associations** since, being massive, they are short-lived and therefore don't survive long enough to move far from their place of origin. B-type **supergiants**, such as **Rigel**, have masses up to 25 M_{sun} and luminosities up to 250,000 L_{sun}.

Baade, (Wilhelm Heinrich) Walter (1893–1960)

A German-born American astronomer who classified stars into two distinct **population** types, found each had a distinct kind of **Cepheid variable** (a crucial cosmic distance indicator), and deduced from this that the universe was roughly twice as big and old as previously thought. Having earned his Ph.D. at Göttingen, Baade began his career at the Hamburg Observatory but moved to the United States in 1931 in search of bigger telescopes. The rest of his career, until his retirement in 1958, was spent at the Mount Wilson and Palomar Observatories. In the 1940s, as an enemy alien, Baade was able to continue observing while other astronomers were away on war duty and to take advantage of the unusually dark skies provided by the Los Angeles blackout. Under these favorable conditions he began a survey of the **Andromeda Galaxy**, resolving, for the first time, stars in the central region. This enabled him to distinguish two types of stars: young, hot **Population I** objects concentrated in the spiral arms of the galaxy, and an older stellar **Population II** in the central galactic region. He later found that the Cepheid variables of these two populations differed in their **period-luminosity relations**, forcing a reevaluation of the distance to Andromeda and other external galaxies. In 1952, he published results showing that the universe was more than double the size and age of earlier estimates (a relief to geologists who could now sleep at night knowing that Earth was younger than the cosmos).

Baade produced other important results, particularly through collaboration with Fritz **Zwicky** and Rudolf **Minkowski**. He and Zwicky proposed, in 1934, that **supernovae** could produce **cosmic rays** and **neutron stars**, and Baade made extensive studies of the **Crab Nebula** and its central star. He distinguished two types of **supernovae** and, with Minkowski, identified optical counterparts of many of the first-discovered radio sources, including **Cygnus A** and **Cassiopeia A**. Minkowski, however, disagreed with Baade's suggestion that Cygnus A was two galaxies in collision and a bottle of whiskey was bet on the outcome. Weeks later, after obtaining a spectrum of Cygnus A, Minkowski conceded the bet–but left Baade disappointed. "For me," Baade said, "a bottle is a quart. What Minkowski brought was a hip flask. . . . Two days later . . . Minkowski visited me in order to show me something, saw the flask, and emptied it." (With hindsight, the drink should probably have been split: Cygnus A is now known to be a single elliptical galaxy–not two colliding spirals as Baade had proposed–but almost certainly the product of a past collision.) Baade is also remembered for his discovery of the minor planets (1920) **Hidalgo** and (1949) **Icarus**.

Baade's Window

A small clearing in the dust clouds of **Sagittarius**, near the globular cluster NGC 6522, through which astronomers can view stars in the **galactic bulge**. The window lies 3.9° south of the galactic center, corresponding to a line of sight that passes within 1,800 light-years of the Milky Way's heart. It is named after Walter **Baade** who used it to observe **RR Lyrae stars** in the bulge region.

Baade-Wesselink method

A way of calculating the sizes and other properties of certain kinds of **pulsating variables**, including **Cepheid variables**. Measurements of a star's color and light output at two different times, t_1 and t_2, in the pulsation cycle are used to find the ratio of the star's radii, $R(t_2)/R(t_1)$, at these times. Spectra of the star throughout its pulsation period are then used to find the **radial velocity** of its surface. Given how fast the star's surface is moving, the difference $R(t_2) - R(t_1)$ is found by adding the products of velocity and time during the interval t_1 to t_2. From the values of $R(t_2)/R(t_1)$ and $R(t_2) - R(t_1)$, it is easy to solve for the radii. The method is named after Walter **Baade** and the Dutch astronomer Adriaan Jan Wesselink (1909–1995).

Babcock, Harold Delos (1882–1968), and Horace Welcome (1912–)

American astrophysicists, father and son, who did pioneering research on the magnetic fields of the Sun and other stars. Harold Babcock was one of the first staff members of the **Mount Wilson Observatory**, remaining there from 1909 until 1948. Harold carried out solar research alongside George **Hale** and made precise laboratory measurements of atomic spectra that helped others identify the first **forbidden lines** in the laboratory. With C. E. St. John he obtained precise wavelengths for some 22,000 lines in the solar spectrum and extended measurements into the ultraviolet and infrared. His son Horace worked at **Lick Observatory** (1938–1939), then at the **Yerkes Observatory** and **McDonald Observatory** (1939–1941), before joining Harold at Mount Wilson after World War II. The two collaborated, developing the solar **magnetograph** in 1948 and using it to measure the distribution of magnetic fields over the solar surface. Previously, only the powerful fields in the vicinity of **sunspots** had been known. By the late 1940s the Babcocks were able to report the presence of weak general magnetic fields on the Sun, restricted to latitudes greater than 55°, and the discovery of a magnetic field associated with another star–78 Virginis. By 1958 they had established the presence of magnetic fields in some 89 stars. Horace served from 1964 until his retirement in 1978 as director of the Mount Wilson and Palomar Observatories (known, from 1969 on, as the Hale Observatories).

Babylonian astronomy

Babylonian astronomy goes back at least as far as 1800 B.C. and centered mainly on the problem of establishing an accurate **calendar**, hence the emphasis was on recording and calculating the motions of the Sun and the Moon. Early on, observation played an important role, but this gave way later to analyzing records of ancient observations, which in turn led to the mathematical prediction of current and future astronomical events. The continuity of civilization in this part of the world enabled records to be kept over a long enough period for features such as the **precession** of the equinoxes and the regularity of eclipses to be recognized. The Babylonians also divided the sky into zones, the most important being that which lay along the celestial equator (ecliptic), the apparent path followed by the Sun, the Moon, and the planets across the backdrop of the sky. The Latin names of the **signs of the zodiac** as we know them today are translations of the old Babylonian constellations.

In connection with the planets, the Babylonians appear to have been motivated by religious-philosophical reasons to take note only of isolated events, such as a planet's first and last appearances in the sky. Such occurrences were taken to have astrological significance: they might foretell human fate. There is no evidence that the Babylonians, unlike the Greeks, came up with any geometrical model of the cosmos. Even so, at the height of its creativity, in the so-called Seleucid era, around 600 B.C., their astronomy could predict planetary motions with surprising accuracy. The Babylonians made careful observations from ancient times using a powerful mathematical tool, the sexagesimal system of numbers–a place-value system based on 60 that we still use. Babylon became part of the Persian empire, and its glory dimmed for a while. However, after Alexander conquered the Persian empire, Babylon's culture and science had a significant influence on the Greeks. See also **Egyptian astronomy, Chinese astronomy, Indian astronomy**, and **Greek astronomy**.

Bailey, Solon Irving (1854–1931)

An American astronomer who joined the **Harvard College Observatory** staff in 1879 and was for many years in charge of the Harvard southern station at Arequipa, Peru. His studies of **globular clusters** led to the discovery of "cluster variables," now known as **RR Lyrae stars.**

Bailey types

A classification of **RR Lyrae stars** according to the shape and amplitude of their **light curves**, devised by Solon **Bailey** in 1899. Types *a, b,* and *c* were defined originally, but types *a* and *b* are usually combined today. RR*a* stars show a sharp rise to maximum followed by a slow fall to minimum; RR*c* stars have a smaller amplitude and a rise and fall of equal length.

Baily's beads

Small "beads" of sunlight that shine through the valleys on the limb of the Moon in the instant before (or after) totality in a **solar eclipse**. They are named after the English astronomer Francis Baily (1774–1844) who first drew attention to them in 1836. See also **diamond ring effect**.

Balmer series

A series of emission or absorption lines in the visible part of the **hydrogen spectrum** that is due to transitions between the second (or first excited) state and higher energy states of the hydrogen atom; they are named after their discoverer, the Swiss physicist Johann Balmer (1825–1898). The transition from the third level to the second level yields the red **H-alpha** (Hα) emission line at 6563 Å; Hβ is at 4861 Å, Hγ at 4342 Å, and Hδ at 4101 Å. The *Balmer jump* is the relatively abrupt decrease in a **continuous spectrum** at about 3650 Å caused by hydrogen absorption lines in the Balmer series crowding to their series limit.

Bamberga (minor planet 324)

A main-belt **asteroid** that has one of the darkest known surfaces in the solar system. It was discovered by Johann **Palisa** in 1892. Diameter 228 km, spectral class C, albedo less than 0.05, rotational period 8 hours. Orbit: semimajor axis 2.683 AU, perihelion 1.78 AU, aphelion 3.59 AU, eccentricity 0.338, inclination 11.10°, period 4.40 years.

band spectrum

A spectrum in which various series of very closely spaced **absorption lines**, due to **absorption** by molecules, appear as bands. Each line represents an increment of energy due to a change in the rotational state of a molecule. Bands caused by titanium oxide, zirconium, and carbon compounds are characteristic of low-temperature stars.

bandpass filter

A filter that transmits a continuous range (band) of wavelengths but reflects or absorbs wavelengths above and below this range. Bandpass filters are specified by the *full width at half maximum* (FWHM) of their transmission peak: wideband (greater than 60 nm FWHM), medium band (20 to 60 nm FWHM), and narrowband (less than 20 nm FWHM).

bandwidth

In **radio astronomy**, the range of frequencies occupied by a signal or to which a detector is sensitive.

Banneker, Benjamin (1731–1806)

An astronomer often considered the first African American man of science. He was born in Baltimore County, Maryland, the son of a freed slave and of Mary Banneker, daughter of a white indentured English servant and her husband Bannka, also a freed slave. Banneker received no formal schooling except for a few weeks in a one-room Quaker schoolhouse near his father's farm. Although he read widely his first love was mathematics, which led him to devise and construct a wooden striking clock around 1753. He raised tobacco, grew vegetables, and cultivated orchards and bees until the late 1780s, when he was forced to retire from farming due to poor health. Thereafter, he developed an interest in astronomy. By borrowing instruments and texts, he taught himself enough mathematics and astronomy to make celestial observations and calculations. In 1791, he was employed for three months as a surveyor on the project commissioned by President George Washington to lay out a national capital. A year later, he wrote to then Secretary of State Thomas Jefferson, submitting a copy of his tabular calculation (or ephemeris) showing the positions of celestial bodies. This manuscript, he hoped, would serve as an example of intellectual achievement by an African American and might motivate Jefferson and others to work toward ending slavery. Jefferson replied, "Nobody wishes more than I do to see such proofs as you exhibit, that nature has given to our black brethren, talents equal to those of other colors of men, and that the appearance of a want of them is owing to the degraded condition of their existence." The ephemeris was published in Baltimore under the title, *Benjamin Banneker's Pennsylvania, Delaware, Maryland, and Virginia Almanack and Ephermeris for the Year of Our Lord 1792.* Banneker's almanacs appeared annually until 1797, were promoted by abolitionist societies, and were distributed widely in the United States and England.

Bappu, (Manali Kallat) Vainu (1927–1982)

A prominent Indian astronomer who did much to set up new observatories and astronomical research centers in his native country. His father was an assistant at the Nizamiah Observatory in Hyderabad, and Bappu was exposed to astronomy at an early age. He won a scholarship to Harvard and codiscovered a comet, named Bappu-Bok-Newkirk, shortly after his arrival. Later he went to the Palomar Observatory where he and Colin Wilson discovered a relationship, now known as the *Wilson-Bappu effect,* between the luminosity of certain kinds of stars and some of their spectral characteristics. In 1953, he returned to India and worked at the Uttar Pradesh State Observatory before becoming director of the observatory at Kodaikanal, the oldest in India. He helped establish the Indian Institute of Astrophysics at Bangalore and set up the largest telescope in India, a 2.34 m reflector at the Kavalur Observatory–an instrument named in his honor in 1986. See also **Indian astronomy**.

barium star

A giant star of **spectral type** G2 to K4 whose spectrum contains unusually strong **absorption lines** of barium (notably the 4554 Å line) and other heavy elements. Also known as *heavy-metal stars* or *Ba II stars,* barium stars are thought to have been contaminated with by-products of nuclear fusion by companions that have already far evolved and are now **white dwarfs**. An example is Altarf (Beta Cancri).

Barlow lens

A diverging lens, placed in front of the **focal point** of a telescope, that increases both **magnification** and **eye relief** when used in conjunction with an eyepiece. The magnification is typically doubled, though at the cost of a fainter image. Invented in 1834 by the English engineer and mathematician Peter Barlow (1776–1862), the lens is commonly used by amateur astronomers.

Barnard, Edward Emerson (1857–1923)

One of the greatest observational astronomers of his time, and discoverer of Jupiter's moon **Amalthea** (the last satellite to be found without photographic aid) and the nearby star that now bears his name (see **Barnard's Star**). Born into poverty in Nashville, Tennessee, Barnard began work in a photographic studio when he was only nine. He became a brilliant amateur astronomer, discovering 10 comets before the age of 30. During the 1880s, a wealthy patron of astronomy in Rochester, New York, awarded $200 each time a new comet was found. Barnard soon earned enough to build a "comet house" for his bride. His discoveries and brilliance as an observer drew the attention of other amateur astronomers in Nashville who raised enough money for Barnard to attend Vanderbilt University. In 1887, Barnard joined the staff of **Lick Observatory** and used the new 36-inch (91-cm) Lick refractor to discover Amalthea and the first comet to be found by photography, both in 1892. In 1895 he moved to the University of Chicago's not-yet-completed **Yerkes Observatory** and helped test the great 40-inch (102-cm) refractor following its installation. On May 29, 1897, Barnard narrowly escaped death when, just hours after he had left the observatory's dome, the 37-ton elevating floor, used to lift observers to the level of the telescope's eyepiece, collapsed after a supporting cable broke. Barnard spent 28 years as an astronomer at Yerkes using the giant refractor as well as the 10-inch (25-cm) Bruce wide-field telescope, built specially for him, to measure star positions and to pioneer wide-field photography for studying the structure of the Milky Way. He discovered the star, subsequently named after him, with the largest known proper motion, and numerous dark clouds and **globules**. His *Photographic Atlas of Selected Regions of the Milky Way,* published posthumously in 1927, identifies 349 dark nebulae north of declination –35° that are still known by their Barnard (B) numbers.

Barnard's Galaxy Hubble-X, one of the most active star-forming regions within Barnard's Galaxy (NGC 6822), seen through the Hubble Space Telescope. *NASA/STScI/AURA*

Barnard's Galaxy (NGC 6822)

A dwarf **irregular galaxy** about 1.6 million light-years away and perhaps one-tenth our Galaxy's size; it lies in **Sagittarius** and is a member of the **Local Group**. In the 1920s, Edwin **Hubble** found three star clusters in Barnard's Galaxy that he believed were all very old objects similar to **globular clusters** in the Milky Way Galaxy. However, images taken by the Hubble Space Telescope have shown that the three clusters are of completely different ages. The stars in the cluster called Hubble VII were formed about 15 billion years ago and are about the same age as our own Galaxy and the universe itself. A second cluster known as Hubble VIII contains stars about 1.8 billion years old, while a third cluster, Hubble VI, has stars that are as young as 100 million years. It seems that whereas our Galaxy formed most of its big clusters in the first couple of billion years after the **Big Bang**, Barnard's Galaxy has been generating new massive star clusters all along. The largest currently active star formation region in NGC 6822 is Hubble-X, shown in the photograph. The nearly circular bright cloud at the core of Hubble-X measures about 110 light-years across and contains a central cluster, less than 4 million years old, of many thousands of young stars, the brightest of which can be seen in the Hubble image as numerous bright white dots. To give some idea of the scale of Hubble-X, the tiny cloud barely visible just below it is roughly about the same size and brightness as the **Orion Nebula**.

Barnard's Loop

A huge nebular shell, about 1,600 light-years away and appearing as a semicircular arc 14° across, that envelopes a large part of the **Orion Complex**. Although William **Herschel** may have been the first to see it in 1786, its discovery is generally credited to Edward **Barnard** (who called it the "Orion Loop") around 1900. Barnard's Loop is thought to have been formed by a series of **supernovae** about 3 million years ago and is kept luminous by a group of hot young stars in the Orion OB1 Association.

The ionized shell is part of an even larger hydrogen cloud measuring some 30° across.

Barnard's Star

The fourth nearest star to the Sun after **Proxima Centauri** and **Alpha Centauri** A and B. Barnard's Star is a **red dwarf** in the northernmost part of **Ophiuchus**, west of Cebelrai (Beta Ophiuchi). It was discovered in 1916 by Edward **Barnard** who found that it has the highest **proper motion** of any star–10.4″ per year, or the equivalent of a lunar diameter every 180 years. It is also approaching us at the unusually high rate of 108 km/s, so that every century its distance decreases by 0.036 light-year. By A.D. 11,800, at its point of closest approach, it will be just 3.85 light-years away. If put in place of the Sun, Barnard's Star would appear from Earth only 100 times brighter than the Full Moon and be such a feeble source of heat that our atmosphere would freeze. It appears to be an old **disk star** that formed before the Galaxy became enriched with heavy elements. Although it may already be around 10 billion years old, it will shine for at least another 40 billion years before cooling to become a black dwarf. See **stars, nearest**.

Visual magnitude:	9.54
Absolute magnitude:	13.24
Spectral type:	M3.8Ve
Distance:	5.94 light-years
Luminosity:	0.0005 L_{sun}
Radius:	0.2 R_{sun}
Mass:	0.2 M_{sun}

barred spiral galaxy

A **disk galaxy** with a rectangular or cigar-shaped nucleus; from the ends of this nucleus **spiral arms** extend. Bars are large bodies of gas, dust, and stars that rotate as if they were solid objects. Typically they are 2.5 to 5 times longer than they are wide and may contribute up to one-third of a galaxy's luminosity. By channeling gas and **dust** to the center of the galaxies, they may trigger bursts of star formation, or, alternatively, feed material to a supermassive **black hole** in the galactic core. Bars are believed to represent a temporary stage in the life of some spiral systems. According to one theory, they form spontaneously through global disturbances in disk galaxies. Another theory suggests they come about from interactions with nearby galaxies. Barred galaxies are grouped by their appearance using three criteria: the central bulge and light distribution; the tightness with which the spiral arms are wound; and the degree to which the spiral arms are resolved into stars and nebulae. The three main types are SBa, SBb, and SBc (see **galaxy classification**). (See photo, page 56.)

Barringer, Daniel Moreau (1860–1929)

A successful mining engineer who, in 1902, first heard of the crater in Arizona that would eventually be named after him (see **Barringer Crater**). A Princeton graduate (1879), Barringer came from a well-connected family, went hunting with Theodore Roosevelt, and owned a silver mine. When he learned that beads of iron were mixed in with the rocks of the crater's rim, Barringer immediately concluded that the great pit had been blasted out by an **iron meteorite**, roughly as wide as the crater, that now lay buried under the crater floor. Before even visiting the site, he formed the Standard Iron Company and began securing mining patents. But his excavation was doomed to failure. As we now know, the impacting body, although huge by everyday meteorite standards, was much smaller than Barringer had supposed, and most of it was broken apart and showered over a wide surrounding area. Barringer was wrong about the physics of hypervelocity cratering, and, after more than a quarter of a century of searching, his Meteor Crater Exploration and Mining Company ran out of money and shut down in 1929. He died shortly after, having lost nearly all his fortune. However, Barringer had by then succeeded in convincing most of the scientific community that his impact theory was correct.

Barringer Crater

The best known and best preserved **impact crater** on Earth; named after Daniel **Barringer** and still owned by his family, it is also known as Meteor Crater, Coon Butte, and Canyon Diablo. Measuring 1.2 km across and 175 m deep, with a rim 45 m higher on average than the surrounding plain, it lies 55 km east of Flagstaff, Arizona, at 35° 02′ N, 111° 01′ W. It was formed about 50,000 years ago by the impact of an **iron meteorite**, some 50 m across and weighing several hundred thousand tons. Most of the meteorite was vaporized or melted, leaving only numerous, mostly small fragments of the **octahedrite** type, scattered up to 7 km from the impact site. Only about 30 tons, including a 693-kg sample, are known to have been recovered. (See photo, page 57.)

Bartsch, Jakob (1600–1633)

A physician, professor of mathematics at the University of Strassburg (now Strasbourg), assistant to and son-in-law of Johannes **Kepler**, and author of various astronomical treatises. In his star chart of 1624, based on data by Philipp Müller of Leipzig, he provides the first known reference to three constellations: **Monoceros** ("Unicorni"), **Camelopardalis**, and **Columba**. Bartsch is also credited with naming **Crux**, the Cross, by regrouping four stars from Ptolemy's original Centaurus.

barred spiral galaxy NGC 1365, one of the most prominent barred spiral galaxies in the sky, imaged by the Very Large Telescope. It is a supergiant galaxy with a diameter of about 200,000 light-years, lies some 60 million light-years away in the direction of the southern constellation Fornax, and is a major member of the Fornax Cluster. *European Southern Observatory*

Barwell meteorite

The biggest meteorite to land in Britain in recorded history; a **carbonaceous chondrite**, it fell in pieces on the Leicestershire village of Barwell on Christmas Eve 1965. Fragments variously penetrated 20 cm into a tarmac drive, landed on the hood of a car, and smashed through a factory roof; one tiny piece was even found later in a vase. With the British Museum offering seven shillings and sixpence (the equivalent today of more than $10) per ounce for specimens, Barwell became something of a gold town. When the fragments were put together, the Barwell meteorite proved, appropriately, to be about the size of a Christmas turkey.

Barringer Crater An aerial view of the Barringer Crater in Arizona. *U.S. Geological Survey*

barycenter

The center of mass of a system of massive bodies and the point around which the system orbits. The barycenter of the Earth-Moon system is actually located inside our planet.

baryon

Any of the class of the heaviest subatomic particles that includes **protons**, **neutrons**, and other particles whose eventual decay products include protons. Baryons, which interact via the strong force, form a subclass of the **hadrons** and are subdivided into nucleons and hyperons. The ratio of the number of baryons to the number of photons (light particles) in the universe, the *baryon-to-photon ratio,* which is about 10^{-9}, is thought to have remained unchanged since the **Big Bang**.

baryon degenerate matter

A form of **degenerate matter** in which **electrons** have been forced into the confines of atomic nuclei by the weight of overlying material. Once inside the nucleus, electrons merge with **protons** to form neutrons, which, obeying the **Pauli exclusion principle**, provide the degeneracy pressure to resist further collapse. This pressure can only be overcome in a collapsed star that exceeds the **Oppenheimer-Volkoff limit**.

basalt

A fine-grained, dark-colored rock of volcanic origin composed primarily of **plagioclase feldspar**, and **pyroxine**, together with other minerals, usually including **olivine** and ilmenite (an oxide of iron and titanium). Basalt is the most common extrusive igneous rock on the terrestrial planets and covers about 70% of Earth's surface.

basin

A large, shallow, circular impact structure on the surface of a planet or moon. Basins may show concentric rings and are often filled in by lava from subsequent volcanic activity.

Bautz-Morgan classification

A system for classifying **clusters of galaxies** based on morphology. It recognizes three main types of clusters. Type I has a single supergiant cD galaxy, type III (e.g., **Virgo Cluster**) has no members significantly brighter than the general bright population, and type II (e.g., **Coma Cluster**) is intermediate between the other two.

Bayer, Johann (1572–1625)

A German lawyer and astronomer whose hugely influential star atlas, ***Uranometria*** (1603), was the first to cover the whole sky and introduced the system of labeling stars with Greek letters that remains in use today (see **Bayer designation**). He was less successful, however, in his attempts to reform the names of constellations. His posthumously published *Coelum stellatum christianum* (1627) proposed replacing their heathen labels with biblical ones, but scholars continued to prefer such traditional names as Cassiopeia and Argo to his suggested Mary Magdalen and Noah's Ark.

Bayer designation

The naming system for the brighter stars introduced by Johann **Bayer** in his star atlas ***Uranometria*** (1603), consisting of a Greek letter followed by the genitive of the name of the constellation in which the star lies. In principle, the brightest star of the constellation should be called Alpha, the next brightest Beta, and so on. In practice, there are many cases where the designations are out of order, and there are even cases where a star is named after a different constellation from the one in which it actually lies (according to the modern constellation boundaries). Two stars have double designations: Beta Tauri (Gamma Aurigae) and Alpha Andromedae (Delta Pegasi). Nonetheless, Bayer's system has proved useful and is widely used today. The designation can be written either in full, as in Alpha Canis Majoris or Beta Persei, or as a lowercase Greek letter before the standard three-letter abbreviation of the constellation, as in α CMa or β Per. Although most common Bayer letters are Greek, the system was extended, first by using lowercase Latin letters, and then by using uppercase Latin letters. Most of these are little used, but there are some exceptions such as h Persei (which is actually a star cluster) and P Cygni. Uppercase Latin Bayer designations never went beyond Q, and names such as W Virginis are **variable star** designations. A further complication is the use of numeric superscripts to distinguish between stars with the same Bayer letter. Usually these are double stars (mostly optical doubles rather than true binaries), but there are some exceptions such as the chain of stars π^1, π^2, π^3, π^4, π^5, and π^6 Orionis. See **Flamsteed number**.

Bayer's constellations

Twelve constellations in the southern hemisphere that were first described by Johann **Bayer** in his 1603 star atlas ***Uranometria.*** They are **Apus** (the Bird of Paradise), **Chamaeleon, Dorado** (the Goldfish), **Grus** (the Crane), **Hydrus** (the Lesser Water Snake), **Indus** (the Indian), **Musca** (the Fly), **Pavo** (the Peacock), **Phoenix** (the Firebird), **Triangulum Australe** (the Southern Triangle), **Tucana** (the Toucan), and **Volans** (the Flying Fish).

B-class asteroid

A subcategory of **C-class asteroids** whose members have **albedos** in the range 0.04 to 0.08. Examples include (2) **Pallas** (the second largest asteroid), (379) Huenna (diameter 62 km), and (431) Nephele (diameter 78 km).

Be star

A **B star** that shows **emission lines** of hydrogen, typically with variable and complex line profiles, of which **Achernar** is a well-known example. The emission comes from a circumstellar disk of material that has formed due to a combination of mass loss and rapid rotation. The one-fifth of B stars that fall into the Be category spin unusually fast (even for B stars), with rotational velocities of 250 to 500 km/s. Different models propose disk radii ranging from 2 to 20 times the radius of the central star, which may itself have a radius of 3 to 12 R_{sun}. Be stars vary in brightness and spectrum, on several different timescales. There are variations on long timescales of weeks to decades connected with the formation and dispersal of the disk, on medium timescales of days to weeks often connected with the binary motion of some of these stars, and on short timescales of 0.3 to 2 days that may be due to **nonradial pulsation** or rotation. Some Be stars, such as **Gamma Cassiopeiae**, undergo occasional fadings and show deep, narrow **absorption lines** due to the ejection of a shell. Individual stars can change from a B to a Be to a B-type **shell star** and back to B again. In certain cases, whether an object appears as a Be star or a shell star may depend on the direction from which it is being viewed. O and A stars that show the same behavior are customarily referred to as Be stars, too.

One of the big theoretical problems with Be stars has been to explain why material in the disk is held some distance from the star instead of either being pulled closer or flying away into space. A promising new model, described in 2002, calls for the existence of a magnetic field around Be stars producing what is called a *magnetically torqued disk.* Magnetic field lines channel **stellar wind** material leaving the surface of the star down toward the equatorial plane. A disk then forms in the region where particles have enough angular velocity to counteract gravity. This model suggests that only a narrow range of types of star would form a detectable magnetically torqued disk and be seen as Be stars. Heavier stars would need an unreasonably large magnetic field while lighter stars would produce disks too small to be detected.

beat Cepheid

A **Cepheid variable** in which two or more nearly equal periods of variability pass into and out of phase with each other, producing a "beat." Also known as *double-mode Cepheids,* such stars account for nearly half of all Cepheids with periods of 2 to 4 days. Beat periods are typically about 2 hours.

Becklin-Neugebauer Object (BN Object)

An intense source of **infrared** radiation in the **Orion Nebula** associated with a **dust** cloud that surrounds a newly formed star. Discovered by Eric Becklin (1940–) and Gerry Neugebauer (1932–) at the California Institute of Technology in 1966–one of the first big finds of infrared astronomy–it is the brightest object in the sky (apart from the Sun) at wavelengths of around 10 microns. The BN Object is about the size of the solar system and has a surface temperature of only about 700 K, but deep inside, hidden from view, is a luminous star with a mass of some 15 M_{sun} and a surface temperature of some 26,000 K. Its birthing process involved the hot, bright stars of the **Trapezium** whose radiation exerts pressure on the nebular material around them. In the direction toward the Sun, this pressure has already dispersed the remaining dust and gas from which the Trapezium stars formed. In the direction away from us, however, the radiation pressure continues to compress dust and gas, triggering the collapse of new stars in the region where the luminous nebula and the dark, dense cloud behind it meet: the interface at which the BN Object was formed. Nearby, in the same environment, lies more evidence of stellar birth in the form of the **Kleinmann-Low Nebula**.

Beer, Wilhelm (1797–1850)

A Berlin banker and brother of the composer Meyerbeer, who set up a private observatory equipped with a 9.5-cm refractor. In 1830, in collaboration with Johann von **Mädler**, he produced the first accurate lunar map (*Mappa selenographica*) and a companion volume (*Der Mond*), describing the surface features. Beer and Mädler's map remained the best available for several decades and

Becklin-Neugebauer Object Deeply embedded in the Orion Nebula, the Becklin-Neugebauer (BN) Object (the brightest object in the photo), is seen clearly in this image from the Very Large Telescope. In the immediate vicinity of the BN Object are highly dynamic outflows and cloudlets, glowing in the light of shock-excited molecular hydrogen. *European Southern Observatory*

helped persuade most professional astronomers that the Moon is uninhabited. The two also collaborated in the production of the first systematic chart of the surface of **Mars**.

Belinda

The ninth moon of **Uranus**, also known as Uranus XIV; it was discovered in 1986 on images returned by **Voyager** 2. Belinda has a diameter of 68 km and moves in a nearly circular orbit 75,260 km from (the center of) the planet.

Bell Burnell, (Susan) Jocelyn (1943–)

A British observational astronomer known for her discovery of **pulsars**, for which her graduate adviser, Anthony **Hewish**, won the 1974 Nobel Prize (with Martin **Ryle**). Born in Belfast (her father was architect of the Armagh Observatory) and educated at York, Glasgow, and Cambridge, Bell (later Burnell) used a radio telescope at Cambridge that she helped build to detect a rapid set of pulses occurring at regular intervals. She determined that the unusual radio source lay beyond the solar system and, over the next few months, discovered three more pulsating radio sources–**pulsars**–later found to be rapidly rotating neutron stars. She subsequently carried out research at gamma-ray, X-ray, infrared, and short radio wavelengths, and currently holds the chair of physics at the Open University in England. Although Burnell shared the prestigious Michelson Award with Hewish in 1973, the Nobel Committee did not acknowledge her role in the discovery of pulsars when it awarded Ryle and Hewish the 1974 Nobel Prize in physics "for their pioneering research in radio astrophysics. Ryle for his observations and inventions . . . and Hewish for his decisive role in the discovery of pulsars." Many distinguished astronomers, including Fred Hoyle, Thomas Gold, and Jeremiah Ostriker, argued that Burnell should have shared the Nobel Prize. In 1993, when the award went to the discoverers, Russell Hulse and Joseph Taylor Jr., of the first **binary pulsar**, both student and supervisor were recognized.

Bellatrix (Gamma Orionis, γ Ori)

A blue-white giant **B star** that is one of the hotter stars visible to the naked eye. Its Latin name means "female warrior." Bellatrix was once believed to be part of the physical association of O and B stars that makes up much of **Orion**, but it is now known to lie much closer to the Sun and therefore to be an independent foreground object. Bellatrix also used to be taken as a standard against which astronomers could track the fluctuations of variable stars, but now it seems that Bellatrix is itself slightly variable, changing in brightness by a few percent over an undetermined period.

Visual magnitude:	1.64
Absolute magnitude:	–2.73
Spectral type:	B2III
Temperature:	21,500 K
Luminosity:	2,400 L_{sun}
Mass:	8 to 9 M_{sun}
Distance:	243 light-years

Bennett, Comet (C/1969 Y1)

A **long-period comet** discovered on December 28, 1969, by the South African amateur astronomer John ("Jack") Bennett (1914–1990). Having reached perihelion on March 20, 1970, it put on a spectacular show throughout April 1970, reaching zero magnitude with a tail up to 30° long. Observations made with the fifth Orbiting Geophysical Observatory showed the head of the comet to be surrounded by a vast hydrogen cloud, spanning some 13 million km in the direction parallel with the tail. Orbit: perihelion 0.54 AU; eccentricity 0.996; inclination 90.0°; period about 1,700 years.

Bessel, Friedrich Wilhelm (1784–1846)

A German astronomer and mathematician who made the first measurement of a star's distance by **parallax** (1838) and deduced the existence of unseen companions of **Sirius** and **Procyon** (1844). Bessel became director of the new Königsberg Observatory in 1810. After publishing his *Fundamenta Astronomiae* (1818), a **fundamental catalog** of 3,000 stars, he began a major program of measuring star positions and **proper motions**, and by 1833 had accurate data on 50,000 stars. These he combined with previous observations into a catalog of 63,000 stars, which provided the base for modern **astrometry**. The value he found for the parallax of **61 Cygni** is only 6% higher than the figure accepted today.

Besselian elements

(1) Values used in calculating solar eclipses that establish the position and size of the Moon's shadow relative to Earth. (2) Values used in calculating an occultation of a star or a planet by the Moon.

Besselian year

The basic unit of the *Besselian epoch,* defined as the time taken for the **right ascension** of the **mean sun** to increase by 24 hours starting from when the mean sun's longitude is 280° (chosen because it corresponds roughly to January 1). It is virtually the same as the tropical **year**.

Beta Cephei star (β Cep star)

A **pulsating variable**, of spectral type O9 to B3, with a short period (3.5 to 6 hours) and small magnitude range of 0.1 to 0.3 (less than those of **Cepheid variables**). Beta Cephei stars are confined to a narrow band of the **Hertzsprung-Russell diagram** near the end of core-hydrogen-burning stars of roughly 10 to 20 M_{sun}. Most such stars expand and contract radially, but a few have odd quivering motions with complex and multiple periods that give them more random-looking light curves. The prototype is **Alfirk** (Beta Cephei), although members of the group are also called *Beta Canis Majoris stars,* after another bright example, also known by its proper name **Mirzam**.

Beta Lyrae (β Lyr)

The prototype for a remarkable class of interacting **eclipsing binary** stars (see **Beta Lyrae stars**); its proper name, Sheliak, comes from an Arabic word for "harp," the constellation in which the star lies. β Lyr is a **close binary** of the *semidetached* variety, consisting of a **B star,** which is the visible component, and another much more massive (12 M_{sun}) **main sequence** star that is largely obscured by a thick **accretion disk** of material that has been lost by the smaller star. The plane of the orbit is pitched so that, from our vantage point, each star passes alternately in front of the other. The bigger star completely eclipses its companion every 12.9 hours (giving a primary minimum of magnitude 4.3) and, 6.5 hours later, the companion blocks some of the giant's light (causing a secondary minimum of 4.8). In addition, being massive and very close together, the two stars have pulled each other into an egg-shape so that β Lyr, and systems like it, are not only eclipsing but also **ellipsoidal variables**. β Lyr's light changes, easily visible to the naked eye by comparing the star to others in the constellation, were first noted by John **Goodricke** in 1784.

Visual magnitude:	3.4 to 4.3
Absolute magnitude:	–3.64
Spectral type (primary):	B7Ve
Mass (primary):	$2M_{sun}$
Distance:	881 light-years

Beta Lyrae star (β Lyr star)

A binary star that is both an **ellipsoidal variable** and an **eclipsing binary**, and has a secondary minimum intermediate between that of a **Beta Persei star** and a **W Ursae Majoris star**. Even when the components are not in eclipse, we view them from different angles and see a different amount of light, so that the brightness changes are fairly smooth and continuous, unlike the abrupt changes of a normal eclipsing system. The prototype is **Beta Lyrae**.

beta particle

A fast-moving **electron** or positron (antielectron) that is emitted from a nucleus during the radioactive process known as *beta decay*.

Beta Pictoris (β Pic)

A relatively nearby star that is surrounded by a large **dust** ring seen almost edge-on. The existence of this circumstellar disk, containing gas and dust at a temperature of about 100 K and extending up to 600 AU from the star, was first revealed in 1983 by the **Infrared Astronomy Satellite** (IRAS). An image of the disk was obtained in 1984 by Richard Terrile and Bradford Smith using the 2.5-m telescope at Las Campanas Observatory. Speculation followed about whether it might be a protoplanetary disk; however, this is almost certainly not the case, since β Pic is a **main sequence** star with an age of approximately 100 million years. The dust presently in its disk has a lifetime on the order of 100 *thousand* years and therefore must undergo continuous replenishment. Although the disk is unlikely to be the site of planet formation today, it may be that existing planets move within it. Support for this idea came with the observation of a gap in the disk between radii of 10 to 30 AU. It has been suggested that the dust-free region may have been cleared by a planet of about five Earth masses moving in a near-circular orbit 20 AU from the central star. Spectral analysis has revealed **redshifted** absorption features that may be due to comets falling onto the star's outer atmosphere. If this is the case, it is again consistent with the presence of a planetary system since one or more gas giants might be responsible for the infall events. In 1996, the Hubble Space Telescope revealed a small warp in the disk within 80 AU that, according to one theory, is due to the gravitational influence of a massive planet orbiting at right angles to the disk. In 1998 came the announcement of a larger warp farther out in the disk, some 11 billion km from the star. This might be due to a **brown dwarf** in a wide orbit around the central star.

Visual magnitude:	3.85
Absolute magnitude:	2.42
Spectral type:	A3V
Distance:	63 light-years

Betelgeuse (Alpha Orionis, α Ori)

A red **supergiant** and **irregular variable** that is (usually) the tenth brightest star in the sky and second brightest in **Orion**, although it occasionally outshines **Rigel** (Beta Ori), with which it makes a striking color contrast. At an infrared wavelength of 2 microns, Betelgeuse is the brightest object in the sky. Its name (also written as "Betelgeuze" or "Betelgeux") is a corruption of the Arabic *yad al jauza*, which means the "hand of al-jauza," *al-jauza* being the ancient Arabs' "Central One," a mysterious woman. It is often pronounced "beetle-juice," though "betel-jerze" is closer to the original Arabic. It marks the upper left corner of the figure of the ancient hunter. One of the sky's two first magnitude supergiants (the other is **Antares**), Betelgeuse is among the larger stars on view to the naked eye, with a measured angular size of 0.05″, corresponding to a radius of 2.9 AU at 427 light-years, so that if put in place of the Sun it would extend out as far as the asteroid belt. However, uncertainty about its distance allows a possible range in size of 45 to 70% of Jupiter's orbit, and a corresponding luminosity range (including infrared) of 40,000 to 100,000 L_{sun}. Its disk is so large that it was the first to be reconstructed by the technique of **speckle interferometry** (1975) and the first to be directly imaged, by the Hubble Space Telescope (1995). The Hubble image revealed a huge atmosphere with a mysterious hot spot on the star's surface. The enormous bright spot, more than 10 times the diameter of Earth, is at least 2,000 K hotter than the rest of the star's surface. Measurements by the International Ultraviolet Explorer found a 420-day period, during which the star vibrates like a ringing bell. These oscillations, thought to be caused by turbulence below the surface of the star, may cause the bright spot's position to change over time. Betelgeuse is also surrounded by a huge **dusty** circumstellar shell, extending up to 1 trillion km from the star and composed of matter that has been ejected in the form of a vigorous **stellar wind**. The first image of the cloud was captured in 1998 using the **Multiple Mirror Telescope**. Dust is not the only thing surrounding Betelgeuse: in 1985 it was found to have two close companions. The inner orbits every two years or so at a mean distance of about 5 AU, while the outer lies some 40 to 50 AU away. The chances are that Betelgeuse is now fusing helium in its core. Theory suggests that its mass has now fallen to between 12 and 17 M_{sun}. If at the high end of this range, the core will fuse elements through neon, magnesium, sodium, and silicon all the way to iron. The star will then explode as a **supernova**. If it were to explode today, it would become as bright as a crescent Moon, would cast strong shadows on the ground, and would be easily seen in full daylight. If the star is near or under the lower end of the predicted mass range, then it may eventually become a neon-oxygen **white dwarf** about the size of Earth.

Visual magnitude:	0.0 to 1.3
Absolute magnitude:	–5.14
Spectral type:	M2Iab
Surface temperature:	3,100 K
Total luminosity:	60,000 L_{sun}
Radius:	630 R_{sun}
Distance:	427 light-years

Bethe, Hans Albrecht (1906–)

A German-American physicist who, in 1938, worked out the details of the **proton-proton chain**, the main energy-producing reaction in stars less massive than the Sun. In the same year, with Carl von **Weizsäcker**, he developed the theory of the **carbon-nitrogen-oxygen cycle**, which powers the Sun and more massive stars. Bethe studied at the universities of Frankfurt and Munich, where he earned his Ph.D. under Arnold Sommerfeld (1928). He made important contributions to the theories of electrons in crystals, the negative hydrogen ion, and the passage of charged particles through matter, before leaving Germany in 1933 for England. There he began working in nuclear physics, and he and Rudolph Peierls developed the theoretical model of the deuteron shortly after its discovery. Bethe has been at Cornell University since 1935. His three review articles in the late 1930s (two of them with coauthors) became known as Bethe's Bible and

Betelgeuse The first direct extended image of a star, other than the Sun, made from observations by the Hubble Space Telescope in 1995. Visible is Betelgeuse's huge atmosphere together with a mysterious hot spot on the lower part of the disk. Betelgeuse is the upper left bright star in the photo on the right showing the constellation Orion. *NASA/ESA/Andrea Dupree (Harvard-Smithsonian CfA) and Ronald Gilliland (STScI)*

formed the first textbook of nuclear physics. At the beginning of World War II, Bethe formed a theory of the penetration of armor by projectiles, and, with Edward Teller, produced a major paper on shock waves. From 1943 to 1946 he headed the theoretical group at Los Alamos, where the first nuclear bomb was designed and built. After the war he worked on nuclear matter and meson theory and was the first to explain the Lamb shift (a tiny change in the energy of an orbiting electron caused by interaction of the electron with other particles that come and go quickly in "quantum fluctuations") in the hydrogen atom. He has been an important adviser to government agencies on both defense and energy policies, and has written widely about arms control and the need for new energy policies.

Bianca

The third moon of **Uranus**, also known as Uranus VIII; it was discovered in 1986 on images sent back by **Voyager** 2. Bianca has a diameter of 44 km and moves in a nearly circular orbit 59,170 km from (the center of) the planet.

Biela's comet

A periodic comet named after Baron Wilhelm von Biela (1782–1856), an Austrian army officer and amateur astronomer, who observed it in 1826, although it had been seen first in 1772. Biela's comet broke up on its 1846 return and subsequently gave rise to some spectacular showers of the Andromedids, also known as the *Bielids*.

Big Bang

See article, pages 63–64.

Big Bear Solar Observatory (BBSO)

An observatory built on an artificial island in the middle of Big Bear Lake in the San Bernadino Mountains of south-

(continued on page 65)

Big Bang

The event in which, according to standard modern cosmology, the universe came into existence some 12 to 15 billion years ago. The Big Bang is sometimes described as an "explosion"; however, it is wrong to imagine that matter and energy erupted into a preexisting space. Modern Big Bang theory holds that space and time came into being simultaneously with matter and energy, hard as that may be to imagine. The possible overall forms that space and time could take–closed, open, or flat–are described by three different **cosmological models**.

Creation to Inflation

According to current theory, the first physically distinct period in the universe lasted from "time zero" (the Big Bang itself) to 10^{-43} second later, when the universe was about 100 million trillion times smaller than a proton and had a temperature of 10^{34} K. During this so-called *Planck era,* quantum gravitational effects dominated and there was no distinction between (what would later be) the four fundamental forces of nature–gravity, electromagnetism, and the strong and weak forces. Gravity was the first to split away, at the end of the Planck era, which marks the earliest point at which present science has any real understanding. Physicists have successfully developed a theory that unifies the strong, weak, and electromagnetic forces, called the Grand Unified Theory (GUT). The GUT era lasted until about 10^{-38} second after the Big Bang, at which point the strong force broke away from the others, releasing, in the process, a vast amount of energy that, it is believed, caused the universe to expand at an extraordinary rate. In the brief ensuing interval of so-called *inflation,* the universe grew by a factor of 10^{35} (100 billion trillion trillion) in 10^{-32} second, from being unimaginably smaller than a subatomic particle to about the size of a grapefruit.

Big Bang The initial expansion of the universe as computer-modeled by cosmologist Andre Linde based on his inflationary theory. *Andre Linde, Stanford University*

Postulating this burst of exponential growth helps remove two major problems in cosmology: the *horizon problem* and the *flatness problem.* The horizon problem is to explain how the **cosmic microwave background**–a kind of residual glow of the Big Bang from all parts of the sky–is very nearly isotropic (the same in all directions) despite the fact that the observable universe isn't yet old enough for light, or any other kind of signal, to have traveled from one side of it to the other. The flatness problem is to explain why space, on a cosmic scale, seems to be almost exactly flat, leaving the universe effectively teetering on a knife-edge between eternal expansion and eventual collapse. Both near-isotropy and near-flatness follow directly from the inflationary scenario.

Electroweak Era (10^{-38} to 10^{-10} second)

At the end of the inflationary epoch, the so-called *vacuum energy* of space underwent a phase of transition (similar to when water vapor in the atmosphere condenses as water droplets in a cloud) suddenly giving rise to a seething soup of elementary particles, including photons, gluons, and quarks. At the same time, the expansion of the universe dramatically slowed to the "normal" rate governed by the **Hubble law**. At about 10^{-10} second, the electroweak force separated into the electromagnetic and weak forces, establishing a universe in which the physical laws and the four distinct forces of nature were as we now experience them.

Particle Era (10^{-10} to 1 second)

The biggest chunks of matter, as the universe ended its first trillionth of a second or so, were individual quarks and their **antiparticles**, anti-quarks–the underlying particles out of which future atoms, asteroids, and astronomers would be made. As time went on, quarks and anti-quarks annihilated each other. However, either because of a slight asymmetry in the behavior of the particles or a slight initial excess of particles over antiparticles, the mutual destruction ended with a surplus of quarks. Only because of this (relatively minor) discrepancy do stars, planets, and human beings exist today.

Between 10^{-6} and 10^{-5} second after the beginning of the universe, when the ambient cosmic temperature had fallen to a balmy 10^{15} K, quarks began to combine to form a variety of **hadrons**. All of the short-lived hadrons quickly decayed leaving only the familiar **protons** and **neutrons**, of which the nuclei of atoms-to-come would be made. This *hadron era* was followed by the *lepton era*, during which most of the matter in the universe consisted of **leptons** and their antiparticles. The lepton era drew to a close when the majority of leptons and antileptons annihilated one another, leaving, again, a comparatively small surplus to populate the future universe.

One to 100 Seconds

Up to this stage, neutrons and protons had been rapidly changing into one another through the emission and absorption of **neutrinos**. But, by the age of 1 second, the universe was cool enough for neutron-proton transformations to slow dramatically. A ratio of about seven protons for every neutron ensued. Because only one proton is needed to make a hydrogen nucleus, whereas helium requires two protons and two neutrons, a 7:1 excess of protons over neutrons would lead to a similar excess of hydrogen over helium–which is what is observed today. At about the 100-second mark, with the temperature at a mere billion K, neutrons and protons were able to stick together. The majority of neutrons in the universe wound up in combinations of two protons and two neutrons as helium nuclei. A small proportion of neutrons contributed to making lithium, with three protons and three neutrons, and the leftovers ended up in deuterium–an isotope of hydrogen with one proton and one neutron.

The First 10,000 Years

Most of the action, at the level of particle physics, was compressed into the first couple of minutes after the Big Bang. Thereafter, the universe settled down to a much lengthier period of cooling and expansion in which change was less frenetic. Gradually, more and more matter was created from the high energy radiation that bathed the cosmos. The expansion of the universe, in other words, caused matter to lose less energy than did the radiation, so that an increasing proportion of the cosmic energy density came to be invested in nuclei rather than in massless, or nearly massless, particles (mainly photons). From a situation in which the energy carried by radiation dominated the expansion of space-time, the universe evolved to the point at which matter became the determining factor. Around 10,000 years after the Big Bang, the *radiation era* drew to a close and the *matter era* began.

When the Universe Became Transparent

About 300,000 years after the Big Bang, when the cosmic temperature had dropped to just 3,000 K, the first atoms formed. It was then cool enough to allow protons to capture one electron each and form neutral atoms of hydrogen. While free, the electrons had interacted strongly with light and other forms of electromagnetic radiation, making the universe effectively opaque. But bound up inside atoms, the electrons lost this capacity, matter and energy became decoupled, and, for the first time, light could travel freely across space. This, then, marks the earliest point in time to which we can see back. The cosmic microwave background is the greatly **redshift**ed first burst of light to reach us from the early universe and provides an imprint of what the universe looked like about a third of a million years after the Big Bang. Fluctuations in the nearly uniform density of the infant universe show up as tiny temperature differences in the microwave background from point to point in the sky. These fluctuations are believed to be the seeds from which future galaxies and clusters of galaxies arose. See also **galaxy formation**.

Big Bear Solar Observatory (BBSO)

(continued from page 62)

west California. The island location reduces the image distortion that usually occurs when the Sun heats the ground and produces convection in the air above. Turbulent motions in the air near the observatory are also reduced by the smooth flow of wind across the lake. These conditions, combined with the usually cloudless skies over Big Bear Lake and the clarity of the air at its 2,000-m elevation, make the observatory a premier site for solar observations. The top floor of the observatory contains a single fork mount supporting four main instruments: a 65-cm vacuum reflector, and three smaller refractors for **earthshine** studies. The telescopes are equipped with special filters and cameras that isolate small portions of the visible, near infrared, and near ultraviolet portions of the Sun's spectrum. BBSO was built by the California Institute of Technology in 1969. Management of the observatory, and of an array of solar radio telescopes at **Owens Valley Radio Observatory** was transferred to the New Jersey Institute of Technology in 1997.

Big Crunch

The endpoint of a *closed universe* (see **cosmological model**). If the universe contains enough mass to halt the **Hubble flow** and reverse it, all the matter contained within the universe will eventually be drawn together by the force of its mutual gravitational attraction. This collapse will cause the matter to be concentrated in an ever decreasing volume with densities and temperatures reaching those last seen in the **Big Bang**. In one scenario, since a Big Crunch would match the conditions of the Big Bang, it could mark the start of a new phase of expansion and the universe would begin again. This concept is known as the *oscillating universe.*

Big Dipper

The familiar bright seven-star asterism in **Ursa Major**, known in Britain as the Plough. The five middle stars are all moving through space together as part of a loosely bound group known as the **Ursa Major Moving Cluster**. **Alkaid** and **Dubhe**, however, are moving in different directions, ultimately dooming the Dipper's shape.

binary asteroid

An **asteroid** that consists of two roughly equal parts that revolve around each other at close range. Several examples are known including (90) **Antiope** and (762) Pulcova. One way to explain such twinning is in terms of impacts that reduced the ancestral body to a collection of rubble. A subsequent glancing blow by another asteroid might then have spun the rubble pile, causing it to fly apart into two equal-size piles that still orbited their center of mass. It has also been hypothesized that the nearly 10% of large craters on Earth that are doublets (e.g., Clearwater Lakes Craters) were formed by the impact of binary asteroids (see **Earth impact craters**). This suggests that there could be a substantial number of binaries among the **near-Earth asteroid** population, and, moreover, that Earth may be the cause of these binaries in the first place. A **rubble-pile asteroid** passing close to Earth could be pulled apart by the planet's tidal force, then, at a later time, collide with Earth and create two nearly equal impact craters.

binary pulsar

(1) Two **pulsars** in orbit around their mutual center of gravity. The first such system to be discovered was PSR 1913+16, also known as the **Hulse-Taylor Pulsar**. (2) More generally, a system in which a pulsar orbits any other type of star; examples include the **Black Widow Pulsar** and **Centaurus X-3**.

binary star

Two stars that orbit around a common center of mass. *Physical* **double stars** are binaries, whereas *optical* doubles are not. A **visual binary** is one whose components can be resolved visually or photographically. An **astrometric binary** gives itself away by a regular wobble in the motion of a visible component. In a **spectroscopic binary**, the apparent separation is so small that the presence of two stars can only be deduced from regular changes in the **Doppler shift** of the stars' spectral lines. In a **close binary** the separation is comparable to the diameter of the stars. At least half of the stars in space appear to be members of binary or **multiple star** systems.

binoculars

A pair of low-power prismatic spotting scopes mounted together for three-dimensional viewing. Prisms and/or porro mirrors shorten the instrument size and deliver a corrected image to the viewer. The common, incorrect usage, "a pair of binoculars," really means two sets of twin oculars, or four oculars. Binoculars are specified by numbers such as 10 × 50, the first of which gives the magnification and the second the aperture in millimeters. Good binoculars with a large aperture (50 mm or more) are a useful astronomical tool for comet-hunting and observing large star clusters and nebulae.

Biot, Jean-Baptiste (1774–1862)

A French physicist who demonstrated the cosmic origin of **meteorites**, which had been suggested by the Swiss physicist Marc Pictet (1752–1825). Biot is best known for his investigations of polarized light and optical rotation, and, with Savart in 1820, for establishing a formula for the magnetic field of a long, straight, current-carrying conductor (known as the Biot-Savart law). In 1803 his analysis of specimens from the **l'Aigle meteorite shower** and eyewitness reports enabled him to show beyond doubt that the stones had come from space.

bipolar group

A pair or group (*multipolar*) of **sunspots** in which the leading and following spots (with respect to the Sun's rotation) are of opposite magnetic polarity. Such groups obey **Hale's polarity law**.

bipolar nebula

A gas cloud with two main lobes that lie symmetrically on either side of a central star from which there is a **bipolar outflow**.

bipolar outflow

A stream of matter in two opposing directions from a central object, usually a star. Bipolar outflows represent significant periods of mass loss in a star's life. They tend to occur during the **protostar** and pre-main-sequence phase and, again, during the **red giant** phase just before the production of a **planetary nebula** (e.g., see **Ant Nebula**). It isn't certain if the bipolar flow is caused by a lack of material being ejected at other stellar latitudes or if something, such as an **accretion disk**, blocks the material in the equatorial regions and allows only that which is ejected at the poles to escape. It has also been suggested that magnetic fields may constrain the outflowing material. The outflows carve out cavities in the surrounding interstellar medium and result in the formation of **bipolar nebulae**.

Birkeland, Kristian Olaf Bernhard (1867–1917)

A Norwegian physicist who studied and gave the first accurate explanation of the **aurora** borealis. Although supported both by his own observations and experiments in the laboratory, Birkeland's theories were not widely accepted until confirmed by satellite evidence in the 1960s. To fund his research, Birkeland cofounded a commercial venture to produce artificial fertilizer (Norsk Hydro, Norway's largest company today) and his technique for nitrogen fixation brought him into contention for a Nobel Prize. Among his other inventions were an electromagnetic cannon, an improved design for hearing aids, and a new method of manufacturing margarine. Eccentric, enthusiastic, and absorbed in his work, Birkeland kept notes on scraps of paper that he filed under seat cushions, wore a fez and slippers in his lab, and was hurled through the air by bolts of electricity when his experiments went awry. He arranged to give an important lecture on the morning of his wedding and was obliged to talk quickly in order to reach the ceremony on time. After five years of neglect, Birkeland's wife left him. In 1913, Birkeland traveled to Egypt to study the **zodiacal light** and found himself stranded there after the outbreak of war in Europe. Isolated from friends and colleagues, and already accustomed to drinking heavily and taking large doses of Veronal to cure his insomnia, these two habits now spiraled out of control. Eventually, Birkeland accepted an offer to accompany the Danish consul back to the Baltic through Asia. In 1917, he was found dead in a Tokyo hotel room after taking 20 times the recommended dose of Veronal. The treatise, on which he had been working furiously in the final months of his life, was lost when a ship carrying his belongings back to Europe went down off the coast of Korea.

BL Herculis star (BL Her star)

A type of **W Virginis star** with a period of less than 8 days that also shows a characteristic bump on the fading portion of its **light curve**.

BL Lacertae object (BL Lac object)

A type of **active galactic nucleus** (AGN) characterized by a nearly featureless spectrum (i.e., virtually devoid of spectral lines); rapid, marked variability at radio, infrared, and optical wavelengths (often by several magnitudes over a few days or weeks); and strong, variable **polarization** at radio and optical wavelengths. The prototype, in **Lacerta**, was originally classified as a variable star. BL Lac objects belong to the category of AGN now known as **blazars** and are believed to be essentially **quasars** viewed directly down the axis of their powerful **jets**.

Blaauw, Adriaan (1914–)

A Dutch astronomer whose studies have included the motions of star clusters and associations, **runaway stars**, star formation, and the determination of the cosmic distance scale. Blaauw has figured prominently in international collaborations, including construction of the **European Southern Observatory** in Chile, the merger of several European astronomical journals into *Astronomy and Astrophysics,* and development of the **Hipparcos** astrometric satellite. Educated at the universities of Leiden and Groningen in the Netherlands, he has worked at Leiden, Yerkes, and Groningen, where he rebuilt the Kapteyn Astronomical Laboratory into a major research center on galactic structure.

Blaauw mechanism

A mathematical explanation, provided by Adriaan **Blaauw**, of the disruption of a binary star system as one component throws off a shell of gas and the resulting loss of gravitational attraction changes the orbit of, or ejects altogether, the companion star.

black dwarf

A degenerate star, either a **white dwarf** or a **neutron star**, that has cooled to the point at which it is no longer visible. The universe is probably not yet old enough to harbor any such objects.

Black Eye Galaxy (M64, NGC 4826)

A relatively nearby **spiral galaxy** with a conspicuous dark feature to one side of the bright nucleus. Discovered by Johan **Bode** in 1779, it is also called the *Sleeping Beauty Galaxy*. Although M64 can be glimpsed with good binoculars, the oval eye only starts to show in telescopes of 10- to 15-cm aperture. M64 has two counter-rotating systems of stars and gas in its disk: an inner zone, about 3,000 light-years in radius, that rubs along the inner edge of an outer disk, which rotates in the opposite direction at about 300 km/s and extends out to at least 40,000 light-years. This rubbing may explain the vigorous burst of star formation that is currently taking place in the galaxy and is visible as blue knots embedded in the huge dust lane. The strange disk and dust lane, according to one theory, may be the result of material from a former companion galaxy that has been accreted but has yet to settle into the orbital plane of the disk. Another suggestion is that M64 may be the prototype for a class of galaxies called ESWAG, or *evolved second wave activity galaxy*. According to this idea, the main spiral pattern consists of an intermediate-aged stellar **population**. Star formation first evolved outside the current "black eye" region, following the density gradient, manufacturing stars as long as there was enough interstellar matter available, and then slowly died out. As matter was re-released into space from the evolved stars, by way of stellar winds, supernovae, and planetary nebulae, more and more interstellar matter accumulated again, until finally there was enough to enable a new wave of star formation to begin. This second wave, the theory maintains, has now reached the region where the dark dust lane appears.

Visual magnitude:	8.5
Apparent size:	9.3′
Diameter:	51,000 light-years
Distance:	19 million light-years

black hole

See article, page 68.

Black Widow Pulsar

A **binary pulsar**, discovered in 1988, in which a pulsar, with a pulse period of 1.6 milliseconds, is in an eclipsing orbit with a normal stellar companion. The signal from the pulsar is delayed, causing the period to lengthen, a few minutes before eclipse and for about 20 minutes after the pulsar reappears. This delay arises because the signal must travel through the ionized plasma surrounding the companion. It is thought that most of the companion's material lies well beyond its **Roche lobe**, and that this outer material is being blown off the companion by the neutron star. Calculations suggest that the Black Widow will evaporate its companion in about 1 billion years.

blackbody

A theoretical object that is both a perfect absorber and a perfect radiator of **electromagnetic radiation**. A blackbody absorbs all the radiation that falls on it, converts it into internal energy (heat), and then reradiates this energy into the surroundings. The reradiated thermal energy, known as *blackbody radiation,* has a **continuous spectrum** governed solely by the body's temperature. For any given temperature, there is a specific wavelength at which radiation emission is greatest.

The *effective temperature* (T_e), or *blackbody temperature,* is the surface temperature that a star, or other object, would have if it were a blackbody that radiated the same amount of energy per unit area. This is a useful and widely employed measure of stellar surface temperature. T_e can be calculated from the *Stefan-Boltzmann law,* which states that the total energy radiated by a blackbody varies as the fourth power of its absolute temperature. This law leads to the formula:

$$L = 4\pi R^2\, \sigma T_e^4$$

where L is the luminosity of the body, R is its radius, and σ (= 5.67×10^{-8} W/m²/K⁴) is the Stefan-Boltzmann constant. Substituting solar values for L and R gives a value for the effective temperature of the Sun of about 5,780 K.

Blagg, Mary Adela (1858–1944)

An English amateur astronomer who played an important role in standardizing the names of features on the Moon. Blagg lived her whole life in Cheadle, Staffordshire, where she was born. She taught herself math from her brother's textbooks and became an expert at solving equations in harmonic analysis. In middle age, she attended a university extension course in astronomy and proved so keen to continue with original work that her tutor suggested she look at the problem of developing a uniform lunar nomenclature. At the time, there were big discrepancies between different maps of the Moon. In 1905, the International Association of Academies set up a committee to address this issue and Blagg was appointed to collate the names of all the lunar formations. Her *Collated List* was published in 1913. In 1920, she was appointed to the Lunar Commission of the newly formed **International**

(continued on page 69)

black hole

An object whose gravitational field is so intense that its escape velocity exceeds the speed of light. In theory, any sufficiently compressed mass would become a black hole. The Sun would suffer this fate if it were shrunk down to a ball about 2.5 km in diameter. In practice, a stellar black hole is only likely to result from a heavyweight star whose remnant core exceeds the **Oppenheimer-Volkoff limit** following a **supernova** explosion. More than two dozen stellar black holes have been tentatively identified in the Milky Way, all of them part of binary systems in which the other component is a visible star. Observations of highly variable X-ray emission from the **accretion disk** surrounding the dark companion together with a mass determined from observations of the visible star enable a black hole characterization to be made. Among the best stellar black hole candidates are **Cygnus X-1**, V404 Cygni, and several **microquasars**. One of the latter, an object known as GRS 1915+105, is the heaviest stellar black hole found to date, with a mass of 14 M_{sun}. Given that massive stars lose a significant fraction of their content through violent **stellar winds** toward the end of their lives, and that interaction between the members of a binary system can further increase the mass loss of the heavier star, it is a challenge to theorists to explain how any star could retain enough matter to form a black hole as heavy as that of GRS 1915+105.

Supermassive black holes are known almost certainly to exist at the center of many large galaxies, and to be the ultimate source of the energy behind the phenomenon of the **active galactic nucleus** (AGN). At the other end of the scale, it has been hypothesized that countless numbers of mini black holes may populate the universe, having been formed in the early stages of the **Big Bang**; however, there is yet no observational evidence for them. In 2002, astronomers found a missing link between stellar-mass black holes and the supermassive variety in the form of middleweight black holes at the center of some large **globular clusters**. The giant G1 cluster in the **Andromeda Galaxy** appears to contain a black hole of some 20,000 M_{sun}. Another globular cluster, 32,000 light-years away within our own Milky Way, apparently harbors a similar object weighing 4,000 M_{sun}. Interestingly, the ratio of the black hole's mass to the total mass of the host cluster appears constant, at about 0.5%. This proportion matches that of a typical supermassive black hole at a galaxy's center, compared to the total galactic mass. If this result turns out to be true for many more cluster black holes, it will suggest some profound link between the way the two types of black holes form. It is possible that supermassive black holes form when clusters deposit their middleweight black hole cargoes in the galactic centers, and they merge together.

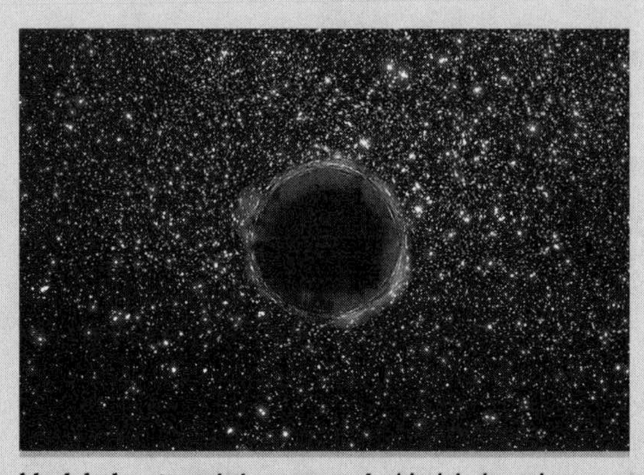

black hole An artist's concept of a black hole at the center of a globular cluster. *NASA/G. Bacon*

According to the **general theory of relativity**, the material inside a black hole is squashed inside an infinitely dense point, known as a **singularity**. This is surrounded by a surface called the *event horizon* at which the escape velocity equals the speed of light and that thus marks the outer boundary of a black hole. Nothing from within the event horizon can travel back into the outside universe; on the other hand, matter and energy can pass through this surface-of-no-return from outside and travel deeper into the black hole. For a nonrotating black hole, the event horizon is a spherical surface, with a radius equal to the **Schwarzschild radius**, centered on the singularity at the black hole's heart. For a spinning black hole (a much more likely contingency in reality), the event horizon is distorted—in effect, caused to bulge at the equator by the rotation. Within the event horizon, objects and information can only move inward, quickly reaching the singularity. A technical exception is **Hawking radiation**, a quantum mechanical process that is unimaginably weak for massive black holes but that would tend to cause the mini variety to explode.

Black Eye Galaxy Dark dust clouds around the nucleus give the Black Eye Galaxy its characteristic appearance. *Isaac Newton Group of Telescopes*

Blagg, Mary Adela (1858–1944)

(continued from page 67)

Astronomical Union and asked to continue her standardization work, together with Karl Muller of Vienna. The resulting *Named Lunar Formations,* published in 1935, became the standard reference on lunar nomenclature.

blazar

A class of **active galaxy** that includes **BL Lacertae objects** and **optically violent variables** (OVVs). Blazars are characterized by extreme and rapid variability across a wide range of wavelengths, and are believed to be **quasars** seen with their **jets** pointed straight at us. In the early

1990s, the **Compton Gamma-Ray Observatory** was used to discover a new class of blazar known as a *gamma-ray blazar,* whose members appear to have gamma-ray–emitting jets that are more tightly bound than the usual radio jets. These gamma-ray–loud blazars may also be a major source of the cosmic diffuse background above energies of about 100 MeV.

Blaze Star

The popular name for the recurrent **nova** T Coronae Borealis of which two major outbursts have been observed: in 1866 and 1946, when it rose rapidly to the second and third magnitude, respectively, before slowly returning to its quiescent magnitude of 10.8. Other, smaller increases in brightness have been recorded at ultraviolet wavelengths. The star is a **spectroscopic binary** with a red giant component and a period of 227.6 days.

Blazhko effect

A secondary variation of the amplitude and period of some **RR Lyrae stars** and related pulsating variables, named for its discoverer, the Russian astronomer Sergei Blazhko (1870–1956). The effect is displayed, for example, by the RR Lyrae star XZ Cygni, which has a primary pulsation period of about 0.466 days but displays a further modulation with a 57-day cycle. Although the Blazhko effect is not properly understood, it is thought to involve double-mode pulsation or, in some cases, an **oblique rotator**.

Blinking Nebula (NGC 6826)

A **planetary nebula** in **Cygnus** whose nickname stems from the fact that, when looked at directly through a small telescope, it seems to disappear, leaving only the central star–one of the brightest of any planetary–still in view. Using **averted vision**, the nebula winks back into view. NGC 6826's eyelike appearance is notable for its prominent red **FLIERs**.

Visual magnitude:	9.0
Angular diameter:	25′
Distance:	2,200 light-years
Position:	R.A. 19h 44.8m, Dec. +50° 31′

blue compact dwarf galaxy (BCD)

A small galaxy, about one-tenth the size of a typical large spiral such as the Milky Way, that appears blue by virtue of containing large clusters of hot, massive stars, which ionize the surrounding interstellar gas with their intense ultraviolet radiation. These massive blue stars are very young by stellar standards–under 3 million years old. They were created in a huge starburst, a violent episode of star formation that in some cases engulfs an entire galaxy. Observations of the young universe show a much larger population of blue galaxies at early epochs, suggesting that they might play a key role in the early history of **galaxy formation**. Local BCDs, such as Haro 3 (NGC 3353), may be the galactic equivalent of the coelacanth (the famous "fossil" fish)–primeval structures surviving into a mature cosmos that form laboratories for the study of galactic evolution. See also **galaxy formation**.

blue giant

A massive, giant star of **spectral type** O or B. Blue giants typically have a luminosity of 10,000 L_{sun} and a surface temperature of 30,000 K. They have exhausted their core supply of hydrogen and are evolving to the stage at which they will expand further and cool to become **red giants**.

blue horizontal-branch star

A kind of **Population II** star, found in the **galactic halo** and especially in **globular clusters**, that has evolved past the **red giant** stage and is now burning helium in its core. Blue horizontal-branch stars are typically of spectral type B3 to A0, and have spectra characterized by strong, sharp hydrogen lines, a large Balmer jump, and very weak lines of other elements (see **Balmer series**).

blue moon

(1) The second full moon in a month that has two full moons. (2) The third full moon in a season of the year that has four full moons (usually each season has only three full moons).

Blue Planetary (NGC 3918)

A **planetary nebula** in **Centaurus**, 2.5° northwest of Delta Crux, that was discovered and named by John **Herschel** in 1834. In a 25-cm telescope, at low to medium power (75 to 120×), it appears as a small pale blue disk similar to that of Neptune.

Visual magnitude:	8.1
Angular size:	12″
Distance:	2,600 light-years
Position:	R.A. 11h 50.3m, Dec. –57° 11′

Blue Snowball Nebula (NGC 7662)

A **planetary nebula** in the west of **Andromeda**, 2.5° west-southwest of the star Iota Andromedae. One of the easiest planetaries to observe with small instruments, it appears blue-green and has a very faint central star that is variable with a magnitude range of 12 to 16.

Visual magnitude:	8.3
Angular size:	32″ × 28″
Distance:	2,200 light-years
Position:	R.A. 23h 25m 54s, Dec. +42° 37′

blue straggler

A hot, bright star in a **globular cluster** or, occasionally, an **open cluster**, that lies close to the cluster's extrapolated **main sequence** but a few magnitudes above its **turnoff point**. Various theories have been put forward to explain why such stars have not already evolved to become **red giants**. The most favored of these involve mass transfer from, or coalescence with, a binary companion.

blueshift

A general displacement of **spectral lines** to shorter wavelengths that happens when a source of electromagnetic radiation moves toward the observer: the greater the velocity of approach along the line of sight, the greater the blueshift. Individual stars within the Milky Way Galaxy frequently show blueshifts. However, with the exception of a few galaxies within the **Local Group**, including the **Andromeda Galaxy**, most extragalactic objects exhibit **redshifts**, due to the overall expansion of the universe.

Bode, Johann Elert (1747–1826)

A German mathematician and astronomer best known for his popularization of an empirical mathematical rule giving the relative mean distances between the Sun and planets. Often referred to simply as Bode's law, this rule had been discovered earlier by Johann Titius (1729–1796) of Wittenberg and so is more properly called Titius's law or the **Titius-Bode Law**. Bode founded the *Berliner Astronomisches Jahrbuch* (Astronomic Yearbook of Berlin) in 1774, and went on to compile and issue 51 yearly volumes of it. From 1772 to 1825 he was astronomer of the Academy of Science, Berlin, and from 1786, director of the Berlin Observatory. His most important contribution to astronomy was the *Uranographia* (1801), a collection of star maps and a catalog of 17,240 stars and nebulae–12,000 more than had appeared in earlier charts.

Bok, Bartholomeus ("Bart") Jan (1906–1983)

A Dutch astronomer famous for his studies of star-forming regions and of the structure of the Galaxy. Educated at the universities of Leiden and Groningen, he worked at Harvard University from 1929 to 1957, and for the next nine years directed the Mount Stromlo Observatory in Australia. In the early 1940s he helped set up the National Observatory of Mexico at Tonantzintla, and a decade later did a similar job for Harvard's southern station in South Africa. In Australia he helped establish the Siding Spring Observatory. The latter part of his career was spent at the University of Arizona, where he directed the **Steward Observatory** from 1966 to 1970. Working closely with his wife, Priscilla Fairfield Bok, he studied the structure and evolution of star clusters and the Galaxy, mapping the spiral arms of the Milky Way, especially in the **Carina** region, and the **Magellanic Clouds**. Bok initiated radio astronomy at Harvard and promoted it elsewhere. His investigations of interstellar gas and dust led to studies of star formation, and he became known for his work on small dark nebulae now called Bok **globules**. He was an important teacher, writer, leader, and popularizer of astronomy.

bolide

A very bright **meteor**, or fireball, that breaks apart or explodes in the air and produces a sonic boom.

bolometer

In astronomy, a device for measuring the amount of radiant energy received from a celestial object. Most bolometers are sensitive to **infrared** radiation longer than about 6 microns.

bolometric magnitude (M_{bol})

The **magnitude** of a star measured across all wavelengths, so that it takes into account the total amount of energy radiated. If a star is a strong infrared or ultraviolet emitter, its bolometric magnitude will differ greatly from its **visual magnitude**. The absolute bolometric magnitude (see **absolute magnitude**) of a star is a measure of its total energy emission per second, or **luminosity**. The *bolometric correction* is the visual magnitude of an object minus its bolometric magnitude.

Bol'shoi Teleskop Azimultal'nyi (BTA)

A 6-m telescope at an altitude of 2,070 m on the slopes of Mount Pastukhov on the northern side of the Caucasus range in Russia. Opened in 1976, it was the world's largest optical instrument, beating out the **Hale Telescope**, until the first of the telescopes at the **Keck Observatory** went into operation in 1993. Few detailed results from the BTA have appeared in the West and it is widely recognized that, although impressive in size, it suffers from a number of problems, including less-than-ideal climate, suspect optics, and a lack of state-of-the-art digital instrumentation.

Bond, William Cranch (1789–1859), and George Phillips (1825–1865)

Father and son American astronomers. William became an expert maker of chronometers and by 1812 was fashioning most of the superior ones used by ships sailing out of Boston. He developed a passion for astronomy and turned part of his home into an observatory. In 1815, he was sent by Harvard College to Europe to visit existing observatories and gather data preliminary to the building of the **Harvard College Observatory** (HCO). Bond supervised the HCO's construction and, in 1839, became its first director. In 1847, a 15-inch (37.5-cm) telescope was installed, matched in size by only one other in the world. With it, Bond studied **sunspots**, the **Orion Nebula**, and **Saturn.**

George became his father's assistant and, in 1859, succeeded him as director of the HCO. Much of his work was done in cooperation with his father. While they were studying Saturn together in 1848, George discovered its eighth satellite, **Hyperion** (found independently in the same year by William **Lassell**). His observations led him to reject the previously held theory that the rings of Saturn were solid, although his hypothesis of their being in a fluid state was in turn soon discarded. His memoir on **Donati's Comet** of 1858 in the *Annals of the Harvard College Observatory,* vol. III, remains one of the most complete descriptions of a great comet ever written. With his father, he developed the chronograph for automatically recording the position of stars and was a pioneer in the use of the chronometer and the telegraph for determining longitude. Together, they applied Daguerre's photographic process to astronomy for the first time in America.

Bonner Dürchmusterung (BD)

A star atlas and catalog listing the positions and magnitudes of 324,188 stars in the northern hemisphere to Dec. 2° S and ninth magnitude. It was compiled by Friedrich **Argelander** in Bonn between 1859 and 1862, and extended by Eduard Schönfeld with a supplement of 133,659 southern stars to Dec. –23° S in 1886. Stars are numbered and designated in declination zones, for example, "BD +12° 1438." Later came two more large southern star catalogs: the ***Córdoba Dürchmusterung*** produced between 1892 and 1932, and the ***Cape Photographic Dürchmusterung*** between 1895 and 1900.

Boötes (abbr. Boo; gen. Boötis)

The Herdsman; a large, kite-shaped **constellation** of uncertain mythological provenance that is dominated by **Arcturus**, the brightest star in the northern hemisphere. Boötes borders on Canes Venatici and Coma Berenices to the west; Virgo to the south; Serpens, Corona Borealis, and Hercules to the east; Draco to the north; and Ursa Major to the northwest. It is home to the prototype **Lambda Boötis star** and also contains Xi Boötis, an easily seen binary with yellow and orange components of magnitude 5 and 7, respectively, that is also a **BY Draconis star**. (See table, "Boötes: Stars Brighter than Magnitude 4.0," and star chart 18.)

Borrelly, Comet (19P/Borrelly)

A **short-period comet** discovered by the French astronomer Alphonse Borrelly (1842–1926). It became the second comet to have its nucleus photographed by a space probe when **Deep Space 1** passed it at a distance of only 2,170 km on September 22, 2001. Images showed an 8-km-long by 4-km-wide nucleus shaped like a bowling pin, spraying material into space along a handful of tightly collimated jets. Scientists were surprised when the probe revealed that, although the **solar wind** flows symmetrically around the coma of Comet Borrelly, the nucleus lies to one side, shooting out a great jet of material that forms the cloud that makes the comet visible

Boötes: Stars Brighter than Magnitude 4.0

Star	Magnitude		Spectral	Distance	Position	
	Visual	Absolute	Type	(light-yr)	R.A. (h m s)	Dec. (° ′ ″)
α **Arcturus**	–0.05	–0.30	K2IIIbCN	37	14 15 40	+19 10 57
ε **Izar** (Pulcherimma)	2.35	–1.69	K0II-III+A2V	210	14 44 59	+27 04 27
η Muphrid	2.68	2.41	G0IV	37	13 54 41	+18 23 51
γ Seginus	3.04	0.95	A7III	85	14 32 05	+38 13 30
δ	3.46	0.69	G8IIICN	117	15 15 30	+33 18 53
β Nekkar	3.49	–0.65	G8IIIa:Ba0.4	219	15 01 57	+40 23 26
ρ	3.57	0.27	K3III	148	14 31 50	+30 22 17
ζ	3.78	0.06	A2III+A2III	181	14 41 09	+13 43 42

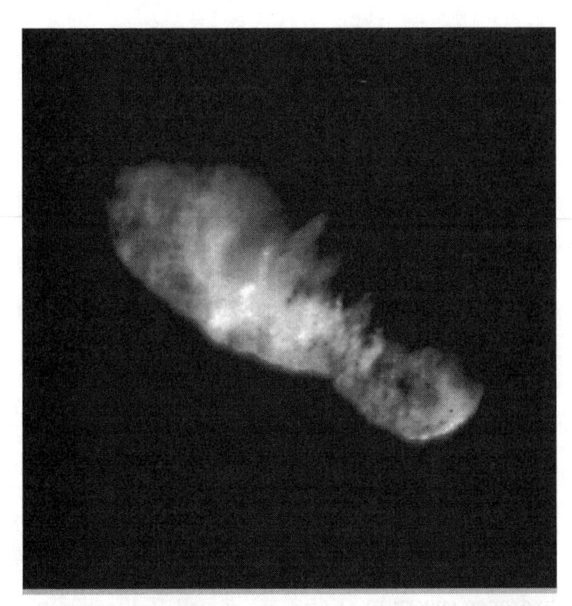

Comet Borrelly The highest resolution view obtained of the 8-km-long nucleus of Comet Borrelly. It was taken from a distance of 3,417 km, just 160 seconds before Deep Space 1's closest approach. A variety of terrain and surface textures, mountains and fault structures, and darkened material are visible over the surface of the nucleus. Smooth, rolling plains containing brighter regions are present in the middle of the nucleus and seem to be the source of dust jets seen in the coma. *NASA/JPL*

from Earth. Around the most active regions are a series of flat-topped, steep-sided hills, or mesas, that probably formed much like terrestrial mesas. The top of each mesa has a thick insulating layer of dust, but the steep sides expose the underlying ice-rich comet material. Ices sublimate from the sides, undercutting the thick, insulating layer and causing sections of it to collapse on the valley floor. Borrelly seems to be broken into two pieces, canted at about 15°, that appear to chaff against each other, raising what look like compressional ridges at the boundary of the two sections. Orbit: semimajor axis 3.61 AU, perihelion 1.358 AU, eccentricity 0.624, inclination 30.3°, period 6.86 years.

boson

A particle, such as a **photon** (a particle of light), that does not obey the **Pauli exclusion principle**. This means there is no limit to the number of bosons that can occupy the same point in space at the same time; so, for example, there is no limit to the brightness of a beam of light. Bosons have an integer (intrinsic) spin quantum number. Mesons have spin 0, photons spin 1, and the as yet-undetected graviton, associated with the gravitational force between masses, spin 2.

Boss General Catalogue *(GC)*

The *General Catalogue of 33,342 Stars,* compiled by the American astronomer Benjamin Boss (1880–1970) and published by the Carnegie Institution of Washington, D.C., in 1936. It lists the position, **proper motion, magnitude**, and **spectral type** of all stars brighter than seventh magnitude over the whole sky, plus some fainter stars for which accurate proper motions were available. It succeeded the *Preliminary General Catalogue* of 6,188 stars, published by Benjamin's father Lewis Boss (1846–1912), who also initiated the larger work.

Bouvard, Alexis (1767–1843)

A French astronomer and mathematician, who was born in a hut in Chamonix and grew up a shepherd boy. He went to Paris, taught himself mathematics, and was appointed assistant to Pierre **Laplace**. He discovered several comets and drew up tables of the planets, which led him to notice that **Uranus** moved in a way that couldn't be completely explained as a result of the gravitational pulls of the then-known planets. Bouvard left it to his successors, however, to decide if the mystery was due to inaccurate observations or to a genuine eighth planet. See also John **Adams** and Urbain **Leverrier**.

bow shock

The interface that forms between a supersonic fluid and an obstacle, such as a denser medium. Many examples are found in astronomy. A bow shock forms at the outermost part of a planetary **magnetosphere**, where the high-speed flow of the **solar wind** is suddenly slowed to subsonic speed by the planetary magnetic field. A similar boundary layer separates the **heliosphere** from interstellar space. Bow shocks also form around hot young stars where vigorous **stellar winds** slam into the surrounding **interstellar medium**.

Bowen, Ira ("Ike") Sprague (1898–1973)

An American astronomer who first explained the **spectral lines** seen in nebulae. His investigation of the ul-traviolet spectra of highly ionized atoms led to his identification, in 1927, of the mysterious "nebulium" spectral lines of gaseous nebulae as **forbidden lines** of ionized oxygen and nitrogen. As director of the Mount Wilson Observatory from 1946 to 1948 and of the combined Mt. Wilson and Palomar Observatories from 1948 to 1964, he directed the completion of the **Hale Telescope** and the 48-inch Schmidt telescope and designed many of their instruments, including a novel **spectrograph**.

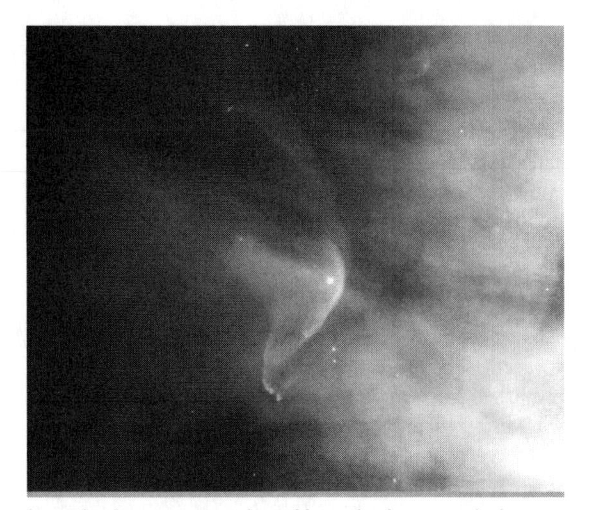

bow shock A crescent-shaped bow shock surrounds the very young star LL Ori. It results from a collision between the vigorous stellar wind, emitted by LL Ori, and slower-moving gas that is evaporating away from the center of the Orion Nebula, located to the lower right in this image. A second, fainter bow shock can be seen around a star near the upper right corner. *NASA/STScI/AURA*

Box, the (Hickson 61)

A very tight grouping of four galaxies in **Coma Berenices** that form a near perfect rectangle. Spanning just 3.8′ of the sky, they all fit into the same medium-power field of view; particularly striking is how the two most elongated galaxies line up, forming virtually one long streak. The brightest member, NGC 4169, is a magnitude 13.3 **lenticular galaxy**. NGC 4174 is a nearly edge-on magnitude 14.5 spiral. NGC 4175 is a magnitude 14.3 edge-on spiral that shows evidence of interaction with its apparent neighbor, NGC 4173, which is the largest of the four but the most difficult to see because of its low surface brightness. Three of the galaxies have recession velocities of around 2,900 km/s, but, at 1,100 km/s, NGC 4173 is receding much more slowly and is not, despite appearances, physically associated with the others.

Bp star

A **B star** whose spectrum is chemically "peculiar" (hence the "p"), usually with regard to the concentration of helium, which may be low, high, or variable (see **helium star**). Bp stars can be considered to be a hot subgroup of the **Ap stars**.

brachinite

A small subgroup of primitive (little altered since their formation) **achondrites**, named for the Brachina meteorite that was found in Australia in 1974. Originally, the **olivine**-rich Brachina was thought to be a second **chassignite**–a Mars meteorite that contains primarily olivine. However, further research showed Brachina to have a distinct trace-element pattern and a unique oxygen isotopic composition. With only a handful of known representatives, brachinites are composed mainly of small, equigranular olivine grains, among which are scattered small amounts of clinopyroxene, **orthopyroxene**, and **plagioclase feldspar**. Recent studies of the olivine compositions of different asteroids suggest that (289) Nenetta may be the parent body of this group.

Brackett series

A series of absorption or emission lines due to **electron** jumps between the fourth and higher energy levels of the hydrogen atom. These lines lie in the **infrared** with wavelengths from 4.05 microns (Brackett-alpha) to 1.46 microns (the series limit), and are named after the American physicist Frederick Brackett (1896–1980). See also **hydrogen spectrum**.

Bradley, James (1693–1762)

An English astronomer who, in searching for stellar **parallax**es, discovered, in 1728, the **aberration of starlight**–an apparent change in the positions of stars caused by Earth's annual motion. This provided the first direct evidence for the revolution of our planet around the Sun and enabled him to calculate the speed of light to within 2% of its correct value. Bradley became the third **Astronomer Royal** (1742) and subsequently discovered the periodic shift in Earth's axis known as **nutation**.

Brahe, Tycho (1546–1601)

A Danish pre-telescopic astronomer and nobleman whose accurate astronomical observations, especially of Mars, in the last quarter of the sixteenth century, gave the empirical basis for **Kepler's laws of planetary motion**; born Tyge Ottesen, he is always referred to by his Latinized name. While studying law at the University of Leipzig (1562–1565), Tycho secretly carried on his celestial studies. In 1566, he famously lost his nose in a dual with Manderup Parsberg, his third cousin and a fellow student, over who was the better mathematician. Subsequently, he wore a prosthetic nose piece, purportedly made of pure gold, but which, upon examination by Czech scholars following Tycho's disinterment in 1901, proved to be an alloy of gold, silver, and copper. During his studies in chemistry at Augustburg (1570–1572), Tycho persuaded his maternal uncle to install a laboratory in his castle at Herritzwad. There, on November 11, 1572, Tycho caught sight of the famous "new star" in Cassiopeia (a supernova), which became the subject of his first book, *De nova stella,* in 1573. In the same year, his family relations became strained when he married a peasant girl. He had intended to settle in Basel, but Fredrik II, King of Denmark, bestowed upon him for life the island of Hveen, along with a substantial pension, which enabled Tycho to set up his famous observatory of Uraniborg. To achieve maximum accuracy, Tycho made his instruments as large as possible without sacrificing mechanical rigidity. One of the largest was the great mural **quadrant**, permanently mounted in the plane of the meridian and with an arc radius of slightly over 1.8 m. Because the mural quadrant measured only meridian altitudes, Tycho also used two other large quadrants of the **azimuth**al type. Both had circles approximately 1.8 m in radius, angular scales divided to 10″, and parallax-free sights. Among other major items at Uraniborg were several large equatorial **armillary spheres**. Although used in China at least as early as 1260, the equatorial armillary had been unknown in Europe, and Tycho considered himself its inventor. The need to measure the angular distance between any two objects in the sky prompted Tycho to develop the sextant: the largest at Uraniborg had a radius of 1.7 m. Tycho also encouraged the entry of women into science. Sponsored by her brother, Sofie Brahe (1556–1643) made critical new observations that would later allow **Kepler** to establish the elliptical orbits of the planets. Sofie worked with her brother during the whole of his tenure at Uraniborg, and later compiled her observations into an unpublished, but preserved, memoir. Although Tycho incorrectly believed that Earth was at the center of the universe and that the Sun and the stars revolved around Earth, he did accept part of the Copernican system, that the other planets orbit the Sun. His principal work, *Astronomiae Instauratae Propgrymnasmata* (Exercises toward a Restored Astronomy) was published posthumously in two volumes (1602–1603) and edited by Kepler, who had worked alongside Brahe for the last few years of the Dane's life. The first volume gave the positions of 777 stars and formed the basis of Kepler's *Rudolphine Tables* (1627), while the second volume dealt primarily with the comet of 1577. Tycho died, not of a burst bladder following a gastronomic orgy, as legend suggests, but from high levels of mercury (which he may have taken as medication after falling ill from the infamous meal) in his blood.

Brahmagupta (c. a.d. 598–c. 665)

The last and most accomplished of the ancient Indian astronomers, who applied algebraic methods to astronomical problems. He set forth the Hindu astronomical system in verse form in the *Brahma-sphuṯa-siddhaṉta,* a standard early work on the subject. See also **Indian astronomy**.

Braille (minor planet 9969)

A small **asteroid** (2.2 × 1 km), discovered in 1992 by American astronomers Eleanor Helin and Kenneth Lawrence and named after the Frenchman Louis Braille (1809–1852), inventor of the most widely used reading system for the blind. NASA's **Deep Space 1** probe flew past it, on July 29, 1999, at a distance of 26 km. The spacecraft's infrared sensor confirmed that Braille is similar in composition to **Vesta**, a rare type of **asteroid** and one of the largest bodies in the main asteroid belt. It remains uncertain whether Braille is a piece that broke off Vesta, or whether the two asteroids both came from a larger body that was destroyed long ago. It is estimated that in about 4,000 years, Braille will join the hundreds of other asteroids that drift in and out of Earth's orbit.

breccia

A coarse **sedimentary** or **igneous** rock composed of angular fragments (clasts) of older rocks that are cemented together in a fine matrix. Sources of breccias include scree slopes (taluses), faulting, volcanic eruptions, and meteorite impacts.

bremsstrahlung

Electromagnetic radiation given off by fast-moving charged particles, such as **electron**s, when they are slowed down by passing through matter. *Bremsstrahlung* is German for "braking radiation."

bridge, intergalactic

A stream of matter connecting two interacting galaxies that have passed close to each other. Material is torn from one galaxy, funnels down onto the other, is compressed, and often gives rise to a starburst–a sudden spate of star formation. One of the most dramatic examples is seen in the **Antennae**.

bright nebula

A luminous interstellar cloud, which may be an **emission nebula** (giving off its own light) or a **reflection nebula** (reflecting the light of nearby stars).

Bright Star Catalogue

A star catalog published by Yale University Observatory and known formally as the *Yale Catalog of Bright Stars;* it lists all stars brighter than magnitude 6.5 contained in the *Harvard Revised Photometry Catalogue.* The first edition appeared in 1930 and the most recent (fifth), in electronic format, in 1991. A *Supplement* listing a further 2,603 stars down to magnitude 7.1 was published in 1983. Although much smaller than the ***Smithsonian Astrophysical Observatory Catalog***, with 258,857 stars, or the ***Guide Star Catalog,*** with 18 million, the *Bright Star Catalogue* provides a wealth of interesting information on each of its objects, including **Bayer designations** and **Flamsteed numbers**, **proper motion**, **parallax**, **magnitude**, and miscellaneous comments.

brightness

The amount of light, or other radiation coming from an object; it can be quantified in several different ways. **Luminosity** is a measure of how bright something is intrinsically. It can be measured in watts, or watts per hertz to describe the luminosity at a specific frequency. For historical reasons, **absolute magnitude** is used in optical astronomy, while radio astronomers quote the power, which is the luminosity per steradian (multiply by 4π to get back to luminosity). **Flux density** is a measure of how bright an object seems to an observer on Earth. It depends both on the luminosity and the distance, since distant objects appear fainter. Optical astronomers measure this as **apparent magnitude**, while radio astronomers use a unit called the **jansky**. **Intensity** (or surface brightness) is a measure of the flux density received, not from the object as a whole, but from each unit area of the sky (technically, *solid angle*). The flux density of an object is thus the product of its intensity times its solid angle.

brightness temperature

The temperature that a **blackbody** would need to have in order to emit radiation of the observed intensity at a given wavelength.

bright-rimmed cloud (BRC)

A type of star-forming cloud that, unlike a dark **globule** (which is entirely self-gravitating), is triggered to collapse by the pressure of ultraviolet light from nearby young stars.

broadband photometry

Photometry carried out using filters with a bandwidth of 30 to 100 nanometers. Various systems have been developed including Johnson photometry, Krons-Cousins RI photometry, and RGU photometry. It is used instead of narrowband photometry or intermediate-band photometry in studies, such as star counts, in which it is important to observe fainter stars.

brown dwarf

See article, pages 77–78.

Bruno, Giordano (1548–1600)

A scholastic philosopher and lapsed Dominican monk who had a dangerous passion for unorthodox cosmological beliefs, which included support for the heliocentric model of Copernicus. He was one of the first early Renaissance thinkers to espouse the view that there might be many inhabited worlds. However, he was eventually burned at the stake in Rome not principally for his astronomical views but for his denial of Christ's divinity.

Bubble Nebula (NGC 7635)

The smallest of three bubbles surrounding the massive star BD +60° 2522 in **Cassiopeia** (R.A. 23h 20.7m Dec. +61° 12′), and part of the gigantic bubble network S162, created with the help of other massive stars. As fast-moving gas spreads out from BD +60° 2522, it pushes surrounding sparse gas into a shell. The energetic starlight then ionizes the shell, causing it to glow. The nebula is about 6 light-years across and visible even with a small telescope.

Bug Nebula (NGC 6302)

A **planetary nebula** in **Scorpius** with a double-lobed structure that has resulted from a **bipolar outflow**. The extremely high rate of expansion (nearly 400 km/s) and the high concentration of ionized iron in the nebular gas suggests that the material making up the Bug was ejected with unusual violence. The Bug also stands out for being one of two objects (the other is the **Red Spider Nebula**) in which carbonates that have no links with liquid water were first found. The discovery, using the **Infrared Space Observatory**, of large amounts of calcite and dolomite in these clouds, cast off by dying stars, breaks the automatic association between these minerals and an aqueous environment.

Visual magnitude:	9.6
Apparent size:	50″
Distance:	6,500 light-years
Position:	R.A. 17h 13.7m, Dec. –37° 6′

bulge population

See **population** and **galactic bulge**.

bump Cepheid

A **Delta Cephei star** (classical Cepheid) that has a bump in its **light curve**. The light curves of Delta Cephei stars are asymmetrical, with a rapid rise to maximum light and a slower fall. The form of the light curve changes with period in a systematic way known as the *Hertzsprung progression.* A bump appears on the descending branch of the light curve of stars with periods of about a week and is found at earlier phases in stars of successively longer periods so that the bump is near maximum light in stars of 10-day periods, which may show a double maximum. The bump falls on the rising branch in stars of longer periods. Stars of the shortest or longest periods have

(continued on page 78)

brown dwarf

The dimmest and least massive kind of star. Brown dwarfs (which are not really brown, but very dull red) fall below the mass limit at which normal hydrogen **fusion** can take place in a stellar core. Theory indicates this limit is about 0.084 M_{sun}, or 88 times the mass of Jupiter (M_{Jup}), though there is some uncertainty. The lower mass limit for brown dwarfs is also fuzzy because there is no clear transition point between a heavyweight planet and a lightweight brown dwarf, but it is generally taken to be about 0.013 M_{sun} or 13–14 M_{Jup}. A more significant distinction between planets and brown dwarfs is in the way they form: brown dwarfs form in the same way as stars, as condensations in an interstellar gas cloud, whereas planets accrete from material in a circumstellar disk. The first brown dwarf to be confirmed, in 1995, through a combination of mass determination, spectroscopic studies, and direct imaging was Gliese 229B, a 50-M_{Jup} companion of the **red dwarf** Gliese 229. Since then, many more have been found, including a number in the **Pleiades**, the **Sigma Orionis star cluster**, and the **Trapezium** star cluster. While some brown dwarfs, like Gliese 229B, are part of binary systems, others have been found floating around on their own. PP1 15 in the Pleiades is a binary system in which *both* components are brown dwarfs. S Ori 47, in the Sigma Orionis cluster, holds the record for the smallest known brown dwarf mass–a mere 0.015 M_{sun}. The nearest known brown dwarf is in the **Epsilon Indi** system, less than 12 light-years away.

brown dwarf An artist's impression of a brown dwarf, showing cloud belts in the star's atmosphere similar to those in a gas giant such as Jupiter. *Douglas Pierce-Price, Joint Astronomy Centre*

Although brown dwarfs haven't enough mass to ignite normal hydrogen fusion in their cores, they glow dully and give off a substantial amount of infrared radiation as a result of slow gravitational contraction and small-scale **deuterium** fusion. At the high-mass, high-temperature end of the brown dwarf scale are the **ultra-cool stars**, with dusty atmospheres and a **spectral type** of M7 or later. Late M-types, with surface temperatures as low as 2,200 K, have water and strong oxide features in their spectra and may be red dwarfs or brown dwarfs, depending on their mass. For brown dwarfs smaller and cooler than type M, a new spectral category has been allocated–type L. With temperatures of about 1,500 to 2,200 K, L dwarfs have spectra characterized by strong metal hydride bands and even prominent water bands. Cooler still are the so-called T-types, or *methane dwarfs*, with surface temperatures ranging from about 1,500 to 1,000 K or even 800 K, and spectra that show strong absorption by methane and water.

Brown dwarfs do not glow, even dully, for very long. As soon as they have used up their meager supply of deuterium, which takes about 10 million years, they fade from dim dark red to black. However, there are stars that start out as ordinary hydrogen-fusing red dwarfs and then get whittled away to brown dwarf size. The binary systems LL Andromedae and EF Eridani both contain **white dwarf** primaries that have looted material from their partners and reduced them to 40-M_{Jup} objects, with surface temperatures of about 1,300 and 1,650 K, respectively.

Brown Dwarfs with Partners, Alone, and (Perhaps) with Planets

A surprisingly high proportion of brown dwarfs have been found as companions to low-mass (red dwarf or other brown dwarf) stars, and, within these systems, the separation between the two components is typically very small, averaging about 4 AU. This goes against the prediction by some theorists that most very low-mass stars and brown dwarfs are solo objects, wandering through space alone after being ejected from their stellar nurseries during the star formation process. Very few brown dwarf companions of larger, Sun-like stars have been found inside 5 AU, a deficiency that has been dubbed the "brown dwarf desert"; however, there is no such desert associated with low-mass stars. The observations to date strongly support the idea that low-mass binaries form in a process similar to that of more massive binaries, and that the percentage of binary systems is similar for bodies spanning the range from one solar mass to as little as 0.05 M_{sun}.

Yet, there are also many lone brown dwarfs, such as KELU-1, discovered in 1997. At a distance of only 33 light-years from the Sun, it was one of the closest brown dwarfs known at that time. Such solitary dwarfs could be *ejected stellar embryos*—small infant stars that were still accreting material when they were kicked out of the nest by gravitational interactions with more massive siblings in multiple stellar systems. On the other hand, observations of some brown dwarfs in the Orion Nebula, which show an excess of near-infrared radiation, point to the presence of dusty disks around these objects. Not only does this suggest a normal stellar formation process but also the possibility that brown dwarfs might develop planetary systems. Whether life could ever emerge on such worlds whose sun is so cool and dim is a question for the future.

bump Cepheid

(continued from page 76)

smooth light curves. The amplitude of the pulsation increases slowly with periods up to about 10 days, where there is a drop in amplitude; it then increases more rapidly to longer periods. The bumps may represent an echo of the surface pulsation from the deep interior; alternatively, they may result from a resonance when the second overtone period is about one-half of the fundamental period.

brown dwarf About 50 brown dwarfs have been identified in this infrared Hubble Space Telescope view of the region of the Orion Nebula around the Trapezium Cluster (the bright quartet of stars in the center). *NASA/K. L. Luhman (Harvard-Smithsonian CfA) and G. Schneider, et al. (Steward Observatory)*

Burbidge, (Eleanor) Margaret (1919–), and Geoffrey Ronald (1925–)

British-born wife and husband astrophysicists, who have been based in the United States for many years. Margaret née Peachey, was educated at the University of London, where she remained until 1951. She was director of the **Royal Greenwich Observatory**, worked at **Yerkes Observatory** and the California Institute of Technology, and has been at the University of California, San Diego, since 1962, becoming the first director of its Center for Astrophysics and Space Sciences. In 1957 she, husband Geoffrey Burbidge, William **Fowler**, and Fred **Hoyle** showed how all of the elements except the very lightest are produced by nuclear reactions in stellar interiors. She is also well known for her spectroscopic studies of **quasars** and has played a major role in developing instrumentation for the **Hubble Space Telescope**.

Geoff Burbidge was educated at the universities of Bristol and London, where he earned his Ph.D. in theoretical physics in 1951. He has said that he got into astronomy by marrying an astronomer. He has worked at Cambridge University, the California Institute of Technology, and Yerkes Observatory, and has been at the University of California, San Diego since 1962, aside from six years as director of the **Kitt Peak National Observatory**. He has done research on the physics of nonthermal radiation processes, and the structures and masses of galaxies and quasars.

Burnham, Sherburne Wesley (1838–1921)

A gifted American observational astronomer who specialized in double star work. He started as an amateur astronomer, then joined the staff of **Yerkes Observatory** (1897–1914), although he never gave up his other job as a court reporter in Chicago. He discovered 1,290 double stars and, in 1906, published his *General Catalogue of Double Stars,* which listed all 13,665 known double stars in the northern hemisphere.

Burnham's Nebula

The popular name for the small nebula that surrounds the star **T Tauri**. It was first reported by Sherburne **Burnham** in 1890, nearly 40 years after the discovery of T Tauri itself. T Tauri was then shining feebly at fourteenth magnitude–the limit of most telescopes of the time. The condensed nebula appeared to be about 4″ in its longest dimension. T Tauri has been no brighter than tenth magnitude since the early twentieth century making this feature very hard to detect. Burnham's Nebula can be discerned in an overexposed image of the star as a bulge extending about 10″ from the variable. Unlike **Hind's Variable Nebula**, it is not believed to be a reflection nebula.

burst, solar

Suddenly enhanced nonthermal radio emission from the high solar **corona** immediately following a **solar flare**, probably due to energetic electrons trapped in the coronal magnetic field. Bursts are divided into several types, depending on their time frequency characteristics (type III is the most common). They are classified on a scale of importance ranging from –1 (least important) to +3. Bursts are generally attributed to a sudden acceleration of 10^{35} to 10^{36} electrons to energies greater than 100 keV in less than 1 second.

Butcher-Oemler effect

The tendency of **clusters of galaxies** at great distances (**redshifts** of 0.4 or more) to have a higher fraction of optically blue galaxies (mostly bright spirals and irregulars) than clusters of galaxies nearby. Although some astronomers suspect this may be only a selection effect, most believe the phenomenon is real and is probably related to the greater availability in the past of material from which new stars could form. The effect was first reported in 1978 by the American astronomers Harvey Raymond Butcher (1947–) and Augustus Oemler Jr. (1954–).

Butterfly Cluster (M6, NGC 6405)

An **open cluster** in **Scorpius**, 4° northwest of M7, that contains about 80 stars roughly in the shape of a butterfly. Visible to the naked eye, it was first noted by Giovanni **Hodierna** before 1654. Its most conspicuous star is an orange giant **K star** of magnitude 6.2.

Visual magnitude:	5.3
Angular size:	25′
Diameter:	20 light-years
Age:	About 50 million years
Distance:	2,000 light-years
Position:	R.A. 17h 40.1m, Dec. –32° 13′

butterfly diagram

A plot of heliographic latitude of **sunspots** versus time, developed by E. Walter **Maunder** in 1904 to illustrate the **solar cycle**. It shows how spots move from high latitudes (30° to 40° N or S) toward the equator throughout each cycle, in accordance with **Spörer's law**. The shape of the distributions, when plotted for both solar hemispheres, resembles the wings of a butterfly.

Butterfly Nebula (M2-9)

A **planetary nebula** with a dramatic bipolar structure; it lies 2,100 light-years away in **Ophiuchus** and was discovered by Rudolph **Minkowski** in 1947. Many of the unusual features of the Butterfly can be understood in terms of the central star, which is a close, mass-exchanging binary system, like a **Z Andromedae star** (or *symbiotic nova*). One of the stars is pulling matter off its partner and spinning it into a giant disk, about 10 times the diameter of **Pluto**'s orbit. Hydrodynamic models of the type used to design jet engines show that such a disk can successfully account for the jet-exhaustlike appearance of M2-9. The high-speed wind from one of the stars rams into the surrounding disk, which serves as a nozzle. The wind is deflected in a perpendicular direction and forms the pair of **jets** seen in the nebula's image. This is much the same process that takes place in a jet engine: the burning and expanding gases are deflected by the engine walls through a nozzle to form long, collimated jets of hot air at high speeds.

Bw star

A **B star** with weak helium lines; in other words, a B star which, if classified according to its color, would have helium lines too weak for the classification, and which, if classified according to its helium lines, would have colors too blue for its spectral type. Also known as a *helium-weak star.*

BY Draconis star (BY Dra star)

A type of **rotating variable** dwarf star, of spectral type G, K, or M, that shows quasiperiodic light changes, ranging from a few hundredths to 0.5 magnitude, with a period from a few hours up to 120 days. The variations are due to surface features, such as starspots, passing in and out of view, as the star spins on its axis. Some of these stars also show flares, similar to **UV Ceti stars**, in which case they belong to both types of variable star. BY Draconis is a **close binary** consisting of a K6V dwarf and a M0V dwarf with an orbital period of 5.9 days and a mean separation of 0.05 AU. The M star exhibits flares that increase its luminosity periodically by a factor of 2 to 4, and also shows smaller variability keyed to its orbit and chromospheric activity.

Byurakan Astrophysical Observatory

A facility of the Armenian Academy of Sciences, founded in 1946 through the efforts of Viktor **Ambartsumian.** Located at an altitude of 1,500 m on Mount Aragatz, 40 km north of Yerevan (in sight of Mount Arafat), its main instrument is a 2.6-m (100-inch) reflector.

C

Cacciatore, Niccolò (1780–1841)

An Italian astronomer who was Guiseppe **Piazzi**'s assistant at Palermo Observatory before taking over the directorship himself in 1817. He is best remembered for being the only person to name two stars after himself: Sualocin (Alpha Delphini) and Rotanev (Beta Delphini). Read backwards, these become Nicolaus Venator, which is the Latinized form of Niccolò Cacciatore (in English, Nicholas Hunter). The names first appeared in the Palermo star catalog of 1814 and it took some detective work by the English astronomer Thomas Webb to unravel their origin.

Caelum (abbr. Cae; gen. Caeli)

The Sculptor's Chisel; a very small and obscure southern **constellation** that lies between Columba and the southern part of Eridanus. Its brightest star is Alpha Cae (visual magnitude 4.44, absolute magnitude 2.92, spectral type F2V, distance 66 light-years); a small telescope reveals Gamma to be a binary, composed of a red giant (magnitude 4.5) and a white giant (magnitude 6.3). (See star chart 3.)

Calabash Nebula (OH231.8+4.2)

A **protoplanetary nebula** that lies about 5,000 light-years away in **Puppis** and is also known as the *Rotten Egg Nebula* because of its unusually large abundance of sulfur compounds (hydrogen sulfide being the culprit for the smell of bad eggs). The Calabash is made up of gas ejected by the central star and subsequently accelerated in opposite directions, at speeds of up to 1.5 million km/hr, to form a bipolar structure. Observations by the **Hubble Space Telescope** have shown how the gas stream rams into the surrounding interstellar material–an interaction that is thought to be preeminent in the formation of **planetary nebulae**. Due to the high speed of the gas, shock fronts are formed on impact, which heat the surrounding gas. Much of the gas flow seen today seems to stem from a sudden acceleration that took place about 800 years ago. Over the next 1,000 years or so, the Calabash is expected to evolve into a fully fledged planetary nebula.

calcium aluminum inclusions (CAIs)

White, millimeter-sized objects found, often together with **chondrules**, in the most primitive kinds of **chondrite** meteorites, notably some types of **carbonaceous chondrites**. They consist of high-temperature minerals, including silicates and oxides of calcium, aluminum, and titanium. In 2002, an international team of scientists accurately dated CAIs at 4.57 billion years, making them the oldest known objects in the solar system. The same team found that chondrules, another of the earliest relics of the solar system, are 2 to 3 million years younger than CAIs. Both types of object formed when dusty regions of the solar nebula were heated to high temperatures. The **dust** melted and then crystallized, forming first CAIs and then chondrules. Larger objects, like asteroids and planets, formed 10 to 50 million years later.

caldera

A large (wider than 1.5 km) basin-shaped depression formed by the collapse or explosion of a volcanic vent; the word is Spanish for "cauldron." Crater Lake in the American Cascades is a good example. Some of the largest calderas in the solar system are found on **Mars**, notably that of Olympus Mons.

calendar

A practical system for fixing the length and beginning of years and their subdivision into months and days; the word "calendar" comes from the Latin *calendarium* for "account book." Only the familiar **Gregorian calendar** is used on a worldwide basis today. It superceded the **Julian calendar** in the sixteenth century; however, because it took from 1582 to 1918 for the Julian system to die out completely, astronomers have to specify to which calendar any given date in this interval refers. Both the Gregorian and Julian systems are solar calendars, tied to the annual motion of Earth around the Sun. The Persian calendar, too, is solar. On the other hand, the Islamic calendar is lunar (based on the waxing and waning of the Moon), while both the Hebrew and the Chinese calendars are lunisolar, with months that keep in step with the phases of the Moon and years that line up with the seasons.

Caliban (97U1)

The sixteenth moon of **Uranus**, also known as Uranus XVI; it was discovered with **Sycorax** in 1997 by Brett Gladman, Phil Nicholson, Joseph Burns, and J. J. Kavelaars using the 5-m Hale Telescope. Caliban has a diameter of about 80 km and is an **irregular moon** with a highly inclined, retrograde orbit at a mean distance of some 7.2 million km from the planet. Like Sycorax, it has a reddish hue, suggesting it may have come from the **Kuiper Belt**.

California Nebula (NGC 1499)

A large but visually faint **H II region**, about 1,000 light-years away in **Perseus**, probably ionized by the hot O star Xi Persei. It was discovered by Edward **Barnard** in 1884 and is named for its resemblance to the state of California.

Callisto

The darkest, least dense, and outermost of the **Galilean moons** of **Jupiter** and the third largest moon in the solar system, after **Ganymede** and **Titan**; it is also known as Jupiter IV. Callisto has one of the oldest, most heavily cratered surfaces of any body yet observed and appears to have changed little, apart from an occasional impact and some surface ice flow, over the past 4 billion years. Unlike Ganymede, with its complex, varied terrains, Callisto, while sharing similar bulk properties with its Galilean neighbor, shows little evidence of **tectonic** activity. Also unlike Ganymede, Callisto seems to have little internal structure, although data from the Galileo spacecraft suggests partial differentiation of Callisto's interior, with the percentage of rock increasing toward the center. The different geologic histories of these giant moons, though not fully understood, appears intimately connected with the different orbital and tidal evolution of the two bodies. The Galileo probe's detection of a weak magnetic field around Callisto led to the suggestion that this field may arise from an interaction between Jupiter's magnetic field and a subsurface layer of salty fluid on the moon–possibly a saltwater ocean of the type thought to exist on **Europa** and, less certainly, on Ganymede.

Callisto's ancient craters have collapsed due to the plastic flow of their icy materials and so lack the high mountain rings, radial rays, and central depressions common to craters on the Moon and Mercury. Images from Galileo show that in some areas small craters have been mostly obliterated, and Callisto's **multiringed basins**, the largest of which, at almost 3,000 km wide, is Valhalla, have been smoothed out by eons of slow ice movement. Another interesting feature is a long, linear series of impact craters known as Gipul Catena. This was probably caused by an object, similar to Comet **Shoemaker-Levy 9**, that was tidally disrupted as it passed close to Jupiter and then crashed into Callisto. (See photos, page 83.)

Diameter:	4,821 km
Density:	1.7 g/cm^3
Composition:	40% ice, 60% rock/iron
Albedo:	0.15
Rotational period:	16.7 days
Orbit	
Mean distance from Jupiter:	1,882,700 km
Eccentricity:	0.0075
Inclination:	0.3°
Period:	16.7 days

Caltech Submillimeter Observatory (CSO)

A 10.4-m-wide radio dish of hexagonally segmented design, located near the summit of Mauna Kea, Hawaii, and operated by the California Institute for Technology. One of the most advanced instruments for submillimeter astronomy, it is also the world's only professional observatory without an operator: the observing astronomer is given full control of the telescope.

Calypso

The eleventh moon of **Saturn**, also known as Saturn XIII; it was discovered in 1980 by Daniel Pascu, P. Kenneth Seidelmann, William Baum, and Douglas Currie from ground-based observations with prototype cameras destined for the Hubble Space Telescope. Calypso measures 34 × 22 × 22 km and is **co-orbital** with **Tethys** and **Telesto**, lying at Tethys's trailing **Lagrangian point** (i.e., 60° behind the larger moon).

Camelopardus (abbr. Cam; gen. Camelopardalis)

The Giraffe; a large, barren **constellation** near the north celestial pole, between Ursa Major and Cassiopeia. Only three of its stars, Beta, 7 Cam, and Alpha, reach fourth magnitude. Alpha is a notably remote and intrinsically luminous **O star** (visual magnitude 4.26, absolute magnitude –7.38, spectral type O9Ia, distance 6,940 light-years). (See table, "Camelopardus: Objects of Interest," and star charts 2 and 17.)

Campbell, William Wallace (1862–1938)

An American astronomer who, while director of **Lick Observatory** (1900–1930), headed a vast program to measure the **radial velocities** of stars. He studied civil engineering at the University of Michigan and then taught astronomy there from 1888 to 1891. In 1890 he worked as a summer volunteer at Lick Observatory where he learned **spectroscopy** under James **Keeler** and subsequently carried out important spectroscopic studies of Mars, hot stars, nebulae, and Nova Aurigae. He discov-

Callisto An area within the Valhalla multiringed structure on Callisto. A prominent fault scarp is visible, together with several smaller parallel ridges. The images that form this mosaic were obtained by the Galileo spacecraft on November 4, 1996, cover an area some 33 km across, and show detail down to a resolution of about 160 meters. *NASA/JPL*

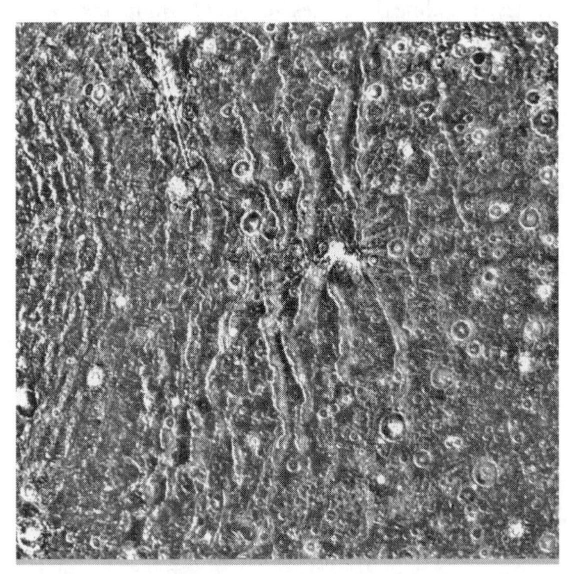

Callisto The concentric rings surrounding Valhalla are the most distinctive geological feature on Callisto. This Voyager 1 image shows a segment of the ridged terrain. The presence of impact craters indicates that the rings formed early in Callisto's history; however, the density of craters is less here than on other parts of the satellite, where the surface is even older. *NASA/JPL*

ered numerous **spectroscopic binaries**, determined the motion of the solar system with respect to surrounding stars, led a number of eclipse expeditions (one of which, in 1921, confirmed the Einstein deflection of light by the Sun), and founded the Lick southern station in Chile.

Canada-France-Hawaii Telescope (CFHT)

A 3.6-m optical and infrared telescope located above 4,000 m on Mauna Kea, Hawaii. The CFHT, which opened in 1979, is operated by the CFHT Corporation under a joint agreement between the National Research Council of Canada, the Centre National de la Recherche Scientifique of France, and the University of Hawaii. The addition of the world's largest digital camera, called MegaPrime, in 2003 has restored the aging telescope to the forefront of astronomical research.

Cancer (abbr. Cnc; gen. Cancri)

The Crab; in Greek mythology, sent by Juno to save the Hydra in its battle with Hercules but unfortunately crushed underfoot during the proceedings. A dim northern **constellation** (shaped a bit like Orion) and the fourth **sign of the zodiac**, located between Castor and Pollux of Gemini and Regulus of Leo. In former times the Sun used to lie in Cancer when it reached its point farthest north–the time of the summer solstice–but today, because of

Camelopardus: Objects of Interest

Object	Notes
Star	
Z Cam	The prototype **Z Camelopardus star**.
Planetary nebula	
IC 3568	A small planetary just south of 32 Cam that looks like a star slightly out of focus. Magnitude 11.6; diameter 18″; distance 9,000 light-years; R.A. 12h 33.11m, Dec. +82° 33.8′.
Open cluster	
NGC 1502	A tiny but striking object containing about 15 stars. The brighter members are in a triangular arrangement while the fainter ones lie in a circular background. Magnitude 5.7; diameter 5.7′; R.A. 4h 7.7m, Dec. +62° 20′.
Galaxy	
NGC 2403	A spiral galaxy readily seen in small telescopes at low power and even visible with good binoculars as a hazy glow. In larger instruments it appears as a fat cigar shape. It is an outlying member of the M81 Group. Magnitude 8.4; size 18′ × 11′; distance 12 million light-years; R.A. 7h 37′, Dec. 65° 36′.

precession, the Sun passes through Cancer from late July to early August. X Cnc is a semiregular variable with a striking red color and an easy object to spot with binoculars (magnitude range 5.6 to 7.5; period 195 days; R.A. 8h 55.4m, Dec. +17° 14′). Cancer also contains the open clusters **Praesepe** (M44) and M67 (NGC 2682). The latter has at least 200 stars and is best seen through larger telescopes (magnitude 6.9; diameter 30′; R.A. 8h 50.4m, Dec. +11° 49′). (See table, "Cancer: Stars Brighter than Magnitude 4.0," and star chart 12.)

Canes Venatici (abbr. CVn; gen. Canum Venaticorum)

The Hunting Dogs; a small, faint northern **constellation**, surrounded by Ursa Major to the north, Boötes to the east, and Coma Berenices to the south. Its only star brighter than magnitude 4.0 is α^2 CVn, or **Cor Caroli** (visual magnitude 2.89; absolute magnitude 0.89; spectral type A0Vp; distance 82 light-years; R.A. 12h 56m 2s, Dec. +38° 19′ 03″). Canes Venatici is particularly well-endowed with galaxies. (See table, "Canes Venatici: Objects of Interest." See also star chart 17.)

Canes Venatici Cloud

One of the clouds of galaxies into which the **Local Supercluster** is considered to be divided. The Canes Venatici Cloud contains the **Local Group**, and a dozen others, including the **Canes Venatici I Group**, the **Canes Venatici II Group**, the **Sculptor Group**, the **M81** Group, the **M101** Group, and the Coma I Group. It takes the form of a flat disk, 650,000 light-years thick and 46 million light-years across.

Canes Venatici I Group

A loose and sparse group of galaxies centered on NGC 4214, 4244, and 4395, about 15 million light-years away. It contains a large number of medium-sized dwarf systems. The spiral galaxy M94 also probably lies at the back of this group.

Canes Venatici II Group

A group with a large collection of spiral galaxies, the dominant members of which are probably M106, NGC 4096, and NGC 4490. It lies at an average distance of 22 million light-years.

Cancer: Stars Brighter than Magnitude 4.0

Star	Magnitude		Spectral	Distance	Position	
	Visual	Absolute	Type	(light-yr)	R.A. (h m s)	Dec. (° ′ ″)
β Altarf	3.53	0.14	K4IIIBa0.5	290	08 16 31	+09 11 08
δ Asellus Australis	3.94	0.83	K0III-IIIb	136	08 44 41	+18 09 15

Canes Venatici: Objects of Interest

Object	Notes
Star	
Y CVn **La Superba**	See separate entry.
Globular cluster	
M3 (NGC 5272)	One of the brightest of its class, it looks like a small comet through binoculars. Magnitude 6.4; diameter 10′; R.A. 13h 41m, Dec. +28° 32′.
Galaxies	
Whirlpool Galaxy	M51, NGC 5272. See separate entry.
Sunflower Galaxy	M63, NGC 5055. See separate entry.
M94 (NGC 4736)	A face-on Sab-type spiral with a very bright core. Magnitude 8.2; size 11.0′ × 9.1′; R.A. 12h 50m, Dec. +41° 17′.
Clusters of galaxies	
Canes Venatici I Group	See separate entry.
Canes Venatici II Group	See separate entry.

Canis Major (abbr. CMa; gen. Canis Majoris)

The Great Dog, the larger of Orion's two hunting dogs. A prominent southern **constellation**, it lies to the east and a bit south of Orion and is easily found by way of **Sirius**, its brightest star and the brightest star in the entire sky. (See tables, "Canis Major: Stars Brighter than Magnitude 4.0," and "Canis Major: Other Objects of Interest." See also star chart 3.)

Canis Minor (abbr. CMi; gen. Canis Minoris)

The Little Dog, the lesser of Orion's hunting dogs; a small northern **constellation** to the east of Gemini. Its brightest star, **Procyon**, forms the northeast apex of the Winter Triangle. (See table, "Canis Minor: Stars Brighter than Magnitude 4.0," and star chart 3.)

Cannon, Annie Jump (1863–1941)

An American astronomer who compiled the enormously influential ***Henry Draper Catalogue***. Cannon was introduced to astronomy by her mother, studied physics at Wellesley College, and specialized in astronomy at Radcliffe College. In 1896 she was appointed by Edward Pickering to the staff at **Harvard College Observatory**, where she began to study variable stars and stellar spectra. As

Canis Major: Stars Brighter than Magnitude 4.0

	Magnitude		Spectral	Distance	Position	
Star	Visual	Absolute	Type	(light-yr)	R.A. (h m s)	Dec. (° ′ ″)
α **Sirius**	−1.44	1.46	A0Vm	8.6	06 45 09	−16 42 58
ε **Adhara**	1.50	−4.11	B2II	430	06 58 38	−28 58 20
δ **Wezen**	1.83	−6.87	F8Ia	1,790	07 08 23	−26 23 36
β **Mirzam**	1.98v	−3.95	B1II-III	500	06 22 42	−17 57 22
η **Aludra**	2.45	−7.51	B5Ia	3,200	07 24 06	−29 18 11
ζ Phurad	3.02	−2.05	B2.5V	336	06 20 19	−30 03 48
o^2	3.02	−6.46	B3Iab	2,570	07 03 01	−23 50 00
κ	3.50	−3.42	B1.5IVne	790	06 49 50	−32 30 31
o^1	3.87	−4.37	K2Iab	1,220	06 54 08	−24 11 02
ω^1	3.89	−5.03	K3Ia	1,980	07 14 49	−26 46 22
ν^2	3.95	2.46	K1III	65	06 36 41	−19 15 22

Canis Major: Other Objects of Interest

Object	Notes
Star	
UW CMa	A **Beta Lyrae star** (eclipsing variable) in the same low-power field as NGC 2362. The components have masses of about 23 M_{sun} and 19 M_{sun} and a combined lum
Open clusters	
M41 (NGC 2287)	More than 50 stars of about eighth magnitude with a bright orange-red star near the center, partly resolved with binoculars. Magnitude 4.5; diameter 38′; R.A. 6h 47.0m, Dec. –20° 44′ (about 4° SW of Sirius).
NGC 2362	A young cluster, about 3,500 light-years away, centered on the hot, luminous star Tau CMa. Magnitude 4; diameter 8′; R.A. 7h 17.8m, Dec. –24° 57′.

color film was not yet available, classification was completed, tediously and with great difficulty, by eye. In 1901, she published a catalog of 1,122 southern stars, which represented a sequence of continuous change from the blue-white stars of types O and B, with their strong helium lines, through types A, F, G, and K, characterized by hydrogen and various metal lines, to the red stars of type M, with their spectral bands of titanium and carbon oxides. It was not known at the time that this was a temperature sequence. In 1910, her scheme was adopted by observatories worldwide. Then came her most famous work, the *Henry Draper Catalogue*, published by the Harvard Observatory between 1918 and 1924, which lists the spectral types, magnitudes, and positions of 225,300 stars, down to ninth magnitude–every one classified personally by Cannon. In 1922, her system was adopted by the International Astronomical Union as the official system for the classification of stellar spectra. She then continued the study down to eleventh magnitude and, in the process, discovered 277 variable stars and 5 novae. Among her many international accolades, she was the first woman to be granted an honorary Ph.D. by Oxford University.

Canopus (Alpha Carinae, α Car)

The brightest star in **Carina** and the second brightest star in the sky. Its name is of uncertain origin; one theory is that it comes from the Coptic or Egyptian *kahi nub* meaning "golden earth." Canopus lies almost due south of first-ranked **Sirius** and nearly 53° south of the celestial equator, so that it can't be seen from latitudes above 37° N. However, for the southern half of the United States, Canopus and Sirius make a marvelous winter sight, as they do all summer in the Southern Hemisphere. Of the pair, Canopus is very much more remote and luminous–a rare class F yellow-white **supergiant**, big enough to stretch three-fourths of the way across Mercury's orbit. It may once have been a **red giant**, or it may become one yet, before ending its days as a massive **white dwarf**.

Visual magnitude:	–0.62
Absolute magnitude:	–5.53
Spectral type:	F0Ib
Surface temperature:	7,800 K
Luminosity:	15,000 L_{sun}
Radius:	65 R_{sun}
Mass:	8 to 9 M_{sun}
Distance:	313 light-years

Cape Photographic Dürchmusterung (CPD)

A Southern Hemisphere extension of the ***Bonner Dürchmusterung*** and the first star catalog produced from pho-

Canis Minor: Stars Brighter than Magnitude 4.0

	Magnitude		Spectral	Distance	Position	
Star	Visual	Absolute	Type	(light-yr)	R.A. (h m s)	Dec. (° ′ ″)
α **Procyon**	0.40	2.68	F5IV-V	11.4	07 39 18	+05 13 30
β Gomeisa	2.89	–0.70	B8Ve	170	07 27 09	+08 17 21

tographic measurements of the sky. The photographs were taken by the Scottish astronomer David Gill (1843–1914) at the Cape Observatory, South Africa, between 1895 and 1900, and the star positions on them were measured by Jacobus **Kapteyn** in the Netherlands. The three-volume catalog, containing 454,877 stars down to tenth magnitude, between Dec. –19° and –90° (the south celestial pole), was published from 1896 to 1900.

Cape York meteorite

A huge **iron meteorite** (a type IIIA **octahedrite**) that fell as the largest known shower more than 1,000 years ago in Cape York, western Greenland. Its existence became widely known when European explorers encountered members of an Inuit tribe in northwestern Greenland in 1818 and were astonished to find that they had knife blades, harpoon points, and engraving tools made of iron. The area has no natural metal deposits, yet the plentiful supply of meteoric iron enabled the polar hunters to develop Iron Age technology to help them survive. Five expeditions from 1818 to 1883 failed to find the source of the iron until Robert Peary was led by a local guide to the site on Saviksoah Island off northern Greenland's Cape York in 1894. The meteorite was in three main chunks, known from Inuit folklore as Ahnighito (the Tent), weighing 31 tons, the Woman (2.5 tons), and the Dog (0.5 ton). Over the next three years, Peary's men built Greenland's only railway, specifically to transport the metal, and managed to load the pieces onto ships despite severe weather and engineering problems. Upon arrival in New York City, the source of Greenland's Iron Age was sold to the American Museum of Natural History (AMNH) for $40,000. The 3.4 × 2.1 × 1.7-m Ahnighito is now on display in the Arthur Ross Hall of the AMNH–the largest meteorite in any museum in the world.

Capella (Alpha Aurigae, α Aur)

A yellow giant **G star**, the brightest star in **Auriga** and the sixth brightest in the whole sky; its Latin name means "the she-goat." Capella is actually a quadruple system in which the dominant components are two post-main-sequence G-type stars, each with roughly the same temperature as the Sun, a radius of about 10 R_{sun}, and a mass of 2.5 M_{sun}; one has a luminosity of 50 L_{sun}, the other of 80 L_{sun}. This pair, separable only by spectroscope, lies less than 100 million km apart and orbit each other every 104.02 days. Capella is a source of X rays, probably because of surface magnetic activity similar to that seen on the Sun. Orbiting the main pair, at a distance of about 0.2 light-year, is a much fainter pair of **red dwarfs** of about 0.4 M_{sun} and 0.1 M_{sun}.

Visual magnitude:	0.08
Absolute magnitude:	–0.48
Spectral types:	G0III + G5III + M5V
Distance:	42 light-years

Caph (Beta Cassiopeiae, β Cas)

A giant or subgiant **F star** that is the second brightest star in **Cassiopeia**. Its name comes from an Arabic phrase meaning "the stained hand"–the W-shape of this constellation being interpreted in Arab lore as the splayed fingers of a hand that has been colored by henna. In terms of its evolution, Caph is currently in the **Hertzsprung Gap**, a relatively short-lived stage (lasting a few tens of millions of years), in which its core will shrink while its outer layers expand to become a **red giant**. It is also the brightest **Delta Scuti star**, varying in brightness by about 0.3 magnitude over a 2.4-hour period. It has a small companion, about which little is known, in a 27-day orbit.

Visual magnitude:	2.28
Absolute magnitude:	1.16
Spectral type:	F2III
Surface temperature:	6,700 K
Luminosity:	28 L_{sun}
Radius:	4 R_{sun}
Mass:	2 M_{sun}
Distance:	54 light-years

Capricornus (abbr. Cap; gen. Capricorni)

The Sea Goat; an extensive but dull southern **constellation** and the tenth **sign of the zodiac**. It lies to the north of Microscopium and Pisces Austrinus, to the west and south of Aquarius, and to the southeast of Aquila. In ancient times the Sun was in Capricorn at the winter **solstice** but, because of **precession**, is now in neighboring Sagittarius at this time of the year. Capricorn appears directly overhead the latitude known as the *Tropic of Capricorn.* Among the constellation's deep sky objects are the globular cluster M30 (NGC 7099), which is easily seen with a small telescope but difficult to resolve into individual stars (magnitude 7.7; diameter 11′; distance 41,000 light-years; R.A. 21h 40.4m, Dec. 23° 11′) and the **Capricornus Dwarf**. (See table, "Capricornus: Stars Brighter than Magnitude 4.0," and star chart 14.)

Capricornus Dwarf (Palomar 12)

An object originally classified as a dwarf spheroidal galaxy by Fritz **Zwicky** (though he cataloged it with the wrong sign in declination) and presumed to be a member of the **Local Group**. Later it was found actually to be a remote

Capricornus: Stars Brighter than Magnitude 4.0

Star	Magnitude		Spectral	Distance	Position	
	Visual	Absolute	Type	(light-yr)	R.A. (h m s)	Dec. (° ′ ″)
δ Deneb Algedi	2.85	2.48	A5IVmv	39	21 47 02	–16 07 38
β Dabih	3.05	–1.87	F8V+A0V	314	20 21 01	–14 46 53
α^2 Algedi (Giedi)	3.58	0.96	G6IIIb	109	20 18 03	–12 32 42
γ Nashira	3.69	0.54	A7IIIp	139	21 40 05	–16 39 45
ζ Yen	3.77	–1.67	G4Ib	398	21 26 40	–22 24 41

globular cluster within our own Galaxy, and cataloged as the twelfth object in the Palomar catalog of globular clusters. Palomar 12 lies about 62,000 light-years from the Sun and some 49,000 light-years from the galactic center at R.A. 21h 46.6m, Dec. –21° 15′. According to one hypothesis, it may have escaped from the **Sagittarius Dwarf Elliptical Galaxy** to become part of the Milky Way's halo.

carbon (C)

The element, of **atomic number** six, that is the basis of all terrestrial life. Carbon is produced during **helium burning** in **red giant**s and is ejected into the Galaxy when these stars form **planetary nebula**e. Some carbon also comes from high-mass stars that explode as **supernovae**.

carbon burning

The stage at which a star fuses **carbon** inside its core, making heavier elements such as neon and magnesium. Carbon burning eventually occurs in all stars that start out with greater than about eight solar masses.

carbon dioxide (CO_2)

A gas that makes up a minor but essential component of Earth's atmosphere (required by plants), and the major component of **Venus**'s thick atmosphere and **Mars**'s thin atmosphere. It is an important greenhouse gas (see **greenhouse effect**). Below –78°C it forms a solid, commonly known as dry ice, and is found in this state in the Martian polar ice caps and in comets.

carbon star

A **red giant** whose spectrum is dominated by strong **absorption band**s of **carbon**-containing molecules; the **Swan bands** of C_2 are especially prominent, with absorption by CN, CH, C_3, SiC_2, and Ca II present to varying degrees, often with a strong sodium **D line**. Carbon stars, also known as *C stars,* have carbon/oxygen ratios that are typically four to five times higher than those of normal red giants and show little trace of the light metal oxide bands that are the usual red giant hallmark. They resemble **S stars** in their relative proportion of heavy and light metals, but contain far more carbon in their upper layers. Carbon stars were previously classified as stars of spectral type R (hotter, with surface temperatures of 4,000 to 5,000 K) and N (up to 10 times more luminous but cooler, with a temperature of about 3,000 K). They are typically associated with some circumstellar material in the form of dusty shells, disks, or clouds. See also **CH star** and **CN star**.

carbonaceous chondrite

The most primitive and unaltered type of **meteorite** known, with an elemental composition probably similar to that of the **solar nebula** from which the solar system formed. Carbonaceous **chondrite**s, a type of **stony meteorite**, contain silicates, oxides, and sulfides, but most distinctively contain water (chemically bonded as water of hydration) or minerals that have been altered in the presence of water, together with large amounts of carbon, including organic compounds. The most pristine carbonaceous chondrites have never been heated above 50°C. Different groups of carbonaceous chondrites have been identified that came from parent bodies in different parts of the solar nebula.

CI chondrites, only a handful of which are known, are named after the Ivuna meteorite that fell in Tanzania in 1938 and are among the most primitive, friable (crumbly), and interesting of all meteorites. They have undergone extensive aqueous alteration, and as a result, they lack relict **chondrule**s but contain up to 20% water, as well as various minerals altered in the presence of water, such as hydrous phyllosilicates (similar to terrestrial clays), oxidized iron in the form of magnetite, and **olivine** crystals sparsely scattered in a black matrix. They also contain organic matter, including **polycyclic aromatic hydrocarbons** (PAHs) and **amino acids**, which makes them important in the search for clues to the origin of **life in the universe**. CIs have never been heated above 50°C, indicating that they came from the outer part of the **solar nebula**.

CM chondrites, named after the Mighei meteorite that fell in Ukraine in 1889, have about half the water content of CI specimens and show less aqueous alteration and some well-preserved chondrules. Like CIs, however, they harbor a wealth of organic material–more than 230 different amino acids, in the case of the **Murchison meteorite**. Comparisons of spectra point to the asteroid (19) Fortuna or, possibly, the largest asteroid, **Ceres**, as candidate parent bodies.

CV chondites, named after the Vigarano meteorite that fell in Italy in 1910, more closely resemble **ordinary chondrites**. They have large, well-defined chondrules of magnesium-rich olivine, often surrounded by iron sulfide, in a dark-gray matrix of mainly iron-rich olivine. They also contain **calcium aluminum inclusions** (CAIs)–the most ancient minerals known in the solar system–that typically make up more than 5% of the meteorite.

CO chondrites, named after the Ornans meteorite that fell in France in 1868, are related in chemistry and composition to the CV chondrites and may, with them, represent a distinct clan of carbonaceous chondrites that formed in the same region of the early solar system. However, COs are mostly blacker than CVs and have much smaller chondrules that are densely packed within the matrix and represent over 70% of the meteorite. As in the CV group, CAIs are present but are commonly much smaller and spread more sparsely in the matrix. Also typical of COs are small inclusions of free metal, mostly nickel-iron, that appear as tiny flakes on the polished surfaces of fresh, unweathered samples.

CK chondrites, named for the Karoonda meteorite that fell in Australia in 1930, were initially regarded as members of the CV group. However, they are now grouped separately since they differ in some respect from all other carbonaceous chondrites. They appear dark-gray or black due to a high percentage of **magnetite** that is dispersed in a matrix of dark silicates, consisting of iron-rich olivine and pyroxene. This shows they formed under oxidizing conditions, yet there is no sign of aqueous alteration or phyllosilicates. Elemental abundances and oxygen isotopic signatures suggest that CKs are closely related to CO and CV types. Most CK chondrites contain large CAIs and some show shock veins that point to a violent impact history.

CR chondrites, named for the Renazzo meteorite that fell in Italy in 1824, are similar to CMs in that they contain hydrosilicates, magnetite, and traces of water. The main difference is that CRs contain reduced metal in the form of nickel-iron and iron sulfide that occurs in the black matrix as well as in the large chondrules that make up about 50% of the meteorites. A possible parent body is **Pallas**, the second largest asteroid.

CH chondrites, named for their high metal content, contain up to 15% nickel-iron and are chemically very close to the CRs and CBs (description follows). They also show many fragmented chondrules, most of which, along with the less abundant CAIs, are very small. As with the CRs, the CHs contain some phyllosilicates and other traces of aqueous alteration. According to one theory, the CHs formed at an early stage from the hot primordial nebula inside what is today the orbit of Mercury, and later were transported to outer, cooler regions of the nebula where they have been more or less preserved to this day. Mercury may have formed from similar, metal-rich material, which would explain its high density and extraordinary large metal core.

CB chondrites, also known as *bencubbites,* are named after the prototype found near Bencubbin, Australia, in 1930. Only a handful of these strange meteorites are known, all composed of more than 50% nickel-iron, together with highly reduced silicates, and chondrules similar to those found in members of the CR group. The CH and CB chondrites are so closely related to the CRs that all three groups may have come from the same parent or at least from the same region of the solar nebula.

C ungrouped chondrites (or C UNGRs) fall outside the other groups and probably represent other parent bodies of carbonaceous chondrites or source regions of the primordial solar nebula.

carbon-nitrogen-oxygen cycle (CNO cycle)

A series of **fusion** reactions in which hydrogen is converted to helium; the main energy-producing process in stars whose core temperature exceeds 18 million K. This is typically the case with stars that are more than twice as massive than the Sun. At lower temperatures (and stellar masses), the **proton-proton chain** becomes increasingly important. The CNO cycle is so named because nuclei of carbon, nitrogen, and oxygen serve as catalysts. In simplified form, the reaction steps are:

$^{12}\text{C} + \text{proton } (^{1}\text{H}) \rightarrow {}^{13}\text{N}$

$^{13}\text{N} - \text{electron} \rightarrow {}^{13}\text{C}$

$^{13}\text{C} + \text{proton} \rightarrow {}^{14}\text{N}$

$^{14}\text{N} + \text{proton} \rightarrow {}^{15}\text{O}$

$^{15}\text{O} - \text{electron} \rightarrow {}^{15}\text{N}$

$^{15}\text{N} + \text{proton} \rightarrow {}^{12}\text{C} + {}^{4}\text{He}$

Carina (abbr. Car; gen. Carinae)

The Keel (part of mythical Jason's ship, the Argo); a major southern **constellation** that is home to the second brightest

Carina: Stars Brighter than Magnitude 4.0

Star	Magnitude		Spectral	Distance	Position	
	Visual	Absolute	Type	(light-yr)	R.A. (h m s)	Dec. (° ′ ″)
α **Canopus**	–0.62	–5.53	F0Ib	313	06 23 57	–52 41 44
β **Miaplacidus**	1.67	–1.00	A2III	111	09 13 12	–69 43 02
ε **Avior**	1.86	–4.58	K3III+B2V	632	08 22 31	–59 30 34
ι Tureis (Aspidiske)	2.21	–4.43	A8Ib	693	09 17 05	–59 16 31
θ	2.74	–2.91	B0V	439	10 42 57	–64 23 39
υ	2.92	–5.57	A7Ib+B7III	1,620	09 47 06	–65 04 18
ω	3.29	–1.99	B8IIIe	370	10 13 44	–70 02 16
q	3.38	–3.38	K3IIa	736	10 17 05	–61 19 56
a	3.43	–2.11	B2IVe	419	09 10 58	–58 58 01
χ	3.46	–1.91	B3IVp	387	07 56 47	–52 58 56
x	3.93	–7.37	G0Ia	5,930	11 08 35	–58 58 30

star in the sky, **Canopus**. It lies to the south of Vela and the southwest of Centaurus. The Milky Way passing through Carina and Vela makes for a field rich in **open clusters**. (See tables, "Carina: Stars Brighter than Magnitude 4.0," and "Carina: Other Objects of Interest." See also star chart 4.)

Carina Dwarf Galaxy

A **dwarf spheroidal galaxy** that orbits the Milky Way; it was discovered in 1977 and lies 330,000 light-years from the galactic center. The Carina Dwarf appears to have formed several billion years after the other satellite galax-

Carina: Other Objects of Interest

Object	Notes
Planetary nebula	
NGC 2867	An object just within binocular range, lying between Iota and Chi Car. Magnitude 9.7; diameter 11″; R.A. 9h 21.4m, Dec. –58° 19′.
Diffuse nebula	
Eta Carinae Nebula	NGC 3372. See separate entry.
Open clusters	
NGC 2516	A large cluster best seen with binoculars. It contains about 100 stars, including a fifth magnitude red giant. Magnitude 3.8; diameter 30′; R.A. 7h 58.3m, Dec. –60° 52′.
NGC 3532	About 150 stars of sixth magnitude and fainter. Small telescopes show it well and reveal an elliptical shape. A magnitude 4 orange star at one edge is not a true member but is much more distant. Magnitude 3.0; diameter 55′; distance 1,300 light-years; R.A. 11h 6.4m, Dec. –58° 40′.
Southern Pleiades	IC 2602. See separate entry.
Carina OB1 and OB2	See separate entry.
Globular cluster	
NGC 2808	A faint naked-eye object. Magnitude 6.3; diameter 14′; R.A. 9h 12.0m, Dec. –64 52′.
Galaxy	
Carina Dwarf Galaxy	See separate entry.

ies of the Milky Way, its oldest stars dating back no more than about 7 billion years. See also **Local group**.

Carina OB1 and OB2

Two rich **OB associations** that lie about 10,000 light-years away, near **Eta Carinae**. They contain some exceptional objects, including Star 12 in Carina OB2, which, along with Eta Carinae, is one of the brightest and most massive stars known, and the **Wolf-Rayet star** HD 93162 in Carina OB1, which has one of the largest known stellar X-ray luminosities.

Carme

One of the outer moons of **Jupiter**, also known as Jupiter XI; it was discovered in 1938 by Seth **Nicholson**. Carme has a diameter of 46 km and moves in a **retrograde** orbit at a mean distance of 23.4 million km from the planet.

Carnegie Observatories

The astronomy department of the Carnegie Institution of Washington, D.C. It operates the **Las Campanas Observatory** in Chile, the chief observatory for American astronomers in the Southern Hemisphere. The Carnegie Observatories has its main office in Pasadena, California, and a smaller office, serving the Las Campanas Observatory, in the Chilean coastal city of La Serena.

Carrington, Richard Christopher (1826–1875)

An English amateur astronomer who, from a close study of **sunspots** from 1853 to 1861, discovered the latitude drift of **sunspots** during the 11-year **solar cycle**. Unfortunately, his astronomical work, carried out from his observatory at Redhill, Surrey, came to an abrupt end when he had to take over his family's brewing business and his important discovery is commonly attributed to Gustav **Spörer**, who analyzed the drift in some detail.

Carte du Ciel

"Chart of the Sky," an ambitious program to photograph every part of the sky, showing stars as faint as fourteenth magnitude, and to compile a catalog, now known as the ***Astrographic Catalogue***, listing the positions of stars down to eleventh magnitude. It was conceived by the Scottish astronomer David Gill (1843–1914) and Admiral Mouchez, director of the Paris Observatory, who called an international conference in 1887 to initiate the project. Eighteen observatories agreed to cooperate and to adopt the 13-inch refractor as a standard design for a photographic telescope, developed for the Paris Observatory by the brothers Paul and Prosper Henry. The sky was divided into zones and allocated to the various observatories. Although parts of the program were never completed, the plates obtained are of great value for comparison with fields rephotographed many years later.

Cartwheel Galaxy (ESO350-40)

A galaxy, lying about 500 million light-years away in **Sculptor**, that has been tidally distorted by an encounter with another galaxy into a ring-and-hub structure. The striking cartwheel appearance is the result of a smaller intruder galaxy having careened through the core of the larger system, which was probably once a spiral similar to the Milky Way. Like a pebble tossed in a lake, the collision sent a ripple of energy into space, plowing gas and dust in front of it. Expanding at a rate of more than 300,000 km/hr, this cosmic tsunami left a burst of new star creations in its wake. Images taken by the **Hubble Space Telescope** have resolved bright blue knots that are gigantic clusters of newborn stars and immense loops and bubbles blown into space by **supernovae**. The intruder was almost certainly one of two small galaxies seen near the Cartwheel. The bluer of this pair is disrupted and has new star formation, which strongly suggests it is the interloper.

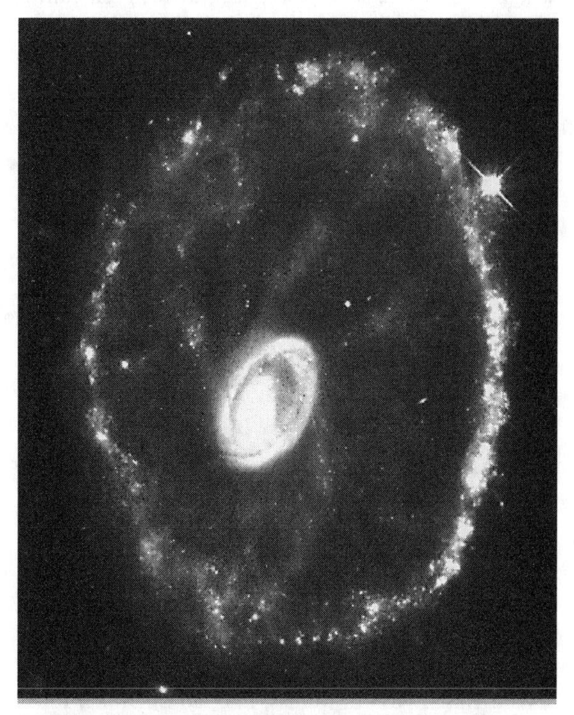

Cartwheel Galaxy A view of this remarkable ring galaxy by the Hubble Space Telescope in 1994. Bright knots in the expanding ring of the Cartwheel are huge clusters of newborn stars and immense loops and bubbles blown into space by supernovae going off like a string of firecrackers. *NASA/Kirk Borne (STScI)*

However, the smoother looking companion has no gas, which is consistent with the idea that gas was stripped out of it during passage through the Cartwheel. The Cartwheel's original spiral structure is beginning to reemerge, as seen in the faint arms or spokes between the outer ring and bull's-eye–shaped nucleus. The ring contains at least several billion new stars that would not normally have formed in such a short time span and is so large–150,000 light-years across–that our entire Milky Way Galaxy would fit inside it.

Cassegrain telescope

A type of **reflecting telescope** with a folded optical path achieved by two mirrors–a large concave paraboloidal primary with a central hole and a small hyperboloidal convex mirror mounted on the large front **corrector plate**. Light strikes the primary mirror, which reflects the image back to the smaller convex secondary mirror, which in turn reflects the magnified image through the center hole and on to the eyepiece. The design was conceived in about 1672 by the Frenchman Guillaume Cassegrain, about whom little is known. In an alternate scheme, called a *modified Cassegrain,* a small flat mirror placed immediately in front of the primary brings the light out to the side of the telescope tube and eliminates the need for a perforated primary. Like the Gregorian and Newtonian telescopes, the paraboloid-hyperboloid combination of the Cassegrain is free from **spherical aberration**. The *Dall-Kirkham telescope* is a variant of the Cassegrain that uses a concave ellipsoidal primary mirror and a convex spherical secondary mirror. Invented independently by the English amateur telescope maker Horace Dall (1901–1986) and the American Alan Kirkham, it is particularly suited to planetary observation where good **resolution** is more important than wide **field of view**. See also **Schmidt-Cassegrain telescope**.

Cassini (spacecraft)

A **Saturn** probe that was built by NASA and the European Space Agency and named for Giovanni **Cassini**. The spacecraft has two main parts: the Cassini orbiter and the Huygens probe, which will be released into the atmosphere of **Titan**. Cassini will enter orbit around the sixth planet on July 1, 2004, to begin a complex four-year sequence of orbits designed to let it observe Saturn's near-polar atmosphere and magnetic field, and carry out several close flybys of Mimas, Enceladus, Dione, Rhea, and Iapetus, and multiple flybys of Titan.

Cassini, Giovanni Domenico (1625–1712)

An Italian-born French-naturalized astronomer (known also by his Gallicized name, Jean Dominique) who discovered four of Saturn's moons–**Iapetus** (1671), **Rhea** (1672), **Tethys**, and **Dione** (both 1684), and the major division in its rings (1675), now named after him. Cassini became the first director of the Paris Observatory (1669), found the rotational periods of Mars and Jupiter, and the distance to Mars (1672) by triangulation with the help of observations by Jean Richer. His refinement of the scale of the solar system led to a value of the astronomical unit only 7% short of that accepted today. The improved tables he drew up for Jupiter's satellites were important in Ole Rømer's determination of the speed of light. Cassini's son, Jacques (1677–1756), grandson Cesar Francois (1714–1784), and great-grandson Jean Dominique IV (1748–1845) all became successful astronomers, the first two succeeding the elder Cassini as director of the Paris Observatory.

Cassiopeia (abbr. Cas; gen. Cassiopeiae)

The Queen, mother of Andromeda; a large northern **constellation** with a distinctive "W" shape and with the Milky Way as a backdrop. A binocular sweep of the region is rewarding (especially around Delta Cas), and small instruments reveal an interesting assortment of **nebula**e and **open clusters**. (See tables, "Cassiopeia: Stars Brighter than Magnitude 4.0," and "Cassiopeia: Other Interesting Objects." See also star chart 1.)

Cassiopeia A (3C 461)

The strongest radio source in the sky, apart from the Sun; it is associated with the youngest known **supernova remnant** in the Milky Way Galaxy. Cas A lies about 10,000 light-years away in **Cassiopeia** (R.A. 23h 23.4m, Dec. +58° 8.9′) and coincides with a shell-type remnant of a Type II **supernova** whose light reached Earth in about 1667. Optically, it is a faint nebula, 10 light-years across, with an expansion velocity of about 800 km/s and a mass of a few solar masses. It is also an extended source of soft X rays.

Cassiopeia Dwarf (Cas dSph)

A recently discovered **dwarf spheroidal galaxy** that is also a **satellite galaxy** of the **Andromeda Galaxy** (M31), and hence also a member of the **Local Group**. The Cassiopaeia Dwarf was found in 1998, together with another small satellite of M31, the **Pegasus Dwarf** (Peg dSph), by a team of astronomers in Russia and the Ukraine. At a distance of 2.6 million light-years from the Sun, Cas dSph and Peg dSph are farther from Andromeda than its other known companion galaxies, yet still appear bound to it by gravity. Neither galaxy contains any young, massive stars or shows traces of recent star formation.

Cassiopeia: Stars Brighter than Magnitude 4.0

Star	Magnitude		Spectral	Distance	Position	
	Visual	Absolute	Type	(light-yr)	R.A. (h m s)	Dec. (° ′ ″)
α **Shedar**	2.24	−1.99	K0IIa	229	00 40 30	+56 32 15
γ **Gamma Cassiopeiae**	2.15v	−4.22	B0IVe	613	00 56 43	+60 43 00
β **Caph**	2.28	1.16	F2III-IV	54	00 09 11	+59 08 59
δ Ruchbah	2.66	0.24	A5III-IVv	99	01 25 49	+60 14 07
ε Segin	3.35	−2.31	B3III	442	01 54 24	+63 40 13
η Achird	3.46	−4.73	G0V+dM0	1,410	00 49 06	+57 48 58
ζ	3.69	−2.63	B2IV	597	00 36 58	+53 53 49

Instead, both seem dominated by very old stars, with ages up to 10 billion years.

Castalia (minor planet 4769)

An Earth-crossing and **potentially hazardous asteroid** that is a member of the **Apollo group**, discovered by Eleanor Helin (Caltech) and colleagues. Castalia became the first asteroid for which astronomers were able to generate an image. In August 1989, when it passed 5.6 million km from Earth (11 times the distance of the Moon), Steven Ostro and his group, at the Jet Propulsion Laboratory, bounced radar beams off Castalia using the Arecibo radio telescope. The three-dimensional model built from the resulting data was of a double-lobed object–a peanut shape–nearly 1.8 km across at its longest. This suggests that Castalia consists of two 800-m pieces resting together: the first known contact-binary object in the solar system. Radar echoes from other **near-Earth asteroids** have since shown that this configuration is quite common.

Cassiopeia: Other Interesting Objects

Object	Notes
Planetary nebula	
Bubble Nebula	NGC 7635. See separate entry.
Supernova remnant	
Cassiopeia A	See separate entry.
Diffuse nebulae	
I 59	Nebulosity associated with Gamma Cas, consisting of two fans pointing northwest. Magnitude 9.9; diameter 10′; R.A. 0h 57.5m, Dec. +61° 9′.
NGC 281	One degree west of Shedir. Magnitude 7; diameter 35′; R.A. 0h 52.8m, Dec. +56° 37′.
Open clusters	
M52 (NGC 7654)	Kidney shaped, containing over 100 stars, with a prominent magnitude 8 star at one edge. A good object for binoculars. Magnitude 6.9; diameter 13′; R.A. 23h 24.2m, Dec. +61° 35′.
M103 (NGC 581)	A fan-shaped cluster with at least 40 stars. Magnitude 7.4, diameter 6′; R.A. 1h 33.2m, Dec. +60° 42′.
NGC 457	One of the brightest open clusters in the sky and an attractive object for small telescopes. Its stars appear to be arranged in chains. Magnitude 6.4; diameter 13′; R.A. 1h 19.1m, Dec. +58° 20′.
Galaxy	
Cassiopeia Dwarf	See separate entry.

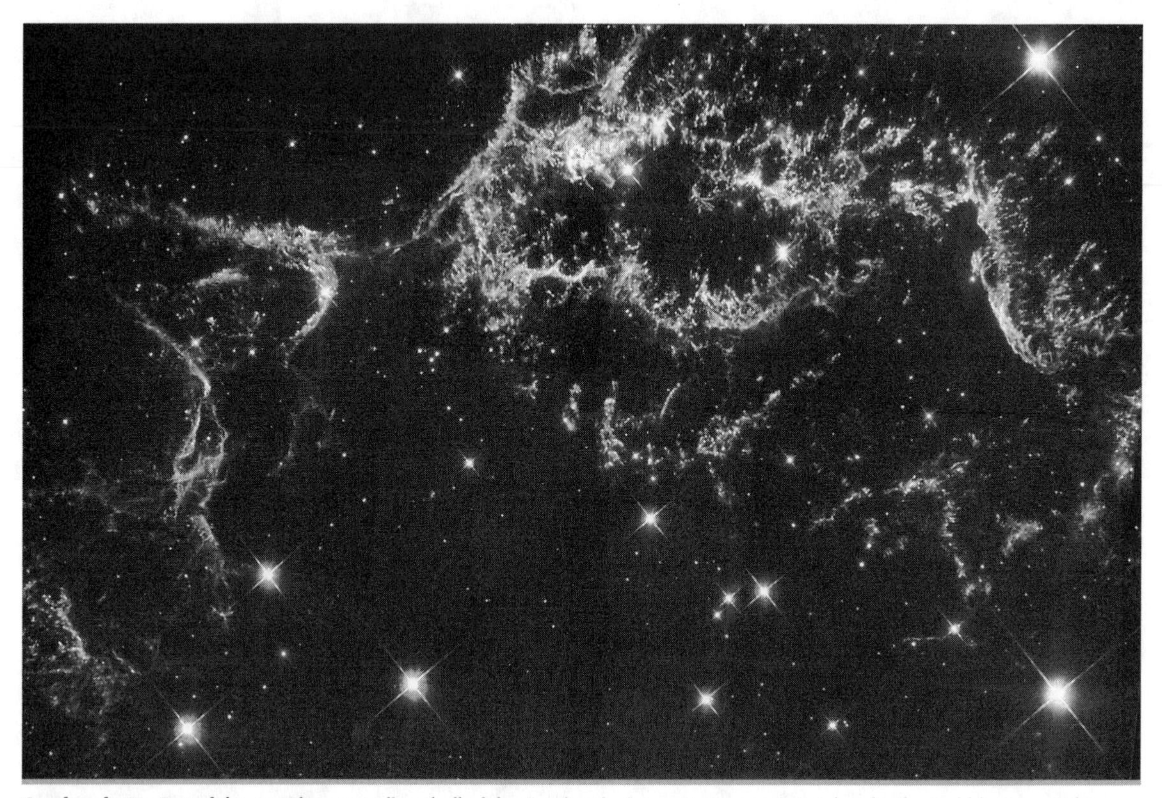

Cassiopeia A Part of the complex expanding shell of the Cassiopeia A supernova remnant, taken by the Hubble Space Telescope. The distance from left to right across this image corresponds to about 15 light-years. *NASA/STScI/AURA*

Diameter:	1.8 × 0.8 km
Density:	2.1 g/cm^3
Spectral class:	S
Rotational period:	4 hours
Orbit	
Semimajor axis:	1.063 AU
Eccentricity:	0.4831
Inclination:	8.89°
Period:	1.10 years

Castor (Alpha Geminorum, α Gem)

The second brightest star in **Gemini** and, in mythology, the twin of **Pollux**. Though appearing as a single white star to the naked eye, Castor is really a remarkable multiple system, consisting of three stellar pairs. A modest telescope shows two similar **A stars**, of magnitude 2.0 and 2.8, a couple of arcseconds apart, which orbit each other every 380 years. About 73′ to the south lies a ninth magnitude companion that orbits the bright pair at a distance of about 1,200 AU. A spectrograph shows that each of the two bright components, A and B, is a double. Castor A consists of almost identical stars, with a combined mass of 3.2 M_{sun}, orbiting each other every 9.22 days. Castor B's twin stars orbit even faster, making their round trip in a mere 2.9 days. The faint, distant star, Castor C, is also double, consisting of nearly identical, low-mass **M stars**, with temperatures of about 4,000 K, some 2.5 million km apart with an orbital period of 19.5 hours. One or both are **flare stars**. Castor is part of a group of widely dispersed stars, known as the **Castor Moving Group**.

Visual magnitude:	1.58
Absolute magnitude:	0.58
Spectral type:	A1V + A2V
Distance:	52 light-years

Castor Moving Group

An extensive but physically related group of more than 50 stars, including **Castor**, **Vega**, **Formalhaut**, **Alderamin**, and Zubenelgenubi, that has a common motion through space. The properties of the group suggest its members are 100 to 300 million years old.

cataclysmic binary

A **close binary** system consisting of a **white dwarf** primary and an orange or red (K- or M-type) main-sequence or giant secondary, in which matter flows from the secondary toward the primary. In most cases, the stripped hydrogen-rich gas enters an **accretion disk** around the white dwarf and may subsequently fall onto the condensed star to ignite violently in a fusion reaction. Cataclysmic variables fall into two groups. The nonmagnetic group, which is by far the most populous, includes **U Geminorum stars** (dwarf novae) and classical and recurrent **novae**. The much rarer, magnetic group includes **AM Herculis stars** and **DQ Herculis stars** and is distinguished by the presence of a powerful magnetic field around the primary that dramatically affects the accretion flow.

cataclysmic variable

(1) A **variable star** that shows a sudden and dramatic change in brightness, including any type of **cataclysmic binary**, some types of **symbiotic star**, and all **supernovae**. (2) More restrictively, a cataclysmic binary.

catadioptric telescope

A telescope that uses a combination of mirrors and lenses to bring light to a focus. The best known types are the **Schmidt camera**, the **Schmidt-Cassegrain telescope**, and the **Maksutov telescope.**

Cat's Eye Nebula (NGC 6543)

A young **planetary nebula** in **Draco**, midway between Delta and Zeta Dra. One of the brightest objects of its type, its resemblance to a cat's eye is due to a series of gas loops that have been ejected by the central star over the past 1,000 years or so. The great complexity of the Cat's Eye's structure, with its concentric gas shells, **jets** of high-speed gas, and unusual shock-induced knots of gas, has led astronomers to suggest that the central star may be a binary system. According to this idea, a fast **stellar wind** from the central star created the elongated shell of dense, glowing gas visible in images of the nebula. This structure is embedded within two larger gas lobes, previously blown off by the star, that are pinched by a ring of denser gas, presumably ejected along the orbital plane of the binary companion. The suspected companion star also might be responsible for a pair of high-speed jets of gas that lie at right angles to the equatorial ring. If the companion were pulling in material from a neighboring star, jets escaping along the companion's rotation axis could be produced. These jets would explain several puzzling features along the periphery of the gas lobes. The fact that the twin jets are now pointing in different directions than these features suggests that the jets are wobbling, or precessing, and turning on and off episodically. Historically, NGC 6543 was the first planetary to be observed with a spectroscope and to reveal the presence of emission lines. This started the controversy about whether planetaries consist of numerous stars or, as turned out to be the case, clouds of diffuse gas. An 8-cm telescope will show a foggy blue-green disk; more powerful instruments are needed to reveal the internal structure.

Magnitude:	8.8
Diameter:	18″ × 350″
Distance:	3,000 light-years
Position:	R.A. 7h 58.7m, Dec. +66° 38′

CCD (charge-coupled device)

A small photoelectronic imaging device (typically 1.5 cm square) made from a crystal of silicon in which numerous (at least 250,000) individual light-sensitive picture elements (pixels) have been fabricated. Each pixel (less than 0.03 mm) is capable of storing electronic charges created by the absorption of light. The name derives from the method of extracting the locally stored charges from each pixel, which is done by transferring or "coupling" charges from one pixel to the next by the controlled collapse and growth of adjacent storage sites or potential wells. Each well is formed inside the silicon crystal by the electric field generated by voltages applied to tiny, semitransparent metallic electrodes on the CCD surface. CCDs, placed at the receiving end of telescopes to take pictures of very faint astronomical objects, have almost completely superceded photographic plates for professional use. A *CCD spectrometer* is an instrument that uses a modified CCD to obtain images and spectra of X-ray sources.

C-class asteroid

A very dark and nonreflective **asteroid** with a composition believed to be similar to that of **carbonaceous chondrites**. C-class asteroids are the commonest type known and dominate the outer part of the main **asteroid belt.** They have an **albedo** of 0.03 to 0.09 and a reflectance spectrum that is flat at wavelengths longer than 0.4 micron but shows a feature shorter than 0.4 micron thought to be due to water molecules bound to crystals. Examples include (10) **Hygeia** and (253) **Mathilde**. Subclasses include B, F, and G (see separate entries for each).

celestial coordinates

Any system of coordinates that can be used to give the position of an object, such as a star, on the **celestial sphere**. A particular system is defined by the chosen point of observation (the origin), the plane of reference, and whether the coordinates are spherical or rectangular.

As an origin, *geocentric coordinates* use the center of Earth, *topocentric coordinates* use a specific point on Earth's surface, and *heliocentric coordinates* use the center of the Sun. This choice of origins can be combined with a variety of reference planes. (1) **Equatorial coordinates**, the most commonly used in astronomy, may be geocentric or topocentric, and use the celestial equator as their reference plane. (2) **Ecliptic coordinates**, may be geocentric or heliocentric, and refer to the **ecliptic**. (3) **Horizontal coordinates** are topocentric and refer to the observer's horizon. (4) **Galactic coordinates** are specified relative to the plane of the Galaxy. In all these systems, **spherical coordinates** dominate but **rectangular coordinates** are occasionally used for some special purposes.

celestial mechanics

The branch of **astronomy** concerned with the motion of celestial objects and, particularly, the determination of orbits.

celestial sphere

The immense imaginary sphere used as a background for ascribing the positions of celestial objects; it is centered on the origin of the chosen system of **celestial coordinates**. The *celestial axis* is the projection of Earth's rotation axis, north and south, onto the celestial sphere. The *celestial poles* are the two points where an extension of Earth's axis intersects the celestial sphere and about which the celestial sphere appears to rotate daily. As a result of **precession**, the celestial poles complete a circle around the ecliptic poles every 25,800 years. The *celestial meridian* is the great circle on the celestial sphere that passes through the celestial poles and the zenith of the observer. The *celestial equator* is the great circle on the celestial sphere that divides the northern and southern hemispheres and serves as the zero-mark for **declination**; it is the projection into space of Earth's equatorial plane. The *celestial latitude* is the angular distance on the celestial sphere measured north or south of the ecliptic along the great circle passing through the poles of the ecliptic and the celestial object. The *celestial longitude* is the angular distance along the ecliptic from the vernal **equinox** eastward.

Centaur

A minor planet whose orbit around the Sun lies typically between the orbits of Jupiter and Neptune (5 to 30 AU), though it may extend inward almost as far as Mars (as in the case of Damocles) or outward beyond Neptune (as in the case of 1995SN55, which ranges out to about 40 AU). Because Centaurs cross the orbits of one or more of the giant outer planets, they are also known as *outer planet crossers*. The orbits of the Centaurs are dynamically unstable due to interactions with the **giant planets**, so they must be objects in transition from a large, outer reservoir of small bodies, believed to be the **Kuiper Belt**, to potentially active, cometlike inner solar system objects. Known sizes range from a few tens to a few hundreds of kilometers across, though there are doubtless many smaller ones awaiting discovery. Their composition is probably intermediate between that of **comets** and ordinary **asteroids**; indeed, the first object to be called a Centaur, **Chiron**, is now also classified as a comet following the discovery of a **coma** around it. In the same way that Centaurs, together with **trans-Neptunian objects** (TNOs), may be considered to be protocomets, some comets, such as 29P/Schwassmann-Wachmann 1 and 39P/Oterma, have orbits that would allow them to be called Centaurs. Classification of Centaurs is often difficult, and there is clearly a continuum of types that includes Centaurs, comets, asteroids, Kuiper Belt objects, and other entities such as **cubewanos** and **Scattered Disk** objects. **Pholus**, which orbits from Saturn to past Neptune, is a member of a subgroup known as *red Centaurs*, remarkable for their surface coloration, which is believed to be due to a coating of organic materials. If any Centaur were to be perturbed into an orbit that approaches the Sun, it would become a truly spectacular comet, many times brighter than any seen in historic times.

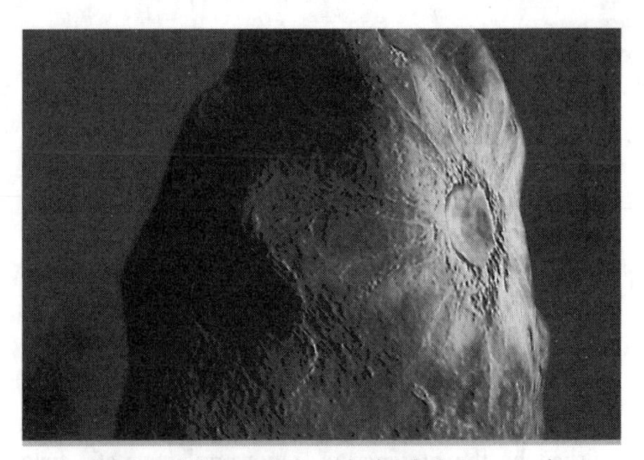

Centaur An artist's impression of a bright feature on the surface of 8405 Asbolus, an 80-km-wide Centaur that lies between Saturn and Uranus. Infrared observations by the Hubble Space Telescope led to the discovery of this feature, which may be a fresh crater less than 10 million years old, in and around which some form of ice has been exposed. This finding is important because it shows that Centaurs do not have a simple uniform surface composition. *NASA/Greg Bacon (STScI/AVL)*

Centaurus (abbr. Cen; gen. Centauri)

The Centaur; a large, bright southern **constellation**. It is most famous for containing **Alpha Centauri**, the nearest star to the Sun. A line from Alpha through Beta Cen (the Pointers) leads to Crux, the Southern Cross. (See tables, "Centaurus: Stars Brighter than Magnitude 4.0," and "Centaurus: Other Objects of Interest." See also star chart 5.)

Centaurus A (NGC 5128)

The second strongest extragalactic radio source (after **Cygnus A**) and the nearest **radio galaxy** to the Milky Way; it lies 13 to 15 million light-years away in the M83 group of galaxies. Cen A's optical counterpart, discovered by James Dunlop in 1826, is an unusual **elliptical galaxy** (NGC 5128) bisected by a dark circumgalactic dust belt that appears to be debris from a merger between the elliptical and at least one spiral galaxy over the past few billion years. At radio wavelengths, Cen A shows two vast lobes of radio emission that extend thousands of light-years in opposite directions along the polar axis of the disk of NGC 5128. The most active radio emission, however, is associated with Cen A's compact core, which is the foremost example of a radio-loud **active galactic nucleus** and, at only 10 light-days across, the smallest known extragalactic radio source. Infrared measurements have re-vealed high-speed motions in this core that indicate a fast-spinning disk containing some 200 million M_{sun} of material. These data confirm a previous suspicion that the active nucleus of Cen A is powered by a supermassive **black hole** with a mass of about 100 million M_{sun}.

Centaurus Cluster

A cluster of hundreds of galaxies whose center lies only 170 million light-years away. Like other large clusters of galaxies, it is filled with gas at temperatures of 10 million K, making it an extended, luminous source of X rays. The brightest individual member is the eleventh magnitude elliptical NGC 4696. Together with the IC 4329 Cluster and Hydra I, the Centaurus Cluster is part of the Hydra-Centaurus Supercluster.

Centaurus: Stars Brighter than Magnitude 4.0

Star	Magnitude		Spectral	Distance	Position	
	Visual	Absolute	Type	(light-yr)	R.A. (h m s)	Dec. (° ′ ″)
α **Alpha Centauri**	–0.27	4.08		4.25		
α^1	–0.01	4.40	G2V		14 39 37	–60 50 02
α^2	1.33	5.70	K1V		14 39 35	–60 50 13
β **Hadar** (Agena)	0.61	–5.43	B1III	525	14 03 49	–60 22 22
θ **Menkent**	2.06	0.70	K0IIIb	61	14 06 41	–36 22 12
γ Muhlifain	2.20	–0.81	A1IV	130	12 41 31	–48 57 34
ε	2.29	–3.02	B1III	376	13 39 53	–53 27 58
η	2.33	–2.55	B1.5Vne	309	14 35 30	–42 09 28
ζ Al Nair al Kentaurus	2.55	–2.81	B2.5IV	385	13 55 32	–47 17 17
δ	2.58	–2.73	B2IVne	376	12 08 22	–50 43 20
ι	2.75	1.48	A2V	59	13 20 36	–36 42 44
κ Ke Kwan	3.13	–2.96	B2IV	539	14 59 10	–42 06 15
λ	3.13	–2.39	B9II	410	11 35 47	–63 01 11
ν	3.41	–2.41	B2IV	475	13 49 30	–41 41 16
μ	3.47	–2.57	B2IVnpe	527	13 49 37	–42 28 25
ϕ	3.83	–1.94	B2IV	465	13 58 16	–42 06 02
τ	3.85	0.82	A2V	132	12 37 42	–48 32 28
υ^1	3.87	–1.67	B2IV-V	418	13 58 41	–44 48 13
π	3.90	–1.07	B5Vn	321	11 21 00	–54 29 27
σ	3.91	–1.76	B3V	443	12 28 02	–50 13 51

Centaurus: Other Objects of Interest

Object	Notes
Planetary nebula	
NGC 3918	**Blue Planetary**. See separate entry.
Open cluster	
NGC 3766	A cluster of about 60 stars that is visible to the naked eye. Small telescopes show a pattern of loops containing orange, yellow, white, and bluish stars. Magnitude 5.3; diameter 12′; distance 1,700 light-years; R.A. 11h 36.1m, Dec. –61° 37′.
NGC 5460	A scattered object well suited for small telescopes. Magnitude 5.6; diameter 25′; distance 2,700 light-years; R.A. 14h 7.6m, Dec. –48° 19′.
Globular cluster	
Omega Centauri	NGC 5139. See separate entry.
Dark cloud	
Coalsack	See separate entry.
Galaxies	
NGC 4945	An edgewise open spiral galaxy, similar to our own, that appears as a long narrow luminous haze. It is a member of a small group of galaxies that includes NGC 5128 and M83. Magnitude 9.0; diameter 20′; R.A. 13h 5m, Dec. –49° 28′.
Centaurus A	NGC 5128. See separate entry.

Centaurus X-3

A pulsating X-ray source, roughly 30,000 light-years away in **Centaurus**, that appears to consist of a **pulsar** (a neutron star) of mass 0.65 to 0.83 M_{sun} in orbit around a giant or supergiant **B star** known as V779 Cen, or Krzeminski's Star (after the Polish astronomer W. Krzeminski). The pulsar spins on its axis once every 4.8 seconds giving rise to regular X-ray variations with this period. It also orbits the giant star every 2.087 days, during which time it is occulted (blocked from our view), so that no X rays are detected, for 0.488 days. In addition to these variations, the X-ray emission from Cen X-3 flickers at rates of between 100 and 2,000 times per second–the fastest known X-ray variations of any collapsed object in the universe. This discovery, made using the Rossi X-ray Timing Explorer satellite in 1997, provides strong support for an earlier theory about the X-ray behavior of a neutron star that is part of a binary system. According to this theory, matter sucked off the companion is channeled by the neutron star's intense magnetic field onto the polar caps; moving at one-third light-speed, it converts its energy into intense radiation. The rain of hot matter and radiation onto the polar cap of the neutron star is like a super-violent version of Earth's **aurorae**. The radiation creates a strong pressure near the surface of the neutron star and pushes the infalling matter aside, poking holes in the matter and creating empty bubbles that fill with 100-million-degree radiation. These bubbles of X-ray light, known as *photon bubbles,* rise like hot fingers to a few kilometers above the surface of the neutron star only to fall and disintegrate, releasing their energy in about a thousandth of a second. Computer simulations predict that the photon bubble "fingers" release radiation in a more or less regular fashion, causing the neutron star to flicker with a quasiregular period of about a millisecond–exactly as seen in the case of Cen X-3.

center of gravity

The point from which the gravitational attraction of a body appears to act. In a uniform gravitational field this is coincident with the body's **center of mass**.

center of mass

The point at which a system of masses would balance if placed on a pivot. In a binary star system, the center of mass would lie nearer to the more massive component. In the case of a single mass, the **center of gravity**, for the sake of simplicity of calculations, is the point at which it can be assumed all the mass is concentrated.

central meridian

An imaginary line bisecting the apparent disk of a planet or the Sun and joining the poles of rotation. The passage of a surface feature across the central meridian, due to the

Centaurus A Large numbers of massive, luminous stars are resolved individually to the upper right and lower left of the broad dust swathe in this image of Centaurus A taken by the Very Large Telescope in 2000. *European Southern Observatory*

axial rotation of a planet, is known as a *central meridian transit.* The timing of central meridian transits is an important means of determining the planetographic longitudes of planetary surface features.

centripetal force

The force that any mass m, moving at a velocity v in a circle of radius r, experiences toward the center of the circle; its value is given by mv^2/r. In the case of one body in orbit around another, the centripetal force to remain in orbit is supplied by the force of gravity.

Cepheid variable

A yellow giant or supergiant **pulsating variable** whose period of pulsation is directly related to its luminosity: the longer the period, the greater the mean intrinsic

Centaurus A A close-up of the central region of Centaurus A, showing individual stars in more detail. *European Southern Observatory*

brightness. This strict **period-luminosity relation**ship makes Cepheids important cosmic distance indicators, since, by measuring a Cepheid's period and comparing the intrinsic brightness that the period yields, with the apparent brightness, the star's distance can be worked out. In addition, since Cepheids are so bright that they can be seen in other galaxies, they help establish a distance scale well beyond the Milky Way (see **cosmological distance ladder**). Cepheids fall into two distinct categories. Type I, also known as **Delta Cephei stars** or *classical Cepheids,* are extreme **Population I** objects, found in the spiral arms of galaxies. They typically have periods of 5 to 10 days, and are more metal rich and about four times more luminous than Type II Cepheids. The Type IIs, or **W Virginis stars**, are **Population II** objects with characteristic periods of 10 to 30 days, found primarily in globular clusters, galactic halos, and elliptical galaxies. Although the luminosities of all Cepheids are proportional to their periods, there is a different relationship for each type. The **light curves** of Cepheid variables have a characteristic shark-fin shape, with a rapid rise to maximum, a brief stay at peak brightness, and a smooth, slow decline to minimum. In evolutionary terms, Cepheids have left the main sequence and lie on the **instability strip** of the **Hertzsprung-Russell diagram**. Over 1,000 have been cataloged, of which the first to be found was **Eta Aquilae** and the best known is **Polaris**, the North Star.

Cepheus (abbr. Cep; gen. Cephei)

The King, father of Andromeda; a northern **constellation**, mostly circumpolar from as far south as latitude 30° N, with a noticeable quadrilateral shape made from Alpha, Beta, Iota, and Zeta. Cepheus contains several variable stars that are prototypes of their classes, the most important of which is **Delta Cephei**. VV Cep is a huge eclipsing binary of the **Zeta Aurigae** type, consisting of a red **supergiant** and a smaller blue companion in a 20.3-year orbit (magnitude range 4.7 to 5.4; R.A. 21h 56.7m, Dec. +63° 38′). There are also two interesting **open clusters**. NGC 6939 is a fine cluster of about 80 stars in a rich field that includes NGC 6946, a face-on spiral galaxy. It lies 2.5° south of Theta Cep and 2° southwest of Eta Cep (magnitude 7.8; diameter 7.8′; R.A. 20h 31.4m, Dec. +60° 38′). NGC 188 contains about 120 stars and is possibly the oldest known open cluster, with an age of about 5 billion years (magnitude 8.1; diameter 14′; distance 5,000

Cepheus: Stars Brighter than Magnitude 4.0

	Magnitude		Spectral	Distance	Position	
Star	Visual	Absolute	Type	(light-yr)	R.A. (h m s)	Dec. (° ′ ″)
α **Alderamin**	2.45	1.57	A7IV	49	21 18 35	+62 35 08
γ Alrai	3.21	2.51	K1IV	45	23 39 21	+77 37 57
β **Alfirk**	3.23v	–3.08	B2III	595	21 28 39	+70 33 39
ζ	3.39	–3.35	K1.5Ib	726	22 10 51	+58 12 05
η	3.41	2.62	K0IV	47	20 45 17	+61 50 20
ι	3.50	0.76	K0III	115	22 49 41	+66 12 02
μ **Garnet Star**	3.4max	–6.81	M2Iae	5,260	21 43 30	+58 46 48
δ **Delta Cephei**	3.5max	–4.07	F5Ib-G2Ib	950	22 29 10	+58 24 55

Cepheid variable Cross-hairs identify five Cepheid variables in this image of part of the nearby spiral galaxy NGC 300 taken by the 2.2-meter telescope at La Silla Observatory in 2002. When fully studied, the data on these and more than 100 other known Cepheids in NGC 300 will enable a very accurate distance to be determined, making this galaxy important in the future recalibration of the cosmic distance scale. *European Southern Observatory*

light-years; R.A. 0h 44.5m, Dec. +85° 20′). (See table, "Cepheus: Stars Brighter than Magnitude 4.0." See also star charts 7 and 9.)

Ceres (minor planet 1)

The largest **asteroid** (though exceeded in size by some **Kuiper Belt objects**) and the first to be discovered, by Giuseppe **Piazzi** on January 1, 1801; it contains about a third of the mass of all the asteroids combined. Probably similar in composition to a **carbonaceous chondrite**, Ceres is so dark that if placed at the same distance from Earth as **Vesta** (the third largest asteroid, which has only 55% of Ceres's diameter but an albedo of 0.35) it would appear the fainter of the two. Even so, Ceres reaches magnitude 7.4 at **opposition**, second only to Vesta.

For a while after its discovery, Ceres was lost. The observations made by Piazzi before his newly found object entered the daytime sky were too few to calculate an orbit accurate enough to predict where the object would reappear when it moved back into the night sky. But Ceres lay at the heliocentric distance predicted by the **Titius-Bode law** and therefore seemed as if it might be the missing planet that Johann **Bode** argued must exist between Mars and Jupiter. The discovery of Uranus in 1781 at a distance that closely fit the one predicted by the Titius-Bode law had already bolstered confidence that the law was valid. So convinced were some astronomers that Ceres was Bode's missing planet that, during an astronomical conference in 1796, they agreed to undertake a systematic search. In 1801, this led Carl **Gauss** to develop, a method for computing the orbit of an asteroid from only a few observations–a technique that hasn't been significantly improved since. Using Gauss's predictions, the German astronomer Franz von Zach recovered Bode's mystery world on January 1, 1802. Piazzi named it Ceres after the Roman goddess of grain and patron goddess of Sicily, thereby starting a tradition that continues to this day: asteroids are named by their discoverers (in contrast to comets, which are named *for* their discoverers).

Diameter:	933 km
Density:	2.7 g/cm^3
Class:	G
Albedo:	0.09
Rotational period:	9.08 hours
Orbit	
Semimajor axis:	2.767 AU
Perihelion:	2.550 AU
Aphelion:	2.982 AU
Period:	4.60 years
Eccentricity:	0.078
Inclination:	10.58°

Cerro Tololo Inter-American Observatory (CTIO)

A complex of astronomical telescopes and instruments located 70 km east of La Serena, Chile, at an altitude of about 2,200 m on Cerro Tololo mountain. Founded in 1963 and part of the U.S. National Optical Astronomy Observatories, its main instrument is the 4-m Blanco Telescope, opened in 1976, that is the southern twin of the 4-m instrument at **Kitt Peak National Observatory**. On a peak nearby is the giant **Gemini** South telescope.

Cetus (abbr. Cet; gen. Ceti)

The Whale or Sea Monster; a very large **constellation** on the celestial equator, west of Eridanus and south of Aries and Pisces. Cetus contains the important variables **Mira** and **UV Ceti**. It is also home to two particularly interesting galaxies. M77 (NGC 1068) is a **Seyfert galaxy**, seen almost face on, that lies about 50 million light-years away and 1° southeast of Delta Cet (magnitude 8.8; size 6.9′ × 5.9′; R.A. 2h 42.7m, Dec. –0° 1′). NGC 247 is a large, fairly bright spiral galaxy with a compact nucleus (magnitude 8.8; size 20′ × 7.4′; R.A. 0h 47.1m, Dec. –20° 46′). (See table, "Cetus: Stars Brighter than Magnitude 4.0," and star chart 6.)

Cetus: Stars Brighter than Magnitude 4.0

Star	Magnitude		Spectral	Distance	Position	
	Visual	Absolute	Type	(light-yr)	R.A. (h m s)	Dec. (° ′ ″)
β **Deneb Kaitos**	2.04	−0.30	KOIIICH	96	00 43 35	−17 59 12
α Menkar	2.54	−1.61	M1.5IIIa	220	03 02 17	+04 05 23
η Haratan	3.45	0.67	K1.5IIICN	118	01 08 35	−10 10 56
γ Kaffaljidhma	3.47	1.47	A3V	82	02 43 18	+03 14 09
τ Tau Ceti	3.49	5.68	G8V	11.9	01 44 04	−15 56 15
ι Baten Kaitos Shemali	3.56	−1.18	K1.5III	290	00 19 26	−08 49 26
θ	3.60	0.87	KOIIIb	115	01 24 01	−08 11 01
ζ Baten Kaitos	3.74	−0.76	K2IIIBa	259	01 51 27	−10 20 06

CH Cygni (CH Cyg)

The brightest and one of the most unusual known symbiotic stars; it lies 2° south-southwest of Iota Cygni at a distance of about 870 light-years. CH Cyg has ranged in brightness from magnitude 5.6 (visible to the naked eye) to 10.5. It was originally thought to be a semiregular variable, with a 90- to 100-day period and an amplitude of about 1 magnitude. However, in 1976 a hot blue continuum appeared in the spectrum, together with **emission lines** of hydrogen, helium, calcium, and iron; at the same time, CH Cyg grew brighter than it had ever been seen before and showed rapid fluctuations, especially in ultraviolet light. This superposition of two distinct spectra–a hot blue continuum with emission lines on top of a late-type absorption spectrum–led to CH Cyg's reclassification as a symbiotic star. It is thought to consist of an M-type **red giant** and a much hotter, smaller star, possibly a **white dwarf**, in a 2.07-year orbit. The pair are embedded in a **dusty** cloud of material that has been pulled off the giant by the hot dwarf; there also appears to be a third star in a wider, 14.5-year orbit. CH Cyg remained in its "blue outburst" state from 1976 to 1986, then faded abruptly by about 2.5 magnitudes. Subsequently it continued to fade, reaching an all-time low magnitude of 10.5 in 1997. However, the decline wasn't smooth and there were semiperiodic oscillations with each minimum fainter than the last. Since 1997, CH Cyg has varied between magnitudes 7.7 and 9.1. In 1984, radio emission was detected in the CH Cyg system, and further studies revealed a bipolar jetlike structure. Although the physical mechanism of **jet** formation in symbiotic systems is still poorly understood, one possibility is that CH Cyg's jet is ejected from an **accretion disk** around the hot companion.

CH star

A type of **carbon star** in which the molecular bands of CH (methylidyne: a carbon-hydrogen molecule) are very strong. CH stars are also typically high-velocity **Population II** objects found in the **galactic halo** and in **globular clusters**. They were first identified as a distinct group by Philip Keenan (1908–2000) in 1942, a year after he and William **Morgan** developed their unified carbon star classification scheme.

Chameleon (abbr. Cha; gen. Chameleontis)

A small, faint **constellation** near the south celestial pole. None of the stars in Chameleon are brighter than fourth magnitude. However, in 1999, an unusual cluster of about a dozen very young stars (about 8 million years old) was found, about 316 light-years away, centered on the reasonably bright star Eta Cha. The brightest members of the Chameleon cluster may be the only infant stars that can be seen from Earth with just binoculars or a small telescope. The cluster is unusual in that it is the first of over a thousand known **open clusters** discovered because of the X-ray emission of its member stars and it is the nearest cluster to Earth found in the past 100 years. It is also odd because it appears isolated in space, instead if being in or near clouds of gas and dust that provide the raw material for star-making. The Eta Cha stars may have formed in the **Scorpius-Centaurus Association** and then had their gas cloud stripped away by powerful **stellar winds** in the Sco-Cen cloud. (See star chart 4.)

Chandler wobble

One of several wobbling motions that Earth undergoes as it spins on its axis; it is named after Seth Chandler Jr. (1846–1913), an American businessman turned astronomer, who discovered it in 1891. The Chandler wobble has a slightly variable period of 416 to 433 days and causes the location of the poles to wander by as much as 15 m from their mean positions. Calculations show that the Chandler wobble would be damped out completely in just 68 years unless some force were constantly acting to reinvigorate it. The nature of this force remained a mystery for over a century until 2001, when Richard Gross, a geophysicist at the Jet Propulsion Laboratory in Pasadena, California, put the matter to rest. Using computer simulations he showed that about two-thirds of the Chandler wobble is due to fluctuating pressure on the bottom of the ocean, caused by temperature and salinity changes and wind-driven changes in the circulation of the oceans. He concluded that the remaining one-third comes from fluctuations in atmospheric pressure.

Chandra X-ray Observatory

A large and very sensitive X-ray telescope, launched in July 1999 into a highly elliptical orbit that carries it as far as 139,000 km from Earth and enables it to make long uninterrupted observations. Chandra's telescope uses four pairs of nearly cylindrical mirrors with diameters of 0.68 to 1.4 m to direct X rays in the energy range 0.1 to 10 keV onto the observatory's four science experiments. The spacecraft is one of NASA's four **Great Observatories** and was named in honor of Subrahmanyan **Chandrasekhar**.

Chandrasekhar, Subrahmanyan (1910–1995)

An Indian-born American astrophysicist renowned for his theoretical work on compact celestial objects, notably **white dwarfs** and **neutron stars**. He determined that the maximum mass of a white dwarf is about 1.4 M_{sun}, now known as the **Chandrasekhar limit**, and that above this mass the star must collapse further to become a neutron star or a **black hole**. Known for his love of mathematical beauty and precision, he investigated and wrote important books on stellar structure and evolution, dynamical properties of star clusters and galaxies, radiative transfer of energy, hydrodynamic and hydromagnetic stability, the stability of ellipsoidal figures of equilibrium, and the mathematical theory of black holes. He edited the *Astrophysical Journal* for nearly 20 years and won the Nobel prize for physics in 1983. His last book was *Newton's Principia for the Common Reader.* Chandrasekhar received a B.A. at Madras University and a Ph.D. at Cambridge, and spent the rest of his career at the University of Chicago (1937–1995). The **Chandra X-ray Observatory** was named in his honor.

Chandrasekhar limit

The highest possible mass of a **white dwarf**–about 1.44 M_{sun}–for a star with no hydrogen. Above this mass, electron degeneracy pressure is insufficient to prevent gravity from collapsing the star further to become either a **neutron star** or, if the **Oppenheimer-Volkoff limit** is also exceeded, a **black hole**.

chaotic orbit

An orbit whose evolution is so sensitive to minor changes in the orbiting object's position and/or velocity with respect to other gravitating bodies that it is essentially unpredictable. Saturn's moons **Pandora** and **Prometheus** have chaotic orbits. The two satellites give each other a gravitational kick each time Pandora passes inside Prometheus, about every 28 days. Because neither's orbit is quite circular, the distance between them on these occasions–hence the strength of the kick–varies. The perturbations lead to changes in motion that are not periodic or predictable.

chaotic terrain

Part of a planetary surface where features such as ridges, cracks, smooth plains, and tilted slopes are jumbled together. Regions of this type occur, for example, in several places on **Mars** and on much of Jupiter's moon **Europa**. Martian chaotic terrain, such as Gorgonum Chaos, is often found near major outwash channels, suggesting it may be the result of water having run out from porous spaces underground and caused the ground above to become unstable. On Europa, vast regions of chaotic terrain occur where the surface ice layer has shattered into small blocks. These blocks appear to have rotated and tilted in a sea of soft liquid or slush, which then refroze and locked them in their new positions. Such activity is thought to be due to melt-through events, in which either localized **tidal heating** in the ice layer itself or an underlying volcanic heat source completely (although briefly) melted through the ice layer.

Charon

The only known moon of **Pluto**; it was discovered in 1978 by **U.S. Naval Observatory** astronomer James Christy (1938–). With just over half the diameter of Pluto and one-seventh the mass, Charon is easily the largest satellite relative to its planet in the solar system. During the 1980s, Earth crossed the orbital plane of Charon so that, from our vantage point, Charon and Pluto alternately passed in front of each other. These eclipses enabled the size of the two objects and other valuable data on the Pluto-Charon system to be collected. **Albedo** measurements suggest that Charon is covered mainly with water-ice, while Pluto has a coating of frozen nitrogen. Also Charon's density (about 1.2 g/cm^3) is significantly lower than that of Pluto (about 2.0 g/cm^3) suggesting that the moon has very little rock and adding to the controversy of how Charon formed. Set against the favored view that Charon stemmed from the collision between Pluto and another large object in the early days of the solar system, is the suggestion that Pluto and Charon formed independently.

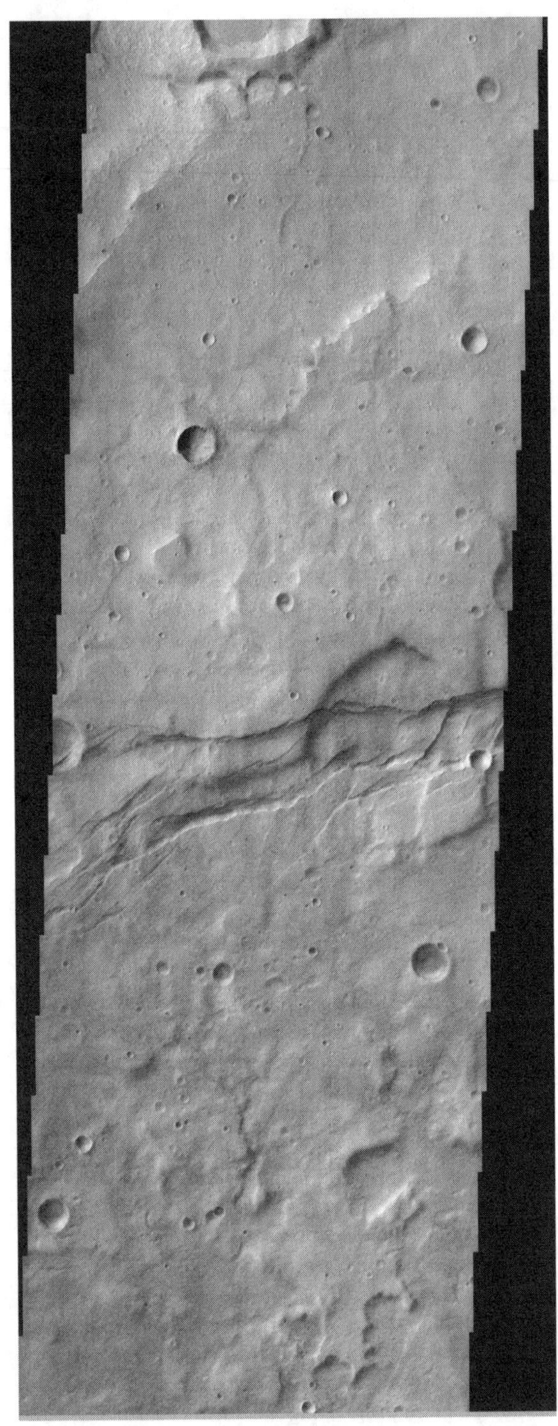

chaotic terrain Part of the region on Mars known as Gorgonum Chaos. This 2001 Mars Odyssey image shows long parallel to semiparallel fault features, called graben, running across the center from left to right. *NASA/JPL/ Arizona State University*

Diameter:	1,190 km
Mass (Earth = 1):	0.00032
Density:	1.2 g/cm^3
Mean distance from Pluto:	19,640 km
Rotational period:	6.387 days
Orbit	
Period:	6.387 days
Eccentricity:	0.00
Inclination:	98.80°
Escape velocity:	0.610 km/s
Visual albedo:	0.5

chassignite

One of the types of *SNC meteorites* believed to have come from Mars (see **Mars meteorites**). The group is named for its only known member, a meteorite that was seen to fall in Chassigny, France, in 1815; its subsequent recovery led to it being one of the first meteorites to be recognized as a genuine rock from space. Chassigny resembles a terrestrial dunite–a coarse-grained, deep-seated igneous rock–and consists of about 91% iron-rich **olivine**, 5% clino**pyroxene**, 1.7% **plagioclase feldspar**, 1.4% chromite, 0.3% melt inclusions, and other minerals. Cracks within Chassigny are filled with carbonate and sulfate salts that point to chemical alteration by water before its arrival on Earth. Its crystallization age of 1.36 billion years and its composition, suggest a close relationship with the **nakhlites** and an origin in the same parent magma on Mars. However, Chassigny contains noble gas values that are entirely different from those found in other Mars meteorites or in the Martian atmosphere. If these gases came from the Martian mantle, as suspected, Chassigny must have originated within a magma pluton deep inside the Martian crust.

Chicxulub Basin

A multiringed **impact basin**, 170 km across, that is buried under several hundred meters of sediment on Mexico's Yucatan Peninsula (21° 20′ N, 89° 30′ W); it is named after a village located near its center. The Chicxulub Basin is believed to have been formed 64.98 million years ago by the impact of an asteroid 10 to 20 km in diameter that also triggered the **mass extinction** at the end of the Cretaceous Period (see **K-T boundary**). The asteroid hit a geologically unique, sulfur-rich region, kicking billions of tons of sulfur and other materials into the atmosphere. This would have caused global temperatures to plunge to near freezing for perhaps several months–long enough to wipe out half the species on Earth, including the last of the dinosaurs.

Chinese astronomy

Astronomical ideas played an important part in the culture of China. An Anyang bone inscription records a lunar eclipse "on the fifteenth day of the twelfth moon of the twenty-ninth year of King Wu-Ting (November 23, 1311 B.C.)." In 1137 B.C., the Shang Emporer Ti-hsin is recorded as ordering a sacrifice because the predicted lunar eclipse fell on the wrong day. By 700 B.C. the Chinese observed the shadows of the Sun from special towers, which remained in continuous use for determining calendars for 1,500 years. By 350 B.C. the length of the solar year was known to be 365¼ days, based on the helical rising of **Spica**. Cosmological ideas were interwoven with the Chinese world concept, which embodied the knowledge of the **celestial sphere** with its daily rotation around a pole. The horizon played a role, but although the motions of the planets were noticed, they were not considered important. See **Egyptian astronomy**, **Babylonian astronomy**, **Indian astronomy**, and **Greek astronomy**.

Chiron (minor planet 2060)

An object, discovered by the American astronomer Charles Kowal (1940–) in 1977, that was originally classified as an asteroid but is now considered to be a **Centaur** or a weakly active comet (95P/Chiron). Its eccentric orbit brings it within the orbit of Saturn at perihelion and almost as far away as the orbit of Uranus at aphelion. Its cometary status was suspected following the 1989 discovery of a dusty **coma** and confirmed in 1991 by the detection of cyanogen (CN) radicals, a known constituent of cometary gas comas. Because Chiron moves in a chaotic, planet-crossing orbit, it is likely in time to collide with a planet or be permanently ejected from the solar system. Diameter 180 km, spectral class B, rotational period 5.9 hours. Orbit: semimajor axis 13.63 AU, perihelion 8.43 AU, aphelion 18.8 AU, eccentricity 0.380, inclination 6.94°, period 50.7 years.

chondrite

A type of **stony meteorite** made mostly of iron- and magnesium-bearing silicate minerals that remains little changed since the dawn of the solar system, over 4.5 billion years ago. Chondrites are the commonest kind of meteorite, accounting for about 86% of falls. Because they come from asteroids that never melted or underwent **differentiation**, they have the same elemental composition, except for the lightest elements, hydrogen and helium as the original **solar nebula**. Chondrites are so named because they nearly all contain **chondrule**s–little round droplets of **olivine** and **pyroxene** that apparently condensed and crystallized in the solar nebula and then accreted with other material to form a matrix. The variation in chemical composition among chondrites reflects the fact that their parent bodies formed in different regions of the solar nebula and were further subjected to different thermal and chemical processes as well as to impacts with other asteroids. The main groups are H, L, and LL, known collectively as **ordinary chondrites**; E, known as **enstatite**s; and C, known as **carbonaceous chondrites**. Other, rarer groups, include the **rumurutiites** (group R), the **kakangariite**s (group K), and the **forsterites** (group F). Each group is further subdivided into petrologic types 1 to 7. Types 1 and 2, which are represented only by carbonaceous chondrites, show evidence of having been chemically altered by water to the extent that chondrites are either absent (type 1) or rare (type 2). Petrologic types 3 to 7 have been exposed to varying degrees of thermal metamorphism, reflected in an alteration of the chondrules. Type 3 contain plentiful, unaltered and distinct chondrules, while the chondrules of types 4 to 6 are increasingly indistinct due to heat-caused chemical changes and recrystallization. Type 7 chondrites, which lack chondrules, are transitional between chondrites and primitive **achondrites**.

chondrule

A roughly spherical object, large numbers of which are found in all **chondrite**s except for the CI group of **carbonaceous chondrites**. Chondrules are typically 0.5 to 2 mm in diameter and are usually composed of iron, aluminum, or magnesium silicates in the form of the minerals **olivine** and **pyroxene**, with smaller amounts of glass and iron-nickel. Together with **calcium aluminium inclusions**, which predate them by a couple of million years, they are among the oldest objects in the solar system with an age of about 4.57 billion years. They formed when dusty regions of the **solar nebula** were heated to very high temperatures, became molten, and then resolidified as tiny droplets.

Christmas Tree Cluster (NGC 2264)

A dozen or so stars in **Monoceros**, engulfed in nebulosity, that form the nearly perfect outline of a fir tree. The fifth magnitude star 15 Monocerotis (also known as S Mon) lies at the base of the Cluster, while another reasonably bright star, immediately to the north of the **Cone Nebula**, marks the apex. The Christmas Tree Cluster and the Cone Nebula were both discovered by William **Herschel** in 1783.

chromatic aberration

A defect in a lens in which the various colors of the spectrum are not brought to the same focus. Blue light, for example, is refracted (bends) more than red light when it passes through a lens and hence comes to a focus inside that of red light. The result is "fringing"–the formation of a colored halo around the image. This problem seriously affected the performance of **refracting telescopes** for centuries and was the reason that so many refractors were built with large **focal ratios**: longer focal length lenses show less chromatic error. A better solution came with the introduction of corrective elements, using at least two different types of glass to make a compound lens. An **achromatic lens** corrects for red and blue light, whereas an **apochromatic** lens corrects for at least red, blue, and green. Reflecting telescopes are free from this type of aberration. See **aberration, optical**.

chromosphere

A narrow region of a star's atmosphere that lies between its **photosphere** and its **corona**. The Sun's chromosphere starts at a temperature of about 6,000 K at the top of the photosphere. At about 500 km above the photosphere, there is a **temperature minimum** of about 4,300 K and, above this, for about another 2,000 km, an increase in temperature to about 20,000 K. From this point, up to the base of the corona, about 10,000 km above the photosphere, is the **transition region** in which the temperature climbs swiftly to over 1 million K. The composition and density of the chromosphere also changes markedly. While the lower 1,500 km or so is more or less continuous, the upper chromosphere features jagged **spicules**. At the same time, the density drops by a factor of about a million from bottom to top. Immediately before and after a total eclipse, the light of the chromosphere ("color sphere") is dominated by the red glow of **H-alpha**. At other times it can be observed using H-alpha and calcium K-line **filtergrams** and, from space, in ultraviolet emission lines.

chromospheric network

An ever-present network of long thin sinuous chains extending over much of the solar disk in **H-alpha** and consisting of tiny points that are brighter than the surrounding **chromosphere**. These points, called **filigree** (also found in **plages**), often have darker **spicules** or short **fibrils** sticking out of, or running over them (part of the fine disk detail known as the dark mottles), making the actual network harder to see. The chromospheric network is formed by magnetic field lines related to **supergranulation cells** in the **photosphere** that lies below the chromosphere.

Circinus (abbr. Cir; gen. Circini)

The Compasses; a very small, faint southern **constellation**. Its brightest star, Alpha (visual magnitude 3.18; absolute magnitude 2.10; spectral type F1V; distance 53 light-years; R.A. 14h 42m 28s, Dec. –64° 58′ 43″), and next two brightest stars, form a narrow triangle that lies east of **Alpha Centauri**. Two deep sky objects of interest are **Circinus X-1** and the **Circinus Galaxy**. (See star chart 5.)

Circinus Galaxy

A spiral **active galaxy** that lies only 15 million light-years away but went unnoticed until 1975 because it lies just 4° from the plane of the Milky Way and is thus heavily obscured. It is the scene of tumultuous gas motions, most of them concentrated in two rings. One, about 700 light-years from the center, appears to be undergoing tremendous bursts of star formation. An inner ring, only 130 light-years from the center, almost certainly encircles a massive **black hole**.

Circinus X-1 (Cir X-1)

An **X-ray binary** in which a **neutron star**, slightly more massive than the Sun, is in orbit around a normal main sequence star; it lies about 20,000 light-years away in **Circinus**. Although Cir X-1 was discovered in 1971, many of its properties remained mysterious because it lies in the galactic plane where obscuring dust and gas block its effective study at many wavelengths. X rays, however, shine through the interstellar medium relatively unaffected and reveal that the collapsed component is surrounded by a fast-spinning, multimillion-degree **accretion disk**, composed of matter looted from the second star. In 2002, the **Chandra X-ray Observatory** detected, for the first time in X rays, a stellar fingerprint known as a **P Cygni profile**–the distinctive spectral signature of a powerful **stellar wind**. Chandra's data reveal that the radiation and rotational forces in the Cir X-1 disk are blasting some of the inward-spiraling gas back into space at speeds of up to 7.2 million km/hr. Cir X-1's high-velocity wind gives this small two-star system a striking similarity to broad-absorption-line **quasars**–a type of active galaxy with an accretion disk circling its black hole plus very powerful winds created when radiation pushes material off the disk and into space. Astronomers hope that X-ray P Cygni profiles will turn out to be fairly commonly in X-ray binaries containing neutron stars and black holes and thus help to shed new light on these objects.

circular orbit

Any orbit that has an **eccentricity** of zero–an ideal condition that is seldom (if ever) actually achieved. The velocity needed to maintain such an orbit, known as the circular velocity, is given by

$$G(M + m)/r$$

where G is the gravitational constant, M and m are the masses of the central body and orbiting object, respectively, and r is the radius of the orbit.

circumpolar star

A star that never sets as seen from a given latitude. At either of Earth's poles all stars are circumpolar, whereas none are at the equator.

circumstellar maser

Maser emission from molecules in the circumstellar envelopes of **red giants**, and also of some young stars, of **spectral types** M, S, and C. The kinds of masers found in circumstellar space closely correlate with the spectral type of the associated stars. In **M stars**, where oxygen is more plentiful than carbon, masers from oxygen-bearing molecules, such as OH (hydroxyl), H_2O (water), and SiO (silicon oxide), are found. OH 1612-MHz masers occur at the outer edge of circumstellar envelopes, where interstellar ultraviolet radiation produces OH from the dissociation of H_2O. The line profile of OH 1612-MHz masers has a characteristic double peak, interpreted as emission from the approaching and receding parts of the expanding shell. H_2O and SiO masers occur closer to the star and show intensity variations on a timescale of a month to a year. In **carbon stars**, where carbon is more abundant than oxygen, masers from HCN (hydrogen cyanide) and SiS (silicon sulfide) molecules are found. HCN masers serve as a useful probe of the inner region of the envelope in carbon stars, whereas SiS masers provide data on the outer envelope. In evolved (red giant) S stars, where oxygen is as abundant as carbon, SiO masers are frequently found. They have also been discovered, however, in several star-forming regions, including the **Orion Complex**, and the W51 and **Sagittarius B2** molecular clouds.

Clark, Alvan (1804–1887)

A renowned American maker of some of the world's largest and best lenses for **refracting telescopes**. Together with his sons, George Bassett Clark (1827–1891) and Alvan Graham Clark (1832–1897), he founded Alvan Clark & Sons at Cambridgeport, Massachusetts, in 1846, and proceeded five times to make the objective lenses for the largest refracting telescopes in the world. These included the 26-inch (66-cm) lens at the **U.S. Naval Observatory** (the first **achromatic lens** produced in the United States), the 36-inch (91-cm) lens at **Lick Observatory**, and the 40-inch (102-cm) lens at **Yerkes Observatory**, the largest ever built. The optical work of Clark & Sons was then recognized as unsurpassed anywhere and represented the first significant American contribution to astronomical instrument-making; prior to this, American telescopes had never compared with those of European manufacture. Alvan Graham Clark is also remembered particularly for his discovery of the (white dwarf) companion of **Sirius** in 1862.

Clementine

A small lunar probe, sponsored jointly by NASA and the Department of Defense and launched in January 1994, that orbited the Moon for several months and sent back the first evidence of ice under the surface near the lunar poles. Further, but still inclusive, evidence for this ice came from **Lunar Prospector**.

close binary

A **binary star** system in which the separation of the components is comparable to the diameter of the stars. There are three main types, distinguished by the extent to which each star fills its **Roche lobe**. In a *detached binary,* neither star fills its Roche lobe, so that there is no significant **mass transfer** between the components. In a *semidetached binary,* one of the stars fills its Roche lobe, which results in this star losing material in a stream that either falls directly onto its companion, or, as is more usual, enters an **accretion disk**. In a *contact binary* both components fill their Roche lobes or, more often, overflow them so that there is a common convective envelope. The proximity of stars in close binaries typically stretches at least one component into an **elliptical variable**, and often produces an **eclipsing binary** as well.

Cloverleaf Quasar (H1413+117)

A **quasar**, discovered in 1988 and located about 8 billion light-years away in **Leo**, that appears as four different images because of a **gravitational lens** effect caused by a foreground elliptical galaxy. Molecular gas (notably CO) detected in the host galaxy associated with the quasar is the oldest molecular material known and provides evidence of large-scale star formation in the early universe.

cluster of galaxies

A collection of galaxies, containing a few to a few thousand members, which may or may not be held together by its own gravity. Our **Local Group** is a smallish, loose cluster of 40 or so galaxies, a few million light-years across. At the other end of the scale are **rich clusters of galaxies**, such as the **Virgo Cluster** and **Coma Cluster**, that contain hundreds or thousands of systems in a volume of space comparable to that of a loose group. Relatively nearby rich clusters were first systematically cataloged by George Abell and are now known as **Abell clusters**. Among the methods used to describe galaxy clusters are the **Bautz-Morgan classification** and the Rood-Sastry classification. Between the galaxies within clusters is a super-hot, super-dilute gruel of X-ray emitting gas known as the **intra-cluster medium**.

CN star

A giant **K star** or **G star** whose spectrum contains strong **cyanogen bands** (CN) and stronger metallic lines than those found in normal giants.

cluster of galaxies A peculiar cluster of galaxies lying in the same sky field as the quasar PB5763. Among the objects on view are an interesting spindle-shaped galaxy with what may be an equatorial ring, a fine spiral galaxy, and many fainter galaxies. These may be dwarf members of the cluster or they may lie in the background at even greater distances. *European Southern Observatory*

CNO star

A late **O star** or early **B star** (O8 to B4) in whose spectrum the lines of some of the elements carbon, nitrogen, and oxygen are weaker or stronger than is usual in stars of these spectral types.

Coalsack

A prominent **dark nebula** in **Crux**, readily visible to the naked eye against the background of the Milky Way. It lies in the galactic plane, about 550 light-years away, spans some 60 light-years (7° × 5° of the sky), and has a mass of about 3,500 M_{sun}.

Coathanger (Collinder 399)

An **open cluster** of about 40 stars in **Vulpecula**. Six of its brighter stars, of sixth and seventh magnitude, are lined up in a nearly perfect row, from the center of which four stars form a hook to complete the coathanger shape. The cluster was first described by **Al-Sufi** in A.D. 964 and rediscovered independently by Giovanni **Hodierna**. It was included in Per Collinder's 1931 catalog of open clusters and is also known as *Brocchi's Cluster,* named after the American amateur astronomer D. F. Brocchi who created a map of it in the 1920s for calibrating **photometers**. In the 1980s the Coathanger was found to share roughly the same motion with about 10 other clusters, including the **Pleiades**, NGC 6633, 6709, 6882, and 6885, and IC 4665. The Coathanger is best viewed through binoculars; to the naked eye, it appears as an unresolved patch at the border of Vulpecula and Sagitta.

Visual magnitude:	3.6
Apparent diameter:	About 1°
Distance:	420 light-years
Position:	R.A. 19h 25.4m, Dec. +20° 11′

COBE (Cosmic Background Explorer)

A NASA satellite, launched in November 1989, that was designed to study the residual radiation of the **Big Bang**. Its historic discovery of "cosmic ripples"–tiny fluctuations in the temperature of the **cosmic microwave background**–was announced in 1992. COBE also mapped interstellar and interplanetary dust clouds.

Cocoon Nebula (IC 5146)

A small round **emission nebula** centered near the apex of several dark lanes in the otherwise rich star fields of **Cygnus**. The Cocoon is embedded in an **open cluster** of about 80 young stars.

Visual magnitude:	7.0
Angular diameter:	12′
Distance:	3,000 light-years
Position:	R.A. 21h 51m, Dec. +47° 02′

cocoon star

A star that is a strong source of **infrared** but is invisible at optical wavelengths because it is shrouded by a dense cloud of gas and dust. Some such stars are less than a million years old and are hidden by material left over from their formation. Others are **red giants** or supergiants, in an advanced stage of evolution, that have ejected the material that makes their cocoon. Some of these aging cocoon stars are OH-IR sources.

Coddington's Nebula (IC 2574)

A dwarf spiral galaxy that is an outlying member of the **M81** Group. A broadband photometric study of IC 2574 indicates that 90% of its mass is in the form of **dark matter**, a figure in keeping with that of larger systems and with some other dwarf galaxies.

Visual magnitude:	10.6
Angular size:	12.3′ × 5.9′
Distance:	11,700 light-years
Position:	10h 28.4m, Dec. +68° 25′

coelostat

Pronounced "seelostat," a device that uses two moveable mirrors to reflect sunlight along a particular path–for instance, down a fixed telescope tube–so that the image at the other end of the telescope doesn't appear to move or rotate when the Sun moves across the sky. One of the mirrors rotates around an axis parallel to Earth's axis of rotation, at half the solar rate. The orientation of the other mirror is adjustable in all directions. A coelostat is similar to a **heliostat**, but has a more complex design and, unlike a heliostat, gives an image in a fixed orientation.

color

A subjective quality that depends partly on the vagaries of the human visual system. Because the eye is poor at detecting color at low light levels, most stars in the night sky appear white. Blues, reds, and yellows are only evident in some of the brighter stars (and planets), or with the aid of binoculars or a telescope. In any event, a quantitative system is needed for scientific purposes and for this the **color index** is used.

color excess

The difference between the observed **color index** of a star and the intrinsic color index corresponding to its **spectral type**. It indicates the amount of **interstellar reddening** suffered by the light from the star when it passes through **dust** in space.

color index

The difference between the **apparent magnitude** of a star at two different wavelengths (always the shorter-wavelength magnitude minus the longer-wavelength magnitude) to give a quantification of the star's color. In the widely used *Johnson system,* a star's brightness is measured through a U filter, which transmits ultraviolet and violet light; a B filter, which lets through blue light; and a V ("visual") filter, which lets through green-yellow light (to which the human eye is most sensitive). If a star's observed magnitudes, corrected for absorption in Earth's atmosphere, are U, B, and V, then the color indices are the differences U – B and B – V. The system is defined so that for a star of **spectral type** A0 (equivalent to a surface temperature of 9,200 K), B – V = U – B = 0. Hotter (bluer) stars have slightly negative color indices, while cooler (yellow, orange, and red) stars have increasingly positive color indices. The full range runs from about –0.4 (blue) to +2.0 (red **carbon stars**). Color indices can also be defined that are sensitive to spectral lines or bands. By comparing several different color indices it is possible to deduce such important quantities as a star's temperature and intrinsic **luminosity**, and the extent to which its light has been absorbed by interstellar dust (see **intrinsic color index**). Color index is often referred to simply as *color.*

color temperature

The temperature of a **blackbody** that has the same **color index** as a given star.

color-magnitude diagram

A plot of **color index** against **absolute magnitude** for a group of stars; also known as a *color-luminosity diagram.* For stars intrinsically fainter than the Sun, it defines the **main sequence** of the **Hertzsprung-Russell diagram.**

Columba (abbr. Col; gen. Columbae)

The Dove (as released by Noah in the Bible); a small southern **constellation,** the main part of which is a noticeable triangle of stars set into a fairly blank area of sky. Mu Col is a blue star (type O9.5, magnitude 5.16) that is one of three **runaway stars** diverging from a point in Orion with a speed of about 100 km/s. The other two are AE Aurigae and 53 Arietis. According to one theory, these stars once belonged to a quadruple system, of which the fourth member exploded in a supernova, about 3 million years ago, pushing the other three stars away in different directions. Columba also contains the **globular cluster** NGC 1851 (magnitude 7.1; diameter 11′; distance 40,000 light-years; R.A. 5h 14.1m, Dec. –40° 2.8′). (See table, "Columba: Stars Brighter than Magnitude 4.0," and star chart 3.)

colure

A **great circle** that passes through the celestial poles and meets the **ecliptic** at either the **solstice** points (the *solstitial colure*) or the **equinox** points (the *equinoctial colure*).

coma, cometary

A glowing envelope of gas and dust that surrounds a **comet**'s nucleus when the comet comes within a few AU of the Sun (although **Chiron** has been seen to develop a coma at 11 AU). At perihelion, the coma is typically about 100,000 km across, and shaped into a teardrop by the **solar wind.** Together, the coma and the nucleus form the comet's head.

coma, optical

A defect of an objective mirror or lens in which rays of light, striking the objective away from the optical axis are not brought to focus in the same image plane. The result is that star images toward the edge of the **field of view** appear to have cometlike tails spreading radially out from the optical axis (*negative coma*) or in toward the axis (*positive coma*). See **aberration, optical.**

Coma Berenices (abbr. Com; gen. Comae Berenices)

A very faint northern **constellation** considered to be part of Leo by the Greeks but made separate by Gerardus Mercator in 1551. It lies between Canes Venatici to the north, Virgo to the south, Leo to the west, and Boötes to the east, and contains the north galactic pole. Although devoid of bright stars, Coma Berenices abounds in galaxies and other deep sky objects. (See table, "Coma Berenices: Objects of Interest," and star chart 18.)

Coma Cluster (Abell 1656)

The nearest massive **cluster of galaxies;** it is roughly spherical, about 20 million light-years in diameter, contains more than 3,000 galaxies, and lies about 280 million light-years away in **Coma Berenices.** Its location close to the north galactic pole makes it ideally placed for observation free from the effects of galactic obscuration. As is usual for clusters of this richness, the largest member galaxies are overwhelmingly elliptical or lenticular, with only a few spirals (most of these probably near the outskirts of the cluster). The central region is dominated by two giant galaxies: NGC 4889 (an elliptical) and NGC 4874 (an S0 type). The Coma Cluster contains hot intracluster gas that is an extended source of X rays, about 0.5° across, known as Coma X-1.

Coma Star Cluster (Melotte 111)

A large, conspicuous **open cluster** of roughly 100 stars, some as bright as fifth magnitude, arranged in a roughly triangular shape. It lies below Gamma Com and shows up well with binoculars.

Columba: Stars Brighter than Magnitude 4.0

	Magnitude		Spectral	Distance	Position	
Star	Visual	Absolute	Type	(light-yr)	R.A. (h m s)	Dec. (° ′ ″)
α Phakt	2.65	−1.93	B7IVe	268	05 39 39	−34 04 27
β Wasn	3.12	1.01	K1III	86	05 50 58	−35 46 06
δ	3.85	−0.46	G7II	237	06 22 07	−33 26 11
ε	3.86	−0.79	K1IIa	277	05 31 13	−35 28 15

Coma Berenices: Objects of Interest

Object	Notes
Open cluster	
Coma Star Cluster	See separate entry.
Globular cluster	
M53 (NGC 5024)	Appears as a misty patch in small telescopes, close to Alpha Com. Magnitude 7.7; diameter 12.6′; R.A. 13h 12.9m, Dec. +18° 10′.
Galaxies	
Black Eye Galaxy	M64 (NGC 4826). See separate entry.
M88 (NGC 4501)	SBb galaxy. Magnitude 9.5; diameter 6.9′ × 3.9′; R.A. 12h 32.0m, Dec. +14° 25′.
M98 (NGC 4192)	Sb galaxy. Magnitude 10.1; diameter 9.5′ × 3.2′; R.A. 12h 13.8m, Dec. +14° 45′.
M99 (NGC 4254)	Sc galaxy. Magnitude 9.8; diameter 5.4′ × 4.8′; R.A. 12h 18.8m, Dec. +14° 25′.
M100 (NGC 4321)	Sc galaxy. Magnitude 9.4; diameter 6.9′ × 6.2′; R.A. 12h 22.9m, Dec. +15° 49′.
Cluster of galaxies	
Coma Cluster	See separate entry.

Visual magnitude:	1.8
Angular size:	275′
Distance:	288 light-years
Position:	R.A. 12h 25m, Dec. +26° 23′

Coma Cluster The elliptical galaxy NGC 4881, which, along with the bright spiral galaxy to the left, belongs to the Coma Cluster. Most of the objects in this 1994 Hubble Space Telescope image are background galaxies that lie at much greater distances. *NASA/ESA/STScI*

combined magnitude

The total brightness of two or more celestial objects, such as the stars of a binary system that appear as a single point to the naked eye. Unfortunately, this can't be found simply by adding the separate **magnitudes**, m_1 and m_2, because the magnitude scale is logarithmic. Instead the combined magnitude, m, has to be calculated from the formula:

$$m = m_1 - 2.5 \log \{1 + \text{antilog}\,[-0.4\,(m_2 - m_1)]\}.$$

comet

See article, page 112.

cometary globule

A **reflection nebula** with a dusty head and a fan-shaped tail that bears a superficial resemblance to a comet. Cometary globules are often the birthplaces of stars, and many show very young stars in their heads. Classical examples of the heads of cometary nebulae are R Monocerotis in **Hubble's Variable Nebula**, R Coronae Australis, and RY Tauri. All have A0 to G0 type spectra that resemble the spectrum of a **T Tauri star**, and show brightness variations from year to year.

common-envelope star

A **close binary** system in which a relatively massive donor star transfers material faster than it can be digested by a less massive accreting star. As a result, the transferred material eventually overflows the latter's **Roche lobe**, forming a common envelope around both stars. Frictional drag between the common envelope and the stars within it causes the two stars to spiral inward and the common

(continued on page 114)

comet

A small body, typically a few kilometers across, containing icy chunks and frozen gases with bits of embedded rock and dust, and possibly a rocky core, aptly described as a "dirty snowball." Comets are leftovers from the formation of the solar system and are believed to exist in vast numbers in the **Oort Cloud** and the **Kuiper Belt**. From these regions they can be perturbed by the gravitational influence of passing stars or interstellar clouds and thrown into new, highly elliptical orbits that bring them into the inner solar system. As a comet draws nearer to the Sun, solar radiation causes the comet's frozen gases to sublime (turn directly from a solid into a gas) and be released, so that, in addition to the frozen *nucleus*, several new features develop. These include a *coma*, a luminous cloud of water vapor, carbon dioxide, and other neutral gases driven off the nucleus; a *hydrogen cloud*, a huge (millions of km in diameter), tenuous envelope of neutral hydrogen; a *dust tail* (the most prominent part of a comet to the unaided eye) up to 10 million km long, composed of smoke-sized dust particles driven off the nucleus by escaping gases; and an *ion tail*, as much as several hundred million km long, composed of plasma and laced with rays and streamers caused by interactions with the **solar wind**.

A comet that has been deflected into the inner solar system from the Oort Cloud will have an incredibly elongated path, perhaps with a perihelion of 1 AU and an aphelion of 10,000 AU, so that it would take many centuries to return again; it might even approach on a hyperbolic path that will eventually carry it into interstellar space. Many comets, however, are eventually nudged by close passages of the giant planets into smaller orbits which, although still elongated, bring them close to the Sun on a more regular basis. Of the two dozen or so comets that are seen each year through telescopes, the majority are **short-period comets**, with periods of less than 200 years, that are either new discoveries or are returning as predicted on well-known orbits, the most famous being **Halley's comet**. The rest are **long-period comets**, seen for the first time, with periods of more than 200 years; among the brightest have been several **Great comets**, the **Daylight comet of 1910**, and **Donati's comet**. When a newly discovered comet is confirmed, the International Astronomical Union assigns an interim designation consisting of the year of discovery followed by a lowercase letter in order of discovery for that year. Often the discoverer's name precedes the designation; for example, Comet Bennett 1969i. If a reliable orbit is later established, the comet is given a permanent designation consisting of the year of perihelion passage followed by a roman numeral in order of perihelion passage; for example, Comet Bennett 1970 II. If the comet is periodic, the letter P followed by the discoverer's (or computer's) name is used, as in comet 1910 II P/Halley.

Apart from their rocky content, comets are made mostly of water-ice, plus frozen methane, ammonia,

comet Comet Hale-Bopp and its spectacular ion tail photographed by an 8-inch wide-field telescope at Lowell Observatory on April 9, 1997, when the comet was 139 million km (0.925 AU) from the Sun. The image covers a region of the sky corresponding to 9.7 million km by 11.6 million km at the distance of the comet from Earth. Various detailed structures visible in the tail are caused by the changing solar wind. A faint, diffuse dust tail is also visible, caused by sunlight reflected by dust grains. *David Schleicher and Tony Farnham (Lowell Observatory)*

comet An artist's impression of a comet nucleus at a distance of about 1.5 AU from the Sun. *Dale Rutter/NASA*

carbon monoxide, and carbon dioxide. They also contain an interesting collection of organic chemicals, including formaldehyde (H_2CO), hydrogen cyanide (HCN), and methyl cyanide (CH_3CN), which are believed to have formed in interstellar space and then become incorporated into comets in the early development of the **solar nebula**. These organic substances, and possibly others of greater complexity, would have been delivered to Earth's surface in significant amounts by impacting comets between 3.5 and 4.5 billion years ago. Astrobiologists have conjectured that such material could have played an important role in the early development of terrestrial life. A comet also contains dust particles which, after being released around the time of perihelion, contribute to the **zodiacal cloud** and also spread out along a comet's orbit, giving rise to annual **meteor showers**. As well as the many individual comets on record, two distinct *comet families* are known: the **Jupiter-family comets**, whose members have different origins but have all been captured by Jupiter into similar orbits, and the *Kreutz family* of **sungrazing-comets**, which may be fragments of a single parent body.

common-envelope star

(continued from page 111)

envelope to be ejected. In this way it is possible for the components of a binary to reduce their separation from many AU (corresponding to an orbital period of tens of years) to a few stellar radii (corresponding to an orbital period of hours). One possible example of a star in the process of ejecting its common envelope is **Eta Carinae**.

compact galaxy

A galaxy with an apparent diameter of 2″ to 5″ and a high surface brightness that appears only barely larger than a starlike point on a sky survey photograph. A compact galaxy is similar to an **N galaxy** but without any sign of a disk or a nebulous background.

compact source

A radio source of small angular extent that is strongest at shorter wavelengths. Compare with **extended source**.

companion

Either one of the stars in a binary system (although usually the less massive), sometimes only detectable by spectroscopy.

comparator

A device used to look for **parallax** motion, **proper motion**, asteroids, or variable stars by quickly alternating between views of two photographic plates taken at different times; also known as a *blink microscope*. Changes in the configuration of the objects on the plates, such as the movement of a suspected planet, are detectable by their relative motion, or new objects become apparent by their alternate appearance and disappearance. A similar instrument is the *stereocomparator*, which highlights objects that have changed position by their stereoscopic appearance.

Compton Gamma-ray Observatory (CGRO)

A large satellite, named after the American physicist Arthur Holly Compton (1892–1962) and launched in April 1991, that revolutionized understanding of the universe in **gamma rays**. It carried four instruments that covered an unprecedented six orders of magnitude in energy, from 30 keV to 30 GeV, improving sensitivity over previous missions by a factor of 10. The second of NASA's **Great Observatories**, it was intentionally de-orbited in June 2000 following the failure of one of its three gyroscopes.

Compton effect

The increase in wavelength and change in direction of an X-ray or gamma-ray **photon** when it collides with a charged particle, usually an **electron**; also known as *Compton scattering* and named after the American physicist Arthur Holly Compton (1892–1962). The opposite effect, in which a photon gains energy (and therefore decreases in wavelength) from a fast-moving electron, is known as the *inverse Compton effect.*

concave

Curved inwards, or having negative curvature, as in certain mirrors and lenses. A *biconcave lens* has two concave surfaces, not necessarily with the same radii of curvature. Compare with **convex**.

Cone Nebula

A **diffuse nebula** and associated **open cluster**, the **Christmas Tree Cluster** (NGC 2264), in **Monoceros**; William **Herschel** discovered both, in 1785 and 1784, respectively. The distinctive cone shape is a dusty pillar about 7 light-years long; its edges are bathed in ultraviolet light from nearby hot stars, releasing gas into the relatively empty region of surrounding space. There, further irradiation makes the hydrogen gas glow, pro-

Cone Nebula The upper 2.5 light-years of the Cone Nebula as seen by the Hubble Space Telescope. Radiation from hot, young stars (located beyond the top of the image) has slowly eroded the Cone over millions of years. Ultraviolet light heats the edges of the dark cloud, releasing gas into the relatively empty region of surrounding space. There, additional ultraviolet radiation causes the hydrogen gas to glow, which produces the halo of light seen around the pillar. A similar process occurs, on a much smaller scale, to gas surrounding a single star, forming the bow-shaped arc seen near the upper left side of the Cone. *NASA/STScI/H. Ford (JHU) et al.*

ducing the red halo and tendrils of light seen around the pillar. The Cone belongs to a much larger complex that is the site of active star formation.

Visual magnitude:	3.9
Angular size:	20′
Distance:	2,700 light-years
Position:	R.A. 6 hr 41.1 min; Dec. +9 hr 53 min

conic section
A curve formed by slicing through a cone. Conic sections include circles, **ellipse**s, **parabola**s, and hyperbolas. These are also the possible orbits that an object can follow around a larger gravitating mass.

conjunction
An **inferior planet** is in *inferior conjunction* when it is directly between Earth and the Sun. It is in *superior conjunction* when it is on the opposite side of the Sun from Earth. A **superior planet** is in conjunction when it is on the opposite side of the Sun from Earth. A superior planet cannot have an inferior conjunction. When Earth is at inferior conjunction with respect to a superior planet, that planet is in *opposition* from Earth's perspective.

constellation
(1) Any of the 88 unequal regions into which the **celestial sphere** is divided by international agreement. (2) A

(continued on page 118)

The 88 Modern Constellations

Name	Genitive	Abbr.	Area (sq. deg.)	Rank (size)	Origin*	Position R.A. range (h m)	Position Dec. range (°)
Andromeda	Andromedae	And	722	19	1	23 00 to 02 40	+21 to +53
Antlia	Antliae	Ant	239	62	6	09 25 to 11 05	–24 to –40
Apus	Apodis	Aps	206	67	3	13 50 to 18 05	–67 to –83
Aquarius	Aquarii	Aqr	980	10	1	20 40 to 00 00	+3 to –24
Aquila	Aquilae	Aql	652	22	1	19 00 to 20 30	+10 to –10
Ara	Arae	Ara	237	63	1	16 35 to 18 10	–55 to –68
Aries	Arietis	Ari	441	39	1	01 40 to 03 30	+10 to +30
Auriga	Aurigae	Aur	657	21	1	04 40 to 07 30	+28 to +55
Boötes	Boötis	Boo	907	13	1	13 40 to 15 50	+8 to +55
Caelum	Caeli	Cae	125	81	6	04 20 to 05 10	–27 to –49
Camelopardalis	Camelopardalis	Cam	757	18	4	03 10 to 14 30	+52 to +87
Cancer	Cancri	Cnc	506	31	1	07 50 to 09 20	+7 to +33
Canes Venatici	Canum Venaticorum	CVn	465	38	5	12 10 to 14 10	+28 to +53
Canis Major	Canis Majoris	CMa	380	43	1	06 10 to 07 30	–11 to –33
Canis Minor	Canis Minoris	CMi	183	71	1	07 05 to 08 10	0 to +12
Capricornus	Capricorni	Cap	414	40	1	20 10 to 22 00	–9 to 27
Carina	Carinae	Car	494	34	6	06 05 to 11 20	–51 to –75
Cassiopeia	Cassiopeiae	Cas	598	25	1	23 00 to 03 00	+50 to +60
Centaurus	Centauri	Cen	1,060	9	1	11 05 to 15 00	–30 to –65
Cepheus	Cephei	Cep	588	27	1	20 05 to 00 00	+53 to +87
Cetus	Ceti	Cet	1,231	4	1	00 00 to 03 25	+10 to –25
Chamaeleon	Chamaeleontis	Cha	132	79	3	07 30 to 13 50	+74 to +83
Circinus	Circini	Cir	93	85	6	13 45 to 15 25	–54 to –70
Columba	Columbae	Col	270	54	4	05 05 to 06 40	–27 to –43
Coma Berenices	Comae Berenices	Com	386	42	2	12 00 to 13 53	+14 to +34
Corona Australis	Coronae Australis	CrA	128	80	1	18 00 to 19 20	–37 to –45
Corona Borealis	Coronae Borealis	CrB	179	738	1	15 15 to 16 25	+26 to +40
Corvus	Corvi	Crv	184	70	1	11 55 to 13 00	–11 to –25
Crater	Crateris	Crt	282	53	1	10 50 to 11 55	–6 to –25
Crux	Crucis	Cru	68	88	4	12 00 to 13 00	–56 to –65
Cygnus	Cygni	Cyg	804	16	1	19 10 to 22 00	+28 to +60
Delphinus	Delphini	Del	189	69	1	20 10 to 21 05	+2 to +21
Dorado	Doradus	Dor	179	72	3	03 50 to 06 40	–49 to –85
Draco	Draconis	Dra	1,083	8	1	10 00 to 20 00	+50 to +80
Equuleus	Equulei	Equ	72	87	1	20 50 to 21 25	+2 to +13
Eridanus	Eridani	Eri	1,138	6	1	01 20 to 05 10	0 to –58
Fornax	Fornacis	For	398	41	6	01 45 to 03 50	–24 to –40

The 88 Modern Constellations

Name	Genitive	Abbr.	Area (sq. deg.)	Rank (size)	Origin*	Position R.A. range (h m)	Position Dec. range (°)
Gemini	Geminorum	Gem	514	30	1	06 00 to 08 05	+10 to +35
Grus	Gruis	Grin	366	45	3	21 30 to 23 30	−37 to −57
Hercules	Herculis	Her	1,225	5	1	15 50 to 19 00	+4 to +50
Horologium	Horologii	Hor	249	58	6	02 10 to 04 20	−40 to −67
Hydra	Hydrae	Hya	1,303	1	1	08 05 to 15 00	−22 to −65
Hydrus	Hydri	Hyi	243	61	3	01 25 to 04 30	−58 to −90
Indus	Indi	Ind	294	49	3	20 30 to 23 30	−45 to −75
Lacerta	Lacertae	Lac	201	68	5	21 55 to 22 55	33 to 57
Leo	Leonis	Leo	947	12	1	09 20 to 11 55	−6 to +33
Leo Minor	Leonis Minoris	LMi	232	64	5	09 15 to 11 05	+23 to +42
Lepus	Leporis	Lep	290	51	1	04 55 to 06 10	−11 to −27
Libra	Librae	Lib	538	29	1	14 20 to 16 00	0 to −30
Lupus	Lupi	Lup	334	46	1	14 15 to 16 05	−30 to −55
Lynx	Lyncis	Lyn	545	28	6	06 20 to 09 40	+34 to +62
Lyra	Lyrae	Lyr	286	52	1	18 10 to 19 30	+26 to +48
Mensa	Mensae	Men	153	75	6	03 30 to 07 40	−70 to −85
Microscopium	Microscopii	Mic	210	66	6	20 25 to 21 25	−28 to −45
Monoceros	Monocerotis	Mon	482	35	4	06 00 to 08 10	−11 to +12
Musca	Muscae	Mus	138	77	3	11 20 to 13 50	−64 to −74
Norma	Normae	Nor	165	74	6	15 25 to 16 35	−42 to −60
Octans	Octantis	Oct	291	50	6	00 00 to 24 00	−75 to −90
Ophiuchus	Ophiuchi	Oph	948	11	1	16 00 to 18 40	+14 to −30
Orion	Orionis	Ori	594	26	1	04 40 to 06 20	+8 to +23
Pavo	Pavonis	Pav	378	44	3	17 40 to 21 30	−57 to −75
Pegasus	Pegasi	Peg	1,121	7	1	21 05 to 00 15	+2 to +37
Perseus	Persei	Per	615	24	1	01 30 to 04 50	+31 to +59
Phoenix	Phoenicis	Phe	469	37	3	23 20 to 02 25	−40 to −59
Pictor	Pictoris	Pic	247	59	6	04 35 to 06 55	−43 to −64
Pisces	Piscium	Psc	889	14	1	22 50 to 02 10	−5 to +34
Piscis Austrinus	Piscis Austrini	PsA	245	60	1	21 25 to 23 05	−25 to −36
Puppis	Puppis	Pup	673	20	6	06 00 to 08 30	−12 to −51
Pyxis	Pyxidis	Pyx	221	65	6	08 25 to 09 30	−17 to −38
Reticulum	Reticuli	Ret	114	82	6	03 15 to 04 40	+53 to +67
Sagitta	Sagittae	Sge	80	86	1	18 55 to 20 20	+17 to +22
Sagittarius	Sagittarii	Sgr	867	15	1	18 00 to 20 25	−12 to −46
Scorpius	Scorpii	Sco	497	33	1	15 45 to 17 55	−8 to −45
Sculptor	Sculptoris	Scl	475	36	6	23 05 to 01 45	−25 to −59
Scutum	Scuti	Sct	109	84	5	18 15 to 18 55	−4 to −16

(continued)

The 88 Modern Constellations *(continued)*

Name	Genitive	Abbr.	Area (sq. deg.)	Rank (size)	Origin*	Position R.A. range (h m)	Dec. range (°)
Serpens	Serpentis	Ser	637	23	1	15 10 to 16 20 and 17 15 to 18 55	–4 to +20 and –15 to +6
Sextans	Sextantis	Sex	314	47	5	09 65 to 10 50	–11 to +7
Taurus	Tauri	Tau	797	17	1	03 20 to 06 00	+10 to +30
Telescopium	Telescopii	Tel	252	57	6	18 10 to 20 30	–46 to –57
Triangulum	Trianguli	Tri	132	78	1	01 30 to 02 50	26 to 37
Triangulum Australe	Trianguli Australis	TrA	110	83	3	15 00 to 17 00	–60 to –70
Tucana	Tucana	Tuc	295	48	3	22 10 to 01 20	56 to 75
Ursa Major	Ursae Majoris	UMa	1,280	3	1	08 35 to 14 30	29 to 73
Ursa Minor	Ursae Minoris	UMi	256	56	1	00 00 to 24 00	66 to 90
Vela	Velorum	Vel	500	32	6	08 00 to 11 05	–40 to –57
Virgo	Virginis	Vir	1,294	2	1	11 35 to 15 10	–22 to 15
Volans	Volantis	Vol	141	76	3	06 30 to 09 00	–64 to –75
Vulpecula	Vulpeculae	Vul	268	55	5	19 00 to 21 30	20° and 30°

*1 = Ptolemy, 2 = Mercator, 3 = Keyser/de Houtman, 4 = Plancius, 5 = Hevelius, 6 = Lacaille

constellation

(continued from page 115)

grouping of stars within such a region, derived from some mythical or pictorial association.

The present constellation boundaries, originally drawn up by Eugène **Delaporte**, were adopted in 1930 by the **International Astronomical Union** (IAU). They are similar to the rectangular borders of some American states that run exactly north/south or east/west. The constellation boundaries run along lines of declination and right ascension for the **epoch** 1875 (chosen because this epoch had already been used by Benjamin **Gould** in defining boundaries for the southern constellations).

Forty-eight of our present-day constellations were identified by **Ptolemy** in his great *Almagest* of the second century A.D. Coma Berenices was invented by Gerardus Mercator in 1551 by taking a few stars from Leo. A further 12, in the region around the celestial south pole, were staked out by the Dutch explorers Pieter Keyser and Frederick de Houtman and included on a globe of the sky made by Petrus **Plancius** in 1598 and then in Johann **Bayer**'s *Uranometria* star atlas in 1603. Jakob **Bartsch** added three constellations in spaces between existing patterns and is also credited with naming Crux, the Cross, by regrouping four stars from Ptolemy's original Centaurus. Johannes **Hevelius** contributed seven more in his star atlas of 1687. Finally, in 1750, during a trip to the Cape of Good Hope, Nicolas de **Lacaille** penciled in the last 14 of our modern constellations to fill in some star-poor regions between existing groups. (See table, "The 88 Modern Constellations.")

contact

Any of the four stages that mark the start or end of an **eclipse, occultation,** or **transit.** *First contact* is the instant when an eclipse, occultation, or transit begins. *Second contact* is the instant that totality begins in a total solar or total lunar eclipse. *Third contact* is the instant that totality ends. *Fourth contact* is the instant at which an eclipse, occultation, or transit ends.

continuous spectrum

An unbroken **emission spectrum** spanning a range of wavelengths. It is produced by electrons undergoing **free-bound transitions** in a hot gas.

contrast

The ratio of the brightness of two parts of an image. The relationship between contrast as measured by a photocell or a light meter and contrast as reported by visual observers is complicated and poorly understood.

convection

A form of energy transport in which the material containing the energy moves. The hotter material moves toward the cooler area and the cooler material toward the hotter area. On Earth, it can be seen in a pan of boiling water (where hot water moves up and cooler water moves down), and also in a thunderstorm (where warmer, moist air moves up and forms clouds).

convective envelope

A region inside a star where energy is transported mainly by **convection**. In the Sun, the convective envelope begins just below the **photosphere** and extends down for about 180,000 km (about a fifth of a solar radius); it accounts for about two-thirds of the Sun's volume but only about one-sixtieth of the Sun's mass. The temperature inside this layer is thought to vary between 2 million and 6,500 K, and the density between 100 times more and 4,000 times less than that of air at Earth's surface. In stars less massive than the Sun, the convective envelope extends down to comparatively much greater depths, and in stars of less than 0.4 M_{sun} energy transport is entirely convective. In high mass stars, however, the convective envelope is much smaller or is entirely replaced by a *radiative envelope.*

convective overshoot

The penetration of **convection** currents into an otherwise nonconvective zone of a star from an underlying or overlying adjacent region. In evolved stars, it may be significant in moving heavy elements from the stellar interior to outer regions that they might not otherwise reach. Evidence of convective overshoot in the Sun appears in the form of **granulation** at the surface.

convergent point

The point on the **celestial sphere** toward which the **proper motions** of stars in a **moving cluster** appear to converge.

convex

Curved outward, or having positive curvature, as in certain mirrors and lenses. A *biconvex lens* has two convex surfaces, not necessarily with the same radii of curvature. Compare with **concave**.

cooling timescale

The characteristic timescale it takes for an **interstellar cloud** to cool. This determines whether a cloud will free fall without pressure support or if it will heat the nebula adiabatically (see **adiabatic process**) as it collapses.

co-orbital

Sharing the same orbit, as when objects, known as **Trojans**, move in the same orbit as a planet, at the **Lagrangian points**. A similar example is when two or more small moons occupy more or less the same orbit as a larger moon; this is the case with Saturn's moons **Calypso** and **Telesto**, which are co-orbital with **Tethys**.

Copernican system

A heliocentric model of the solar system introduced by Nicolaus **Copernicus** in the first half of the sixteenth century. Despite causing little stir at the time, it laid the foundation for the revolution in astronomy that was realized through the work of Johannes **Kepler** and **Galileo**. Although modern in its reversal of roles for the Sun and Earth, the Copernican system owed much to the geocentric **Ptolemaic system** that it opposed. It still involved **epicycles** and circular orbits, and incorporated ideas from variations to the Ptolemaic model proposed by Arab astronomers. Its importance lay not in its improved accuracy–Kepler's elliptical orbits would be needed for that–but in its challenge to the orthodox view that Earth was at the center of the universe. A much earlier heliocentric scheme had been proposed by **Aristarchus of Samos** in the third century B.C., a fact known to Copernicus but long ignored by others prior to him.

Copernicus, Nicolaus (1473–1543)

A Polish astronomer who was the first in Renaissance Europe to advance the heliocentric theory, namely, that Earth and other planets revolve around the Sun. The son of a merchant, Copernicus was born in Torun and educated at Cracow University and at various Italian universities where he studied medicine and law. On his return to Poland in 1506, he served as physician and secretary to his uncle Lucas, Bishop of Ermland. On his uncle's death in 1512, Copernicus took up the post of canon of Frauenburg Cathedral. By this time he had already abandoned the **Ptolemaic system** and had begun to formulate the revolutionary system with which his name has been associated (see **Copernican system**). The new system was first described in his *Commentariolus,* a brief tract completed sometime before 1514 and circulated in manuscript form to interested scholars. Thereafter he worked out the details of the new system in his *De revolutionibus orbium coelestium* (1543). Although it was a complete manuscript by 1530 Copernicus seemed, for reasons that are unclear, reluctant to publish it. In fact, it was not until **Rheticus** arrived in Frauenburg in 1539 and intervened that Copernicus reluctantly allowed its publication. The work finally appeared just in time, according to popular legend, for it to be shown to Copernicus on his deathbed.

Cor Caroli (Alpha^2 Canum Venaticorum, α^2 CVn)

The brightest star in **Canes Venatici**; its name, meaning "Charles's Heart," was given by Edmund **Halley** in 1725 in honor of England's King Charles II. Cor Caroli is actually a binary system consisting of α^1, an undistinguished fifth magnitude star, and α^2, a far more interesting component. α^2 has one of the strongest known magnetic fields among otherwise normal **main sequence** stars–some 1,500 times stronger than the Sun's. Variations in the strength of the field and in light output, by 0.03 magnitude, occur with a period of 5.5 days. The star also has a strange chemical composition in which elements such as silicon and mercury, and rarer ones such as europium, are greatly enhanced. It is thought that the magnetic field plays a key role in redistributing the elements in the star's atmosphere, appearing to enrich some and to deplete others. α^2 is the prototype **Alpha² Canum Venaticorum star**.

Visual magnitude:	2.89
Absolute magnitude:	0.89
Spectral type:	A0VpSiEuHg
Surface temperature:	9,500 K
Luminosity:	50 L_{sun}
Distance:	82 light-years

Cordelia

The innermost moon of **Uranus**, also known as Uranus VI; it was discovered in 1986 from images sent back by **Voyager** 2. Cordelia has a diameter of 26 km and moves in a nearly circular orbit 49,750 km from (the center of) the planet. It appears to be the inner **shepherd moon** of Uranus's Epsilon ring.

Córdoba Dürchmusterung (CD)

A southern extension of the ***Bonner Dürchmusterung***, compiled from observations at the Cordoba Observatory, Argentina, by J. M. Thome and published in the *Resultados del Observatorio Nacional Argentino* from 1892 onward. It includes 613,959 stars down to about tenth magnitude, south of Dec. –22°. Objects are referred to in the form "CD –37° 2853."

core, planetary

The central part of a planet, large moon, or large asteroid that is denser than, and compositionally distinct from, the layers that surround it. Earth has a solid inner core with a radius of about 1,300 km below a fluid outer core some 2,300 km thick. Both regions of the core consist largely of iron and nickel that sank to the center of the planet while it was still molten. Circulating currents in the core give rise to Earth's magnetic field.

core, stellar

The dense, hot central part of a star in which **hydrogen burning** takes place while the star is on the **main sequence** and in which the fusion of heavier elements occurs later on.

Coriolis effect

An apparent deflection of the path of an object that moves within a rotating coordinate system; the object doesn't actually deviate from its path, but it appears to do so because of the motion of the coordinate system. The Coriolis effect explains the directions of the trade winds in equatorial regions on Earth and, in general, plays a prominent part in studies of atmospheric dynamics of planets with dense atmospheres. It also figures in some aspects of astrophysics and stellar dynamics; for example, it is a controlling factor in the direction of rotation of **sunspots**. The effect, also known as the *Coriolis force,* is named after the French physicist Gustave de Coriolis (1792–1843).

corona

(1) The tenuous uppermost level of the Sun's (or another star's) atmosphere; lying immediately above the **chromosphere**, it consists of hot (1 to 4 million K), low-density (about 10^{-16} g/cm^3) gas that extends for millions of km from the Sun's surface. The *white-light corona,* visible during a total eclipse or with a **coronograph**, has three components. The faint *E corona* (*emission corona*) results from emission lines, including **forbidden lines** of calcium, iron, and some other elements. The *K corona* (or *continuum corona*), which is the innermost part of the corona closest to the photosphere, extending to about two solar radii, is caused by sunlight scattering off electrons. The *F corona* (*Fraunhofer corona*), lying outermost, is caused by sunlight scattering or reflecting off dust in interplanetary space. X rays, too, come from the corona (as well as from solar flares) but not from all parts equally. Movies made from X-ray pictures show that the corona is extraordinarily dynamic with an appearance that changes, not only daily but over the course of a **solar cycle**. At solar maximum the dominant features are **coronal loops** and streamers associated with active regions, but at minimum these give way to coronal holes at each pole and a sheetlike structure near the equator. (2) A circular to elongate feature (pl. coronae) on the surface of a planet or moon surrounded by multiple concentric ridges. Coronae are thought to be formed by hot spots. (3) A region of very hot, tenuous gas that stretches out of the galactic plane in spiral galaxies such as the Milky Way; also known as the *galactic corona.*

Corona Australis (abbr. CrA; gen. Coronae Australis)

The Southern Crown, also known (less commonly but more correctly) as Corona Austrini; a small, faint southern **constellation**. It lies at the southeastern edge of Sagittarius, north of Telescopium. Its principal stars (Theta, Eta^2, Gamma, Delta, Beta, Alpha, and Gamma) form an arc, similar to but fainter than that of **Corona Borealis**. Located at the edge of the Milky Way, Corona Australis contains some interesting deep sky objects. Among these are the variable **diffuse nebula** NGC 6729, which surrounds the erratic variable R Coronae Australis (R.A. 19h 1.9m, Dec. –36° 57′), and the **globular cluster** NGC 6541, an interesting object seen with binoculars or a small telescope, that lies between Theta CrA and Theta Scorpii (magnitude 6.6; diameter 13.1′; distance 14,000 light-years; R.A. 18h 8m, Dec. –43° 42′). See star chart 14.)

Corona Borealis (abbr. CrB; gen. Coronae Borealis)

The Northern Crown, given by Bacchus to Ariadne, daughter of King Minos of Crete; a small but prominent northern **constellation**. It lies east of Arcturus between Boötes and Hercules, and comprises a distinctive arc formed by the stars Theta, Beta, Alpha, Gamma, Delta, Epsilon, and Iota. Also within its boundaries are the prototype **R Coronae Borealis star** and T CrB, popularly known as the **Blaze Star**. (See table, "Corona Borealis: Stars Brighter than Magnitude 4.0," and star chart 11.)

coronagraph

A device for studying the solar **corona** at any time of the day by creating an artificial eclipse, invented by Bernard Lyot in 1930. It consists of a high-quality refractor in whose focal plane a small disk occults the image of the Sun. The diameter of the disk exactly equals the diameter of the solar image, so that only the faint light from the surrounding corona reaches the end of the instrument where a camera is mounted. For good results, the objective lens must be superbly polished and be entirely free of internal defects such as striations or bubbles. Great care is taken to reduce scattered light by mounting a series of diaphragm stops inside the coronagraph tube. Observations need to be carried out under the most favorable atmospheric conditions–clean air being absolutely essential–so that coronagraphs are normally installed at high-altitude stations.

coronal condensation

An area of the Sun's **corona** that is hotter (up to 4 million K) and denser than its surroundings. Coronal condensations are seen at the Sun's limb above **sunspot** groups and show loop structures that outline magnetic field lines.

coronal hole

A region of the Sun's **corona** that appears dark in pictures taken with a **coronagraph** or during a total solar eclipse, and that shows up as a void in X-ray and extreme ultraviolet images. Coronal holes are of very low density (typically 100 times lower than the rest of the corona) and have an open magnetic field structure; in other words, magnetic field lines emerging from the holes extend indefinitely into space rather than looping back into the **photosphere**. This open structure allows charged particles to escape from the Sun and results in coronal holes being the primary source of the **solar wind** and the exclusive source of its high-speed component. During the minimum years of the solar cycle, coronal holes are confined to the Sun's polar regions, while at solar maximum they can open up at any latitudes. See also **helmet streamer**.

coronal lines

Strong **emission lines** in the spectrum of the Sun's **corona** caused by very high excitation metal ions, especially those of iron. The strongest of all is the green coronal green line [Fe XIV] (due to the Fe^{13+} ion) at 5303 Å. Other prominent ones are [Fe X] at 6375 Å and [Fe XI] at 7892 Å.

coronal loop

A feature in the Sun's **corona** visible at X-ray, ultraviolet, and white-light wavelengths, consisting of an arch, extending upward from the **photosphere** for tens or hundreds of thousands of kilometers. Bright coronal loops,

Corona Borealis: Stars Brighter than Magnitude 4.0

Star	Magnitude		Spectral	Distance	Position	
	Visual	Absolute	Type	(light-yr)	R.A. (h m s)	Dec. (° ′ ″)
α **Alphekka**	2.22	0.42	A0V	75	15 34 41	+26 42 53
β Nusakan	3.66	0.94	F0Vp	114	15 27 50	+29 06 21
γ	3.81	0.57	B9IV+A3V	145	15 42 45	+26 17 44

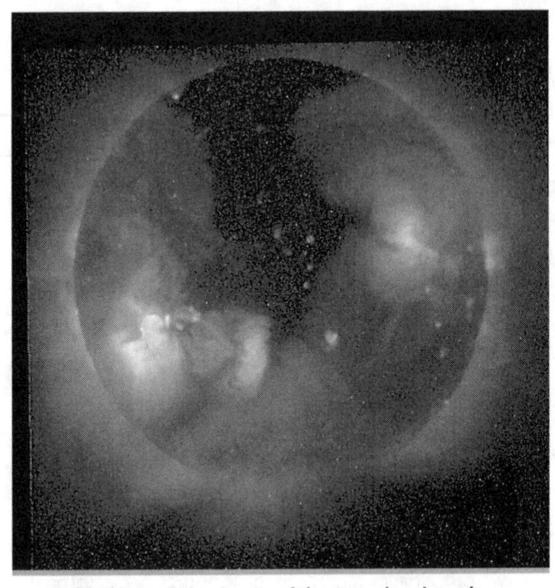

coronal hole An X-ray image of the Sun, showing a large coronal hole (the dark region), taken by the ACE (Advanced Composition Explorer) satellite. *NASA*

in the form of **coronal condensation**s and bright spots, are common around the time of solar maximum. Larger faint ones, lasting days or weeks, are more typical of the quiet corona, when solar activity is low. The two ends of a loop, known as *footprints,* lie in regions of the photosphere of opposite magnetic polarity to each other. Until recently, researchers had suspected that coronal loops were essentially static, plasma-filled structures. However, movies made from observations by the TRACE (Transition Region and Coronal Explorer) spacecraft show bright blobs of plasma racing up and down the coronal loops. **SOHO** (Solar and Heliospheric Observatory) data confirmed that these plasma blobs were moving at tremendous speeds, leading to the new view that coronal loops are hypervelocity currents of plasma blasted from the solar surface and squirted between the magnetic structures in the corona. Rather than being tubes of plasma enclosed within a magnetic container, the loops are jets of hot plasma flowing along in the alleys between the strong coronal magnetic fields. If coronal loops are indeed currents of plasma being propelled against solar gravity, they would have about the same density along their entire height, like an arc of water from a fountain. Plasma flows are seen in roughly half of all coronal loops visible by TRACE; flows may be present in the remainder but may be too faint for TRACE to detect. The plasma current that forms a coronal loop is probably caused by uneven heating at the bases of the loop, with plasma racing from the hotter end to the cooler end. The bases of a coronal loop are separated by many thousands of km, and there is no reason to assume that the environment at one end will be exactly the same, and input exactly the same amount of heat, as the environment at the other end. Although it isn't clear what causes coronal-loop heating in the first place, these new discoveries may help uncover the mechanism, shedding light on the long-standing mystery of why the corona is hundreds of times hotter than the solar surface.

coronal mass ejection (CME)

A huge eruption of material from the Sun's **corona** into interplanetary space. CMEs are the most energetic of solar explosions and result in the ejection, over the course of several hours, of up to 100 billion kg of multimillion-degree **plasma** at speeds ranging from 10 to 2,000 km/s. They often look like bubbles and, when seen close to the Sun, can appear bigger than the Sun itself, though their density is extremely low. In contrast to the steady state **solar wind,** CMEs originate in regions where the magnetic field is closed and result from the catastrophic disruption of large-scale coronal magnetic structures, such as **coronal streamers**. CMEs can occur at any time during the **solar cycle,** but increase in daily frequency from about 0.5 during minimum years to about 2.5 around solar maximum. Fast CMEs–those which outpace the ambient solar wind–give rise to large geomagnetic storms when they encounter Earth's **magnetosphere**. Such storms, which can disrupt power grids, damage satellite systems, and threaten the safety of astronauts, can result from the passage either of the CME itself or of the shock created by the fast CME's interaction with the slower-moving solar wind.

coronal rain

Material that condenses in the Sun's **corona** and falls along curved paths onto the **chromosphere**. Observed in **H-alpha** light at the solar limb above strong **sunspots**, coronal rain consists of gas ejected by a **loop prominence** that returns, several hours later, along the outline of the now invisible loop.

coronal streamer

A wisplike stream of particles traveling through the Sun's **corona**, visible in images taken with a **coronagraph** or during a total solar eclipse. Coronal streamers are thought to be associated with **active region**s and/or **prominence**s and are most impressive near the maximum of the solar cycle. Although they can be longer than the diameter of the Sun, they are very tenuous; the

Corvus: Stars Brighter than Magnitude 4.0

Star	Magnitude		Spectral	Distance	Position	
	Visual	Absolute	Type	(light-yr)	R.A. (h m s)	Dec. (° ′ ″)
γ Minkar	2.58	–0.94	B8IIIpHgMn	165	12 15 48	–17 32 31
β Kraz	2.65	–0.51	G5II	140	12 34 23	–23 23 48
δ Algorab	2.94	0.78	B9.5V	88	12 29 52	–16 30 55
ε	3.02	–1.82	K2.5IIIaBa0.2	303	12 10 07	–22 37 11

material in them gradually moves away from the Sun and becomes part of the **solar wind**.

COROT (Convection, Rotation, and Planetary Transits)

A French-led mission, scheduled for launch in 2005, one of the objects of which will be to search for **extrasolar planets** by **photometry**. It will use a 30-cm telescope, equipped with **CCDs**, to monitor selected stars for tiny, regular changes in brightness that might be due to planets in **transit**. COROT will also be able to detect the light variations caused by seismic disturbances inside stars and so provide data helpful in determining stellar mass, age, and chemical composition (see **astroseismology**).

co-rotation

The situation in which at least one of two bodies that are orbiting around each other, has the same **axial period** as its **rotation period** so that it always keeps the same face toward its partner. The co-rotation of the Moon means that the lunar farside is never visible from Earth; the co-rotation of both Pluto and Charon ensures that both objects never turn their backs on each other.

corrector plate

A thin lens, or combination of lenses, placed at the front of a **catadioptric telescope** to correct the **spherical aberration** of the primary mirror; also known as *correcting lens.*

Corvus (abbr. Crv; gen. Corvi)

The Crow; a small but fairly conspicuous southern **constellation** whose four main stars make up an easily distinguished quadrilateral. It lies south and west of Virgo and east of Crater. Among its premier deep sky objects are the planetary nebula NGC 4361, which is large and reminiscent of the **Owl Nebula**, with an easily seen central star (magnitude 10.3; diameter 45″; R.A. 12h 24.5m, Dec. –18° 47.1′), and the extraordinary interacting galaxies known as the **Antennae**. (See table, "Corvus: Stars Brighter than Magnitude 4.0," and star chart 5.)

cosmic abundance

The relative proportions of the chemical **elements** in the universe as a whole. Hydrogen and helium dominate, heavy element production in stars having had very little effect, even after more than 12 billion years, on the overall balance. But although the relative amount of elements heavier than helium is still tiny, these substances are critical for planet formation and, ultimately, for the evolution of life. (See table, "The Most Common Elements in the Universe.")

cosmic background radiation

Diffuse electromagnetic radiation coming from all parts of the sky. It is made up of the **cosmic microwave background** (CMB), **cosmic X-ray background** (CXB),

The Most Common Elements in the Universe

Element	Number of Atoms per 10 Million Atoms of Hydrogen
Hydrogen	10,000,000
Helium	1,400,000
Oxygen	6,800
Carbon	3,000
Neon	2,800
Nitrogen	910
Magnesium	290
Silicon	250
Sulfur	95
Iron	80
Argon	42
Aluminum	19
Sodium	17
Calcium	17

cosmic infrared background (CIB), and cosmic optical background (COB). Of these components, the CMB is by far the largest, with a total intensity of 996 nW/m^2/steradian (nW = nanowatt, or 10^{-9} W; "steradian" is the solid angle whose opening is one radian). The cosmic far-infrared background has a total intensity of 34 nW/m^2/sr, while the cosmic near-infrared background and COB combined have a total intensity of slightly less than 60 nW/m^2/sr. Together the infrared-optical backgrounds add up to about 9% of the CMB's intensity.

cosmic infrared background (CIB)

The total of the **redshift**ed and reprocessed radiation from the era of **galaxy formation**. The intensity and spectrum of the CIB provides information about the history of star and galaxy formation, and the presence or absence of dust in early galaxies. Measurements of the CIB by the 2MASS telescope, published in 2002, confirmed earlier estimates by COBE (Cosmic Background Explorer) that the CIB is two to three times brighter than expected based on extrapolations of observed galaxies. The brightness indicates that there was an incredible burst of star formation during the universe's youth. It also supports the idea that a great deal of **dark matter** was present, pulling gas together shortly after the universe's birth, and enabling the first stars and galaxies to form.

cosmic microwave background (CMB)

Diffuse electromagnetic radiation, most intense around a wavelength of 1 mm, that fills the universe; it had been predicted theoretically but was first observed in 1964 by the American telecommunications engineers Arno Penzias and Robert Wilson. The strongest component of the **cosmic background radiation,** it is believed to have orig-

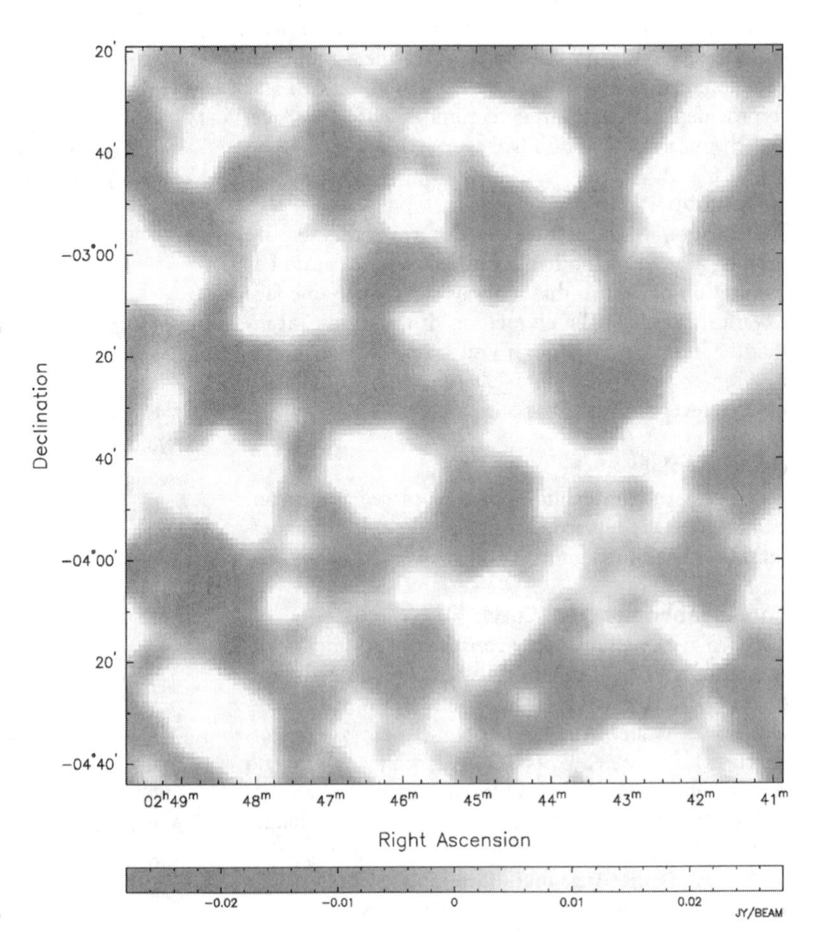

cosmic microwave background One of the sharpest images ever obtained of the cosmic microwave background; it was taken by the Cosmic Background Imager and has a resolution of 7 minutes of arc. Dark regions indicate cool spots; bright patches correspond to hot spots. The fluctuations in intensity are a mere ten-thousandth of a kelvin about the average temperature of 2.73 K. The image is of the surface of the "last scattering," from which photons were emitted about 14 billion years ago when the universe first became transparent. The fluctuations in temperature correspond to tiny enhancements in the density of the hot primordial plasma. As the universe expanded, the gravity of the dark matter within these clumps made them collapse into clusters of galaxies. *CBI/Caltech/NSF*

inated in the *decoupling era,* about 300,000 years ago, when radiation was first able to travel freely over great distances without being absorbed by ordinary matter. The temperature of the universe at this time was about 3,000 K, but the expansion of the universe has **redshift**ed the relict radiation into the microwave region of the spectrum so that it now appears as if it has come from a **blackbody** with a temperature of just 2.73 K. Sensitive measurements of the microwave background by the **Cosmic Background Explorer** (COBE) and other spacecraft have shown slight variations in temperature of the CMB with direction; these are taken to indicate slight fluctuations in the density of matter in the early universe, which would have been critical to the formation of the first galaxies.

cosmic rays

See article, page 126.

cosmic string

Hypothetical material left over from the **Big Bang** in the form of filaments of energy, as narrow as an atomic nucleus but possibly billions of light-years long. Cosmic string, not to be confused with the subatomic strings postulated by particle physicists, may be thought of as tubular samples of the universe from 10^{-35} second after the beginning of time. There is no observational evidence for its existence.

cosmic X-ray background (CXB)

An X-ray and gamma-ray glow that comes from all parts of the sky. First revealed in the 1970s by early X-ray satellites, its origin proved something of a mystery. Beginning in 2000, however, more powerful orbiting X-ray instruments, such as the **XMM-Newton Observatory**, have resolved some of the enigmatic background into many faint individual sources. Their X-ray characteristics point to huge amounts of material falling into massive **black holes** in very distant galaxies. These results add to the growing consensus that massive black holes hold court at the center of all large galaxies and that, from across the universe, X rays produced as matter that feeds these black holes account for the bulk of the CXB.

cosmic year

The time the Sun takes to complete one orbit around the center of the Galaxy–about 225 million years.

cosmic-ray exposure age

The period over which a **meteoroid** has been exposed to radiation in space, or a rock has been exposed on the surface of a body.

cosmological constant (λ)

Originally, a "fudge factor" introduced by Einstein into the equations of **general theory of relativity** to produce a static model for the universe; this was in 1917, before observations showed that the universe was expanding. Einstein later called it "my greatest blunder," since without the cosmological constant he could have theoretically predicted that the universe must be changing in size. The cosmological constant can be interpreted as a repulsive force–a kind of antigravity. Recently it has become important again, first in the context of the **inflationary model** and second because compelling evidence has been found that the expansion of the universe is accelerating, which implies there is some additional force at work that opposes gravity.

cosmological distance ladder

The chain of overlapping methods by which astronomers establish a distance scale for objects in the universe, from nearby planets to the most remote quasars and galaxies. At every step of the distance ladder, errors and uncertainties creep in. Each step inherits all the problems of the ones below it, and also the errors intrinsic to each step tend to get larger for the more distant objects; thus the spectacular precision at the base of the ladder degenerates into much greater uncertainty at the very top.

Distances within the solar system are known to extreme accuracy by a variety of methods, including the motions of the planets in the sky, radar, and timing of signals from interplanetary probes. Distances to stars within a couple of thousand light-years come from various geometrical methods; the most accurate values are those based on measurements of the annual **parallax** of about 10,000 nearby stars made by the **Hipparcos** satellite. The **moving cluster** method can be applied over a similar range, while **main-sequence fitting** works with **open clusters** out to a distance of about 60,000 light-years.

Beyond the Milky Way Galaxy, distances can be established most reliably using the **period-luminosity relation** of **Cepheid variables**, backed by similar observations of other bright stars whose intrinsic brightness is reasonably well-known, including **RR Lyrae stars** and **novae**. This method can be applied out to the limit at which Cepheids and other individual stars can be distinguished inside their host galaxies–up to about 100 million light-years. For more distant galaxies, standard candles brighter than Cepheids are needed. These include **globular clusters** and Type Ia **supernovae**, which can be calibrated as distance indicators using Cepheids in relatively nearby galaxies and then applied farther afield–up to about 200 million light-years for globulars andout to at least 3 billion light-years for supernovae. At the farthest limits, only whole galaxies

cosmic rays

Subatomic particles that move through space close to the speed of light; their origin is one of the major unsolved mysteries of astrophysics, although researchers are moving closer to a solution. Cosmic rays consist of about 85% **protons**, 14% **alpha particles** (helium nuclei), about 1% **electrons** and other elementary particles, and a tiny smattering of nuclei heavier than helium. Their energies range from 10 million eV to 1 million trillion eV–equivalent, at the high end, to the punch of a major league baseball pitch. Three categories of cosmic rays are recognized: solar, galactic, and extragalactic. *Solar cosmic rays,* with energies of 10^7 to 10^{10} eV, are ejected by the Sun during **solar flares**. *Galactic cosmic rays,* with energies of 10^{10} to 10^{15} eV and *extragalactic cosmic rays,* with energies up to 10^{18} eV, come from all parts of the sky and, at lower energies, have their original directions partially scrambled by the galactic magnetic field.

Cosmic rays in space are known as *primary cosmic rays* and can be detected directly only by instruments above Earth's atmosphere. When they collide with atoms and molecules in the upper atmosphere, they generate showers (known as *cosmic-ray showers* or *air showers*) of *secondary cosmic rays.* The initial collision produces pions, which quickly decay into muons, some of which decay further into electrons, positrons, and neutrinos. Deceleration of the electrons and positrons in the atmosphere produces a flash of light that can be observed from the ground with special telescopes; however, most of the secondary cosmic-ray particles that reach sea-level are undecayed muons. Observations of these muons and of the aerial light-flashes from electrons and positrons provide information on the primary cosmic ray that caused the cascade. Increasingly powerful equipment in space, in the atmosphere, and on the ground is helping unravel the enigma of galactic and extragalactic cosmic rays.

It had been suspected that many of the galactic variety are generated by shock waves from **supernovae**. Researchers had already shown that supernova remnants can accelerate electrons to cosmic-ray energies, but there was no evidence that protons are accelerated by the same mechanism. Then in 2002, a Japanese team reported that it had, for the first time, associated a supernova remnant with cosmic-ray protons. On several occasions, the team detected light showers due to protons coming from a patch of sky that contains a supernova remnant called RX J1713. 7-3946.

The origin of *extragalactic cosmic rays,* the most energetic particles known, is also becoming clearer. Analyzing data from high-energy cosmic-ray detectors in Japan and England, researchers announced in 2002 that they had traced the trajectories of several cosmic rays to four galaxies known to surround dead or dormant **quasars**, which almost certainly contain supermassive **black holes**. The finding fits with a scenario in which a spinning, supermassive black hole acts like a giant battery. Magnetic field lines in close contact with the rotating hole generate a billion trillion volts, which accelerate charged particles to ultrahigh energies. In this theory, the quasar must be dormant. If the cosmic rays revved up by the black hole were to collide with intense radiation from an active quasar, their energy would be drained away.

are detectable, so methods such as the **Tully-Fisher relation** and **Faber-Jackson relation** are used, which link measurable properties of galaxies, or clusters of galaxies, to their luminosity. Extragalactic distance indicators enable estimates to be made of the **Hubble constant**, a measure of the rate at which the universe as a whole is expanding. Observation of the **redshift** of a remote galaxy or quasar then supplies the object's distance. This method, however, carries a lot of uncertainty, because not only do estimates of the Hubble constant vary widely (between about 50 and 100 km/s/Mpc) but it appears the constant itself may vary over time!

cosmological model

A description of the universe in terms of its space-time configuration. Assuming that matter, overall, is distributed evenly and the same in all directions, there are three basic possibilities. (1) A *closed universe* is positively curved

cosmological distance ladder The magnificent spiral galaxy NGC 4603–the most distant galaxy in which Cepheid variables have been identified. NGC 4603 lies in the enormous Centaurus cluster. The discovery of Cepheids within it is an important step toward establishing a more reliable cosmological distance ladder. *NASA/Jeffrey Newman (Univ. of California at Berkeley)*

like the surface of a ball and is finite in extent; such a universe grows to a certain size before collapsing again. (2) An *open universe* is negatively curved like a saddle and is infinite in extent. (3) A flat *universe* has zero curvature and is infinite in extent. The factor determining which kind of space-time we actually live in is the average density of matter in the universe. Depending on whether the average density is greater than, less than, or equal to the so-called **critical density**, the cosmos is closed, open, or flat, respectively. Current observations suggest a value tantalizingly close to that needed for space-time to be flat. See **Big Bang**.

cosmological principle

The assumption that, ignoring local irregularities, the universe looks pretty much the same to every observer within it; in other words, there is no special cosmic vantage point.

cosmological redshift

The **redshift** produced by the expansion of the universe and the reason most galaxies in the universe have redshifts. Contrary to popular belief, this is *not* a Doppler shift. A Doppler redshift arises when an object moves away from us. Most galaxies move away from us, but this is *not* the cause of their redshifts. Instead, as a light wave travels through the fabric of space-time, the universe expands and the light wave gets stretched and therefore redshifted. This is a subtle but important difference. The farther away a galaxy is, the longer its light waves have traveled through space and the more redshifted they have become.

cosmology

The study of the origin, structure, and evolution of the universe on the largest possible scale.

coudé focus

A telescopic focus used primarily for **spectroscopy**. In this arrangement, light from the primary mirror is reflected along the polar axis to focus at a fixed place separate from the moving parts of the telescope, where large pieces of equipment can be fitted without interfering with the telescope's balance. The word comes from a French word meaning "bent like an elbow," not from a person's name.

Crab Nebula (M1, NGC 1952)

The most famous and conspicuous known **supernova remnant**. First noted by Chinese astronomers on July 4, 1054, it is the centuries-old wreckage of a stellar explosion that reached a peak magnitude of –6 (about four times brighter than Venus). According to the Chinese records, it was visible in daylight for 23 days and in the night sky to the naked eye for 653 days. Petroglyphs found in Navaho Canyon and White Mesa (both in Arizona) and in the Chaco Canyon National Park (New Mexico) appear to be depictions of the event by Anasazi Indian artists. The Crab lies about 6,300 light-years away in **Taurus**, measures roughly 10 light-years across, and is expanding at an average speed of 1,800 km/s (some 0.2″ per year). (See photo, page 129.) Surprisingly, its expansion rate seems to be accelerating, driven by radiation from the central pulsar (see below). Its luminosity at visible wavelengths exceeds 1,000 L_{sun} and comes from two major contributions, revealed spectroscopically by Roscoe Frank Sanford in 1919 and confirmed photographically by Walter **Baade** and Rudolph **Minkowski** in 1930. First, a reddish component, with an **emission line** spectrum like that of emission nebula, forms a far-flung chaotic web of bright filaments. Second, a bluish component, with a **continuous spectrum** due to highly polarized **synchrotron radiation,** supplies a diffuse background in the central region of the Crab. This synchrotron component is emitted by high-speed electrons spiraling in a strong magnetic field. In 1948, the Crab was identified as a strong source of radio waves, and cataloged as Taurus A and, later, as 3C 144. In 1964, it was also found to be a bright X-ray source (Taurus X-1), emitting 100 times more energy in X rays than it does in optical wavelengths. The collapsed remnant of the star that produced the nebula, the Crab pulsar, was first detected in 1968.

The Crab Nebula was discovered in 1731 by the British physician and amateur astronomer John Bevis (1695–1771). Charles **Messier** independently found it on August 28, 1758, when looking for Halley's comet on its first predicted return and initially thought it was a comet. It inspired Messier to begin his famous catalog and formed its first entry, M1. It was christened the "Crab" on the basis of a drawing made by the Earl of **Rosse** (William Parsons) in about 1844. The first photo of M1 was obtained in 1892 and the first serious investigation of its spectrum was carried out from 1913 to 1915 by Vesto **Slipher**.

The Crab can be found starting from Zeta Tauri, a third-magnitude star east-northeast of **Aldebaran**. Lying about 1° N and 1° W of Zeta, the Crab is visible as a dim patch in 7 × 50 or 10 × 50 binoculars. With more magnification, it is seen as a nebulous oval patch, surrounded by haze. In telescopes larger than 10 cm, some detail in its shape becomes apparent, with the suggestion of mottled or streaky structure in the inner part of the nebula.

Crab Pulsar (PSR 0531+21, NP0532)

The **pulsar** associated with the Crab Nebula has a period of 33.085 milliseconds, corresponding to a spin rate of 30 revolutions per second, and is one of the few pulsars detected at optical wavelengths. It was discovered on November 9, 1968, as a pulsating radio source by astronomers at the Arecibo Observatory, and later observed to be flashing at visible wavelengths by observers at Steward Observatory. As an optical pulsar it is sometimes also referred to by the Crab's variable star designation, CM Tauri. At visible wavelengths it shines about as brightly as the Sun–despite the fact that it measures less than 30 km across.

Crab Nebula
See article, this page.

Crater (abbr. Crt; gen. Crateris)
The Cup; a small, faint southern **constellation** that lies next to Hydra. Its only star above fourth magnitude is Delta Crt (visual magnitude 3.56, absolute magnitude –0.32, spectral type K0III, distance 195 light-years). (See star charts 5 and 12.)

crater
A bowl-like depression (Latin for "cup") on the surface of a planet, moon, or asteroid. Craters range in size from a few centimeters to over 1,000 km across, and are mostly

Crab Nebula The most famous star wreck in the sky, seen by the Very Large Telescope. *European Southern Observatory*

caused by impact (see **impact crater** and **impact basin**) or by volcanic activity (see **volcanism**), though some are due to **ice volcanism**.

crater chain

Several **craters** along a general line that may be overlapping, touching, or detached from one another. They are typically the result of either secondary impacts or volcanic activity. A secondary impact crater chain is usually aligned with a much larger impact crater, whereas a volcanic crater chain is associated with a volcanic fissure and may be due to explosions or collapse. Several unusual crater chains have been found on **Ganymede** and **Callisto** that appear to be the impact scars of tidally disrupted **comets** or **asteroids**. These features serve to record the characteristics of comets and support the "rubble-pile" model for comet nuclei and some asteroids, in which these objects are formed of many small, loosely bound fragments. A couple of crater chains have been found on Earth, including the Aorounga Craters (see **Earth impact craters**). A crater chain with a curved form is known as a *crater arc.*

Crescent Nebula (NGC 6888)

A rapidly expanding, luminous shell of gas that surrounds the **Wolf-Rayet star** WR 136 in **Cygnus**. The Crescent Nebula started to form about 250,000 years ago as the central star began shedding its outer envelope in a strong **stellar wind.** As this wind rammed into the surrounding interstellar medium, it compacted it and sent shock waves into it, resulting in a series of shells, which ultraviolet radiation from WR 136 subsequently ionized and made visible.

Angular diameter:	20′
Distance:	4,700 light-years
Position:	R.A. 20h 12.0m, Dec. +38° 21′

Cressida

The fourth moon of **Uranus**, also known as Uranus IX; it was discovered in 1986 by **Voyager** 2. Cressida has a diameter of 66 km and moves in a nearly circular orbit 61,770 km from (the center of) the planet.

critical density

The exact average density of matter in the universe today that would be needed to halt the cosmic expansion at some point in the future. A universe that has precisely the same density as the critical density is said to be *flat* or *Euclidean.* If the density of the universe is greater than the critical density, then not only will the expansion be stopped but there will be a collapse of the universe in the distant future. In this closed universe scenario, the universe will eventually implode under its own gravitational pull, leading to an event known as the **Big Crunch**. If the density of the universe is less than the critical density, an open universe scenario plays out in which the cosmic expansion will continue forever. The critical density is calculated to be about (1 to 2) $\times 10^{-26}$ kg/m^3–about 100 times greater than the average density inferred from all the known visible matter in the form of galaxies. When the inferred presence of **dark matter** is taken into account, the universe appears to be pretty close to the density called for by the flat scenario. However, the recent discovery that the rate of cosmic expansion is accelerating suggests that large amounts of **dark energy** exist, which are more than counteracting the in-pull of gravity.

cross-cutting

An interruption of one geologic feature by another, which can give an indication of the relative ages of these geologic features/events; for example, a fault cutting across an impact crater would be younger than the crater.

crown glass

Soda-lime glass used to make lenses and prisms. It has a lower index of refraction and less dispersion than flint glass, but is more durable. See **achromatic lens.**

Cruithne (minor planet 3753)

A 5-km-wide **near-Earth asteroid** that is co-orbital (shares the same orbit) with our planet. Its path is highly inclined and horseshoe-shaped with respect to Earth, causing Cruithne to alternatively move closer to and then much farther away from us. At its closest approach, which happens every 100,000 years, Cruithne comes within about 15 million km of Earth. It was discovered in 1986 by D. Waldron at Siding Spring Observatory, Coonabarabran, Australia, and named for the first Celtic tribal group that settled in the British Isles. Two other near-Earth asteroids are known to be currently in resonant states similar to that of Cruithne: 1998 UP1 and 2000 PH5.

crust

The solid surface layer of a planet or moon. Earth's crust, which forms the upper part of the **lithosphere**, ranges in depth from about 10 km under the oceans to about 40 km under the continents. As in the case of other planetary bodies in the inner solar system, it consists of lighter rocks that rose to the surface when molten, leaving denser materials to form the **mantle** and **core**. Large solid bodies in the outer solar system typically have crusts made of water-ice.

Crux (abbr. Cru; gen. Crucis)

The Cross; a tiny but brilliant southern **constellation** invisible to most of the populated Northern Hemisphere. Surrounded on three sides by Centaurus, with Musca to the south, it is well seen only south of the tropic of Cancer. Crux contains the most famous of dark nebulae, the **Coalsack**, and several bright **open clusters**, including the **Jewel Box**, NGC 4349 (magnitude 7.4; diameter 16′; R.A. 12h 24.5m, Dec. –61° 54′), and NGC 4463 (magnitude 7.2; diameter 5′; R.A. 12h 30.0m, Dec. –64° 48′). (See table, "Crux: Stars Brighter than Magnitude 4.0," and star chart 4.)

cryovolcanism

The eruption of water and other liquid or vapor-phase volatiles onto the frigid surfaces of the icy satellites of the

Crux: Stars Brighter than Magnitude 4.0

Star	Magnitude		Spectral	Distance	Position	
	Visual	Absolute	Type	(light-yr)	R.A. (h m s)	Dec. (° ′ ″)
α **Acrux**	0.77	–4.19		321		
α^1	1.58	–3.9	B0.5IV		12 26 36	–63 05 56
α^2	2.09	–3.4	B1V		12 26 37	–63 05 58
β **Mimosa**	1.25	–3.92	B0.5III	353	12 47 43	–59 41 19
γ **Gacrux**	1.59	–0.56	M3.5III	88	12 31 10	–57 06 47
δ	2.79	–2.45	B2IV	364	12 15 09	–58 44 55
ε	3.59	–0.63	K3III	228	12 21 22	–60 24 04

giant planets. It is known to occur on only one of these worlds: **Triton**, a moon of distant Neptune, where geyser-like plumes of nitrogen were discovered during the explorations of **Voyager** 2. There is indirect evidence that cryovolcanic processes may have taken place elsewhere in the outer solar system and might even be active today.

cryptovolcano

A volcano-like feature on a planet or moon, that may or may not be an actual volcano. Examples have been found on the surface of **Ganymede**.

cubewano

An object that orbits within the main **Kuiper Belt**, between 42 and 46 AU from the Sun, beyond the distance at which bodies can be controlled by resonances with the outer planets. The first cubewano discovered was 1992 QB1 (nicknamed "Smiley").

culmination

The instant at which a celestial object reaches its highest altitude above the horizon as Earth's rotation carries it across the sky; in other words, the moment the object crosses the observer's meridian (the north-south line in the sky). *Upper culmination* (also known as *culmination above pole* for circumpolar stars and the Moon) is the crossing closer to the observer's zenith; *lower culmination* (or *culmination below pole*) is the crossing farther from the zenith. Alternative names for culmination are *meridian passage* and *transit.*

Curtis, Heber Doust (1872–1942)

An American astronomer who started out as a student and then a professor of Latin and Greek, but went on to play an important role in establishing the nature of external galaxies. Having earned his Ph.D. in astronomy (1902) from the University of Virginia he joined the staff of the **Lick Observatory** where he remained until 1920. He then became director of the University of Pittsburgh's Allegheny Observatory and, finally, in 1930, was appointed director of the University of Michigan's observatory. Curtis's early work involved measuring the **radial velocities** of stars. In 1910, however, he began investigating spiral nebulae and became convinced that they were independent star systems. In 1917 he argued that the observed brightness of **novae**, found by him and by George Ritchey on photographs of spiral nebulae, indicated that the nebulae lay well beyond our Galaxy. He also maintained that extremely bright novae, later identified as **supernovae**, could not be included with the novae as distance indicators. He estimated the Andromeda Nebula (now known as the **Andromeda Galaxy**) to be 500,000 light-years away, a view opposed by many, including Harlow **Shapley** who proposed that the Milky Way Galaxy was 300,000 light-years in diameter–far larger than previously assumed–and that the spiral nebulae lay within it. In 1920, at a meeting of the National Academy of Sciences, Curtis engaged in a famous debate with Shapley over the size of the Galaxy and the distance of the spiral nebulae. The matter lay unresolved, however, until 1924 when Edwin **Hubble** redetermined the distance of the Andromeda Nebula and demonstrated that it was a galaxy in its own right.

curvature of field

The apparent bending of the **field of view** of a telescope; also known as *field curvature.* It can be compensated for by changing the focus for the edge areas.

CW Leonis (CW Leo, IRC 10°216)

The second brightest extrasolar object in the sky (after **Eta Carinae**) at an **infrared** wavelength of 10 microns. Visually, an eighteenth magnitude long-period **pulsating variable,** CW Leo is an **asymptotic giant branch** (AGB) star about 650 light-years away. A luminous giant star, with a surface temperature of 2,330 K and a radius of about 500 R_{sun}, it is cocooned within a shell of gas and dust in which carbon and dozens of different types of molecules have been detected. This shell has been formed from material lost in the form of a **stellar wind** that is blowing at a speed of about 14.5 km/s and carrying away about 3×10^{-5} M_{sun} per year. CW Leo is believed to be a rare case of a **protoplanetary nebula**–a system in the early stages of evolving to the **planetary nebula** stage.

cyanogen bands (CN bands)

Molecular bands found in the spectra of stars of type G0 and later. Cyanogen absorption is an important **luminosity** criterion, and is more pronounced in giants than in dwarfs of the same spectral type.

Cybele group

A group of **asteroids** in the outer part of the main asteroid belt at a mean distance from the Sun of 3.4 AU, corresponding roughly to a 4:7 resonance with Jupiter. Along with the Jupiter **Trojans**, the **Hilda group**, and **Thule**, the Cybeles are thought to be remains of the original asteroid population that have remained in more-or-less the same orbital locations since the dawn of the solar system. The prototype, (65) Cybele, of spectral class P and diameter 237 km, was discovered in 1861 by the German astronomer Wilhelm **Tempel**. Its orbital details are: semimajor axis 3.433 AU, perihelion 3.077 AU, aphelion 3.789 AU, eccentricity 0.104, inclination 3.55°.

cyclotron radiation

Electromagnetic radiation given off by charged particles that are spiraling along magnetic field lines. The *cyclotron frequency* (also known as the *gyrofrequency*), which is the number of times per second that a particle orbits a magnetic field line, is completely determined by the strength of the field and the particle's charge-to-mass ratio. The cyclotron frequency is twice the Larmor frequency of **precession**. See **pitch angle**.

61 Cygni (61 Cyg)

A nearby binary star system that consists of two orange-**red dwarfs** moving in a stretched-out orbit with a mean separation of about double the distance from Pluto to the Sun. It lies in **Cygnus**, southeast of Sadr (Gamma Cygni) and can just be glimpsed with the naked eye. 61 Cygni was christened the "Flying Star" in 1792 by Guiseppe **Piazzi** for its unusually large **proper motion**. In 1838, it became the first star (other than the Sun) to have its distance to Earth successfully calculated, by Friedrich **Bessel**, using trigonometric **parallax**. Both components appear to be slightly variable.

Visual magnitude:	5.20
Absolute magnitude:	7.49
Spectral type:	K5Ve (A), K7Ve (B)
Luminosity:	0.06 L_{sun} (A), 0.04 L_{sun} (B)
Separation (mean)	
Apparent:	24.4"
Actual:	85.2 AU
Orbit	
Period:	659 years
Eccentricity:	0.48
Inclination:	126°
Distance:	11.43 light-years

Cygnus (abbr. Cyg; gen. Cygni)

The Swan; a conspicuous northern **constellation** that lies east of Lyra and north of Vulpecula. The main stars form a cruciform asterism known as the Northern Cross–a familiar sight overhead in northern winter evening skies. (See tables, "Cygnus: Stars Brighter than Magnitude 4.0," and "Cygnus: Other Objects of Interest." See also star chart 7.)

Cygnus Loop A small portion of the Cygnus Loop, captured by the Hubble Space Telescope. The image shows a supernova blast wave overrunning dense clumps of gas. A ribbon of light, stretching left to right across the picture, may be a knot of gas ejected by the supernova that is just catching up with the shock front, which has slowed down by plowing into interstellar material. *NASA/ESA/STScI*

Cygnus: Stars Brighter than Magnitude 4.0

Star	Magnitude		Spectral	Distance	Position	
	Visual	Absolute	Type	(light-yr)	R.A. (h m s)	Dec. (° ′ ″)
α **Deneb**	1.25	–8.73	A2Iae	3,230	20 41 26	+45 16 49
γ **Sadr**	2.23	–6.12	F8Ib	1,520	20 22 14	+40 15 24
ε Gienah	2.58	–0.94	K0III	165	20 46 13	+33 58 13
δ	2.86	–0.74	B9.5IV+F1V	171	19 44 58	+45 07 51
β **Albireo**	3.05	–2.32	K3II+B0.5V	386	19 30 43	+27 57 35
ζ	3.21	–0.12	G8III-IIIaBa	151	21 12 56	+30 13 37
ξ	3.72	–4.07	K5Ib-II	1,180	21 04 56	+43 55 40
τ	3.73	2.13	F1IV	68	21 14 47	+38 02 44
ι	3.76	0.88	A5Vn	122	19 29 42	+51 43 47
κ	3.80	0.91	K0III	123	19 17 06	+53 22 07
η	3.89	0.73	K0III	139	19 56 18	+35 05 00
ν	3.94	–1.25	A1Vn	356	20 57 10	+41 10 02

Cygnus: Other Objects of Interest

Name	Notes
Stars	
61 Cyg	See separate entry.
P Cyg	The prototype **P Cygni star**.
SS Cyg	The prototype **SS Cygni star**.
Cygnus X-1	See separate entry.
Planetary nebulae	
Blinking Nebula	NGC 6826. See separate entry.
NGC 7072	An object with an irregular shape and four bright condensations. Magnitude 13.9; diameter 0.8′; R.A. 21h 30.6m, Dec. –43° 9′.
Diffuse nebulae	
NGC 6960, 6992	The western and eastern parts, respectively, of the **Veil Nebula**.
North American Nebula	NGC 7000. See separate entry.
Pelican Nebula	IC 5067, 5070. See separate entry.
Dark Nebula	
Great Rift	See separate entry.
Open clusters	
M29 (NGC 6913)	A cluster of about 50 stars near Gamma Cyg. Magnitude 6.6; diameter 7′; R.A. 20h 23.9m, Dec. +48° 26′.
M39 (NGC 7092)	About 50 stars in a loose grouping about 9° east and slightly north of Deneb. Magnitude 4.6; diameter 32′; R.A. 21h 32.2m, Dec. +44° 20′.
Galaxies	
Cygnus A	See separate entry.

Cygnus A (Cyg A, 3C 405)

The strongest extragalactic radio source and third strongest radio source in the sky (after the Sun and **Cassiopeiae A**). At one time believed to be the result of a collision of two galaxies, it has now been identified as a classic double-lobed **radio galaxy** that, at optical wavelengths, is a fifteenth magnitude cD galaxy (a supergiant elliptical) lying about 1 billion light-years away.

Cygnus Loop

A large nebula in **Cygnus**, some 80 light-years across, lying about 2,500 light-years away and 330 light-years above the galactic plane, that is the remains of a **supernova** explosion inside another supernova explosion. The most prominent part of it at visible wavelengths is the **Veil Nebula**. The initial supernova took place roughly 18,000 years ago and sent out powerful shock fronts that propagated into the interstellar medium, creating a giant cavity in the surrounding clouds. This was followed by the explosion of a second massive star, at least 5,000 years ago: its shock waves are now interacting with the original cavity walls. The result of this interaction is ragged, sweeping filaments of dense interstellar matter that have been energized to the point at which they radiate at visible wavelengths. These form together with smoother, more delicate filaments, at the outer edges of the expanding shock fronts, where the density of interstellar matter is much lower.

Cygnus X-1 (Cyg X-1 3U 1956+35)

A powerful, variable X-ray source, lying about 8,000 light-years away in **Cygnus**, that is one of the best known stellar **black hole** candidates. It corresponds with a binary system whose visible component is a ninth-magnitude 20-M_{sun} supergiant **O star** (spectral type O9.7Iab) cataloged as HDE 226868. The invisible companion has a mass of at least 6 M_{sun}, well above the limit at which a collapsed object must become a black hole. The two components have a very close orbit with a period of just 5.6 days. The telltale signs that a black hole is at work are the rapidly flickering X rays that come from superhot material, stolen from the bright supergiant, whizzing around an **accretion disk** centered on the dark star.

D

D galaxy

A luminous giant **elliptical galaxy** typically found near the center of a **cluster of galaxies**. D galaxies are similar in appearance to ordinary ellipticals, but their brightness declines more slowly with distance from the center, making them look as if they have diffuse halos (the "D" stands for "diffuse"). A *cD galaxy* is even bigger and brighter than a D galaxy and is found centrally located in a **rich cluster of galaxies**. cD galaxies are believed to have grown so large through **galaxy mergers**. Indeed, several cD galaxies are observed to have multiple galactic nuclei, as if they are still in the process of cannibalizing smaller systems.

D lines

Two close prominent lines in the yellow region of the **spectrum** of neutral sodium at wavelengths of 5890 Å and 5896 Å. They were labeled as feature D in the solar spectrum by Joseph von Fraunhofer and are conspicuous in the light of stars similar in temperature to the Sun (see **Fraunhofer lines**). They also show up in the spectra of remote stars due to absorption by sodium atoms in the **interstellar medium**.

Dactyl

The first satellite of an **asteroid** to be discovered. Dactyl was found in orbit around **Ida** by the **Galileo** probe when it flew by the asteroid in 1993. The little moon, measuring $1.6 \times 1.4 \times 1.2$ km, was about 100 km from its much larger companion when photographed. Interestingly, the spectra of Ida (an **S-class asteroid**) and Dactyl reveal that the compositions of the two objects, though similar, are not identical; Dactyl is not simply a bit of Ida that broke off. Instead, it is thought that the binary system may have formed during the collision and breakup that created the **Koronis family** of asteroids. What seems certain is that the moon is a *captured object*–something created completely separately from Ida that happened to wander near the asteroid and be caught by its gravitational field. According to the laws of celestial mechanics, such an event would deflect the smaller object, but it would not be captured into orbit unless a third force of some kind slowed it down. It is named after the Dactyli, a group of mythological beings who lived on Mount Ida. The Dactyli protected the infant Zeus after the nymph Ida hid and raised the god on the mountain.

d'Alembert, Jean le Rond (1717–1783)

A French mathematician (named after the church of St. Jean le Rond upon whose steps he was found as a baby) who, together with his compatriots Joseph **Lagrange**, Pierre **Laplace**, Leonhard **Euler**, and Alexis Clairault, applied calculus to **celestial mechanics** in order to tackle the **three-body problem**–the question of how three mutually gravitating objects move. In 1754, his efforts enabled him to mathematically explain Newton's discovery of the **precession** of the equinoxes, and also **perturbations** in the orbits of the planets. He was persuaded by his friend Denis Diderot to contribute scientific articles to the monumental and influential *Encyclopédie* (1751) but pulled out of the project when the first volume was heavily criticized by the Church. Shortly before his death, d'Alembert almost had a nonexistent moon of Venus named after him by its "discoverer," the German astronomer and mathematician Johann Lambert (1728–1777). However, he declined the offer when his own calculations led him to doubt the object was real.

Damocloid family

An unusual collection of **asteroids** with highly elliptical and, sometimes, highly inclined orbits that resemble those of **short-period comets**. Because of this similarity, it is thought that Damocloids may be the dark remains of old comets. The prototype of the group, (5335) Damocles, has a diameter of 15 to 20 km. It was discovered in 1991 by Robert McNaught (1956–) at the Anglo-Australian Observatory and named for a courtier of the ancient tyrant Dionysus I (the Elder) of Syracuse who lived in the fourth century B.C.; semimajor axis 11.88 AU, perihelion 1.58 AU, aphelion 22.18 AU, inclination 61.9°, period 40.6 years. Another suspected Damocloid, asteroid 2001 OG108, discovered in July 2001, has an orbit that takes it from just inside Earth's orbit out as far as Uranus. If its estimated diameter of 15 km is correct, this would make it one of the biggest Earth-crossing asteroids known.

dark adaptation

Heightened sensitivity to light when the eye is subjected to darkness for an extended period. Chemical changes (involving the buildup of rhodopsin) take place in the retina, mostly in the first 20 minutes in darkness but continuing for up to 2 hours, that greatly improve the observer's ability to see faint objects. However, they can be canceled quickly by a sudden exposure to light. Amateur astronomers carry red-filtered flashlights into the field for use for reading star charts, setting circles, and telescope controls since dim red light least affects the eye's dark adaptation. See also **averted vision**.

dark energy

A property of empty space, allowed by Einstein's **general theory of relativity**, that acts as large, negative pressure and pushes the universe apart at an increasingly fast rate. Its existence has been hypothesized to explain the fact that the expansion of the universe seems be accelerating. Observations of distant quasars that have been subject to a gravitational lens effect by intervening galaxies suggest that two-thirds of the cosmos may consist of dark energy. Various types of dark energy have been proposed, including a cosmic field associated with inflation; a different, low-energy field dubbed "quintessence"; and the **cosmological constant**, or vacuum energy of empty space. Unlike Einstein's famous fudge factor, the cosmological constant in its present incarnation doesn't delicately (and artificially) balance gravity in order to maintain a static universe; instead, it has "negative pressure" that causes expansion to accelerate.

dark halo

The massive outer region of the **Milky Way Galaxy** that surrounds the disk and stellar halo. The dark halo consists mostly of **dark matter**, whose form is unknown. Though it emits almost no light, the dark halo outweighs the rest of the Galaxy. Similar halos of nonluminous matter surround and envelope other large galaxies.

dark halo crater

A **crater**, possibly volcanic in origin, that is surrounded by an ejecta blanket darker than the adjacent landscape.

dark matter

Matter that cannot be detected directly, but whose existence can be inferred on the basis of dynamical studies. Within a **spiral galaxy**, the stars move as if large quantities of dark matter exist around the galaxy's disk; similarly, within **clusters of galaxies**, the individual galaxies move as if 10 times more matter is present than that visible in the form of stars and interstellar gas and dust. Cosmologists suspect that dark matter may account for most of the so-called **missing mass** needed to make the average cosmic density fit that predicted by the **inflationary model**. Although the nature of dark matter remains unknown, it could take two possible forms: *baryonic matter* or *nonbaryonic matter* (see **baryon**). The former is "ordinary" matter, the type that makes up the luminous portions of the universe, cast in the guise of objects that are difficult to detect, such as planets, **brown dwarfs**, and **black holes**. Nonbaryonic matter, on the other hand, would exist as exotic particles, predicted by certain grand unified theories. There are two possible subgroups of exotic dark matter: *hot dark matter* (HDM) and *cold dark matter* (CDM). HDM would be composed of particles such as **neutrinos**, described as hot because they travel at or very close to the speed of light. Neutrinos are a prime candidate, especially in view of recent evidence that they have mass. CDM would be composed of *weakly interacting massive particles* (WIMPs). These particles have relatively large masses, travel relatively slowly, and interact only weakly with normal baryonic material; hence, they are difficult to detect.

The currently favored view is that CDM makes up the bulk of dark matter in the universe, with perhaps a minor contribution of 10 to 20% from HDM. A key prediction of the CDM model, that large galaxies are accompanied by huge retinues of **dwarf galaxies** (gravitationally drawn in by the CDM halos), won observational support in 2002. A study of large galaxies that act as **gravitational lens**es revealed that about 2% of their mass must be in the form of dwarf galaxies–equivalent to thousands of these small systems–to explain the lensing effects.

dark nebula

A relatively dense (up to 10^4 particles/cm^3) cloud of interstellar matter whose **dust** particles obscure the light from stars beyond it and give the cloud the appearance of a starless region.

d'Arrest, Heinrich Ludwig (1822–1875)

A Danish astronomer who assisted Johann **Galle** with the first observations of **Neptune**. After receiving its predicted position from Urbain **Leverrier**, Galle and d'Arrest began searching. With Galle at the eyepiece and d'Arrest reading the chart, they scanned the sky and checked that each star seen was actually on the chart. Just a few minutes after their search began, d'Arrest cried out, "That star is not on the map!" and earned his place in history. D'Arrest also discovered 342 NGC objects, mainly with a 28-cm refractor.

Darwin

A European Space Agency mission, expected to be launched by about 2015, that will consist of a flotilla of eight spacecraft flying in formation and combining their observations to detect the light from Earth-like planets around other stars. By analyzing that light, astronomers will be able to deduce the chemical compositions of distant planets' atmospheres and search for the telltale chemicals related to life. Darwin will build on the work of the **Eddington** mission.

dark matter An X-ray image of the cluster of galaxies EMSS 1358+6245, about 4 billion light-years away in the constellation Draco, obtained by the Chandra X-ray Observatory. When combined with Chandra's X-ray spectrum, this image has allowed scientists to determine that the mass of dark matter in the cluster is about four times that of normal matter. *NASA/MIT/J. Arabadjis et al.*

Davida (minor planet 511)

The seventh largest known **asteroid**, discovered in 1903 by the American astronomer Raymond Smith Dugan (1878–1940). Diameter 323 km, spectral class C, rotation period 5.17 hours. Orbit: semimajor axis 3.171 AU, perihelion 2.591 AU, aphelion 3.750 AU, eccentricity 0.183, inclination 15.94°.

Davis, Ray(mond) Jr. (1914–)

An American physicist who was the first to detect solar **neutrinos**, the signature of nuclear fusion reactions occurring in the Sun's core. In research from 1967 to 1985 (for which he shared the 2002 Nobel prize in physics), using chlorine detectors located deep underground in a South Dakota gold mine, Davis found only one-third of the neutrinos that standard theories predicted. His results threw the field of astrophysics into an uproar, and, for nearly three decades, physicists tried to resolve the so-called solar neutrino puzzle. Experiments in the 1990s using different detectors around the world eventually confirmed the solar neutrino discrepancy. Davis's lower-than-expected neutrino detection rate is now accepted as evidence that neutrinos have the ability to change into one of three known neutrino forms. This property, called *neutrino oscillation,* implies that the neutrino has mass. Davis's detector was sensitive to only one form of the neutrino, so he observed less than the expected number of solar neutrinos. Davis retired from

Brookhaven in 1984, but remained a research collaborator in the university's chemistry department.

Dawes' limit

An empirical formula derived by the English amateur astronomer William Dawes (1799–1868) that gives the smallest **angular separation** of two stars in which each is still observable with a telescope of given **aperture**. The Dawes' limit in arcseconds is $11.6/D$, where D is the aperture of the telescope in cm.

dawn

The first light in the sky before sunrise, equivalent to morning astronomical **twilight**. The *dawn side* of a planet or moon is the hemisphere that lies nearest the morning **terminator**.

Dawn

A NASA probe designed to study the two largest asteroids, **Ceres** and **Vesta**, scheduled for launch in May 2006. Dawn will arrive at Vesta in 2010 and orbit it for nine months, then move on to Ceres in 2014, for a further nine-month orbital stint. The probe's instrument will measure the exact mass, shape, and spin of the two asteroids from orbits 100 to 800 km high, record their magnetization and composition, use gravity and magnetic data to determine the size of any metallic core, and use infrared and gamma-ray spectroscopy to search for water-bearing minerals. Flybys of more than a dozen other asteroids are planned along the way.

day

The time it takes Earth to spin once around on its axis relative to some external reference. The two main references are the Sun, which leads to the *solar day,* and the stars, which leads to the *sidereal day.*

The *apparent solar day* is the interval between two consecutive **upper transits** of the Sun, i.e., the period between one passage of the Sun across the observer's **meridian** and the next. The apparent solar day varies with the time of year because the Sun moves in the **ecliptic** instead of along the celestial equator, and also because the Sun moves along the ecliptic at a variable rate (due to the varying distance of Earth from Sun during the year). The *mean solar day* is the average of the apparent solar day over a whole year or–what amounts to the same thing–the length of day reckoned according to the **mean sun**.

The *equinoctial sidereal day* is the interval between two successive meridian transits of the vernal **equinox** (equal to 23h 56m 4.091s). Because of **precession**, the sidereal day is about 0.0084 second shorter than the *true sidereal day,* which is the period of Earth's rotation relative to a fixed direction, i.e., the interval between two successive upper transits of a star from a fixed point on Earth's surface.

Daylight comet of 1910

The brightest comet of the twentieth century. It was independently found by so many people in the Southern Hemisphere that no single original discoverer could be named, though the first astronomer to see it appears to have been Robert **Innes** on January 17, 1910, at the Cape Observatory in South Africa. Most observers judged the comet to be brighter than Venus, giving it a magnitude of about –5.

daylight stream

A **meteor** stream that is active and above the horizon at the same time as the Sun. Daylight streams can only be observed by radar and radio-echo techniques.

db galaxy

One of a small number of dumbbell-shaped **radio galaxies**. They are sometimes called D systems with double nuclei, in which two elliptical nuclei share a common extended envelope.

D-class asteroid

A very dark and nonreflective **asteroid**, reddish in color, probably due to the surface presence of organic materials. Rare in the main belt, D-class asteroids crop up with increasing regularity beyond about 3.3 AU from the Sun. They have **albedos** of 0.02 to 0.05 and appear to be made of some of the most primitive material in the solar system. Examples are found among many of Jupiter **Trojan**'s, including the largest, **Hektor**. Also the Martian moons **Phobos** and **Deimos** may well be D-class asteroids and the Tagish Lake meteorite has been confirmed as being a likely D-class fragment.

decaying orbit

One in which the orbiting object is slowly spiraling toward the primary. This is the case with **Phobos**, the larger Martian moon, which is doomed to collide with Mars within the next few tens of millions of years.

declination (Dec.; sym. δ)

The angular distance of an object north or south of the celestial equator (the projection of Earth's equator onto the **celestial sphere**); it is the equivalent of latitude on Earth. Together with **right ascension**, it can be used to locate any position in the sky. See **equatorial coordinates**.

Deep Impact

A NASA mission to collide with a **comet** nucleus and study the material thrown out by the impact. If launched as planned in December 2004, Deep Impact will encounter comet Tempel I in July 2005. The spacecraft consists of a flyby probe and a smart impactor that will separate from the main craft 24 hours before collision. The 500-kg cylindrical copper impactor will approach the sunlit side of the nucleus, sending back pictures as it closes in, at a relative velocity of 10 km/s. The collision will create a new crater, larger than a football field and deeper than a seven-story building. Two optical imaging systems on the flyby craft will record the impact events and the subsurface cometary structure, while two near-infrared imaging spectrometers will determine the composition of the cometary material.

deep sky object

A term used by amateur astronomers to indicate any object outside the solar system, with the possible exception of double stars and variable stars. More specifically, an object contained in the ***Messier Catalogue*** and/or the ***New General Catalogue (NGC)***.

deep space

Originally, any region of space beyond the Moon's orbit. Now more generally used to indicate any region beyond the orbit of Mars.

Deep Space 1

A experimental NASA probe, designed to test a new ion engine and a range of other advanced technologies. It was launched in October 1998 and flew past asteroid **Braille** in July 1999 and Comet **Borrelly** in September 2001. Deep Space 1 came within 2,171 km of Borrelly and snapped 30 superb black and white photos of the comet's nucleus.

degenerate matter

Matter at such high density that quantum effects dominate its behavior. In the present universe, two specific forms of degenerate matter are known to be important: **electron degenerate matter**, found in **white dwarfs**, and **neutron degenerate matter**, found in **neutron stars**.

degenerate star

A dead star made of **degenerate matter**.

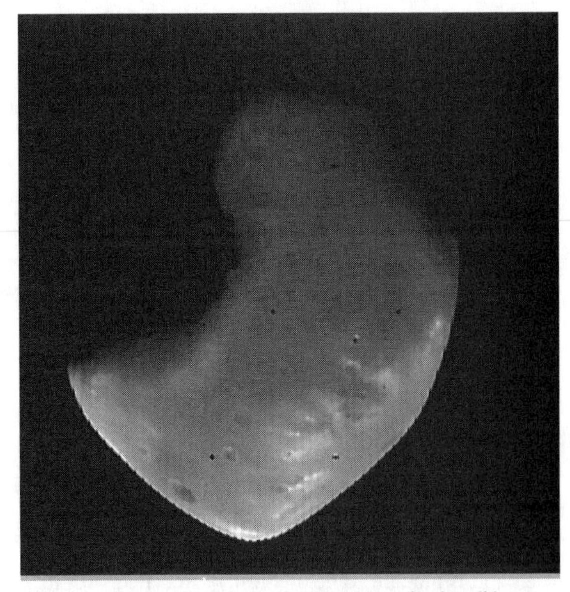

Deimos The smaller moon of Mars as seen by the Viking 2 Orbiter in 1977 from a distance of 1,400 km. *NASA/JPL*

Deimos

The outer and smaller of the two tiny satellites of **Mars**, discovered in 1897 by Asaph **Hall**. Its name is Greek for "terror." Like its sister moon **Phobos**, it is locked in synchronous rotation with the planet, irregular in shape, and almost certainly a captured asteroid.

Delaporte, Eugène Joseph (1882–1955)

A Belgian astronomer who, at the request of the newly formed **International Astronomical Union**, drew up the **constellation** boundaries as they are known today, publishing his results in 1930. Two years later, he discovered the prototype of the **Amor group** of asteroids.

Delphinus (abbr. Del; gen. Delphini)

The Dolphin; a small northern **constellation**, lying just north of the celestial equator between Pegasus and Aquila. Its four main stars, none brighter than third magnitude, form an asterism known as Job's Coffin. Gamma Del is a wide double of magnitudes 4.5 and 5.5 that can be

Delphinus: Stars Brighter than Magnitude 4.0

	Magnitude				Position	
Star	Visual	Absolute	Spectral Type	Distance (light-yr)	R.A. (h m s)	Dec. (° ′ ″)
β Rotanev	3.64	1.26	F5IV	97	20 37 33	+14 35 43
α Sualocin	3.77	–0.57	B9V	241	20 39 38	+15 54 43

resolved by binoculars. The region around Delphinus has been an unusually rich hunting ground for **novae**. (See table, "Delphinus: Stars Brighter than Magnitude 4.0," and star chart 7.)

Delta Cephei (Delta Cep, δ Cep)

The second **Cepheid variable** to be identified and the prototype for the category known as *Type I Cepheids,* or **Delta Cephei stars**. With a change in apparent magnitude of 3.5 to 4.4 over a period of 5.366 days, Delta Cep's entire range of variability can be followed with the naked eye. At a declination of +58°, Delta makes a perfect target for northern observers, especially those who see Cepheus as a circumpolar constellation. Also, it lies conveniently between two bright stars that shine close to each end of its range of variability: Zeta Cep (magnitude 3.6) and Epsilon Cep (magnitude 4.2). For those observing with binoculars or a modest telescope, Delta Cep's companion stars can be discerned. A seventh magnitude mate is located 41″ from the variable itself and is thought to be physically associated with it. A second nearby star of thirteenth magnitude, located 20.9″ from Delta, is most likely a line-of-sight object.

Visual magnitude:	4.07 (mean)
Absolute magnitude:	–3.26
Spectral type:	G2Ib
Distance:	950 light-years

Delta Cephei star (δ Cep star)

A Type I **Cepheid variable**, also known as a *classical Cepheid,* and named after the prototype, **Delta Cephei**. Such stars are massive (5 to 20 M_{sun}) **Population I** objects of high luminosity that undergo very regular pulsations, with periods of 1 to 135 days. Accompanying these pulsations are changes in brightness, with amplitudes of 0.1 to 2 magnitudes, and in **spectral type**, typically from F at maximum to G or K at minimum. Delta Cephei stars occupy a region of thze **Hertzsprung-Russell diagram** known as the **instability strip**; having started life as hydrogen-burning B stars they are now in a post-main-sequence phase marked by instability to **radial pulsation**. A star may pass through such a phase several times after leaving the main sequence: once during shell hydrogen burning, twice more during core helium burning, and twice again during shell helium burning.

Delta Scuti star (δ Sct star)

A type of **pulsating variable** that shows small, regular light variations, with an amplitude of 0.003 to 0.9 magnitude and a period of 0.25 to 5 hours. Delta Scuti stars are **Population I** objects of spectral type A0 to F5 and lie in the lower part of the **instability strip**, either on the main sequence or among the **subgiants** and **giants**. Although superficially like low-luminosity **Delta Cephei stars**, and sometimes still known by the old name *dwarf Cepheid,* they display both **radial pulsation** and **nonradial pulsation** modes simultaneously. They are closely related to both **SX Phoenicis stars** and **AI Velorum stars**.

Dembowska (minor planet 349)

An **asteroid** that is very red in color and of a unique type (R); it was discovered in 1892 by the French astronomer August Charlois. Dembowska is highly irregular in shape, spins around 5 times a day, and has a surface made up largely of the bright green mineral, **olivine**. Its composition seems to closely match that of the pyroxene-olivine **achondrite** meteorites. Diameter 160 km, spectral class R. Orbit: semimajor axis 2.925 AU, eccentricity 0.0856, inclination 8.24°, period 5.00 years.

Deneb (Alpha Cygni, α Cyg)

The brightest star in **Cygnus**, the nineteenth brightest star in the sky, and one of the most luminous known stars in the Milky Way. Its name, meaning "tail" in Arabic, comes from its position in the Swan. It also makes the western apex of the **Summer Triangle** and the tip of the Northern Cross. Deneb is an extremely luminous supergiant **A star**, although its exact luminosity and evolutionary state are undetermined because of uncertainty about its distance.

Visual magnitude:	1.25
Absolute magnitude:	–8.73
Spectral type:	A2Ia
Luminosity:	200,000 L_{sun}
Radius:	200 R_{sun}
Mass:	25 M_{sun}
Distance:	3,230 light-years

Deneb Kaitos (Beta Ceti, β Cet)

The brightest star in **Cetus**; it stands out in an otherwise barren area south of the Square of Pegasus and northeast of **Fomalhaut**. Its Arabic name means the "whale's tail," while an alternative name "Diphda" (or "Dipda"), also Arabic, refers to the "first frog" (and Formalhaut was the

Visual magnitude:	2.04
Absolute magnitude:	–0.30
Spectral type:	K0III
Surface temperature:	4,800 K
Luminosity:	145 L_{sun}
Radius:	17 R_{sun}
Mass:	3 M_{sun}
Distance:	96 light-years

"second frog"). A giant **K star** (almost type G), Deneb Kaitos is one of the brightest X-ray stars in the solar neighborhood, which is something of a puzzle. This high-energy radiation comes from a magnetically heated **corona**, similar to that of the Sun, and is expected to be related to rotation through a dynamo action; yet the star is spinning quite slowly. The unusual X-ray activity also suggests that the star has only recently left the main sequence; yet the chemical composition revealed by its spectrum argues that it is already fusing helium into carbon.

Denebola (Beta Leonis, β Leo)

The second brightest star in **Leo** and the easternmost of a prominent triangle of stars to the east of **Regulus**. Denebola marks the Lion's tail and is named for the Arabic phrase *al dhanab al asad* for that anatomical part. It is a main-sequence **A star**, a **Delta Scuti star**, and, most interestingly, like **Vega**, seems to be encircled by a disk of infrared-emitting dust. Viewed through a telescope, it also appears to have an orange companion, but the two stars are not physically connected.

Visual magnitude:	2.14
Absolute magnitude:	1.91
Spectral type:	A3V
Surface temperature:	8,500 K
Luminosity:	12 L_{sun}
Distance:	36 light-years

density

The ratio between the mass of an object and its volume. In the metric system, density is measured in grams per cubic centimeter (or kilograms per liter). (See table, "Densities of Familiar Substances and Objects.")

Densities of Familiar Substances and Objects

Object/Substance	Density (g/cm³)
water	1.0
iron	7.89
Sun	1.41
white dwarf	2×10^6
neutron star	7×10^{14}
Venus	5.2
Earth	5.52
Mars	3.94
Jupiter	1.31
Saturn	0.69
Uranus	1.29
Neptune	1.64
Pluto	2.03

density parameter (Ω)

The ratio of the actual mean density of matter in the universe to the **critical density**. A value of Ω greater than 1 implies that the universe will eventually collapse; a value less than 1 suggests eternal expansion. The actual value remains to be determined but is thought to be very close to 1.

density wave

A sound wave, or any other kind of material wave, that produces a series of alternate condensations and rarefactions of the material through which it passes. The *density wave theory,* first put forward by Bertil **Lindblad** in about 1925, is a possible explanation for the **spiral arms** of spiral galaxies. According to this idea, the arms represent regions of enhanced density (density waves) that rotate more slowly than the galaxy's stars and gases. As gas enters a density wave, it gets squeezed and makes new stars, some of which are short-lived blue stars that light the arms.

density wave theory The winding arms of the Whirlpool Galaxy, shown here, like those of all spiral galaxies, are believed to mark the position of density waves that trigger the collapse of gas and dust into a new generation of stars.
NASA/STScI/AURA

Subsequent work in the 1960s, by Chia Lin and Frank Shu explained the large-scale structure of spirals in terms of a small-amplitude wave propagating with fixed angular velocity. As the compression wave goes through, it triggers star formation on the leading edge of the spiral arms.

depth of field

In an imaging system, the distance in object space over which the system delivers an acceptably sharp image. Depth of field increases with increasing f-number.

descending node

The point in an orbit where the orbiting body crosses a reference plane, such as the **ecliptic** or the celestial equator, going from north to south.

Desdemona

The fifth moon of **Uranus**, also known as Uranus X; it was discovered in 1986 by **Voyager** 2. Desdemona has a diameter of 58 km and moves in a nearly circular orbit 62,660 km from (the center of) the planet.

Despina

The third moon of **Neptune**, also known as Neptune IV; it was discovered in 1989 by **Voyager** 2. Despina has a diameter of 148 km and orbits at a mean distance of 52,530 km from (the center of) the planet.

deuterium

A heavy isotope of **hydrogen** (^2H), with one proton and one neutron in its nucleus (compared with ordinary hydrogen's single proton). The abundance of deuterium in interstellar space is about 1.4×10^{-5} that of hydrogen. Because deuterium is difficult to manufacture and is quickly destroyed in stellar nuclear reactions, it is generally accepted that most of the deuterium found in the universe today was formed in the **Big Bang**.

diagonal

The small flat mirror used near the upper end of a **Newtonian telescope** to direct the converging beam of light over to the side of the tube where the eyepiece is located. See also **star diagonal**.

diamond ring effect

A dazzling burst of light that happens a few seconds before and after totality during a **solar eclipse**. The effect is caused by the last bit of sunlight shining through valleys on the edge of the Moon. See **Bailey's beads**.

Dicke, Robert Henry (1916–1997)

An American physicist who established the importance of the measurements of Arno Penzias and Robert Wilson in terms of the **cosmic microwave background**. In 1964, unaware that he was repeating a line of thought pursued earlier (in 1948) by George **Gamow**, Ralph **Alpher**, and Robert C. Herman, Dicke began to think about the consequences of a **Big Bang** origin of the universe. He deduced that the glow of the primordial fireball in which the universe was born ought today to be still visible as feeble **blackbody** radiation coming from all parts of the sky. At Dicke's instigation, his colleague P. James Peebles calculated that this remnant radiation should now have a temperature of about 10 K, later corrected to about 3 K. At this temperature a blackbody should radiate a weak signal at microwave wavelengths from 0.05 mm to 50 cm with a peak at about 2 mm; further, the signal should be constant throughout the universe. Dicke began to organize a search for such radiation and had begun to install an antenna on his laboratory roof when he heard from Penzias and Wilson that they had detected background microwave radiation at a wavelength of 7 cm. It was this confluence of theory, calculation, and observation that helped establish the Big Bang theory. Dicke graduated in 1939 from Princeton University and obtained a Ph.D. in 1941 from the University of Rochester. He spent the war at the radiation laboratory of the Massachusetts Institute of Technology, joining the Princeton faculty in 1946. In 1957 he was appointed professor of physics and served from 1975 to 1984 as Albert Einstein Professor of Science. In 1984 he was appointed Albert Einstein Emeritus Professor of Science.

differential rotation

(1) The variable rotation rate of a gassy body, such as a star or gas giant planet with latitude; gas near the equator spins around faster than gas near the poles. (2) The variable rotation in a disk-shaped structure, such as a galaxy, according to distance from its center. Stars near the center take less time to complete one rotation than those farther away.

differentiation

A process that takes place in a molten or partially molten planetary body by which heavier materials sink to the center and lighter ones rise to the surface. In Earth's case, differentiation resulted in an iron-nickel core and a crust and mantle rich in silicate rocks.

diffraction

A wavelike property of light that allows it to bend around obstacles whose size is about that of the light's wavelength. The disturbed waves then interfere to produce ripple patterns. Diffraction effects are classified into either Fresnel or Fraunhofer types. *Fresnel diffraction* is concerned mainly with what happens to light in the immediate neighborhood of a diffracting object or aperture, so is only of interest when the illumination source is close by. *Fraunhofer diffraction* is the light-spreading effect of an aperture when the aperture (or object) is lit by plane waves, i.e., waves that effectively come from a source that is infinitely far away.

diffraction grating

An optical device, ruled with thousands of fine parallel grooves, which produces interference patterns in a way that separates all the components of the light into a spectrum. A diffraction grating can be used as the main dispersing element in a **spectrograph**. The **diffraction pattern** produced by the grating is described by the equation

$$m\lambda = d \sin \theta$$

where m is the order number, λ is a selected wavelength, d is the spacing of the grooves, and θ is the angle of incidence of light. A *transmission grating* has grooves ruled onto a transparent material, such as glass or Perspex, so that a beam of light passed through the grating is partly split into sets, or *orders,* with spectra on either side; the blue light is diffracted the least and the red light the most in each order. The orders of spectra increase in dispersion and faintness with distance from the direct beam. A *reflection grating* has grooves ruled onto a reflective coating on a surface that may be plane or concave, the latter being able to focus light. Its advantage over a transmission grating is that it produces a spectrum extending from ultraviolet to infrared, since the light doesn't pass through the grating material.

diffraction pattern

In general, the interference pattern that results when diffracted waves are superposed. In astronomy, the image of a star formed by an optically perfect telescope. In a refractor, this consists of a small bright disk surrounded by concentric rings of light. In a reflector, the disk will have (typically) four spikes caused by diffraction from the secondary mirror supports.

diffraction-limited

Said of an optical system that meets the **Rayleigh criterion**, or, put another way, that can bring light to a focus at the theoretical minimum of the **Airy disk**. To be diffraction-limited there must be no more than ¼ wavelength of light between wave fronts at the focal point. Modern optical systems can exceed this theoretical design limit, but it remains a standard.

diffuse galactic light

Scattered, integrated starlight; a small contribution to the background glow by starlight reflected and scattered by interstellar **dust** near the galactic plane. Its measurement requires exceptionally dark sky conditions.

diffuse interstellar bands (DIBs)

A series of diffuse **absorption** bands, of interstellar origin, first recorded on photographic plates in the early 1900s. Well over 100 such bands have been identified in the ultraviolet, visible, and near infrared regions of the spectrum. According to some estimates, up to 10% of cosmic carbon may be in the molecules that cause these features. Identifying the carriers of DIBs has become one of the classic astrophysical spectroscopic problems. Recent work suggests they are caused by **polycyclic aromatic hydrocarbons** (PAHs), or, most likely, their cations, since PAH ions of all sizes absorb light in the visible and near infrared, and such molecules are expected to be ionized by the intense ultraviolet field present in much of the **interstellar medium**.

diffuse nebula

An irregularly shaped cloud of interstellar gas or dust whose spectrum may contain **emission lines** (emission nebula) or **absorption lines** characteristic of the spectrum of nearby illuminating stars (reflection nebula).

Digges, Leonard (c. 1520–c. 1559), and Thomas (1546–1595)

English father and son who pioneered the construction of the telescope (Leonard) and its use (Thomas). Leonard Digges was educated at Oxford and made his name as a mathematician, a surveyor, and an author of several books. He invented a **reflecting telescope** a century before Isaac **Newton**, and may also have built a **refracting telescope**. However, he had little chance to use them. In 1554 he was sentenced to death for his part in a rebellion, and although this was commuted to seizure of his estates, he spent the rest of his life trying to regain his property. Thomas was only 13 when his father died but had John Dee, a mathematician, as his guardian. In 1571, Thomas published a mathematical work of his own and a posthumous book, *Pantometria,* by his father in which Leonard's invention of the telescope is discussed. Thomas's observations of the supernova of 1572 were used by Tycho **Brahe** in his analysis of this event.

Digital Sky Survey (DSS)

An extremely large dataset consisting of the scanned images from the ***Palomar Observatory Sky Survey*** and several other sky surveys, compressed to fit on 100 CD-ROMs. *RealSky* is a version of the *DSS* that has been compressed still further, allowing it to be stored on fewer CDs at the cost of some image quality.

diogenite

A class of **achondrite** composed mostly of magnesium-rich ortho**pyroxene**, with minor amounts of **olivine** and **plagioclase feldspar**. The pyroxene is usually coarse-grained, suggesting an origin for the diogenites in magma chambers within the deeper regions of **Vesta**'s crust. They are intrusive igneous rocks similar to plutonic rocks found on Earth, which cooled slowly and allowed their pyroxene to form sizable crystals. This is especially true for the Tatahouine meteorite, a unique diogenite that fell in Tunisia in 1931 and is renowned for its green, centimeter-sized pyroxene crystals. Diogenites belong to the **HED group**, the same family of meteorites as **howardite**s and **eucrites**. They are named after the Greek philosopher Diogenes of Apollonia, of the fifth century B.C., who was the first to suggest that meteorites come from space (a realization that was subsequently forgotten for the next 2,000 years).

Dione

The fourth largest and twelfth nearest moon of **Saturn** with a diameter of 1,120 km, also known as Saturn IV; it was discovered by Giovanni **Cassini** in 1684. Dione is the densest of Saturn's moons (1.43 g/cm^3), except for **Titan**, suggesting that it contains a rocky core, making up about one-third of its mass, surrounded by water-ice. In composition, albedo, and terrain, it is similar to **Rhea**; both moons have strikingly different leading and trailing hemispheres. On Dione's trailing hemisphere, which is lightly cratered, bright wispy features overlie a darker background. Most of these features radiate from the large crater Amata but don't look like crater rays. According to one idea, they may be surface deposits of ice erupted from a network of fractures. The leading hemisphere is much more heavily cratered. This is the reverse of the normal situation and suggests that Dione may have been spun around by a colliding asteroid. The small moon **Helene** is located at Dione's leading **Lagrangian point.**

direct motion

Rotation or orbital motion in a counterclockwise direction when viewed looking down from above the Sun's north pole, or the movement of a body from west to east on the celestial sphere. Also known as *prograde,* it is the normal direction of orbital and axial movement in the solar system; its opposite is **retrograde**.

Dione Saturn's moon Dione is seen to be well peppered with impact craters in this mosaic of images taken by Voyager 1 from a range of 162,000 km on November 12, 1980. The largest crater is less than 100 km in diameter and shows a well-developed central peak. Bright rays represent material ejected from other impact craters. Sinuous valleys probably formed by faults breaking the moon's icy crust. *NASA/JPL*

discordant redshift

A discrepancy between the **redshifts** of galaxies that lie close together in the sky and that appear to be interacting or to be connected by luminous bridges. The nature of the discordant redshift of members of compact groups of galaxies, such as **Stefan's Quintet** and **Seyfert's Sextet**, has been a subject of debate for many years. If the frequency of discordant galaxies is greater than the statistics of chance projection allow, or if apparent connections between discordant galaxies are genuine bridges of matter, as has been argued by Halton **Arp**, the **Burbidges**, and others, it challenges the almost universally held belief that galactic redshifts are due to cosmic expansion and might signify the need for new physical theories. However, although the situation has not been completely settled, evidence now strongly suggests that discordant galaxies do not occur more often than can be explained by line-of-sight coincidences; also, there is independent physical evidence that discordant galaxies all have physical properties consistent with a cosmological distance; for example, those with higher redshift tend to be smaller and fainter than other members of the group.

discordant redshift The spiral galaxy NGC 4319 (center) and the quasar Markarian 205 (upper right) appear to be neighbors in this image taken by the Hubble Space Telescope. In reality, their very different redshifts reveal them to be widely separated in space and time. NGC 4319 is 80 million light-years from Earth; Markarian 205 is more than 14 times farther away, at a distance of 1 billion light-years. *NASA/STScI/AURA*

discrete radio source

An object that can be observationally separated at radio wavelengths from its local background; it may range from a planet to the **jet** of an active galaxy.

disk

(1) A flattened, circular region of gas, dust, and/or stars. It may refer to material surrounding a newly formed star, material accreting onto a black hole or neutron star, or the large region of a spiral galaxy containing the spiral arms. (2) The apparent circular shape of the Sun, a planet, or the Moon when seen with the naked eye or through a telescope.

disk galaxy

A galaxy with a prominent wide, thin disk of stars. This description applies to **spiral galaxies** (including barred spirals) and **lenticular galaxies**, but not to elliptical galaxies, dwarf spheroidals, and some peculiar galaxies.

disk star

A star that orbits within the **galactic disk** of the Milky Way or that of some other **disk galaxy**. See also **population**.

dispersion

The separation of a beam of light into the individual wavelengths of which it is composed by means of **refraction** or **diffraction**.

distance modulus

The difference between the **apparent magnitude** (m) of a star or galaxy and its **absolute magnitude** (M); it is given by $m - M = 5 \log d - 5$, where d is the distance in parsecs.

distortion

A type of optical **aberration** that happens when there is a variation in magnification across the visual field. *Barrel distortion* results when a lens magnifies slightly more at its thickest part (along the optical axis) than it does at its edges; the image of a square shape thus appears to have bulged-out sides. Barrel distortion is associated with wide-angle lenses and only occurs at the wide end of a zoom lens. *Pincushion distortion* comes about when magnification increases away from the optical axis causing image lines, both vertical and horizontal, to bend in toward the center of the image. Neither kind of distortion affects **resolution**; it simply means that the image shape doesn't correspond exactly to the shape of the object.

diurnal

Daily. *Diurnal motion* is the apparent daily motion of celestial bodies across the sky from east to west, including the rising and setting of the Sun, caused by Earth's rotation. A *diurnal circle* is the apparent curving track of a celestial object relative to the horizon. For *diurnal aberration* see **aberration of starlight**, for *diurnal libration* see **libration**, and for *diurnal parallax* see **parallax**.

Dobsonian telescope

A **Newtonian telescope** with a low **altazimuth mounting**, typically using Teflon pads as bearings on a flat board, in which the telescope is supported by V-shaped cuts in the side of a wooden box. Named for the Californian "father of sidewalk astronomy" John Dobson (1915–), who advocated simplicity in equipment, the basic Dobsonian was designed so that amateur astronomers could easily create their own large-aperture reflecting telescope mounts in home workshops.

Dollfus, Audouin Charles (1924–)

A French astronomer who, in 1966, discovered what was, at the time, the innermost known moon of Saturn, **Janus**. Before the first planetary probes, Dollfus sought evidence for tenuous atmospheres around Mercury and the Moon, and made the first ascent in a stratospheric balloon in France in order to carry out detailed studies of Mars. Before the **Viking** landings, he correctly concluded, from **polarization** experiments with hundreds of different terrestrial minerals and polarization observations of the Red Planet, that the Martian deserts are strewn with iron oxide (Fe_2O_3). Since 1946, he has been at the Meudon Observatory outside Paris.

Dollond, John (1706–1761)

An English optician who introduced **achromatic lens**es for telescopes and microscopes. A London-born son of immigrant Huguenots from France, Dollond took up the weaving profession of his father, but at the age of 46 joined his son Peter in the optical business. In 1752 he became involved in a controversy over the question of whether Isaac **Newton** had been correct in asserting that it was not possible to combine two lenses of different materials to obtain an achromatic lens (a lens that focuses all colors of light equally). Even though Dollond was supposedly an uneducated optical worker, he stood up to Leonhard **Euler**, one of Europe's most learned mathematicians. And when he did experiments with prisms of different materials he proved Newton wrong by demonstrating that a prism combination could be found that gave **refraction** without **dispersion**. This opened the door to making greatly improved telescopes. By combining two lenses with the proper curvatures, one of **crown glass** and the other of **flint glass**, he and his son could make telescope objectives that didn't have the usual blurring that results from a variation of the focal length with color. There was a controversial claim that, in 1729, an English barrister named Chester Moor Hall had designed the first achromatic lens and that Dollond may have known about it. However, Hall never submitted a claim for his invention nor did he attempt to publicize it, and there is no doubt that Dollond was the first to make successful achromatic telescopes in quantity and offer them commercially. The firm of Dollond & Son became the preeminent English telescope maker and telescopes made by them went to all parts of the world. After John died in 1761 the business flourished through four more generations of Dollonds. In 1871 the business went out of the family but continues to this day as Dollond & Aitchison.

Dominian Astrophysical Observatory (DAO)

An optical observatory, founded in 1917, operated by the Herzberg Institute of Astrophysics, and located in Victoria, British Columbia, Canada. Its main instrument is the 72-inch (1.85-m) Plaskett reflector, named after the observatory's first director, John Plaskett, which was equipped with a new mirror in 1974. The DAO also houses the 1.22-m McKellar reflector, inaugurated in 1962, and the Canadian Astronomy Data Centre.

Donati, Giovanni Battista (1826–1873)

An Italian astronomer who carried out early spectroscopic studies of the Sun and stars and was the first to obtain and analyze the spectrum of a **comet**, concluding that comets are, at least in part, gaseous. Between 1854 and 1864 he discovered six new comets, the brightest of which, found in 1858, became known as **Donati's comet**. His spectroscopic observations of Comet Tempel (1864 II)–the first ever of a comet–showed three prominent lines which Donati named alpha, beta, and gamma. These same lines were seen in an 1866 comet by Angelo **Secchi**, and shown by William **Huggins** in 1868 to be due to the presence of carbon. After graduating from the university in his native city of Pisa, Donati joined the staff of the Florence Observatory in 1852 and was appointed director in 1864. He died from bubonic plague.

Donati's comet (C/1858 L1)

One of the brightest and most visible **comets** of the nineteenth century; it was discovered by Giovanni **Donati** on June 2, 1858, brightened to magnitude –1 in September 1858, reached perihelion on September 30, came closest to Earth (0.5 AU) on October 9, and was last seen on March 4, 1859. Around the time of closest approach to Earth, the comet developed a prominent dust tail, up to

60° long and curved like a scimitar, for which it is best remembered. Perihelion 0.58 AU, eccentricity 0.996, inclination 117.0°, period about 2,000 years.

Doppler boosting

An effect, predicted by the special theory of relativity, that enhances the radiation from material that is moving toward the observer at nearly the speed of light, and hides material moving in the opposite direction at such speeds. For example, if there are many **quasars** with radio **jets** moving at relativistic speeds, the strongest radio sources will be those that are most nearly pointed in our direction. This is important in understanding **blazars** and **superluminal radio sources**.

Doppler broadening

The broadening of **spectral lines** caused by the thermal, turbulent, or mass motions of atoms along the line of sight.

Doppler shift

The apparent change in wavelength of sound or light caused by the motion of the source, the observer, or both. Waves emitted by a moving object as received by an observer will be **blueshifted** (compressed) if approaching and **redshifted** (stretched) if receding. The frequency of change depends on the speed of the object moving toward or away from the receiver. Named after Christian Andreas Doppler (1803–1853).

Dorado (abbr. Dor; gen. Doradus)

The Goldfish, or Swordfish; a southern **constellation** that is distinguished only by containing most of the **Large Magellanic Cloud** and having, in Beta Dor, one of the brightest **Cepheid variables** (magnitude range 3.5 to 4.1, period 9.83 days). (See table, "Dorado: Stars Brighter than Magnitude 4.0," and star chart 8.)

Dorado: Stars Brighter than Magnitude 4.0

Star	Magnitude: Visual	Magnitude: Absolute	Spectral Type	Distance (light-yr)	Position: R.A. (h m s)	Position: Dec. (° ′ ″)
α	3.30	−0.36	A0IIISi	176	04 34 00	−55 02 42
β	3.5 max	−3.76	F6Ia	1,040	05 33 38	−62 29 24

Double Cluster (h and Chi Persei; NGC 869 and NGC 884)

Two young **open clusters** that lie close together in the sky and are visible to the naked eye on a clear, dark night as a hazy patch of light. Both clusters lie in the Perseus OB1 association and contain many **B stars** and M **supergiants**.

Double Cluster Two splendid open clusters lying just a few hundred light-years apart. *A. Steere*

Visual magnitude:	4.3 (h), 4.4 (Chi)
Age (millions of years, est.):	5.6 (h), 3.2 (Chi)
Distance (light-years):	7,100 (h), 7,400 (Chi)

double planet

Two objects of comparable size and of planetary mass orbiting each other. There is controversy over where to draw the line between a double planet and a system consisting of a planet and a moon. In most cases, satellites are of much smaller mass than their primary. However, there are two known examples of moon/planet mass ratios much greater than average: those of Earth and the Moon, and of Pluto and **Charon**. A commonly accepted criterion of a double planet is when the center of gravity of the two objects is not located inside either body. By this rule, Pluto and Charon count as a double planet while the Earth-Moon system does not. The question of whether Pluto is a planet at all or should instead be regarded as a large, wayward **Kuiper Belt object** is a separate matter.

Double Quasar (Q0957+561A/B)

The first confirmed case of **gravitational lensing**, an effect long predicted by the **general theory of relativity**.

It consists of two images, 5.7″ apart, of the same **quasar**, and results from the gravitational effect of an intervening **cluster of galaxies** that lies about 3.5 billion light-years away. The Double Quasar was discovered in 1979 and lies in **Ursa Major** at R.A. 10h 1.3m, Dec. +55° 54′.

double radio source

A **radio galaxy** in which the bulk of radio emission comes from two sources on opposite sides of the visual galaxy. The radiation is due to the violent expulsion of high-speed energetic particles, in two opposite directions, from the nucleus of the parent galaxy. About one-third of all known radio galaxies are double sources.

double star

Two stars that lie close together on the **celestial sphere**. If they are physically linked by gravity so that they orbit around each another, they are said to be a *physical double* or a **binary star**. If their apparent proximity is due only to a chance alignment, they are referred to as an *optical double*. Some writers use "double star" to mean only the latter.

double-mode variable

A **pulsating variable** that oscillates in the **fundamental mode** plus one overtone of this mode. First overtone pulsations are seen, for example, in **beat Cepheids**. The **Blazhko effect**, shown by some **RR Lyrae stars**, may be explainable in terms of a third overtone.

doublet

A compound lens made from two elements. The elements may be cemented together (forming a *cemented doublet*) or separated by an air space (forming a *separated doublet*). The most common doublet is the **achromatic lens**, consisting of a **crown glass** and a **flint glass** element chosen so that the indices of refraction of two separated wavelengths (one red, one blue) are equal.

DQ Herculis star (DQ Her star)

Also known as an *intermediate polar*, a **magnetic cataclysmic variable** in which the **white dwarf** component generates a magnetic field on the order of 10^6 to 10^7 gauss–extremely powerful but weaker than the field associated with an **AM Herculis star**. The type star, known by its variable star name of DQ Herculis, was seen in outburst as **Nova Herculis 1934**. In a DQ Her system, an **accretion disk** forms from material stripped from the **red dwarf** secondary but is disrupted close to the white dwarf primary by this star's magnetic field. As the material in the accretion disk approaches the white dwarf, it is swept up by the magnetic field of the white dwarf to form *accretion curtains* that follow the magnetic lines of force. The **magnetosphere** of the white dwarf is tilted with respect to the orbital plane, causing the accretion curtain to split, and the material to follow the shortest route along the magnetosphere. It then crashes onto the magnetic poles of the white dwarf, causing intense emission. Although the white dwarf's magnetosphere is strong, it is not strong enough to synchronize the orbits of the rotating white dwarf with the orbital period of the system as in AM Her stars.

Draco (abbr. Dra; gen. Draconis)

The Dragon; a large **constellation** that coils around the north celestial pole, appearing to encircle Ursa Minor. It

Draco: Stars Brighter than Magnitude 4.0

	Magnitude				Position	
Star	Visual	Absolute	Spectral Type	Distance (light-yr)	R.A. (h m s)	Dec. (° ′ ″)
γ **Eltanin**	2.24	–1.04	K5III	148	17 56 36	+51 29 20
η Aldhibain	2.73	–6.27	G8IIIab	2,050	16 23 59	+61 30 50
β Rastaban	2.79	–2.44	G2Ib-IIa	361	17 30 26	+52 18 05
δ Taïs	3.07	0.63	G9III	100	19 12 33	+67 39 41
ζ Aldhibah	3.17	–1.90	B6III	320	17 08 47	+65 42 53
ι Edasich	3.29	–1.92	B6III	339	15 24 56	+58 57 58
χ	3.55	4.01	F7V	26	18 21 03	+72 43 58
α **Thuban**	3.67	–1.21	A0III	308	14 04 23	+64 22 33
ξ Juza	3.73	1.06	K2III	111	17 53 32	+56 52 21
ε Tyl	3.84	0.59	G8IIIbCN	146	19 48 10	+70 16 04
λ Giausar	3.82	–1.24	M0IIICa	334	11 31 24	+69 19 52
κ	3.85	–2.07	B6IIIpe	498	12 33 29	+69 47 17

is one of the few star patterns to somewhat resemble the object after which it is named. The four stars forming the dragon's head–Beta, Gamma, Xi, and Nu Dra–make up a conspicuous asterism known as the Lozenge. Among Draco's interesting deep sky objects are the **Cat's Eye Nebula** (NGC 6543), the **Draco Dwarf**, and 3C 351, a quasar lying at a distance of about 7 billion light-years (magnitude 15.3; R.A. 17h, Dec. +60°). (See table, "Draco: Stars Brighter than Magnitude 4.0," and star chart 9.)

Draco Dwarf

A **dwarf elliptical galaxy** that orbits the Milky Way at a distance of about 270,000 light-years from the galactic center. The Draco Dwarf is the least luminous galaxy known, with an absolute magnitude of –8.6 and a diameter of about 3,000 light-years. It was discovered in 1954 at the same time as another Milky Way satellite, the **Ursa Minor Dwarf**. See also **Local Group**.

DRAGN (double radio source associated with a galactic nucleus)

An acronym coined by Patrick Leahy in 1993 (in *Jets in Extragalactic Radio Sources*, eds. H.-J. Röser, & K. Meisenheimer; Berlin: Springer-Verlag) that has entered the astronomical vernacular.

Draper, Henry (1837–1882)

An American pioneer of astronomical **spectroscopy** who established the observing techniques and program for the work that would bear his name when published, seven years after his death. The son of John William **Draper**, he trained to be a medical doctor but, because he completed all of his medical courses at New York University by the age of 20, he traveled in Europe for a year until he was old enough to graduate. In Ireland he visited, and was greatly influenced by, the Third Earl of **Ross**, William Parsons. Subsequently, he wove his interests in telescope-making and photography, developed during his travels, into his professional career. When he returned from Europe, Draper began preparing his own glass mirror, which he installed in his new observatory on his father's estate at Hastings on Hudson, New York. He started his astronomical research career by making preliminary studies of the spectra of the more common elements and photographing the solar spectrum. By 1873 he had produced a **spectrograph** that was similar to the visual spectroscope of William **Huggins**; he clarified the spectral lines by using a slit and incorporating a reference spectra so that elements could be identified more easily. The spectroscopic studies of Huggins and Norman **Lockyer** in Europe stimulated Draper's research, and during the last years of his life he worked toward acquiring high-quality spectra of celestial objects. After his untimely death, from pleurisy, his widow established a fund to further support a spectral studies program. In 1886 a team at **Harvard College Observatory** began the program to establish a useful classification scheme for stars and a catalog of spectra. The Harvard project, named the ***Henry Draper Catalogue***, was completed in 1897 and resulted in the first comprehensive classification of stars according to their spectra.

Draper, John William (1811–1882)

A distinguished English physician and chemist, and father of Henry **Draper**. Educated at University College, London, he emigrated to the United States in 1833 and studied medicine at the University of Pennsylvania in 1836. After a short period of teaching in Virginia he moved to New York University (1838) where he taught chemistry and in 1841 helped to start the medical school of which he became president in 1850. Most of his chemical work was done in the field of photochemistry. He was one of the first scientists to use Louis Daguerre's new invention of photography and took the first photograph of the Moon in 1840. In the same year he took a photograph of his sister, Dorothy, which is the oldest surviving photographic study of the human face. In 1843 he obtained the first photographic plate of the solar spectrum. On the theoretical level, Draper was one of the earliest to grasp that only those rays that are absorbed produce chemical change and that not all rays are equally powerful in their effect. He also, in a series of papers, showed that the amount of chemical change is proportional to the intensity of the absorbed radiation multiplied by the time it has to act. Draper's work was continued and largely confirmed by the work of Robert Bunsen and Henry Roscoe in 1857.

Dreyer, John Louis Emil (1852–1926)

A Danish astronomer (also known as Johan Ludvig Emil) who compiled the ***New General Catalogue (NGC)*** and its supplements, the ***Index Catalogue (IC)***. In 1874, he was appointed as an assistant of the Third Earl of **Rosse** at Birr Castle, and worked with what was at that time the world's largest telescope. He began observing and surveying deep sky objects (star clusters, nebulae, and galaxies), and in 1878 published a supplement to John **Herschel**'s *General Catalogue*, containing about 1,000 new "nebulae." Also in 1878, he went to Dublin and worked as assistant at Dunsink Observatory from 1878 to 1882 before he became director of Armagh Observatory (1882–1916) in Northern Ireland. Armagh Observatory was badly funded in those days and mainly had old, smaller instruments. So Dreyer concentrated on compiling new catalogs from older observations, including the NGC and its supplements. Among his other works was a biography of his childhood hero, Tycho **Brahe**.

Dschubba (Delta Scorpii, δ Sco)

The fifth brightest star in **Scorpius** and the middle star of three that make the head of the Scorpion. Its name stems from the Arabic *al-jabbah*, meaning "the forehead" of the Scorpion, which was originally applied to the whole line–Graffias, Dschubba, and the fainter Pi Sco. Dschubba is a multiple star system. The main component, a subgiant **B star**, recently brightened, in July 2000, from magnitude 2.3 to 1.9, and developed the spectrum of a **Be star**. Since then it has brightened further, to near magnitude 1.6. A rapid rotator, it has evidently flung off a hot mass of gas from its equator and may be going through a phase like that which **Gamma Cassiopeiae** began in 1937. It is accompanied by a cooler B-type companion in a 20-day orbit. A third component lies farther away, at least at Saturn's distance from the Sun, and takes more than 10 years to make the round trip. Twice as far away again lies the fourth and faintest member of the quartet. Dschubba is part of the **Scorpius-Centaurus Association**.

Visual magnitude:	2.29
Absolute magnitude:	–3.16
Spectral type:	B0.3IV
Luminosity:	14,000 L_{sun}
Radius (mainstar):	5 R_{sun}
Distance:	402 light-years

Dubhe (Alpha Ursae Majoris, α UMa)

The second brightest star in **Ursa Major**; it lies at the front of the Big Dipper's bowl and with **Merak** (Beta UMa) makes the famed Pointers. Its name comes from the Arabic phrase *thur al Dubbal Akbar*, "the back of the Great Bear." Unlike the middle five stars of the Big Dipper, Dubhe is not part of the **Ursa Major Moving Cluster,** and, as a giant **K star** with a temperature of 4,500 K, it is much cooler than its A-type constellation neighbors, more evolved (in the core helium-burning stage), and noticeably orange in color. Dubhe is orbited every 44 years at a distance of about 23 AU by a warmer, much dimmer, and less massive F star. Over 400 times farther away is another F star that also has a companion (with a 6-day period), making Dubhe a quadruple system.

Visual magnitude:	1.81
Absolute magnitude:	–1.09
Spectral type:	K0III + F0V
Luminosity:	300 L_{sun}
Radius:	30 R_{sun}
Distance:	124 light-years

Dumbbell Nebula (M27, NGC 6853)

One of the brightest and best known of **planetary nebulae**, named after its double-lobed appearance. It lies in **Vulpecula**, was the first planetary to be discovered by Charles **Messier** in 1764, and is the easiest object of its type to see with a small telescope. The bright portion of the nebula is expanding at a rate of 6.8″ per century, leading to an estimated age of 3,000 to 4,000 years. It is seen roughly along its equatorial plane; from near one pole, it would probably have the shape of a ring and perhaps appear similar to the **Ring Nebula**. The central star of M27 is quite bright at magnitude 13.5 and may have a faint (magnitude 17) yellow companion 6.5″ away.

Visual magnitude:	7.4
Angular diameter:	6′ (bright portion), 15′ (including faint halo)
Distance:	1,200 light-years
Position:	R.A. 19h 59.6m, Dec. +22° 43′

Dürchmusterung (DM)

See ***Bonner Dürchmusterung***, ***Córdoba Dürchmusterung***, and ***Cape Photographic Dürchmusterung***.

dust

Microscopic grains of matter found in space, the composition, size, and other properties of which vary from one location to another. Dust grains in dense **interstellar clouds**, for example, are larger than those in the general **interstellar medium**, while even larger particles are found in circumstellar disks. Most of the dust in space comes from **red giants** and red **supergiants** whose extended atmospheres are rich in silicon, oxygen, and carbon–elements that were manufactured in the stellar core but have been dredged to the surface by **convection** currents. Depending on its life history, a red giant may have surface layers that are rich in either carbon or oxygen. A **carbon star** gives rise to a dense pall of carbon particles in the form of graphite flakes or amorphous lumps, each about 0.01 micron across. In the case of an oxygen-rich star, the oxygen atoms react with silicon and

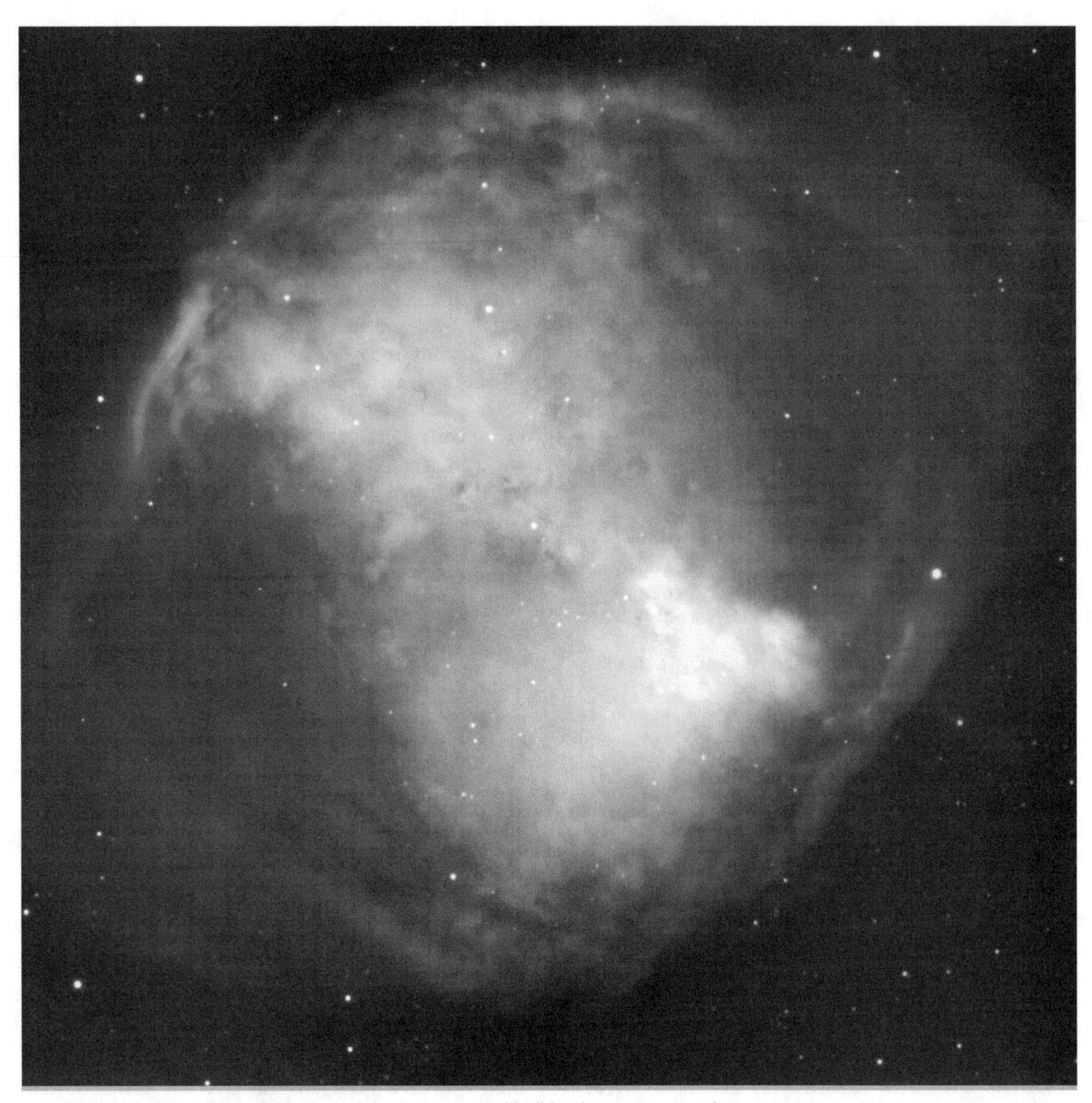

Dumbbell Nebula A spectacular view of the famous Dumbbell by the Very Large Telescope. *European Southern Observatory*

any metal atoms in the star's atmosphere to form silicate grains roughly 1 micron across. As the grains are blown away from the star by radiation pressure or by mechanical means and their temperature falls, they begin acquiring additional atoms of hydrogen, oxygen, nitrogen, and sulfur that have also escaped from the parent star. These accreted materials build up into icy mantles of water-ice and solid ammonia, methane, and carbon dioxide. Other substances may then be added to the mantle ices, including small molecules such as carbon monoxide and hydrogen sulfide. Bombardment by ultraviolet radiation from local hot stars or more remote stars triggers reactions between the different chemical species on a grain's surface and leads to the formation of simple organic substances. Dust grains that drift into the general interstellar medium find their way into clouds where the density is sufficiently high for more complex organic synthesis to take place. How far up the scale of prebiotic synthesis

dust Huge clouds of dust and gas extend along, as well as far above, the disk of NGC 4013–a galaxy seen perfectly edge-on and lying some 55 million light-years away in the direction of Ursa Major. The dark dusty band, about 500 light-years thick, absorbs the light of background stars and gives the illusion of cutting the galaxy in two. *NASA/STScI/AURA*

such interstellar cookery can lead has yet to be determined, but it certainly extends as far as simple **amino acids** and sugars.

Dust grains are the repository of most of the interstellar medium's elements heavier than helium. Typical interstellar dust grains are about 0.5 micron in diameter, but can grow to several microns or more in the deep, dark cores of collapsing dust clouds. Eventually these dust grains stick together to form larger bodies all the way up to asteroids and planets.

Interstellar dust particles strongly absorb, scatter, and polarize visible light at wavelengths comparable to their size, reemitting the light in the far-infrared region of the spectrum. The amount of visual **interstellar extinction** is wavelength-dependent and leads to both a dimming and a reddening of starlight, as blue wavelengths tend to be scattered the most. Views along the plane of the Milky Way are severely limited by the dust which congregates there. **Elliptical galaxies** have less dust than our galaxy (but are not dust-free), while some galaxies are experiencing such tremendous episodes of star formation that the dust in them converts nearly all the visible light into infrared, resulting in an **ultra-luminous infrared galaxy**.

dwarf Cepheid

An old and somewhat misleading name for what is now more commonly called either a **Delta Scuti star** or an **AI Velorum star**.

dwarf elliptical galaxy (abbr. dE)

A dim, gas-poor **dwarf galaxy**, with low surface brightness, of which large numbers are found in **clusters of galaxies**, especially in the vicinity of large galaxies. Many of the brighter ones, classified as dE.N, have a well-defined nucleus. Despite their name, dwarf ellipticals are not really fainter versions of true **elliptical galaxies**, but are structurally distinct. Typical dE's have masses of about 1 billion M_{sun}, or about 1/1000 that of a typical giant galaxy. They contain very little or no gas, which makes them different from **dwarf irregular galaxies**. Three relatively bright dE's are known in the **Local Group**: NGC 147, 185, and 205, all companions of the **Andromeda Galaxy**. Hundreds of similar galaxies exist in the relatively nearby **Virgo Cluster** and even more in the large **Coma Cluster**. Dwarf ellipticals are of special interest because it is possible that they are very similar to the fragments from which larger galaxies formed. In addition to the brighter dE galaxies, there are much less luminous examples known as **dwarf spheroidal galaxies**.

dwarf galaxy

A small, faint galaxy, of which there are two main types. **Dwarf irregular galaxies** tend to contain a lot of gas and usually show strong signs of ongoing star formation, whereas **dwarf elliptical galaxies** (of which **dwarf spheroidal galaxies** are a subcategory) tend to be gas-poor, lower in mass and luminosity, and quiescent. Dwarf galaxies are the commonest variety in the **Local Group** and, almost certainly, in the universe as a whole. In fact, evidence of very extensive halos of thousands of small dwarf systems around large galaxies has been uncovered in a study of gravitational lensing (see **dark matter** for details).

Observations by the Chandra X-ray Observatory support the idea that dwarf systems might be responsible for most of the **heavy elements** in intergalactic space. The orbiting observatory found huge amounts of oxygen and other heavy elements escaping in hot bubbles of gas, thousands of light-years in diameter, from NGC 1569, a dwarf galaxy about 7 million light-years away. For the past 10 to 20 million years, NGC 1569 has been undergoing a burst of star formation and **supernova** explosions, which have driven heavy elements at high speed into the gas in the galaxy, heating it to millions of degrees and, because of the dwarf's relatively weak gravitational field, causing it to escape from the galaxy altogether. It has been speculated that heavy elements escaping from dwarf galaxies in the early universe may have played a dominant role in enriching the intergalactic gas from which other galaxies form. Enriched gas cools more quickly, so the rate and manner of formation of new galaxies in the early universe would have been strongly affected by this process.

dwarf irregular galaxy (abbr. dI)

A **dwarf galaxy** that lacks any apparent structure or uniformity of shape. Dwarf irregulars are becoming increasingly important in understanding the evolution of galaxies in general, because, with many examples nearby, they enable detailed study of important issues such as the occurrence of galactic winds, the chemical enrichment of the interstellar and intergalactic media, and the photometric evolution of galaxies. In addition, their low level of evolution, as implied by their low **metallicity** and high gas content, makes these systems the most similar to primeval galaxies and, therefore, the most useful to infer the primordial galaxy conditions. It has also been suggested that dwarf irregulars represent the local counterpart of faint blue galaxies found in excess in deep galaxy counts.

dwarf molecular cloud (DMC)

A **molecular cloud** that is much smaller and denser than a **giant molecular cloud** but may still be the scene of star formation. Because DMCs are not restricted to the spiral arms of the Galaxy, they can probably survive one or more galactic rotations and thus have lifetimes of 100 million years or more.

dwarf spheroidal galaxy (abbr. dSph)

A low-luminosity (less than absolute visual magnitude –14) **dwarf elliptical galaxy** of low surface brightness and spherelike shape that lacks a nucleus. Although often thought in the past to be merely large, low-density **globular clusters**, recent studies have shown that dSph galaxies have a more complex stellar **population** than that found in globulars. Most important, dwarf spheroidals show evidence of star formation over extended periods, even though they show no sign of current or recent star formation and have no detectable interstellar matter. The stellar populations of dwarf spheroidals consist of two basic components: an old metal-poor population similar to that of globular clusters, and an intermediate-age population, whose ages range from 1 to 10 billion years. The mass-to-light ratios of dSph's is also higher than that of globular clusters, indicating that these galaxies contain a large amount of **dark matter**. Thirteen dwarf spheroidals are known in the **Local Group**, all with masses of 10 to 100 million M_{sun}; nine are satellite galaxies of the **Milky Way Galaxy** and three are companions of the **Andromeda Galaxy**.

dwarf star

(1) A star on the **main sequence**; in other words, a star that is fusing hydrogen into helium in its core. The Sun, for example, is a yellow dwarf. (2) A type of degenerate star known as a **white dwarf**. (3) A star that is not massive enough to fuse hydrogen in its core, known as a **brown dwarf**.

Dwingeloo 1

A **barred spiral galaxy** and member of the nearby **Maffei 1 Group** of galaxies. Despite its considerable size (about a third that of our Galaxy) and proximity (about 10 million light-years away), it remained undiscovered until 1994 because more than 99% of its visible light is absorbed by **dust** in the plane of the Milky Way. It was the first system to be found in the Dwingeloo Obscured Galaxy Survey carried out by astronomers at the Dwingeloo Radio Observatory in the Netherlands.

dynamical method

A method of measuring the mass of an object, such as a star, a galaxy, or a cluster of galaxies, that makes use of the gravitational interactions of two or more bodies.

dynamical time

The timescale used in calculations of orbital motions within the solar system. The most widely used form of it, known as *Terrestrial Time* (TT), uses 86,400 Systeme Internationale seconds (1 day) as its fundamental unit, and is related to **International Atomic Time** (TAI), by TT = TAI + 32.184 seconds.

dynamical timescale

The characteristic time it takes a cloud to collapse and release its gravitational energy; also known as the *freefall timescale.*

Dyson, Frank Watson (1868–1939)

The ninth **Astronomer Royal** (1910–1933); the son of a minister, he studied mathematics and astronomy at Trinity College, Cambridge, and then spent his entire career, except for five years in Edinburgh, at the **Royal**

Greenwich Observatory. There, he directed measurements of terrestrial magnetism, latitude, and time, and he initiated the radio broadcast of time. He determined proper motions of northern stars and completed his portion of the international ***Carte du Ciel*** project of photographing the entire sky. Dyson is best known for directing, with Arthur **Eddington**, the 1919 eclipse expedition, which confirmed the bending of starlight by the Sun's gravity, as predicted by Einstein's **general theory of relativity**.

E

Eagle Nebula (IC 4703)

A diffuse nebula in **Serpens**, discovered by Philippe Loys de Chéseaux in 1745–1746. The Eagle is an active region of star formation that gave birth to the open cluster M16 (NGC 611) about 5.5 million years ago and that is now illuminated by the hot stars of this cluster. One of the most stunning photos taken by the **Hubble Space Telescope** is of the so-called "pillars of creation" in the Eagle–dark columns of gas and dust, also known as *elephant trunks,* reaching into the brighter parts of the nebula. At the end of each pillar, the intense ultraviolet light of newborn stars is photovaporizing some of the hydrogen gas and shaping structures called EGGs (evaporating gaseous globules).

Visual magnitude:	6.4
Angular size:	7°
Linear size:	15 light-years
Distance:	7,000 light-years
Position:	R.A. 18h 18.8 m, Dec. –13° 47′

early-type galaxy

A now generally disused term that stems from the idea that type S0 (lenticular galaxies) and SB0 galaxies might have evolved from E0 and other ellipticals and might also be evolutionary precursors to other spiral and barred spiral galaxies.

early-type star

Stars of **spectral types** O, B, A, and F0 to F5; the name derives from a former, mistaken belief that hot stars such as these are at an early stage of their evolution. The word "early" is also used to describe any star that is at the hotter end of its spectral class: for example, a G2 star is earlier than a G5 star.

Earth

The third planet from the Sun and the only world known to harbor life. Its surface is 70.8% covered by water. It has a relatively thick **atmosphere** composed of 76% nitrogen, 21% oxygen, 1% argon, plus traces of other gases including carbon dioxide and water vapor. The interior of Earth is, like that of the other terrestrial planets, divided into an outer silicate-rich solid crust, with a highly viscous mantle, an outer liquid core that is less viscous than the mantle, and a solid inner nickel-iron core. Convection currents in the material of the outer core give rise to a weak magnetic field. Earth is unusual in the solar system in having a terrestrial planetlike satellite, the **Moon**, that is about one-quarter of Earth's diameter. By coincidence, the Moon is just far enough away to have, when seen from Earth, the same apparent angular size as the Sun, enabling total solar eclipses to be seen.

Diameter	
Polar:	12,713.8 km
Equatorial:	12,756.3 km
Area	
Total:	510 million km^2
Land:	149 million km^2
Water:	361 million km^2
Mass:	5.97×10^{27} g
Mean density:	5.517 g/cm^3
Surface gravity:	9.80 m/s^2
Escape velocity:	11.19 km/s
Rotation period:	23h 56m 4s
Obliquity:	23° 26′ 34″
Mean albedo:	0.39
Atmospheric pressure (at sea level):	1,013 millibars
Orbit	
Mean distance from Sun:	149,598,500 km (8.3 light-minutes)
Perihelion:	147,100,000 km
Aphelion:	152,100,000 km
Eccentricity:	0.0167
Period:	365.256 days

Earth impact craters

See article, pages 156–157.

Earth-crossing asteroid

An asteroid whose orbit crosses that of Earth. Included in this category are **Apollo asteroids** and **Aten asteroids**.

earthshine

Light reflected from Earth and its atmosphere onto the dark part of the Moon when the Moon is not full.

eccentricity

A value that defines the shape of an **ellipse**, such as a planetary orbit. The eccentricity of an ellipse is the ratio

(continued on page 157)

Earth impact craters

Large **meteorites** and **asteroids** occasionally strike the Earth, as they do other worlds in the solar system, and create **impact craters**. However, intense erosion and tectonic processes on our planet quickly (on geological timescales) take their toll on these structures, wiping out all traces of them. Of those of which visible evidence remains, the best known, best preserved, and mostly recently formed is the **Barringer Crater**. One of the largest, and most important in terms of its biological effects, is the **Chicxculub Basin**. This is similar in size, but much younger than and probably different in origin to, the **Sudbury Crater**. Some others follow.

Aorounga Craters

A group of three impact craters formed about 360 million years ago in the Sahara Desert in northern Chad Africa, at 19° 6′ N, 19° 15′ E. The first to be found, the 17-km-wide crater now known as Aorounga South, had been buried by sediments, which were then partially eroded to reveal the current ringlike structure. Two further circular features, both about 16 km across, known as Aorounga North and Aorounga Central, were detected in radar images taken from the space shuttle in 1994. The Aorounga craters are only the second suspected **crater chain** known on Earth; the first consists of eight round depressions, 3 to 17 km wide, stretching along a 700-km line from southern Illinois to Kansas.

Bosumtwi Crater

An impact crater, 10.5 km in diameter and about 103 million years old, located in Ghana in crystalline bedrocks of the West African Shield (6° 32′ N, 1° 25′ W). It is almost entirely filled by Lake Bosumtwi. Chemical, isotopic, and age studies show that the crater is probably the source of the Ivory Coast **tektites**, which are found on land in central Africa and as microtektites in nearby ocean sediments.

Clearwater Lakes Craters

Twin lake-filled impact craters in crystalline bedrocks of the Canadian shield in Quebec; they were formed simultaneously, about 290 million years ago, probably by either a binary or a dumbbell-shaped asteroid. The larger crater, Clearwater Lake West (56° 13′ N, 74° 30′ W), has a diameter of 32 km and a prominent ring of islands some 10 km across that are covered with impact melts. The central peak of the 22-km-wide Clearwater Lake East (56° 05′ N, 74° 07′ W) is submerged.

Deep Bay Crater

An impact crater in Saskatchewan, Canada (56° 24′ N, 102° 59′ W), formed about 100 million years ago in Precambrian metamorphic rock, that appears as a nearly circular bay, about 5 km wide and 220 m deep, in the otherwise shallow Reindeer Lake. The round shoreline, with a diameter of 11 km, is partially surrounded by a ridge that rises to 100 m above the lake surface. The diameter of this ridge, at some 13 km, probably represents the outer rim of the impact structure. Much of Deep Bay's complex structure, including its central uplift, is underwater.

Kara-Kul Crater

A remarkable impact crater, partly filled by the 25-km diameter Kara-Kul Lake (the highest saltwater lake in the world), almost 6,000 m above sea level in the

Earth impact craters/Manicougan impact structure A photo of the Manicougan crater in Quebec taken on June 1, 2001, by the Terra satellite. *NASA*

Pamir Mountain Range in Tajikstan, near the Afghan border (38° 57′ N, 73° 24′ E). The crater, whose rim has a diameter of 52 km, was formed less than 5 million years ago.

Earth impact craters A winter view of the Mistastin Crater in Newfoundland from the space shuttle. *NASA/LPI*

Manicougan Impact Structure

One of the largest preserved impact craters on Earth, located in Quebec, Canada, at 51° 23′ N, 68° 42′ W; it was formed about 212 million years ago. An ice-covered lake, 70 km across, which serves as a hydroelectric reservoir, now lies within a ring of rock showing clear signs of having been melted and altered by a violent collision. The original rim of the crater, now eroded away, is thought to have had a diameter of about 100 km.

Wolfe Creek Crater

A well-preserved crater, formed some 300,000 years ago, that is partly buried under wind-blown sand in the flat desert plains of north-central Australia at 19° 18′ S, 127° 46′ E. Its circular rim is 885 m in diameter and rises 25 m above the surrounding plains and 60 m above the crater floor. Oxidized remnants of iron meteoritic material, impact glass, and signs of stress fractures have been found in the vicinity.

eccentricity

(continued from page 155)

of the distance between the foci and the **major axis.** Equivalently, the eccentricity of an orbit is $(r_a - r_p)/(r_a + r_p)$ where r_a is the **apoapsis** distance and r_p is the **periapsis** distance.

échelle grating

A type of **diffraction grating** characterized by a relatively wide spacing (typically 10 to 20 microns) and by grooves that have a steplike profile; the name is French for "ladder." The light to be dispersed is allowed to fall on to the grating at right angles to the faces of the grooves, resulting in a set of many overlapping spectra and a high resolution. See **échelle spectrograph.**

échelle spectrograph

A **spectrograph** that uses an **échelle grating** to achieve high spectral resolution and wide wavelength coverage. The overlapping spectra produced by the grating are separated by a prism or a grism before reaching the detector. The sensitivity brought to modern échelle spectrographs by **CCD** detectors has made them indispensable in such fields as **astroseismology** and **extrasolar planet** searches.

E-class asteroid

A fairly rare, very reflective **asteroid** that is slightly red in color, probably due to the presence of some surface organic chemicals. E-class asteroids dominate the **Hungaria family** of the main belt, though some, including the two largest, (44) Nysa (see **Nysa-Polana family**) and (64) **Angelina**, are found scattered farther out. Their flat reflectance spectra in the 0.3 to 1.0 micron region and high **albedos** (usually above 0.3) suggest a connection with the **aubrite** (otherwise known as *enstatite achondrite,* hence the "E") category of meteorite. However, this link is problematic because aubrites are clearly of igneous origin, whereas some E-class asteroids (including Nysa) show an absorption feature at 3 microns that points to water- and/or hydroxyl-bearing minerals and a nonigneous history.

eclipse

The total or partial concealment of one celestial body by another, a phenomenon seen in **solar eclipses, lunar eclipses,** and **eclipsing binary** stars. Compare with **occultation** and **transit.**

eclipse season

A period in a given year during which the Earth-Sun line falls near the line of **nodes** of the Moon's orbit, so that lunar and solar eclipses can take place. Eclipse seasons recur every 173.31 days and last about 24 days for lunar eclipses and about 37 days for solar eclipses. Two eclipse seasons make one eclipse **year**.

eclipsing binary

A **binary star** in which, as seen from Earth, at least one of the two bodies regularly passes in front of the other. The primary minimum occurs when the component with the higher surface luminosity is eclipsed by its fainter companion. Three main types of eclipsing binaries are distinguished on the basis of their **light curves: Algol stars, Beta Lyrae stars**, and **W Ursae Majoris stars**. Eclipses may also occur in some kinds of **cataclysmic binary** systems, including dwarf novae, novae, and symbiotic stars.

ecliptic

The **great circle** cut in the **celestial sphere** by an extension of the plane of Earth's orbit; equivalently, the apparent annual path of the Sun against the background of the stars. Because of Earth's axial tilt, the ecliptic is inclined at about 23.4° to the celestial equator, an angle known as the *obliquity of the ecliptic*. The ecliptic intersects the celestial equator at the **equinoxes**. The *ecliptic poles* are the two points on the celestial sphere that lie 90° north and south of the plane of the ecliptic.

ecliptic coordinates

A system that uses the plane of the **ecliptic** as a reference plane. Ecliptic coordinates of objects with respect to the center of Earth, known as *geocentric ecliptic coordinates*, are given in terms of celestial latitude (the perpendicular distance of the orbit from the ecliptic in angular measure) and celestial longitude (the angular distance along the ecliptic between the plane through the object and the vernal equinox). Ecliptic coordinates that refer to the center of the Sun, known as *heliocentric ecliptic coordinates*, are given in terms of heliocentric latitude and heliocentric longitude.

Eddington, Arthur Stanley (1882–1944)

An English astrophysicist who pioneered the study of stellar structure. Educated at Manchester and Cambridge, he spent seven years as chief assistant at the Royal Greenwich Observatory before becoming Plumian Professor of Astronomy at Cambridge in 1913. He made important investigations of stellar dynamics, and was an influential supporter of the view that "spiral nebulae" were external galaxies. He contributed much to the introduction of Einstein's **general theory of relativity** into cosmology, writing books on the new theory for both his fellow scientists and the public, and led one of the two 1919 solar eclipse expeditions that confirmed the predicted bending of starlight by gravity. Eddington's greatest contributions concerned the astrophysics of stars. He dealt with the importance of radiation pressure, the transfer of energy by radiation, the **mass-luminosity relation**, pulsations in **Cepheid variables**, and the very high densities of **white dwarfs**, and was among the first to argue that subatomic reactions must power the stars.

Eddington (spacecraft)

A European Space Agency (ESA) mission to detect seismic vibrations in the surfaces of stars and to search for Earthlike planets using precise **photometry**. Named in honor of Arthur **Eddington**, it is expected to be launched after 2008 and will orbit at the second **Lagrangian point**, beyond the Moon's orbit, on a 5-year primary mission. For the first 2 years, it will look at about 50,000 individual stars for signs of tiny oscillations that betray the stars' internal compositions and allow their ages to be deduced (see **astroseismology**). For the next 3 years, it will watch simultaneously about 500,000 stars in a single patch of the sky, looking for regular light dips that might indicate the passage (transit) across the face of a star of a planet as small as Mars. Eddington's mission is similar to that of NASA's **Kepler** and will be a forerunner to that of ESA's **Gaia**.

Eddington limit

The theoretical limit at which the **radiation pressure** of a light-emitting body would exceed the body's gravitational attraction. A star emitting radiation at greater than the Eddington limit would break up. This would happen to a star of more than about 120 M_{sun}, or to the Sun if its luminosity were increased by a factor of 30,000. The Eddington limit, named after Arthur **Eddington**, is given as

$$L = 4\pi G M m_p c/\sigma_T$$

where G is the gravitational pressure, M is the mass of the luminous object, m_p is the mass of a proton, c is the speed of light, and σ_T is the effective area of an electron when it is illuminated by radiation.

effective radius

The distance from the center of a galaxy from within which half the galaxy's **luminosity** is emitted. See also **Holmberg radius**.

Effelsberg Radio Observatory

The observatory of the Max Planck Institut für Radioastronomie. Located in the Eifel mountains, 40 km southwest of Bonn, Germany, it is home to a 100-m radio telescope, opened in 1971, which was the largest fully steerable dish in the world until the completion of the **Green Bank Telescope**.

Egg Nebula (CRL 2688)

A **protoplanetary nebula** in **Cygnus**. The central star of the Egg was a **red giant** until a few hundred years ago. It then began shedding its outer layers, which today are visible as a cloud of matter about 0.6 light-year across. Observations by the Hubble Space Telescope have shown bright arcs of matter within the cloud, almost like tree rings, that reveal the way in which the rate of mass ejection from the central star has varied throughout its recent period of mass loss. Hubble has also shown starlight escaping in narrow, oppositely directed beams through holes in the circumstellar cocoon. The beams may result from shadows cast by blobs of material distributed within the region of ringlike holes that are carved out by a wobbling, high-speed stream of matter. Alternatively, they may be due to starlight reflected off fine jetlike streams of matter being ejected from the center and confined to the walls of a conical region around the symmetry axis. Both theories call for the ejection of high-speed material in a narrow beam by some mechanism that isn't properly understood. Similar fine **jets** have been seen in the **Cat's Eye Nebula**.

Egg Nebula A Hubble Space Telescope view of the Egg Nebula (CRL 2688), showing a pair of mysterious searchlight beams emerging from a hidden star and crisscrossed by numerous bright arcs. A dense cocoon of dust (the dark band in the center) enshrouds the star and hides it from view.
NASA/Raghvendra Sahai and John Trauger (JPL)

Visual magnitude:	13.5
Angular diameter:	19″ × 7″
Distance:	3,000 light-years
Position:	R.A. 23h 2.3m, Dec. +36° 42′

Egyptian astronomy

The ancient Egyptians kept a complicated **calendar** with two different, mutually shifting years and festivals. They had a 365-day (36 decans: 36 × 10 = 360, plus 5 festival days) calendar for New Year rites and a Sothis (Sirius) calendar of 365¼ days for agriculture (sowing and harvesting), measured by watching the **helical rising** of Sirius. The latter was called the Canicular Year (Sirius is known as the Dog Star, and *canis* is Latin for "dog"). The Egyptians knew that the two calendars would be out of step by one year after 1,461 years, known as the **Sothic cycle** (because 1,460 × 365¼ = 1461 × 365 = 533,265).

In the Egyptian language the word "sky" is feminine. Thus for the Egyptians, unlike most other peoples, the sky was a goddess (Nut or Hathor) represented either as a cow with four hooves planted on Earth or as a woman whose body bends in an arc so that her toes and fingertips touch Earth. She gives birth every day to the Sun, which similarly has different names depending on whether it is the rising Sun (Khepri), the Sun at zenith (Ra), or the setting Sun (Atum). The Moon's various names, Aah, Thoth, and Khons, correspond with its depiction as a dog-headed ape, an ibis, and the left eye of a great celestial hawk. These identifications predate the development of hieroglyphic writing. The oldest megalithic site with an astronomical orientation was found in Nabta in 1998. See **Babylonian astronomy**, **Chinese astronomy**, **Indian astronomy**, and **Greek astronomy**.

Eight-burst Nebula (NGC 3132)

A **planetary nebula** in **Vela**, also known as the *Southern Ring Nebula.* The simple nebular oval evident to an observer at the eyepiece of a telescope, transforms into several superimposed glowing rings on long photographic exposures. Neither the unusual shape of the peripheral shell nor the structure and placements of the filamentary dust lanes running across the nebula are well understood.

Visual magnitude:	8
Angular diameter:	0.8′
Linear diameter:	0.4 light-year
Distance:	2,000 light-years
Position:	R.A. 10h 7m; Dec. –40° 26′

Einstein Cross (G2237+030)

The most spectacular of the 20 or so well-attested instances of **gravitational lens**es that involve a **quasar** and foreground galaxy. The Einstein Cross consists of four symmetrically placed images of the same quasar, which lies 8 billion light-years away and almost directly behind the nucleus of a galaxy that is only 500 million light-years away. The alignment is so close that if the lens effect could be removed, the quasar would appear within 0.05″ of the galaxy's nucleus. The system was discovered in the course of a **redshift** survey by the Smithsonian Center for Astrophysics by John Huchra (1948–), and is sometimes referred to as the *Huchra Lens.* Einstein crosses form when the source and lens are in alignment with the observer but the lensing mass is unevenly distributed. When there is alignment plus even distribution of the lensing mass the result is an **Einstein ring**.

Einstein ring

A **gravitational lens** effect in which the image of a remote background object, such as a **quasar**, is distorted into a ring by a foreground galaxy. A perfect ring will only result if the source, the lensing object, and the observer are exactly lined up, and, in addition, the mass of the lensing object is evenly distributed (see **Einstein Cross**). A few good approximations to Einstein rings have been found, such as MG1131+0546 and B1938+666. In other cases, where the alignment is not perfect, the gravitational lens produces one or more arcs.

ejecta

Fractured and/or molten rocky debris thrown out of a **crater** during an impact event, or, alternatively, material, including volcanic ash, pyroclasts, and bombs (rounded, congealed lumps of magma), discharged by a volcano. An *ejecta blanket* is a generally symmetrical apron of ejecta that surrounds a crater; it is thickly layered at the crater's rim and varies from thin to discontinuous at the blanket's outer edge.

Elara

The thirteenth moon of **Jupiter**, also known as Jupiter VII; it was discovered in 1905 by Charles Perrine. Elara has a diameter of about 78 km and orbits at a mean distance of 11.74 million km from the planet.

electromagnetic radiation

A form of radiation in which energy is carried by vibrating electric and magnetic fields that move together and travel through space at the speed of light. Electromagnetic radiation may also be regarded as a stream of particles known as **photon**s.

electromagnetic spectrum

The complete range of **electromagnetic radiation**, from very short-wavelength (high-frequency) **gamma rays**, through **X rays** and **ultraviolet** light, to the small range of visible light, and further to **infrared** radiation, **microwaves**, and the comparatively long-wavelength/low-frequency **radio waves**.

electron

A fundamental particle of matter that carries a negative charge. Electrons occur in atoms and are arranged in energy levels.

electron degenerate matter

A form of **degenerate matter** in which the weight of overlying material tries to force all of the **electron**s surrounding the atomic nucleus into the lowest energy quantum state. The electrons resist, because of the **Pauli exclusion principle**, and so exert a pressure, known as *electron degeneracy pressure,* that halts further collapse. This pressure is sufficient up to the **Chandrasekar limit** of about 1.4 M_{sun}. However, beyond this critical mass, gravity overwhelms the degeneracy pressure and results in further collapse. See also **baryon degenerate matter**.

electron temperature

The temperature determined by the mean kinetic energy of free **electron**s in a **plasma**; also known as the *kinetic temperature.* Inside stars, electron temperature is likely to match the temperature as measured by other methods, but in the rarefied environment of a nebula or the outer atmospheres of stars, it may be quite different.

element, chemical

A substance made up entirely of atoms that each contain the same number of protons in their nuclei (and therefore the same number of electrons). Most atoms may have different numbers of neutrons in their nuclei, leading to different **isotope**s of the same element. There are 92 naturally occurring elements, ranging from the lightest, hydrogen, to the heaviest, uranium.

Ellerman bomb

A small, bright structure, also known as a *moustache,* that appears in a solar **active region**, most often in an emerging flux region or on the edge of a **sunspot** where the magnetic field is breaking through the surface. Ellerman bombs usually last less than 5 minutes and are best seen in the wings of **H-alpha**; they are named after the American astronomer Ferdinand Ellerman (1869–1940).

ellipse

A flattened circle or oval: the shape of most **orbits**. Johannes **Kepler** first established that the orbits of the planets are ellipses, not circles, based on the careful observations by Tycho **Brahe**.

ellipsoid

A body or surface of which every cross-section that passes through the center is a circle or an **ellipse**.

ellipsoidal variable

A **close binary** in which the stars are so near together (though not actually in contact) that one or both are drawn out into the shape of an **ellipsoid**. As the components orbit each other, they are seen from different angles and therefore present a continuously varying total luminous area. The observed magnitude variations are small, typically much less than 0.2 magnitude, because there is a limit to how far a star can be stretched before it is torn apart. However, if, as in the case of **Beta Lyrae stars**, the components also eclipse each another, the brightness changes are much greater.

elliptical galaxy

A **galaxy** that has a superficially smooth, featureless appearance and an ellipsoidal shape. *Giant ellipticals* may contain over 10 trillion M_{sun} in the form of stars, are among the largest of galaxies, and are often found at the heart of rich clusters of galaxies (see **D galaxy**). *Dwarf ellipticals,* on the other hand, may have masses as low as 10 million M_{sun} and lie at the bottom end of the galactic size range. Normal ellipticals are quite rare throughout much of space, but form the majority of ellipticals found in big clusters.

Many old ideas about ellipticals have been debunked by discoveries made over the past couple of decades. Among these is the notion that ellipticals contain hardly any interstellar matter. When looked at closely, about a third of them show features due to absorption by **dust**. The dust seems to be distributed in rings or disks, in some cases aligned with either the major or minor axes and in other cases warped. Ellipticals contain modest amounts of cool and warm gas, although not as much as found in spiral galaxies and not usually enough to support much star formation. A few have extended disks of neutral hydrogen. X-ray observations show that many ellipticals contain between 1 billion and 10 billion M_{sun} of gas at temperatures of 10 million K, typically in the form of a pressure-supported atmosphere around the galaxy. The surface brightness of ellipticals doesn't always decline smoothly with radius. When a smooth luminosity profile is subtracted from the actual surface brightness, concentric shells or ripples are often seen. These shells are somewhat bluer than the rest of the galaxy and appear to be composed of younger stars.

The orbits of stars in ellipticals are in random directions, and often very elongated, taking them close to the center and then far into the outskirts. It was once thought that ellipticals were *oblate spheroids*–shaped like a hamburger bun–and that they kept this shape because they were slowly rotating. It is now know that they are often *prolate,* like a rugby ball, or even *triaxial,* meaning they have different diameters in all three directions. They hardly rotate at all; instead the shape is maintained because the stars move faster in the direction of the long axis, and so can travel farther from the center before gravity turns them back.

ellipticity

A measure of the amount by which an object, such as a planet or a galaxy, deviates from a perfect sphere; also known as *oblateness.* It is found by subtracting the polar diameter from the equatorial diameter, then dividing by the equatorial diameter. Ellipticity is typically an indication of how fast a body is rotating. The ellipticity of Earth is about 1/299. The ellipticity of elliptical galaxies increases, going from type E0 to E7.

elongation

(1) The geocentric angle between a planet and the Sun, measured in the plane of the planet, Earth, and Sun. Planetary elongations range from 0° to 180°, east or west of the Sun. *Greatest elongation* is the instant at which the geocentric angular distance of Mercury or Venus from the Sun is a maximum. (2) The angle between a planet and one of its satellites as seen from Earth, measured from 0° east or west of the planet, in the plane of the planet, satellite, and Earth.

Eltanin (Gamma Draconis, γ Dra)

The brightest star in **Draco**; its name (also given as Etamin) comes from the Arabic *al Ras al Tinnin,* which means "the Dragon's Head." Eltanin was also formerly known as the Zenith Star since it passes directly overhead as seen from the Royal Greenwich Observatory in London. In 1728, James Bradley sought to measure the **parallax** of Eltanin and, in so doing, discovered the **aberration of starlight**, caused by Earth's motion through space. Eltanin is approaching the sun and will make its closest pass of 28 light-years about 1.5 million years from now, when it will be the brightest star in the sky and a rival to our current Sirius. Physically, Eltanin is an orange giant **K star**, a bit over half the size of Mercury's orbit, that shows a slight variability of about 0.05 magnitude.

elliptical galaxy The peculiar giant elliptical galaxy NGC 1316, located the in Fornax Cluster, is seen to have extensive dust lanes in this photo by the Very Large Telescope. *European Southern Observatory*

Visual magnitude:	2.24
Absolute magnitude:	–1.04
Spectral type:	K5III
Surface temperature:	4,000 K
Luminosity:	600 L_{sun}
Radius:	50 R_{sun}
Mass:	1.7 M_{sun}
Distance:	148 light-years

emission

The release of a **photon** of specific energy from an atom when an **electron** jumps from an outer orbit to an inner orbit around the nucleus. The energy radiated, as line emission, corresponds to the lost energy of the electron.

emission line

A bright line in a **spectrum** caused by the **emission** of light at a specific wavelength. Emission lines may appear on their own, as in the spectrum of a nebula energized by a radiation from a nearby hot star, or they may be superimposed on an **absorption spectrum**, as happens when a star is surrounded by hot gas.

emission nebula

A nebula that displays an **emission spectrum** because of energy that has been absorbed from one or more hot, luminous stars and reemitted by the nebular gas at specific wavelengths. This is different from a **reflection nebula**, where the light from the nebula is simply reflected light from the central star. The difference becomes clear from looking at the spectrum of the nebula and comparing it to that of the stars providing the initial energy. If it is a reflection nebula, then the spectra will match; if it is an emission nebula, it will show emission lines of its own.

emission spectrum

A **spectrum** that consists predominantly or solely of **emission lines**. It indicates the presence of hot gas and a nearby source of energy, as found, for example, in **planetary nebula**e and **quasar**s.

En Henduanna (fl. 2350 B.C.)

The earliest female known to have had a connection with astronomy. She was the daughter of Sargon of Akkad who established the Sargonian Dynasty in Babylon (see **Babylonian astronomy**). He appointed her the chief priestess of the moon goddess of the city–a position of great power and prestige. Only through the auspices of the high priestess could a leader achieve a legitimate claim to rule. Of her written work, translations of 48 of her poems survive. To put her in perspective, astronomy began with the priests and priestesses in Sumeria and Babylon. As early as 3000 B.C., these sacred temples in Sumer were complex structures that directed every essential activity of life including trade, farming, and crafts. The priests and priestesses established a network of observatories to monitor the movements of the stars. The calendar they created is still used to date certain religious events like Easter and Passover.

Enceladus

The eighth moon of **Saturn**, also known as Saturn II; it was discovered by William **Herschel** in 1789. Enceladus has a diameter of 498 km and orbits at a mean distance of 238,020 km from the planet's center, within the tenuous E-ring. It is covered by pristine ice which gives it the highest **albedo** (more than 0.9) of any body in the solar system–an extreme reflectivity that explains the moon's surface temperature of –201°C, which is lower than that of its satellite neighbors. Where **impact crater**s do occur they are flattened and of fairly low density, while a large portion of the surface is crater-free and covered by smooth plains separated from the cratered areas by long sinuous ridges. Evidence suggests that at least part of the surface is **young** and that Enceladus has been geologically active in the past 100 million years and may still be active today. Perhaps some sort of **ice volcanism** is at work driven by **tidal heating** caused by the gravitational tugging of Saturn and of the moon **Dione** with which Enceladus is locked in a 1:2 resonance. Eruptions on Enceladus may be the source of material in the E-ring.

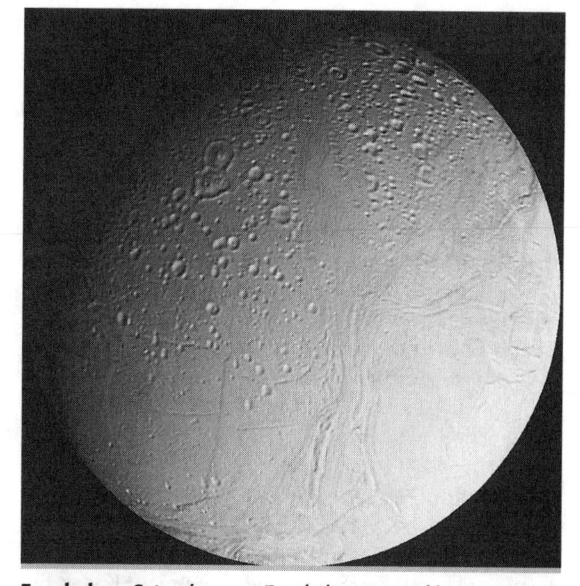

Enceladus Saturn's moon Enceladus captured in a mosaic of images taken by Voyager 2. Enceladus would fit within the state boundaries of Arizona, yet despite its small size it reveals one of the most interesting surfaces of all the icy satellites. *NASA/JPL*

Encke, Johann Franz (1791–1865)

A German astronomer and mathematician who calculated the orbit of a short-period comet, now named after him (see **Encke's comet**). In 1838, he discovered a division between the A- and F-ring of **Saturn**, also named in his honor. In 1811, Encke began his study of mathematics at Göttingen under Carl **Gauss**. He moved to the Seeberg Observatory, Switzerland, in 1816, to work as an observer and, in 1822, was appointed its director. In 1825 he was invited to become director of the Berlin Observatory and in 1844 he took up a professorship at the University of Berlin. He is best known for figuring out, at the suggestion of Jean **Pons**, the orbital elements of a comet found by Pons in 1818. Pons suspected that one of the three comets discovered in 1818 had already been discovered by Encke in 1805. This comet was found to have a period of 3.3 years and Encke predicted its return for 1822. In 1846, Johann **Galle** discovered Neptune with the help of star charts edited by Encke.

Encke's comet

A small, relatively inactive comet with the shortest period–just over three years–and the smallest aphelion of any known comet. It has been observed at every **apparition** since its discovery in 1818. Named after Johann **Encke**, who computed its orbit, it is the parent body of the Taurids **meteor shower**. Semimajor axis 2.21 AU, aphelion 4.1 AU, eccentricity 0.847, inclination 12.4°, period 3.30 years.

Enif (Epsilon Pegasi, ε Peg)

The brightest star in **Pegasus**; its name comes from the Arabic *al anf* for "the nose" and refers to the muzzle of the winged horse. Enif is an orange **supergiant K star**, big enough, if put in place of the Sun, to engulf the orbit of Venus. It seems to have connections with two other supergiants–the Alpha and Beta stars, Sadalmelik and Sadalsuud, of nearby Aquarius. This stellar trio, with similar luminosities and distances (Sadalmelik at 760 light-years, Sadalsuud at 610) may have formed together in the same association and, over the past 15 million years or so, have drifted apart. Most odd about Enif is its erratic variability, which can take it from magnitude 3.5 to 0.7, although it generally hovers around 2.4. On September 26 and 27, 1972, an observer in Florida saw Enif briefly rival **Altair**–becoming five times brighter than normal–after which it faded. Probably a colossal flare was responsible, one far more luminous than those seen on the Sun. Only two dozen such stellar outbursts are on record and they are poorly understood.

Visual magnitude:	2.38
Absolute magnitude:	–4.19
Spectral type:	K2Ib
Surface temperature:	4,460 K
Luminosity:	6,700 L_{sun}
Radius:	150 R_{sun}
Mass:	10 M_{sun}
Distance:	673 light-years

Ensisheim Meteorite

The oldest **meteorite** whose fall can be dated precisely. On November 7, 1492, near noon, a loud explosion preceded the arrival of a 127-kg stone meteorite in a wheat field near the village of Ensisheim in the province of Alsace, France, which at the time was part of Germany. An old woodcut depicting the scene shows the fall watched by two people emerging from a forest. In fact, a young boy was the only eyewitness, and he led the local populace to the field where the meteorite lay in a 1-m hole. After it was retrieved, the townsfolk, believing the object to be of supernatural origin, began to chip off bits for souvenirs until stopped by the local magistrate. Many of these fragments ended up in museums around the world. The remaining specimen, a rounded gray mass weighing only 55 kg and nearly without any **fusion crust**, can be seen today at Ensisheim resting in an elegant case in the middle of the main hall of the Regency Palace built in 1535 by Emperor Ferdinand of Austria.

enstatite

The rarest of the three main types of **chondrite**, accounting for about 1.5% of falls in this category of meteorite. The name derives from the most abundant mineral present, a **pyroxene** known as enstatite, $Mg_2Si_2O_6$. Whereas virtually all pyroxenes found on Earth, the Moon, and in other meteorites contain both magnesium (Mg) and iron (Fe), enstatite is almost iron-free. The iron found in enstatite chondrites, which accounts for 15 to 25% of their mass, is in a metallic state. Stranger still, many elements that normally reside in silicate (rocky) minerals are found instead in sulfide minerals with the result that enstatites contain a wide assortment of uncommon sulfides. Even the metallic iron has up to several percent silicon. Moving silicon to metallic iron requires a very reducing, oxygen-depleted environment. Some researchers believe the parent body may be an asteroid inside the orbit of Venus or even Mercury (see **vulcanoid**), while others suggest that a formation in the inner part of the main asteroid belt would have provided the same conditions in the early solar system. Enstatite chondrites are similar to ordinary chondrites and have been further subdivided based on their content of total iron: members of the EL group contain less iron than members of the EH group. Despite these differences, most researchers believe that both subgroups originated on the same asteroid, most probably representing different layers of the parent body. A comparison of the reflectance spectra of different asteroids to the spectrum of the EH chondrite Abee suggests that the main belt asteroid (16) Psyche may be the common parent for the enstatite chondrites.

entrainment

An effect in which the material from the medium through which an astrophysical **jet** is traveling is pulled into the jet itself. Interstellar gas may be entrained by the jets in **active galaxies**. Similarly, **neutrons** in the upper layers of the Sun's atmosphere may be entrained by the **solar wind**.

entrance pupil
The optical aperture of a telescope, eyepiece, binoculars, or other optical system through which light initially enters.

envelope
A cloud of gas and dust that surrounds one or more stars or some other astronomical object. Young, hot stars often eject envelopes or produce them by ionizing nearby material. Old stars, in their **red giant** phase, shed their outer layers and produce cool envelopes rich in molecules and dust. When the core of the dying star is exposed, ultraviolet light from the core makes the envelope luminesce as a **planetary nebula**.

Eos family
A family of **asteroids**, first detected by the Japanese astronomer Kiyotsugu Hirayama (1874–1943) in 1918, that moves in the outer asteroid belt at a mean distance of 3.02 AU from the Sun. Most Eos asteroids appear to belong to class C (carbonaceous) or S (silicaceous); spectral studies have linked some Eos family members specifically with **carbonaceous chondrites** of the CO and CV type, while others have yet to be matched with known meteorite counterparts. This compositional variation suggests that the Eos parent body was partially differentiated. The family is named after (221) Eos, a 112-km-wide object of spectral class S, discovered in 1812 by Johann **Palisa**; semimajor axis 3.014 AU, perihelion 2.72 AU, aphelion 3.31 AU, eccentricity 0.077, inclination 10.9°, period 5.23 years.

Ep galaxy
In Morgan's classification (see **galaxy classification**), an **elliptical galaxy** that shows some peculiarity in appearance. This might involve patches or lanes of dust; some hot, young stars; high-energy phenomena associated with the nucleus; or an untypical fall-off in luminosity with increased distance from the center of the galaxy. Among the best-known Ep galaxies are **M87**, which has a prominent jet, and **M32**, a satellite of the Andromeda Galaxy that appears to have been denuded of its outer regions by its massive neighbor.

ephemeral region (EFR)
An area of the **Sun** where a magnetic dipole, or flux tube, surfaces on the disk and eventually produces a bipolar **sunspot** group. In **H-alpha**, an EFR usually appears as a small oval area of bright **plage** (typically about 7,000 km across) and often contains a series of short-lived narrow **fibrils**, known as an *arch filament system,* running roughly from one end of the dipole to the other. Each pole of an EFR is often marked by **pores** or small developing sunspots. Surges or even small **solar flares** can sometimes occur in EFRs.

ephemeris (pl. ephemerides)
A tabulation of the predicted positions of a celestial object in an orderly sequence for a number of dates. An ephemeris may also list distances of an object from the Sun and/or Earth, as well as magnitudes. Sets of ephemerides are published annually in **almanacs**.

epicycle
A small orbit within a larger orbit that was used (mistakenly) to describe the movements of celestial objects in the **Ptolemaic system** (about A.D. 150). In Ptolemy's model of the solar system, the Sun, the Moon, or a planet moved in an epicycle, the center of which traveled along a bigger circular orbit, known as the *deferent,* whose center was offset from Earth. Many layers of epicycles were needed to approximate real orbits with their retrograde motion. Nicolaus **Copernicus** also used epicycles in his heliocentric (Sun-centered) representation of the solar system in the mid-1500s, and they were only superceded following Johann **Kepler**'s discovery of the elliptical nature of orbits in the early 1600s.

Epimetheus
The fifth moon of **Saturn**, also known as Saturn XI; it is nearly co-orbital (i.e., shares its orbit) with **Janus** and is located between Saturn's F- and G-rings at a mean distance of 151,422 km from the planet's center. It measures 144 × 108 × 98 km and has several craters larger than 30 km across, as well as a variety of ridges and grooves. Epimetheus was first observed by Richard Walker in 1966 but the situation was confused since Janus is in a similar orbit. So Walker officially shares the discovery of Epimetheus with John Fountain and Stephen Larson who showed in 1977 that there were two satellites involved. The situation was clarified in 1980 by **Voyager** 1.

epoch
The time and date at which an astronomical observation is made (*epoch of observation*), or the date for which **orbital elements** (*epoch of elements*) or the positions of celestial objects are calculated. Specifying the epoch is important because the apparent positions of objects in the sky change gradually due to **precession** and **nutation**, while orbital elements change due to the gravitational effects of the planets. Data given in star catalogs and ephemerides are referred to a *standard epoch,* also known as a *fundamental epoch.* Prior to 1984, coordinates of star catalogs were commonly referred to as **Besselian epochs**. However, from 1984 on, **Julian Dates** have been used, denoted by the prefix J. The current standard epoch is J2000.0. It will be superceded in half a century by J2050.0.

Epsilon Aurigae (ε Aur)

One of the strangest and least understood stars in the sky: a binary system in which an enormous yellow-white **supergiant** is periodically eclipsed by an object that is vastly larger. Its occasionally used Arabic name, Almaaz, means "he-goat." The bright component of Epsilon Aurigae is a hot-end supergiant **F star**, slightly more than 1 AU in diameter. Despite its size, every 27.1 years the bright star is eclipsed for *two years* by something of truly colossal proportions. The prevailing idea is that the mysterious dark component is a star surrounded by a thick ring of obscuring dust set nearly on edge. The supergiant we see and the mystery star are perhaps 30 AU apart, and the dust ring about the secondary star is some 20 AU in diameter. The ring has some sort of gap in the middle, as Epsilon Aur brightens a bit at mid-eclipse. We have little idea what lies at the center the dusty ring. One theoretical model predicts an object with a mass of 4 M_{sun}, another with a mass 15 M_{sun}. It could be one star that has generated a disk through a fierce out-flowing wind or, as more commonly believed, a pair of class B stars that are themselves in tight orbit. The last eclipse took place from 1982 to 1984. The next will be from 2009 to 2011, when a new generation of telescopes will be trained on this stellar enigma in an effort to unlock its mysteries.

Data for the Bright Component of Epsilon Aurigae	
Visual magnitude:	3.03
Spectral type:	F0Ia
Surface temperature:	7,800 K
Radius:	100 R_{sun}
Luminosity:	47,000 L_{sun}
Mass:	15 to 19 M_{sun}
Distance:	2,040 light-years

Epsilon Eridani (ε Eri)

A nearby star, in the northeastern part of **Eridanus**, that is known to have at least one planet. Somewhat smaller, cooler, and less luminous than the Sun, Epsilon Eri is the third closest star visible to the naked eye. An orange-**red dwarf**, perhaps only 500 million to 1 billion years old, it spins relatively fast, with a rotational period of 11 days (compared with the Sun's 27 days). This quick rotation generates a comparatively strong magnetic field, which results in a lot of chromospheric activity, including large starspots, and a variable spectrum characteristic of a **BY Draconis star**. In 1998, a cold dust disk was found around Epsilon Eri at about where the **Kuiper Belt** of the Solar System would be–from inside Neptune's orbit to twice Pluto's average distance from the Sun. Most of the dust lies in a ring from 35 to 75 AU around the star, peaking at a radius of about 60 AU, with an estimated mass of about 0.014 to 0.4 M_{sun}. In August 2000, astronomers announced the discovery of a Jupiter-like planet with a mass of about 1.2 $M_{Jupiter}$ in a highly elliptical orbit (eccentricity 0.6) of mean radius 3.3 AU and period 6.8 years. Although Epsilon Eri has been monitored extensively by SETI (Search for Extraterrestrial Intelligence) projects, including the first, Project Ozma, its young age makes it an unlikely host for advanced life. The presence of at least one planet, however, will make this stellar neighbor a prominent target for future planet searches and efforts to detect primitive life signs. See **stars, nearest**.

Visual magnitude:	3.72
Absolute magnitude:	6.18
Spectral type:	K2V
Luminosity:	0.34 L_{sun}
Mass:	0.85 M_{sun}
Radius:	0.88 R_{sun}
Distance:	10.50 light-years

Epsilon Indi (ε Ind)

One of the nearest stars to the Sun, a dwarf **K star**, somewhat cooler than and about one-seventh as luminous as the Sun. Around it, at an average distance of 1.46 AU (220 million km), orbits our nearest known **brown dwarf**. Discovered in 2003, this brown dwarf has a mass of 40 to 60 $M_{Jupiter}$ and a surface temperature of about 1,260 K. See **stars, nearest**.

Visual magnitude:	4.69
Absolute magnitude:	7.00
Spectral type:	K5Ve
Luminosity:	0.14 M_{sun}
Mass:	0.77 M_{sun}
Radius:	0.76 R_{sun}
Distance:	11.8 light-years

Epsilon Lyrae (ε Lyr)

The fifth brightest star in **Lyra** and one of the most famous multiple stars in the sky: a naked-eye double (on a very clear, moonless night) that a 6-cm telescope with high magnification shows to be a quadruple system made up of two pairs of stars; it is commonly known as the *Double Double*. $Epsilon^1$ consists of two stars, of magnitude 4.7 and 6.2, that orbit about their center of gravity every 1,200 years. $Epsilon^2$ consists of a magnitude 5.1 and 5.5 pair with an orbital period of 585 years. In addition $Epsilon^1$ and $Epsilon^2$ orbit around each other in a huge orbit, with a separation of 0.16 light-year and a period measured in hundreds of thousands of years.

Visual magnitude:	4.67 (ε^1), 4.59 (ε^2)
Absolute magnitude:	1.18 (ε^1), 1.10 (ε^2)
Spectral type:	F1V (ε^1), A8V (ε^2)
Distance:	162 light-years

equation of light

A correction to allow for the time it takes light to cross Earth's orbit. It is important when recording events such as stellar **occultations**, since the timing could be off by as much as 17 minutes depending on Earth's orbital position. For some events within the solar system, such as eclipses of planet's satellites, account must also be taken of the position of the planet and its moons.

equation of the center

In elliptical motion, the true **anomaly** minus the mean anomaly. In other words, it is the difference between the actual angular position in the elliptic orbit and the position the body would have if its angular motion were uniform.

equation of time

The apparent **solar time** minus the mean solar time. It varies throughout the year, and slightly from year to year. At present, it reaches extremes of about –14 minutes in February, and about +16 minutes in November.

equator

The great circle on the surface of a body formed by the intersection of the surface with the plane passing through the center of the body perpendicular to the axis of rotation.

equatorial coordinates

The most common system of **celestial coordinates** used by astronomers. Because it uses the plane of the celestial equator (an extension into space of Earth's equatorial plane) as a reference plane, it is the sky equivalent of longitude and latitude. Celestial longitude is called **right ascension** (R.A.) and is measured eastward along the celestial equator, in hours, minutes, and seconds of **sidereal time**, or, alternatively, in degrees, starting from 0h (or 0°) at the vernal **equinox**. Celestial latitude is known as **declination** (Dec.) and is the angular distance of a body north or south of the celestial equator, reckoned positive when north and negative when south. Sometimes, instead of R.A. and Dec., **hour angle** and **polar distance** are used.

equatorial mounting

See article, page 168.

equinox

(1) Either of the two points on the **celestial sphere** where the celestial equator intersects the **ecliptic**. (2) Either of the times at which the center of the Sun's disk passes through these points. The *vernal equinox* (or *spring equinox*), when the Sun reaches its **ascending node** (i.e., crosses the celestial equator moving northward), falls on or around March 21 and marks the start of spring. The *autumnal equinox,* when the Sun reaches its **descending node**, occurs on or around September 22 and marks the start of autumn. These are the two days of the year on which, everywhere on Earth, day and night are of equal duration: hence the name. The equinoxes drift very slightly across the sky because of **precession**. For example, in the time of Hipparchus, about 2,100 years ago, the vernal equinox lay in Aries and it is still referred to as the *First Point in Aries,* even though it has now moved into Pisces. Likewise, the autumnal equinox, also called the *First Point in Pisces,* now lies in Virgo.

equipotential surface

A surface surrounding a body, or group of bodies, over which the gravitational field is of constant strength and, at all points, is directed perpendicular to the surface. The concept of the *critical equipotential surface,* popularly known as the **Roche lobe**, is often misunderstood and incorrectly applied. This equipotential surface arises from the solution of the restricted **three-body problem**; it describes the gravitational field experienced by a massless body in the vicinity of two point masses orbiting each other in a perfect circle, in the absence of any other force such as **radiation pressure** or **stellar wind**. Unless the restricted three-body conditions are met, the surface doesn't exist mathematically. Since there probably aren't any binaries with perfect circular orbits, the restricted three-body solutions are at best approximations to real binaries. The use of the term Roche lobe encourages the idea of a kind of impenetrable container from which material can only escape through the gravitationally neutral inner-**Lagrangian point**. But this is a faulty notion. Earth, for instance, has an infinite number of equipotential surfaces, yet air carries molecules through those surfaces freely. However, the concept of the critical Roche equipotential surface can be useful in understanding the evolutionary processes in **close binary** stars. It defines an *approximate* boundary within which the atmosphere of the evolving star can expand while still remaining a distinctly separate object. If that rough boundary is exceeded, the expanding atmosphere will have access to the domain containing both stars.

equivalent width

A measure of the strength of a **spectral line**. On a plot of intensity against wavelength, a spectral line appears as

(continued on page 169)

equatorial mounting

A telescope **mounting** with one axis, called the *polar axis,* that is parallel to Earth's rotation axis, and the other, called the *declination axis,* at right angles to it. The telescope moves north-south about the declination axis and east-west about the polar axis. To *point* at a target requires moving the telescope about both axes. To *track* a target, however, requires movement about the polar axis only, at the same rate that Earth spins. This is the chief advantage of the equatorial mounting: the N–S position doesn't change, and a single drive can regulate the E–W tracking. The main drawback is that the polar axis is difficult to orientate with respect to the ground, and it is different for every observatory. Positioning the mounting at odd angles creates difficulties that increase rapidly as the size and mass of the telescope increase. For this reason, computer-controlled **altazimuth mountings** are now used exclusively for new large instruments. Several designs of equatorial mounting have been used extensively over the years.

German Mounting

In this approach, the declination axis is at the end of the polar axis, which is on top of a pier to raise the telescope to a convenient height. This arrangement can point to any part of the sky, but it results in a lot of mechanical stress because the weight of the telescope has to be held at the end of the axis. A counterweight is used to balance the telescope on the declination axis, but this doubles the weight that the polar axis must support and limits the German mounting to relatively small, light telescopes.

Fork Mounting

In this scheme, the polar axis branches into a fork, while the declination axis holding the telescope is anchored on both ends by the two sides of the fork. This provides much stronger declination support for the telescope, but there is still no support for the weight at the end of the polar axis. In addition, there is limited room for the telescope at the bottom of the fork. If the telescope has only a small tube, it can be used to view all parts of the sky, but if it has a large tube it cannot be used to view along the polar axis. The fork mounting is practical for telescopes with mirror diameters of up to 2.5 m.

English Mounting

In this design, the polar axis is supported at the top and the bottom on vertical piers. This relieves the stress on the polar axis but transfers it to the declination axis. The English mounting isn't restricted to just part of the sky, as in the case with the fork mounting, but it is convenient for only half of the sky at one time. For one half of the sky the telescope is slung beneath the polar axis, making it easy to reach the focus for observing. For the other side of the sky, however, the telescope is on top of the polar axis, making it difficult to reach the focus, except by standing on a tall ladder. The difficulty can be solved by reversing the side of the polar axis on which the telescope is attached, but this only swaps the sides of the sky that the telescope can view easily.

Horseshoe Mounting

A solution to the problem of supporting telescopes larger than 2.5 m; it was introduced with the **Hale Telescope** in the late 1940s and used for all the 3- to 5-m telescopes built from 1950 to 1970. The **Canada-France-Hawaii Telescope**, finished in the early 1970s, was one of the last telescopes built using this plan. The forklike polar axis is mounted on the top end of a horseshoe-shaped support. The fork provides support for both ends of the declination axis, and the horseshoe mount provides support at the top end of the polar axis. Because the horseshoe is open, the telescope can be tilted all the way down to see objects that lie directly along the polar axis in the sky. For every mount, the piers on which the polar axis rest are separate from the floor and the remainder of the observatory building. The piers extend down through the building and into the ground, all the way to the level of solid rock. In this way, the telescope is isolated from any vibrations in the observatory building.

equivalent width

(continued from page 167)

curve with a shape defined by the **line profile**. The equivalent width is the width of a rectangle centered on a spectral line that, on a plot of intensity against wavelength, has the same area as the line.

Equuleus (abbr. Equ; gen. Equulei)

The Little Horse; an ancient but obscure northern **constellation** that represents a small horse's head and shoulders. It lies between Pegasus and Delphinus and ranks ahead only of Crux in size. Its only star brighter than magnitude 4 is Kitalpha (Alpha Equ), a binary star (visual magnitude 3.92, absolute magnitude 0.13, spectral type G0III+A5V, distance 186 light-years).

ER Ursae Majoris star (ER UMa star)

A variety of **SU Ursae Majoris star** (a type of dwarf nova) in which the interval between **super-outbursts** is unusually short. ER UMa stars typically spend a third to a half their time in super-outburst, with a super-cycle (the interval between super-outbusts) of only 20 to 50 days. When not in super-outburst these stars show frequent normal outbursts–about one every 4 days.

Eratosthenes of Cyrene (c. 276–c. 196 B.C.)

A Greek scholar who was the first person to determine the circumference of Earth. He compared the midsummer's noon shadow in deep wells in Cyrene (now Aswan on the Nile in Egypt) and Alexandria. Then, correctly assuming that the Sun's rays are virtually parallel (since the Sun is so far away) and knowing the distance between the two locations, he worked out the Earth's circumference to be 250,000 stadia. The exact length of a stadium is not known, so his accuracy is uncertain, but he wasn't far off the mark. Among his many other accomplishments, he accurately measured the tilt of Earth's axis and the distance to the Sun and Moon, and devised a method for finding all the prime numbers up to a given number (the Sieve of Eratosthenes).

Eridanus (abbr. Eri; gen. Eridani)

The River; a very large southern **constellation** that starts near the southwest corner of Orion and meanders down to the region of the South Pole. Although one of Ptolemy's original constellations, its southernmost extension, including its brightest star, was added later. The two Omicrons, o^1 (Beid) and o^2 (Keid), are not physically connected. Beid is a giant **F star** 125 light-years distant, while Keid is a multiple system, a mere 16 light-years away. Keid contains an easy, wide binary, the secondary of which is itself a binary made up of a red dwarf and the brightest (in terms of apparent magnitude) **white dwarf** in the sky. (See table, "Eridanus: Stars Brighter than Magnitude 4.0," and star chart 10.)

Eros (minor planet 433)

The first **near-Earth asteroid** to be found and the second-largest one known with dimensions of 35 × 13 × 13 km.

Eridanus: Stars Brighter than Magnitude 4.0

Star	Magnitude		Spectral	Distance	Position	
	Visual	Absolute	Type	(light-yr)	R.A. (h m s)	Dec. (° ′ ″)
α Achernar	0.45	−2.77	B3Vpe	144	01 37 43	−57 14 12
β Cursa	2.78	0.60	A3III	89	05 07 51	−05 05 11
θ Acamar	2.88	−0.59	A4III+A1V	161	02 58 16	−40 18 17
γ Zaurak	2.97	−1.19	M0.5IIICa-ICr	221	03 58 02	−13 30 31
δ Rana	3.52	3.74	K0IV	29	03 43 15	−09 45 48
υ^4	3.55	−0.15	B9V	178	04 17 54	−33 47 54
φ	3.56	0.17	B8IV	155	02 16 31	−51 30 44
τ^4 Angetenar	3.70	−0.79	M3.5IIIaCa	258	03 19 31	−21 45 28
χ	3.69	2.47	G8IIIbCNIV	57	01 55 58	−51 36 32
ε Epsilon Eridani	3.72	6.18	K2V	10.5	03 32 56	−09 27 30
υ^2 Theemini	3.81	−0.22	G8III	209	04 35 33	−30 33 45
53 Sceptrum	3.86	1.23	K1IIIb	109	04 38 11	−14 18 15
η Azha	3.89	0.83	K1IIIbBa0.2	133	02 56 26	−08 53 54

Eros A view looking down on the north polar region of Eros made from six images taken on February 29, 2000, by NEAR-Shoemaker from an orbital altitude of about 200 km. This vantage point highlights the major physiographic features of the northern hemisphere: the saddle seen at the bottom, the 5.3-km-diameter crater at the top, and a major ridge system running between the two features that spans at least one-third of the asteroid's circumference. *NASA/Johns Hopkins University*

It was discovered in 1898 by the German astronomer Gustav Witt (1866–1946), director of the Urania Observatory in Berlin, and independently on the same day by the French astronomer Auguste Charlois (1864–1910) in Nice. As a member of the **Amor group**, Eros moves in an orbit that crosses the orbit of Mars but does not intersect that of Earth, so there is no danger of it colliding with us. It came about as close as it ever gets on January 23, 1975, when its distance was about 0.15 AU (22 million km). A 90-kg person on Earth would weigh about 60 grams on Eros, and a rock thrown from the surface at 10 m/s (about a quarter the speed of a top pitcher's fast-ball) would escape into space. Although gravity on Eros is very weak it is strong enough to hold a spacecraft in orbit, as demonstrated when **NEAR-Shoemaker** entered orbit in February 2000. NEAR-Shoemaker spent a year circling around the asteroid, sending back 160,000 pictures, and spotting more than 100,000 craters, about a million house-sized (or bigger) boulders, and a layer of debris resulting from a long history of impacts. Finally, on February 12, 2001, the little probe–never designed to land–descended to the surface, returning its last image from a height of just 120 m, before touching down. Density 2.7 g/cm^3, spectral class S, albedo 0.16, rotational period 5.27 hours; orbit: semimajor axis 1.458 AU, perihelion 1.13 AU, aphelion 1.78, eccentricity 0.223, inclination 10.8°, period 1.76 years.

eruptive variable

A **variable star** that undergoes sudden and marked changes in brightness due to activity in its **chromosphere** or **corona**. The brightness variations may also be accompanied by an enhanced **stellar wind** or the ejection of shells of matter. Among the types of eruptive variable are **BY Draconis stars**, **FU Orionis stars**, **Gamma Cassiopeiae stars**, **irregular variables**, **nebular variables**, **R Coronae Borealis stars**, **RS Canum Venaticorum stars**, **S Doradus stars**, **T Tauri stars**, **UV Ceti stars**, and **Wolf-Rayet stars**. Eruptive variables should not be confused with **cataclysmic variables**.

escape velocity

The velocity an object needs to attain a **parabolic orbit**–the lowest-energy open orbit–and thus escape from a given gravitational field. For an object leaving Earth's surface this is 11.2 km/s. An *escape orbit* is any orbit whose **apoapsis** lies at infinity. This includes parabolic orbits and **hyperbolic orbits**.

Eskimo Nebula (Clown-face Nebula, NGC 2392)

A **planetary nebula** in **Gemini** that looks like a face peering out of the fur-lined hood of a parka; it was discovered by William **Herschel** in 1787. The parka is really a disk of material embellished with a ring of comet-shaped objects, with their tails streaming away from the central star. The Eskimo's face is formed from a bubble of material being blown into space by the central star's intense wind of high-speed material. The nebula is composed of two elliptically shaped lobes of matter streaming above and below the dying star. Scientists believe that a ring of dense material around the star's equator, ejected during its **red giant** phase, created the nebula's shape. This dense waste of material is plodding along at 115,000 km/hr, preventing high-velocity **stellar winds** from pushing matter along the equator. Instead, the 1.5-million-km/hour winds are sweeping the material above and below the star, creating the elongated bubbles. The bubbles are not smooth like balloons but have filaments of denser matter. Each bubble is about 1 light-year long and about 0.5 light-year wide. One possible explanation is that these objects formed from a collision of slow- and fast-moving gases.

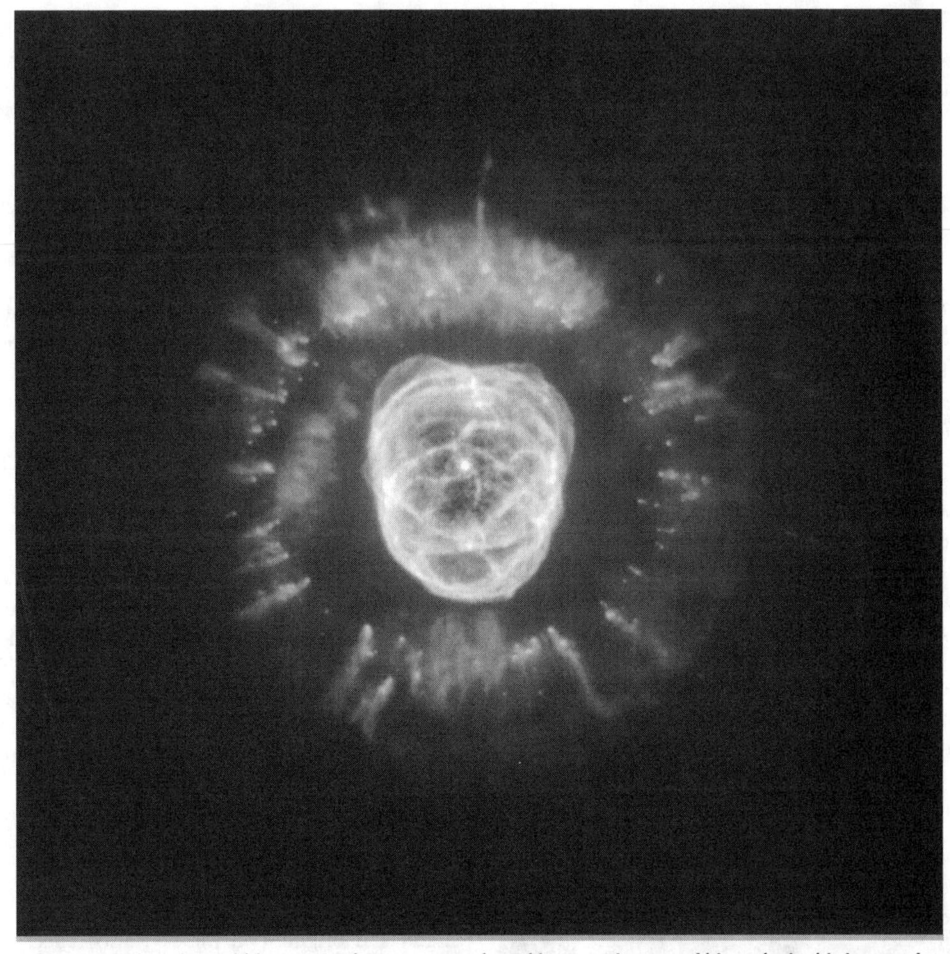

Eskimo Nebula The Hubble Space Telescope spots the Eskimo peering out of his parka in this image of the well-known planetary nebula taken in 2000. *NASA/STScI*

Visual magnitude	
Nebula:	8.5
Central star:	11
Angular size:	0.8′ × 0.7′
Age:	about 10,000 years
Distance:	3,000 light-years
Position:	R.A. 7h 29m; Dec. 20° 55′

Eta Aquarids

A strong annual **meteor shower,** best seen from the southern hemisphere, that ranges from April 19 to May 28 and reaches a peak **zenithal hourly rate** of about 60 meteors per hour on May 5 to 6. The Eta Aquarids consist of fast-moving (66 km/s) debris from **Halley's comet.** Radiant: R.A. 22h 32m; Dec. 0°.

Eta Aquilae (η Aql)

One of the sky's most prominent **Cepheid variables**; its variability was discovered in 1784 by Edward **Pigott**. Eta Aql lies 8° south of **Altair** and near one of **Aquila**'s (the Eagle's) talons. Eta also represents the head of the now-defunct constellation Antinous, who was honored in the sky by the Roman emperor Hadrian and depicted as being carried by Aquila. Eta, like other Type I Cepheids is a luminous yellow-white supergiant. It changes brightness by 0.8 magnitude and back again with a precise period of 7d 4h 14m 22s. As it dims, it dips to spectral class G and from a maximum temperature of 6,200 to 5,300 K.

Visual magnitude:	3.6 to 4.4
Absolute magnitude:	–3.91
Spectral type (at max.):	F6Ib
Surface temperature (max.):	6,200 K
Luminosity:	3,100 L_{sun}
Radius:	60 R_{sun}
Mass:	7 M_{sun}
Distance:	1,170 light-years

Eta Carinae (η Car)

One of the most massive and remarkable known stars. Surrounded by the largest diffuse nebula in the sky, the **Eta Carinae Nebula**, it is an **S Doradus star** with a mass of over 100 M_{sun} and a luminosity of about 4 million L_{sun}, putting it close to the theoretical limit of stellar stability. Only by shedding matter at the prodigious rate of 0.1 M_{sun} per year has it managed to stay in one piece. Its variability is extraordinary. In 1843, it reached a visual magnitude of –1, making it, briefly, the second brightest star after Sirius, despite its distance of some 7,500 light-years. Accompanying this visual brightening was an expulsion of 2 to 3 M_{sun} of material from the star's polar regions. This material, spewed from the star at speeds close to 700 km/s, formed two large, grayish, bipolar lobes, nicknamed the *Homonculus Nebula,* that have been photographed in spectacular detail by the Hubble Space Telescope. Each lobe currently expands at a rate of 2.4 million km/hr and spans about 6.4 trillion km. After the great eruption of the mid-nineteenth century, Eta Car faded in spurts to below naked-eye visibility, settling at about seventh magnitude. Recently, it has started to brighten again, gaining a full magnitude from 1950 to 1992 and is continuing its ascent. Eta Car emits powerfully across a range of wavelengths. At some infrared wavelengths, the star and its nebula are the brightest objects in the sky beyond the solar system. The mid-infrared emission originates in dust ejected by the star during giant mass-loss events within the past several hundred years. X rays come from an outer, horseshoe-shaped ring with an electron temperature of about 3 million K, which is about 2 light-years in diameter and was probably caused by an outburst that happened more than a thousand years ago. Regular, small-scale variations in the star's ultraviolet and X-ray output, with a period of 5.5 years, have led to the suggestion that Eta Car is actually a binary star. According to this theory, previous eruptions may have been due to the orbital interactions of the two stars. As for the star's powerful X-ray emission, most astronomers agree that this is the result of the collision of two dense **stellar winds**, but whether these emanate from the two stars of a close interacting binary system or from the fast and slow stellar winds of a single star remains unclear. One thing seems certain: Eta Carina is doomed to explode as a **supernova** in the not-too-distant future.

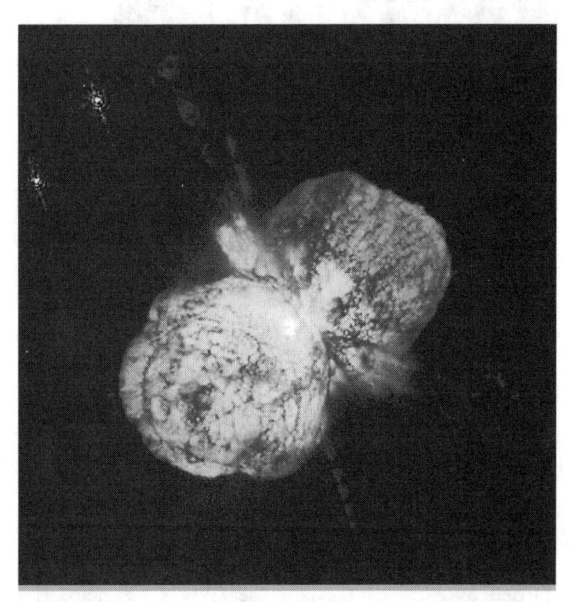

Eta Carinae A huge, billowing pair of gas and dust clouds are captured in this stunning image of the supermassive star Eta Carinae by the Hubble Space Telescope. *NASA/STScI/ Jon Morse*

Eta Carinae Nebula (NGC 3372)

A spectacular **diffuse nebula** in **Carina**–one of the largest and brightest in the sky; also called simply the *Carina Nebula.* The only reason it is not better known is that it cannot be seen from most of the northern hemisphere. In addition to **Eta Carinae** itself, it contains several stars that are among the hottest and most massive known, each about 10 times as hot, and 100 times as massive, as our Sun. The *Keyhole Nebula* (NGC 3324) is a dark cloud, whose round portion is some 7 light-years across, that is silhouetted against the brighter background near Eta Carinae.

Visual magnitude:	1.0
Angular diameter:	2°
Linear diameter:	200 light-years
Distance:	7,500 light-years
Position:	R.A. 10h 44m; Dec. –59° 52′

eucrite

The most common class of **achondrite** meteorite and a member of the **HED group** for which the parent body is believed to be the asteroid **Vesta**. Eucrites are basalts–

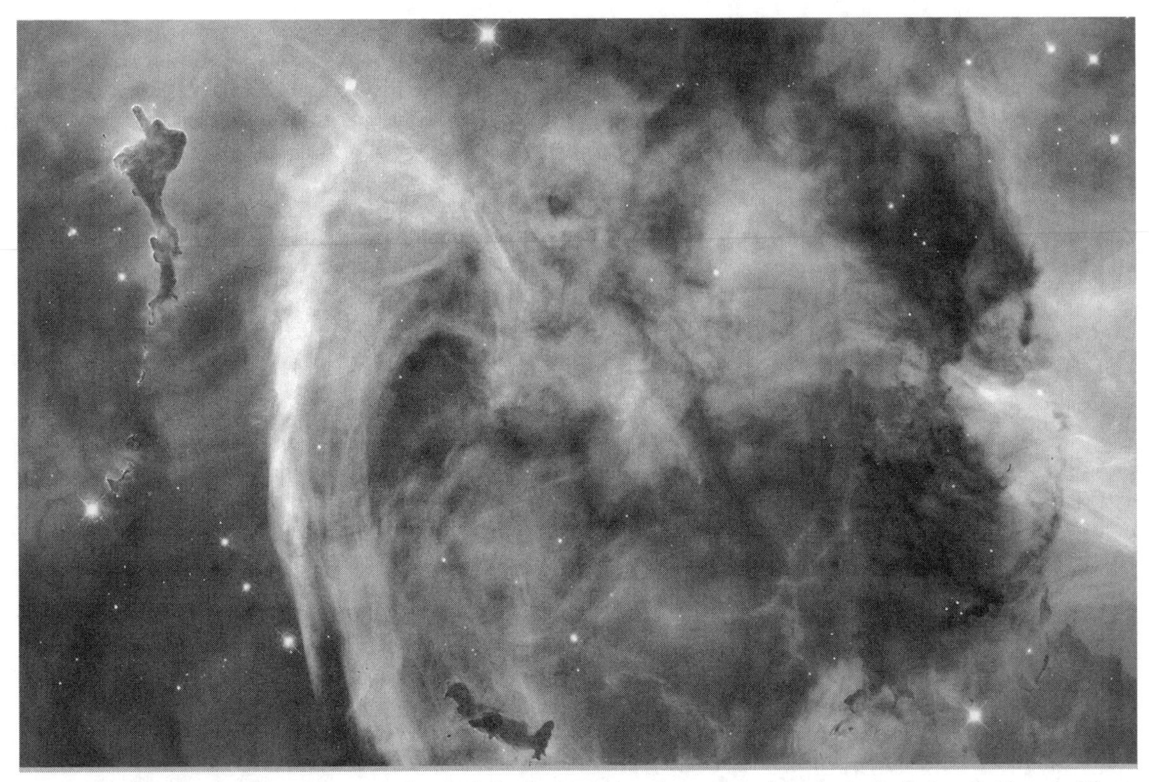

Eta Carinae Nebula A region of the Eta Carinae Nebula dominated by a large, approximately circular feature that is part of the Keyhole Nebula and lies adjacent to the extraordinary star Eta Carinae. *NASA/STScI/AURA*

volcanic rocks of magmatic origin–composed primarily of the calcium-poor **pyroxene** pigeonite and the calcium-rich **plagioclase feldspar** anorthite. Based on mineralogical and chemical differences, they have been placed into three distinct subgroups. The *non-cumulate eucrites* represent the upper crust of Vesta that solidified on a magma ocean after the core and the mantle had already been formed. The rare *cumulate eucrites* are the products of the gravitational settling of crystallized minerals, mainly pyroxene and plagioclase, within magma chambers trapped below Vesta's early crust. Finally, the *polymict eucrites* are breccias that contain more than 90% eucritic material and less than 10% diogenitic clasts. This 9:1 ratio is an arbitrary criterion to discriminate between polymict eucrites and the closely related **howardites**, the latter showing a more even distribution of eucritic and diogenitic clasts.

Eudoxus of Cnidus (c. 400–c. 347 B.C.)

A Greek mathematician and astronomer who devised a scheme of the heavens in which the Sun, Moon, and planets traveled around Earth on 27 geocentric spheres, rotating at different speeds and inclinations. This concept of planetary motion, refined by Callipus and ultimately by **Ptolemy**, provided the standard model of the cosmos for 2,000 years.

Eugenia (minor planet 45)

A main-belt **asteroid** around which, in 1998, a satellite was discovered–the first such discovery using a ground-based telescope. The little moon, named Petit-Prince, is about 13 km in diameter, and travels around Eugenia in a 5-day circular orbit at a distance of about 1,190 km. The orbit of the moon yields the mass of the main asteroid and hence its density, which turns out to be surprisingly low–only about 1.2 times that of water. Two possible explanations exist for this low density: either the asteroid is made mostly of ice, or it is made of rocks with big gaps in between. Because most asteroids, including Eugenia, are very dark, they cannot have icy surfaces. Eugenia might, however, be the burned-out remains of a comet, with a dark coating and an icy interior. The more popular view, though, is that it is a **rubble-pile asteroid**–a highly porous jumble of fairly loose rocks. Diameter 226 km, spectral class FC, rotational period 5.699 hours; orbit: semimajor axis 2.721 AU, eccentricity 0.083, inclination 6.61°, period 4.49 years.

Euler, Leonhard (1707–1783)

A Swiss mathematician who, along with Joseph **Lagrange**, Pierre **Laplace**, and others, helped develop the science of **celestial mechanics**. He applied powerful new mathematical techniques to problems of cometary orbits, planetary **perturbations**, and the tides. He also refined the theory of the Moon's motion and calculated more accurate orbits for Jupiter and Saturn.

Eunomia family

A small family of S- and C-class **asteroids** that orbit in the middle of the main asteroid belt at a mean distance from the Sun of 2.6 to 2.7 AU and a mean inclination of 12°. The prototype and largest member is (15) Eunomia, discovered in 1851 by Annibale de Gasparis (1819–1892) in Naples; diameter 272 km, spectral class S, semimajor axis 2.644 AU, perihelion 2.151 AU, aphelion 3.136 AU, eccentricity 0.186, inclination 11.75°, period 4.30 years.

Europa

See article, pages 175–176.

Europa (minor planet 52)

The seventh largest known **asteroid**; it was discovered by Hermann Goldschmidt in 1858. Diameter 303 km, albedo 0.06, semimajor axis 3.098 AU, perihelion 2.786 AU, aphelion 3.410 AU, eccentricity 0.101, inclination 7.47°.

European Northern Observatory (ENO)

The name by which the Instituto de Astrofisica de Canarias (IAC) and its observatories, the Teide Observatory, on Tenerife, and the **Roque de los Muchachos Observatory**, on La Palma, are collectively known.

European Southern Observatory (ESO)

An intergovernmental organization with 10 European member states (Belgium, Denmark, France, Germany, Italy, the Netherlands, Portugal, Sweden, Switzerland, and United Kingdom) that operates two major observatories in Chile, **La Silla Observatory** and the **Very Large Telescope** at Paranal Observatory, and has its headquarters in Garching, near Munich, Germany. ESO also has a major role in the **Atacama Large Millimeter Array** (ALMA) project and is carrying out a concept study for the Overwhelmingly Large Telescope (see **extremely large telescope**).

EUVE (Extreme Ultraviolet Explorer)

A NASA satellite, launched in June 1992, that carried out the first all-sky survey in the extreme **ultraviolet**, below the Lyman limit and into the soft X-ray region (roughly 7 to 76 nm). It also conducted a deep survey of a strip of the sky along the ecliptic with very high sensitivity and follow-up spectroscopic observations on bright UV sources.

event horizon

The boundary surrounding a **black hole** from within which it is impossible for matter or energy to escape the black hole's gravitational pull. In other words, the escape velocity at the event horizon is greater than the speed of light.

Evershed effect

The horizontal outward flow of gas in the **penumbra**e of a **sunspot**; the effect is named after its discoverer, the English astronomer John Evershed (1864–1956). The maximum outflow velocity is about 2 km/s.

excitation

The lifting of **electrons** from lower energy levels in atoms to higher energy levels by the injection of energy. This energy can come from two main sources. In *radiative excitation,* a **photon** is absorbed whose energy is equal to the difference between the energy levels. In *collisional excitation,* the energy needed for a particular electron jump comes from the impact between the atom and another particle. In both cases, when electrons fall back to lower energy levels they give off photons at specific wavelengths. The *excitation temperature* of a gas or plasma is the temperature determined by the proportions of atoms or ions in the **ground state** and in excited states.

exit pupil

The minimum diameter of the light beam leaving an **eyepiece** through which all of the light from the eyepiece passes. It can be calculated in either of two ways: by dividing the diameter of the telescope's primary objective by the magnification, or by dividing the **focal length** of the eyepiece by the **focal ratio** (f-number) of the telescope. For example, a telescope with a 20-cm (8-inch) primary mirror using a magnification 38× results in an exit pupil of 5.3 mm. The observer' eye should be located at the exit pupil to see the fullest and brightest field of view.

EXOSAT (European X-ray Observing Satellite)

A European Space Agency satellite for X-ray astronomy, launched in May 1983, that carried out observations of variable X-ray sources in the energy range 0.05 to 20 keV. It was equipped with three instruments: an imaging telescope, a proportional counter, and a gas scintillation proportional counter array. EXOSAT made 1,780 observations of a wide variety of objects, including active galactic nuclei, stellar coronae, cataclysmic variables, X-ray binaries, clusters of galaxies, and supernova remnants.

exosphere

The extremely tenuous, outermost layer of Earth's **atmosphere**; it lies above the **ionosphere** and extends

(continued on page 176)

Europa

The second closest and smallest of the **Galilean moons** of Jupiter, the sixth largest moon in the solar system (slightly smaller than our own Moon), and an object of tremendous interest because of the strong possibility that it may have a subsurface watery ocean. Europa's bright icy surface is unlike any other in the solar system, mostly flat with few features more than a few hundred meters high and very few craters, none of them large, suggesting that the surface is **young**, no more than 30 million years old. It has two main types of terrain: one is mottled, brown or gray in color, and consists of mainly small hills, while the other consists of large smooth plains crisscrossed by numerous cracks, some curved, some straight, extending up to thousands of kilometers. The cracked surface strongly resembles pack-ice on polar seas during spring thaws on Earth. Beneath this icy crust, which may be 12 to 15 kilometers thick, may lie a layer kept liquid by heat generated tidally due to a gravitational tug-of-war with Jupiter and the other Galilean satellites. The notion that this layer is made of saltwater has been strengthened by the observation, by the **Galileo** spacecraft, that Europa's weak magnetic field varies periodically as Europa passes through Jupiter's powerful magnetosphere. This points to the existence

Europa A high-resolution image of Europa showing crustal plates ranging up to 13 km across, which have been broken apart and rafted into new positions, superficially resembling the disruption of pack-ice on polar seas during spring thaws on Earth. The size and geometry of these features suggest that motion was enabled by ice-crusted water or soft ice close to the surface at the time of disruption. The area shown is about 34 km by 42 km, and the resolution is 54 meters. This picture was taken by the Galileo spacecraft on February 20, 1997, from a distance of 5,340 km. *NASA/JPL*

Europa A close-up of the icy surface of Europa taken by the Galileo spacecraft on December 20, 1996, during its fourth orbit of Jupiter. The view covers an area of about 11 km by 16 km and has a resolution of 26 meters. A flat smooth area about 3.2 km across, seen in the left part of the picture, has resulted from flooding by a fluid which erupted onto the surface and buried sets of ridges and grooves. The smooth area contrasts with a rugged patch of terrain farther east, to the right of the prominent ridge system running down the middle of the picture. The rugged patch of terrain is 4 km across and represents localized disruption of the complex network of ridges in the area. *NASA/JPL*

of conducting material beneath Europa's surface. The possible presence of a vast amount of liquid water has led to much speculation about the prospects for Europan life. Reddish spots on the moon's surface may indicate pockets of warmer ice rising from below in a "lava lamp" action. The red coloration is probably a result of organic chemicals that have been brought up from the subterranean sea. Europa appears to have a layered structure with a small iron or iron-sulfur core, surrounded by a mantle of silicate rock. A tenuous atmosphere of oxygen has been detected.

Diameter:	3,122 km
Mass (Earth = 1):	0.0083
Density:	3.01 g/cm^3
Surface gravity (Earth = 1):	0.135
Escape velocity:	2.02 km/s
Visual albedo:	0.64
Overall composition:	40% ice, 60% rock/iron
Rotational period:	3.551 days
Orbit	
Mean distance from Jupiter:	670,900 km
Eccentricity:	0.009
Inclination:	0.47°
Period:	3.551 days

exosphere

(continued from page 174)

from the so-called *exobase,* at a height of about 500 km, to the edge of interplanetary space. At or below the exobase, the atmosphere is sufficiently dense that collisions dominate the motion of gas molecules and atoms. Above the exobase, on the other hand, collisions are so infrequent that atoms moving with sufficient velocity have a high probability of escaping from Earth's gravitational field into interplanetary space. Escaping atoms make up only a portion of the exospheric hydrogen, however. There is also a gravitationally bound component that consists both of atoms following ballistic trajectories and of satellite atoms that orbit the Earth for some time before returning to the denser atmosphere. The density and structure of the exosphere are influenced by a number of factors, including variations in the temperature and density of the atmosphere below the exobase, **photoionization** and ionization by impact with solar wind particles, charge exchange with the plasma of the plasmasphere, and **radiation pressure** exerted by solar far-ultraviolet photons.

exposure age

The interval during which a **meteorite** was an independent body in space; in other words, the time between when a meteoroid was broken off its parent body (such as an asteroid) and its arrival on Earth as a meteorite. It is also known as the *cosmic-ray exposure age* because it is estimated from the observed effects on the meteorite of bombardment by **cosmic rays** from the Sun and the rest of the Galaxy. As cosmic rays strike the meteorite, they produce characteristic **isotopes**, both radioactive, such as helium-3, neon-21, and argon-38, and stable. The longer a meteorite was exposed to cosmic rays, the more of these new isotopes are found to be present. Further dating information comes from an analysis of the fission tracks (thin trails left in a substance by a fast-moving atomic nucleus) that cosmic rays cause. Exposure ages range typically from a few million to a few hundred million years. To obtain the total age of a meteorite, including its time on Earth, the exposure age must be added to the **terrestrial age**.

extended source

(1) A source whose angular size exceeds the **resolution** of the instrument being used to observe it. Extended sources, such as many galaxies and nebulae, can be resolved by telescopes, whereas most stars cannot and are therefore classified as **point sources**. (2) In radio astronomy, formerly a source whose angular extent could be measured, as distinguished from a point source. Now, a source that has a large angular extent and is strongest at longer wavelengths, as distinguished from a **compact source**.

extragalactic background light (EBL)

The faint diffuse light of the night sky that comes from distant but unresolved sources outside the Milky Way. EBL at optical, ultraviolet, and near-infrared (less than about 5 microns) wavelengths is thought to consist mainly of **redshift**ed starlight from unresolved galaxies, with possible additional contributions from stars or gas in intergalactic space, and from decaying elementary particles. In the mid- and far-infrared, the main contribution is thought to be redshifted emission from **dust** particles,

extrasolar planet

Scores of planets have been discovered in orbit around other stars within about 200 light-years of the Sun. (See illustration, page 178.) Curiously, the very first extrasolar planets to be confirmed were not among these relatively nearby worlds but were oddball **pulsar planets** circling a couple of neutron stars, thousands of light-years away. The first extrasolar planet around an ordinary, sunlike star was found by the Swiss astronomers Michel Mayor and Didier Queloz as a companion to 51 Pegasi in 1995. Other similar discoveries were quickly announced. All involved giant planets, typically with several times the mass of Jupiter and presumably **gas giants**, in either very tiny orbits (closer than Mercury is to the Sun) or in highly elliptical orbits. The idea soon took hold that these planets could not have formed in their present locations but must have formed farther out from their host stars and then, in their infancy, been deflected inward by some process such as the gravitational interaction with another massive planet. The discovery, early on, of many "hot Jupiters" in small or elongated orbits is a selection effect that stems from the main method used in planet-hunting, which involves looking for the tiny wobbling motion produced in stars by any planets that are going around them. The more massive the planet and the smaller its orbit, the bigger the wobbling motion that it creates. As planet searches have become more sensitive and extensive, so they have begun to reveal the presence of Jupiter-like worlds in more Jupiterlike locations—fairly circular orbits that are several AU from the central stars. We are now entering a phase in which the general properties of planetary systems can be compared to those of the solar system and in which much smaller planets, ultimately with masses as low as that of Earth, can be detected. The next 10 to 15 years will reveal how often planetary systems have an architecture like ours, with terrestrial, rocky planets close in and gaseous giant planets far out. This has a strong bearing on the quest for **life in the universe.** (See table, "Known Extrasolar Planets," pages 179–181.)

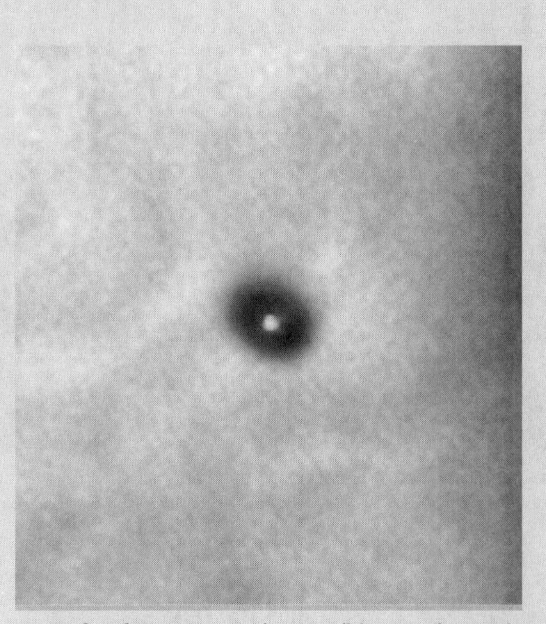

extrasolar planet A protoplanetary disk—an early stage in the process of planet formation—around a young star in the Orion Nebula, photographed by the Hubble Space Telescope in 1995. The dark disk shows up because it is silhouetted against the backdrop of the Orion Nebula's luminous gas. The glow in the center of the disk is a newly formed star, barely 1 million years old. *NASA/ESA/Mark McCaughrean (Max-Planck-Institute for Astronomy) and C. Robert O'Dell (Rice University)*

heated by starlight in galaxies. Studies of the EBL spectrum can serve as important tracers of the history of the formation of stars and galaxies.

extragalactic reference frame

A reference frame tied to remote extragalactic objects, such as **quasars**. Such a coordinate system forms a true inertial reference frame because the reference objects are at such great distances that their motions across the sky are undetectable. The extragalactic reference frame that has been adopted by the astronomical community is known as the **International Celestial Reference Frame** (ICRF).

extrasolar planet

See article, this page.

extrasolar planet An artist's conception of an extrasolar planet. *NASA/Greg Bacon (STScI)*

extreme horizontal branch star (EHB star)

The hottest variety of stars on the horizontal branch of the **Hertzsprung-Russell diagram**, with temperatures of about 25,000 K; also known as *extended horizontal branch stars*. EHB stars are believed to account for the faint **ultraviolet** glow shown by many **elliptical galaxies**, which otherwise consist of old, cool stars. **Subdwarf** B (sdB) stars are low-mass evolved extreme horizontal branch stars in which core **helium-burning** has just started.

extremely large telescope

A ground-based optical telescope, with a mirror 30 to 100 m in diameter and a resolution of a few milliarcseconds, that would complement and match the performance of equivalent instruments at other wavelengths, notably the **Next Generation Space Telescope** and the **Atacama Large Millimeter Array**. Among the candidates under study for such a telescope are the University of California and Caltech's 30-m California Extremely Large Telescope (CELT), the **National Optical Astronomy Observatory**'s 30-m Giant Segmented Mirror Telescope (GSMT), the 50-m Euro50 Telescope (proposed by a consortium of institutes and groups in Finland, Ireland, Spain, Sweden and the United Kingdom), and the **European Southern Observatory**'s Overwhelmingly Large Telescope (OWL). OWL is the most ambitious, with a mirror 100 m in diameter supported by a structure 200 m tall and weighing 40,000 tons. It would have 10 times more light-gathering power than all the other telescopes that have ever been built combined, and could be completed by 2015 for a cost of less than $1 billion. Among the many uses of these colossal instruments would be obtaining spectra of the atmospheres of **extrasolar planets** to allow astronomers to look for signatures of life.

extrinsic variable

A **variable star**, the light fluctuations of which are not due to changes in the star itself but to an external effect such as eclipses, obscuration, rotation, or tidal distortion.

(continued on page 181)

Known Extrasolar Planets

	Host Star		Planet					
	Distance (light-yr)	Spectral Type	Mass ($M_{Jupiter}$)	Period (d)	Semimajor Axis (AU)	Eccentricity	Year of Discovery
OGLE-TR-56	4,900	G	0.90	1.2	0.023	–	2003
HD 73256	119	G9V	1.85	2.549	0.037	0.04	2003
HD 83443	142	K0V	0.35	2.986	0.038	0.00	2000
HD 46375	109	K11V	0.25	3.024	0.041	0.02	2000
HD 187123	163	G5V	0.54	3.097	0.042	0.01	1998
HD 179949	88	F8V	0.93	3.092	0.045	0.00	2000
Tau Boo	51	F7V	4.14	3.313	0.047	0.04	1997
BD-10 3166	–	G4V	0.48	3.487	0.046	0.05	2000
HD 75289	94	G0V	0.46	3.508	0.047	0.01	2000
HD 209458	153	F8V	0.63	3.524	0.046	0.02	2000
HD 76700	195	G6V	0.19	3.971	0.049	0.00	2002
51 Peg	50	G5V	0.46	4.231	0.052	0.01	1995
Ups And	44	F8V	0.69b	4.617	0.059	0.01	1997
HD 49674	133	G5V	0.12	4.948	0.057	0.00	2002
HD 68988	192	G2V	1.90	6.276	0.071	0.14	2001
HD 168746	141	G5V	0.24	6.400	0.066	0.00	2002
HD 217107	64	G8IV	1.29	7.130	0.072	0.14	1998
HD 162020	102	K2V	13.73	8.420	0.072	0.28	2002
HD 130322	98	K0V	1.15	10.72	0.092	0.05	2000
HD 108147	126	G0V	0.41	10.90	0.079	0.20	2002
HD 385296	138	G4IV	0.78b	14.31	0.129	0.28	2001
55 Cnc	41	G8IV	0.84b	14.65	0.115	0.02	1997
G J 86	36	K1V	4.23	15.80	0.117	0.04	2000
HD 195019	13.8	G4IV	3.55	18.20	0.136	0.02	2001
HD 6434	132	G3IV	0.48	22.09	0.154	0.30	2000
Gliese 876	15.3	M4V	0.56c	30.12	0.130	0.27	2001
Rho CrB	57	G2V	0.99b	39.81	0.224	0.07	1997
55 Cnc	41	G8IV	0.21c	44.28	0.241	0.34	2002
HD 74156	211	G0V	1.55b	51.60	0.276	0.65	2002
HD 168443	126	G5IV	7.64b	58.10	0.295	0.53	1999
Gliese 876	15.3	M4V	1.89b	61.02	0.207	0.10	2001
HD 3651	145	K0V	0.20	62.23	0.284	0.63	2003
HD 121504	145	G2V	0.89	64.62	0.317	0.13	2000
HD 178911	125	G5V	6.46	71.50	0.326	0.14	2001
HD 16141	117	G5V	0.22	75.80	0.351	0.00	2000
HD 114762	132	F9V	10.96	84.03	0.351	0.33	1989
HD 80606	190	G5V	3.43	111.8	0.438	0.93	2001
HD 219542B	178	G7V	0.30	112.1	0.460	0.32	2003
70 Vir	59	G5V	7.41	116.7	0.482	0.40	1996

(continued)

Known Extrasolar Planets *(continued)*

	Host Star		Planet				
	Distance (light-yr)	Spectral Type	Mass ($M_{Jupiter}$)	Period (d)	Semimajor Axis (AU)	Eccentricity	Year of Discovery
HD 52265	91	G0V	1.14	119.0	0.493	0.29	2000
HD 1237	57	G6V	3.45	133.8	0.505	0.51	2000
HD 37124	108	G4V	0.86	153.0b	0.543	0.20	2000
HD 73526	309	G7V	3.63	188.0	0.647	0.52	2002
HD 82943	90	G0V	0.88c	221.6	0.728	0.54	2000
HD 8574	144	F8V	2.08	228.5	0.770	0.30	2002
HD 169830	118	F8V	2.95	230.4	0.823	0.34	2001
Ups And	44	F8V	1.90c	241.3	0.829	0.28	1999
HD 12661	121	G6V	2.30b	263.3	0.823	0.35	2000
HD 89744	127	F7V	7.17	256.0	0.883	0.70	2000
HD 202206	151	G6V	14.68	258.9	0.768	0.42	2002
HD 40979	108	F8V	3.16	260.0	0.818	0.26	2002
HD 150706	89	G0V	1.00	264.9	0.820	0.38	2002
HD 134987	82	G5V	1.63	265.0	0.821	0.37	2000
Iota Hor	51	G0V	2.12	312.0	0.909	0.15	2000
HD 92788	105	G5V	3.88	337.0	0.969	0.28	2001
HD 142	76	G11V	1.36	338.0	0.980	0.37	2002
HD 28185	129	G5V	5.70	383.0	1.03	0.07	2001
HD 177830	192	K0	1.24	391.0	1.10	0.40	2000
HD 108874	223	G5V	1.65	401.0	1.07	0.20	2002
HD 4203	254	G5V	1.64	406.0	1.09	0.53	2001
HD 128311	54	K0V	2.63	414.0	1.01	0.21	2002
HD 27442	59	K2IV	1.32	415.0	1.16	0.06	2003
HD 210277	72	G0V	1.29	436.6	1.12	0.45	1999
HD 82943	90	G0V	1.63b	444.6	1.16	0.41	2001
HD 19994	73	F8V	1.66	454.0	1.19	0.20	2001
HD 20367	88	G0V	1.12	500.0	1.28	0.23	2003
HD 114783	67	K2V	0.99	501.0	1.20	0.10	2001
HD 147513	42	G4V	1.00	540.4	1.26	0.52	2003
Iota Dra	98	K2III	8.68	550.0	1.34	0.71	2000
HD 222582	137	G5V	5.20	577.1	1.36	0.76	2000
HD 23079	113	F9V	2.76	628.0	1.48	0.14	2002
HD 160691	50	G3IV/V	1.74	637.3	1.48	0.31	2002
HD 141937	109	G2V	9.67	658.8	1.48	0.40	2002
16 Cyg	70	G5V	1.68b	798.4	1.69	0.68	1997
HD 4208	107	G5V	0.81	829.0	1.69	0.04	2001
HD 114386	91	K3V	0.99	872.0	1.62	0.28	2002
HD 213240	133	G4IV	4.49	951.0	2.02	0.45	2001
HD 10697	98	G5IV	6.08	1074	2.12	0.11	1999

Known Extrasolar Planets

	Host Star		Planet				
	Distance (light-yr)	Spectral Type	Mass ($M_{Jupiter}$)	Period (d)	Semimajor Axis (AU)	Eccentricity	Year of Discovery
47 UMa	46	G0V	2.56	1091	2.09	0.06	1996
HD 190228	216	G5IV	3.44	1112	1.98	0.52	2000
HD 114729	114	G0V	0.88	1136	2.08	0.33	2002
HD 2039	293	G2IV/V	5.10	1190	2.20	0.69	2002
HD 136118	171	F9V	11.91	1209	2.39	0.37	2002
HD 50554	102	F8V	3.72	1254	2.32	0.51	2002
HD 216437	86	G2IV	2.09	1294	2.38	0.34	2002
Ups And	44	F8V	3.75d	1284	2.52	0.27	1999
HD 196050	153	G4V	2.81	1300	2.41	0.20	2002
HD 216435	109	G0V	1.23	1326	2.60	0.14	2002
HD 106252	122	G0V	6.79	1503	2.53	0.57	2002
HD 12661	121	G6V	1.56c	1445	2.56	0.20	2002?
HD 23596	170	F8V	8.00	1558	2.87	0.31	2002
HD 30177	178	G8V	7.64	1620	2.65	0.21	2002
HD 168443	126	G5IV	16.96c	1770	2.87	0.20	2001
Her 14	59	K0V	3.09	1775	2.87	0.37	2000
HD 37124	108	G4V	1.00c	1550	2.50	0.40	2002
HD 38529	138	G4IV	12.78c	2207	3.71	0.33	2001
HD 72659	168	G0V	2.54	2185	3.24	0.18	2002
HD 39091	59	G1V	10.39	2280	3.50	0.63	2002
HD 74156	211	G0V	7.46c	2300	3.47	0.40	2002
HD 33636	94	G0V	9.30	2404	3.50	0.52	2001
Epsilon Eri	11.4	K2V	0.92	2550	3.39	0.43	2000
GJ 777A	52	G6IV	1.15	2613	3.65	0.00	2002
47 UMa	46	G0V	0.76c	2640	3.78	0.00	2002
55 Cnc	41	G8V	4.05d	5360	5.90	0.16	1999

extrinsic variable

(continued from page 178)

The main categories are **eclipsing binary**, which includes **Algol stars, Beta Lyrae stars**, and **W Ursae Majoris stars**, and **rotating variables**, which includes **Alpha² Canum Venaticorum stars, BY Draconis stars, ellipsoidal variables, FK Comae Berenices stars**, and **SX Arietis stars**. Compare with **intrinsic variable**.

eye relief

The distance from the observer's cornea to the nearest optical surface of the **eyepiece**; this is the distance at which the entire (or largest) **field of view** is visible.

eyepiece

See article, pages 182–183.

eyepoint

The position in which the eye is placed to see the full circumference of the image. See **eye relief**.

Eyes, the (NGC 4435 and 4438)

A pair of adjacent galaxies in the **Markarian Chain**, 20′ east of M86 in the central core of the **Virgo Cluster**, which were nicknamed by the nineteenth-century observer L. S. Copeland. The likeness to a pair of eyes is emphasized

(continued on page 183)

eyepiece

A combination of lenses, also known as an *ocular,* used to magnify the image formed by the **objective** of a telescope. In practice, eyepieces contain at least two lenses: the *field lens,* which faces the objective and collects the light from it, and the *eyelens,* which faces the observer and magnifies the image. A *field stop* (a circular aperture) inside the eyepiece limits the **field of view**, helping to give it a sharp edge. There are various types of eyepiece designs, from very simple **achromatic lens**es to complex eyepieces with many optical elements. Usually, the more complex an eyepiece, the more optical corrections, and the better **eye relief** and wider field of view it has. On the other hand, adding many glass surfaces dims the image and may also increase internal reflections, or "ghosting."

Huygenian Eyepiece

One of the earliest compound lenses, introduced by Christian **Huygens** in 1664. It uses two plano-**convex** lenses, both mounted with the convex surface toward the objective. Huygenians offer good eye relief but suffer badly from **spherical aberration** and various other defects. They are often included with the least expensive and least effective amateur telescopes but are only useful at **focal ratios** faster than about f/12.

Ramsden Eyepiece

Another old design, invented by the English instrument-maker Jesse Ramsden (1735–1800) in 1782. In this case, the two plano-convex lenses both have their flat sides facing outward. Ramsdens are less prone to spherical aberration than are Huygenians and are free of **coma**, but fare badly with regard to **chromatic aberration** and have poor eye relief and prominent ghosts. Their apparent field of view is quite small, as little as 30°, and they don't perform well at focal ratios shorter than about f/9. However, good ones are surprisingly effective at longer focal ratios for lunar, planetary, and double-star work.

Kellner Eyepiece

A three-element, fairly aberration-free improvement on the Ramsden thanks to the addition of an achromat for the eyelens. Sometimes called *achromatic Ramsdens,* Kellners, developed by Carl Kellner in 1849, have a fairly wide field of view, up to 50°, and good eye relief, but need excellent internal coatings to minimize ghosting. They work best in long focal length telescopes and offer a good balance between performance and economy.

Orthoscopic Eyepiece

A four-element eyepiece (typically with a simple lens nearest the eye and a cemented triplet farther away), more expensive than a Kellner, that is considered among the best eyepieces for lunar, planetary, and double-star work, and a good choice for eyeglass wearers because of their long eye relief. Although an excellent all-around performer offering crisp images, the orthoscopic has a somewhat restricted field of view, in the 35 to 50° range. It was invented by the German physicist Ernst Abbe (1840–1905) in 1880.

Plossl Eyepiece

A four- or five-element design (consisting of two doublets, sometimes with an intermediate lens) offering excellent performance, especially in the 15- to 30-mm size range. Plossls, invented by the Austrian optician G.S. Plossl in 1860, provide good color correction, flat field, adequate eye relief (except for 10 mm and shorter lenses), and a moderate field of view (40 to 50°); some **astigmatism** is present, especially at the edge of the field.

Wide-Field Eyepieces

Eyepieces that offer a field of view of 60 to 80°, providing spectacular panoramas of the night sky, but at some cost, both financially and observationally. They often won't work in certain telescope drawtubes and with particular telescope designs; instruments lacking good optical performance around the edge of field will have their worst traits emphasized; and at high magnifications with some telescopes, some ultra-wide eyepieces are unusable.

Erfle eyepiece. The earliest type of wide-field eyepiece, developed in 1917 by Heinrich Erfle (1884–1923).

Erfles use three lenses, at least two of which are doublets, to give a field of view of up to 68°. Although they suffer from ghost images and astigmatism at the edges of the field, which makes them unsuitable for planetary viewing, they are relatively inexpensive.

König eyepiece. A short-focal-length version of the Erfle, giving high magnification and a field of view of up to 70°. Königs may contain anywhere from four to seven simple lenses, grouped into various combinations of cemented doublets and singlets. They are named after the German optician Albert König (1871–1946).

Nagler eyepiece. A seven-element design, introduced in 1982 by the American optician Al Nagler (1944–), with a remarkable 82° field of view. Naglers come in 2-in. barrel size only, and are heavy–up to 1 kg. Although expensive, they are an excellent choice for observers with very fast focal ratio telescopes (up to f/4).

Eyes, the (NGC 4435 and 4438)

(continued from page 181)

by the fact that NGC 4435 and NGC 4438 are both elongated in a SSW–NNE direction. There is debate about whether NGC 4435 and NGC 4438 are interacting, with most authorities doubting it. NGC 4435 (magnitude 17.5; R.A. 12h 27.7m, Dec. +13° 5′) is oval with a bright core and appears completely free of any tidal disturbances due to its sky neighbor. NGC 4438 (magnitude 17.5; R.A. 12h 27.8m, Dec. +13° 1′) looks as if it is the result of a merger between two galaxies, its convulsed appearance probably being due to this rather than any gravitational effect of the other "eye."

F

F star

A yellow-white star of **spectral type** F and a surface temperature of 6,000 to 7,400 K. F stars on the **main sequence**, such as **Procyon**, are typically several times more luminous than the Sun and have masses of 1.2 to 1.6 M_{sun}. F-type **supergiants**, on the other hand, such as **Canopus** and **Polaris**, can shine as brilliantly as 30,000 Suns and weigh up to 12 M_{sun}. In the spectra of F stars, going from F9 (cooler) to F0 (hotter), the Balmer lines of hydrogen strengthen and the **H and K lines** of calcium weaken along with the lines of other metals.

Faber-Jackson relation

An observed correlation between the random speeds of stars in the center of an **elliptical galaxy** and the intrinsic **luminosity** of the galaxy–the higher the random speeds, the more luminous the galaxy. Since the speeds of stars can be directly measured by the **Doppler shift** in their spectra, the Faber-Jackson relation gives a way of estimating an elliptical galaxy's intrinsic luminosity. By comparing this with the *observed* brightness, the distance to the galaxy may be inferred. It is named after the American astronomers Sandra Moore Faber (1944–) and Robert Earl Jackson (1949–). See also the **Tully-Fisher relation**, which is applied to spiral galaxies.

Fabricius, David (1564–1617), and Johannes (1587–1616)

German father and son astronomers. David was a Lutheran pastor in Osteel, East Frisia (northwest Germany), who corresponded with Johannes **Kepler** and discovered the first known variable star, **Mira**, in 1596. Early in 1611, his son Johannes, then at university, returned from the Netherlands with a telescope. On March 9, Johannes turned the telescope at the rising Sun and saw several dark spots on it. He and his father continued making solar observations, quickly switching, however, to the projection method by means of a camera obscura. Over the next several months they tracked spots as they moved across the Sun's face and found that, a dozen or so days after they had disappeared from the western edge of the Sun, they reappeared on the eastern edge. In a tract dedicated June 1611, Johannes wrote on **sunspots**, *De Maculis in Sole Observatis, et Apparente earum cum Sole Conversione Narratio* (Narration on Spots Observed on the Sun and their Apparent Rotation with the Sun), giving his opinion that they were on the Sun and that the Sun therefore probably rotated on its axis (a notion already suggested by Giordano **Bruno** and Kepler). Because of the lack of a powerful patron who might have publicized the book, it drew little attention, and by the time Kepler became aware of its existence it was eclipsed by Christoph Scheiner's first publication on sunspots (January 1612). Johannes's diffidence may have been caused by a disagreement with his father about the nature of sunspots. In December 1611, David Fabricius wrote to Michael **Maestlin** (Kepler's old teacher) that he didn't believe the spots were on the Sun's body, though the center of their motions clearly lay in the Sun. Little else is known about Johannes Fabricius, except that he died at the age of 29. A year later his father was killed when an irate peasant, whom he had accused of stealing a goose, hit him over the head with a shovel.

Fabry, Marie Paul Auguste Charles (1867–1945)

A French physicist who coinvented the **Fabry-Pérot interferometer**, with Alfred Pérot (1863–1925), and suggested the use of the **Fabry lens**. He is also known for his research into light in connection with astronomical phenomena, and as the discoverer of the **ozone layer** in the upper atmosphere. He became a professor at the University of Marseilles (1904) and the Sorbonne (1920).

Fabry lens

A lens that forms an image of the telescope's primary mirror onto the photocathode of a photomultiplier tube and, in this way, serves as a key element of a **photometer**. A mask, placed in front of it, isolates the object whose light intensity is to be measured. The use of such a lens was first proposed by the French physicist Charles **Fabry**.

Fabry-Pérot interferometer

A type of optical **interferometer** that, when placed in front of the **diffraction grating** in a **spectrograph**, can reveal fine details in the spectrum of a galaxy, nebula, or other extended object that are beyond the scope of a grating alone. Light from an object is passed through an *etalon*–a pair of partially silvered, parallel glass plates separated by an adjustable air space–that uses interference to transmit only a narrow range of wavelengths. By changing the gap between the plates in steps, it is possible to scan the spectral region of interest and produce an image of the object at each chosen wavelength. The design was conceived by French physicists Charles **Fabry** and Alfred Pérot (1863–1925) in the late nineteenth century.

facula (pl. faculae)

(1) A bright area on the face of the Sun, commonly seen near an **active region**, such as a **sunspot**, or where such a region is about to form. Faculae, which last on average about 15 days, are best seen in blue light and are not visible at all in **H-alpha**. They were named by Johannes **Hevelius** and are thought to be caused by luminous hydrogen clouds close to the **photosphere**. (2) A bright spot on the surface of a planet or moon.

fall

A **meteorite** that was seen to fall and was recovered. Because this type of meteorite is usually collected soon after falling, weathering and other terrestrial processes do not have an opportunity to degrade the sample.

Fanaroff-Riley classification

A way of categorizing extragalactic radio sources, especially **radio galaxies** and radio-loud **quasars**, in terms of the separation between the brightest parts of their radio-emitting lobes. It is named after the South African astronomer Bernard Lewis Fanaroff (1947–) and the British astronomer Julia Margaret Riley (1947–) who published their results in 1974. Fanaroff and Riley noticed that the relative positions of regions of high and low surface brightness in the lobes of extragalactic radio sources are correlated with their radio luminosity. Their conclusion was based on a set of 57 radio galaxies and quasars that were clearly resolved at 1.4 GHz or 5 GHz into two or more components. Fanaroff and Riley divided this sample into two classes using the ratio, R_{FR}, of the distance between the regions of highest surface brightness on opposite sides of the central galaxy or quasar to the total extent of the source up to the lowest brightness contour in the map. Sources with $R_{FR} < 0.5$ were placed in Class I and sources with $R_{FR} > 0.5$ in Class II. Various properties of sources in the two classes are different, which indicates a direct link between luminosity and the way in which energy is transported from the central region and converted to radio emission in the outer parts. *Fanaroff-Riley Class I* (*FR-I*) sources have their low brightness regions farther from the central galaxy or quasar than their high brightness regions. The sources become fainter toward the outer extremities of the lobes and the spectra here are steepest, indicating that the radiating particles have aged the most. **Jets** are detected in 80% of FR-I galaxies, which also tend to be bright, large galaxies (D or cD) that have a flatter light distribution than an average elliptical galaxy and are often located in rich clusters with extreme X-ray emitting gas. As the galaxy moves through the cluster, the gas can sweep back and distort the radio structure through ram pressure, which explains why narrow-angle-tail or wide-angle-tail sources appear to be derived from the FR-I class of objects. *Fanaroff-Riley Class II* (*FR-II*) is made up of luminous radio sources with hotspots in their lobes at distances from the center such that $R_{FR} > 0.5$. These sources are said to be *edge-darkened*, a particularly apt terminology when the angular resolution and dynamic range used in observing the classical sources weren't always good enough to reveal the hotspots as distinct structures. In keeping with the overall high luminosity of this type of source, the cores and jets in them are also brighter than those in FR-I galaxies in absolute terms, but relative to the lobes these features are much fainter in FR-II galaxies. Jets are detected in less than 10% of luminous radio galaxies, but in nearly all quasars. FR-II sources are generally associated with galaxies that appear normal, except that they have nuclear and extended emission line regions. The galaxies are giant ellipticals, but not first-ranked cluster galaxies.

Faraday rotation

A magneto-optic effect, also known as the *Faraday effect*, in which the plane of **polarization** of an electromagnetic wave is rotated under the influence of a magnetic field parallel to the direction of propagation; it is named after the English physicist Michael Faraday (1791–1867), who first observed the effect in 1845. The amount of rotation, in radians, is given by $R_m \lambda^2$, where λ is the wavelength of the radiation and R_m is a factor known as the *rotation measure*. Faraday rotation is displayed by **radio waves** as they travel through the **interstellar medium**. Observing it in the radiation from **pulsars** is among the most important ways of studying the **galactic magnetic field**.

fault

A fracture or zone of fractures in a planet's crust, accompanied by displacement of the opposing sides.

Faye's comet

Originally a **long-period comet** that was deflected by a close passage of Jupiter into a much smaller orbit with a period of 7.34 years, just before its discovery in 1843. The solar heating of Faye's nucleus is such that gas and dust are not emitted from the subsolar point, that is, where it is noon on the comet, but from a spot on the afternoon section. The jet effect that results gradually changes the comet's orbit. Moreover, the jet itself seems to be shifting position, because of **precession** of the comet's spin axis, which modifies the orbit still further.

F-class asteroid
A dark **asteroid** belonging to a subcategory of **C-class asteroids** distinguished by having a weak to absent ultraviolet absorption feature. F-class asteroids account for most of the Polana faction of the **Nysa-Polana family**. They have **albedos** of 0.03 to 0.07 and featureless reflectance spectra across the range of 0.3 to 1.1 microns.

feldspar
Any of a group of aluminosilicate (containing aluminum and silicon) minerals that also contain calcium, sodium, or potassium. Feldspars make up about 60% of Earth's crust and are the major component in nearly all **igneous** rocks found on Earth, on the Moon, and in some meteorites. They also are common in **metamorphic** and some **sedimentary** rocks. Their complex chemical and structural properties make them useful for interpreting the origins of rocks. Natural feldspars can be divided into alkali and **plagioclase feldspars**.

Fermi acceleration
A mechanism, first suggested by the Italian-American physicist Enrico Fermi (1901–1954) in 1949, to explain the origin of **cosmic rays**. It involves charged particles being reflected by the moving interstellar magnetic field and either gaining or losing energy, depending on whether the "magnetic mirror" is approaching or receding. In a typical environment, Fermi argued, the probability of a head-on collision is greater than a head-tail collision, so particles would, on average, be accelerated. This random process is now called *second-order Fermi acceleration,* because the mean energy gain per bounce depends on the mirror velocity squared. In 1977, theorists showed that Fermi acceleration by **supernova remnant** shocks is particularly efficient, because the motions are not random. A charged particle ahead of the shock front can pass through the shock and then be scattered by magnetic inhomogeneities behind the shock. The particle gains energy from this bounce and flies back across the shock, where it can be scattered by magnetic inhomogeneities ahead of the shock. This enables the particle to bounce back and forth again and again, gaining energy each time. Because the mean energy gain depends only linearly on the shock velocity, this process is now called *first-order Fermi acceleration.*

fermion
An elementary particle that obeys the **Pauli exclusion principle;** familiar examples include the **proton, neutron,** and **electron.** All fermions have half-integer intrinsic angular momentum (spin). They are named after the Italian-American physicist Enrico Fermi (1901–1954). Compare with **boson.**

FG Sagittae (FG Sag)
One of the most extraordinary variable stars known; a **supergiant,** it lies at the center of a young **planetary nebula** (He 1–5), which it ejected a few thousand years ago, and has evolved significantly within a human lifetime. Its **spectral type** changed from B4Ia in 1955 to A5Ia in 1967 to F6Ia in 1972 and is currently K2Ib. Its magnitude, which was 13 in 1900, brightened to 9 over the next 60 years, before fading again to 14 in 1992 and a deep minimum of 16 in 1996. Since the early 1990s, its variability has been like that of an **R Coronae Borealis star,** dominated by dimming events and partial recoveries most likely produced by obscurations of the star by dust. The consensus view is that FG Sagittae is a **final helium flash object,** similar to **Sakurai's Object,** about to make a rapid transition to the **white dwarf** stage.

fibril
A fine dark line seen in **H-alpha** light in the low **chromosphere** of the Sun. Fibrils form whirls near strong **sunspots** and **plages** or in **filament** channels. They connect spots and plages of opposite polarities, running along magnetic field lines, and surround large, mature spots in a radial pattern. Individually, they are about 10,000 km long and last for 10 to 20 minutes.

field curvature
A form of optical distortion in which the focus changes from the center to the edge of the **field of view.** In the presence of **astigmatism,** this problem is compounded because there are two separate astigmatic focal surfaces. Field curvature varies with the square of field angle or the square of image height. Therefore, by reducing the field angle by one-half, it is possible to reduce the blur from field curvature to a value of 0.25 of its original size. Positive lens elements usually have inward curving fields, and negative lenses have outward curving fields. Field curvature can thus be corrected to some extent by combining positive and negative lens elements. Lenses with virtually no field curvature are called flat-field lenses.

field galaxy
An isolated galaxy that doesn't belong to any **cluster of galaxies.** The ratio of galaxies in clusters to field galaxies is about 23:1.

field lens
A lens placed in or near the focal plane of a telescope to create an image of the primary mirror inside the instrument. In an **eyepiece,** it is the lens that receives the image from the objective and relays it on the eyelens (the lens nearest the eye).

field of view

(1) The angular size of the actual patch of sky being viewed by an instrument (usually less than 1°). (2) The apparent size of the **field stop** as seen through an **eyepiece** (20 to 90°, depending on the eyepiece). The *actual field of view* is the angle, measured on the sky, from one edge of the eyepiece field of view to the other; approximately equal to the apparent field of view divided by the magnification. The *apparent field of view* is the field of view that is written on an eyepiece. To get the true field of view, the apparent field must be divided by the magnification of the telescope.

field rotation

The rotation of a star field about the center that occurs in a telescope with an **altazimuth mounting** because the motion is not about the polar axis. Field rotation is seen in photographs or **CCD** images as stars trailed around the center of the picture. If a telescope has an **equatorial mounting**, an attached camera correctly tracks the subject across the sky, since the telescope and Earth now rotate about the same axis. By placing an altazimuth telescope on a wedge, field rotation can be eliminated.

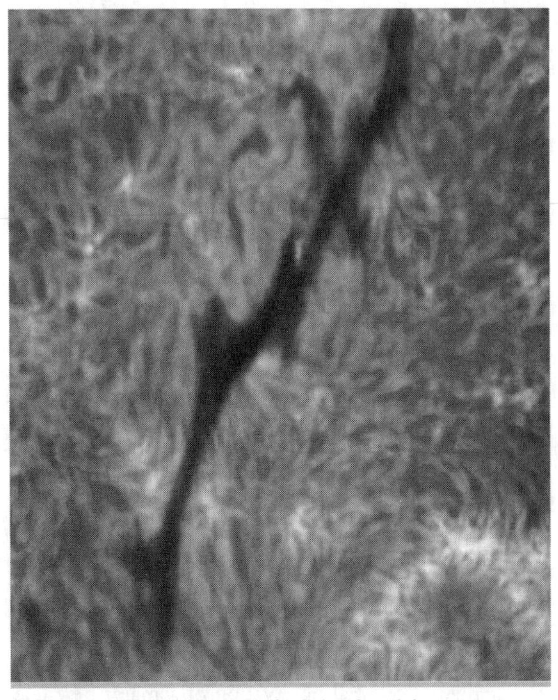

filament A filament seen in the light of H-alpha. *Tenerife VTT/MSDP*

field star

A star that lies along roughly the same line of sight as a star cluster but is not actually part of the cluster.

field stop

A round aperture in an **eyepiece** that limits the **field of view**.

figuring

The final stage of grinding a mirror or lens to give it the exact shape required. After figuring, a mirror is ready to be polished and coated.

filament

A strand of relatively cool gas suspended by magnetic fields over the solar **photosphere** so that it appears as a dark line over the Sun's disk. A filament on the limb of the Sun seen in emission against the dark sky is called a **prominence**. Filaments often mark areas of magnetic shearing and can be seen only in the centers of strong spectral lines, such as **H-alpha** or the **H and K lines** of calcium. A *filament channel* is a broad pattern of **fibrils** in the **chromosphere**, marking where a filament may soon form or where a filament recently disappeared.

filigree

A string of bright points on the Sun's **photosphere** that are sometimes visible in **intergranular lanes** in continuum images; the smallest points are only about 150 km wide and last for less than half an hour. Filigrees are thought to be places where **flux tubes** penetrate the photosphere.

filter

An accessory used with an optical instrument or detector of electromagnetic radiation to either narrow down the wavelength band or to reduce the total intensity passing into the instrument. Filters are often valuable in improving **contrast** and delineating detail that would otherwise be less visible.

filtergram

A photograph (usually of the Sun) taken in the light of a specific, narrow region of the spectrum, typically centered on a prominent **Fraunhofer line** such as **H-alpha**.

final helium flash object

An evolved star that, in descending the **white dwarf** cooling track, undergoes a final thermal pulse when its helium shell ignites causing it to expand rapidly to high luminosity and become a born-again **planetary nebula**. This phase of stellar evolution is extremely short, lasting only a few decades to a few centuries and is rarely observed. A handful of final helium flash object candidates have been identified, including **Sakurai's Object**, **FG Sagittae**, and V605 Aquilae.

find

A **meteorite** that was not seen to fall, but was found at some later date. For example, many finds from Antarctica fell 10,000 to 700,000 years ago.

fine structure

Closely spaced lines in part of the **spectrum** of a given element; a familiar example is the **D lines** of sodium. Fine structure results from the interaction between **electrons** in an atom and the atom's own magnetic field, which splits the various energy levels into a number of near-spaced sublevels.

fireball

A **meteor** that is brighter than any planet or star, i.e., with an apparent magnitude of –5 or greater.

first light

The inaugural celestial observation of a major new telescope–an emotional moment for those involved in the project, and a critical one, since it is the first test of the optical system on a "live" object.

first lunar meridian

The **meridian** of lunar longitude that passes through the point in Sinus Medii that marks 0° longitude; it is identical in function to the **prime meridian** on Earth.

First Point of Aries

Another name for the vernal **equinox**.

First Point of Libra

Another name for the autumnal **equinox**.

fixed star

An ancient term once used to distinguish the stars that maintained the same relative position from the "wanderers," or planets.

FK Comae Berenices star (FK Com star)

A fast-spinning giant star of spectral type G or K (about as warm or slightly cooler than the Sun but much larger) whose light variations have the same period as the rotation–typically a few days. Changes in brightness of a few tenths of a magnitude are caused by a surface that is more luminous in some areas than others. FK Com stars may represent a late stage in the evolution of **common-envelope stars** at which the stellar cores have coalesced.

Flame Nebula (NGC 2024)

A beautiful emission nebula in **Orion** that is ionized and caused to luminesce by the easternmost star in Orion's Belt, **Alnitak** (Zeta Orionis). The Flame Nebula glows in a variety of colors, from yellow to orange, though the predominant hue is shell-pink.

Flaming Star Nebula (IC 405)

A diffuse nebula in **Auriga** (R.A. 5h 13m, Dec. +34° 16′) that mainly surrounds the star AE Auriga and gives the impression that the star is burning, hence its name. AE Aur is an erratic variable and one of the **runaway stars** whose proper motion can be traced back to the area of **Orion's Belt**.

Flammarion, (Nicolas) Camille (1842–1925), and Gabrielle (nee Renaudot) (1876–1962)

French husband and wife astronomers. The author of more than 70 books, Camille did more to encourage public interest in astronomy than anyone else of his day, although many of his scientific and philosophical arguments were eccentric. He served for some years at the Paris Observatory and at the Bureau of Longitudes, but in 1883 he set up a private observatory at Juvisy (near Paris) and continued his studies, especially of double and multiple stars and of the Moon and Mars. His first book, *La pluralité des mondes habités* (The Plurality of Inhabited Worlds), originally published in 1862, secured his reputation as both a great popularizer and an advocate of life on many worlds. By 1882, it had gone through 33 editions, and continued to be translated and reprinted well into the twentieth century. Flammarion's best-selling work, his epic *Astronomie populaire* (1880), translated as *Popular Astronomy* (1894), is filled with speculation about extraterrestrial life. An entire chapter argues the case for lunar life, while he considers Mars "an earth almost similar to ours [with] water, air . . . showers, brooks, fountains. . . . This is certainly a place little different from that which we inhabit." His later studies were on psychical research, on which he wrote many works, among them *Death and Its Mystery* (3 vol., 1920–1921). Flammarion earned the amorous attention of a French countess who died prematurely of tuberculosis. Although they never met, the young woman made an unusual request to her doctor, that when she died he would cut a large piece of skin from her back, bring it to Flammarion, and ask that he have it tanned and used to bind a copy of his next book. (The woman also had a picture of Flammarion tattooed on herself!) Flammarion's first copy of *Terres du Ciel* (Lands of the sky) was bound thus, with an inscription in gold on the front cover: "Pious fulfillment of an anonymous wish/Binding in human skin (woman) 1882."

In 1919, Camille married Gabrielle Renaudot and for six years they worked side by side to promote astronomy in France. After Camille died, Gabrielle continued to maintain Juvisy Observatory and even made arrangements for work to continue after her death. She is buried next to her husband in the observatory park.

Flamsteed, John (1646–1719)

An English astronomer, founder of the **Royal Greenwich Observatory**, and the first **Astronomer Royal**. He was appointed by King Charles II in response to the need to find a way to accurately measure longitude at sea. Flamsteed got the job having recommended that the solution was to produce better tables of the movements of the Moon and the positions of the stars. His *Historia coelestis Britannica,* listing 2,935 stars, was the first major star catalog compiled with telescopic aid. A preliminary version of it, published by Edmond **Halley** and Isaac **Newton** in 1712 without Flamsteed's approval, introduced the method of designating stars now known as **Flamsteed numbers**.

Flamsteed number

A combination of number and constellation name (e.g., 61 Cygni or 36 Ophiuchi) used to identify naked-eye stars that was adopted by John **Flamsteed** in his stellar catalog. The numbers were originally assigned in order of increasing **right ascension** (west to east) within each constellation, but due to the effects of **precession** they are now slightly out of order in some places. The designations gained popularity throughout the eighteenth century and are now commonly used when no **Bayer designation** exists. Not all naked-eye stars in the Northern Hemisphere have a Flamsteed number and, because Flamsteed's catalog included only the stars visible from London, most stars in the far Southern Hemisphere are not covered by the system.

flare star

An orange or **red dwarf** star that experiences sudden, intense outbursts of energy. Although the term is often used synonymously with **UV Ceti star**, some **BY Draconis stars** also show flare activity.

flash spectrum

An **emission spectrum** of the solar **chromosphere**, obtained by placing an objective prism in front of the telescopic lens the instant before or after totality in a **solar eclipse**.

flat

A mirror with a flat surface used as the **secondary mirror** in a **Newtonian telescope**.

flat field

A lens or optical system that corrects **spherical aberration** and gives uniform focus across the **field of view**.

Fleming, Williamina Paton Stevens (1857–1911)

A Scottish astronomer who, in the United States, did important work on stellar classification (see **spectral type**). Fleming initially served as a maid in the home of Edward **Pickering**, then was offered part-time work as a (human!) computer at **Harvard College Observatory**, of which Pickering was director. Although positions in computing were poorly paid, they were the only way at the time that women could get into astronomical work–university courses and paid work as assistants at major observatories being largely a male preserve. Fleming's talent was such, however, that in 1881 she was made a permanent member of the staff. She developed an empirical star classification scheme consisting of 17 categories (a huge advance upon Angelo **Secchi**'s seminal scheme) and went on to classify 10,351 stars based on their photographed spectra. Her work, published as a catalog in 1890, was further refined by Annie Jump **Cannon**. While engaged in her monumental task, Fleming discovered 10 novae, about 60 new nebulae, and more than 300 variable stars.

flickering

Fast, irregular changes in the brightness of a stellar-sized object, over seconds, minutes, or hours, of up to several tenths of a magnitude. Flickering is displayed by a variety of **interacting binary** systems, including **AM Herculis stars** and some **symbiotic stars**.

FLIERs (fast, low ionization emission regions)

Blobs of material that form a symmetrical pattern in some **planetary nebulae**, such as the **Blinking Nebula** and **Saturn Nebula**. Their origin is not well understood. Some of the observed characteristics of FLIERs suggest that they are like sparks flung outward from the central star within the past thousand years or so. Yet they can also be interpreted as essentially stationary structures, from the surface of which gas is scraped by the outward flow of other gas.

flint glass

Any highly refractive lead-containing glass used to make lenses and prisms. Because it absorbs most **ultraviolet** light but comparatively little visible light, it is also used for telescope lenses. Flint glass typically has a much higher **refractive index** than does **crown glass**, and exhibits much higher dispersion, however it is less resistant to damage. Flint glass is used in conjunction with crown glass to make **achromatic lenses**.

Flora family

A collection of **asteroids** near the inner edge of the **asteroid belt** whose orbits have semimajor axes of 2.12 to 2.25 AU and mean inclinations of about 5°. All the members of the Flora family probably came from the breakup of a single parent body; however, other asteroids in this region with similar orbital characteristics appear to have different origins. The prototype is (8) Flora, which has a diameter of 162 km and a spectral class of C, and was discovered by John **Hind** in 1847. Its orbital details are: semimajor axis 2.202 AU, period 3.27 years, perihelion 1.86 AU, aphelion 2.55 AU, eccentricity 0.156, inclination 5.89°.

flux

The rate of flow of a physical quantity through a reference surface. *Luminous flux* is the rate of flow of **electromagnetic radiation** and is measured in lumens. *Particle flux* is the number of particles passing through a unit area per second.

flux collector

An imprecise term generally applied to a large **reflecting telescope** that is designed to capture as much radiation from a celestial source as possible. Also known as *light buckets,* flux collectors may sacrifice some accuracy of form for large aperture and are usually designed for use at infrared and submillimeter wavelengths.

flux density

The radiant power that falls per unit surface area of a detector per unit bandwidth of the radiation. It thus measures both the *spatial* (surface) density and a *spectral* density of the flux received from an astronomical source.

flux tube

A tubelike space in which there is a strong magnetic field. Small flux tubes at the surface of the Sun, with a diameter of less than about 300 km, form **filigrees** and **plages** and may appear bright. Bigger flux tubes, up to about 2,500 km diameter, appear dark and are called **pores** or **sunspots**.

focal length

The distance between a mirror or a lens and its **focal point**. The focal length of a compound optical system, given as if it were a single optical element, is called the *equivalent focal length.* The focal length at which an optical system seems to be working in a given situation is known as the *effective focal length.* For example, if an **eyepiece** projects an image onto a film, the image is magnified as if it were made by a telescope of several times the telescope's equivalent focal length. The *back focal length* is the distance between the last surface of a compound optical system and the focal plane of the system and may be quite different from the actual focal length.

focal plane

The flat plane at right angles to the **optical axis** onto which a lens will focus an image. **Curvature of field** results in an extended image forming on a curved surface known as the *focal surface.* A *focal plane scale* is the relationship between angles on the sky, in arcseconds, and millimeters of size at the focus of the telescope, i.e., the number of arcseconds per millimeter.

focal point

The convergence point, or sharpest point of focus, for an image; also known simply as the *focus.* For a positive (converging) lens or mirror, the focal points of the element are the points at which a ray, approaching the element parallel to the **optical axis**, crosses the optical axis on the opposite side. For a negative (diverging) lens, the focal points are the points at which the backward extension of the diverging ray intersects the optical axis.

focal ratio

The ratio of the **focal length** (F) of a mirror or lens to its diameter (D) expressed as a number, $f/\# = F/D$. Borrowing the language of photography, small focal ratios, below about *f*/6, are said to be *fast* and result in a brighter image for a given aperture. Large focal ratios, equal to or above f/8, are said to be *slow.*

focal reducer

An auxiliary lens, used mainly for photographic or CCD-imaging purposes. It reduces a telescope's effective **focal length**, allowing images to be taken much faster, and also gives a wider **field of view**. Some focal reducers may be used visually to obtain a wider field of view, but others are intended for photographic/CCD use only.

focus, orbital (pl. foci)

One of two points on the **major axis** of an **ellipse** whose separation from the other focus determines the shape of an elliptical orbit. At one focus is the body being orbited; for example, the Sun lies at the focus of Earth's orbit. The other focus is unoccupied. If the orbit is parabolic or hyperbolic, there is only one focus and it is occupied by the orbited body.

Fokker-Planck equation

A formula that, by making some simplifying assumptions to sidestep the **many-body problem**, enables the general evolution of the orbits of stars in **globular clusters** or in galaxies to be worked out. It is named after the Dutch physicist Adriaan Fokker (1887–1972) and the German physicist Max Planck (1858–1947).

following

A term used to describe the side of an object, or member of a group of objects, that comes into view later in the motion across the sky or across the face of a rotating body. Examples include the following **limb** of a planet, the following spot in a group of **sunspots**, and the following component of a binary star. What comes into view earlier is said to be *preceding.*

Fomalhaut (Alpha Piscis Austrini, α PsA)

The brightest star in **Piscis Austrinis**; its name (pronounced "fo-ma-low") comes from the Arabic *fum al hut* meaning "the fish's mouth." A main sequence **A star**, it is known to have a dust disk, similar to that around **Vega**, four times the size the Sun's planetary system. Observations of Fomalhaut's disk have revealed a central clearing and a warp farther out that may be caused by a planet with roughly the mass of Saturn. Fomalhaut is only about 200 million years old and may offer a glimpse of what our own solar system was like some 4.3 billion years ago.

Visual magnitude:	1.17
Absolute magnitude:	1.74
Spectral type:	A3V
Surface temperature:	8,500 K
Luminosity:	16 L_{sun}
Distance:	25 light-years

forbidden line

An **emission line** found in the spectrum of a rarefied gas under special conditions, such as those found in some **nebulae**, the solar **corona**, and parts of **active galactic nuclei**. Although not seen on Earth–hence their name–forbidden lines may account for 90% or more of the total visual brightness of an object such as a **planetary nebula**. A forbidden line arises when an **electron** in an excited (energized) atom jumps from a *metastable state* to a lower energy level. Under normal circumstances, when particle densities are higher (greater than about 10^8 per cm^3), such an electron would almost immediately be knocked out of its metastable state by collision and not be given time to emit a photon. But in an environment like that of a planetary nebula, the time between collisions averages 10 to 10,000 seconds. Consequently, when ions such as O^+, O^{2+} (singly and doubly ionized oxygen), or N^+ (singly ionized nitrogen) go into metastable states by allowed transitions from higher states, they remain there undisturbed until they radiate spontaneously. A large fraction of the more highly excited ions eventually drop into these states and, in a nebular environment, practically every ion goes from the metastable states to the **ground states** by forbidden radiation. Forbidden lines are denoted by enclosing them in square brackets. The strongest are two lines of doubly ionized oxygen [O III], in the green part of the spectrum (to which the human eye is most sensitive) at 4,959 and 5,007 Å. When these lines were first seen, in the spectra of planetary nebulae in the 1860s, their true nature wasn't recognized and it was thought they might be due to a new element, which was dubbed "nebulium." More than half a century passed before Ira **Bowen** provided the right explanation. Besides forbidden lines of ionized oxygen, others of neon, nitrogen, and other relatively abundant elements are seen making up the light of nebulae, as well as the ordinary, permitted lines of hydrogen and helium.

formation age

The time since a rock was hot enough to allow movement of chemical elements between different minerals.

formation interval

The time between the formation of the atoms of a chemical element and their incorporation into a planetary body.

Fornax (abbr. For; gen. Fornacis)

The Chemical Furnace, a faint southern **constellation** created by **Lacaille** in 1751–1752 made up of several dim stars previously in **Eridanus** and named by him in honor of the chemist Antoine Lavoisier, who was guillotined in the French Revolution. Its only star brighter than magnitude 4.0 is Alpha For, or Fornacis (visual magnitude 3.80; absolute magnitude 3.05; spectral type F8V; distance 46 light-years; R.A. 3h 12m 4s, Dec. −28° 59′ 13″). Extragalactic objects of interest include **Fornax A** (NGC 1316), the **Fornax Dwarf**, and the **Fornax Cluster**. (See star chart 6.)

Fornax A (NGC 1316)

An unusual giant **elliptical galaxy** in **Fornax** that appears to have merged, or be in the process of merging, with a spiral galaxy (see **galaxy merger**). Evidence for this includes a dark central lane of dust, arcs and plumes of stars in the outer parts of the galaxy, and an enormous outpouring of radio waves, powered ultimately by material falling onto a central supermassive **black hole**. The designation Fornax A means that this is the brightest radio source in its constellation; in fact, it is one of the strongest and largest radio sources in the sky, with radio lobes extending over several degrees.

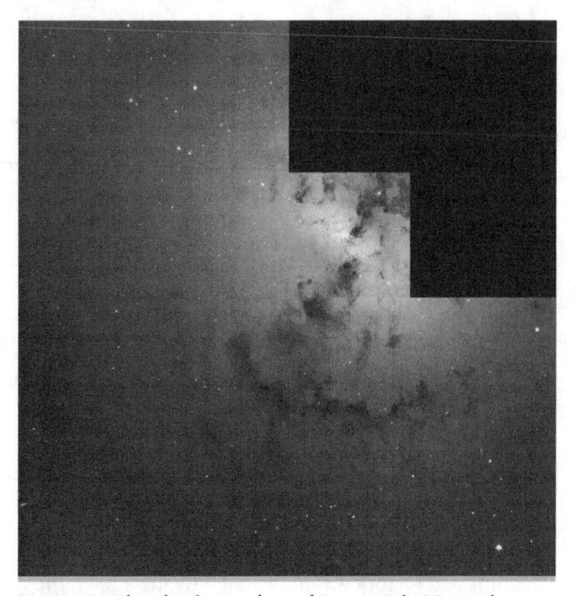

Fornax A The glowing nucleus of Fornax A (NGC 1316), marked by the silhouette of dark dust clouds. This Hubble image reveals what appears to be the aftermath of a 100-million-year-old cosmic collision between the elliptical and a smaller companion galaxy. *NASA/Carl Grillmair (California Institute of Technology)*

Fornax Cluster

The second richest **cluster of galaxies** within 100 million light-years, although it is much smaller than the ultrarich **Virgo Cluster**. There are actually two neighboring clusters, Fornax I and Fornax II (also known as the Eridanus Group), which contain 30 to 40 large galaxies each.

Fornax Dwarf

A **dwarf spheroidal galaxy** that orbits the Milky Way at a distance of about 450,000 light-years from the galactic center. Discovered in 1938, at the same time as the **Sculptor Dwarf**, it has a mass of about 20 million M_{sun} and has at least six **globular cluster**s orbiting it. See also **Local Group**.

forsterite

A member of a rare group (also known as the F group) of **chondrite** meteorites, named for the presence in them of the magnesium-rich type of **olivine** known as forsterite. In terms of their chemical and mineralogical makeup, forsterites are intermediate between the H group of ordinary chondrites and the E chondrites. They are thought to have derived from a small and primitive asteroid of F chondritic composition that collided with the **aubrite** parent body shortly after the forsterite's formation in the early solar system.

forward scattering

Scattering of **photon**s by a medium in a such way that most of the photons end up traveling in roughly the direction they started. Forward scattering is always accompanied by some **backscattering**, the proportion of the former increasing with particle size.

Foucault, Jean Bernard Léon (1819–1868)

A French physicist who was the first to show how a pendulum can track Earth's rotation (see **Foucault pendulum**). He also invented the gyroscope (1852), showed that light travels more slowly in water than in air (1850), and improved the mirrors of reflecting telescopes. More important even than his perfection of the silver-on-glass process was his development of a simple technique, now known as the **Foucault knife-edge test**, for determining the exact figure of a mirror.

Foucault knife-edge test

A test of optical surfaces, developed by Léon **Foucault**, that uses reflection and geometrical optical principles to amplify shadows of defects on a telescope mirror so that they are made easily visible. Like John **Hadley** some 200 years earlier, Foucault placed a pinhole source at the mirror's center of curvature and arranged the image to be formed alongside the source. However, unlike Hadley, Foucault examined the rays converging to a focus by placing his eye behind a knife-edge, which he then slowly introduced into the image. If the surface of the mirror darkened uniformly, he knew the mirror was spherical; if it didn't, he was able to deduce where and by how much the mirror surface deviated from sphericity. This technique, now known as the *Foucault knife-test,* is incredibly sensitive: bulges or hollows in a mirror surface with a relief as little as one millionth of an inch are easily detectable. Armed with his knife-edge, Foucault was able to produce mirrors with an accuracy of figure never before achieved. His technique is still much used today.

Foucault pendulum

A simple pendulum (a weight on a long string attached to a support) that shows the rotation of Earth. As the pendulum swings, Earth rotates under the pendulum, so the pendulum seems to rotate. It was first demonstrated by Léon **Foucault**, in 1851 at the Paris World's Fair.

Fowler, William ("Willy") Alfred (1911–1995)

An American astrophysicist who studied the nuclear reactions believed to occur in stellar interiors. In 1938, he showed that the **proton-proton chain**, proposed by Hans **Bethe**, was a viable means of producing energy in stars. In 1957, he coauthored, with Margaret and Geoffrey **Burbidge** and Fred **Hoyle**, a famous paper (referred to as B^2FH) titled "Synthesis of the Elements in Stars," which showed how the cosmic abundances of essentially all but the lightest nuclides could be explained as the result of stellar **nucleosynthesis**. Together with colleagues at the California Institute of Technology's Kellogg Radiation Laboratory, Fowler measured the rates of numerous nuclear reactions of astrophysical interest. He earned his B.S. in engineering physics at Ohio State University and his Ph.D. in nuclear physics at the California Institute of Technology, where he remained for the rest of his life, aside from frequent visits to the University of Cambridge.

Fp star

An **F star** whose spectrum reveals unusual chemical abundances of some metals. Fp stars are essentially an extension of **Ap stars** to cooler temperatures and spectral type F2, and include the group known as chromium-europium-strontium stars.

Franklin-Adams charts

An early photographic atlas of the sky by the English amateur astronomer John Franklin-Adams (1843–1912), based on plates taken at Johannesburg, South Africa, and Godalming, England, and published posthumously in 1913–1914. There are 206 charts covering the whole sky, each 15° square, and showing stars as faint as seventeenth magnitude.

Fraunhofer, Joseph von (1787–1826)

A Bavarian physicist and optician who devised the first **spectrometer** and used it to identify dark lines in the Sun's spectrum now known as **Fraunhofer lines**. A skilled lens-grinder and instrument-maker, he built the 9-inch (24-cm) Dorpat refractor, installed at the Pulkovo Observatory in Russia around 1825, and the 16-cm heliometer used by Bessel to measure the parallax of **61 Cygni**. He also developed a type of **equatorial mounting** that became known as the German mount.

Fraunhofer lines

Absorption lines in the spectrum of the Sun, or of another star, first studied and named by Joseph von **Fraunhofer** in 1814. The nine most prominent he labeled with capital letters A to K, starting at the red end. The A and B bands are now known to be caused by absorption in Earth's atmosphere, while the rest are due to absorption in the Sun's photosphere. C and F are now better known as **H-alpha** and H-beta; the **D lines** are of sodium, the **H and K lines** of calcium, and the G band of neutral iron and the CH molecule. All these features occur generally in stars of spectral types F, G, and K. (See table, "A Selection of Fraunhofer Lines.")

A Selection of Fraunhofer Lines

Lines	Due To	Wavelengths (Å)
A band	O_2 (molecular oxygen in Earth's atmosphere)	7594–7621
B band	O_2 (molecular oxygen in Earth's atmosphere)	6867–6884
C (H-alpha)	H (hydrogen)	6563
a band	O_2 (molecular oxygen in Earth's atmosphere)	6276–6287
D_1 & D_2	Na (sodium)	5896 & 5890
E	Fe (iron)	5270
b_1, b_2, b_3, b_4	Mg (magnesium)	5184, 5173, 5169, 5167
c	Fe (iron)	4958
F (H-beta)	H (hydrogen)	4861
d	Fe (iron)	4668
e	Fe (iron)	4384
f	H (hydrogen)	4340
G	Fe (iron)	4308
g	Ca (calcium)	4227
h (H-delta)	H (hydrogen)	4102
H & K	Ca (calcium)	3968 & 3934

free-bound transition
The capture of a free **electron** by an **ion**–a process known as *recombination,* in which a **photon** is emitted. The energy of the photon radiated is the difference between the original energy of the electron and its new energy (at the particular level of the atom it ends up in). Since this difference can have any value, the result of many free-bound transitions is a **continuous spectrum**.

free-floating planet
See **rogue planet**.

free-free transition
The emission or absorption of radiation by free **electrons** when they are accelerated or decelerated, respectively.

frequency
The number of peaks (often called crests) of a propagating wave that cross a given point in a unit of time. For example, if 1,000 peaks cross a given point in one second, the frequency is 1,000 cycles per second or 1,000 hertz.

Friedman, Herbert (1916–2000)
An American space scientist and astrophysicist who played an important role in the development of X-ray astronomy. After earning a Ph.D. from Johns Hopkins University (1940), he spent most of his career at the U.S. Naval Research Laboratory. He pioneered observations of the X-ray sky using rocket-borne instruments. Although X rays from the Sun were first detected in 1948 by T. R. Burnright, they were studied systematically from 1949 by Friedman and his colleagues, who observed X-ray activity throughout a full **solar cycle** of 11 years. Friedman also studied solar ultraviolet radiation and in 1960 produced the first X-ray and ultraviolet photographs of the Sun. In 1965, his observations of an occultation of the **Crab Nebula** by the Moon proved that Tau X-1, the second X-ray source to be detected beyond the solar system, coincided with the Crab.

Friedmann, Aleksandr Aleksandrovich (1888–1925)
A Russian mathematician and physicist who formulated an early **Big Bang** theory. Born the son of a composer in St. Petersburg, Friedmann was educated at the university there. He began his scientific career in 1913 at Pavlovsk Observatory in St. Petersburg and in 1918, after war service, was appointed professor of theoretical mechanics at Perm University. In 1920 he returned to St. Petersburg Observatory where he became director shortly before his death from typhoid fever. Friedmann established an early reputation for his work on atmospheric and meteorological physics. However, he is best known for his 1922 paper on the expanding universe in which he applied the equations of the **general theory of relativity** to cosmology. Friedmann developed a theoretical model of the universe using Einstein's theory, in which the average mass density is constant and space has a constant curvature. Different cosmological models are possible depending on whether the curvature is zero, negative, or positive. Such models are called *Friedmann universes.*

fringing
A **chromatic aberration** caused by the failure of a lens system to focus all colors at the same point, so a red and violet image, for example, are slightly offset from one another.

FU Orionis star (FU Ori star)
A pre-main-sequence **eruptive variable**. FU Orionis stars appear to be a stage in the development of **T Tauri stars**. They gradually brighten by about six magnitudes over several months, during which time matter is ejected, then remain almost steady or slowly decline by a magnitude or two over a period of years. All known FU Ori stars (commonly known as *fuors*) are associated with **reflection nebulae**. The prototype is a cF5- to G3 Ia-type star, presently near the top of its **Hayashi track**. In 1936 it suddenly appeared in the middle of a dark cloud, and rose by 6 magnitudes in the photographic band. Its lithium abundance is 80 times that of the Sun and it has developed a reflection nebula.

fundamental catalog
A reference catalog listing precisely measured star positions and **proper motions**, obtained mostly from **meridian** observations over many years. Among the most important are the ***Fundamental Katalog*s** and the ***Boss General Catalogue***.

Fundamental Katalog (FK)
A series of **fundamental catalogs** published in Germany. The first, in 1879, provided reference positions for the ***Astronomische Gesellschaft Katalog***. There followed the second FK (NFK) in 1907, FK3 in 1937–1938, and FK4 in 1963. FK5, published in 1988, contained, like its predecessor, 1,535 stars, brighter than about magnitude 7.5, distributed rather evenly across the entire sky. An extension to FK5, listing a further 3,117 stars, down to magnitude 9.5, appeared in 1991.

fundamental star
A star whose position and **proper motion** is so well known that it has been adopted for use as part of a frame of reference for positional observations of other objects.

FUSE (Far Ultraviolet Spectroscopic Explorer)

A NASA ultraviolet astronomy satellite, launched in June 1999. FUSE carries four 0.35-m UV telescopes, each with a high resolution **spectrograph**, that cover the ultraviolet band from the hydrogen ionization edge at 91 to 119 nm, just short of the **Lyman-alpha** line. The observations are used to measure the abundance of **deuterium** in the universe, as well as study helium absorption in the **intergalactic medium**, hot gas in the **galactic halo**, and cold gas in **molecular clouds** from molecular hydrogen lines. NASA has recommended a two-year extension beyond the three-year primary science mission.

fusion, nuclear

A process in which several small nuclei combine to make a larger one whose mass is slightly smaller than the sum of the small ones. The difference in mass is converted to energy according to Einstein's equivalence $E = mc^2$. In the center of most stars, **hydrogen** fuses into **helium**. The energy emitted by fusion prevents the star from collapsing in on itself and causes the star to glow.

fusion crust

A charred layer on a **meteorite** formed by the freezing of surface melt at the end of a meteorite's high-speed atmospheric flight. When a meteorite tears through the atmosphere at 12 to 72 km/s, friction with the air can raise its surface temperature to 4,800 K. At these temperatures the surface minerals melt and flow backward over the surface. As the meteorite slows and the fireball is extinguished, the molten minerals begin to cool and fuse together to form a thin, glassy skin that envelopes the whole meteorite. Typically, the fusion crust is black or bluish-black, although it can be light colored and transparent. It often helps to make the meteorite stand out against the background of terrestrial rocks.

G

G star

A yellowish star of **spectral type** G, with a surface temperature of 4,900 to 6,000 K. G stars have spectra in which the **H and K lines** of ionized calcium are particularly strong and there are numerous lines of both neutral and ionized metals, particularly iron; the Balmer lines of hydrogen, although weaker, are still recognizable. Familiar examples include Sun, **Alpha Centauri** A, and **Capella**.

gabbro

A dense, dark, course-grained **igneous** rock consisting largely of **plagioclase feldspar**, **pyroxene**, and **olivine**. It is the **intrusive** equivalent of **basalt**.

Gacrux (Gamma Crucis, γ Cru)

The third brightest star in **Crux**, contrasting with its three bright, blue-white constellation neighbors by being a **red giant** M star. If put in place of the Sun, it would extend over halfway to Earth's orbit. A dim A-type companion, about 2′ away, happens to be along the same line of sight. The fact that Gacrux is a mild **barium star,** however, does suggest that its surface has been contaminated by material from a binary companion that evolved first and has now become a white dwarf. Gacrux also fluctuates by a few tenths of a magnitude and is classified as a **semiregular variable**. A fairly strong **stellar wind** blows from its surface. If, as has been suggested, it is in the act of helium-burning, it may be about to become a red giant for the second time in its career–a speculation reinforced by its variability.

Visual magnitude:	1.59
Absolute magnitude:	–0.56
Spectral type:	M3.5III
Surface temperature:	3,400 K
Luminosity:	1,500 M_{sun}
Radius:	113 R_{sun}
Mass:	3 M_{sun}
Distance:	88 light-years

Gaia

A European Space Agency mission, scheduled for launch no later than 2012, that will survey the nearest 1 billion stars to provide the most precise positional and brightness data ever. This will be invaluable to every aspect of astronomy, from nearby asteroid searches to cosmology. Gaia is also expected to find as many as 10,000 new **extrasolar planets** by detecting tiny brightness changes as planets pass across the face of their host stars and also by the way stars wobble in response to their planets' gravitational pulls. In this respect it will be a successor to ESA's **Eddington** and a forerunner of the **Darwin** mission.

galactic anticenter

The point in the galactic plane that lies directly opposite the **galactic center**. It lies in **Auriga** at approximately R.A. 5h 46m, Dec. +28° 56′; the nearest bright star to it is **Alnath** in Taurus.

galactic bulge

The spheroidal mass of stars that forms the central hub of spiral and lenticular galaxies–the yoke, if such galaxies are

galactic bulge The spiral galaxy NGC 4013, some 55 million light-years away in the direction of Ursa Major, seen exactly edge-on in this Hubble Space Telescope image. The central bulge rises to either side of the dark band that marks the galaxy's disk. (The bright spot superimposed on the disk is a foreground star in our own Galaxy.) *NASA/STScI/AURA*

imagined to resemble fried eggs. The bulge diminishes in size relative to the **galactic disk** in the sequence of spirals Sa to Sd. Stars that populate the bulge are normally old, dating back to their galaxy's earliest period. Studying bulges can therefore tell astronomers about how galaxies formed and evolved. According to current theory, a **spiral galaxy** begins as a giant, roughly spherical rotating mass of gas and dust, that gradually flattens out at the edges to create the disk. The original spherical shape lives on in the outermost region of a galaxy, known as the **galactic halo**, and, to a lesser extent, in the bulge. However, this view is challenged by observations of some bulges. The bulge of the **Triangulum Galaxy** (M33), for example, contains young- and intermediate-age stars, and has a star distribution that suggests the disk goes all the way to the center. This raises questions about how M33 as a whole formed and what triggered the birth of the relatively youthful stars in its bulge.

galactic center
The center of the Milky Way or any other **galaxy**. It is the point about which the disk of a spiral galaxy rotates and it lies within the **galactic nucleus**.

galactic coordinates
Coordinates based on the plane of the **Milky Way Galaxy**, which is inclined about 63° to the celestial equator, and centered on the Sun, with the zero point of longitude and latitude pointing directly at the **galactic center**. Before 1958, the zero point of galactic latitude and longitude was taken to lie at R.A. 17h 45.6m, Dec. –28° 56.2′ (in Sagittarius). Galactic latitude (*b*) is measured from the galactic equator north (+) or south (–); galactic longitude (*l*) is measured eastward along the galactic plane from the galactic center. In 1958, because of increased precision in determining the location of the galactic center, based on observations of the **21-centimeter line**, a new system of galactic coordinates was adopted with the origin at the galactic center in Sagittarius at R.A. 17h 42.4m, Dec. –28° 55′ (epoch 1950). The new system is designated by a superior Roman numeral II (i.e., b^{II}, l^{II}) and the old system by a superior Roman numeral I.

galactic disk
The plate-shaped component of a **spiral galaxy** in which the **spiral arms** are found. It contains primarily young stars, gas, and dust clouds.

galactic equator
The imaginary great circle on the celestial sphere that lies 90° from both **galactic poles**; it closely follows the middle of the visible Milky Way.

galactic halo
A spherical region surrounding the center of a **galaxy**. It may extend beyond the luminous boundaries of the galaxy and contain a significant fraction of the galaxy's mass. In the case of the Milky Way, most of the *stellar halo* lies closer to the galactic center than the Sun does, while most of the *dark halo* lies farther from the galactic center than the Sun.

galactic magnetic field
A weak and largely disordered magnetic field, with a strength of about 5×10^{-10} tesla, that pervades the disk of the **Milky Way Galaxy** and controls the alignment of interstellar **dust** particles. Similar fields exist in other **disk galaxies**; those of **elliptical galaxies** are more difficult to estimate because of the lack of interstellar matter.

galactic noise
A diffuse radio signal that comes from the **synchrotron radiation** of electrons spiraling in the galaxy's magnetic field. Also known as *galactic radiation* or *galactic radio waves*.

galactic nucleus
The small, central, high-density region of a **galaxy**. If the nucleus gives off considerably more energy than can be accounted for by its stellar contents, it is said to be an **active galactic nucleus** (AGN). For details of our own galactic nucleus, see **Milky Way Galaxy**.

galactic plane
The plane in which the disk of the **Milky Way Galaxy** lies. By definition, one direction perpendicular to this plane is called *above* or *north*, and the opposite direction, also perpendicular to the galactic plane, is called *below* or *south*. From Earth, due galactic north is marked by the *north galactic pole*, which lies in **Coma Berenices**, near the bright star Arcturus, and due galactic south is marked by the *south galactic pole*, which lies in **Sculptor**.

galactic rotation
The systematic, **differential rotation** of stars and interstellar matter in a galaxy about an axis that passes through the **galactic center**. Spiral and lenticular galaxies have a well-defined rotational structure whereas the individual stars in elliptical galaxies tend to move in random orbits.

galactic window
A region of sky near the plane of the **Milky Way** that is unusually free of dust and gas and thus permits relatively unobscured views of distant galaxies. The best known example is **Baade's Window**; others occur in **Circinus**.

galactocentric distance

A star's distance from the center of the **Milky Way Galaxy**. The Sun's galactocentric distance is about 27,700 light-years.

Galatea

The fourth moon of **Neptune**, also known as Neptune VI; it was discovered in 1989 by **Voyager** 2. Galatea has a diameter of 158 km and orbits at a mean distance of 61,950 km from (the center of) the planet.

Galaxy, the

See **Milky Way Galaxy**.

galaxy

A large system of stars, together with interstellar material and (at least in some cases) **dark matter**, held together by gravity. There are three basic types: the **spiral galaxy**, the **elliptical galaxy**, and the **irregular galaxy**. A **lenticular galaxy** is midway in form between a spiral and an elliptical. Galaxies range in size from the smallest **dwarf galaxies** only a few hundred light-years across with just a few million stars, through normal galaxies like our own **Milky Way Galaxy**, with a few hundred billion stars, to giant ellipticals spanning over hundreds of thousands of light-years and containing several trillion stars. Various schemes have been devised to categorize them (see **galaxy classification**). Most **galaxy formation** is thought to have taken place in the early universe, beginning less than half a billion years after the **Big Bang**. Galaxies may be solitary, or in small groups like our **Local Group**, or in larger **clusters of galaxies**. In **rich clusters of galaxies**, the brightest systems tend to be ellipticals and lenticulars, with spirals making up only 5 to 10% of the population. However, the proportion of spirals in these clusters was probably higher in the past, **galaxy cannibalism** and **galaxy mergers** having turned them into the more amorphous types. In low-density environments, spirals account for about 80% of the bright galaxy count. Some galaxies display unusual activity in their cores associated with **active galactic nuclei**.

galaxy cannibalism

The swallowing of a smaller galaxy by a much larger one, in contrast to a **galactic merger**, which involves two galaxies of similar size. Most examples of cannibalism involve giant or supergiant ellipticals, at the heart of **rich clusters of galaxies**, drawing in and ingesting other members of their fold. Telltale signs of a big elliptical having recently (in cosmic terms) swallowed a disk galaxy, for example, include significant amounts of dust and young stars, loops and shells of luminous matter, and various other phenomena, such as powerful radio emission, that suggest the central supermassive **black hole** of the swallowing galaxy has just received a fresh consignment of matter. On a less dramatic scale, dwarf galaxies are frequently cannibalized by their larger neighbors. The **Sagittarius Dwarf Galaxy**, for example, is currently being devoured by the Milky Way.

galaxy classification

Various schemes have been devised to bring order to the galactic zoo by pigeonholing galaxies according to one or more properties, including shape, spectrum, and luminosity. A few are listed below but there are many others, including ones that specialize in radio galaxies or unusual systems.

Hubble Classification of Galaxies

The best known and most often used general scheme, in which galaxies are grouped according to their appearance. Devised by Edwin **Hubble**, it splits galaxies into ellipticals, spirals (normal and barred), and irregulars, and is represented by the familiar tuning-fork diagram. Elliptical galaxies are graded from E0 (spherical) to E7 (very elongated) in terms of increasing eccentricity. Normal spirals range from Sa (arms tightly wound around the nucleus) to Sc (arms widely spread from the nucleus), and, similarly, barred spirals from SBa (arms tightly wound) to SBc (arms widely spaced). Irregulars are designated Irr. To this original scheme, Alan Sandage (1926–) added another category, S0, to describe lenticular systems with a nucleus surrounded by a disklike structure that lacks spiral arms. Galaxies are often said to be "early" (E and S0) or "late" (Sb,Sc, Irr) in type, a remnant of early notions that galaxies physically evolve along the Hubble sequence. Unfortunately, this nomenclature is opposite to that of the dominant stellar **population** in these types.

Morgan Classification of Galaxies

A scheme invented by William **Morgan** that uses the integrated spectrum of the stars of a galaxy together with its shape (real and apparent) and its degree of central concentration. It specifies the galactic spectral type, a, af, f, fg, g, gk, or k (corresponding to the integrated stellar types); the form, type S (spiral), B (barred spiral), E (elliptical), I (irregular), Ep (elliptical with dust absorption), D (rotational symmetry without pronounced spiral or elliptical structure), L (low surface brightness), or N (small bright nucleus); and the inclination to the line of sight, from 1 (face-on) to 7 (edge-on). For example, the Andromeda Galaxy is classified as kS5.

de Vaucouleurs-Sandage Classification of Spiral Galaxies

Specifies SA (ordinary spirals) or SB (barred spirals): then in parentheses a lower case s (for S-shaped spirals) or r (for the ringed type). Finally, several transitional stages have been added between the SA or SB spirals and the Magellanic irregulars Im. In this classification, the Andromeda Galaxy is SA(s)b.

DDO (or van den Bergh) Classification of Galaxies

This contains two parameters: (1) the galactic type (Sa, Sb, Sc, Ir) and (2) the luminosity class (I, II, III, IV, V), similar to the **Morgan-Keenan classification** system of stellar luminosity class. The notations S^- and S^+ are used to denote subgiant species with low and high resolution, respectively. The notation S(B) has been introduced to denote objects intermediate between true spirals and barred spirals.

galaxy formation

The sequence of events by which **galaxies** took shape. Most galaxy formation is thought to have happened in the early universe following the **recombination** era that ended about 300,000 years after the **Big Bang**. Small fluctuations present in the **cosmic microwave background** at this time, first detected by the **Cosmic Background Explorer** (COBE), provide evidence of matter clumping together–an essential prerequisite for galaxies to assemble in a cosmos that is generally flying apart. Islands of matter had to have been able to come together at this stage, since there would have been no chance later on when the cosmic contents had become more dilute. The specifics of galaxy formation are unknown, though there's no shortage of theories. It seems certain that **dark matter** played a crucial role in providing a gravitational anchor for normal matter to condense into galaxies, but the details depend on whether dark matter is hot or cold, and this is undetermined. (Cold dark matter makes galaxy formation easier to understand because the hot variety would tend to smear out, by rapid particle motions, any density enhancements before they had a chance to condense further.) The basic idea is that the first lumps of matter to break free of the universe's expansion were mostly dark matter together with some neutral hydrogen and a dash of helium. As these lumps condensed further under their own gravity, the dark matter and the ordinary matter would have separated because the latter can dissipate energy. As the atoms in the hydrogen/helium gas came closer together, during gravitational collapse, they collided more often, heated up, and got rid of this heat as infrared radiation, allowing the collapse to continue. Dark matter doesn't interact in this way and would have continued to orbit in the halo of the galaxy-to-be–exactly where it is found in galaxies today. As the hydrogen/helium gas in the protogalaxy lost energy, its density rose, and gas clouds formed. When two clouds collided, the gas was compressed into a shock front, triggering a burst of star formation. With the production of its first light by its first stars, the protogalaxy became a primeval galaxy.

According to one scenario, the newborn galaxies were small, perhaps no bigger than globular clusters, and lay along filaments that threaded the dawn universe like a spider web. Moving in random directions these infant star systems (or perhaps they were still at the pre-stellar stage) occasionally came close and merged. Repeating the process many times over caused increasingly large galaxies to form. Over billions of years, the filaments were replaced by **clusters of galaxies** connected by bridges–the remains of the largest of the original filaments. This spider web theory of galaxy formation gained ground in 2001 thanks to observations by the **Very Large Telescope** that show, according to the team involved, a string of dense clumps of hydrogen in the early universe, glowing because of hot young stars inside them. The researchers claim the clumps are protogalaxies lying within a tubular region of space–a filament stuffed with galaxies in the making. Some galaxy formation may even be taking place in the modern universe. One of these late-bloomers is POX 186, a **blue compact** dwarf galaxy, less than 100 million years old that inhabits a cosmic **void** some 68 million light-years away.

galaxy interaction

An encounter between two (or more) galaxies that is close enough for the mutual gravitational attraction to alter the appearance and, in some cases, the composition, of one or both systems. Classic examples include NGC 5195, the little, tidally distorted companion of the **Whirlpool**

galaxy interaction A pipeline of material is seen flowing between two battered galaxies that bumped into each other about 100 million years ago in this image from the Hubble Space Telescope. The dark string of matter begins in NGC 1410 (the galaxy at left), crosses over 20,000 light-years of intergalactic space, and wraps around NGC 1409 (the companion galaxy to the right). The pair lie about 300 million light-years away in the direction of Taurus. *NASA/William Keel (University of Alabama)*

Galaxy, and the **Antennae**. Extreme cases of interaction result in **galaxy mergers** or **galaxy cannibalism**.

galaxy merger

The process that takes place when two galaxies collide with one another. If the galaxies become caught in each other's gravitational field they do not simply make a close pass, undergo **galaxy interaction**, and continue their separate ways into intergalactic space. Instead they spiral into one another and form a single galaxy at the end of the interaction. It is thought that giant **elliptical galaxies** are the product of mergers between **spiral galaxies**. Since the spaces between individual stars in galaxies are so large, even though the galaxies appear to collide, the stars contained within them do not. There's enough space for the two galaxies to virtually pass through one another like interlinking fingers. The huge gravitational fields of the galaxies, however, distort their shapes and the close stellar encounters swing the stars into randomly orientated orbits. Often the interstellar clouds collapse during the merger and many new stars are formed in a process called a *starburst.* As the galaxies become increasingly close, long tails of stars are strung out in their wake. Sometimes these tail fragments contain enough matter to be thought of as dwarf galaxies in their own right.

Galilean moons

Jupiter's four largest moons, **Io**, **Europa**, **Ganymede**, and **Callisto**, discovered independently by **Galileo** Galilei

galaxy merger The merging galaxy system known as ESO202-G23 as seen by the Very Large Telescope. At least one of the two nuclei involved is an active galactic nucleus whose partially collimated ultraviolet radiation is exciting the surrounding gas to the north. Also visible is a blue star-forming complex to the south of the center and a complicated pattern of gas emission due to the combination of arms resulting from the merger. The arc of points in the lower part of the image are more distant galaxies. *European Southern Observatory*

and Simon **Marius** in 1610. Galileo proposed that they be named the *Medicean stars,* in honor of his patron Cosimo II de Medici; however, the present names were chosen by Marius. All four are easily visible in a small telescope or binoculars. Io, Europa, and Ganymede periodically line up with the result that their gravitational interactions during the alignments force the three moons into noncircular orbits; the moons' varying distances from the planet lead to tidal distortions, and subsequent **tidal heating**, caused by Jupiter's gravity. The heating, in turn, fuels extreme volcanism on the innermost of the Galilean moons Io and, evidence suggests, helps maintain a liquid sea beneath the frozen crust of the second Galilean moon, Europa, and possibly also that of Ganymede.

Galilean telescope

A **refracting telescope** of the type used famously by **Galileo**, with a convex objective lens and a concave lens as an eyepiece, separated by the difference of their focal lengths. Galilean telescopes are shorter than equivalent **Keplerian telescope**s and give bright, upright images, but they suffer from a limited **field of view**, **spherical** and **chromatic aberration**, and, except at low magnification (about 30× or less), poor **eye relief**.

Galileo Galilei (1564–1642)

A great Italian astronomer and physicist, renowned for his epoch-making contributions to physics, astronomy, and scientific philosophy. In 1610, he was among the first to use a telescope to study the heavens and with it discovered the four big moons of Jupiter, the phases of Venus, the mountains of the Moon, and the starry nature of the Milky Way, breakthroughs that he announced the same year in his *Siderius Nuncius* (Starry Messenger). His defense of the Sun-centered **Copernican system** in *Dialogue on the Two Chief World Systems* (1632) brought severe censure from the Church and he was forced to recant before, at the age of 69, being sentenced to life imprisonment (commuted to house arrest); he was not formally exonerated by the Catholic Church until 1992. Having heard, in 1609, of the invention of the telescope, but lacking a detailed description, he set about learning the principles of the instrument himself and, within a matter of weeks, had produced his first simple "optik tube," which he immediately directed to the skies. Thus began a new and exciting era of observational astronomy that continues to this day. Galileo made a number of telescopes ranging up to 5 cm in aperture and 170 cm in focal length, and with magnifications from about 8 to 30.

Galileo (spacecraft)

A NASA spacecraft that carried out the first studies of **Jupiter**'s atmosphere, satellites, and magnetosphere from orbit around the planet. The mission was named after **Galileo Galilei** and launched in October 1989. En route, Galileo made the first and second flybys of asteroids–**Gaspra** in October 1991 and **Ida** in August 1993. It was also the only vehicle in a position to obtain images of the far side of Jupiter when Comet **Shoemaker-Levy 9** plunged into Jupiter's atmosphere in July 1994. On December 5, 1995, as Galileo approached the giant planet, it released a small probe that descended into Jupiter's atmosphere. After this, the main spacecraft went into orbit around Jupiter to begin an exploration of the planet and its moons that continued until the end of 2002.

Galilean moons A family portrait of the Galilean satellites–the four largest moons of Jupiter. Io, Europa, Ganymede, and Callisto are shown to scale and in order, from left to right, of increasing distance from Jupiter. *NASA/JPL*

Galle, Johann Gottfried (1812–1910)

A German astronomer who, with Heinrich Louis **d'Arrest**, made the first observation of **Neptune** (1846) based on calculations by **Leverrier**. Though Galle was the first to observe Neptune, its discovery is usually credited to John **Adams** (who made an earlier calculation) and Leverrier. Galle also discovered the crêpe ring of **Saturn** (1838).

Gamma Cassiopeiae (γ Cas)

The central star of the "W" of **Cassiopeia** and the first known **Be star** (identified by Angelo **Secchi** in 1866); it is among the brightest stars in the sky to have no proper Western name, though the ancient Chinese knew it as Tsih (the whip). γ Cas is an **eruptive variable** and has given its name to a category of similar objects (see **Gamma Cassiopeiae star**). In 1937 γ Cas brightened almost to first magnitude, and it has been as faint as third. Its high rotational speed (at least 300 km/s at the equator) and high luminosity conspire to drive mass from the star into a surrounding disk that radiates the emission lines, while at the same time dimming the direct light from the star. γ Cas is a member of a visual double system and is also a member of a spectroscopic binary with a period of 203.59 days, an eccentricity of 0.26, and a companion with about the same mass as the Sun. The companion could be a normal star, or it might be a white dwarf or neutron star. If it were a collapsed object, this would explain why γ Cas is also an X-ray source: gas from the Be star accreting onto a compact companion would release gravitational energy, which would be transformed into thermal energy and then into X rays.

Visual magnitude:	2.15
Absolute magnitude:	–4.22
Spectral type:	B0.5IVe
Surface temperature:	25,000 K
Luminosity:	70,000 L_{sun}
Distance:	613 light-years

Gamma Cassiopeiae star (γ Cas star)

A fast-spinning **eruptive variable** with a Be III to Be IV spectrum that, at irregular intervals, shows a series of strong, sharp **absorption lines** accompanied by a fading due to the ejection of matter from the equatorial region of the star. The ejected material may settle into a ring (in a variable **Be star**) or form a shell (in a **shell star**), and be accompanied by a fading of up to 1.5 magnitude. Well-known examples are **Gamma Cassiopeiae** itself, the brightest of its type, and **Pleione** in the Pleiades.

gamma rays

Electromagnetic radiation with a wavelength less than about 1 Å (10^{-10} m). Gamma-ray energies as high as 1 TeV (trillion eV) have been recorded from cosmic sources.

Gamma Velorum (γ Vel)

A remarkable group of five stars. Also known by the proper name Regor, γ^2 is the brightest star in **Vela** and one of the intrinsically brightest stars in the night sky. In fact, it is a **spectroscopic binary**. The brighter of the pair is a giant **O star**, while its partner is the visually brightest **Wolf-Rayet star** in the sky, the two separated by about 1 AU in a 78.5-day orbit. A third component, known as γ^1, is a **B star** that appears to be physically unconnected to γ^2 and has two wide companions. The O-type member of γ^2 has a surface temperature of 35,000 K, a total luminosity of 200,000 L_{sun}, and a mass of 30 M_{sun}, and appears to be at the core **helium-burning** stage. Its partner, of the carbon-rich Wolf-Rayet variety, is about half as luminous and contains less than 10 M_{sun}, but is perhaps as hot as 60,000 K. Because massive stars always evolve faster than lighter ones, the Wolf-Rayet star must once have been heavier than its O-type companion. It probably started out with somewhere around 40 M_{sun} and has now stripped itself to a quarter of this. Though only a few million years old, the Wolf-Rayet star is on the verge of blowing up as a **supernova**.

Visual magnitude:	1.75
Absolute magnitude:	–5.31
Distance:	840 light-years

gamma-ray burst (GRB)

The most violent and powerful event in the universe: a sudden, intense burst of **gamma rays** lasting from as little as a hundredth of a second to as much as a few minutes. GRBs are typically registered once or twice a day, from random directions in space, by spacecraft in orbit and were first detected, by accident, in the 1960s by military satellites looking for evidence of nuclear explosions on Earth. Until recently, there were two competing theories to explain GRBs: hypernovae–exceptionally violent **supernovae** in which **black holes** are formed, and mergers between **neutron stars** or black holes. The hypernova theory has now gained the upper hand. Beginning in 1997, astronomers began tracking GRBs so closely that they could train optical telescopes on the site of a burst almost immediately after it had happened. This allowed them to detect the afterglows from the blast. It was quickly determined that the

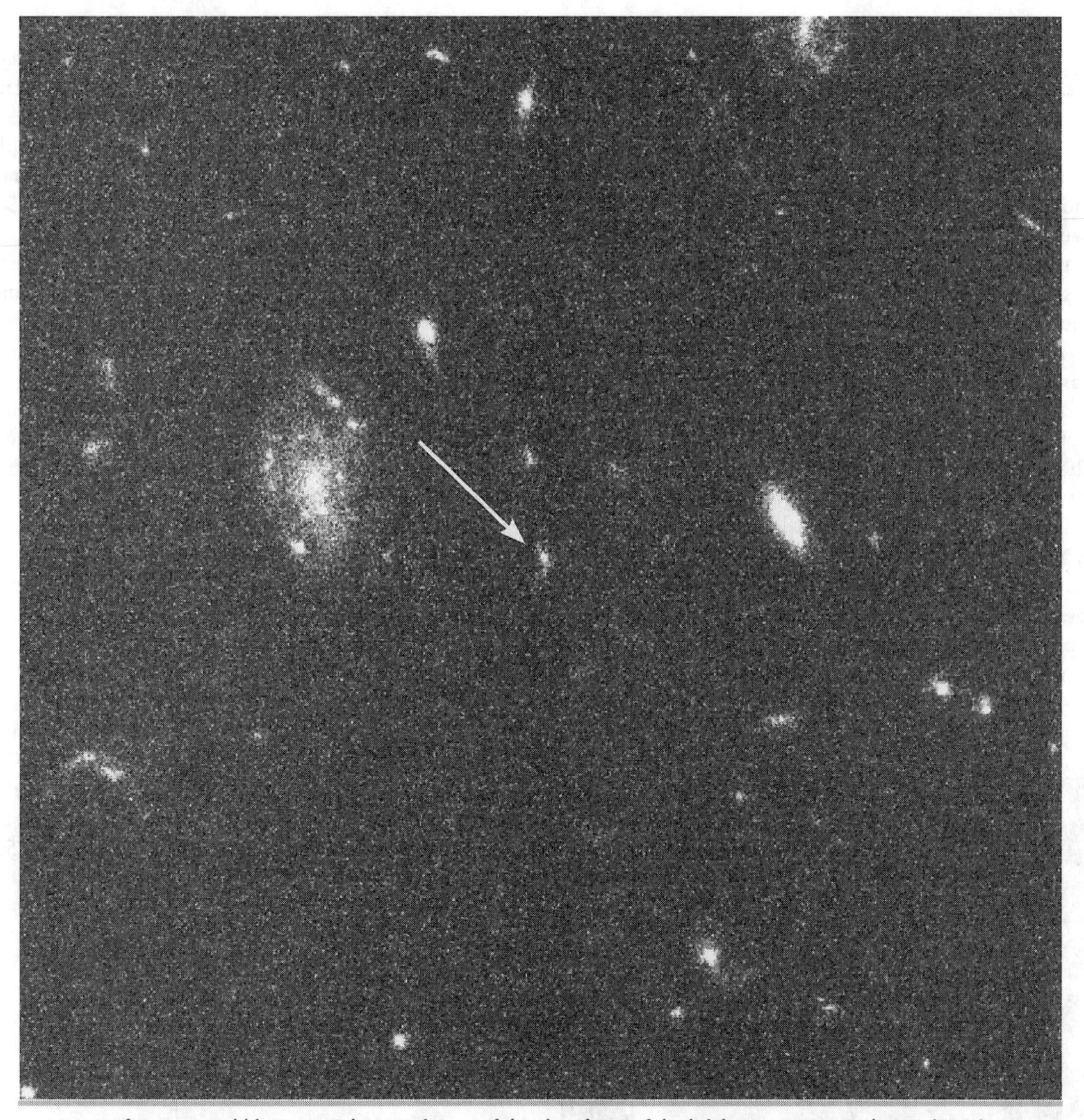

gamma-ray burst A Hubble Space Telescope image of the site of one of the brightest gamma-ray bursts (GRBs) ever recorded, GRB 971214, about four months after the burst took place in 1998. The faint object marked with an arrow is the host galaxy of the burst, an object that lies about 12 billion light-years from Earth. *NASA/S. R. Kulkarni and S. G. Djorgovski (Caltech)*

explosions occur in remote galaxies. A burst in 1998 was tracked to a region near the center of its host galaxy where star birth is occurring at a rapid rate, supporting the idea that gamma-ray bursts come from the death explosions of very young, massive stars. Then, in 2001, observations of the X-ray afterglow of a GRB by the **XMM-Newton Observatory** provided what may be the clincher. Spectral analysis of the site of the burst, in a galaxy some 10 billion light-years away, showed traces of the elements magnesium, silicon, sulfur, argon, and calcium racing toward us at one-tenth of the speed of light. These light elements and such an expansion rate would not be expected from a neutron star collision, but they are exactly the signatures predicted for a violent supernova explosion in which a black hole is created. In October 2002, scientists were able to study the afterglow of a GRB just nine minutes after its onset, thanks to precise coordination and fast slewing of ground-based telescopes upon detection of the burst by the **HETE** (High-Energy Transient Explorer) satellite.

Gamow, George (1904–1968)

A Ukrainian-American nuclear physicist, cosmologist, and writer who formulated an early theory of the hot **Big Bang** with Ralph **Alpher** in 1948, and worked on quantum mechanics, stellar evolution, and genetic theory (proposing the existence of DNA in 1954). Gamow's popular books included the *Mr. Tomkins* series (1939–1967), *One, Two, Three . . . Infinity* (1947), and *The Creation of the Universe* (1952; revised 1961).

Ganymede

The largest satellite of **Jupiter** and of the entire solar system, slightly bigger even than Mercury (but with only half its mass); it was discovered by **Galileo** and Simon **Marius** in 1610. Ganymede appears to have a differentiated internal structure, with a small molten iron or iron-sulfur core surrounded by a rocky silicate mantle with an icy shell on top. Its surface is a roughly equal blend of ancient, densely cratered dark terrain, and younger, lighter regions that are marked with a network of grooves and ridges. This network is evidently **tectonic** in nature, though, unlike on Earth, there are no signs of recent tectonic activity. Evidence for a very thin oxygen atmosphere on Ganymede, like that of **Europa,** has been found by the Hubble Space Telescope. Similar ridge and groove terrain is seen on **Enceladus, Miranda,** and **Ariel.** The dark regions are similar to the surface of **Callisto.** Extensive cratering is seen on both types of terrain, with a density that points to a surface age of 3 to 3.5 billion years, similar to that of the Moon. Craters both overlay and are cross-cut by the groove systems showing that the grooves are quite ancient, too. Relatively young craters with rays of ejecta are also visible. However, these craters are quite flat, lacking the ring mountains and central depressions common to craters on the Moon and Mercury. This is due to the plastic nature of Ganymede's icy crust, which can flow over millions of years and thereby smooth out any surface features. Ganymede has its own **magnetosphere,** embedded within Jupiter's huge one, that is probably generated in a similar way to Earth's, as a result of motion of conducting material in the interior. *Galileo Regio* is a prominent, dark, oval region, about 3,200 km across, centered approximately at latitude +36° and longitude 138°. *Gilgamesh* is the largest preserved **impact basin** on Ganymede, with a rim 580 km in diameter surrounding a roughly 300-km-wide smooth central depression. An **ejecta** blanket and secondary **crater chains** blasted out from the basin are visible out to approximately 500 km from the basin center.

Ganymede Jupiter's largest moon as seen by the Galileo spacecraft during its first encounter with the satellite. North is to the top of the picture and the Sun illuminates the surface from the right. The dark areas are the older, more heavily cratered regions, whereas the light areas are younger and more tectonically deformed. Bright spots are geologically recent impact craters and their ejecta. The smallest details visible in this image are 13.4 km across. *NASA/JPL*

Diameter:	5,262 km
Mass (Earth = 1):	0.0247
Surface gravity (Earth = 1):	0.145
Mean distance from Jupiter:	1,070,000 km
Orbital period:	7.1546 days
Rotational period (days):	7.1546 days
Density:	1.94 gm/cm3
Eccentricity:	0.002
Inclination:	0.18°
Escape velocity:	2.74 km/s
Visual albedo:	0.43
Temperature	
Subsolar	156 K
Equatorial Subsurface	117 K

gardening

A process in which countless small impacts create a **regolith** on a planetary surface and then continually keep

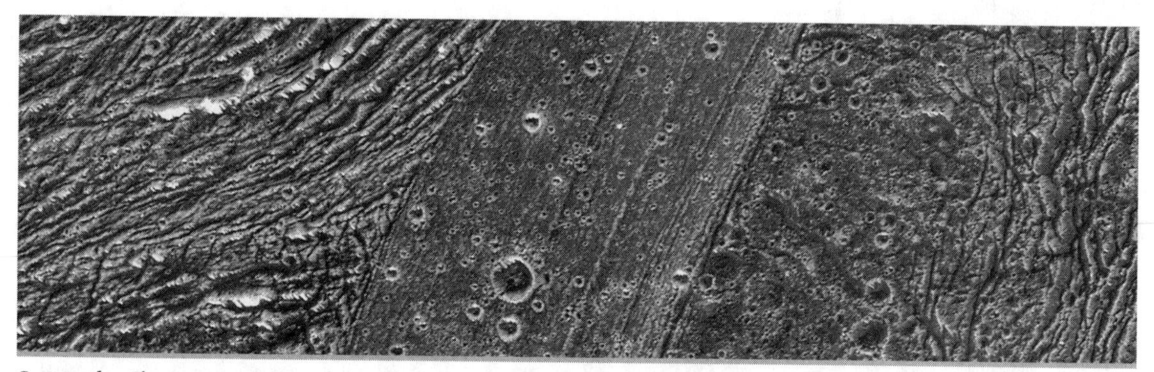

Ganymede The areas on Ganymede known as Nicholson Regio and Arbela Sulcus, photographed by Galileo on May 20, 2000, from a range of 3,350 km. The bright terrain of Arbela Sulcus, finely striated and relatively lightly cratered, is the youngest shown here, slicing north-south across the picture. To the east (right) is the oldest terrain on view, rolling and relatively densely cratered Nicholson Regio. To the west (left) is a region of highly deformed grooved terrain of intermediate age. The image corresponds to an area of about 89 km by 26 km. *NASA/JPL*

it mixed, broken up, and turned over; much like the act of cultivating a garden.

Garnet Star (Mu Cephei, μ Cep)

A very luminous red **supergiant,** one of the largest and brightest stars visible not only to the naked eye but in the entire Galaxy. Even at its great (uncertain) distance, Mu Cephei is big enough to show a measurable angular diameter, of around 0.02″, equivalent to 2.4 billion km across, so that if put in place of the Sun, it would extend midway between the orbits of Jupiter and Saturn. In common with most large supergiants, the Garnet Star pulsates, varying in brightness by about 1.67 magnitude in a semi-irregular way over a period of 2 to 2.5 years, the average magnitude varying over a period of a decade or so and dropping as low as magnitude 5.1. At the same time it is losing mass through a strong wind. Although the Garnet has certainly left the main sequence, its exact evolutionary status is uncertain; most likely, it is in the core **helium**-burning stage and is destined to explode as a supernova in the next million years or so.

Visual magnitude:	4.23
Absolute magnitude:	–6.81
Spectral type:	M2Iae
Surface temperature:	3,500 K
Luminosity:	250,000 L_{sun}
Radius:	1,500 R_{sun}
Mass:	20 to 25 M_{sun}
Distance:	5,260 light-years

gas giant

A large planet composed mainly of hydrogen and helium, but with a substantial metal-rock core around which the lighter materials that form the bulk of the object have accreted. There are four gas giants in the solar system: **Jupiter**, **Saturn**, **Uranus**, and **Neptune**. The example of Jupiter reveals why the term "gas giant" is somewhat of a misnomer. Apart from having a sizeable solid core, Jupiter also has much of its hydrogen in either liquid or, at greater depths, solid, quasi-metallic form. Most of the **extrasolar planets** discovered so far also appear to be gas giants, which is a selection effect due to the fact that current detection methods are heavily biased toward finding massive planets. The upper mass limit of gas giants is thought to be around 13 times the mass of Jupiter. Above this, theory suggests that the pressure inside an object is sufficient to induce a modest level of **deuterium** fusion and enough heat for the object to glow dully as a **brown dwarf.**

gas-bounded nebula

An **interstellar cloud** that is kept more-or-less completely ionized by one or more hot, luminous stars within it. The visible extent of the nebula is thus limited by the actual size of the cloud and the quantity of material it contains, rather than in the case of a **radiation-bounded nebula,** where the limiting factor is the amount of ionizing radiation available.

gaseous nebula

An **H II region,** a **supernova remnant,** or a **planetary nebula.** H II regions have an **emission-line** optical spectrum, and a thermal **continuous spectrum** declining in intensity as the wavelength increases (from maximum in the ultraviolet) through infrared and radio. Supernova remnants have an emission-line optical spectrum and a nonthermal radio spectrum. Temperatures of planetary nebulae are much higher than those of H II regions.

Gaspra (minor planet 951)

The first **asteroid** to be seen close-up when the **Galileo** spacecraft flew past it at a distance of 1,600 km, on October 29, 1991, en route to Jupiter. Gaspra is a member of the **Flora family,** elongated and irregular in shape, and lacking any large craters, which suggests that it has a comparatively recent origin, most likely from the collisional breakup of a larger body some 300 to 500 million years ago. It was discovered by Grigoriy N. Neujamin in 1916 and named after a Russian resort and spa near Yalta, Crimea, that was visited by contemporaries such as Tolstoy and Gorky. Diameter 19 × 12 × 11 km, spectral class S, albedo 0.2, rotational period 7.04 hours; orbit: semimajor axis 2.209 AU, perihelion 1.828 AU, aphelion 2.592 AU, eccentricity 0.174 AU, inclination 4.10°, period 3.29 years.

Gauss, Carl Friedrich (1777–1855)

A German mathematician–one of the greatest ever known–who devised a method for working out the orbit of a body from just three observations. Gauss turned his attention to mathematical applications for astronomy at about the same time that Guiseppe **Piazzi** discovered the first asteroid, **Ceres**, in 1801. By 1807 Gauss had become director of the Göttingen Observatory and in 1809 published his quick method for calculating the asteroid's orbit. This enabled astronomers to recover Ceres after it had become "lost" behind the Sun following its discovery. Gauss also worked out the theories of **perturbations** that were eventually used by Urbain **Leverrier** and John **Adams** in their independent calculations that led to the discovery of Neptune. After 1817 Gauss did no further work in theoretical astronomy, although he continued to work in positional astronomy for the rest of his life. The 1,001st asteroid to be discovered was named in his honor.

G-class asteroid

A dark **asteroid** belonging to a subcategory of **C-class asteroids**, distinguished by having a very strong ultraviolet absorption feature at wavelengths shorter than 0.4 microns due to water of hydration in the surface rocks. G-class asteroids, of which the best known example is **Ceres,** have albedos of 0.05 to 0.09.

gegenschein

A faint round or oval spot of light, about 20° across, in the midnight sky near the **ecliptic** and opposite the Sun, best seen under clear, dark conditions in September and October; also called the *counterglow* (its meaning in German).

Geminga

A **neutron star** in **Gemini** that is the second brightest source of high-energy **gamma rays** in the sky. Discovered in 1972, by the SAS-2 satellite, its name is both a contraction of "Gemini gamma-ray source" and an expression in Milanese dialect meaning "it's not there." For nearly 20 years, the nature of Geminga was unknown, since it did not seem to show up at any other wavelengths. Then, in 1991, an exceptionally regular periodicity of 0.237 second was detected by the **ROSAT** satellite in soft X-ray emission, indicating that Geminga is almost certainly a **pulsar** (a flickering neutron star). Its invisibility at radio wavelengths may be because its beams of radio radiation don't sweep past Earth. It has also been identified optically with an extremely dim blue star, 100 million times fainter than anything visible to the naked eye. A comparison of images of the suspected optical counterpart taken over an eight-year period shows a **proper motion** that is consistent with a distance to Geminga of about 330 light-years. Geminga is believed to be the stellar remains of a **supernova** that took place some 300,000 years ago and which is, at least partly, responsible for clearing out a low-density cavity in the interstellar medium in the vicinity of the solar system, known as the **Local Bubble.**

Gemini (abbr. Gem; gen. Geminorum)

The Twins from Greek mythology: divine Pollux fathered by Zeus, and mortal Castor, both placed in the sky to allow them to be together for all time; a prominent **constellation** of the northern hemisphere and the third (and northernmost) of the **zodiac.** It lies south and east of Auriga, west of Cancer, and north and east of Orion. (See tables, "Gemini: Stars Brighter than Magnitude 4.0," and "Gemini: Other Objects of Interest." See also star chart 2.)

Gemini Observatory

An international collaboration, involving the United States, Britain, Canada, Chile, Australia, Argentina, and Brazil, that has built two identical 8.1-m telescopes located at Mauna Kea, Hawaii (Gemini North) and Cerro Pachón in central Chile (Gemini South) to provide full coverage of both hemispheres of the sky. Both telescopes incorporate new technologies that allow large, relatively thin mirrors under active control to collect and focus optical and infrared radiation. Gemini North was dedicated in 1998 and Gemini South in 2001.

Geminids

One of the three most prominent annual **meteor showers**, with activity that ranges from December 7 to 17 and peaks on December 14. The maximum **zenithal hourly rate** is normally around 60 meteors per hour but can reach as high as 120 meteors per hour. The Geminids

Gemini: Stars Brighter than Magnitude 4.0

Star	Magnitude		Spectral Type	Distance (light-yr)	Position	
	Visual	Absolute			R.A. (h m s)	Dec. (° ′ ″)
β **Pollux**	1.16	1.08	K0IIIb	34	07 45 19	+28 01 34
α **Castor**	1.58	0.58	A2Vm+A1V	52	07 34 36	+31 53 18
γ **Alhena**	1.93	–0.61	A0IV	105	06 37 43	+16 23 57
μ Tejat posterior	2.87	–1.39	M3IIIab	232	06 22 58	+23 30 49
ε Mebsuta	3.06	–4.15	G8Ib	904	06 43 56	+25 07 52
η Tejat prior	3.31	–1.84	M3IIIab	349	06 22 58	+22 30 49
ξ Alzirr	3.35	2.13	F5IV	57	06 45 17	+12 53 44
δ Wasat	3.50	2.22	F0IV	59	07 20 07	+21 58 56
κ	3.57	0.35	G8IIIa	144	07 44 27	+24 23 52
λ	3.58	1.27	A3V	94	07 18 06	+16 32 25
θ	3.60	–0.30	A3III	197	06 52 47	+33 57 40
ζ **Mekbuta**	4.01v	–3.77	F7-G3Ib	1,170	07 04 07	+20 34 13
ι	3.78	0.84	G9IIIb	126	07 25 44	+27 47 53

move relatively slowly (about 35 km/s) and consist of debris that is spread out along the orbit of **Phaethon.** Radiant: R.A. 7h 28m, Dec. +32° (near *Castor*).

General Catalogue of Variable Stars (GCVS)

A list of all known **variable stars**, first published by the Russian Academy of Sciences in 1948. The fourth edition, identifying 28,435 stars, came out in three volumes in 1985–1987. A companion catalog of suspected variables is also produced, the most recent edition being the *New Catalogue of Suspected Variable Stars,* with 14,810 objects, in 1982.

general precession

The sum of the lunisolar and the planetary **precession.** It causes the **ecliptic** longitude to increase at a constant rate (50.27″ per year) but has no effect on ecliptic latitude.

general theory of relativity

The theory of gravitation developed by Albert Einstein; its fundamental principle is the equivalence of gravitational and inertial forces. General relativity describes a gravitational field in terms of the curvature of **spacetime,** or, as the American physicist John Wheeler put it: "space tells matter how to move; matter tells space how to curve." Among its predictions, borne out by observation, are the **advance of perihelion** of Mercury, the bending of light in a gravitational field (including **gravitational lenses**), and the spin-down of **pulsars** (due to the emission of **gravitational waves**, which have yet to be detected directly).

Genesis

A NASA spacecraft, launched in August 2001, that is designed to collect 10 to 20 micrograms of particles from

Gemini: Other Objects of Interest

Objects	Notes
Star	
Zeta (ζ) Gem	One of the brightest **Cepheid variables.** Magnitude range 3.7 to 4.1; period 10.15 days. See table, "Gemini: Stars Brighter than Magnitude 4.0" for further details.
U Gem	The prototype **U Geminorum star.**
Open cluster	
M35 (NGC 2168)	An outstanding cluster of about 200 stars, visible as a hazy patch with binoculars or a small telescope. Magnitude 5; diameter 28′; R.A. 6h 8.9m, Dec. +24° 20′.
Planetary nebula	
Eskimo Nebula	NGC 2392. See separate entry.

Gemini Observatory The Gemini North Observatory on Mauna Kea, Hawaii. *Gemini Observatory*

the **solar wind** using wafers of aerogel (the lightest solid material known, with a density only three times that of air) set in winglike arrays. For two years, Genesis will orbit around the first **Lagrangian point** of the Earth-Sun system before returning to enable the recovery of its 210-kg sample capsule in September 2004. As the capsule descends by parachute it will be caught by a helicopter over the Utah desert. Scientists know that the solar system evolved a little under 5 billion years ago from an interstellar cloud of gas, dust, and ice, but the exact composition of this cloud remains unknown. As its name suggests, Genesis will help unravel this mystery by recovering material that has been shot out of the upper layers of the Sun–material that has not been modified by nuclear reactions in the Sun's core and is thus representative of the composition of the original **solar nebula.**

geocentric

Earth-centered. *Geocentric coordinates* are a system of **celestial coordinates** with its origin at the center of Earth. For *geocentric parallax* see **parallax.**

geocorona

A halolike extension of the **exosphere** out to several Earth radii, consisting of very tenuous hydrogen atoms, off which solar far-ultraviolet light is reflected.

geodetic precession

A small, relativistic, direct motion of the **equinox** along the **ecliptic,** amounting to 1.915″ per century.

Geographos (minor planet 1620)

An oddly shaped member of the **Apollo group** of Earth-crossing asteroids, discovered by Rudolph **Minkowski** and his American colleague Albert George Wilson (1918–) at Palomar Observatory in 1951. Geographos shows the most extreme variations in its **light curve** of any object in the solar system: the amount of light it reflects varies by a factor of 6.5 over the course of an **axial rotation.** This indicates that Geographos is either very elongated–a cigar-shaped object viewed along a line perpendicular to its spin axis–or is a pair of objects nearly in contact that orbit each other and around their center of mass. Possibly it acquired its strange form after being splintered off a larger body or by stretching through **tidal forces** produced during a past close encounter with Earth. The asteroid's name was chosen to honor the National Geographic Society for its support of the ***Palomar Observatory Sky Survey.*** Diameter 5.1 × 1.8 km, spectral class S, semimajor axis 1.246 AU, perihelion 0.828 AU, aphelion 1.663 AU, inclination 13.34°, period 1.39 years.

geoid

The hypothetical shape of Earth's globe used in geodetic calculations; the surface of the geoid is approximately the same as *mean sea level* envisaged as extending over the entire surface of the globe.

geomagnetic field

The magnetic field observed in and around Earth. The intensity of the magnetic field at Earth's surface is approximately 0.32 gauss at the equator and 0.62 gauss at the North Pole.

geomagnetic storm

A worldwide disturbance of Earth's magnetic field that follows 2 to 3 days after violent activity on the Sun, such as **solar flares** or **coronal mass ejections**. A geomagnetic storm's main phase, which can last as long as 2.5 days in the case of a severe storm, is characterized by multiple intense **substorms** with attendant **auroral** and geomagnetic effects.

ghost crater

A **crater** that has been largely buried under deposits of other material, such as a lava flow or ejecta from later impacts. All that may be visible is the circular outline of parts of the crater rim and perhaps the peak of a central mountain. In other cases, the original crater may have been completely submerged but later subsidence has formed wrinkle ridges, which trace out the old position of the rim.

Ghost Head Nebula (NGC 2080)

One of a chain of star-forming regions lying south of the 30 Doradus Nebula in the **Large Magellanic Cloud**. Two bright regions (the eyes of the ghost), named A1 and A2, are very hot, glowing blobs of hydrogen and oxygen. The bubble in A1 is produced by the hot, intense radiation and powerful **stellar wind** from a single massive star. A2 has a more complex appearance due to the presence of more dust, and it contains several hidden, massive stars. The massive stars in A1 and A2 must have formed within the last 10,000 years since their natal gas shrouds are not yet disrupted by the powerful radiation of the newly born stars.

ghost image

(1) In optics: a faint image caused by reflections at uncoated or antireflection-coated surfaces. A prime example is the faint reflection from the uncoated front surface of a rear-surface mirror. Using a front-surface mirror eliminates the ghost image (but may cause other problems). (2) In spectroscopy: a false image of a **spectral line** caused by irregularities and imperfections in the ruling of a **diffraction grating.**

Ghost of Jupiter (NGC 3242)

A **planetary nebula** in **Hydra,** just south of Mu Hydrae. Although round and similar in apparent size to Jupiter, it is much dimmer and an unmistakable blue-green in color. Even a small telescope will show it well; larger instruments reveal its inner torus and central star.

Visual magnitude:	9
Angular diameter:	20.8′
Distance:	1,400 light-years
Position:	R.A. 10h 24.8m, Dec. −18° 38′

Giacconi, Riccardo (1931–)

An Italian-born astrophysicist and pioneer of X-ray astronomy. Giacconi earned a Ph.D. in cosmic-ray physics at the University of Milan before joining American Science and Engineering, a Massachusetts research firm, in 1959 to begin his X-ray work. He and his colleagues developed **grazing-incidence telescopes** and launched them on rockets. In 1962 they discovered **Scorpius X-1,** the first known X-ray source outside the solar system. They then built the Uhuru X-ray satellite and made the first surveys of the X-ray sky. Joining the Harvard-Smithsonian Center for Astrophysics in 1973, Giacconi led the construction and successful operation of the powerful X-ray observatory, HEAO-2, also known as *Einstein,* which made detailed images of X-ray sources. Giacconi was the first director of the Space Telescope Science Institute from 1981 to 1993, and he directed the **European Southern Observatory** for the next six years. In 1999 he became president of Associated Universities, the operator of the National Radio Astronomy Observatory.

Giacobini-Zinner, Comet (Comet 21P/)

A **short-period comet** that is the parent body of the Draconids (also known as the *Giacobinids*) **meteor shower**; it became the first comet to be visited by a spacecraft when the International Cometary Explorer (ICE) flew past it at a distance of 7,800 km on September 11, 1985. Every alternate appearance of the comet is favorable for observers on Earth and can result in a visual magnitude as high as about 7. The 1946 appearance was especially noteworthy as the comet passed only 0.26 AU from Earth in late September; it was then near magnitude 7, although it experienced an unexpected outburst that caused it to reach magnitude 6 during the first days of October. It was discovered by the French astronomer Michel Giacobini (1873–1938) in 1900 and rediscovered two returns later, in 1913, by the German astronomer Ernst Zinner (1886–1970). Semimajor axis 3.52 AU, perihelion 1.034 AU, eccentricity 0.706, inclination 31.8°, period 6.61 years.

giant branch

A conspicuous sequence of red stars with large radii in the **Hetzsprung-Russell diagram** of a typical **globular cluster**, which extends upward from the main-sequence **turnoff point** to the **red giant tip**.

giant molecular cloud (GMC)

A huge complex of interstellar gas and dust, composed mostly of molecular hydrogen but also containing many other types of **interstellar molecules**. GMCs are the coolest (10 to 20 K) and densest (10^6 to 10^{10} particles/cm^3) portions of the **interstellar medium**, and are one of two recognized types of **molecular cloud.** Stretching typically over 150 light-years and containing several hundred thousand solar masses of material, they are the largest gravitationally bound objects in the Galaxy and, in fact, the largest known objects in the universe made of molecular material. Molecular clouds are the only places where star formation (and planet formation) is known to occur (see **stellar evolution**). The other types of interstellar clouds, in which hydrogen is atomic, are too warm and diffuse to allow stars to form. Since star formation occurs when deeply embedded clumps of interstellar gas and dust collapse, stars that are newborn or in the very process of forming are always obscured from direct optical view, and the only source of information from inside these clumps is provided by longer-wavelength **radio waves** and **infrared** emitted by molecules.

What happens when stars begin forming in GMCs depends on the environment. Under normal conditions in the **Milky Way Galaxy** and in most other present-day spiral galaxies, star birth stops after a relatively small number of stars have been born because the stellar nursery is blown away by some of the newly formed stars. The

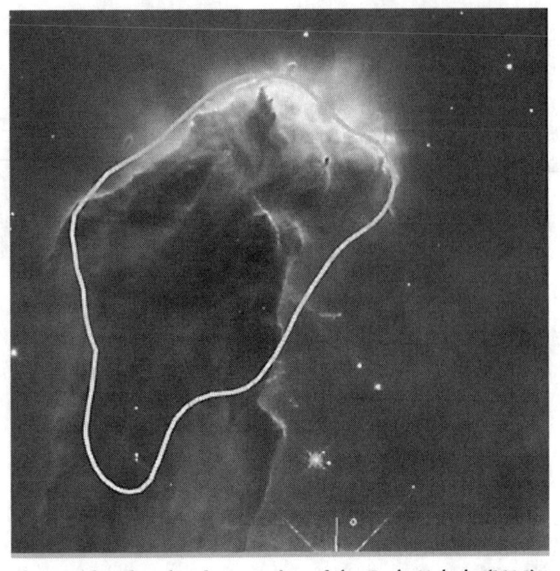

giant molecular cloud A portion of the Eagle Nebula (M16) with the location of the associated molecular cloud outlined.
NASA/ESA/STScI

hottest of these heat the surrounding molecular gas, break up its molecules, and drive the gas away. As the celestial smog of gas and dust clears, the previously hidden young stars become visible, and the molecular cloud and its star-birthing capability cease to exist. GMCs in colliding galaxies may experience a different fate. As the collision crunches the interstellar gas and stars form at an accelerating rate, the gas pressure around the surviving GMCs increases 100- to 1,000-fold. Calculations suggest that the hot surrounding gas can trigger rapid star birth throughout the clouds by driving shock waves into them. The several hundred thousand stars that form from the cold molecular gas in such clouds use up most of the gas before it has time to be heated and dispersed. The result of such violent events is the nearly complete conversion of GMCs into rich star clusters, each containing up to 1 million stars. Observations by the Hubble telescope suggest that many of these newly born star clusters remain bound by their own gravity and evolve into globular clusters, like those observed in the Milky Way's halo.

giant planet

A planet that is much larger and more massive than Earth. In practical terms, this probably means that all giant planets are **gas giants**, since it is hard to see how an exclusively rock-metal planet could acquire enough material in a **protoplanetary nebula** to take on such huge proportions.

giant star

A star that has evolved off the **main sequence,** having exhausted its core supply of hydrogen, and has swollen in size and brightness. A giant has a radius of 5 to 25 R_{sun} and a luminosity from tens to hundreds of L_{sun}. See **red giant** and **blue giant**.

Giotto

A European Space Agency probe, launched in July 1985, that encountered **Halley's comet** the following year. Giotto measured Halley's composition and, on March 16, 1986, returned color images of the comet's nucleus. It was named after the Italian painter, architect, and sculptor Giotto di Bondone (1266–1337) whose fresco, *Adorazione dei Magi* (Adoration of the Magi), painted sometime between 1304 and 1306, features an accurately represented comet above the Nativity stable. The fresco's realistic portrayal strongly suggests that it was based on the artist's first-hand observation of the comet during its appearance in the skies over Europe in October 1301.

GLAST (Gamma-ray Large Area Space Telescope)

An orbiting high-energy astrophysics observatory designed to map the universe in **gamma rays** with up to 100 times the sensitivity, resolution, and coverage of previous missions. It will carry two instruments: the Large Area Telescope, sensitive to gamma rays in the 200 MeV to 300 GeV range, and the Burst Monitor, for detecting **gamma-ray bursts**. Following a one-year sky survey, GLAST will be available to scientists for particular research projects. An international mission led by NASA, GLAST is scheduled for launch in March 2006.

Gliese Catalogue

The usual name given to any of three catalogs of nearby stars compiled by Wilhelm Gliese (1915–1993) and, later, also by Hartmut Jahreiss, in 1957, 1969, and 1993; listed objects are prefixed by "Gl" or "GJ." The *Gliese Catalogue* attempts to list all stars within 25 parsecs (81.5 light-years) of Earth. Numbers in the range 1.0 to 965.0 are from the second edition, *Catalogue of Nearby Stars* (1969). The integers represent stars that were in the first edition, while the numbers with a decimal point are used to insert new stars for the second edition without destroying the original order. Numbers in the range 9001 to 9850 are from the supplement *Extension of the Gliese Catalogue* (1970). Numbers in the ranges 1000 to 1294 and 2001 to 2159 are from the supplement *Nearby Star Data Published 1969–1978.* The range 1000 to 1294 represents nearby stars, while 2001 to 2159 represents suspected nearby stars. Numbers in the range 3001 to 4388 are from *Preliminary Version of the Third Catalogue of Nearby Stars* (1991). Although this version of the catalog was termed preliminary, it is still the current one as of mid-2003. Most of the 3,803 stars it lists already had GJ numbers, but there were also 1,388 that were not numbered. The need to give these 1,388 *some* name has resulted in them being numbered 3001 to 4388.

glitch

A sudden change in a **pulsar**'s period and spin-down rate believed to be caused by a *starquake*–the abrupt release of stress energy either in the crust of the star or between the crust and the superfluid interior. Glitches tend to occur in young pulsars, such as the **Crab pulsar** and **Vela pulsar**, whose spin rate is decreasing most rapidly.

Global Oscillation Network Group (GONG)

A network of six observatories around the world that provides almost continuous, highly sensitive monitoring of the Sun's pulsations. The data collected is used for detailed studies of solar internal structure and dynamics (see **helioseismology**). GONG began operation in 1995, is managed by the **National Solar Observatory**, and involves sites in the Canary Islands (Teide Observatory), Western Australia (Learmonth Solar Observatory), California (**Big Bear Solar Observatory**), Hawaii (Mauna Loa Solar Observatory), India (Udaipur Solar Observatory), and Chile (**Cerro Tololo Inter-American Observatory**).

globular cluster

See article, page 212.

globule

A small, dense, rounded cloud of dust and gas that absorbs radiation and appears as a black blob when seen against the background of a bright nebula or star field; also known as a *Bok globule* after Bart **Bok** who first drew attention to them and suspected they might be **protostars**. Their exact status, however, remains unclear. While it is generally assumed they are the cores of fragments of **molecular clouds** that are in the process of collapsing to form individual stars, they could also be fragments that have become hung up by a combination of internal magnetic and turbulent pressure and are failed star-forming cores. The interiors of globules are the coldest known places in nature, with temperatures as low as a few degrees above absolute zero.

Gold, Thomas (1920–)

An Austrian-American astrophysicist who, in 1948, together with Hermann Bondi and Fred **Hoyle**, proposed the **steady-state theory** of the universe. In the late 1960s he correctly interpreted the newly discovered **pulsars** in terms of rotating **neutron stars** (a proposal made independently by Franco Pacini). More recently, Gold has again been at the center of a controversy with his notion that petroleum deposits formed nonbiologically at great depths. This would explain, Gold believes, the surprisingly high concentration of helium found in petroleum since as the liquid rose to the surface it would collect helium from the many kilometers of rock through which it percolated. A few kilometers underground, as Gold sees it, the upwelling petroleum acts as a nutrient for deep-dwelling microorganisms that are the source of the biological molecules found in crude oil. Gold argues that a number of planets, moons, and even asteroids in the solar system might have internal hydrocarbon reserves, which could provide an energy source for subterranean microbes. He considers **Titan**, in particular, which has methane and ethane in its atmosphere, a promising place to look for subsurface life. Born in Vienna, Gold became a refugee from the Austrian Anschluss and gained his BA in 1942 from Cambridge University, England. He lectured there in physics from 1948 to 1952 before joining the Royal Greenwich Observatory as chief assistant to the Astronomer Royal. He moved to the United States in 1956, founding the Center for Radiophysics and Space Research at Cornell University and serving as its first director from 1959 to 1981, and as professor of astronomy from 1971 to 1986.

Goodricke, John (1764–1786)

A Dutch-born English amateur astronomer who, despite a short life and the handicap of being deaf and mute, did important early work on **variable stars**. Encouraged in this field by his neighbor, Edward **Pigott**, Goodricke determined the period of **Algol** and explained, in 1782, its behavior in terms of an **eclipsing binary** system–a suggestion that was confirmed a century later. Goodricke discovered the variability of **Beta Lyrae** and of **Delta Cephei** in 1784, but then died at the age of 21 after a bout with pneumonia that he reportedly caught while observing Delta Cep.

Gould, Benjamin Apthorp (1824–1896)

An American astronomer who founded the *Astronomical Journal* and later established Argentina's National Observatory. Gould graduated from Harvard in 1844, then, having studied for a year at Berlin, he obtained his Ph.D. from Göttingen University in 1848 under Karl **Gauss**. On his return to the United States he served as head of the longitude department of the U.S. Coast Survey from 1852 to 1867, pioneering the use of the telegraph in measuring longitude. At the same time he founded the *Astronomical Journal* (1849) and edited it until 1861 when its publication was halted by the Civil War. He was also connected with the Dudley Observatory, Albany, from 1855

(continued on page 213)

globular cluster

A spherically symmetric collection of old stars that share a common origin. Globular clusters contain from tens of thousands to millions of stars and measure from 100 to 300 light-years across. Some have been shown, in all likelihood, to contain middleweight **black holes** in their cores. Unlike **open clusters** and **stellar associations**, which are held together only weakly by gravity and contain **Population I** objects, globulars are tightly gravitationally bound collections of **Population II** stars. They inhabit the **galactic halo** and **galactic bulge**, and show significant concentration toward the galactic center. The **Milky Way Galaxy** has about 150 globular clusters, all of them in highly elongated orbits. They are divided into two broad types, known as *Oosterhoff groups* after the Dutch astronomer Pieter Oosterhoff (1904–1978) who first identified them. The main difference between them is that group I clusters have *slightly* weak metal lines whereas group II clusters have *very* weak metal lines.

A topic of great interest is the origin of globular clusters. It had been assumed that the majority of globulars in the Milky Way were ancient native systems that formed around the same time as the rest of the Galaxy some 10 billion years ago, though it was known that nearly two dozen group II clusters are prisoners-of-war captured from other, smaller galaxies in the **Local Group**. Recent research, however, has shown that even some group II systems–the metal-poor variety–have been looted from outside. These presumed aborigines have, it turns out, come from the **Large Magellanic Cloud** and are a billion years younger than had been previously estimated. As a result, astronomers have been forced to rethink their ideas about how the Milky Way evolved and, because globulars are used as a cosmic yardstick, to recheck the distances between galaxies, which might be in error by as much as 7%. Further evidence that globular clusters are not necessarily relics of the earliest generations of stars in a galaxy comes from observations of **galaxy interactions**. These can give rise to **ultra-luminous clusters**, which appear to be globular clusters in the making.

Palomar 5, on the other hand, is a globular cluster in its death throes–being torn apart by tidal forces of the Milky Way. This ancient object, on the outskirts of our Galaxy, some 75,000 light-years from the Sun, has only about 10,000 stars left. All the rest have been stripped away and now lie in two incredibly long tails, strung across 13,000 light-years of space. In fact, the tails of Palomar 5 delineate the orbital path of this cluster and thus provide a unique opportunity to determine its motion around the Galaxy. Together with the so-called Sagittarius Stream, which emerges from the **Sagittarius Dwarf Elliptical Galaxy**, there are now two different examples of extended stream-like structures in the galactic halo. The geometry and the velocities of those tidal streams will become important tools for determining the mass of the **dark matter** halo that surrounds the Milky Way.

globular cluster The globular cluster M15 as seen by the Hubble Space Telescope. M15 lies some 40,000 light-years away in the constellation Pegasus. The extended, bloblike object to the upper left in this image is Kuestner 648–the first planetary nebula to be identified in a globular cluster. *NASA/STScI/AURA*

globule Barnard 68, one of the nearest globules to Earth at a distance of only 410 light-years. It is about 2 trillion km (12,500 AU) across, which is similar in the size to the Oort Cloud of dormant comets that surrounds the solar system. The temperature of Barnard 68 is 16 K (–257°C) and the pressure at its boundary is about 10 times higher than in the interstellar medium. The total mass of the cloud is about twice that of the Sun. *European Southern Observatory*

Gould, Benjamin Apthorp (1824–1896)

(continued from page 211)

and served as its director briefly in 1858 before being forced to get out of town the following year. After his traumatic expulsion from Albany he handled his father's business for some time. He set up a private observatory in Cambridge, Massachusetts, financed by his wife, and in 1862 produced a star catalog that brought together measurements made at various observatories. He left for Argentina in 1870. The 15 years spent in Cordoba were by far the most productive of Gould's career. He established the Argentine National Observatory and began the first major survey of the southern skies. The Observatory's first survey of naked-eye stars was published as the

Uranometria Argentina (1879). This was followed by the fuller recording, published in 1884, of 73,160 stars from 23° S to 80° S and in 1886 by the publication of the *Catàlago General* (General Catalogue) containing the more accurate recording of 32,448 stellar coordinates. This important work was continued by Gould's successor, Juan Thomé. An extended band of young stars, cloud, and dust that forms a spur off one of the spiral arms of our Galaxy, revealed by the southern surveys, was subsequently named **Gould's Belt**. In 1885 Gould returned to Massachusetts where he restarted the *Astronomical Journal* and worked on the 1,000 photographic plates of star clusters he brought back with him from Cordoba.

Gould's Belt
A partial ring or disk of hot, young stars, of types O and B, with a diameter of about 3,000 light-years, inclined to the **galactic plane** at about 20°, within which the Sun lies. Among its most prominent components are the bright stars in Orion, Canis Major, Puppis, Carina, Centaurus, and Scorpius, including the **Scorpius-Centaurus Association**. It is thought to be 30 to 50 million years old but its origin is unclear. It is named after Benjamin **Gould**, who established its existence in 1879.

graben
A long, relatively narrow area of rock that has slid down between parallel faults. Unless erosion wears the sides down to the level of the graben, the result is a **rift valley**.

Gran Telescopio Canarias (GTC)
A high-performance, segmented 10.4-m telescope to be installed at one of the best sites of the Northern Hemisphere: the **Roque de los Muchachos Observatory** on La Palma in the Canary Islands. When it opened in September 2003 it became the largest single optical/infrared telescope in the world. The GTC project is a Spanish initiative, led by the IAC (Instituto de Astrofisica de Canarias) with the aim of becoming an international project.

granulation
The mottled appearance of the solar surface when seen at high spatial resolution; it is caused by hot gases rising from the interior of the Sun. It looks a bit like rice pudding, with individual bright "rice grains," known as **granules**, separated from each other by a connected network of dark paths, called *intergranular lanes*.

granule
A convective cell, about 1,000 km in diameter, in the solar **photosphere**. Each granule lasts on average about 5 minutes and represents a temperature roughly 300° higher than the surrounding dark areas. At any one time, granules cover about one-third of the solar photosphere and are collectively referred to as **granulation**.

gravitational collapse
The infall of a massive body under its own weight. **Interstellar clouds**, for example, collapse gradually to become stars, at which point the onset of nuclear **fusion** arrests any further infall. The cores of massive stars, by contrast, collapse very suddenly when all viable sources of nuclear fusion are exhausted and the outward **radiation pressure** is no longer sufficient to balance the inward gravitational pressure. In gravitational collapse there is a sudden, catastrophic release of great quantities of gravitational potential energy, which is the underlying cause of **supernovae**, **neutron stars**, and **black holes**.

gravitational equilibrium
The state in a celestial body, such as a star, when gravitational forces pulling inward on a particle are balanced by some outward pressure, such as **radiation pressure** or electron degeneracy (see **electron degenerate matter**), so that no vertical motion results.

gravitational instability
A situation in which an object's self-gravity exceeds opposing forces such as internal gas pressure or material rigidity, and the object collapses. For a gas, gravitational instability sets in when the mass is greater than a certain critical value known as the **Jeans mass**. In the early universe, instabilities were large enough to produce galaxies and clusters of galaxies.

gravitational lens
The gravitational equivalent of a magnifying glass: the focusing or distorting effect produced when a concentrated mass lies along the line of sight from an observer to a more distance light source. Typically, the distant source is a **quasar** and the lensing mass is a galaxy lying between the quasar and Earth. Gravitational lenses arise from the warping of **space-time** around a massive object, as described by Einstein's **general theory of relativity**. Light rays, in following the shortest path through the curved region of space-time, are deflected. But the effect is not as simple as that produced by an ordinary glass lens; instead it is comparable to the distortion produced by looking through the base of a wineglass. Depending on the geometry of the gravitational lens, the resulting image of the lensed object may be an arc, a complete circle (known as an **Einstein ring**), or a series of multiple images. More than 20 examples of cosmic-scale gravitational lenses have been discovered since the first in 1979,

gravitational lens Four images (two of them partly merged in the largest blob) of the quasar MG0414+0534A are a gravitational lens effect caused by an intervening galaxy. *European Southern Observatory*

including the remarkable **Einstein Cross**. Gravitational lensing also occurs on a smaller scale, known as **microlensing**, when a dark object in our Galaxy passes directly in front of a more distant star (e.g., one of the stars in the **Magellanic Clouds**) and makes its image brighten briefly.

gravitational waves

Disturbances or ripples in **space-time** predicted by the **general theory of relativity** due to changing distributions of mass, such as the spin-down of a **binary pulsar** or the implosion of a star during a **supernova**. There have been no confirmed direct detections so far because the equipment used has not been sufficiently sensitive.

gravity

One of the basic forces in nature. Gravity operates between any two objects that have mass with an attractive force that is directly proportional to the masses of the objects and is inversely proportional to the square of the separation distance. In Einstein's **general theory of relativity**, gravity is viewed as a consequence of the curvature of **space-time** induced by the presence of a massive object. In quantum mechanics, the gravitational field is seen as being conveyed by quanta called *gravitons*.

grazing-incidence telescope

A telescope used in extreme ultraviolet, X-ray, and gamma-ray astronomy, which utilizes the fact that electromagnetic radiation at these very short wavelengths behaves like ordinary light rays if it strikes surfaces at a shallow enough angle. Instead of the bowl of a paraboloid, the mirror of a grazing-incidence telescope consists of a ring higher up on the wall of the paraboloid so that the incoming rays meet it at acute angles and are deflected, rather than reflected, onto the detector.

Great Annihilator

A very powerful high energy X-ray source near the center of the **Milky Way Galaxy**; it appears to be a massive **black hole**, 10 to 1,000 times the mass of the Sun, that is producing an enormous flux of positrons (antielectrons). These antiparticles are interacting with ordinary matter nearby to produce a strong flux of electron **annihilation radiation** at a characteristic energy of 511 keV.

Great Attractor

A hypothesized large mass, about 150 million light-years away in the direction of the Virgo-Hydra-Centaurus supercluster, that seems to be affecting the motions of many nearby galaxies. The amount of mass proposed amounts to tens of thousands of galaxies. However, the nature and even the existence of the Great Attractor are

Great Attractor A region of the sky in the direction of the Great Attractor. This region covers a field of 0.5° × 0.5° in the southern constellation Norma and is at an angular distance of about 7° from the main plane of the Milky Way. The foreground stars, in our own Galaxy, appear mostly as whitish spots (the crosses around some of the brighter stars are caused by reflections in the telescope optics). Many background galaxies are also seen. They form a huge cluster (ACO 3627) with a number of bright galaxies near the center that stand out by their larger size and yellowish color. *European Southern Observatory*

subjects of debate. Some astronomers believe that the Great Attractor may be centered on the rich cluster known as Abell 3627. Others propose that, while there may be a cluster of galaxies in the area of the Great Attractor, the large-scale movement of so many superclusters is probably due to the gravitational pull of all clusters combined. One argument against the existence of the Great Attractor is that no one has detected signs of infalling galaxies *behind* the Attractor. On the other hand, it is possible that this infall may be counteracted by the mass of the more distant **Shapley Concentration** whose galaxies may be tugging the Great Attractor galaxies the other way.

great circle

The line of intersection of the surface of a sphere and any plane that passes through the center of the sphere.

Great comets

Long-period comets of the past couple of centuries that achieved unusual brilliance in the nighttime (and, in some cases, even the daytime) sky. The *Great comet of 1811* reached peak brightness during September and October and could be seen with the unaided eye for 9 months–the longest period of naked-eye visibility for a comet on record. It was referred to by Leo Tolstoy in his *War and Peace* and was popularly associated with the turmoil of this period. The *Great March comet of 1843*–visible only from southern latitudes for a few hours on February 28–at its greatest brilliance outshone any comet seen in the previous seven centuries. The tail of the comet holds the record for actual length: an estimated 300 million km. The *Great September comet of 1882* (C/1882 R1) was discovered a few days before reaching perihelion and grew 10 times brighter each day as it plunged toward the Sun. On September 16 and 17, it was an impressive naked-eye object next to the midday Sun. See also **Tebbutt, Comet (C/1861 J1)** and **Daylight comet of 1910**.

Great Observatories

A series of four major NASA orbiting observatories covering different regions of the **electromagnetic spectrum**. They are the **Hubble Space Telescope**, **Compton Gamma-ray Observatory**, **Chandra X-ray Observatory**, and **Space Infrared Telescope Facility**.

Great Rift

A dark divide in the Milky Way between **Cygnus** and **Sagittarius** caused by a succession of large, overlapping clouds of interstellar **dust** in the equatorial plane of the Galaxy.

Great Wall

A sheet of galaxies measuring some 200 million by 600 million light-years in area but with a thickness of only about 20 million light-years. One of the largest known structures in the universe, it lies 300 to 400 million light-years from the Milky Way and appears to connect the Coma and Hercules superclusters. It was discovered in 1988–1989 by Margarite Geller and John Huchra at the Center for Astrophysics while investigating data from a survey of 1,060 galaxies. The Great Wall stretches from R.A. 8h to 17h and from Dec. +26° to +32°, and influences the motions of galaxies with recession velocities (due to the expansion of the universe) of 7,000 to 10,000 km/s.

greatest elongation

The maximum angular separation of an **inferior planet** (Venus or Mercury) from the Sun at a given **apparition**.

Greek astronomy

Unlike the Babylonians and Egyptians, who observed the heavens primarily to keep track of the seasons, the Greeks approached astronomy from a much more fundamental and theoretical viewpoint: they wanted to know the basic nature and makeup of the universe. Early theories about Earth, the Moon, the Sun, the planets, and the stars were put forward, beginning in the sixth century B.C., by **Thales**, **Anaximander**, **Anaximenes**, **Pythagoras**, and **Anaxagoras**. By the time of **Eudoxus** (408–355 B.C.), a standard Greek cosmological model was in place with Earth at the center of the universe surrounded by a series of concentric spheres holding, in order of increasing distance, the Moon, Mercury, Venus, the Sun, Mars, Jupiter, Saturn, and the stars. Eudoxus, **Heraclides**, and Callipus developed this idea further. It was opposed by **Aristarchus**, who put forward the first genuine heliocentric scheme; however, the Sun-centered cosmos was never embraced by the Greeks, partly because the great philosopher **Aristotle**, about 50 years earlier, had "proved" by many philosophical arguments that Earth must be central. Important observational work was done by Aristarchus, **Erastothenes**, and **Hipparchus**, which began to reveal the scale of the solar system and to accurately catalog the positions of stars. Around A.D. 125, **Ptolemy**, a theorist living in Alexandrian Egypt, summarized Greek astronomical knowledge, including the geocentric theory, in a book called (by the Arabs) the *Almagest.* After Ptolemy's generation, the ancient civilizations began to decline. Astronomy came to a halt in

Green Bank Telescope The Green Bank Telescope at dusk.
NRAO

Europe with the fall of Rome and the Ptolemaic system would not be seriously challenged for another 1,400 years. See **Egyptian astronomy, Babylonian astronomy, Chinese astronomy,** and **Indian astronomy**.

Green Bank Telescope

A fully steerable **radio telescope** with an elliptically shaped, 100 × 110 m dish at the National Radio Astronomy Observatory site at Green Bank, West Virginia.

green flash

A rare optical effect in which the tip of the Sun's disk suddenly and briefly changes to a vivid green, either as the last remnant of the setting Sun vanishes or just at the moment of sunrise. Caused by the preferential refraction of light, it requires a distant, sharply defined low horizon (preferably the sea) and special atmospheric conditions, including cool weather. The period of visibility tends to increase in summer months with increasing latitude, i.e., as the angle of descent of the Sun decreases. In the Antarctic, it has been observed for as long as 30 minutes. Occasionally it takes the form of a white flash followed by a deep blue one. A similar phenomenon, the *red flash,* is sometimes seen as the lower edge of the Sun emerges from a dark cloud near the horizon.

greenhouse effect

An increase in a planet's surface temperature caused by the absorption of **infrared** radiation by gases in the atmosphere, including **carbon dioxide, methane,** and water vapor. Incoming short wavelength radiation passes through, but longer wavelengths reradiated from the surface are blocked by the greenhouse gases. Earth's atmosphere is about 35 K warmer, on average, than if there were no atmospheric greenhouse effect. On Venus a *runaway greenhouse effect* massively increases temperatures by about 500 K over what they would be if the atmosphere were completely transparent. By contrast, the thin carbon dioxide atmosphere on Mars contributes only a 5 K rise in surface temperature.

Gregorian calendar

The **calendar** established under the auspices of Pope Gregory XIII in 1582 to replace the **Julian calendar**. Correcting at a stroke the 10-day accumulated error of the Julian calendar, the main innovation was that century-years were discounted as **leap years** unless they were divisible by 400. The Gregorian calendar is now used as the civil calendar in most countries. Historical or astronomical dates based on the Gregorian calendar are said to be *New Style dates* as distinct from *Old Style dates* that refer to the Julian calendar.

Gregorian telescope

An early type of **reflecting telescope**, designed by the Scottish mathematician and astronomer James **Gregory** (1638–1675), that uses two curved mirrors: a concave paraboloidal primary and a concave ellipsoidal secondary. The primary collects and brings light to a focus, while the secondary, positioned a little way beyond the primary's focal plane, reflects the beam, diverging from the focus, back through a hole in the center of the primary and out the bottom end of the instrument. Although the Gregorian is free from **chromatic** and **spherical aberration**, it requires a long telescope tube. It was rendered obsolete by the **Cassegrain telescope**.

Grimaldi, Francesco Maria (1618–1663)

An Italian Jesuit physicist and astronomer who was the first to describe the **diffraction** of light (in a work published posthumously in 1665) and the first to attempt a wave theory of light. He verified **Galileo**'s law of the uniform acceleration of falling bodies, drew a detailed map of the Moon, and began the practice of naming lunar features after astronomers and physicists.

Groombridge Catalogue

The name by which *A Catalog of Circumpolar Stars,* compiled by the English astronomer Stephen Groombridge (1755–1832), is commonly known. In 1806, Groombridge began observations at Blackheath, London, and retired from the West Indian trade in 1815 to devote his full attention to the project. His catalog, listing 4,243 stars located within 50° of the north celestial pole and having apparent magnitudes greater than 9, was published posthumously in 1838.

Grus: Stars Brighter than Magnitude 4.0

Star	Magnitude		Spectral Type	Distance (light-yr)	Position	
	Visual	Absolute			R.A. (h m s)	Dec. (° ′ ″)
α **Alnair**	1.73	–0.74	B7IV	101	22 08 14	–46 57 40
β	2.07v	–1.52	M5III	170	22 42 40	–46 53 05
γ Al Dhanab	3.00	–0.97	B8III	203	21 53 56	–37 21 54
ε	3.49	0.49	A3V	130	22 48 33	–51 19 01
ι	3.88	0.11	K0III	185	23 10 21	–45 14 48
δ^1	3.97	–0.82	G6III	296	22 29 16	–43 29 45

grooved terrain

A region on the surface of a planet or moon marked by a series of almost parallel grooves; the origin may be tectonic or associated with secondary impact craters. Grooved terrain was originally identified on **Phobos**, but has also been seen on **Ganymede** and **Mars**.

ground state

The condition of an atom, ion, or molecule, when all of its **electrons** are in their lowest possible energy levels, i.e., not excited.

ground-level event (GLE)

A sharp increase in the ground-level count of **cosmic rays** by at least 10% above background, associated with solar **protons** of energies greater than 500 MeV. GLEs are relatively rare, occurring only a few times each **solar cycle**.

Grus (abbr. Gru; gen. Gruis)

The Crane; a small southern **constellation**. It lies south of Pisces Austinus between R.A. 21h 30m and 23h 25m and Dec. –36° and –56°. (See table, "Grus: Stars Brighter than Magnitude 4.0.")

Grus Quartet (NGC 7552, NGC 7582, NGC 7590, and NGC 7599)

Four large **spiral galaxies** that are physically very close together and strongly interacting. The high starburst activity (see **starburst galaxy**) of two of the members, NGC 7552 and NGC 7582, is also thought to arise from tidal galaxy-galaxy interactions and subsequent formation of a bar in the disk. Several **tidal tails** are visible extending from NGC 7582, one pointing toward the neighbors in the east and the other toward NGC 7552, which lies at a projected distance of approximately 30′ to the northwest. See also **galaxy interaction**.

Guide Star Catalog (GSC)

A catalog of stars, galaxies, and other celestial objects, based on a digitization of photographic sky survey plates from the Palomar and United Kingdom Schmidt telescopes. There are two versions. *GSC I* is used for the control and target acquisition of the **Hubble Space Telescope** and is an all-sky optical catalog of positions and magnitudes of approximately 19 million stars and other objects in the sixth to fifteenth magnitude range. *GSC II* contains almost a billion objects down to about twentieth magnitude and is used to support operations of the **Very Large Telescope**, the **Gemini Observatory**, and Hubble.

Gum Nebula

A vast, almost circular **emission nebula** that sprawls across nearly 40° of the southern constellations **Vela** and **Puppis**. It is so large, yet faint against a bright and complex background, that it is hard to distinguish; its front edge is as close to us as 450 light-years, while its back edge lies about 1,500 light-years away. The Gum Nebula is believed to be the hugely expanded (and still-expanding, at 20 km/s) remains of a **supernova** that took place about a million years ago. It contains the Vela OB2 Association, the hot stars of which (among them, **Naos**) cause it to shine. Also embedded in the Gum are the **Vela Supernova Remnant** and the **Vela Pulsar**, though both these are very much younger. It is named after its discoverer, the Australian astronomer Colin Stanley Gum (1924–1960), who published his findings in 1955.

gyrofrequency

The frequency of rotation for an **electron** (or other charged particle) as it spirals in a magnetic field.

H

H and K lines

Prominent **absorption lines** in the spectra of stars like the Sun and also cooler stars due to singly ionized calcium (Ca II). Named by Joseph von Fraunhofer, they occur in the near-ultraviolet at wavelengths of 3969 and 3934 Å, respectively (see **Fraunhofer lines**). Emission in the H and K lines is also common in some kinds of eruptive variables, such as **flare stars** and **RS Canum Venaticorum stars**.

H I region

An **interstellar cloud** of neutral **hydrogen**, detectable by its radio wave emission in the **21-centimeter line**. H I regions have a density of 1 to 10 atoms/cm^3 and a temperature of about 125 K (the **spin temperature** of neutral hydrogen). At least 95% of interstellar hydrogen is H I.

H II region

A region of ionized interstellar **hydrogen**. Most H II regions arise from hot blue O and B stars, whose ultraviolet light can ionize all the hydrogen for dozens or even hundreds of light-years in every direction. They have typical **kinetic temperatures** of 10,000 to 20,000 K, and a density of about 10 atoms/cm^3. The most famous H II region is the **Orion Nebula**.

Hadar (Beta Centauri, β Cen)

The second brightest star in **Centaurus** and the eleventh brightest star in the sky; the origin of its name is unclear, though its alternative name "Agena" means the "knee" of the Centaur. Hadar-A consists of a pair of almost identical, giant **B stars**, each some 55,000 times more luminous and 15 times more massive than the Sun, separated by about 3 AU with an orbital period of just under a year. Just 1.3″ away, and hard to isolate optically, is main sequence Hadar-B, a star that is an impressive star in its own right, with a mass of 5 M_{sun} and a luminosity of 1,500 L_{sun}. Hadar-B orbits the main pair at a minimum distance of 210 AU with a period of at least 600 years. The twins that make up Hadar-A are some 12 million years old and due to quickly expand to become red **supergiants** prior to exploding as **supernovae**. One of them is also a **Beta Cephei star**, varying slightly with multiple periods of less than a day.

Visual magnitude:	0.61
Absolute magnitude:	–5.43
Spectral type:	B1III
Surface temperature:	25,500 K
Luminosity:	112,00 L_{sun}
Distance:	525 light-years

Hadley, John H. (1682–1744)

An English scientist and mathematician who made improvements in **reflecting telescope** design and invented the reflecting quadrant–a forerunner of the mariner's sextant. Hadley was the first to devise a laboratory method for testing the parabolic figure of a mirror. He placed a tiny illuminated pinhole at the mirror's center of curvature and examined the reflected cone of light in the vicinity of the image. From the appearance of this cone, he could infer the state of the mirror's surface and was thus able to pass, by successive polishings, from a spherical to a paraboloidal figure.

Hadley circulation

An average thermal circulation in a planet's atmosphere due to warm air rising at lower latitudes, moving to higher latitudes, then descending and moving back to lower latitudes nearer the surface. On Earth, such circulations, known as *Hadley cells,* exist between the equator and 30° N/S (the most prominent and permanent cells), 30° N/S and 60° N/S, and 60° N/S and the poles. On Venus, a single cell extends from the equator to the poles. On Mars, Hadley circulation is strongest in the northern winter, especially when the atmosphere is dusty, and weakest at the equinoxes. During solstice seasons, it consists of a strong cross-equatorial surface flow with rising motion in the summer hemisphere, return flow at high altitude, and descent in mid-latitudes of the winter hemisphere. The effect was first discussed in 1735 by the English meteorologist George Hadley (1685–1768).

hadron

A fundamental particle that can interact via the strong force. The hadron family consists of **baryons**, which are **quark** triplets and include the familiar **proton** and **neutron**, and **mesons**, which are quark doublets.

HALCA (Highly Advanced Laboratory for Communications and Astronomy)

A Japanese satellite, launched in February 1997, that forms the first space-borne element of the VSOP (VLBI Space Observatory Program). HALCA's main instrument is an 8-m-diameter radio telescope. The satellite's highly elliptical orbit enables **very long baseline interferometry** (VLBI) observations, at 1.6 GHz, 5 GHz, and 22 GHz, with baselines up to three times longer than those available on Earth.

Hale, George Ellery (1868–1938)

An American astronomer who founded the **Yerkes Observatory** (1897), **Mount Wilson Observatory** (1904), and **Palomar Observatory** (1934), and made important contributions to solar astronomy. Hale began studying the solar spectrum as a wealthy teenager in Chicago, encouraged by Sherburne **Burnham**. As an undergraduate at the Massachusetts Institute of Technology he invented the **spectroheliograph**, which he used to study **prominences** and other features on the Sun. He worked in his private Kenwood observatory, funded by his father, before being appointed professor of astronomy at the University of Chicago in 1892. Once there he set about establishing a new observatory for the faculty and persuaded the Chicago streetcar magnate Charles T. Yerkes to donate a large sum toward its construction and that of the 40-in. (102-cm) Yerkes refractor. Realizing the importance of a mountain location for future observatories, Hale pressed the Carnegie Institution of Washington to establish a solar laboratory on Mount Wilson, California. From there, he made observations showing that **sunspots** were cooler than the surrounding photosphere (1905) and that they were strongly magnetic (1908). Hale also sought benefactors to build a 60-inch (1.52-m) reflector (completed in 1908), using a mirror blank his father bought, and negotiated with J. D. Hooker, a Los Angeles businessman, to finance the construction of Mount Wilson's 100-inch (2.54-m) reflector (completed in 1918). As well as solar and stellar astrophysics, Hale encouraged research in galactic and extragalactic astronomy and hired Harlow **Shapley** and Edwin **Hubble** as soon as they finished their doctorates. He founded, with James **Keeler**, the *Astrophysical Journal* (1895), played a leading role in setting up the **International Astronomical Union** and the **American Astronomical Society**, and was instrumental in transforming the Throop Institute of Pasadena into the California Institute of Technology. He suffered seriously from headaches, insomnia, and frequent episodes of mental illness, but the often-repeated story that he believed he was frequently visited by an elf who advised him on technical matters is a myth stemming from a misunderstanding by one of his biographers. His last major contribution to science was to instigate and obtain funding (from the Rockefeller Foundation) for the observatory on Mount Palomar and its great 200-inch (5-m) telescope. Although he died before the realization of his greatest dream, the giant Palomar telescope, which revolutionized astronomy, was completed and dedicated to him in 1948.

Hale cycle

The approximate 22-year cycle in which the magnetic polarity of **sunspot** pairs reverses and then returns to its original state; during half the cycle, for example, the leading spot in every pair will have a positive polarity but during the other half, the leading spot will be negative. It is named after George **Hale** who discovered it in 1925.

Hale Observatories

The name by which the combined facilities of **Mount Wilson Observatory, Palomar Observatory, Big Bear Solar Observatory,** and **Las Campanas Observatory** were known from 1970 to 1980, during which time they were jointly owned and operated by the Carnegie Institution and the California Institute of Technology.

Hale Telescope

The 200-inch (5-m) reflector at **Palomar Observatory**. It was opened in 1948 and named in honor of George **Hale**.

Hale-Bopp, Comet (C/1995 O1)

The brightest **comet** seen in recent years, even though it came no closer to Earth than 1.32 AU (197,000,000 km). Discovered independently by amateur astronomers Alan Hale (of New Mexico) and Thomas Bopp (of Arizona) on July 22, 1995, it reached perihelion on April 1, 1997, and was visible to the naked eye for many months. Its nucleus seems to be very large–about 40 km across–and spins around once every 11.4 hours. Last seen in 1997, Hale-Bopp has an orbital period of 2,380 years and won't be seen again until A.D. 4377.

Hale's polarity law

An observational rule governing the direction of regions of strong magnetic fields on the **Sun**, which are grouped in pairs of opposite polarities. At any given time, the ordering of positive/negative regions with respect to the east-west direction (the direction of rotation) is the same in a given hemisphere, but is reversed from northern to southern hemispheres. This rule is named after George **Hale** who first established it. Shortly after his discovery, ongoing studies of the magnetic polarities of **sunspot** pairs by Hale and his collaborators revealed yet another intriguing pattern: from one sunspot cycle to the next, the magnetic polarities of sunspot pairs undergo a reversal in each hemisphere.

Hall, Asaph (1829–1907)

An American astronomer who discovered the two moons of Mars, **Deimos** and **Phobos**. Born in Goshen, Connecticut, Hall left school at 13 to support his family as a carpenter, following the death of his father. He educated himself, and became talented enough in astronomy for George **Bond** to employ him as his assistant at Harvard in 1857. In 1863 Hall became professor of mathematics at the **U.S. Naval Observatory**. In 1877, when Mars was in **opposition**, Hall searched for Martian satellites using the Naval Observatory's 26-in. (66-cm) refractor. On August 11, he was rewarded with the discovery of a tiny moon (Deimos, as it turned out) but was then compelled to wait a further six nights for mist to clear from over the Potomac before he could confirm his sighting and discover another satellite. Much of Hall's later work involved the moons of the outer planets and studies of binary stars. In 1895, he returned to Harvard as professor of astronomy.

Halley, Edmund (1656–1742)

A highly influential and well-liked English astronomer and physicist best known for the comet named after him (see **Halley's comet**). In 1679 he published the first accurate southern sky catalog based on his two years of telescopic observations on St. Helena. His interest in comets was sparked by the Great Comet of 1682, which prompted him to work out the orbits of 24 known comets. Noting that the orbits of comets seen in 1456, 1531, 1607, and 1682 were very similar, he concluded that they were of one and the same object and predicted its return in 1758 (though he died before his prediction could be verified). In 1695, after a decade of lunar studies, he proposed the **secular acceleration** of the Moon. In 1718, by noting that Sirius, Procyon, and Arcturus had changed position since the time of Ptolemy's ***Almagest***, he discovered stellar **proper motion**. After succeeding John **Flamsteed** to become the second **Astronomer Royal** in 1720, he began a series of lunar and solar observations that spanned the 19-year lunar cycle, or **Saros**, and confirmed the Moon's secular acceleration. His celebrated friendship with Isaac **Newton** enabled him to persuade Newton to publish the *Principia* through the Royal Society, of which Halley served as clerk and editor; when the Society was unable to pay for the book's printing, Halley financed it himself. Among his other achievements, he proposed a way of determining the Earth-Sun distance by measuring **transits** of Venus (a method successfully used after his death), he realized that the **aurora** borealis was magnetic in origin, improved understanding of the optics of rainbows, raised the problem of what became known as **Olbers' Paradox**, and devised and personally used the first diving bell.

Halley's comet (1P/Halley)

The most famous of **comets**, named for Edmund **Halley** who first computed its orbit and predicted its return in 1758; records of Halley's comet were subsequently traced back to 240 B.C. During its last visit to the inner solar system, in 1986, it was observed by five spacecraft; Europe's **Giotto**, Japan's Sakigake and Suisei, and Russia's twin Vega probes. Giotto photographed Halley's nucleus close-up and found it to measure 16 × 8 km, to have an **albedo** of 0.04, and to spin on its axis once every 3.7 days. Aphelion 35.3 AU, perihelion 0.587 AU, eccentricity 0.967, inclination 162.3°, period 76.2 years.

Halley-type comet

A kind of **short-period comet** with a period of 20 to 200 years and an orbit of high inclination; also known as an *intermediate-period comet.* Halley-type comets, named after their celebrated prototype, are believed to have come from

Halley's comet The nucleus of Halley's comet as seen by Giotto on March 13, 1986. This picture is a mosaic of eight images sent back by the probe during its flyby. *European Space Agency*

the **Oort Cloud**. This makes them distinct, in origin and orbital characteristics, from **Jupiter-family comets**.

halo, atmospheric

A phenomenon that can range in appearance from a simple and fairly common circle of light around the Sun or the Moon to a rare event in which the whole sky is webbed by intricate arcs. Atmospheric halos are caused by tiny, flat ice crystals in the atmosphere that refract and reflect incoming light. Round solar halos with a radius of 22° happen more often than rainbows, and in Europe and parts of the United States can be seen on average twice a week. Within a solar halo, on opposite sides of the Sun and at the same altitude, may be two bright spots known as parhelia, sundogs, or mock suns. Lunar halos may contain parselenae (mock moons), but these are usually seen only in polar regions and within five days of a full moon.

halo population

Old, metal-poor stars found in the **galactic halo** of the **Milky Way Galaxy** and of other galaxies. See also **population**.

H-alpha (Hα)

A red **spectral line**, at 6563 Å, emitted by a **hydrogen** atom when its **electron** falls from the third lowest energy level ($n = 3$) to the second lowest energy level ($n = 2$); the same line appears in **absorption** when electrons are raised from $n = 2$ to $n = 3$. Many solar features, such as **prominences**, show up best in H-alpha, so that, for observation of the Sun, filters are often used that allow only light close to the H-alpha wavelength to pass. H-alpha is the first member of the **Balmer series**.

Hamal (Alpha Arietis, α Ari)

Easily the brightest star in **Aries**, an orange giant **K star**. Its Arabic name, meaning "the lamb," stands for the whole constellation of the Ram. Hamal has the distinction of having (along with **Shedar**, Alpha Cassiopeiae) the most accurately measured angular diameter–0.00680″ (the width of a penny seen from 60 km away). This precise measurement also enabled Hamal to become one of the few stars for which **limb darkening** has been detected.

Visual magnitude:	2.01
Absolute magnitude:	0.48
Spectral type:	K2III
Surface temperature:	4,590 K
Luminosity:	90 L_{sun}
Radius:	15 R_{sun}
Mass:	2 M_{sun}
Distance:	66 light-years

Haro, Guillermo (1913–1988)

A Mexican astronomer best known as the codiscoverer of **Herbig-Haro (HH) objects** and for important work on **flare stars**, blue galaxies with line emission (see **Haro galaxy**), stellar evolution and formation, and, most important, for putting Mexico on the astronomical map. Haro grew up amid the turbulence of the Mexican Revolution and began his early studies in law. After discovering astronomy through the motivation of L. E. Erro, Haro left law and became his assistant. From 1943 to 1947 he carried out research in the United States at **Harvard College Observatory**, Case Observatory of the University of Chicago, and **McDonald Observatory**. Subsequently he served as director of the Observatory of Tacubaya, the Institute of Astronomy in Mexico City, and the National Astronomy Observatory. He also started the project to establish the Observatory of San Pedro Mártir, Baja California, in 1968, which now bears his name.

Haro galaxy

A blue galaxy, often elliptical or lenticular in form, whose spectrum shows bright, sharp **emission lines**. The emission may be from an **active galactic nucleus** or from a recent burst of star-formation. These galaxies are named after Guillermo **Haro**, who discovered them in 1956.

Hartmann test

A way of checking the accuracy of the shape of a large mirror (or lens). Light from a point source near the center of curvature is reflected from the mirror (or transmitted through the lens), over which a screen pierced with hundreds of holes is placed. The light then reconverges and is intercepted by a photographic plate located just inside or outside the focus. The image records aberrations and can be analyzed to yield the objective's figure. The test is named after the German astronomer Johannes Hartmann (1865–1936).

Harvard College Observatory (HCO)

The astronomical observatory of Harvard College at Cambridge, Massachusetts, founded in 1839. HCO is a partner in the **Magellan Telescopes** project and owns telescopes at Oak Ridge, Massachusetts, that are operated by the **Smithsonian Astrophysical Observatory** (SAO). HCO and SAO jointly run the **Harvard-Smithsonian Center for Astrophysics**.

Harvard Revised Photometry (HR Photometry)

A catalog of 9,096 stars brighter than magnitude 6.5 (visible to the naked eye), published in 1908 by Edward **Pickering** at **Harvard College Observatory**. These stars are still often referred to by their HR number. The HR

Photometry catalog was the forerunner of Yale's ***Bright Star Catalogue***.

Harvard-Smithsonian Center for Astrophysics (CfA)

A research organization headquartered at Cambridge, Massachusetts, founded in 1973, that combines the research activities of **Harvard College Observatory** with those of the neighboring **Smithsonian Astrophysical Observatory**.

harvest moon

The full Moon that appears closest in time to the autumnal **equinox**, in late September or early October.

Hat Creek Radio Observatory

The radio observatory of the University of California at Berkeley, founded in 1960 and located near Mount Lassen, California. Its main instrument is the BIMA Array, a millimeter-wave radio interferometer that consists of ten 6-m antennae that can be moved to various positions along baselines measuring 1 km north-south and 1 km east-west. The BIMA Array is jointly owned by the University of California at Berkeley and the universities of Illinois and Maryland.

Haute-Provence Observatory (Observatoire de Haute-Provence, OHP)

An optical observatory in southeast France, 100 km north of Marseilles, at an altitude of 660 m in the foothills of the Alps; it was founded in 1937 and is owned by the Centre National de la Recherche Scientifique. The main instruments are a 1.93-m reflector, opened in 1958, and a 1.5-m reflector, opened in 1967. Attached to the larger telescope is the Elodie spectrograph, which has been used to discover a number of **extrasolar planets**.

Hawking, Stephen William (1942–)

An English theoretical astrophysicist and cosmologist famed almost equally for his work on **black holes** and the origin of the universe (propagated to a vast lay audience through his nontechnical books, including *A Brief History of Time* and *The Universe in a Nutshell*) and the fact that he has survived for four decades with motor neuron disease, an affliction with which he was diagnosed as a postgraduate at Oxford. Since 1980 he has held the Lucasian Chair of Mathematics at Cambridge University. He supplied a mathematical proof, along with Brandon Carter, W. Israel and D. Robinson, of John Wheeler's "No-Hair Theorem"–namely, that any black hole is fully described by the three properties of mass, angular momentum, and electric charge. He also showed that black holes, especially tiny ones, can effectively evaporate as a result of the production of particle-antiparticle pairs in their vicinity (see **Hawking radiation**). Together with Ian Moss, he has derived a "no-boundary" solution for the origin of space and time in the **Big Bang**.

Hawking radiation

A process, first theorized by Stephen **Hawking**, by which a **black hole** can apparently evaporate. When virtual particles are produced in the vicinity of a black hole, it is possible for one member of the particle-antiparticle pair to be pulled into the black hole while the other escapes into space. The particle that falls into the black hole negates some of its mass and so the black hole shrinks a little. This makes it look as if the particle that escaped into space has come from the black hole. Hawking radiation would be particularly important in the case of miniature black holes, which might explode in this way.

Hayashi, Chushiro (1920–)

A Japanese astrophysicist at Kyoto University who has done important work on the early evolution of stars and of the universe. In 1950 he was the first to offer a variant of the hot **Big Bang** model as put forward in 1946 by George **Gamow** and colleagues in the famous "Alpher-Bethe-Gamow" theory. He is best remembered, however, for his discovery of the almost temperature-independent early evolution of pre-main-sequence stars along what has become known as the **Hayashi track**.

Hayashi track

A nearly vertical path of stellar evolution on the **Hertzsprung-Russell diagram** down which an infant star progresses on its way to the **main sequence**. While on the Hayashi track, a star is largely or completely in convective equilibrium; as it progresses, its luminosity, initially very high, decreases rapidly with contraction, but its surface temperature remains almost the same. The sequence runs in reverse for stars leaving the main sequence to become giants. The track is named after Chushiro **Hayashi**.

Haystack Observatory

A radio observatory at Westford, Massachusetts, whose main instrument is a 36.6-m dish, housed within a protective dome. Opened in 1964 and originally used for radar observations of the planets, it is now used mostly for radio astronomy. It is owned by the Massachusetts Institute of Technology and operated by a consortium of academic institutions on behalf of MIT.

head-tail galaxy

A **radio galaxy** whose emission has been distorted by the movement of the parent galaxy, a giant elliptical, through the **intra-cluster medium**. The result is strong radio emission coming from a bright "head" and more diffuse emission from an extended "tail" that stretches behind for hundreds of thousands of light-years.

heavy element

In astronomy (unlike in chemistry or physics), any element that is heavier than the two lightest ones–hydrogen and helium. Astronomers also use the terms "heavy element" and "metal" almost interchangeably, even though many elements (such as carbon and oxygen) are nonmetals.

HED group

A group of meteorites that comprises three closely related classes of **achondrites**–**howardites**, **eucrites**, and **diogenites**–whose members all resemble terrestrial igneous rocks, such as **basalts**, and are thought to have originated from the same parent body: the third-largest asteroid, **Vega**. Close similarities in the ratio of their oxygen isotopes and other chemical similarities point to a common heritage. Their ancient crystallization ages of 4.43 to 4.55 billion years suggest that the ancestor was large and differentiated and had become geologically inactive after a brief but intense igneous history. Vesta fits this bill and, most tellingly, has a very similar reflectance spectrum to the HED group members. Vesta also bears the scar of a huge impact event–a 30 km-deep, 460 km-wide depression near its south pole that plunges down through Vesta's basaltic crust to expose the mantle. The theory is that, during the impact event, large chunks of matter were ejected from Vesta to form smaller asteroids of similar composition–the so-called Vestoids. Some of these bits of Vesta subsequently entered near-Earth orbits, and are thought to be the sources of the howardites, eucrites, and diogenites.

Hektor (minor planet 624)

The largest of the Jupiter **Trojans**, discovered at the preceding (L_4) **Lagrangian point** by the German astronomer August Kopff (1882–1960) in 1907. Measuring 150 × 300 km, Hektor is highly elongated and may actually be two objects in contact that are rotating around a common center of mass with a period of 6.93 hours. With a visual magnitude that reaches +14.5, it is the brightest of the Trojans. It is of spectral class D. Semimajor axis 5.172 AU, perihelion 5.05 AU, aphelion 5.29 AU, eccentricity 0.02, inclination 18.2°.

Helene

The thirteenth moon of **Saturn**, also known as Saturn XII; it moves ahead of the much larger **Dione** in the same orbit as the leading **Lagrangian point**. Helene measures 36 × 32 × 30 km and was discovered in 1980 by P. Lacques and J. Lecacheux on Voyager 1 images.

heliacal rising

The first observable rising of a celestial object late in morning twilight after **conjunction** with the Sun. The heliacal rising of **Sirius** heralded the Nile floods and marked an important point in the ancient Egyptian calendar (see **Egyptian astronomy**). *Heliacal setting* is the last visible setting of a celestial object in the evening sky before conjunction.

heliocentric

Sun-centered. A *heliocentric orbit* is an orbit with the Sun at one **focus**.

heliocentric coordinates

A system of **celestial coordinates** that has its origin at the center of the Sun and takes the **ecliptic** as its plane of reference. Heliocentric coordinates are used especially in calculations of the relative positions of the planets and other bodies of the solar system.

heliograph

A device for recording the positions of **sunspots**.

heliographic coordinates

The latitude and longitude of a feature on the Sun's surface. *Heliographic latitude* is an object's angular distance north or south of the solar equator; *heliographic longitude* can be measured east or west of the central solar meridian, or given in terms of the **Carrington** rotation number.

helioseismology

The study of the **Sun**'s interior by measuring oscillations of the surface caused by sound waves that travel through the Sun at different depths. Large numbers of distinct, resonating sound waves can be detected by the **Doppler shift** they cause in **spectral lines** emitted at the Sun's surface. The periods of these waves depend on their propagation speeds and the depths of their resonant cavities (similar to the lengths of organ pipes that produce different notes). Data on many resonant modes, with different cavities, serve as a precise probe of the temperature, chemical composition, and motions, from just below the Sun's surface down to its core. This information, in turn, provides a far more stringent test of stellar structure and evolution theory than that previously available from knowledge of just the global properties of radius, luminosity, mass, composition, and (inferred) age. Comparison of oscillation periods, predicted from solar models, will give an accurate measurement of the Sun's helium abundance, which will put bounds on cosmological models of the early universe. Helioseismology will reveal the nature of stellar **convection** and how rotation changes with depth. For stars like the Sun, almost all the energy released by fusion in the core is brought to the surface by convection currents whose large cells are hidden from view. However, these cells should perturb the oscillation periods in a distinctive way that reveals their properties. Likewise, internal rotation imparts a clear signature–a splitting–to the oscillation periods. Among the largest helioseismology projects is the **Global Oscillation Network Group** (GONG).

heliosphere

The region of space, surrounding the Sun, that is inflated by the **solar wind** and within which the Sun exerts a magnetic influence. Despite its name, the heliosphere is almost certainly greatly elongated due to the movement of the Sun with respect to the **interstellar medium.** The surface that encloses the heliosphere, where the pressure of the solar wind balances that of the interstellar medium, is called the *heliopause.* Between the heliopause and the **bow shock** is presumed to be a layer of interstellar gas known as the *heliosheath.*

heliostat

A moveable flat mirror used to reflect sunlight into a stationary **solar telescope**. If the telescope is aligned with

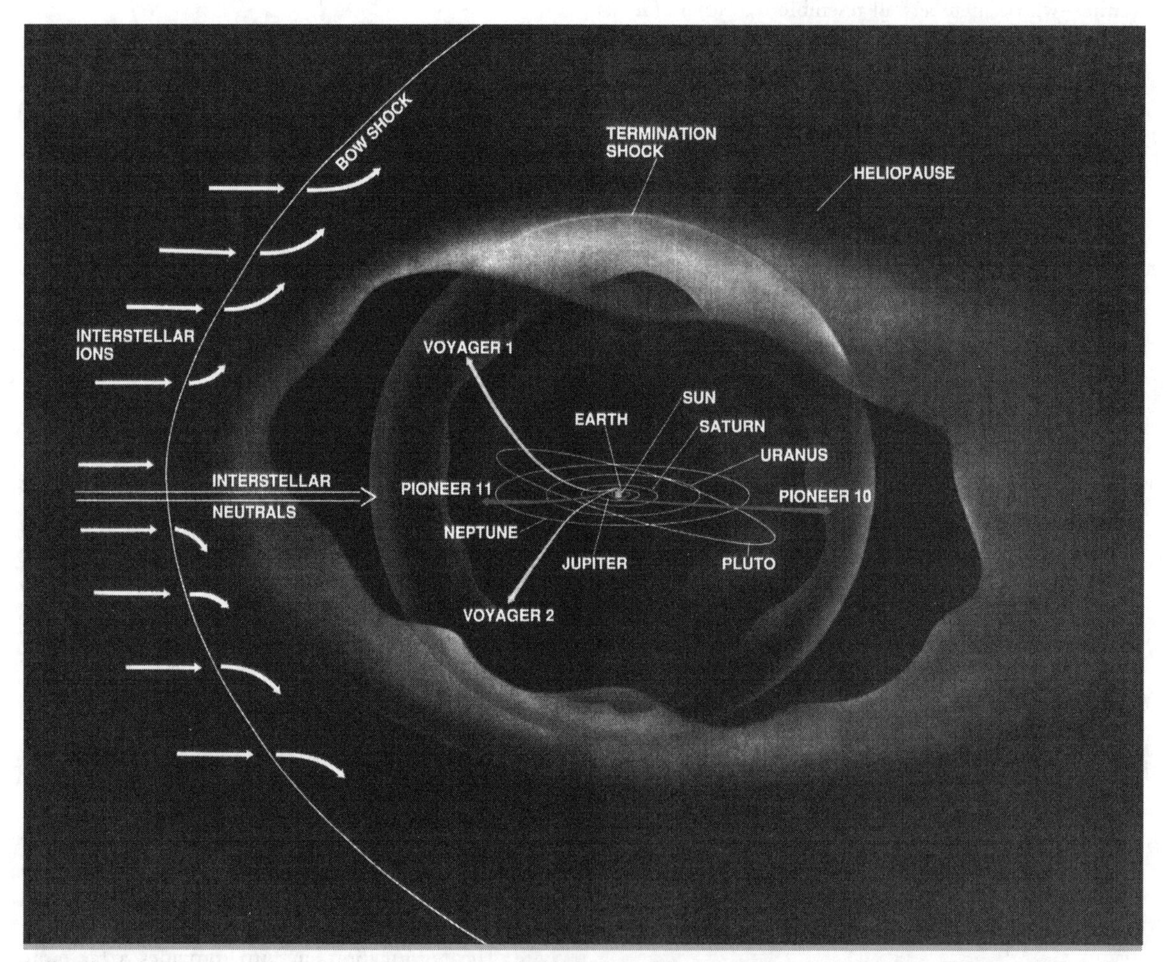

heliosphere A diagram of the main structures associated with the heliosphere, including the heliopause (the edge of the heliosphere), the termination shock, and the bow shock. Also shown are the paths of four probes that are currently leaving the solar system. *NASA*

Earth's axis, only one mirror is needed. A heliostat is similar to a **coelostat**, but has a simpler design and (unlike the coelostat) makes the image of the Sun rotate as the Sun moves along the sky.

helium (He)

The second lightest and second most common element in the universe, after **hydrogen**; most of it was produced immediately after the **Big Bang**, though an additional contribution has come from **hydrogen burning** inside **main sequence** stars. Helium was first discovered, by virtue of its spectral lines, in the Sun. An ordinary helium atom consists of a nucleus of two protons and two neutrons surrounded by two electrons; a nucleus of it, on its own, is called an **alpha particle.**

helium burning

The fusion of helium into carbon by the **triple-alpha process**. All stars born with more than about half a solar mass go through this stage of energy production as they evolve beyond the **main sequence**.

helium flash

The onset of runaway **helium burning** in the core of a low-mass star (such as the Sun). The helium flash happens in the hydrogen-exhausted core of a star that has become a **red giant**. When gravitational pressure has raised the temperature of the dormant helium core to a temperature of about 100 million K, the helium nuclei start to undergo thermonuclear reactions. Once the helium burning has started, the temperature climbs rapidly, without a cooling, stabilizing expansion. The extreme sensitivity of the nuclear reaction rate to temperature causes the helium-burning process to accelerate. This in turn raises the temperature, which further accelerates the helium burning, until a point is reached at which the thermal pressure expands the core and thereby limits the flash. The helium flash can only occur when the helium core is less than the 1.4-M_{sun} **Chandrasekhar limit** and thus it is restricted to fairly low-mass stars.

helium shell flash

One of a series of **helium burning** episodes in the thin helium shell that surrounds the dormant carbon core of an **asymptotic giant branch** star; the helium burning shell does not generate energy at a constant rate but instead produces energy primarily in short flashes. During a flash, the region just outside the helium burning shell becomes unstable to **convection**; the resultant mixing probably leads to the **s-process** as well as to the upward movement of carbon produced by helium burning. The overheating from a flash also causes an expansion of the star's upper layers, followed by a collapse, leading to large-scale pulsations.

helium star

An **O star** or **B star** in which the **absorption lines** of helium are abnormally strong and those of hydrogen are absent or weak. *Extreme helium stars* (also called hydrogen-deficient stars) show no trace of hydrogen, while *intermediate helium-rich stars* have hydrogen lines that are visible but weaker than in normal stars. The loss or depletion of the star's hydrogen envelope, leaving essentially an exposed helium core, may have happened because of a powerful **stellar wind**, as in **Wolf-Rayet stars**, or because of **mass transfer** to a close binary companion. "Helium star" is also an obsolete name for a normal B star. A *helium variable* is a **Bp star** in which the strength of the helium lines varies periodically. At the extreme phases, the object appears as helium-rich, while at other phases helium lines may be very weak or absent.

Helix Galaxy (NGC 2685)

An unusual **lenticular galaxy** in **Ursa Major** that is a member of the very rare class of **polar ring galaxy**; it is also known as the *Pancake Galaxy.* Several thin filamentary strands, consisting of knots of luminous star-forming regions and hydrogen gas, with a diameter comparable to the main disk, form a helical band perpendicular to the main disk and centered on the galactic nucleus. These structures suggest that NGC 2685 once had a companion, perhaps like one of the Milky Way's **Magellanic Clouds**, that was captured into a polar orbit and had its stars eventually merged with those of the larger system, leaving behind the companion's interstellar medium. New generations of stars formed from this material to produce the luminous ring seen today. It is possible that if the Magellanic Clouds had been closer to the Milky Way, they too would have created a polar ring around our Galaxy. Perhaps a foreshadowing of this process can be found in the so-called *Magellanic Stream,* a string of hydrogen clouds trailing the Magellanic Clouds and extending thousands of light-years along their orbit of the Milky Way.

Visual magnitude:	11.9
Angular size:	5.0′ × 1.5′
Position:	R.A. 8h 55.6m, Dec. +58° 44′

Helix Nebula (NGC 7293)

The nearest **planetary nebula** to the Sun; it lies in **Aquarius** and was discovered by Karl Ludwig Harding before 1824. In apparent size (including an outer halo) it

Helix Galaxy Remarkable alternating rings of dust and newly formed star clusters, together with a more diffuse stellar glow, are evident in this image of the Helix Galaxy (NGC 2685). *Isaac Newton Group of Telescopes/Nik Szymanek*

is almost as big as the full Moon. Observations by the Hubble Space Telescope of the inner edge of the main ring have resolved fine details associated with the long-known radial structures, including thousands of comet-like knots, the heads of which are larger than our solar system. Hot, fast moving shells of nebular gas overrunning cooler, denser, slower shells ejected by the star during an earlier expansion may have produced these dropletlike condensations as the two shells intermixed and fragmented. An intriguing possibility is that instead of dissipating over time, these objects might eventually

Visual magnitude:	7.3
Angular size:	28′
Linear diameter:	1.5 light-years
Distance:	450 light-years
Position:	R.A. 22h 29.6m; Dec. –20° 48′

Helix Nebula A close-up view of part of the Helix Nebula, by the Hubble Space Telescope, showing numerous cometary knots—so-called because of their cometlike glowing heads and gossamer tails. The tails, each about 150 billion km long (about 1,000 times the Earth-Sun distance) form a radial pattern around the central star like the spokes on a wagon wheel. *NASA/STScI*

collapse to form Pluto-like bodies. If so, small icy worlds created near the end of a star's life would be numerous in our Galaxy.

helmet streamer

A beautiful structure, seen during a **solar eclipse**, in which million-degree coronal plasma is trapped by closed magnetic field lines. **Prominences** are often situated beneath helmet streamers, and active regions occur beneath streamers near the equator (sometimes called active region streamers). In some regions, the coronal magnetic field cannot confine the plasma, and the plasma expands outward, reaching supersonic velocities. Regions on the Sun with these open magnetic field lines (which stretch far out into the solar system) correspond to **coronal holes** and are the source of the **solar wind**. The electrons in the coronal hole plasma are typically cooler and less dense than streamers, and so they show up as dark regions in both X rays and white light.

Henry Draper Catalogue (HD)

A catalog of the spectral types and positions of 225,300 stars, down to about magnitude 8, compiled by Annie Jump **Cannon** and her coworkers at **Harvard College Observatory** between 1918 and 1924 and named in honor of Henry **Draper**, whose widow donated the money needed to finance it. *HD* numbers are widely used today, especially for stars that have no **Bayer designation** or **Flamsteed number**. Stars numbered 1 to 225,300 are from the original catalog and are numbered in order of **right ascension** for 1900.0. Stars in the range 225,301 to 359,083 are from the *Henry Draper Extension (HDE)*, published in 1925–1936 and 1949. The notation *HDE* is used only for stars in this extension, but even these are usually denoted *HD* as the numbering ensures there can be no ambiguity.

Henyey track

An almost horizontal track of stellar evolution on the **Hertzsprung-Russell diagram**, which traces the path of a star, of mass greater than about 0.4 M_{sun}, that is in radiative equilibrium, as it moves from its **Hayashi track** to the **main sequence**. As a star progresses along the Henyey track–named for the American astrophysicist Louis Henyey (1910–1970)–it contracts and becomes hotter, until its core has become sufficiently small and pressurized for **hydrogen burning** to begin.

Heraclides of Pontus (c. 388–c. 315 B.C.)

A Greek philosopher who was the first to suggest that the rotation of Earth would account for the apparent rotation of the stars. Until fairly recently, it was believed that, although Heraclides argued for a geocentric universe, he suggested that Mercury and Venus orbited the Sun. (Heliocentric theories were rejected at the time of Heraclides because it was believed that Earth's rotation would cause falling bodies to be deflected westward.) However, David C. Lindberg, at the University of Wisconsin at Madison, has produced many references to indicate that Heraclides's theories never espoused heliocentrism. See also **Greek astronomy**.

Herbig Ae/Be star

A still-contracting (pre-main-sequence) star of spectral type A or B, with strong **emission lines** (especially **H-alpha** and the calcium **H and K lines**), associated with fairly bright nebulosity. It is named after the American astronomer George Herbig (1920–) and also known as a *Herbig-Bell star* or a *Herbig emission star.* A well-known example is **R Coronae Borealis**. The number of Herbig Ae/Be stars is smaller than the number of **T Tauri stars** (lower mass pre-main-sequence stars), for two reasons: higher mass stars in general are less common, and also the timescale for the core contraction of Ae/Be stars is smaller than for stars of smaller mass, so they remain deeply embedded until the end of their mass accretion stage.

Herbig-Haro (HH) object

A small bright nebula in a star-forming region, created when fast-moving **jets** of material from a newborn star collide with the **interstellar medium**. As the **bipolar flow** from a young star plows into the surrounding gas, it generates strong shock waves that heat and ionize the gas. In the cooling gas behind the shock front, electrons and ions recombine to give an **emission line** spectrum characteristic of Herbig-Haro objects. They are named after the American astronomer George Herbig (1920–) and Guillermo **Haro** who discovered the first three such objects in 1946–1947 in images of the nebula NGC 1999 in Orion. All known Herbig-Haro objects have been found within the boundaries of **dark clouds** and are strong sources of **infrared**.

Hercules (abbr. Her; gen. Herculis)

A mythical hero famed for his 12 labors; a very large but unspectacular northern **constellation**. It contains the Keystone asterism, named after its shape, which is made by four of the main stars in Hercules–Epsilon, Zeta (**Ruticulus**), Eta, and Pi Her. (See tables, "Hercules: Stars Brighter than Magnitude 4.0," and "Hercules: Other Objects of Interest." See also star chart 11.)

Hercules Cluster (Abell 2151)

A loose unsymmetrical cluster of about 75 bright galaxies, about 500 million light-years away and about 6 million light-years across. It contains an unusually high proportion of spiral galaxies and also galaxies that show evidence of interaction or other peculiar features.

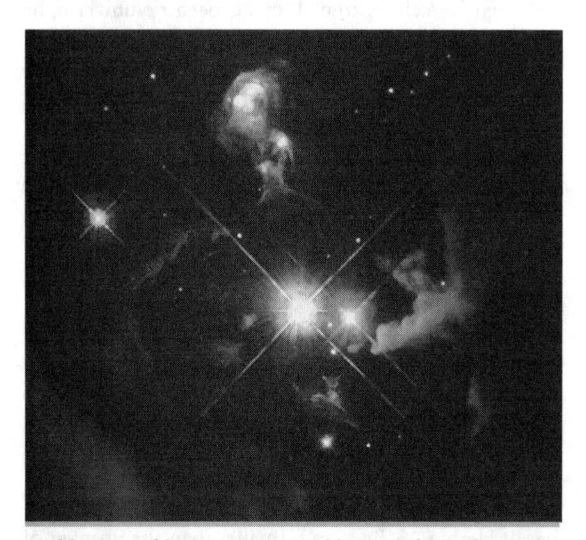

Herbig-Haro object HH 32, a Herbig-Haro object observed by the Hubble Space Telescope in 1994. Lying about 1,000 light-years from Earth, HH32 is old enough that the intense wind of material blowing from the bright, young central star has cleared much of the dust out of the central region, thus exposing the star to direct view. *NASA/STScI*

Hercules Globular Cluster (M13, NGC 6205)

The most prominent **globular cluster** in the northern half of the sky. Discovered by Edmund **Halley** in 1714, who noted that "it shows itself to the naked eye when the sky is serene and the Moon absent," M13 was selected in 1974, by Frank Drake and Carl **Sagan**, as a target for one of the first radio messages addressed to possible extraterrestrial races, sent by the Arecibo Telescope. In small telescopes, M13 appears as a misty patch, denser toward the center; in large scopes it is revealed as a nearly perfect sphere of stars.

Visual magnitude:	5.9
Apparent diameter:	23′
Actual diameter:	165 light-years
Distance:	25,000 light-years
Position:	R.A. 16h 41.7m, Dec. +36° 28′

Hercules X-1 (Her X-1)

An **X-ray pulsar** that is a member of an eclipsing binary system with an orbital period of 1.7 days; it lies about 15,000 light-years away. The visible component has been identified as the blue variable HZ Herculis, whose spectrum varies from late A or early F to B. Her X-1 has a pulsation period of 1.2378 seconds, presumed to be its rotation period, and shows a 35-day quasiperiodicity in the X-ray region (but not in the optical). It is thought to move in a nearly circular orbit, to be accreting matter from HZ Her, and to have a mass of about 0.7 M_{sun}. The orbital period is stable, but the pulsation period is speeding up at a rate of about 1 part in 100,000 per year. The X-ray eclipse lasts 0.24 days.

Hermes

An **asteroid**, discovered by the German astronomer Karl Reinmuth (1892–1979) in 1937, that passed about 780,000 km from Earth; its orbit was never established well enough for it to be located again, and the object is now considered lost.

Herschel, Caroline Lucretia (1750–1848)

One of the first female astronomers to be recognized for her work. Sister of William **Herschel** and aunt of John, she was born in Hanover, Germany, and raised to be the household servant, with little education. Her mother believed that it was her daughter's duty to look after her brothers, while her father, a musician in the Hanoverian Guards, included her in music lessons

Hercules: Stars Brighter than Magnitude 4.0

Star	Magnitude		Spectral Type	Distance (light-yr)	Position	
	Visual	Absolute			R.A. (h m s)	Dec. (° ′ ″)
β Kornephorus	2.78	–0.50	G8IIIa	148	16 30 13	+21 29 22
ζ Rutilicus	2.81	2.64	F9IV	35	16 41 17	+31 36 10
α Rasalgethi	3.00	–0.02	M5Ib-II	382	17 14 39	+14 23 25
δ Sarin	3.12	1.21	A3IV	79	17 15 02	+24 50 21
π	3.16	–2.10	K3IIab	367	17 15 03	+36 48 33
μ	3.42	3.80	G5IV	27	17 46 27	+27 43 15
η	3.48	0.80	G8II	112	16 42 54	+38 55 20
ξ	3.70	0.20	G8III	163	17 57 46	+29 14 52
γ	3.70	0.61	K0III	135	16 21 55	+19 09 11
ι	3.82	–2.09	B3V	495	17 39 28	+46 00 23
ο	3.84	–1.30	B9.5V	347	18 07 32	+28 45 45
109	3.85	0.87	K2.5IIIab	128	18 23 42	+21 46 11
θ	3.86	–2.70	K1IIaCN	670	17 56 15	+37 15 02
τ	3.91	–1.02	B5IV	315	16 19 44	+46 18 48
ε	3.92	0.43	A0V	163	17 00 17	+30 55 35

given to his sons but warned her against any thoughts of marriage because she wasn't good-looking or rich and, therefore, couldn't expect an offer. After her father's death, William persuaded his mother to let Caroline join him in Bath, England. So in 1772, she went, but only after William paid for a servant to replace her at home. Following a brief career as an oratorio singer, she joined William in astronomical work. She became her brother's assistant, pounding and sieving horse manure to make material for telescope mirror molds, and polishing the metal mirrors themselves. She recorded data, prepared catalogs, and wrote papers for publication. In 1783, she began to search for comets and discovered three new nebulae. Between 1786 and 1797, after her discovery of eight comets, she entered the scientific limelight herself. King George III was so impressed that he awarded her 50 pounds a year. When William was made Court Astronomer after his discovery of Uranus in 1781, Caroline was appointed the first official female assistant to this position. She also revised John **Flamsteed**'s star catalog and published it in 1798 as the *Index of Flamsteed's Observations of the Fixed Stars.*

Herschel, John Frederick William (1792–1871)

An English astronomer, only son of F. William **Herschel**, who continued his father's observations of **nebula**e and **double stars**, and, in 1834, began a survey of the southern sky from the Cape of Good Hope. This survey

Hercules: Other Objects of Interest

Name	Notes
Star	
Hercules X-1	See separate entry.
Planetary nebula	
NGC 6210	A bright inner ring surrounded by a faint outer ring. Magnitude 9.7; diameter 13″ × 20″ (inner), 20″ × 43″ (outer); R.A. 16h 44.5m, Dec. +23° 49′.
Globular clusters	
M13 (NGC 6205)	See **Hercules Globular Cluster**.
M92 (NGC 6341)	Nearly a magnitude fainter than M13 but an interesting object for smaller scopes. Magnitude 5.5; diameter 11.2′; R.A. 17h 17.1m, Dec. +43° 8′.

yielded 2,307 nebulae and 2,102 double stars which he listed, together with his father's discoveries, in the *General Catalogue of Nebulae and Clusters*–the forerunner of the *New General Catalogue*. He was the first to notice the variability of **Betelgeuse** (1836), and to measure the **solar constant**, and he was a pioneer of stellar **photometry**.

Herschel, (Frederick) William (1738–1822)

A German-born English astronomer, brother of Caroline **Herschel** and father of John, who discovered **Uranus** (1781), its two largest moons, **Oberon** and **Titania** (1787), and two moons of Saturn, **Mimas** and **Enceladus** (1789). In 1789, with the patronage of King Charles III, he built what was then the world's largest telescope–a Newtonian reflector with a focal length of 12 m (40 ft.) and a mirror diameter of 1.2 m (48 in.). A patient, careful observer, he visually scanned the entire northern sky and cataloged more than 800 double stars and 2,500 nebulae. From counts of stars in different parts of the sky, he reasoned that the Sun is part of a flattened disk, seen from our vantage point in the plane of this disk known as the Milky Way. In 1800, using a thermometer and a prism, he discovered **infrared** radiation.

Herschel Space Observatory

A giant European Space Agency space telescope with a 3.5-m-diameter main mirror (by comparison, Hubble's is only 2.4 m across) designed to observe the universe in unprecedented detail at far infrared and submillimeter wavelengths, from 80 to 670 microns, and to carry out sensitive **photometry** and **spectroscopy**. Herschel will be launched together with **Planck** in early 2007. Once in space, the two satellites will separate and proceed to different orbits around the second **Lagrangian point**, some 1.5 million km from Earth. The observatory is named after William **Herschel**.

Herschelian telescope

A **reflecting telescope**, of a design conceived by William **Herschel** and used in his giant 48-inch (1.22-m) instrument, in which the primary mirror is tilted so that light is focused near one side of the open end of the tube. The eyepiece then picks up this light directly, avoiding light loss from reflection by a secondary mirror. The drawback is **astigmatism**, unless the **focal ratio** is large.

Hertzsprung, Ejnar (1873–1967)

A Danish astronomer best known for his discovery that the variations in the widths of stellar lines, discovered by Antonia Maury, reveal that some stars (giants) are of much lower density than others (main sequence stars) and for publishing the first color-magnitude diagrams (see **Hertzsprung-Russell diagram**). He was the first to calibrate the **period-luminosity relation** for **Cepheid variables**, a relation he used to estimate the distance to the **Small Magellanic Cloud**. He also determined **proper motions**, colors, and magnitudes of many stars in the **Pleiades**, and measured large numbers of photographic positions of binary stars. Hertzsprung studied chemical engineering in Copenhagen, worked as a chemist in St. Petersburg, and studied photochemistry in Leipzig before returning to Denmark in 1901 to become an independent astronomer. In 1909 he was invited to Göttingen to work with Karl **Schwarzschild**, whom he accompanied to the Potsdam Astrophysical Observatory later that year. From 1919 to 1944 he worked at the Leiden Observatory in the Netherlands, the last nine years as director. He then retired to Denmark but continued measuring plates into his nineties.

Hertzsprung Gap

A region of the **Hertzsprung-Russell diagram**, between the high end of the **main sequence** (occupied by stars more massive than the Sun) and the **giant branch**, occupied by very few stars because it corresponds to a relatively very short period in stellar evolution. Among the few types of object that populate this gap are **RR Lyrae stars**.

Hertzsprung-Russell diagram

A plot of stellar color, temperature, or **spectral type** against stellar **luminosity** or **absolute magnitude**. The Hertzsprung-Russell (HR) diagram, named after Ejnar **Hertzsprung** and Henry **Russell**, is dominated by the **main sequence**, which forms a curved, diagonal band from bright blue stars to faint red ones, and contains stars in their core **hydrogen burning** stage, and the **giant branch**, occupied by **red giants**. Other conspicuous regions are represented by the **supergiants** (above the

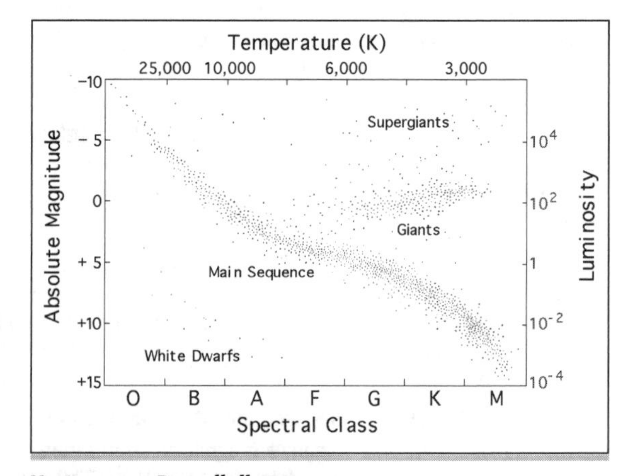

Hertzsprung-Russell diagram

giant branch) and the **white dwarfs** (below the main sequence). The HR diagram can be seen as either a snapshot of the state of a large collection of stars, or a generalization of the evolutionary pathways of stars. See also **asymptotic giant branch**.

Hess, Victor Franz (1883–1964)

An Austrian-American physicist who discovered **cosmic rays**. Beginning in 1912, Hess sent up balloons carrying electroscopes in order to locate the source of background radiation causing ionization in the atmosphere. Hess believed the radiation would decrease at greater altitudes, but found that instead it increased up to eight times. He suggested that the radiation came from space and Robert Millikan proposed the name "cosmic rays." Hess shared the 1936 Nobel Prize in physics with Carl Anderson.

HETE (High Energy Solar Spectroscopic Imager)

An international (United States, France, and Japan) satellite, launched in October 2000, whose main task is a multi-wavelength study of **gamma-ray bursts** using ultraviolet, X-ray, and gamma-ray instruments. A unique feature of the mission is its ability to localize bursts with an accuracy of several arcseconds in near real-time. The positions it identifies are transmitted to the ground and picked up by a global network of primary and secondary ground stations, enabling rapid follow-up studies by radio, optical, and X-ray telescopes. This is crucial because, while the gamma-ray burst itself fades quickly, a lower-energy afterglow may linger for days or weeks. HETE is on an extended mission until 2004.

Hevelius, Johannes (Johann Hewelcke) (1611–1687)

A Polish astronomer, born in Danzig (Gdansk), best remembered for his careful charting of the Moon's surface. A wealthy brewing merchant, he had a well-equipped observatory installed on the roof of his house in 1641, and became one of the most active observers of the seventeenth century. Between 1642 and 1645, he deduced a fairly accurate value for the solar rotation period and gave the first description of the bright markings in the neighborhood of **sunspots**, calling them **"faculae,"** a name still used. He also observed the planets, especially Jupiter and Saturn. In November 1644, he noted that Mercury goes through phases, as had been predicted by Copernicus. In 1647, he published his *Selenographia*, the first comparatively detailed lunar map. He realized that the large, uniform gray regions in the lunar disk were low plains, and that the bright contrasting regions represented higher, mountainous relief. Hevelius also planned a new star catalog of the Northern Hemisphere to supercede that of Tycho **Brahe**. This he began in 1657, but his observatory, with some of his notes, was destroyed by fire in 1679. Nevertheless, his observations enabled him to catalog more than 1,500 stellar positions. The resulting *Uranographia* contains an excellent celestial atlas with 54 plates and introduces several new constellations, but Hevelius's reliance on the naked eye for estimating positions reduced the value of his work. He observed the variable star Omicron Ceti and christened it "**Mira.**" Hevelius discovered four comets, which he called "pseudoplanetae," and suggested (correctly, in some cases) that these bodies orbited in parabolic paths about the Sun. His wife, Elizabeth Margarethe, shared his interest and assisted him greatly. After his death, she edited his famous work, *Prodromus Astronomiae,* and published it in 1690. Although little of Hevelius's lunar nomenclature survives, a large crater, 118 km in diameter, on the western border of the Ocean of Storms, is named after him.

Hewish, Anthony (1924–)

An English radio astronomer who, in collaboration with Martin **Ryle**, published the first four *Cambridge Radiosource Catalogues.* In the 1960s he also developed, at Cambridge's Mullard Observatory, a special radio telescope that was sensitive to rapid fluctuations (scintillation) in radio sources due to disturbances in ionized gas in Earth's atmosphere, within the solar system, and in interstellar space. Using this instrument, his student Jocelyn Bell Burnell discovered the first **pulsar** in 1967. For this discovery and his other work on radio astronomy, Hewish shared with Ryle the 1974 Nobel Prize in physics.

hexahedrite

One of the main types of **iron meteorites**, composed almost entirely of the nickel-iron mineral **kamacite** and named for its cubic (hexahedral) cleavage of alpha-Fe,Ni single crystals. Upon etching, hexahedrites do not display a **Widmanstätten pattern**, but they often exhibit fine, parallel lines called **Neumann lines**, named for their discoverer, Franz Neumann, who first studied them in 1848. These lines represent a shock-induced, structural deformation of the kamacite plates, and they suggest an impact history for the hexahedrite parent body, at least for the hexahedrites related to chemical group IAB.

Hidalgo (minor planet 944)

The **asteroid** with the largest known orbit, the second highest inclination to the ecliptic, and the second highest eccentricity; it was discovered by Walter **Baade** in 1920. Hidalgo, whose perihelion lies at the inner edge of the main asteroid belt and aphelion is just beyond the orbit of Saturn, moves in a **chaotic orbit**, and is considered by some to be a defunct comet. Diameter 30 km, spectral class D, semimajor axis 5.757 AU, perihelion 1.958 AU, aphelion 9.556 AU, eccentricity 0.656, inclination 42.57°, period 13.7 years.

high-velocity cloud (HVC)
A fast-moving **interstellar cloud**, composed mostly of neutral hydrogen, whose velocity is too high to be explained by the ordinary rotation of the Milky Way Galaxy. Many HVCs are thought to be material falling into our Galaxy from the outside; one sign of an extragalactic nature is their **heavy element** content, which appears to be poorer than the Milky Way norm. HVCs may be material left over from the formation of the Galaxy or that of other galaxies in the **Local Group**, or they may consist of matter wrenched away from a satellite system. A spectacular example of tidally displaced debris is the **Magellanic Stream**.

high-velocity star
A star traveling faster than 60 to 100 km/s (definitions vary) relative to the average motion of stars in the solar neighborhood. High-velocity stars belong to the **halo population** and travel around the galactic center in eccentric orbits, often of large inclination to the galactic plane.

Hilda group
A group of **asteroids** that define the outer edge of the main **asteroid belt**, about 4 AU from the Sun. They have become locked into their present orbits by a gravitational resonance with **Jupiter**: a Hilda makes exactly three orbits of the Sun for every two that Jupiter makes. The prototype is the 171-km-wide, P-class (153) Hilda, discovered in 1875 by Johann **Palisa**. Semimajor axis 3.975 AU, perihelion 3.408 AU, aphelion 4.543 AU, eccentricity 0.143, inclination 7.84°, period 9.72 years.

Himalia
The eleventh moon of **Jupiter**, also known as Jupiter VI; it was discovered in 1904 by Charles Perrine. Himalia has a diameter of 184 km and orbits at a mean distance of 11.46 million km from the planet.

Hind, John Russell (1823–1895)
An English astronomer who discovered 11 asteroids and several notable variable stars, including Nova Ophiuchi 1848, **U Geminorum**, R Leporis (also known as **Hind's Crimson Star**), and a mysterious newcomer in the constellation of Taurus. While scanning the sky through the Pleiades and in the direction of the Hyades, he noticed a tenth magnitude star that was missing from the charts he was using. This is the object that became the prototype **T Tauri star**. Hind was in charge of a private observatory in Regents Park, London.

Hind's Crimson Star (R Leporis)
A **long-period variable** in **Lepus**, easily seen with binoculars when at maximum brightness. It was discovered by John Hind who described it as "resembling a blood drop on the background of the sky."

Visual magnitude:	7.7 average; 6th to 10th range
Spectral type:	C6IIe
Period:	About 430 days

Hind's Variable Nebula (NGC 1555)
A **reflection nebula** remarkable for its changes in brightness, which are due to wide variations in the star that illuminates it–**T Tauri**. Discovered by John **Hind** in 1852, it began to fade after 1861 and had disappeared from view to even the largest telescopes of the time by 1868. It was not seen again until 1890 when it was observed by Edward **Barnard** and Sherburne **Burnham**. Since the 1930s it has been gradually brightening but remains a challenge for the amateur observer.

Hipparchus of Nicaea (c.190–c.125 B.C.)
A Greek astronomer and mathematician who compiled an extensive star catalog in which he gave the positions of some 850 stars and also classified them according to their **magnitude** (on a scale of 1 to 6, brightest to faintest). **Ptolemy** later incorporated this information into his ***Almagest***. Hipparchus also found the distance to the Moon using the **parallax** method (for which he had to produce a table of chords–an early example of trigonometric tables), and, independently of Kiddinu of Babylon, discovered the **precession of the equinoxes** by comparing his observations with those of Timocharis 150 years earlier. He extended **Apollonius of Perge**'s work on **epicycles** and eccentrics by offsetting Earth from the center of the planets' orbits in order to explain the different lengths of the seasons. See also **Greek astronomy**.

Hipparcos
A European Space Agency **astrometry** satellite, launched in August 1989, for measuring the position, brightness, and **proper motion** of relatively nearby stars. Its name is an abbreviation of High Precision Parallax Collecting Satellite and was chosen for its (somewhat strained) similarity to that of **Hipparchus of Nicaea**. It resulted in two catalogs: the ***Hipparcos Catalogue*** and the ***Tycho Catalogue***.

Hipparcos Catalogue (abbr. *HIP*)
A catalog of 118,218 stars, mostly down to magnitude 8 but with a sampling of dimmer ones (chosen for their unusual or interesting properties), surveyed by the **Hipparcos** satellite and published in 1997. It contains the

most precise measurements of positions, **parallaxes**, **proper motions**, and magnitudes ever made; although for stars dimmer than the ones surveyed, other catalogs must still be used. The accurate parallaxes have led to a vastly improved knowledge of the distances of stars out to a few hundred light-years from the Sun. Magnitudes given by the Hipparcos dataset were measured using a device sensitive to a wide range of wavelengths (mostly visual). These magnitudes are unique to the detector on the satellite; they correspond roughly, but not exactly, to visual magnitudes, and are called *Hipparcos magnitudes*. See also ***Tycho Catalogue***.

Hirayama family

A group of **asteroids** with similar **orbital elements** (notably, **semimajor axis**, **eccentricity**, and **inclination**) widely believed to have resulted from collisions between larger parent bodies. They are named after the Japanese astronomer Kiyotsugu Hirayama (1874–1943) who first recognized the existences of such families in 1918.

Hoag's Object (PGC 54559)

An unusual **ring galaxy** in **Serpens**, named after the American astronomer Arthur Hoag who discovered it in 1950. Observations have revealed an intricate structure in the bright ring, which has a near-circular projected shape and a diameter of about 100,000 light-years. According to one school of thought, Hoag's Object is a **disk galaxy** and the ring was formed from the end of a central bar that has since dissolved. Countering this is the view that the inner core is an E0 **elliptical galaxy**, not a disk, and that the ring resulted from an accretion event 2 to 3 billion years ago. It has also been proposed that Hoag's Object be considered the prototype of a class called

Hoag's Object A nearly perfect ring of hot, blue-star pinwheels about the nucleus of this unusual galaxy. The Hubble Space Telescope image captures the face-on structure in unprecedented detail. *NASA/STScI*

Hoag-type galaxies, which are neither obviously barred nor obviously inclined disks, and which have outer rings including a significant fraction of the total luminosity.

Visual magnitude:	15.0
Angular size:	54″
Distance:	600 million light-years
Position:	R.A. 15h 17.3m; Dec. +21° 35.2′

Hoba West Meteorite

An **iron meteorite** (of the **ataxite** group) weighing some 55,000 kg–the largest meteorite ever found. It still lies where it was found in 1920 at Hoba Farm, near Grootfontein, Namibia. Surprisingly, there is no crater, perhaps because the great chunk of metal entered our atmosphere at a long, shallow angle and was slowed down considerably by atmospheric drag.

Hobby-Eberly Telescope (HET)

An 11-m (9.2-m effective aperture) reflector at an altitude of 1,980 m on Mount Fowlkes, Texas, opened in 1997, and designed mainly for **spectroscopy** rather than imaging. Its mirror is made from 91 hexagonal segments each 1 m across. Permanently angled at 55°, the telescope is free to move in **azimuth**. The HET, named after two benefactors, is part of the **McDonald Observatory**, and is jointly owned by the University of Texas, Pennsylvania State University, Stanford University, the Ludwig Maximilian University in Munich, and the Georg-August University in Göttongen.

Hodierna, Giovanni Battista (1597–1660)

An Italian astronomer at the court of the Duke of Montechiaro whose *De Admirandis Coeli Caracteribus* (Admirable objects of the sky, 1654) describes some 40 objects found with a simple Galilean refractor of magnification 20. Hodierna discovered the open clusters now known as M6, M36, M37, M38, M41, M47, NGC 2362, NGC 6231, and NGC 6530 (the cluster associated with the **Lagoon Nebula**) and several comets, and made the earliest surviving drawing of the **Orion Nebula**, in which three of the **Trapezium** stars are shown. His astronomy seems to have verged toward astrology, and titles on astrology bulk large in his corpus of work.

Holmberg, Erik Bertil (1908–2000)

A Swedish astronomer known for his studies of external galaxies. In 1941 he published the results of a remarkable simulation to study the effects of interacting galaxies. His experiment consisted of an array of light bulbs to represent, by their location, the stellar distribution in a galaxy and, by their brightness, the local strength of gravity. Using photocells he measured the simulated gravitational forces as, step by step, two galaxies passed close to each other. One of his conclusions was that interacting galaxies often merge into a single larger galaxy. But it was not for another three decades, when fast enough computers became available, that his results could be confirmed. In the late 1940s, he began developing standards for measuring galaxy magnitudes from photographic plates, including the **Holmberg radius**. His work was published in a large catalog of galaxies that has been widely used in extragalactic research, notably in statistical studies of the space density of different types of galaxies. Holmberg found a correlation between galaxy magnitudes and colors (the Holmberg relation), and helped show that elliptical galaxies are generally older than spirals. See **galaxy interaction**.

Holmberg II Dwarf Irregular Galaxy

A much-studied, relatively nearby galaxy in which a great deal of star formation is taking place; it lies 9.8 million light-years away in the **M81** Group. Holmberg II enables astronomers to study star birth in a conveniently close environment that isn't disturbed by **density waves** (as happens in larger galaxies such as the Milky Way) or by deformation caused by the pull of another galaxy. Giant holes in the little galaxy–the largest about 5,500 light-years wide–are regions of old star formation. Waves of energy from mature and dying stars have blown out the surrounding gas and dust, resulting in these great voids, which have held their shape because they have no spiral arms or a massive nucleus to distort them. New star birth is also taking place, but not in the same areas as the holes because these are drained now of gas or dust. The star formation regions in Holmberg II appear as disorganized patches that occupy a relatively large fraction of the disk. They are massive, filled with hundreds of young, blue, **O stars** because the size and luminosity of star-forming regions increases only slowly with the size of the parent galaxy. One region in particular has almost as many young stars as the famous **Tarantula Nebula** of the **Large Magellanic Cloud**. A star formation model known as the "champagne" model has been used to describe the way new stars trigger even newer star birth in Holmberg II. In this scheme the bubble created by the **stellar winds** of a new star expands until it reaches the perimeter of the **molecular cloud** from which the star was born. At this point the gas and dust is being compressed against the outside wall of the molecular cloud. The result is a blistering effect on the wall: these blisters are actually new stars forming on the outside wall of the molecular cloud, resembling champagne bubbles popping out of the parent molecular cloud. The champagne model leads to star formation regions that seem to be connected in chains. As new stars form, they are constantly triggering even newer star birth, which creates a type of domino effect that spreads throughout the galaxy.

Holmberg radius

A criterion, developed by Erik **Holmberg** in 1958, to estimate the size of a galaxy without regard to its orientation in space. It is the radius at which the surface brightness is 26.5 magnitudes per square arcsecond in blue light.

Hooke, Robert (1635–1703)

An extraordinarily prolific English physicist, mathematician, and inventor. Considered the greatest mechanic of his age, he made many improvements in astronomical instruments and clocks, was the first to formulate the theory of planetary movements as a mechanical problem, explored the behavior of elastic materials, and coined the term "cell" in biology. He devised a practicable system of telegraphy, invented the spiral spring in watches, and constructed an early arithmetical machine. He also built the first **Gregorian telescope**, observed that Jupiter revolves on its axis, and made drawings of Mars from which its period of rotation was later found. In 1672 Hooke attempted to prove that Earth moves in an ellipse around the Sun and six years later proposed an inverse square law of gravitation to explain planetary motions. Hooke wrote to Newton in 1679 asking for his opinion "of compounding the celestiall motions of the planetts of a direct motion by the tangent (inertial motion) and an attractive motion towards the centrall body . . . my supposition is that the Attraction always is in a duplicate proportion to the Distance from the Center Reciprocall."

Hooke seemed unable to give mathematical proof of his conjectures. However, he claimed priority over the inverse square law and this led to a bitter dispute with Newton who, as a consequence, removed all references to Hooke from the *Principia.*

horizon

(1) *Cosmological horizon:* the maximum distance from us that light has traveled since the beginning of the universe; objects farther away are invisible to us because there has not been enough time for light to travel from them to Earth. (2) *Astronomical horizon* (or *sensible horizon*): the **great circle** formed by the intersection of the **celestial sphere** with a plane perpendicular to the line from an observer to the zenith; in other words, the great circle whose poles are the nadir and zenith.

horizontal branch

The part of the **Hertzsprung-Russell diagram** of a typical **globular cluster**; it contains stars with masses of 0.6 to 0.8 M_{sun} that are hotter and fainter than those on the **giant branch**. A star appears on the horizontal branch after it has undergone its **helium flash** and begun to burn helium steadily in its core and hydrogen in a surrounding envelope.

horizontal coordinates

A system of **celestial coordinates** that uses the observer's horizon as a reference plane and the observer's latitude and longitude as a point of origin. The position of a star, or other object, on the **celestial sphere**, at any given time, is determined by its *altitude,* the angular distance of the object above the horizon, and its *azimuth,* the angular distance of the object angle measured east from north and parallel to the horizon.

Horologium (abbr. Hor; gen. Horologii)

The Pendulum Clock; a faint southern **constellation**. Its only star brighter than magnitude 4.0 is Alpha Hor (visual magnitude 3.85; absolute magnitude 1.07; spectral type K1III; distance 117 light-years; R.A. 4h 14m, Dec. –42° 18′). One of the few objects of interest to amateur observers is R Horologium, a **Mira star** with a period of 407.6 days and one of the largest magnitude ranges known: 4.7 to 14.3. There are two **globular clusters**, both very faint, one of which, AM1, is the most remote in the Milky Way Galaxy at an incredible distance of 390,000 light-years. (See star chart 18.)

Horrocks, Jeremiah (c. 1618–1641)

An English astronomer and clergyman who applied **Kepler's laws of planetary motion** to the Moon and whose observations of a **transit** of Venus, in 1639, are the earliest on record. Horrocks studied at Cambridge University from 1632 to 1635, then became a tutor in Toxteth, Liverpool, and studied astronomy in his spare time. From Kepler's recently published *Rudolphine Tables* (1627), he worked out that a transit of Venus was due on November, 24, 1639, at 3 P.M. He prepared for the transit by directing the solar image on to a large sheet of paper in a darkened room. However, a late November afternoon in Lancashire is not the best time to observe the Sun. For Horrocks there was another problem. The predicted day was a Sunday, which meant that the puritan curate could well find himself in church at the crucial moment. Horrocks was successful in observing the transit, however, and left an account of the day in his *Venus in Sole Visa* (Venus in the Face of the Sun), published posthumously in 1662. The day was cloudy but at 3:15, "as if by divine interposition," the clouds dispersed. He noted a spot of unusual magnitude on the solar disk and began to trace its path; but, he added, "she was not visible to me longer than half an hour, on account of the Sun quickly setting." With the aid of his observations Horrocks could establish the apparent diameter of Venus as 1′ 12″ compared with the Sun's diameter of 30′, a figure much smaller than the 11′ assigned by Kepler. Before his death, at the age of only 22, Horrocks was working on an *Astronomia Kepleriana* (Astronomy of Kepler), and essays on comets, tides, and the Moon. Unfortunately, none of this was published until many years later. Most of his work was lost in the chaos of the Civil War. Other material sent to a London bookseller was burnt in the Great Fire of 1666. The remainder of his papers were published by John Wallis as *Opera posthuma* (1678; Posthumous Works).

Horsehead Nebula (Barnard 33)

A **dark nebula** in **Orion** that projects into the bright nebula IC 434 south of the star Zeta Orionis. It can only be seen well on long-exposure photographs.

Apparent size:	6′ × 4′
Distance:	1,600 light-years
Position:	R.A. 5 h 40.9 m, Dec. –2° 28′

hot spot

(1) A small, radio-bright region in a lobe of a **radio galaxy**, thought to be caused by high-speed material in a **jet** colliding with the boundary between the lobe and the intergalactic medium. (2) A small, bright region (also known as a *bright spot*) in a **cataclysmic binary**, located either where the matter stream collides with the outer

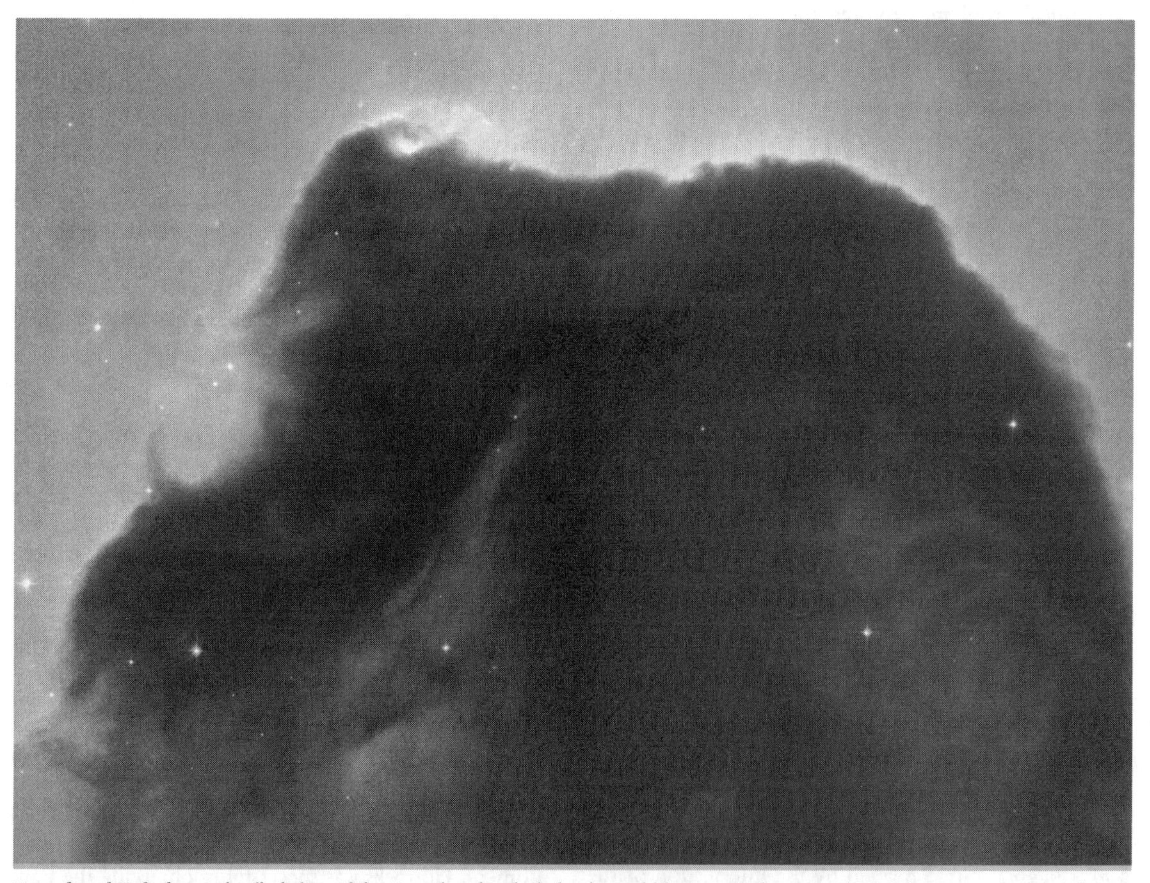

Horsehead Nebula A detailed view of the Horsehead Nebula by the Hubble Space Telescope. The bright area at the top left edge is a young star still embedded in the cloud of gas and dust from which it formed, though radiation from this hot star is eroding the stellar nursery. The top of the nebula also is being sculpted by radiation from a massive star located out of Hubble's field of view. *NASA/STScI*

Types of Hot Subdwarfs

Type		Characteristics of Spectrum
HBB star	(Horizontal Branch B)	Narrow Balmer absorption, He I, Mg
sdB star	(Subdwarf B)	Broader Balmer absorption, weak He I
sdOB star	(Subdwarf OB)	Like sdB, plus He II (at 4686 Å)
sdO star	(Subdwarf O)	He dominated, He II

edge of the **accretion disk**, or at the inner edge of the disk. (3) A center of persistent volcanism, thought to be the surface expression of a rising hot plume in a planetary mantle (as seen, for example, on Earth and on Io).

hot subdwarf

A **helium burning** star, with a helium core mass of about 0.50 M_{sun}, that is covered with a very thin hydrogen shell of about 0.02 M_{sun}. Hot subdwarfs are immediate progenitors of **white dwarfs**. There are several types, recognized by their optical spectra, as listed in the table, "Types of Hot Subdwarfs." Types sdB and sdOB are thought to be extremely blue **horizontal branch** stars that evolved (like HBB stars) from **red giants**. They form a relatively narrow sequence at the blue end of the horizontal branch (HB), and are therefore often referred to as *extended horizontal branch stars* (EHB). They differ from normal HB stars in that they don't evolve to the asymptotic giant branch. By contrast, sdO stars are in a post asymptotic giant branch phase of evolution.

hour

(1) One twenty-fourth of a day. (2) The unit of **right ascension**, equivalent to 15 degrees of arc.

hour angle (HA)

The angle between the **hour circle** of a celestial object and the observer's **meridian**, measured anticlockwise along the celestial equator. It is sometimes used as a coordinate in place of **right ascension**.

hour circle

A **great circle** on the **celestial sphere** that passes through a celestial object and the north and south celestial poles. Everywhere on an hour circle the **right ascension** is the same.

howardite

A type of **achondrite** meteorite and a member of the **HED group** believed to have originated on the asteroid **Vesta**. Howardites are named after the English chemist Edward Howard (1774–1816), one of the pioneers of meteoritics. Consisting mostly of eucritic and diogentic clasts and fragments, howardites are polymict breccias. However, they also contain dark clasts of carbonaceous chondritic matter, other xenolithic inclusions, and **impact melt** clasts, indicating a **regolith** origin. The suspicion is that howardites represent the surface of Vesta, a regolith breccia, consisting of eucritic and diogenitic debris, that was excavated by the large impact that created the enormous crater near Vesta's south pole. These fragments have been mixed with parts of the chondritic impactor, and this mixture has been subsequently pulverized and metamorphosed by smaller impacts and the **solar wind** to form a regolith. Similar regoliths cover the surface of the Moon and as with the howardites, these regolith breccias display high values for noble gases that have been implanted into the rock by the solar wind. Earth never formed any analogous rocks because its atmosphere and magnetic field protect its surface from continuous meteorite bombardment and the destructive radiation of the solar wind.

Hoyle, Fred (1915–2001)

A maverick English astrophysicist and cosmologist, who did important work in **nucleosynthesis** and founded the **steady-state theory** of the universe. Educated in mathematics and theoretical physics at Cambridge University, he returned there after wartime work on radar development and remained from 1945 to 1973, with many long-term visits to the California Institute of Technology and the Mount Wilson and Palomar Observatories. He founded Cambridge's Institute of Theoretical Astronomy and served as its first director. His early work on **stellar evolution** led to his famous collaboration with Margaret and Geoffrey **Burbidge** and William **Fowler** on the synthesis of the elements beyond helium in stars. Hoyle successfully predicted the existence of a resonance in carbon-12 that was essential to **helium burning** in stars. In 1948 Hoyle, Hermann Bondi, and Thomas **Gold** developed the steady-state cosmological model. Hoyle provided a mathematical theory of the model consistent with the **general theory of relativity** and served as the leading spokesman for the new theory, coining the term "Big Bang" for the competing model during a radio lecture. Following his retirement, he became a vigorous proponent of panspermia–the notion that life came to Earth from space–which he developed in collaboration with Chandra Wickramasinghe (see **life in the universe**). The author or coauthor of much science fiction (including *The Black Cloud*), a play, and more than 20 nonfiction books, Hoyle was a leading popularizer of science.

Hubble, Edwin Powell (1889–1953)

An American astronomer whose observations proved that galaxies are "island universes," not nebulae inside our own Galaxy. His greatest discovery was the linear relationship between a galaxy's distance and the speed with which it is moving, now known as the **Hubble law**. Upon graduation from the University of Chicago, Hubble won a Rhodes scholarship and earned a law degree at Oxford University. After obtaining his doctorate he spent his career, aside from army service in both world wars, at **Mount Wilson Observatory**. In 1923 to 1925 he identified **Cepheid variable**s in the "nebulae" NGC 6822, M31 (the **Andromeda Galaxy**), and M33 (the **Triangulum Galaxy**) and proved conclusively that they are outside the Galaxy. His investigation of these and similar objects, which he called extragalactic nebulae and which astronomers today call galaxies, led to his now-standard classification system of elliptical, spiral, and irregular galaxies (see **galaxy classification**), and to the proof that they are distributed uniformly out to great distances. Hubble measured distances to galaxies and, with Milton **Humason**, extended Vesto **Slipher**'s measurements of their **redshifts**. In 1929 he published the velocity-time relation which, taken as evidence of an expanding universe, is the basis of modern cosmology. The **Hubble Space Telescope** is named in his honor.

Hubble constant (H_0)

The present expansion rate of the universe, in units of kilometers per second per megaparsec (km/s/Mpc); it relates the apparent recession velocity of a galaxy to its distance from the Milky Way. The larger the Hubble constant, the younger the universe. Although the precise value of H_0 is unknown, independent measurements have established it to be in the range 50 to 80 km/s/Mpc; in other words, for every megaparsec, an object's cosmological velocity of recession increases by 50 to 80 km/s.

Hubble flow

The general outward movement of galaxies and **clusters of galaxies** resulting from the expansion of the universe. It occurs radially away from the observer and obeys the **Hubble law**. Galaxies can overcome this expansion on scales smaller than that of clusters of galaxies. The clusters, however, are being forever driven apart by the Hubble flow.

Hubble law

The empirical rule that governs the rate of expansion of the universe, discovered by Edwin **Hubble**. Its is simply that the velocity of recession of a galaxy, v, is directly proportional to its distance from the observer, r. The **Hubble constant** is the proportionality factor in this law.

Hubble radius

The distance from the observer at which the cosmological recession velocity of a galaxy would equal the speed of light–in mathematical terms, c/H_0, where c is the velocity of light and H_0 is the **Hubble constant**. Roughly speaking, the Hubble radius is the radius of the observable universe, and is believed to be between 12 and 15 billion light-years.

Hubble Space Telescope (HST)

An orbiting observatory built and operated jointly by NASA and the European Space Agency, launched in April 1990, and named after Edwin **Hubble**. It has a 2.4-m-diameter main mirror and, following repairs and upgrades since its launch, is equipped with the Wide-Field and Planetary Camera 2 (WFPC-2), the Near Infrared Camera and Multi-Object Spectrometer (NICMOS), the Space Telescope Imaging Spectrograph (STIS), and, installed in 2002, the Advanced Camera for Surveys (ACS). Hubble is the visible/ultraviolet/near-infrared element of the **Great Observatories** program. It is expected to remain in service until about 2010 by which time the **Next Generation Space Telescope** (NGST) should be in use.

Hubble's Variable Nebula (NGC 2261)

A **cometary nebula** (a type of **reflection nebula**) whose apex star is the peculiar variable R Monocerotis. The variability of R Mon was discovered at Athens Observatory in 1861, but it was not until 1916 that Edwin **Hubble** discovered that the nebula itself varies, over timescales as short as a few weeks. Carl Lampland at Lowell Observatory carried out a 30-year study of the nebula and found that it changed in apparent size by as much as 1″ in 4 days. These changes are presumed to be caused by shadows cast on the nebulosity by dark material orbiting R Mon. The nebula is a source of CO emission and has a temperature of about 810 K.

Huggins, Margaret Lindsay (neé Murray) (1848–1915)

A self-taught English astronomer who did extensive work in **spectroscopy** and photography. She studied the **Orion Nebula** extensively and was the first, along with her husband (married 1875), William **Huggins**, to realize

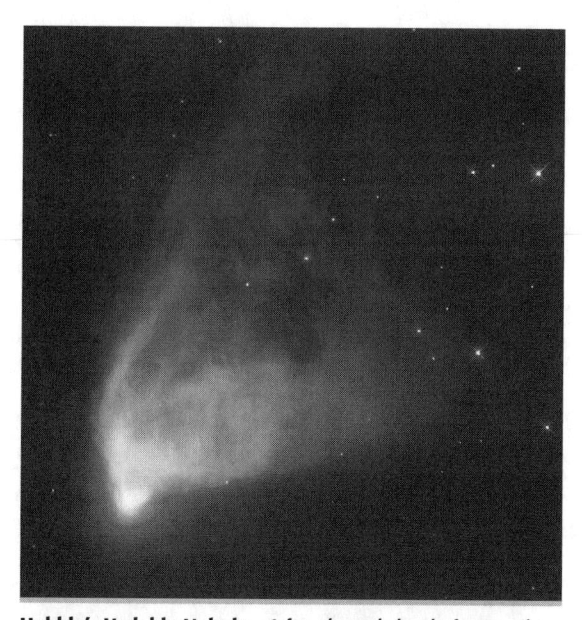

Hubble's Variable Nebula A fan-shaped cloud of gas and dust illuminated by R Monocerotis, the bright star at the bottom left. In this image of Hubble's Variable Nebula, taken, appropriately, by the Hubble Space Telescope, dense condensations of dust near the star can be seen casting shadows out into the nebula. *NASA/STScI*

that some nebulae, such as the Orion Nebula, consisted of amorphous gases rather than clusters of stars as was then generally believed.

Huggins, William (1824–1910)

A wealthy English amateur astronomer who was the first to use **spectroscopy** to determine the compositions of astronomical objects (1861). He determined that the Sun and the stars are made mostly of hydrogen, and together with his wife, Margaret **Huggins**, examined the spectra of nebulae and comets. At the age of 30 he sold the family business and built a private observatory 8 kilometers outside London. After Gustav **Kirchoff** and Robert Bunsen's 1859 discovery that spectral emission and absorption lines could reveal the composition of the source, Huggins took chemicals and batteries into the observatory to compare laboratory spectra with those of stars. First visually and then photographically he explored the spectra of stars, nebulae, and comets. He was the first to show that some nebulae, including the **Orion Nebula**, have pure emission spectra and thus must be truly gaseous, while others, such as that in Andromeda, yield spectra characteristic of stars. However, he attributed the brightest lines in nebular spectra to a new unknown element that he called "nebulium." He was also the first to attempt to measure the **radial velocity** of a star.

Hulse-Taylor Pulsar (PSR 1913+16)

A **binary pulsar** found in 1974 by the American physicists Russell Hulse and Joseph Taylor of Princeton University, a discovery for which they shared the 1993 Nobel Prize in physics. It consists of a pulsar (a neutron star) with a pulsation period of 59 milliseconds (equal to 17 pulses per second) and a companion that move around each other in an elongated orbit (period 7.75 hours, periastron 1.1 R_{sun}, apastron 4.8 R_{sun}). Although the nature of the companion is not known for certain, it is thought to have the same mass as the pulsar (1.4 M_{sun}) and so is probably also a neutron star. The orbit is gradually shrinking, by about 3.1 mm per orbit, because of gravitational waves as predicted by the **general theory of relativity**. This will cause the two stars to merge–about 300 million years from now.

Hulst, Hendrik Christoffel van der (1918–)

A Dutch astronomer who, while a student of Jan **Oort** in 1944, showed that clouds of neutral **hydrogen** in space should emit radio waves with a characteristic wavelength of 21 cm (see **21-centimeter line**). His prediction was borne out in 1951 by American physicists Edward Purcell and Harold Ewen (1922–1997). In the same year, van der Hulst and Oort began using **Doppler shift** measurements of 21-cm emission to map the Galaxy's structure.

Humason, Milton Lasell (1891–1972)

An American astronomer who worked with Edwin **Hubble** on the survey of galaxy **redshifts** that led to the discovery of the expanding universe. Having dropped out of school, he became a mule driver for the pack-trains that traveled the trail between the Sierra Madre and Mount Wilson during construction work on the observatory. In 1911 he married the daughter of the observatory's engineer and became a foreman on a relative's range in La Verne, but in 1917 he joined the staff of **Mount Wilson Observatory** as a janitor and was soon promoted to night assistant. In 1919 George **Hale**, the observatory's director, recognized Humason's unusual ability as an observer and appointed him to the scientific staff. He became involved with Hubble's study of galaxies and personally developed a technique for determining the exposures and plate measurements. From 1930 until his retirement in 1957, Humason measured the redshifts of 620 galaxies, first using the 100-in. (2.5-m) telescope at Mount Wilson and then the 200-inch (5-m) reflector at Palomar. He also applied the techniques he developed for recording spectra of faint objects to the study of **supernovae**, old **novae** that were well past peak brightness, and faint blue stars (including **white dwarfs**). During his studies on galaxies he also discovered comet 1961e, notable for its large perihelion distance.

Humason, Comet (1961e)

A **long-period comet** discovered by Milton **Humason** in 1961. It reached an absolute magnitude of +1.35 (the tenth brightest on record) and is believed to have a period of about 2,900 years.

Hungaria group

A group of **asteroids** near the inner edge of the **asteroid belt** but separated from it by a **Kirkwood gap** at the 4:1 resonance with Jupiter. The Hungarias, of which about 200 are known, have nearly circular orbits (mean eccentricity 0.08) with semimajor axes of 1.81 to 1.99 AU and relatively large inclinations of 22 to 24°. Within the group is at least one family–the *Hungaria family*–believed to have come from the breakup of a single parent body. The prototype of the group, the 11-km-wide, E-class (434) Hungaria, was discovered by Max **Wolf** in 1898; semimajor axis 1.944 AU, perihelion 1.80 AU, aphelion 2.09 AU, inclination 22.5°, period 2.71 years. A few Hungarias have perihelions only slightly farther from the Sun than the aphelion of Mars and so are shallow **Mars-crossers** as well.

hunter's moon

The next full moon after the **harvest moon**; it usually falls in October. Thought of as providing more light for the night hunter, it rises, as does the harvest moon, around the time of sunset for several successive nights in mid- or high-northern latitudes.

Huygens, Christiaan (1629–1695)

A Dutch physicist and astronomer who first described the nature of **Saturn**'s rings and discovered Saturn's largest moon, **Titan**, both in 1655. He also figured in one of the great unsung moments in planetary exploration when, at 7 P.M. on November 28, 1659, he sketched a feature–a dark triangular patch on the surface of **Mars**–that he had just seen through his modest telescope. Eventually, it came to be known as Syrtis Major, the "Great Marsh," and for many years was presumed to be some kind of watery body. The significance of its sighting lay in the fact that whereas the large bright and dark areas on the Moon had been visible for all to see from prehistoric times and **Galileo** had telescopically discovered the Great Red Spot on Jupiter (an atmospheric feature), Syrtis Major was the first *permanent* marking to be glimpsed on the surface of another planet. Huygens quickly used his discovery to show that the Martian day–the time it takes for the planet to spin around once on its axis–is similar in length to our own. With his brother, Constantijn Huygens (1628–1697), he built tubeless telescopes supported by cables of very long focal length to reduce the problem of aberration. He also invented the Huygenian **eyepiece** and the first pendulum clock (1657), and contributed to the wave theory of light (1678) and several other areas of physics.

Hyades

The nearest **open cluster** to the Sun, except for the **Ursa Major Moving Cluster** (which appears spread out as individual stars). The Hyades is centered some 150 light-years away in **Taurus**, and contains more than 200 stars. The central group is about 10 light-years in diameter, while outlying members seem to be spread over a region at least 80 light-years across. Its age is put at 660 million years, which, taken together with its motion through space, suggests it has a common origin with the **Praesepe** cluster. Although the bright red giant **Aldebaran** (Alpha Tauri) lies almost along the same line of sight as the Hyades, it is not a member of the cluster and lies much closer to us (about 60 light-years away). Although visible to the naked eye and therefore known prehistorically, the Hyades appears to have been first cataloged by **Hodierna** in 1654.

Hyakutake, Comet (C/1996 B2)

A **long-period comet**, discovered by the Japanese astronomer Yuji Hyakutake in January 1996. It developed rapidly throughout March, reaching the peak of its display in the northern hemisphere between March 24 and

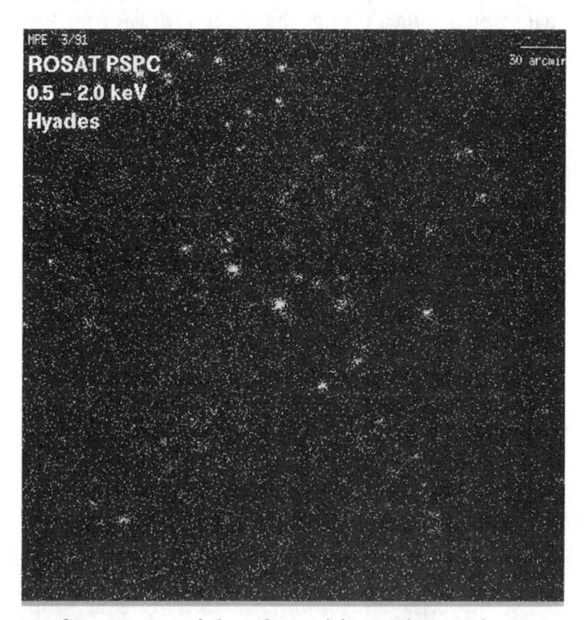

Hyades An unusual view of part of the Hyades star cluster at X-ray wavelengths by the ROSAT X-ray satellite. It shows more than 100 X-ray sources in the central region of the Hyades. Altogether ROSAT has identified over 185 of the 400 or more cluster stars as X-ray emitters. *Max Planck Institut fur extraterrestrische Physik*

Hydra: Stars Brighter than Magnitude 4.0

Star	Magnitude		Spectral Type	Distance (light-yr)	Position	
	Visual	Absolute			R.A. (h m s)	Dec. (° ′ ″)
α Alphard	1.99	−1.69	K3III	177	09 27 35	−08 39 31
γ	2.99	−0.55	G8IIIa	132	13 18 55	−23 10 17
ζ	3.11	−0.22	G8III	151	08 55 24	+05 56 44
ν	3.11	−0.03	K0III	139	10 49 37	−16 11 37
π	3.25	0.75	K2III-IIIbCN	101	14 06 22	−26 40 56
ε	3.38	0.29	G0III	135	08 46 47	+06 25 07
ξ	3.54	0.55	G8III	129	11 33 00	−31 51 27
λ	3.61	0.88	K0IIICN	115	10 10 35	−12 21 15
μ	3.83	−0.58	K4.5III	248	10 26 05	−16 50 11
θ	3.89	0.90	B9.5V	129	09 14 22	+02 18 51

26, when it was visible all night and sported a tail spanning at least 75°. At closest approach to Earth it was only 0.10 AU (15,000,000 km) away.

Hydra (abbr. Hya; gen. Hydrae)

The Water Serpent; the largest **constellation** of all (following the dismantling of **Argo Navis**), stretching across a huge swathe of the southern sky. (See tables, "Hydra: Stars Brighter than Magnitude 4.0," and "Hydra: Other Objects of Interest." See also star charts 12 and 18.)

hydrogen

The lightest and most common element in the universe. Formed in the **Big Bang**, it is the main constituent of stars and, through its fusion into helium, the main source of stellar energy. It comes in three isotopes: ^{1}H (one proton and no neutrons), by far the commonest variety, ^{2}H (one proton and one neutron) or **deuterium**, and ^{3}H (one proton and two neutrons), or tritium, which is radioactive and of no importance astronomically. Hydrogen occurs in space in molecular, atomic (in **H I regions**), and ionic (in **H II regions**) forms.

hydrogen burning

The **fusion** of **hydrogen** into **helium**: the process by which all stars on the **main sequence** generate energy. Every star born with more than about 0.08 M_{sun} burns hydrogen.

hydrogen spectrum

A characteristic pattern of **spectral lines**, either absorption or emission, produced by the **hydrogen** atom. The various series of lines are named according to the lowest energy level involved in the transitions that give rise to the lines. The **Lyman series** involves jumps to or from the ground state (n = 1); the **Balmer series** (in

Hydra: Other Objects of Interest

Object	Notes
Planetary nebula **Ghost of Jupiter**	NGC 3242. See separate entry.
Open cluster M48	On the boundary of Hydra and Monoceros and at the edge of naked-eye visibility. Magnitude 5.8; diameter 54′; R.A. 8h 13.8m, Dec. −5° 48′.
Globular cluster M68	Discovered by Pierre Méchain in 1780. Magnitude 8.2; diameter 12′; distance 39,000 light-years; R.A. 12h 39.5m, Dec. −26° 45′.
Galaxy M83	A face-on type Sc spiral just outside the Local Group. Magnitude 8.2; distance 8.5 million light-years; R.A. 13h 37.0m, Dec. −29° 52′.

which all the lines are in the visible region) corresponds to n = 2, the Paschen series to n = 3, the **Brackett series** to n = 4, and the Pfund series to n = 5.

Hydrus (abbr. Hyi; gen. Hydri)

The Little Water Snake; a small, faint **constellation** of the far southern sky (not to be confused with Hydra), partly sandwiched between the two Magellanic Clouds. (See table, "Hydrus: Stars Brighter than Magnitude 4.0," and star chart 8.)

Hygiea (minor planet 10)

The fourth largest main belt **asteroid**; it was discovered in 1849 by the Italian astronomer Annibale de Gasparis (1819–1892). Diameter 443 km, spectral class C, semimajor axis 3.138 AU, perihelion 2.763 AU, aphelion 3.513 AU, eccentricity 0.119, inclination 3.84°, period 5.56 years.

hyperbolic orbit

An open **orbit** in the shape of a hyperbola, the **eccentricity** of which is greater than 1. A hyperbolic orbit is followed by an object that escapes from the gravitational field of a larger body.

hyperfine structure

The splitting of **spectral lines** into two or more very closely spaced components caused by interaction between the intrinsic spin of **electrons** and the intrinsic spin of the atomic nuclei they orbit. The splitting is much closer than in **fine structure** and only visible at the highest spectral resolutions.

hypergalaxy

A system consisting of a dominant **spiral galaxy** surrounded by a cloud of dwarf satellite galaxies, often ellipticals. Our Galaxy and the Andromeda Galaxy are hypergalaxies.

hypergiant

Precise definitions vary, but all agree that hypergiants are the most massive and luminous type of star. They are on the verge of instability with masses of up to about 100 M_{sun}, strongly developed large-scale atmospheric velocity fields, excessive mass loss, and extended circumstellar envelopes. Hypergiants can survive only a couple of million years or so before exploding as **supernovae** and, theory insists, leaving behind **black holes**. Only about a dozen are known in our Galaxy, including **Eta Carinae**, **P Cygni**, and **Rho Cassiopeiae**. **S Doradus** is a hypergiant in the nearby **Large Magellanic Cloud**.

Hyperion

The sixteenth moon of **Saturn** and the largest nonspherical satellite in the solar system (Neptune's **Proteus** is larger but almost spherical), measuring 410 × 260 × 220 km. Also known as Saturn VII, it was discovered by William and George **Bond** and, independently, by William **Lassell**, in 1848. Possibly it is a fragment of a larger body that was broken long ago by a major impact. Like most of Saturn's moons, Hyperion has a low density, which suggests it is made mostly of water-ice with a small amount of rock. But unlike its satellite neighbors, Hyperion has a low **albedo** (about 0.2), indicating that it is covered by a layer of dark material. This material may have come from **Phoebe** having eluded **Iapetus** on the way. Hyperion moves at a mean distance of 1.48 million km from Saturn and is strongly influenced by nearby **Titan**. Its rotation is chaotic, its axis of rotation wobbling so much that its orientation in space is totally unpredictable. Hyperion is the only known body in the solar system that rotates chaotically, though simulations indicate that other irregular satellites may have behaved this way in the past. Hyperion is unique in combining an irregular shape, a highly eccentric orbit, and a proximity to another large moon. These factors conspire to make rotational instability more likely. The 3:4 orbital resonance between Titan and Hyperion also appears to increase the odds of chaotic rotation. Hyperion's strange spin probably accounts for the fact that Hyperion's surface is more or less uniform, in contrast to many of Saturn's other moons that have distinctly different leading and trailing hemispheres.

Hydrus: Stars Brighter than Magnitude 4.0

	Magnitude				Position	
Star	Visual	Absolute	Spectral Type	Distance (light-yr)	R.A. (h m s)	Dec. (° ′ ″)
β	2.82	3.45	G2IV	24	00 25 46	–77 15 15
α	2.86	1.16	F0V	71	01 58 46	–61 34 12
γ	3.26	–0.83	M2III	214	03 47 15	–74 14 20

hypernova

An usually energetic **supernova** of the type sometimes seen in dense, starburst regions of some irregular galaxies.

hypersensitization

A way of treating photographic emulsions to prevent their sensitivity falling off rapidly when used for long exposures. The commonest method is to soak the film in a chemically reducing gas (an 8% hydrogen, 92% nitrogen mixture is popular for amateur use).

hypersthene

An important rock-forming silicate of magnesium and iron, $(Mg, Fe)SiO_3$, that crystallizes in the orthorhombic system.

hypervelocity impact

A collision of an object into a surface at a speed of at least 1 to 2 km/s, which results in the impact rock behaving like a fluid. Hypervelocity impacts result in **crater**s that are generally circular, as the crater is typically much larger than the impacting body.

HZ Herculis star (HZ Her star)

A low-mass **X-ray pulsar** that is also an optically variable binary in which **mass transfer** is taking place. Such stars consist of a dwarf primary of spectral type B to F, and a neutron star that is an X-ray pulsar and possibly also an optical pulsar. The X-ray emission flips periodically between a *high state* and a *low state,* with peak optical luminosity corresponding to the high state.

I

Iapetus

The third largest moon of **Saturn**, with a diameter of 1,460 km, and seventeenth in order from the planet, at a mean distance of 3,560,000 km; also known as Saturn VIII, it was discovered by Giovanni **Cassini** in 1671. Iapetus has the most extreme variation in **albedo** of any satellite in the solar system, ranging from about 0.05 (as dark as soot) on the leading hemisphere to 0.5 on the trailing side. One possibility is that the leading hemisphere is coated with material knocked off **Phoebe**; however, the color of the leading half of Iapetus and that of Phoebe don't quite match. Alternatively, some internal process on Iapetus may have been responsible. This idea is supported by the fact that the dark material seems to be concentrated in crater floors.

Icarus (minor planet 1566)

A member of the **Apollo group** of asteroids, discovered by Walter **Baade** in 1949. It has a diameter of 1.4 km, belongs to spectral class U, and spins once around on its axis every 2.27 hours. Its orbit carries it well within the orbit of Mercury, but not quite as close to the Sun as 2000 BD19 (0.09AU), 1995 CR (0.12 AU), or 3200 **Phaethon** (0.14 AU). Icarus ranks fifth on the list of **potentially hazardous objects**, although it will not come within several million km of Earth in the foreseeable future. During its last "near miss," on June 14, 1968, it approached us to within 6.5 million km–never a threat, but enough to have one cult retreating to a peak in Colorado to escape the anticipated slide of California into the Pacific. Semimajor axis 1.078 AU, perihelion 0.19 AU, aphelion 1.97 AU, eccentricity 0.827, inclination 22.9°, period 1.12 years.

Iapetus One of the strangest moons in the solar system, revealed in this Voyager 2 image taken on August 22, 1981. The smallest details on Iapetus that can be discerned in this view are about 19 km across. *NASA/JPL*

ice

A term often used by planetary scientists to refer not only to frozen water, but also to frozen **methane** and **ammonia**, which usually occur as solids in the outer solar system. Water-ice is an important geological material on many of the outer planet satellites. Under various conditions ice can have viscosities and flow properties similar to those of different types of molten rock, and therefore can result in the formation of similar geologic structures. Ice that is at near freezing temperatures (273 K) and low pressure (less than 1,000 bars)–Ice I–has different physical properties than ice at lower temperatures and higher pressures. (For example, Ice II is stable at 200 K and 6,000 bars, while Ice VI is stable at 200 K and 15,000 bars.) These phases of ice have different densities and flow properties that may be expressed through tectonic and volcanic processes on icy satellites. When these ices contain small amounts of impurities, such as ammonia, they are known as clathrates, and have a lower melting temperature than pure water-ice. This may be a factor in volcanic processes on certain large outer satellites.

ice volcanism

The eruption of molten **ice** or gas-driven solid fragments onto the surface of a planetary body due to internal heating. Ice volcanism may exist on several large moons in the outer solar system, including **Europa** and **Enceladus**. *Ice-volcanic melt* is the fluid or semifluid material associated with ice volcanism; like molten rock, it can have a wide variety of viscosities and other flow properties.

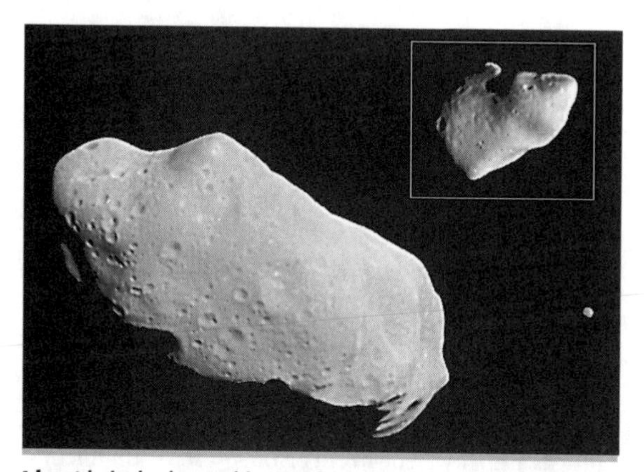

Ida Ida is the large object to the left in this composite picture, taken by the Galileo spacecraft from a range of about 10,500 km, some 14 minutes before its closest approach on August 28, 1993. The tiny object to the right is Ida's moon Dactyl. The inset shows Ida from a different angle. *NASA/JPL*

Ida (minor planet 243)

A member of the **Koronis family** of asteroids; it was discovered by Johann **Palisa** in 1884 and named after the mountain on Crete where Zeus spent his childhood. On August 28, 1993, it became the second asteroid, after **Gaspra**, to be encountered by a spacecraft when **Galileo** flew past it at a distance of 2,400 km and with a relative velocity of 12.4 km/s. The highlight of the flyby was the discovery (made in February 1994 when images stored on the spacecraft's tape recorder were finally transmitted to Earth) that Ida has a satellite, subsequently christened **Dactyl**. Ida's age is a puzzle. Its surface is heavily cratered suggesting that it has existed in its present form for at least a billion years, which predates estimates for the Koronis breakup. Size 56 × 24 × 21 km, density 2.5 g/cm^3, spectral class S, rotational period 4.63 hours, orbit: semimajor axis 2.861 AU, eccentricity 0.045, inclination 1.14°.

igneous

Descriptive of rock or minerals that solidified from molten or partly molten material. Igneous rock is *extrusive* if it formed on the surface of a planetary body and *intrusive* if it solidified deep underground.

igneous age

The period of time since a rock, such as some type of **meteorite**, solidified and crystallized from a molten state. This can be derived from the ratios of the radioactive isotopes of certain elements, such as samarium-neodymium (Sm-Nd) and rubidium-strontium (Rb-Sr).

Ikeya-Seki, Comet (C/1965 S1)

A member of the Kreutz family of **sun-grazing comets** that was visible in broad daylight during October 1965, reaching a visual magnitude of –10, and displayed a tail 60° long at maximum; its nucleus split into three parts at perihelion. It was discovered independently in September 1965 by the Japanese amateur astronomers Kaoru Ikeya (1943–) and Tsutomu Seki (1930–). Perihelion 0.008 AU (1.2 million km), eccentricity 0.9999, inclination 141.9°, period 880 years.

illuminance

Also known as *luminous flux density*, the total amount of visible light falling on a point on a surface from all directions above the surface in a given time. It is equivalent to **irradiance** weighted with the response curve of the human eye. Illuminance is measured in lux (lx), or lumens per square meter (lm/m^2).

image

(1) A figure formed by rays of light or other electromagnetic radiation at a mirror's **focal plane**. (2) A picture obtained by a telescope, camera, or other imaging device.

image tube

A vacuum-filled tube with a photocathode at one end, used to amplify faint images. Electrons, released when light forms an image on the photocathode, are accelerated by magnetic coils around the tube so that they form a second, brighter image when they strike a second phosphor screen. Also known as an *image intensifier tube.*

impact basin

An **impact crater** that has a rim diameter greater than 300 km. Impact basins are produced by such violent impacts that they are always associated with extensive faulting and other deformations of the crust, and very large **ejecta** blankets. Over 40 impact basins have been identified on the Moon, the largest of which is the Imbrium Basin.

impact crater

A **crater** formed by the high-speed impact of a meteoroid, asteroid, or comet with a solid surface. Craters are a common feature on most moons (an exception is *Io*), asteroids, and rocky planets, and range in size from a few cm to over 1,000 km across, in the case of a large **impact basin**. There is a general morphological progression from large to small craters: large craters may have several rings and smooth floors; intermediate craters tend to have a central peak (formed by melting and rebounding of the crust) and smooth floors; small craters have a simple

impact basin The giant impact basin known as Odysseus Basin, on Saturn's moon Tethys, is visible to the lower right in this Voyager 2 image. The basin has been flattened by the flow of softer ice and no longer shows the deep bowl shape characteristic of fresh craters in hard, cold ice or rock. It appears to have been formed early in Tethys's history, at a time when the moon's interior was still relatively warm and soft. *NASA/JPL*

bowl-shaped floor that is rough. Because impact craters degrade at different rates depending on their environment, they are valuable indicators of the age of a surface and the extent to which resurfacing has taken place. On Earth, for example, craters are rapidly degraded and destroyed by weathering processes; about 120 are known, with diameters ranging from 150 m to 180 km. One of the best preserved and most impressive is the **Barringer Crater**, near Winslow in northern Arizona. On Mercury, by contrast, which lacks an atmosphere and is geologically inert, the landscape is peppered with craters dating back over 4 billion years.

impact feature

Any feature on the surface of a planet, a moon, or an asteroid that has resulted from a past collision with another object. Among the types or subtypes identified are **impact craters**, **impact basins**, **multiringed basins**, and **palimpsests**.

impact melt

Rock that has been made temporarily molten as a result of the energy released by the impact of a large colliding body. Impact melts include small particles, known as *impact melt spherules,* that are splashed out of the **impact crater**, and larger pools and sheets of melt that coalesce in low areas within the crater. They are composed predominantly of the target rocks, but can contain a small but measurable amount of the impactor.

inclination (symbol *i*)

Orbital inclination is the angle between the orbital plane of an object and a reference plane that passes through the body being orbited. The inclination of a planetary orbit in the solar system is measured from the ecliptic (the plane of Earth's orbit around the Sun). The inclination of a satellite's orbit is measured from the plane that contains its primary's equator. For binary stars, the inclination is specified relative to the plane of the sky. *Axial inclination* is the angle between a planet's or moon's rotation axis and its orbit; in the case of a star, the reference plane is usually that of the **celestial sphere**.

inclusion

A fragment of older material, including minerals and rock, that has been enclosed within another rock.

Index Catalogue (IC)

Either of two catalogs that served as supplements to the *New General Catalogue (NGC)*; they were compiled, as was the *NGC,* by John **Dreyer**. The first, published in 1895, added 1,529 new star clusters, nebulae, and galaxies; the second, published in 1908, listed a further 3,857 objects.

Indian astronomy

As in other cultures, astronomy grew up in ancient India alongside religion and astrology. The underlying Indian cosmology is based on the idea of the unity of the cosmos and that it should be perceived as an organic whole in which all parts are interdependent. The Sun (Surya) is the center of creation, the point at which the seen and unseen worlds unite, and is the visible source of the world in which we live. Such beliefs may help explain why Indian astronomers, 1,500 years ago, held advanced notions such as the Earth orbits the Sun and that stars are distant suns. In India today, the science of astronomy is called Khagola-shastra, a name possibly deriving from the famous astronomical observatory at the University of Nalanda which was called Khagola. This is where the great fifth-century Indian astronomer **Aryabhata** studied and extended the subject. Only **Brahmagupta**, more than a century later, rivaled him in stature on the subcontinent. See **Egyptian astronomy**, **Babylonian astronomy**, **Greek astronomy**, and **Chinese astronomy**.

Indus (abbr. Ind; gen. Indi)

The Indian; a faint southern **constellation** with very little of interest for the amateur astronomer. (See table, "Indus: Stars Brighter than Magnitude 4.0," and star chart 14.)

inequality

Any irregularity in an object's orbital motion that can't be explained as a result of the mutual gravitational attraction between the body and its primary. Inequalities in a planet's movement around the Sun, for example, are due mainly to **perturbations** by other planets. The *great inequality,* also known as *Laplace's period,* is a slight rocking of both Jupiter and Saturn, as they trek around the **ecliptic**. Each oscillation takes some 840 to 960 years (the precise figure is not known) and is mostly caused by the near 5:2 resonance of their orbits: Jupiter makes five complete trips around the Sun for every two (or thereabouts) of Saturn. The *lunar equality* stems from several factors, which include the perturbing effects of the Sun and other planets, tidal forces, and the nonspherical shapes of Earth and the Moon.

infall velocity

The velocity with which an object falls toward the center of a region of gravitational influence. For example, the infall velocity of an object 1,800 AU away from a star of one solar mass, in the absence of other factors, is about 0.5 km/s. On a larger scale, galaxies and even clusters of galaxies may be drawn toward enormous concentrations of mass. Our own **Local Group** has an infall velocity of about 350 km/s toward the **Virgo Cluster**.

inferior planet

Either of the planets **Mercury** or **Venus**, whose orbits are closer to the Sun than is Earth's orbit. Other planets are called *superior planets.*

inflationary model

A modified version of the **Big Bang** theory of the origin of the universe that allows for a brief period, when the universe was only about 10^{-35} seconds old, in which the expansion of **space-time** took place at a vastly accelerated rate. The huge inflation of space was driven by energy released from a separation of two forces that had previously been one: the strong force and the electroweak force. For around 100 million trillion trillionth of a second, the universe grew many times faster than the speed of light (not in defiance of Einstein's special theory of relativity, however, because nothing was traveling *through* space at a speed greater than that of light) taking it from a size unimaginably smaller than an atomic nucleus to about the size of grapefruit. In the process, tiny isotropic (small in every direction) regions of the early universe were inflated to become larger than the observable universe. The latest version of inflation, known as *chaotic inflationary theory,* posits that the universe grew out of a quantum fluctuation in a preexisting region of spacetime and that other universes could do the same from regions within our universe today. A new universe, or "babyverse," formed by this budding process, would have its own set of physical laws and material particles.

infrared

Electromagnetic radiation of wavelengths longer than the red end of visible light and shorter than microwaves, roughly from 1 micron (10^{-6} m) to 350 microns. Infrared is divided into three spectral regions–near, mid and far infrared–but the boundaries between these are not agreed upon and can vary. The main factor that decides which wavelengths are included in each of the regions is the type of detector technology used. A typical breakdown is shown in the table "The Infrared Spectrum." See **infrared astronomy**.

infrared astronomy

The study of astronomical objects at **infrared** wavelengths. Many objects in the universe that are too cool and faint to be detected in visible light, can be seen in the infrared. These include cool stars, **infrared galaxies**, clouds of particles around stars, nebulae, **interstellar molecules**, **brown dwarfs**, and **extrasolar planets**. In the

Indus: Stars Brighter than Magnitude 4.0

Star	Magnitude		Spectral	Distance	Position	
	Visual	Absolute	Type	(light-yr)	R.A. (h m s)	Dec. (° ′ ″)
α The Persian	3.11	0.65	KOIIICN	101	20 37 34	–47 17 29
β	3.67	–2.67	KOIII	603	20 54 49	–58 27 15

The Infrared Spectrum

Spectral Region	Wavelength Range (microns)	Temperature Range (K)	Objects Seen
Near infrared	1 to 5	740 to 3,000	Cooler red stars and red giants; dust is transparent
Mid infrared	5 to 40	140 to 740	Planets, comets, and asteroids; dust warmed by starlight, protoplanetary disks
Far infrared	40 to 350	11.6 to 140	Emission from cold dust; central regions of galaxies; very cold molecular clouds

case of **dust**, infrared astronomers can either see through it or focus on it, depending on the wavelengths they choose. At near infrared wavelengths, dust is essentially transparent, enabling far-flung views along the dusty plane of the Milky Way that are impossible in visible light. The **galactic bulge**, for example, shows up particularly well in near infrared. On the other hand, at mid infrared wavelengths astronomers pick up radiation generated by dust itself, which is valuable in studying such phenomena as protoplanetary disks and possible extrasolar planets. Work in cosmology also depends crucially on infrared observations. The expansion of the universe means that all of the ultraviolet and much of the visible light from distant sources is shifted into the infrared part of the spectrum.

Infrared Astronomy Satellite (IRAS)

The first spacecraft to be equipped with an **infrared** telescope–a 0.6-m instrument, cooled by liquid helium, that operated successfully for 300 days. Launched in January 1983, a joint project of the United States, the United Kingdom, and the Netherlands, IRAS carried out an all-sky survey at 12, 25, 60, and 100 microns, detecting about 250,000 sources and more than doubling the number previously cataloged. Among its discoveries were warm dust disks around certain stars, including **Vega**, **Formalhaut**, and **Beta Pictoris**, **starburst galaxies**, several new comets and asteroids, and **infrared cirrus**. IRAS also revealed for the first time details of the core of the Milky Way Galaxy.

infrared cirrus

Wispy, filamentary structures seen in **infrared** in all parts of the sky, but particularly well at high galactic latitudes away from all of the other infrared emission associated with the **galactic plane**. First detected by the **Infrared Astronomy Satellite** at wavelengths of 60 and 100 microns, the cirrus is named after its cloudlike appearance and is believed to be due to **dust** grains in otherwise cool (15 to 30 K), diffuse atomic hydrogen clouds that have been warmed slightly by ultraviolet light from nearby stars.

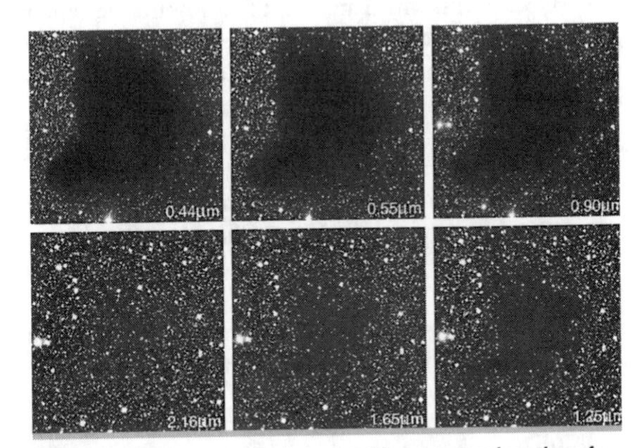

infrared astronomy Six views, at different wavelengths, of the dark cloud Barnard 68 (B68) in front of a rich star field. B68 lies about 500 light-years away in Ophiuchus. The top three photos, taken by the Very Large Telescope at wavelengths of 440 nm (blue), 550 nm (green-yellow), and 900 nm (near infrared), show the cloud as a compact, opaque, and sharply defined object, the central parts of which are so dense that they completely block the light from the stars behind. However, the bottom three images, taken at increasingly long infrared wavelengths (from right to left) of 1,250, 1,650, and 2,160 nm with the SOFI infrared telescope, penetrate the cloud and reveal more and more background stars. Since the outer regions of the cloud are less dense than the inner ones, the apparent size of the cloud also decreases, as more background stars shine through the outer parts. *ESO*

infrared excess

Infrared emission from an object above that expected if the object were radiating solely as a **blackbody**. The excess usually indicates the presence of **dust** in the vicinity that has been warmed as it absorbs shorter wavelength radiation from the object.

infrared galaxy

A dust-laden galaxy that appears bright at **infrared** wavelengths. Many such galaxies are visible *only* in the infrared.

Infrared Space Observatory (ISO)

A European Space Agency orbiting observatory that carried a cryogenically-cooled 60-cm telescope and a suite of deep-**infrared** cameras and spectrometers. ISO, which observed at wavelengths between 2.5 and 240 microns, not only covered a much wider wavelength range than its predecessor, the **Infrared Astronomy Satellite**, but was also thousands of times more sensitive and viewed infrared sources with much greater resolution. It operated from November 1996 to April 1998–three times longer than IRAS–at which point the liquid-helium coolant was exhausted.

infrared star

A star whose radiation is wholly or primarily at **infrared** wavelengths; this indicates that the star is very cool or is embedded in a very dusty environment.

Infra-Red Telescope Facility (IRTF)

A 3.0-m telescope, optimized for **infrared** observations, and located at the summit of Mauna Kea, Hawaii. The observatory is operated and managed for NASA by the University of Hawaii Institute for Astronomy in Honolulu.

Innes, Robert Thorburn Ayton (1861–1933)

A Scottish astronomer who emigrated to Australia (becoming a wine merchant) before moving to South Africa as director of the Observatory at Johannesburg. He measured stellar **proper motions**; introduced the orbital parameters for **double stars** now known as *Thiele-Innes constants,* in conjunction with the Danish astronomer Thorvald Nicolai Thiele (1838–1910); and was the first to use a blink **comparator**–an instrument with which he discovered **Proxima Centauri** in 1915. His *Southern Double Star Catalogue* (1927) included over 1,600 doubles that he found.

Innisfree meteorite

One of the few **meteorites** whose passage through the atmosphere was recorded photographically by cameras at more than one place (others include the **Príbram Meteorite** and **Lost City meteorite**). This is important because it enables a calculation of the object's old orbit and of the area where it is likely to fall. The Innisfree meteorite fell 13 km north of the town of this name, in Alberta, Canada, at 7:17 P.M. on February 5, 1977. Although an immediate search of the area, by light plane and on foot, turned up nothing, photographic records from two stations in the Meteorite Observation and Recording Program (MORP) network allowed a computer projection of the most likely fall area. Eleven days later, the largest piece of the Innisfree rock (2.07 kg) was found only a few hundred meters from the point predicted. Subsequently five other fragments were found, bringing the total mass recovered to 3.79 kg. Analysis showed Innisfree to be an **ordinary chondrite** of the rare LL5 type.

instability strip

A narrow, vertical region in the **Hertzsprung gap** occupied by **pulsating stars** in a post-main-sequence stage of stellar evolution. Stars traverse the instability strip relatively quickly, at least once on their way to their final evolutionary configuration.

INTEGRAL (International Gamma-ray Astrophysics Laboratory)

A European Space Agency satellite, launched in October 2002, that has provided the sharpest pictures of the cosmos in **gamma rays** to date and has identified many new gamma-ray sources. It is designed to carry out fine **spectroscopy** and imaging of cosmic gamma-ray sources in the 15 keV to 10 MeV energy range, with concurrent source monitoring in the X-ray (3 to 35 keV) and optical (V-band, around 550 nanometers) ranges.

integrated magnitude

The **apparent magnitude** that an extended object, such as a nebula or galaxy, would have if all its light were concentrated at a starlike point.

intensity

(1) In **photometry**, a measure of the amount of radiant energy received per unit solid angle, per unit time, per unit area of a surface element orthogonal to the direction of propagation of the radiation. Also known as *specific intensity.* (2) In visual perception, one of the three basic parameters (along with *hue* and *saturation*) that may be used to describe the physical perception of color. Intensity is a measure of the energy of the spectral distribution, at a given point in an image or scene, weighted by the spectral response of the visual system. *Luminance* is the energy of the physical spectrum, but not weighted by the visual response. *Brightness* sometimes is used

synonymously with either term. (3) In **radio astronomy**, a term often used much more loosely to mean either the **flux density** of an unresolved radio source or the **surface brightness** of an extended source.

interacting binary

A binary system in which the two stars orbit around each other so closely that they strongly affect each other's structure and evolution, most notably through **mass transfer** to the primary from the secondary "donor" star. A typical interacting pair consists of a degenerate, or compact, primary–a **black hole**, a **neutron star**, or, most commonly, a **white dwarf**–and a secondary that has either filled its **Roche lobe** or has a large **stellar wind**. Matter is pulled toward the compact object and forms an **accretion disk** around it. Material cannot land directly onto the compact object as it has to get rid of its **angular momentum**–hence, the accretion disk–but when it does, the gravitational potential energy of the material is converted to electromagnetic energy as it falls and/or undergoes explosive fusion on the surface of the compact recipient.

Interamnia (minor planet 704)

The fifth largest main belt **asteroid** with a diameter of 338 km; it was discovered by Vincenzo Cerulli in 1910. Albedo 0.06, spectral class F, rotational period 8.73 hours; orbit: semimajor axis 3.062 AU, perihelion 2.617 AU, aphelion 3.512 AU, eccentricity 0.146, inclination 17.32°, period 5.36 years.

interference

Alternate reinforcement and cancellation of two or more beams of **electromagnetic radiation** from the same source. In constructive interference the two component beams are in phase, and light results. In destructive interference the components are out of phase, and darkness results.

interference pattern

Alternating light and dark bands, known as *fringes*, that are produced by **interference**. The term is also used in **radio astronomy** to describe the pattern that results when the signals picked up by two or more elements of an **interferometer** are combined.

interferometer

A device that combines observations of the same astronomical object made by two or more separate instruments to obtain a higher **resolution** image than would be possible from any of the instruments on its own. The signals from each instrument are put together to form an **interference pattern**. *Radio interferometers* have been used widely in radio astronomy for many years; an example is the **Very Large Array** in New Mexico. More recently, *optical interferometers* have begun to play a major role in observational astronomy at visible wavelengths. The world's most powerful optical interferometer is the European Southern Observatory's **Very Large Telescope**.

intergalactic absorption

The **absorption** of light from distant objects by material in intergalactic space; it is too feeble to notice in the case of nearby galaxies but gives rise to the **Lyman-alpha forest** of spectral lines superimposed on the spectra of **quasars** and other objects billions of light-years away.

intergalactic magnetic field

The weak magnetic field (less than 10^{-12} tesla) that is thought to permeate intergalactic space. Its origin is unknown but may extend back to the origin of the universe.

intergalactic medium (IGM)

Matter in the space between the galaxies. Whether there is a genuine, all-pervading IGM, in the same way that there is an **interstellar medium** within galaxies, is unknown. Certainly, we know, from its X-ray emission, that plenty of hot, tenuous gas resides in **rich clusters of galaxies.** But evidence for the existence of intergalactic gas in the **Local Group** (aside from tidally displaced material such as the **Magellanic Stream**) is controversial and there is no data yet whether, in the vast intercluster spaces, gas just gets increasingly thin as you move from one cluster to the far outskirts of the next, or whether there is a true intercluster medium. The intergalactic presence of **dark matter** is also conjectural at this stage.

Intergalactic Wanderer (NGC 2419)

One of the most remote **globular clusters** of the Milky Way Galaxy, at a distance (both from the Sun and the galactic center) of nearly 300,000 light-years. Although it ranks fourth in intrinsic brightness of our Galaxy's globulars, after **Omega Centauri** (NGC 6388) in Scorpius and M54, its remoteness–nearly double that of the **Large Magellanic Cloud**–leaves it faint in our sky, though within range of moderate-sized amateur telescopes. It lies in **Lynx** in roughly the direction of the galactic anticenter.

Visual magnitude:	10.4
Angular diameter:	4.1′
Distance:	295,000 light-years
Position:	R.A. 7h 38.1m, Dec. +38° 53′

intergranular lane

A dark, relatively cool area between **granules** where material is descending below the surface of the **Sun**.

interferometer The underground interferometric tunnel and the delay lines that connect the main instruments of the European Southern Observatory's Very Large Telescope and allow them to function as the world's most powerful optical interferometer. *European Southern Observatory*

Intergranular lanes form a connected network around and between granulation.

intermediate population star

A star with properties between those of extreme **Population I** stars in the galactic **spiral arms** and extreme **Population II** stars in the **galactic halo.** *Intermediate Population I stars* move in circular orbits but are distributed throughout the **galactic disk** not just in the arms. They are somewhat older than the extreme Population I stars with ages in the range 200 million to 10 billion years. Their metallicites range from 1 to 2% of their overall chemical composition. Our Sun belongs to this class, as do most of the visually observable stars in the disk. *Intermediate Population II stars* are typically at least several billion years old with a metallicity of around 0.8%. They are found in the inner parts of the halo and the **galatic bulge** (though the bulge also contains Population I stars with high metallic ties). Intermediate Population II stars outside of the halo are not confined to circular orbits or to the disk, but instead move in moderately elliptical orbits though still centered on the galactic nucleus.

International Astronomical Union (IAU)

The leading world organization for astronomers, founded in 1919. With over 8,300 individual members and 66 adhering countries, the IAU plays a pivotal role in promoting and coordinating worldwide cooperation in astronomy. It also serves as the internationally recognized authority for assigning designations to celestial bodies and any surface features on them. The scientific and educational activities of the IAU are organized by its 11 scientific divisions and, through them, its 40 specialized commissions covering the full spectrum of astronomy, along with its 70 working and program groups. The long-term policy of the IAU is defined by the general assembly and the focal point of its activities is the permanent IAU Secretariat at Institut d'Astrophysique in Paris, France.

International Atomic Time (Temps Atomique International, TAI)

The most accurate timekeeping system in use. It is based on the Systeme International (SI) second, which is defined as the duration of 9,192,631,770 cycles of radiation corresponding to the transition between two hyperfine levels of the ground state of cesium 133.

International Celestial Reference Frame (ICRF)

The **extragalactic reference frame** adopted in 1998 by the **International Astronomical Union**. It is defined with respect to 618 remote extragalactic radio sources (mostly quasars) whose positions have been accurately determined using **very long baseline interferometry**. Positions of stars in our Galaxy are now tied to the ICRF, as are measurements of Earth's orientation.

International Cometary Explorer (ICE)

A NASA probe that was originally known as ISEE-3 (International Sun-Earth Explorer 3) but was renamed ICE when it was reactivated and diverted, after four years service, in order to pass through the tail of Comet **Giacobini-Zinner** in June 1985 and observe **Halley's Comet** in March 1986.

International Dark-Sky Association (IDA)

An organization, formed in 1988, to combat the adverse environmental impact on dark skies by building awareness of the problem of **light pollution** and of the solutions, and to educate everyone about the value and effectiveness of quality nighttime lighting. The IDA also campaigns against radio interference that hampers work in radio astronomy.

International Ultraviolet Explorer (IUE)

The first satellite totally dedicated to **ultraviolet astronomy**. Launched in January 1978, IUE was a joint mission of NASA and the European Space Agency; it carried a 45-cm-diameter telescope and two **spectrographs** for use in the range 115 to 320 nm. It operated successfully for more than 18 years–the longest-lived astronomical satellite–until it was switched off in 1996.

interplanetary magnetic field (IMF)

Part of the Sun's magnetic field that is carried into interplanetary space by the **solar wind**. At Earth, its strength is about 5×10^{-5} gauss, about 10,000 times weaker than Earth's surface magnetic field. The interplanetary magnetic field lines are effectively frozen in the solar wind plasma; accordingly, because of the Sun's rotation, the IMF, like the solar wind, travels outward in a spiral pattern like the pattern of water sprayed from a rotating lawn sprinkler. The IMF originates in regions on the Sun where the magnetic field is open, where field lines emerging from one region don't return to a conjugate region but extend virtually indefinitely into space. The polarity, or direction, of the field in the Sun's northern hemisphere is opposite that of the field in its southern hemisphere. (The polarities reverse with each **solar cycle**.) Along the plane of the Sun's magnetic equator, the oppositely directed open field lines run parallel to one another and are separated by a thin current sheet known as the *interplanetary current sheet,* or *heliospheric current sheet.* The current sheet is tilted, because of an offset between the Sun's rotational and magnetic axes, and warped, because of a quadrupole moment in the solar magnetic field, and thus has a wavy structure as it extends into interplanetary space. Because Earth is located sometimes above and sometimes below the rotating current sheet, it experiences regular, periodic changes in the polarity of the IMF. These periods of alternating positive (away from the Sun) and negative (toward the Sun) polarity are known as *magnetic sectors.*

interplanetary medium

A tenuous mixture of **dust** and ionized gas, mostly **hydrogen**, that pervades the solar system and carries the expanding solar magnetic field. Dust is present as fragments from collisions in the **asteroid belt** and also as dust from **comets** that break up during their passage through the inner solar system. Charged particles are provided by the Sun in the form of the **solar wind**. There are also neutral gas atoms present that are destroyed close to the Sun but are replenished from the **interstellar medium**.

interstellar absorption

The **absorption** of light from stars and other background objects by gas and **dust** in interstellar space; it is greatest along the **galactic plane** where most of the gas and dust lies. As well as continuous absorption, there are **diffuse interstellar bands** and atomic **absorption lines**, such as the **H and K lines** of calcium and the **D lines** of sodium. Interstellar absorption is a component of **interstellar extinction**, which also includes the effects of scattering from dust.

interstellar extinction

The dimming of light from stars and other distant objects, especially pronounced in the **galactic plane**, due to the combined effects of **interstellar absorption** and scattering of light by **dust** particles. Interstellar extinction increases at shorter (bluer) wavelengths, resulting in **interstellar reddening**. It is least in the radio and infrared region, which makes these wavelengths suitable for seeing across large distances in the galactic plane and, in particular, for probing the nucleus of the Milky Way.

interstellar maser

See article, pages 256–257.

interstellar medium (ISM)

The material between the stars of a galaxy. The ISM of the Milky Way Galaxy consists (by mass) of about 99% gas and 1% **dust**. The gas, mostly **hydrogen**, is a complex soufflé of hot (over 1 million K) and cool (3 to 100 K) components. A pressure balance is maintained, however, because cool regions have a higher density (up to 10^{10} particles/cm^3 in some molecular clouds) while hot regions have a density well below the average density of the ISM of about 1 particle/cm^3. The source of the diversity of the ISM is the continual release of matter and energy from stars in the form of **stellar winds**, **planetary nebulae**, and **supernovae**. This creates thermal instabilities in the ISM resulting in clouds of various temperatures and composition. In **elliptical galaxies** the ISM consists primarily of hot gas (at roughly 10 million K) that extends seamlessly into a gaseous halo surrounding the galaxy.

interstellar molecule

A **molecule** in interstellar space. More than 130 different varieties have been found, ranging in complexity from molecular **hydrogen** (H_2), which is also by far the commonest molecule in space, through other familiar ones like water, hydrogen cyanide (HCN), nitrous oxide or "laughing gas" (N_2O), and ethanol (CH_3CH_2OH), to esoteric carbon chains known as cyano-polyynes, the biggest known of which is $HC_{11}N$. In addition, there is strong evidence for even larger aromatic (carbon-ringed) molecules called **polycyclic aromatic hydrocarbons**, or PAHs. Some molecules, such as HCO^+, were not even known on Earth at the time of their detection in space.

Because chemical bonds holding molecules together are sensitive to high-energy radiation and high temperatures, molecules are found in cool astronomical environments such as the dark interior of dense **interstellar clouds** and in expanding envelopes around dying **red giants** (and also in comets and planetary atmospheres). Most of the molecular material in our Galaxy and elsewhere occurs in **giant molecular clouds** (GMCs). Within these, the greatest concentration and diversity of molecules are found in pockets, known as *hot cores,* near certain recently formed luminous stars. Hot cores are very compact (fractions of a light-year), warm (a few hundred K, compared to 10 K or 20 K for the general interstellar gas), and dense (more than 10^6 hydrogen molecules per cm^3) condensations with remarkably rich millimeter-wave emission-line spectra. They also show up as powerful infrared objects. The **Orion Nebula** contains two hot cores, which, being the nearest such objects, are also the most well-studied.

Interstellar molecules are formed through a complicated network of chemical reactions *in situ,* inside the interstellar or circumstellar clouds where they are found. Crucial to their synthesis is the presence of **dust** grains. The icy surface of these grains both shields molecules from stellar ultraviolet that would otherwise disrupt the chemical bonds and provides a surface on which atoms, radicals, and molecules can congregate and interact. Laboratory studies have shown that simple molecules as well as elaborate molecules (yet to be detected in space) that may be important in the origin of life itself can be made in this way. Among the molecules of biochemical interest already detected in the **Sagittarius B2** cloud, near the center of the Galaxy, is glycolaldehyde ($C_2H_4O_2$), a simple sugar that can combine with other molecules to form the more complex sugars ribose and glucose. Ribose is a building block of nucleic acids, such as RNA and DNA, which carry the genetic code of living organisms. Observations also suggest the presence of glycine, the simplest **amino acid** (a chemical unit of proteins), in Sag B2.

interstellar polarization

The partial **polarization** of starlight produced by elongated **dust** grains aligned by magnetic fields in the **interstellar medium** (a mechanism the details of which are not fully understood). The polarization falls off toward the ultraviolet while the amount of **interstellar extinction** increases.

interstellar reddening

The apparent reddening of light from distant stars and other remote objects, especially in directions along the **galactic plane**, caused by the preferential scattering of short wavelengths (bluer light) by interstellar **dust** particles.

interstellar scintillation

An apparent twinkling of the signals from distant point-like radio sources that is due to changes in the density of the **interstellar medium** through which the signals have passed on their way to Earth.

interstellar wind

A flow of the local **interstellar medium** past the **heliosphere** (the Sun's magnetic sphere of influence) at a

(continued on page 257)

interstellar maser

A **maser** formed through the interaction between high-energy starlight and a nearby region rich in molecules; the first was discovered in 1965. The conditions needed to produce an interstellar maser tend to be found especially with star-forming regions and with late-type stars that are losing mass.

The environs of stellar nurseries contain dense pockets of molecular material rich in hydroxyl (OH), water (H_2O), and methanol (CH_3OH) molecules. Nearby luminous infant O or B stars heat their dusty envelopes with energetic photons, causing the **dust** to reemit in the **infrared**, which in turn excites the molecules into maser action. Within a cloud the maser-emitting region is no more than about 100 billion km across (about 10 times the size of Pluto's orbit). Its power is given in terms of the number of maser photons emitted per second. For typical H_2O masers in star-forming regions, such as those in the Orion molecular cloud, this is about 10^{46} per second; however, the H_2O maser associated with W49, a **Wolf-Rayet star**, puts out 10^{49} per second, making this the strongest maser source in the Galaxy. The intensity of maser emission can vary dramatically on a time scale of a year. One of the most spectacular examples is the H_2O outburst in Orion that happened in late 1979 and lasted for 8 years. The intensity of this one maser component suddenly increased by a factor of 1,000, making it temporarily the brightest H_2O maser source in the sky.

Observational evidence indicates that once a young star begins to radiate through nuclear fusion, OH masers survive only until the ionized region expands to a diameter of about 0.3 light-year, while H_2O masers last for about 100,000 years after the star switches on. OH and H_2O masers have also been found in the nuclei of **active galaxies**, such as NGC 3079 and NGC 1068. The power of these extragalactic masers are stronger (by about a million for OH masers [megamasers] and by a thousand for H_2O masers) than known maser sources in our Galaxy.

Masers are also found near **long-period variables** (LPVs), and arise when the turbulent upper photosphere of a luminous star undergoing mass loss is exposed to the radiation from below. These stars are typically class M (with 10,000 solar luminosities) and have cool surface temperatures, usually around

interstellar maser A strong water maser has been found associated with the infant star cluster, known as GGD 12-15, at the center of this image from the 2MASS project. This cluster, in turn, is part of an active star-forming region located in the Monoceros molecular cloud, about 3,300 light-years away. In addition to the maser, a compact H II region and a bipolar molecular outflow have been found in connection with the cluster—all signs of active ongoing star formation. *The Two Micron All Sky Survey (2MASS), a joint project of the University of Massachusetts and the Infrared Processing and Analysis Center/Caltech, funded by NASA and the NSF*

Molecules That Exhibit Maser Action in Celestial Objects

Molecule			
Formula	**Name**	**Frequency (GHz)**	**Characteristics***
OH	hydroxyl	1.612	O, M
		1.667	O, M
		1.720	O
H_2CO	formaldehyde	4.829	O
CH_3OH	methanol	12.178	O
SiS	silicon sulfide	18.155	C
H_2O	water	22.235	O, M
NH_3	ammonia	23.870	O
SiO	silicon oxide	43.122	M, S, O
		86.243	M, S
HCN	hydrogen cyanide	89.087	C

*O means that the maser emission is frequently found in star-forming regions; M, in M stars; S, in S stars; C, in carbon stars.

2,500 K. Since they are evolved, their photospheres contain appreciable abundances of heavier elements, including silicon and oxygen, which supply an ideal environment in which silicon dioxide (SiO) masers can form. The LPV VY Canis Majoris is a prime example of such a star. VY CMa shows a triple-peaked SiO maser line at 43 GHz. Physically, this has been interpreted as a spherically symmetric circumstellar envelope with an inner maser region at rest relative to the star, and an outer masing region expanding away from the star at around 10 km/s; this gives rise to blue- and redshifted peaks on either side of the one at rest.

Because interstellar maser emission often is very strong and arises from extremely compact regions, it can be observed with interferometric methods, such as **very long baseline interferometry**, that yield high-spatial resolution. Such observations provide detailed information on the physical conditions in the emission regions and their chemical composition, velocity, and magnetic field structure. (See table, "Molecules That Exhibit Maser Action in Celestial Objects.")

interstellar wind

(continued from page 255)

relative speed of about 20 km/s. An interstellar wind surrounds each star's sphere of influence like a river moving around protruding rocks.

intra-cluster medium (ICM)

The hot (tens of millions of K) and extremely tenuous, X-ray emitting gas that exists between galaxies in **clusters of galaxies**. In big clusters, the ICM may contain more material than all the galaxies put together (although both are outweighed 10:1 by **dark matter**). One of the puzzles in astrophysics has been to explain why this gas hasn't cooled down, because of its X-ray emission, and then condensed to form more galaxies. The solution to this now seems tied to **jet**s emitted from massive **black hole**s that lie at the center of active galaxies at the heart of many galaxy clusters. The black holes swallow up any gas coming close to them and liberate enormous amounts of energy in the process. This energy drives very narrow outflows of gas at velocities close to the speed of light, which reheat the intra-cluster gas. Effectively, the black holes act as thermostats. As hot gas in a cluster cools, it flows to the cluster center and is consumed by the black holes inside active galaxies. Some of the energy from this process drives jets into the cluster gas farther out, which heats the remaining gas and drives it away from the cluster center. As the black hole runs out of fuel, it shuts down, ready for the whole cycle to begin again.

intrinsic color index
The **color index** a star would have in the absence of **interstellar extinction**. It is assumed that all stars of the same **spectral type** and **luminosity class** have the same color index.

intrinsic variable
A **variable star** whose apparent brightness changes due to actual changes in the amount of light given off by the star, rather than to due accidents of geometry, such as those that produce an **eclipsing binary**. Intrinsic variables are far more common than **extrinsic variables**.

invariant plane
The plane in the solar system defined by the combined **angular momentum** of the planets and the Sun; it is tilted by about 1° 35′ to the plane of the ecliptic. Also known as the *invariable plane.*

inverse P Cygni profile
A **P Cygni profile** in which the **emission lines** lie on the blue side of the absorption component. It is usually interpreted to mean an infall of matter.

inward orbital migration
See **orbital migration**.

Io
The third largest and fifth closest moon of **Jupiter**, and the innermost and densest **Galilean satellite**; also known as Jupiter I. Although similar in diameter and density to our own Moon, Io is radically difference in appearance. It is the most volcanically active body in the solar system, its yellow and orange sulfur-rich surface peppered with hundreds of active volcanic calderas. Huge eruptions, accompanied by plumes 300 km high, have been observed by the **Voyager** and **Galileo** spacecraft. The energy for all this activity comes from tidal interactions between Io, **Europa**, **Ganymede**, and Jupiter. A slightly oval orbit around Jupiter and the gravitational fields of nearby Europa (which is in a 2:1 orbital resonance with Io) and Ganymede (in a 4:1 resonance), conspire to flex Io's crust and heat it in the same way that repeatedly stretching a rubber band causes it to warm up (see **tidal heating**). Some of the hot spots on Io may reach temperatures as high as 2,000 K, compared with an average surface temperature of about 130 K, and are the principal means by which Io loses its heat. Io's biggest volcano, Loki, is the most powerful volcano in the solar system and consistently gives out more heat than all of Earth's volcanoes put together. Its enormous caldera, larger than the state of Maryland, is continually flooded with lava.

However, not all of Io's mountains are volcanic. Some appear to have been formed by uplift and thrust faulting, and, in some cases, may rise to great heights; Euboea Montes, for example, tops out at an altitude of 13 km. The sheer size and steepness of these peaks argue that the material underlying them is rock and not some form of sulfur. Images of Euboea Montes suggest that it formed from the uplift of a large crustal block, which caused a landslide that left an enormous debris apron at the mountain's base. The

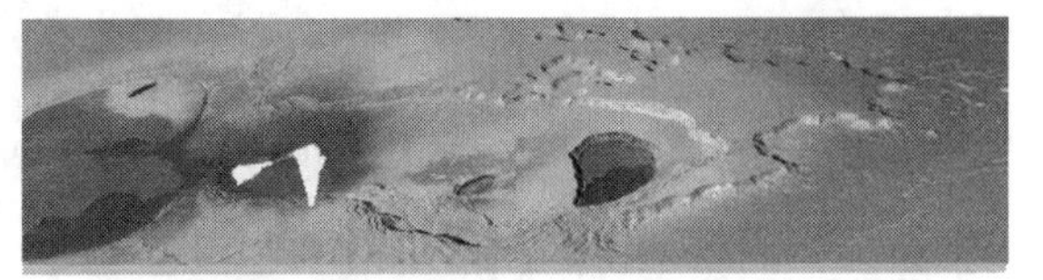

Io A fountain of lava spews above the surface of Io in this image captured by Galileo on November 25, 1999. The active lava is so hot and bright that it overloaded the spacecraft's camera and caused the white blur to the left of center. Most of the hot material is in a wavy line thought to be hot lava shooting more than 1.5 km high out of a fissure on the surface. The scalloped margins of the mesa to the right are typical of a process called sapping, which results from erosion due to a fluid escaping from the base of a cliff. On Io, this liquid is presumed to be pressurized sulfur dioxide. The image covers an area approximately 300 km by 75 km. *NASA/JPL*

Io The highest resolution image of Io obtained by the Galileo spacecraft on July 3, 1999, at a range of about 130,000 km. Most of Io's surface has pastel colors, punctuated by black, brown, green, orange, and red units near the active volcanic centers. North is to the top of the picture and the Sun illuminates the surface from almost directly behind the spacecraft. The resolution is 1.3 km per picture element. *NASA/JPL*

debris flow is 200 km wide and contains an estimated 25,000 cubic km of rock–10,000 times larger than the landslide that accompanied the Mt. St. Helens eruption in 1980. Only on the flanks of Olympus Mons on **Mars** have avalanches of this size been observed elsewhere.

As Io travels around its orbit, it cuts across Jupiter's magnetic field lines, generating an electric current. Although small compared to the tidal heating, this current may carry more than 1 trillion watts. It also strips some material away from Io, which forms a doughnut-shaped cloud of plasma in Io's orbit. This cloud, consisting mostly of oxygen and sulfur ions, emits strongly in the extreme ultraviolet but also radiates at wavelengths long enough to be detected by ground-based telescopes. In 2001, chlorine ions were found in the plasma torus. This discovery, together with the earlier detection in 1974 of sodium in neutral clouds of gas in Io's thin atmosphere, suggested to astronomers that sodium chloride–ordinary table salt–was somehow involved. Sure enough, in 2003, salt was found spectroscopically at high altitudes above Io, presumably having been ejected in gaseous form from volcanoes.

Diameter:	3,643 km
Mean distance from Jupiter:	421,800 km
Mass (Earth = 1):	0.0150
Density:	3.57 g/cm^3
Surface gravity (Earth = 1):	0.183
Escape velocity:	2.56 km/s
Visual albedo:	0.61
Temperature	
Typical subsolar:	135 K
Typical hotspot:	300 K
Rotational period:	1.769 days
Orbit	
Eccentricity:	0.0041
Inclination:	0.04°
Period:	1.769 days

ion

An atom that has either lost or gained one or more **electron**. A positive ion remains if an electron is removed, for example, by being struck by a **photon** or fast charged particle of sufficient energy to overcome the binding energy of that electron.

ionization

The loss or gain by an atom of one or more **electron**, by which process the atom becomes an **ion** and instead of being neutral, has a charge: positive if it has lost an electron, negative if it has gained one. High temperature is particularly conducive to ionization.

ionization equilibrium

The condition in a hot gas when the rate of **ionization** of a particular ion is balanced by the rate of **recombination**.

ionization front

A transition region where interstellar gas changes from a mostly neutral state to a mostly ionized state. Ionization fronts are typically found in the vicinity of hot stars and define the extent to which stellar **ultraviolet** has ionized the material around them.

ionization temperature

The temperature of a gas or **plasma** derived from the relative numbers of neutral atoms and **ions**. Specifically, it is the temperature for which the **Saha equations** would predict these relative numbers, assuming the atoms and ions are in thermodynamic equilibrium.

ionosphere

The part of the **atmosphere** that is kept partially (up to 0.1%) ionized by **ultraviolet** light and **X rays** from the Sun. It lies immediately above the **stratosphere**, roughly between altitudes 50 and 500 km, and is divided into three main layers, D, E, and F, on the basis of radio wave propagation properties. The *D layer,* below about 80 km altitude, mainly absorbs **radio waves**. So-called *sudden ionospheric disturbances* are due to enhancements of the daytime D layer caused by **solar flares**. The *E layer,* between 80 and 120 km, is reflective to shortwave radio and so can be used to bounce signals between distant stations on the ground; however, day-night variations in electron density result in marked variations in reflectivity. The *F layer,* upward of 120 altitude, is also reflective and divides during the day into the *F1* and *F2* regions. The F2 region has the greater electron density, which peaks at midday at an altitude of 250 to 300 km. The F1 region has a smaller peak in electron density, which forms at around 170 km in the daytime. Above the F region is a zone of exponentially decreasing density known as the *topside ionosphere* that extends to an altitude of a few thousand kilometers and, at mid-latitudes, feeds into the **plasmasphere.**

Iota Orionis (ι Ori)

The brightest star in **Orion's Sword**. Its Arabic name, Na'ir al Saif, means simply "the Bright One of the Sword." ι Ori is actually a complex multiple dominated by a 15-M_{sun} **O star**. At distances of 50″ and 11″, respectively, lie an eleventh magnitude A or F star and a seventh magnitude **B star**, with true separations of at least 4,400 and 20,000 AU and orbital periods at least 75,000 and 700,000 years. In addition, the O star has a close B-type companion in a very stretched-out 29-day orbit that varies from 0.8 AU to 0.11 AU from the primary. Colliding **stellar winds** from the two giants generate powerful X rays. ι Ori also has connections with two stars that lie well away from it in our sky: fourth magnitude Mu Columbae, 26° to the south, and AE Aurigae, 40° to the north. These two, almost identical in spectral type to ι Ori, are racing away from each other in opposite directions with velocities of about 100 km/s. They are **runaway stars** and were apparently thrown out of the region of the **Trapezium** Cluster some 2.6 million years ago. One theory is that they used to be part of a pair of tightly knit binary stars. In a close encounter between the two binaries, two of the four stars were ejected, while the remaining two–the close O-B pair of ι Ori–remained, more or less at the site of the violent event, locked in an eccentric embrace.

Visual magnitude:	2.75
Absolute magnitude:	–5.30
Spectral type:	O9III
Luminosity:	12,600 L_{sun}
Distance:	1,325 light-years

IP Pegasi (IP Peg)

A **U Geminorum star** (dwarf nova) that is also an **eclipsing binary** with an orbital period of 3.80 hours. IP Peg is one of the few eclipsing U Gem systems for which the inclination of the orbital plane is large enough for the companion star to eclipse the **white dwarf** primary, the **accretion disk**, *and* the **hot spot** (the place where the gas stream hits the outer edge of the accretion disk). The study of its deep eclipses has provided valuable information on the evolution of accretion disks and the mechanism of outbursts in **cataclysmic variables**.

IRAS-Araki-Alcock, Comet (C/1983 H1)

A **long-period comet** that made an unusually close passage of Earth of just 0.031 AU (4.6 million km) on May 11, 1983. Because of this near approach–bettered only by Comet **Lexell** over the past two centuries–it became quite bright (second magnitude) for a few days even though it was actually a small comet. It was first detected by the **Infrared Astronomy Satellite** on April 25 and found independently by the Japanese amateur astronomer Genich Araki (1954–) and George Alcock (1912–2000) on May 3.

Perihelion:	0.99 AU (on May 21, 1983)
Eccentricity:	0.990
Inclination:	73.3°
Period:	about 1,000 years

Iris (minor planet 7)

One of the brightest **asteroids** as seen from Earth with a maximum visual magnitude of 8.4, outshone only by Vesta, Ceres, and Pallas. An s-class occupant of the main belt with a diameter of 208 km, it was discovered by John **Hind** in 1847. Semimajor axis 2.385 AU, perihelion 1.836 AU, aphelion 2.935 AU, eccentricity 0.231, inclination 5.52°, period 3.68 years.

iron (Fe)

A metallic element, atomic number 26. Created mostly by type Ia **supernovae**, with an additional contribution from type Ib, Ic, and II supernovae, it is, from the point of view of its nucleus, the most stable element. It is also the most abundant element in the metallic cores of the inner planets.

iron meteorite

A **meteorite** composed mainly of iron (Fe) and nickel (Ni) in the form of two nickel-iron alloys, **kamacite** and **taenite**. Due to their metallic makeup and extraordinary weight, iron meteorites are easy to tell from ordinary rocks, even for a layperson. Because they rarely break up in the air and suffer much less from the effects of **ablation** during their passage through the atmosphere, they are usually much larger than stony or stony-iron meteorites. All known iron meteorites together have a mass of more than 500 tons–about 89% of the entire mass of all known meteorites. Yet they are comparatively rare, accounting for just 5.7% of witnessed falls.

There are two ways to classify iron meteorites. The older, structural method is based on characteristic crystalline features that show up when the meteorites are sec-

iron meteorite An iron meteorite found at Derrick Peak, Antarctica. *NASA/JPL*

Relations between Structural and Chemical Groups of Iron Meteorites

Structural Class	Symbol	Kamacite (mm)	Nickel (%)	Related Chemical Groups
Hexahedrites	H	>50	4.5–6.5	IIAB, IIG
Coarsest octahedrites	Ogg	3.3–50	6.5–7.2	IIAB, IIG
Coarse octahedrites	Og	1.3–3.3	6.5–8.5	IAB, IC, IIE, IIIAB, IIIE
Medium octahedrites	Om	0.5–1.3	7.4–10	IAB, IID, IIE, IIIAB, IIIF
Fine octahedrites	Of	0.2–0.5	7.8–13	IID, IIICD, IIIF, IVA
Finest octahedrites	Off	<0.2	7.8–13	IIC, IIICD
Plessitic octahedrites	Opl	<0.2, spindles	9.2–18	IIC, IIF
Ataxites	D	–	>16	IIF, IVB

tioned, etched, and polished. This results in three subdivisions: **hexahedrites** (with 4 to 6% Ni), **octahedrites** (the commonest type, with 6 to 12% Ni), and **ataxites** (with more than 12% Ni). The newer chemical method is far more precise but depends on sophisticated instruments, including electron microprobes and X-ray spectroscopes, to look at the quantities of trace elements such as germanium, gallium, or iridium present. The concentrations of the trace elements are plotted against the overall nickel content on logarithmic scales to resolve well-defined chemical clusters, each representing a distinct chemical group. Fourteen groups, labeled by Roman numbers and letters, such as IAB, have been recognized so far. It is believed that the iron meteorites of each chemical group share the same origin and formed on a common parent body. Iron meteorites come mostly from the cores of small differentiated asteroids that were disrupted by devastating impacts shortly after their formation. (See tables, "Relations between Structural and Chemical Groups of Iron Meteorites," and "Largest Iron Meteorites.")

irradiance

Also known as *radiant flux density*, the total energy at all wavelengths falling on a point on a surface from all directions above the surface in a given time. It is measured in watts per square meter (W/m^2). See **illuminance**.

irradiation

(1) An optical effect of contrast that makes bright objects viewed against a dark background appear to be larger than they really are. (2) Exposure to an intense flux of fast neutrons or to ionizing radiation. The outer layers of a **supernova** are irradiated by neutrons produced as the

Largest Iron Meteorites

Meteorite	Country	Found	Structural Class	Group	Mass (kg)
Hoba	Namibia	1920	ataxite	IVB	60,000
Campo del Cielo	Argentina	1990	octahedrite	IAB	37,000
Cape York (Ahnighito)	Greenland	1894	octahedrite	IIIAB	31,000
Armanty	China	1898	octahedrite	IIIE	23,500
Bacubirito	Mexico	1863	octahedrite	UNG	22,000
Cape York (Agpalilik)	Greenland	1963	octahedrite	IIIAB	20,000
Mbosi	Tanzania	1930	octahedrite	UNG	16,000
Campo del Cielo	Argentina	1570	octahedrite	IAB	15,000
Williamette	USA	1902	octahedrite	IIIAB	14,900
Chupaderos	Mexico	1854	octahedrite	IIIAB	14,100
Mundrabilla	Australia	1911	octahedrite	IIICD	12,000
Morito	Mexico	1600	octahedrite	IIIAB	11,000

core collapses to form a **neutron star**; these neutrons react with atomic nuclei to form heavier elements.

irregular galaxy

A galaxy, with a poorly defined structure, that falls outside the categories of **disk galaxy** or **elliptical galaxy**. Two main types are recognized in the Hubble classification scheme (see **galaxy classification**). Type Irr I galaxies are less massive (10^8 to 10^{10} M_{sun}) than large ellipticals or spirals and often have a high gas content and show evidence of extensive star formation. Type Irr II galaxies make up a catchall group that may, in some cases, include interacting and merging systems.

irregular moon

A satellite whose orbit is very noncircular and/or very inclined with respect to the equatorial plane of its planet. Irregular moons also tend to be very distant from their primary. Examples have been found around all four giant planets in the solar system. Families of irregular moons were probably formed by the collision between former moons and passing comets or asteroids.

irregular variable

A **variable star** with no detectable period in its brightness variations. There are two, very different main types, irregular **eruptive variables** and irregular **pulsating variables**, neither of which is particularly well understood.

Irregular eruptive variables are divided into three categories. Group I variables are split into subgroups IA (spectral types O to A) and IB (spectral types F through M). IN (irregular nebulous) variables, indigenous to star-forming regions, may vary by several magnitudes with rapid changes of up to 1 magnitude in 1 to 10 days, are similarly divided by spectral type into subgroups INA and INB, but with the addition of another subgroup, INT, for **T Tauri stars**, or INT(YY) for **YY Orionis stars**. The third category of eruptive irregulars is the IS stars, which show rapid variations of 0.5 to 1 magnitude in a few hours or days; again, these come in subgroups ISA and ISB.

Irregular pulsating giants or supergiants, all of late spectral types (K, M, C, or S), are classed as type L–LB for giants and LC for supergiants. How many of these are actually **semiregular variables** that simply need more study, remains unclear.

isothermal process

A process during which energy enters or leaves a system in such a way that a constant temperature is maintained. Astrophysical examples include the descent of a **protostar** down the **Hayashi track** and the collapse of an evolved star to become a **white dwarf**. Compare with **adiabatic process**.

isotope

Two nuclei with the same number of **protons** but different numbers of **neutrons** are said to represent the same **element**, but different isotopes. For example, helium-3, with two protons and one neutron in each nucleus, and helium-4, with two protons and two neutrons, are two different isotopes of helium.

isotropy

Having the same value of some physical property (e.g., density) when measured in any direction.

Izar (Epsilon Bootis, ε Boo)

The second brightest star in **Boötes**. Its name comes from an Arabic phrase meaning the "girdle" or "belt." Izar is one of the finest double stars in the sky, consisting of a magnitude 2.35 bright orange giant only 2.9′ from a fifth magnitude white main sequence **A star**. The color contrast is so striking that Friedrich **Struve**, who discovered Izar's dual nature in 1829, was inspired to name the pair Pulcherrima, for "most beautiful." The orange primary (surface temperature 4,500 K, luminosity 400 L_{sun}, mass 4 M_{sun}, radius 33 R_{sun}) and the white secondary (surface temperature 8,700 K, luminosity 27 L_{sun}, mass 2 M_{sun}, radius 2 R_{sun}) orbit each other at a mean distance of 185 AU and with a period of more than 1,000 years.

Visual magnitude:	2.35
Absolute magnitude:	−1.69
Spectral type:	K0II-III + A2V
Distance:	210 light-years

J

James Clerk Maxwell Telescope (JCMT)

The world's largest radio telescope designed specifically to work at submillimeter wavelengths; it covers wavelengths between 0.3 and 2 mm. The 15-m primary dish is made of 276 aluminum panels, each of which is adjustable to keep the surface accurately configured. The JCMT is located near the summit of Mauna Kea, Hawaii, and is operated on behalf of the three partner countries (United Kingdom, Canada, and the Netherlands) by the Joint Astronomy Centre (JAC) in Hilo, Hawaii, which also operates the **United Kingdom Infrared Telescope**.

jansky (Jy)

The unit of **flux density** used in radio astronomy; it was named in 1973 by the International Astronomical Union in honor of Karl **Jansky**. 1 Jy = 10^{-26} watt/hertz/m^2. Pitifully small though this seems, it measures extraordinarily small quantities. Were you to use all the energy ever collected from cosmic radio sources to heat a single cup of coffee, you would still have a cold drink.

Jansky, Karl Guthe (1905–1950)

An American engineer who first identified **radio waves** from beyond the solar system. As a research engineer at the Bell Telephone Company, he was asked to track down and identify the various types of interference from which radio telephony and reception were suffering. The company was particularly concerned with interference at wavelengths of around 15 m (then used for ship-to-shore radio communications), only some of which could be explained by the presence of thunderstorms, nearby electrical equipment, or aircraft. By building a high quality receiver and aerial system that was mounted on wheels and could be rotated in various directions, Jansky was able to identify a new kind of static. He noticed that the background hiss on a loudspeaker attached to the receiver and antenna system peaked every 24 hours. From overhead it seemed to move steadily with the Sun but gained on the Sun by four minutes a day–a period that correlates with the difference of apparent motion, as seen on Earth, between the Sun and the stars. Jansky realized that the source must lie beyond the solar system. By the spring of 1932 he tracked down the source to the direction of Sagittarius–the same direction in which Harlow **Shapley** and Jan **Oort** confirmed the center of the Milky Way Galaxy lay. Jansky published his results in December 1932. In the same month, Bell Laboratories issued a press release on Jansky's discovery to the *New York Times*. It made front-page news–and **radio astronomy** was born. There was no immediate follow-up to Jansky's discovery, however, and it was left to an American amateur astronomer, Grote **Reber**, to pursue the matter. Not until the end of World War II did the importance of radio astronomy begin to be recognized.

Janssen, (Pierre) Jules César (1824–1907)

A French astronomer who discovered an unknown **spectral line** in the Sun in 1868. He forwarded the data to Norman **Lockyer**, who is credited with the discovery of **helium**. Janssen was the first to note the granular appearance of the Sun, and published a monumental solar atlas in 1904 including 6,000 photographs.

Janus

The sixth moon of **Saturn**, also known as Saturn X; it is nearly co-orbital (shares its orbit) with **Epimetheus** and is located between Saturn's F- and G-rings at a mean distance of 151,472 km from the planet's center. Because Janus and Epimetheus differ in their orbital radii by only 50 km, their orbital velocities are very nearly equal and the lower, faster one gradually catches up and overtakes the other. As the moons approach each other they exchange a small amount of momentum that boosts the lower one into a higher orbit while the higher one drops to a lower orbit. This exchange of places happens about once every four years. Audouin **Dollfus** is credited with the discovery of Janus in 1966 but it is not certain whether the object he saw was Janus or Epimetheus and his observations led to a spurious orbit. (Richard Walker discovered it independently but his telegram arrived a few hours after Dollfus's.) Stephen Larson and John Fountain determined in 1978 that there are two moons at about 151,000 km from Saturn–a fact confirmed in 1980 by **Voyager** 1. Janus measures 196 × 192 × 150 km and is extensively cratered, with several craters larger than 30 km, but has few linear features.

Jeans, James Hopwood (1877–1946)

An English mathematician, physicist, and astronomer. He was a professor of applied mathematics at Princeton (1905–1909), later lectured at Cambridge (1910–1912) and Oxford (1922), and was a research associate at **Mount Wilson Observatory** (1923–1944). He devoted himself to mathematical physics and contributed to the dynamical theory of gases and the mathematical theory of electricity and magnetism. Going on to astrophysics and cosmogony, he solved the problem of the behavior of rotating masses of compressible fluids, and addressed the origins of binary stars and the collapse of interstellar clouds (see **Jeans length** and **Jeans mass**). These ideas are presented in *Problems of Cosmogony and Stellar Dynamics* (1919). With Harold Jeffreys (1891–1989) he developed the tidal hypothesis of the origin of Earth. In 1929, Jeans abandoned research and became one of the most outstanding popularizers of science and the philosophy of science. His later works include *The Universe around Us* (1929), *The Mysterious Universe* (1930), and *The Growth of Physical Science* (1947).

Jean's length

The critical size of a gas cloud, of a given temperature and density, above which the cloud must collapse under its own gravity. Below this size, the cloud's internal pressure is enough to resist collapse. The Jean's length (named after James **Jeans**) is given by

$$r = \sqrt{15kT} / 2\pi\rho Gm$$

where k is an important physical constant known as Boltzmann's constant, T is temperature, ρ is density, G is the constant of gravitation, and m is mass.

Jean's mass

The critical mass a volume of space must contain before it will collapse under the force of its own gravity.

jet

A bright, highly directional beam of matter and radiation associated with certain types of stars and **active galaxies**. Jets are observed mainly at radio but sometimes at optical and other wavelengths; in many instances, their existence is presumed, even if they can't be seen directly, because of the existence of giant lobes of material on either side of the central object. They typically come in pairs, with each jet aiming in the opposite direction to the other, and commonly occur where **accretion disks** are present. On a galactic scale, jets are found emerging from many **radio galaxies** and **quasars**. Stellar jets are seen associated particularly with **T Tauri stars** and **FU Orionis** stars.

jet A remarkable jet streaming out from the center of the giant elliptical galaxy M87. In the original color image obtained by the Hubble Space Telescope, the blue of the jet contrasts with the yellow glow from the combined light of billions of unseen stars and the yellow, pointlike globular clusters that make up this galaxy. *NASA/STScI*

Jewel Box (NGC 4755)

An **open cluster** of about 100 stars in **Crux**, also known as the *Kappa Crucis Cluster* after its most prominent member, a blue **supergiant** (visual magnitude 5.89, absolute magnitude –6.10, spectral type B5Ia). Located near Beta Crucis, it was discovered by Nicolas **Lacaille** when he was in South Africa in 1751–1752. One of the youngest clusters known, its age is estimated to be less than 10 million years. Nearby lies the **Coalsack**.

Visual magnitude:	4.2
Angular size:	10′
Distance:	8,150 light-years
Position:	R.A. 12h 53.6m, Dec. –60° 20′

Jilin meteorite

The most massive fall of **stony meteorites** on record; it happened near Jilin in Manchuria, northeastern China (44° 0′ N, 126° 0′ E) on March 8, 1976. Of the four tons of fragments of the type H5 **chondrite** recovered, one piece weighed 1.77 tons, produced an impact pit 6 m deep (only a couple of hundred meters from the nearest house), and is the largest single fragment of stony meteorite ever found.

Jodrell Bank Observatory

The **radio astronomy** observatory of the University of Manchester, located near Macclesfield, Cheshire, England. It is home to the 76-m Lovell Telescope (formerly known as the Mark IA), completed in 1957 under the direction of Bernard **Lovell**, the first director of the

Observatory; this instrument was the first giant steerable radio telescope in the world and remains the third largest of its type. Jodrell Bank is also the operations center for the MERLIN (Multi-element Radio-linked Interferometer Network), a dedicated network of radio telescopes in the United Kingdom for long-baseline **interferometry** with separations of up to 217 km.

Joint Astronomy Centre (JAC)

A facility that incorporates the 15-m **James Clerk Maxwell Telescope** and the 3.8-m **United Kingdom Infrared Telescope** on the summit of Mauna Kea along with the Centre's Hawaii headquarters in Hilo. The facility is operated by the Royal Observatory, Edinburgh on behalf of the Science and Engineering Research Council of the United Kingdom, the Nederlandse Organisatie Voor Wetenschappelijk Onderzoek, and the National Research Council of Canada.

Jovian

(1) Pertaining to the planet Jupiter. (2) Any Jupiter-like (large gas giant) planet.

Joy, Alfred Harrison (1883–1973)

An American astronomer best known for his work on stellar distances, the radial motions of stars, and variable stars. After teaching astronomy for a number of years and spending a year at Yerkes Observatory, Joy came to **Mount Wilson Observatory** in 1915. There he applied Walter **Adams**'s method of spectroscopic **parallax** to determine the distances of thousands of stars. When he retired nearly half of all published **radial velocities** of stars had been found at Mount Wilson, largely through his efforts. His measurements of the radial velocities of **Cepheid variable**s confirmed the distance and direction of the galactic center and the Sun's rate of revolution about it. He also invented the classification of **T Tauri stars** and made extensive studies of them. Although he officially retired in 1948, Joy remained active at Mount Wilson for nearly 66 years.

Julian calendar

A calendar introduced by Julius Caesar in 46 B.C. and created by Caesar's resident expert in such matters, a Greek named Sosigenes, to replace the Roman calendar. Sosigenes set up the months as we now know them and added an extra day in February every fourth (leap) year. This gives an average **year** of 365.25 days, which is pretty close to the period over which the seasons exactly repeat (the tropical year). However, there is an error of about three days every four centuries. By 1582, the calendar was about 10 days out of kilter with the real world. Pope Gregory XIII took two steps to deal with this. First, he decreed that October 4, 1582, would be followed immediately by October 15. Second, to keep the problem from recurring, he decreed that three out of every four century years (those ending in 00) would not be leap years. Thus, 1700, 1800, and 1900 were not leap years, even though they are divisible by 4, but 2000 was a leap year. The result was the **Gregorian calendar** used in most of the world today. Unfortunately, it took from 1582 to 1918 for the Julian calendar to completely die out, so astronomers need to make clear which system is being referred for dates in that interval. Civilian or astronomical dates based on the Julian calendar are said to be *Old Style dates*. Except for some religious purposes, the Julian calendar is now obsolete.

Julian Date (JD)

A timekeeping system used by astronomers to avoid the ambiguities and computational complexities of the civilian calendar; the "Julius" involved is not Julius Caesar, and this system is unrelated to the **Julian calendar**. The Julian date is the number of days that have elapsed since noon (12h **Universal Time**) on January 1, 4713 B.C. For example, January 1, 1970, is JD 2440588. Decimal fractions correspond to fractions of a day so that, for example, an observation made at 15h on June 24, 1962, is given as JD 2437840.63; the whole number part is called the *Julian date number*. The system was proposed by the French scholar Joseph Justus Scaliger (1540–1609) in 1582 and named after his father, Julius Caesar Scaliger. His choice of starting year was based on the convergence in 4713 B.C. of three calendrical cycles, one of which was the 15-year ancient Roman tax cycle of Emperor Constantine, so is of no practical consequence. For convenience, the *Modified Julian Date* (MJD) is sometimes used. This is defined as starting at midnight on November 17, 1958, so that MJD = JD – 2,400,000.5 day.

Juliet

The sixth moon of **Uranus**, also known as Uranus XI; it was discovered in 1986 by **Voyager** 2. Juliet has a diameter of 84 km and moves in a nearly circular orbit 64,360 km from (the center of) the planet.

Juno (minor planet 3)

The ninth largest **asteroid**, with a diameter of 230 × 288 km, and third to be discovered, in 1804, by the German astronomer Karl Harding (1765–1834). Juno belongs to spectral class S, has an albedo of 0.2, and rotates once on its axis in 7.2 hours. Semimajor axis 2.669 AU, perihelion 1.979 AU, aphelion 3.358 AU, eccentricity 0.258, inclination 12.97°, period 4.36 years.

Jupiter

See article, pages 266–270.

Jupiter

The largest planet in the solar system and the fifth in order from the Sun; Jupiter has 11 times the diameter of Earth and is two and half times more massive than all the other planets and satellites combined. Jupiter radiates abour 2½ times more heat than it receives from the Sun, pointing to a substantial source of internal heat, almost certainly gravitational contraction. A **gas giant**, Jupiter has an immense atmosphere that consists (by number of atoms) of about 90% hydrogen and 10% helium (75% and 25%, respectively, by mass), with traces of methane, ammonia, and other light substances, making it similar in composition to the original **solar nebula**. Jupiter probably has a metal-rock core with a mass of 10 to 30 M_{Earth}, a radius of about 1.5 R_{Earth}, and a temperature of some 30,000°C. Above the core lies an extraordinary ocean, perhaps 40,000 km deep, of liquid metallic hydrogen–an unfamiliar form of hydrogen that can exist only at pressures greater than several million times those found at Earth's surface. On top of this is a 21,000-km-thick layer composed mainly of ordinary molecular hydrogen and helium, liquid in the interior and gaseous farther out, capped by a triple-layer cloud-deck believed to consist, from top to bottom, of ammonia ice crystals, ammonium hydrosulfide, and a mixture of ice and water. Results from the atmospheric probe released by the **Galileo** spacecraft didn't exactly confirm this cloud-deck model, showing only faint indications of clouds, very little water, and higher temperatures and densities than expected. However, it now appears that the probe's entry site was unusual and may have been one of the warmest and least cloudy areas on Jupiter at the time. Temperatures generally range from around –130°C at the top of the clouds to 30°C about 70 km below.

Jupiter's upper atmosphere is striated into wide parallel bands at different latitudes because of a combination of the planet's rapid rotation and extensive convection caused by internal heat rising to the surface. Winds of more than 600 km/hr blow in opposite directions in adjacent bands, while slight chemical and temperature differences between the bands are responsible for their different shades of yellow, brown, red, and other hues; the light-colored bands are referred to as zones and the dark ones as belts. The zones are at a slightly higher altitude and about 15°C cooler than the belts. For many years it was thought that the zones, with their pale clouds, were areas of upwelling, partly because clouds on Earth form where air is rising. Because what goes up must come down, the adjacent dark belts were assumed to be where gases descended. However, in 2003, the long-standing assumption was turned on its head by observations by the Cassini spacecraft en route to Saturn. Images returned by Cassini showed that individual complex vortices in the boundary regions between the zones and belts were first seen by **Voyager**. Galileo's descent probe also found turbulence in the Jovian atmosphere, indicating that Jupiter's winds are driven largely by the planet's internal heat rather than by solar radiation as on Earth.

An enormous elliptical region in Jupiter's South Equatorial Belt, known as the Great Red Spot, has been known for more than three centuries, its discovery usually attributed to Giovanni **Cassini** or Robert **Hooke** in the seventeenth century. Measuring about 14,000 km from north to south and about 25,000 to 40,000 km from east to west (big enough to hold a couple of Earths), with a color that changes from bright to dull red and back again over timescales of

Jupiter A Voyager 2 image of Jupiter, taken from 6 million km, showing a region that extends from the equator to the southern polar latitudes in the neighborhood of the Great Red Spot. A white oval, different from the one observed in a similar position at the time of the Voyager 1 encounter, is situated south of the Spot. *NASA/JPL*

Jupiter The Great Red Spot of Jupiter seen through a methane filter that allowed the Galileo spacecraft to peer more clearly through the overlying haze of the giant planet's atmosphere. This picture is a mosaic of six images, taken on June 26, 1996, that have been map-projected to a uniform grid of latitude and longitude. *NASA/JPL*

years, it is thought to be a hurricane-like disturbance caused and maintained by the **Coriolis effect**. Other similar but smaller and less long-lived spots have been known for decades. Infrared observations and the direction of its rotation indicate that the Spot is a high-pressure region whose cloud tops are significantly higher and colder than the surrounding regions. Ultraviolet images of Jupiter's north polar region reveal a swirling dark oval of high-altitude haze similar in size to the Great Red Spot. Intense lightning and powerful aurorae are other features of the Jovian atmosphere.

Measuring longitude values on Jupiter is made difficult by the fact that the planet rotates more rapidly near the equator than it does at the poles. So three systems are used. Jupiter System I is used for features within about 10° of Jupiter's equator, where a full rotation takes about 9h 50.5m. Jupiter System II is used for features north and south of this zone (such as the Great Red Spot), where a rotation takes about 9h 55.7m. There is another system, Jupiter System III, which is based on the rotation of Jupiter's interior; it is used for radio observations, and is not particularly useful for visual observers. This rotation time of 9h 55.5m probably reflects the rate at which the solid core of Jupiter rotates, far below the cloud layers.

Jupiter has a powerful magnetic field of about 4 gauss (the magnetic axis inclined 15° to the rotational axis and about 0.1 Jupiter's radius from the center of the planet) and an immense **magnetosphere** that extends several million kilometers in a Sunward direction and more than 650 million km away from the Sun–past the orbit of Saturn! The Galileo atmospheric probe discovered a new intense radiation belt between Jupiter's ring and the uppermost atmospheric layers that is about 10 times as strong as Earth's **Van Allen belts** and contains high-energy helium ions of unknown origin. In July 1994, Comet **Shoemaker-Levy 9** collided with Jupiter with spectacular results.

Jupiter Facts

Equatorial diameter:	142,980 km
Oblateness:	0.065
Mass (Earth = 1):	318.8
Mean density:	1.33 g/cm3
Surface gravity (Earth = 1):	2.69
Temperature (cloud tops):	120 K
Mean albedo:	0.52
Rotational period	
Equator:	9h 50m
Pole:	9h 55m
Orbit	
Mean distance from Sun:	778,328,000 km (5.203 AU)
Eccentricity:	0.048
Inclination to ecliptic:	3°.1
Period:	11.86 years

Moons of Jupiter

The 61 known moons of Jupiter (as of mid-2003) fall, with one exception (**Themisto**), into four major groups. The inner group of **Metis**, **Adrastea**, **Amalthea**, and **Thebe** are small- to medium-sized, orbit at less than 200,000 km, and were discovered by Voyager. The **Galilean moons**–**Io**, **Europa**, **Ganymede**, and **Callisto**–have orbital radii of 400,000–2 million km and are among the largest satellites in the solar system. The third group includes **Leda**, **Himalia**, **Lysithea**, and **Elara**, which were discovered in the twentieth century but pre-Voyager, and one other moon found recently; all have diameters of less than 200 km and orbits of 11 to 13 million km with inclinations of 26–29°. The fourth group includes four moons–**Ananke**, **Carme**, **Pasiphae**, and **Sinope**–whose twentieth-century discovery predates Voyager, plus 40-odd others found recently; all have diameters of under 50 km and high-inclination, **retrograde** orbits with radii of 19.4 to 28.6 million km. It is thought that the three

Jupiter's Moons

		Orbit			
Name	Distance* (km)	Period (d)	Inclination (°)	Eccentricity	Diameter (km)
Metis	128,100	0.295	0.021	0.001	44
Adrastea	128,900	0.298	0.027	0.002	23 × 20 × 15
Amalthea	181,400	0.498	0.389	0.003	270 × 165 × 150
Thebe	221,900	0.675	1.070	0.018	110 × 90
Io	421,800	1.769	0.036	0.004	3,643
Europa	671,100	3.551	0.467	0.009	3,122
Ganymede	1,070,400	7.155	0.172	0.002	5,262
Callisto	1,882,700	16.689	0.307	0.007	4,821
Themisto	7,507,000	130.0	43.08	0.242	9
Leda	11,165,000	240.9	27.46	0.164	18
Himalia	11,461,000	250.56	27.50	0.162	184
Lysithea	11,717,000	259.22	28.30	0.112	38
Elara	11,741,000	259.65	26.63	0.217	78
S/2000 J11	12,555,000	287.0	28.30	0.248	4
S/2003 J20	17,100,000	456.5	55.1	0.295	3
S/2003 J3	18,340,000	504.0	143.7	0.241	2
S/2003 J12	19,002,480	533.3	145.8	0.376	1
S/2001 J10	19,394,000	553.1	145.8	0.143	2
S/2003 J21	20,600,000	599.0	148.0	0.208	2
S/2003 J18	20,700,000	606.3	146.5	0.119	2
S/2003 J6	20,979,000	617.3	156.1	0.157	4

Jupiter's Moons

Name	Orbit Distance* (km)	Period (d)	Inclination (°)	Eccentricity	Diameter (km)
S/2003 J16	21,000,000	595.4	148.6	0.270	2
S/2001 J7	21,017,000	620.0	148.9	0.230	3
Harpalyeke	21,105,000	623.3	148.6	0.226	4
Praxidike	21,147,000	625.3	149.0	0.230	7
S/2001 J9	21,168,000	623.0	146.0	0.281	2
S/2001 J3	21,252,000	631.9	150.7	0.212	4
Iocaste	21,269,000	631.5	149.4	0.216	5
Ananke	21,276,000	610.5	148.9	0.244	28
S/2001 J2	21,312,000	632.4	148.5	0.228	4
S/2003 J15	22,000,000	668.4	140.8	0.110	2
S/2003 J17	22,000,000	690.3	163.7	0.190	2
S/2003 J11	22,395,000	683.0	163.9	0.223	2
S/2003 J9	22,441,680	683.0	164.5	0.269	1
S/2003 J20	22,800,000	701.3	162.9	0.334	2
S/2001 J6	23,029,000	716.3	165.1	0.267	2
S/2001 J1	23,064,000	715.6	163.1	0.244	3
S/2001 J8	23,124,000	720.9	165.0	0.267	2
Chaldene	23,179,000	723.8	165.2	0.251	4
Isonoe	23,217,000	725.5	165.2	0.246	4
S/2001 J4	23,219,000	720.8	150.4	0.278	3
S/2003 J4	23,258,000	723.2	144.9	0.204	2
Erinome	23,279,000	728.3	164.9	0.266	3
Taygete	23,360,000	732.2	165.2	0.252	5
Carme	23,404,000	702.3	164.9	0.253	46
S/2001 J11	23,547,000	741.0	165.2	0.264	3
Kalyke	23,583,000	743.0	165.2	0.245	5
Pasiphae	23,624,000	708.0	151.4	0.409	58
Megaclite	23,806,000	752.8	152.8	0.421	6
S/2003 J7	23,808,000	748.8	159.4	0.405	4
S/2001 J5	23,808,000	749.1	151.0	0.312	2
Sinope	23,939,000	724.5	158.1	0.250	38
S/2003 J13	24,000,000	737.8	141.0	0.412	2
S/2003 J5	24,084,000	759.7	165.0	0.210	4
Callirrhoe	24,102,000	758.8	147.1	0.283	7
S/2001 J1	24,122,000	756.1	152.4	0.319	4
S/2003 J10	24,250,000	767.0	164.1	0.214	2
S/2003 J8	24,514,000	781.6	152.6	0.264	3
S/2003 J1	24,557,000	781.6	163.4	0.345	4
S/2003 J14	25,000,000	807.8	140.9	0.222	2
S/2003 J2	28,570,000	982.5	151.8	0.38	2

* Mean distance from the center of Jupiter.

groups of smaller moons may each have a common origin, perhaps as a larger moon or captured body that broke up into the existing moons of each group. All Jupiter's moons are tidally locked with the planet so that their rotational periods and orbital periods are the same. (See table, "Jupiter's Moons.")

Rings of Jupiter

Jupiter has a faint ring system with four main components: the *halo ring*, the *main ring*, and the two *gossamer rings;* it was first detected by Voyager 1. The main ring encompasses the orbits of the two innermost moons Adrastea and Metis, and at its inner edge merges into the halo, a broad, faint torus of material extending halfway from the main ring to Jupiter's cloud tops. Just outside the main ring are the broad and extremely faint gossamer rings, one bounded by Amalthea's orbit, the other by Thebe's. In 1996–1997 the Galileo spacecraft recorded events showing how Jupiter's rings are still being formed. Comet and meteor debris, accelerated by Jupiter's powerful gravitational field, smash into the inner four moons flinging dark-red surface material into space. Galileo took pictures of the red dust coming off Amalthea and Thebe, the two moons orbiting in the gossamer ring. The dust travels so fast that it escapes the minute gravitational fields of the tiny moons and goes into orbit. It then enters the gossamer rings and adds to the collection that has been accumulating there over billions of years. A similar process involving Adrastea and Metis is thought to supply the particles for the main and halo rings. (See table, "Jupiter's Rings.")

Jupiter's Rings

	Radius (km)		
Name	**Inner**	**Outer**	**Width (km)**
Halo ring	100,000	122,800	22,800
Main ring	122,800	129,100	6,300
Gossamer ring	129,100	250,000+	121,000+

Jupiter-crosser

A rare type of **asteroid** whose orbit crosses the orbit of Jupiter; the gravitational influence of Jupiter causes such an orbit to be very short-lived.

Jupiter-family comet

A type of **short-period comet** with a period of less than 20 years, a **semimajor axis** shorter or roughly the same as that of **Jupiter**'s orbit, and an orbit of low to moderate inclination, typically less than 40°. Jupiter-family comets are believed to come from the **Kuiper Belt**.

K

K star

A cool orange or red star of **spectral type** K. The spectra of K stars are dominated by the **H and K lines** of calcium and lines of neutral iron and titanium, with molecular bands due to cyanogen (CN) and titanium dioxide (TiO) becoming increasingly prominent at the cooler end of the range. Dwarf K stars have surface temperatures of 3,900 to 5,200 K, luminosities of 0.1 to 0.4 L_{sun}, and masses of 0.5 to 0.8 M_{sun}. Nearby examples include **Alpha Centauri** B and **Epsilon Eridani** (K4). Giant K types are typically 100 to 400 K cooler, and have luminosities of 60 to 300 L_{sun} and masses of 1.1 to 1.2 M_{sun}. Familiar examples include **Arcturus** (K1) and **Aldebaran** (K5).

kakangariite

A member of a rare group of **chondrites**, group K, named for their type specimen, the Kakangari meteorite that fell in Tamil Nadu, India, in 1890. Only three specimens are known with a total mass of less than 0.4 kg, all of which are rich in the iron sulfide, troilite, and show numerous primitive, armored **chondrules**. Unique in their chemical composition and with an oxygen-isotopic signature that distinguishes them from all other chondrites, K chondrites are thought to have originated in a small, primitive parent body that has yet to be identified.

Kakkab (Alpha Lupi, α Lup)

The brightest star in **Lupus**; its (rarely used) ancient name comes from the phrase *kakkab su-gub gud-elim* meaning "the star left of the horned bull." A hot giant **B star**, it belongs to the *Upper Centaurus-Lupus Association* (UCL), which in turn is part of the **Scorpius-Centaurus Association**. From analysis of all its members, UCL lies at an average distance of 450 light-years, well in keeping with Kakkab's individually measured distance. Like many of its spectral type, Kakkab is a **Beta Cephei star**, a variable that pulsates slightly with multiple periods. With a principal oscillation cycle of 0.2598466 days (the value is truly this well known), in which it varies between magnitudes 2.29 and 2.34, Kakkab has one of the longest periods of its class. It is also a soft X-ray source. A thirteenth magnitude star, just 2.76″ away, may be a genuine companion of Kakkab or may simply lie along the same line of sight.

Visual magnitude:	2.30
Absolute magnitude:	–3.83
Spectral type:	B1.5III
Luminosity:	18,000 L_{sun}
Temperature:	21,600 K
Mass:	10 to 11 M_{sun}
Distance:	548 light-years

kamacite

The commoner of the two nickel-iron alloys found in **iron meteorites**; the other is **taenite**. It contains 4 to 7.5% nickel, and forms large crystals that appear like broad bands or beam-like structures on the etched surface of a meteorite; its name comes from the Greek word for "beam."

Kapteyn, Jacobus Cornelius (1851–1922)

A Dutch astronomer who discovered **star streaming**, now known to be the observed effect of the rotation of the Milky Way Galaxy, and compiled the ***Cape Photographic Dürchmusterung*** of almost half a million stars from photographs taken in South Africa by the Scottish astronomer David Gill (1843–1914). Kapteyn studied at the University of Utrecht and served as professor of astronomy at the University of Groningen from 1878 to 1921. After his work on the *Dürchmusterung* he was involved with a program of photographically determining the **parallax**es of 10,000 stars, during which he discovered the high-parallax star now known as **Kapteyn's Star** and the fact that there are two preferred streams of stellar motion. In 1906 he inaugurated a plan to measure the positions, magnitudes, **spectral type**s, and **proper motion**s of stars in selected areas of the sky with a view to determining the shape and structure of the Galaxy (then thought to be the entire universe). *Kapteyn's selected areas,* as they became known, consisted of 206 areas, each about 1° × 1°, uniformly spaced at about 15° intervals over the whole sky, plus 46 others in regions of special importance such as around the galactic poles. From an analysis of the plates, Kapteyn and fellow Dutchman Pieter van Rhijn (1886–1960) came up with a model, known as *Kapteyn's Universe,* which was correct in some details. It described a lens-shaped stellar system whose density decreased away from the center. However, its estimate of the size of the Galaxy (40,000 light-years across) and, in particular, its placement of the Sun only 2,000 light-years from the center, were incorrect because of the lack of knowledge of **interstellar absorption**.

Kapteyn's Star (HD 33793)

A nearby **red dwarf** in **Pictor** that has the second largest **proper motion** (8.72″ per year) of any star. It is also the nearest halo star to the Sun (see **halo population**), the nearest star that orbits the Galaxy backward, and a **high-velocity star** (radial velocity +242 km/s). It is named after Jacobus **Kapteyn**, who discovered its large proper motion in 1897.

Visual magnitude:	8.86
Absolute magnitude:	12.78
Spectral type:	M0V
Luminosity:	0.004 L_{sun}
Distance:	12.78 light-years

Karin Cluster

A family of 13 **asteroids** that appears to have originated in a single asteroid disruption event just 5.8 million years ago; it is named after the largest member, 18-km-long (832) Karin. The relative youth and known age of the Karin Cluster promises to shed light on several important questions about asteroid geology and impact physics. Data from this breakup could be used to validate computer simulations that show the effects of large bodies colliding at high speed. The Karin cluster could also help in our understanding of space weathering. The impacts of energetic particles from the Sun, along with micrometeorite impacts, over time have changed the optical properties of asteroid surfaces. This makes it difficult to identify the kinds of asteroids that produce particular types of **stony meteorite** such as ordinary chondrites. Because objects in the Karin Cluster are young and their formation age is known, further study of their surface properties could give vital clues about the nature and rate at which space weathering alters their surface features. The known age of the Karin Cluster members also could help explain the rate at which asteroids strike one another in the main belt. Because the Karin Cluster asteroids were given blank slates 5.8 million years ago, impact craters formed since that time could be used to estimate the current crater production rate in the main belt. This information could help researchers determine surface ages of asteroids visited by spacecraft. It is also possible that some of the meteorites arriving on Earth today could be traced back to this breakup event. If a firm connection can be made between this event and some class of meteorites found on Earth, laboratory studies of these meteorites could be used to learn more about the nature of asteroids in the Karin Cluster. Results from these studies would be the next best thing to a sample return mission.

Kaus Australis (Epsilon Sagittarii, ε Sag)

The brightest star in **Sagittarius**, despite its lowly Epsilon rating; it marks the southern (Latin *australis*) extremity of the Archer's bow (Arabic *kaus*) and is also the lower right star of the Milk Dipper asterism. Kaus Australis is a giant B star with a fairly high rotational speed of 140 km/s. Deep inside it may have a helium-rich core that is shrinking and heating up to the point when helium burning can begin. A faint (magnitude 12.3) secondary component lies 36″ away.

Visual magnitude:	1.79
Absolute magnitude:	–1.45
Spectral type:	B9.5III
Surface temperature:	9,200 K
Luminosity:	375 L_{sun}
Radius:	7 R_{sun}
Mass:	4 M_{sun}
Distance:	145 light-years

Keck Observatory

At an altitude of 4,160 m on Mauna Kea, Hawaii, the site of the world's largest infrared and optical telescopes, known as Keck 1 and Keck 2. Each telescope has a 10-m primary mirror made up of 36 hexagonal segments, each of which is 1.8 m wide and weighs 400 kg. Keck 1 opened in 1993; Keck 2 opened in 1996. Both telescopes are eight stories tall. The Keck Observatory is jointly run by the California Institute of Technology, the University of California, and NASA.

Keeler, James Edward (1857–1900)

An American astrophysicist, probable discoverer of the dark narrow gap in the outer part of the A-ring of **Saturn**, and the second director of **Lick Observatory**. Keeler was (it seems accidentally) cheated of his rightful fame when the A-ring gap became known as Encke's Division. Johann **Encke** had earlier seen a broad, poor contrast feature in the A-ring (now called the Encke Minimum) which is quite different from the sharp, distinct gap that Keeler recorded on the very first night of observing with the Lick 36-in. (91-cm) refractor. On the other hand, the gap may have been seen even earlier by Francesco

De Vico (1805–1848), William **Lassell**, and William Dawes (1799–1868). In 1895 Keeler made a spectroscopic study of Saturn and its rings in order to explore its period of rotation. He found that the rings did not have a uniform rate of rotation, thus proving for the first time that they were not solid and corroborating James Clerk Maxwell's theory that the rings are composed of meteoritic particles. After 1898 Keeler's attention was focused on the study of the nebulae in William **Herschel**'s hundred-year-old catalog of which he managed to photograph half. During the course of his work he discovered many thousands of new nebulae and revealed their close relationship to stars.

Keenan's System (NGC 5216 and NGC 5218)

A fascinating pair of interacting galaxies in **Ursa Major** (R.A. 12h 30m 30s; Dec. +62° 59′), first recorded by P. C. Keenan who noticed the remarkable filament between the peculiar type spiral NGC 5216 and the globular galaxy NGC 5218. His note on this system seems to have been overlooked by most workers in the field, and the pair was subsequently rediscovered by observers at the Lick and Palomar observatories. While there hovers a mass of luminous debris around and in between the two galaxies, the most singular structures are the concentrated stringlike formation connecting the two systems and the fingerlike extension, or *countertide*, protruding from the globular cluster NGC 518 and starting on the same tangent as the interconnecting filament. Keenan's System lies about 17.3 million light-years away. See **galaxy interaction**.

Kelvin, Lord (William Thomson) (1824–1907)

A Scottish physicist, born in Ireland, who proposed the thermodynamic temperature scale (1848), now measured in **kelvin**, and deduced the *heat death of the universe* based on an extrapolation of the second law of thermodynamics (heat cannot flow spontaneously from a cooler object to a hotter one). The heat death would occur when all the cosmic thermal energy is spread out uniformly through space. Kelvin estimated the age of Earth by calculating how long it would take for an Earth-sized ball of rock to cool from its initial molten state. His value–20 million to 400 million years–was much too low because he knew nothing about the heat still being generated inside our planet by radioactive decay. He also estimated the Sun's age, based on the most efficient energy source he could imagine, which was the slow release of gravitational energy by contraction (see **Kelvin-Helmholtz contraction**). Again, he had no way of knowing that, in nuclear **fusion**, there is a vastly more potent way of generating heat and light.

Kelvin-Helmholtz contraction

The contraction of a ball of gas under gravity, accompanied by the radiation of the lost potential energy as heat. This was proposed by Lord **Kelvin** and Hermann von Helmholtz (1821–1894) as the most efficient means by which the Sun could remain hot for a long period. Unfortunately, it is not long enough. The so-called *Kelvin-Helmholtz timescale* of the Sun is only 20 million to 30 million years. Although stars are now known to shine by nuclear **fusion**, rather than gravitational collapse, Kelvin-Helmholtz contraction is still believed to be a valid description of the way infant stars behave in their pre-main-sequence evolution.

Kepler, Johannes (1571–1630)

A German astronomer and mathematician, considered one of the founders of modern astronomy. Using positional data carefully amassed by Tycho **Brahe**, Kepler formulated his famous three laws (see **Kepler's laws of planetary motion**), including the crucial realization that planetary orbits are **ellipses** not circles. Born in Weil der Stadt, southwest Germany, he studied at the University of Tübingen and, as a graduate, was tutored by Michael **Maestlin** who introduced him to the heliocentric concepts of **Copernicus**. In 1597 he published *The Cosmographic Mystery* in which (revealing his medieval mystical bent) he argued that the distances of the planets from the Sun in the Copernican system were determined by the five regular solids, if one supposed that a planet's orbit was circumscribed about one solid and inscribed in another. Except for Mercury, Kepler's construction gave surprisingly accurate results. Because of the mathematical skills shown in this volume, he was invited by Tycho Brahe to Prague to become his assistant and to calculate new orbits for the planets from Tycho's observations. When Tycho died, in 1601, Kepler was appointed his successor as Imperial Mathematician, the most prestigious job in mathematics in Europe. In Prague, Kepler published *Astronomia pars Optica* (The optical part of astronomy, 1604), in which he dealt with refraction and gave the first modern explanation of the workings of the eye; *De Stella Nova* (Concerning the new star, 1606) on the "new" star that had appeared in 1604 (see **Kepler's Star**); and *Astronomia Nova* (New astronomy, 1609), which contained his first two laws (planets move in elliptical orbits with the Sun at one focus, and a planet sweeps out equal areas in equal times). In 1610 Kepler heard about **Galileo**'s discoveries with the telescope and wrote a long letter of support, which he published as *Dissertatio cum Nuncio Sidereo* (Conversation with the sidereal messenger). Later that year, he presented his own observations of Jupiter's moons. These writings gave tremendous support to Galileo, whose discoveries were being widely doubted and denounced by church authorities. Kepler went on to provide the beginning of a theory of the telescope in his *Dioptrice* (1611), the title being a word he coined himself. A couple of years later he wrote *De Vero Anno quo Aeternus Dei Filius Humanam Naturam in Utero Benedictae Virginis Mariae Assumpsit* (Concerning the true year in which the son of God assumed a human nature in the uterus of the blessed Virgin Mary), arguing that the Christian calendar was out by five years, and that Jesus had been born in 4 B.C. (a conclusion now widely accepted). Between 1617 and 1621 he published *Epitome Astronomiae Copernicanae* (Epitome of Copernican astronomy), which became the most influential introduction to heliocentric astronomy of the time. In his *Harmonice Mundi* (Harmony of the world, 1619), he derived the heliocentric distances of the planets and their periods from considerations of musical harmony, and presented his third law, relating the periods of the planets to their mean orbital radii. His *Tabulae Rudolphinae* (Rudolphine tables, 1627), based on Tycho's observations and calculated according to the elliptical astronomy, were used into the eighteenth century. This was not quite the last of Kepler's published work, however. His *Somnium* (The dream), a precursor of the science fiction novel, appeared posthumously in 1634. The hero of the piece, a young Icelander named Duracotus, travels to the Moon with the aid of his mother, who is an accomplished witch–an arrangement not unfamiliar to Kepler since his own mother was tried, although not convicted, of witchcraft.

Kepler (spacecraft)

A NASA probe, scheduled for launch in 2006, that will search for and characterize Earth-sized **extrasolar planets** by **photometry**. Kepler's main instrument will be a 1-m aperture photometer with a 12° field of view, which will continuously and simultaneously monitor the light from 90,000 main sequence stars in a star field in Cygnus.

Keplerian orbit

An orbit involving two spherical objects and governed by gravitational forces only. Also known as a *Keplerian ellipse.*

Keplerian rotation curve

A plot of rotation speed versus distance from the center of an astronomical system. If most of the mass of the system is concentrated at the center, as in the solar system, then the speed of any orbiting body, such as a planet, is inversely proportional to the square root of its distance from the center.

Keplerian telescope

A telescope or beam expander formed from two positive elements separated by the sum of their **focal lengths**. Keplerian telescopes have a focus between the elements, making them ideal for use with spatial filters.

Kepler's equation

One of the important formulas that enables the position of a body in an elliptical orbit to be calculated at any given time from its **orbital elements**. It relates the mean **anomaly**, M, of the body to its eccentric anomaly, E, by the following:

$$M = E - e \sin E$$

where e is the **eccentricity** of the orbit.

Kepler's laws of planetary motion

The three basic laws of planetary motion, established by Johannes **Kepler**. (1) *Elliptical law:* each planet orbits the Sun in an **ellipse** with the Sun at one focus. (2) *Equal-areas law:* a line directed from the Sun to a planet (the *radius vector*) sweeps out equal areas in equal times as the planet orbits the Sun. (3) *Harmonic law:* the square of the period of a planet's orbit varies as the cube of that planet's **semimajor axis**.

Kepler's Star (SN Oph 1604)

A Type Ia **supernova** whose light reached Earth in 1604; the resulting "new" star, which attained a visual magnitude of about –2.2 in late October, was described by Johannes **Kepler** in his *De Stella Nova* (1606). The visible supernova remnant consists of a few faint filaments and knots in the **galactic halo** at a distance of 30,000 to 40,000 light-years, and 4,000 to 5,000 light-years above the galactic plane. Also known as *Kepler's Supernova.*

Khayyáam, Omar (1048–1122)

A Persian mathematician, astronomer, and poet who, on the accession as sultan of Jalal ad Din Malik Shah, was appointed astronomer royal. Other leading astronomers were also brought to the court observatory in Esfahan and, for 18 years, Khayyáam supervised and produced work of outstanding quality. During this time, Khayyáam led work on compiling astronomical tables and he also contributed to calendar reform in 1079. He measured the length of the year as 365.24219858156 days, which is incredible on two accounts: first, that anyone would have the audacity to claim this degree of accuracy (we know now that the length of the year changes in the sixth decimal place within a lifetime) and second, that it is astonishingly accurate. For comparison, the length of the year at the end of the twentieth century was 365.242190 days. Omar's full name is Ghiyath al-Din Abu'l-Fath Umar ibn Ibrahim Al-Nisaburi al-Khayyami.

kinetic energy

Energy of motion, equivalent to one-half an object's mass multiplied by its velocity squared.

kinetic temperature

The temperature of a gas defined in terms of the average **kinetic energy** of its atoms or molecules. It is given by

$$T = \frac{2}{3}\, 1/k \,\text{avg}(\tfrac{1}{2}\, mv^2)$$

where T is the kinetic temperature, k is an important physical constant known as Boltzmann's constant, m is the particle mass, and v is the particle velocity. In thermal equilibrium, values of kinetic temperature correspond well with those of other measures of temperature, such as **effective temperature**; otherwise, they may bear little resemblance.

Kirchoff, Gustav Robert (1824–1887)

A German physicist who, with the chemist Robert Bunsen (1811–1899), laid the foundations of spectral analysis. In 1859 he realized that the **Fraunhofer lines** in the Sun's spectrum were due to light from the **photosphere** being absorbed at those specific wavelengths by elements in the solar atmosphere. This opened the way for others, such as Angelo **Secchi** and William **Huggins**, to develop astronomical spectroscopy.

Kirkwood, Daniel (1814–1895)

An American astronomer best known for his discovery (1866) of radial gaps in the main **asteroid belt**, now known as **Kirkwood gaps**. Kirkwood was also the first to account for the Cassini and Encke Divisions in **Saturn**'s rings in terms of orbital resonances with the planet's larger moons, and to propose the existence of a group of **sun-grazing comets**.

Kirkwood gaps

Regions in the main **asteroid belt** that have been cleared of asteroids by the perturbing effects of **Jupiter**, named for Daniel **Kirkwood** who discovered them. The Kirkwood gaps are due to **resonances** with Jupiter's orbital period. For example, an asteroid with a **semimajor axis** of 3.3 AU makes two circuits around the Sun in the time it takes Jupiter to make one and is thus said to be in a 2:1 resonance orbit with Jupiter. Once every two orbits, Jupiter and such an asteroid would be in the same relative positions, so that the asteroid would experience a force in a fixed direction. Repeated applications of this force would eventually change the semimajor axes of asteroids in such orbits, creating gaps at that distance. Gaps occur at 4:1, 7:2, 3:1, 5:2, 7:3, and 2:1 resonances, while concentrations occur at the 3:2 (**Hilda group**), 4:3 (**Thule**), and 1:1 (**Trojan** group) resonances. The presence of secular resonances complicates the situation, particularly at the inner edges of the belt. An adequate explanation of why some resonances produce gaps and others produce concentrations has yet to be found.

Kitt Peak National Observatory (KPNO)

An optical observatory at an altitude of 2,120 m on Kitt Peak in the Quinlan Mountains 90 km southwest of Tucson, Arizona. Its main instruments are the 4-m **Mayall Telescope** and the 3.5-m **WIYN Telescope**. Also on Kitt Peak are the **McMath-Pierce Solar Telescope** and the Solar Vacuum Tower, operated by the **National Solar Observatory**, and the **Kitt Peak 12-Meter Telescope**, and various instruments belonging to the **Steward Observatory**.

Kitt Peak 12-Meter Telescope

A dish-type radio telescope used for observations in the millimeter-wave region of the spectrum and operated by the **National Radio Astronomy Observatory**.

Kleinmann-Low Nebula

A cool (less than 600 K), intense, extended **infrared** source that is the most active part of the **Orion Complex**; it lies about 1′ northwest of the **Trapezium** and about 12″ south of the **Becklin-Neugebauer Object** and was discovered in 1967 by the American astronomers Douglas Kleinmann (1942–) and Frank Low (1933–). The Kleinmann-Low Nebula appears to consist of a cluster of young and forming stars embedded in a dusty **molecular cloud**. In visible light, the dust blocks much of the nebula's light, but in infrared, at wavelengths less than 20 microns, the region is bursting with activity. Hot **stellar winds**, flowing off massive young stars, permeate and heat surrounding gas, causing fingerlike intrusions. Near the center of Kleinmann-Low is IRc2, a particularly active star with a mass estimated at more than 30 M_{sun}. A water **maser** and concentrations of carbon monoxide (CO) have also been detected in association with the region.

Kleopatra (minor planet 216)

An M-class (metallic) main belt **asteroid**, discovered in 1880 by Johann **Palisa**, that was the first ever to have its shape imaged by ground-based radar: it resembles a dog bone the size of New Jersey. Kleopatra's metallic composition indicates it came from the core of a large differentiated parent body that was smashed apart. But the origin of its double lobe nature is uncertain. It may have come about through the collision of two objects that had previously been thoroughly fractured and ground into piles of loosely consolidated rubble (see **rubble-pile asteroid**).

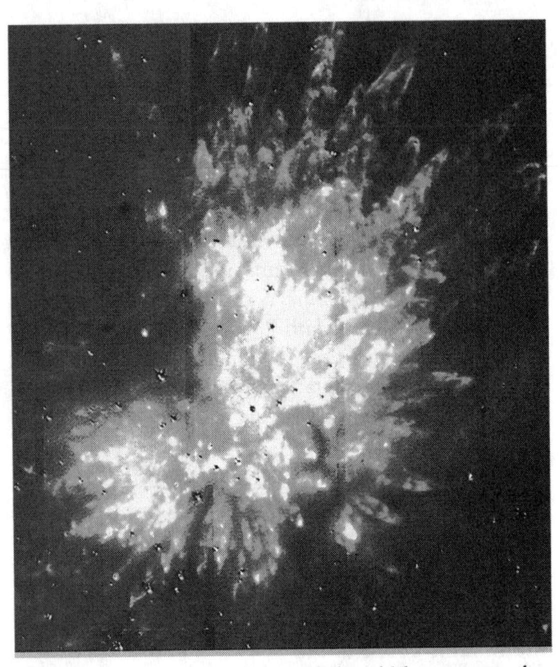

Kleinmann-Low Nebula Infrared light, which penetrates the thick dust that blocks visible light from the region, gives the Kleinmann-Low Nebula the appearance of a colossal explosion. Hot stellar winds flowing off massive young stars in the nebula permeate and heat surrounding gas, causing fingerlike intrusions. Near the center of the nebula is IRc2, a particularly active star thought to be over 30 times more massive than the Sun. *Subaru Telescope, National Astronomical Observatory of Japan*

Or, Kleopatra may once have been two separate lobes in orbit around each other with empty space between them, with subsequent impacts filling in the area between the lobes with debris. Diameter 217 × 94 km, rotational period 5.385h, semimajor axis 2.793 AU, eccentricity 0.254, inclination 13.1°, orbital period 4.67 years.

Klumpke-Roberts, Dorothea (1861–1942)

An American astronomer who became the first woman to make astronomical observations from a balloon. Born in San Francisco, she and her family moved to Europe, where she was educated and spent 50 years of her life. After earning a bachelor's degree at the Sorbonne, she joined the **Paris Observatory** and worked on photographic star charts, then became the first woman to obtain a doctorate from the Sorbonne (1893), for a thesis on **Saturn**'s rings. In 1899, when France, Germany, and Russia launched hot air balloons to observe the Leonids **meteor shower**, Klumpke was invited to ride in one. In 1901, she married a Welsh businessman and amateur astronomer, Isaac Roberts (1829–1904), 30 years her senior, and they moved to London. Three years later, she was widowed and went to live with her sister in France where she continued her astronomical work. In 1929, to commemorate the centenary of her husband's birth, she published an astronomical atlas based on his work. Her views about the importance of women's role in astronomy remained firm and she wrote and lectured on the subject in order to encourage greater female participation in the science. She finally moved back to her native San Francisco in 1930 where, despite ill health, she continued to take an active part in astronomy. Her house became a gathering place for scientists, artists, and musicians.

Kochab (Beta Ursae Minoris, β UMi)

The second brightest star in **Ursa Minor**, marking the top and front of the Little Dipper; its Arabic name (also given as Kocab) is obscure and may simply mean "star." Together with the other bowl star, Pherkad (Gamma UMi), they make the Guardians of the Pole. **Polaris** is only a temporary **pole star** that will get closer to the pole in the next century and then will begin to shift away. About the year 1100 B.C., Kochab made a reasonably close pass to the pole, and there are old references to it being called "Polaris." Kochab has left the main sequence, and is an evolving orange giant probably at the stage of core **helium-burning**. It is also classified as a mild **barium star**.

Visual magnitude:	2.07
Absolute magnitude:	–0.88
Spectral type:	K4IIIBa0.3
Surface temperature:	4,000 K
Luminosity:	500 L_{sun}
Radius:	50 R_{sun}
Distance:	126 light-years

Kohoutek, Comet (C/1973 E1)

A **long-period comet** best-remembered by the public for being such a letdown. It was discovered near Jupiter's orbit, on March 18, 1973, by the Czech astronomer Lubos Kohoutek (1935–), and, based on its unusual brightness at that distance, was predicted to become a splendid naked-eye sight at perihelion. However, although it failed to brighten as much as advertised, it was intensively studied. Its best showing in the night sky was after perihelion when, although it had dimmed to fourth magnitude, it sported a tail up to 25° long together with an **antitail**. Its orbit was found to be hyperbolic, meaning that Kohoutek is on an escape course from the solar system, never to be seen again. Visual magnitude –3 (at perihelion), perihelion 0.14 AU (December 28, 1973), inclination 14.3°.

Kordylewski Clouds
Extremely faint patches of nebulosity, first reported by the Polish astronomer Kazimierz Kordylewski (1903–1981) in the 1950s and hypothesized to be dust that has collected at the stable Earth-Moon **Lagrangian points**. Although confirmed by some observers, the existence and nature of Kordylewski Clouds remains controversial. There is some evidence to suggest that they consist of fine lunar material ejected from the Moon by impacts and that they may be variable, acquiring material on the one hand and losing it through dynamical ejection, on the other. Dust ejected by the Clouds could eventually reach Earth and appear as sporadic meteors.

Koronis family
A **Hirayama family** of **asteroid**s in the outer part of the main asteroid belt orbiting at mean distances of 2.82 to 2.95 AU from the Sun. Mainly of spectral class S (siliceous), with similar colors and albedos, these objects are all believed to have originated in the breakup of a single, highly homogenous parent body, some 100 km in diameter. The family is named after the 36-km-wide, S-class (158) Koronis, the first member to be found, in 1876 by the German astronomer Victor Knorre (1840–1919); semimajor axis 2.871 AU, perihelion 2.72 AU, aphelion 3.02 AU, inclination 1.0°, period 4.86 years. The largest members are (208) Lacrimosa and (167) Urda, with diameters of 48 and 44 km, respectively, and **Ida**.

Kramers' law
A formula, derived in 1923 by the Dutch physicist Henrik Kramers (1894–1952), for the **opacity** of material inside a star. It states that, at high temperature, stellar opacity is proportional to the density divided by the temperature to the power of 3.5. Opacity that follows Kramers' law is mostly due to **free-free transitions** of electrons and is the main source of opacity inside stars below about 1 M_{sun}. In more massive stars, the rule is only applicable to the upper layers, *electron scattering opacity* being more important at greater depths.

KREEP
Lunar **basalt**ic rock, rich in the radioactive elements uranium and thorium; it formed at the base of the **Moon**'s crust and is found in the lunar highlands. The acronym comes from K for potassium, REE for rare earth elements, an P for phosphorus.

Krüger 60
A nearby binary star, 13.07 light-years away in **Cepheus**, consisting of two **red dwarfs** orbiting with a period of 44.5 years. Krüger 60B is also a **flare star**. (See table, "The Krüger 60 Components.")

The Krüger 60 Components

	Krüger 60A	Krüger 60B
Visual magnitude	9.7	11.2
Absolute magnitude	11.9	13.3
Spectral type	M3V	M4V

K-T boundary
The junction between the Cretaceous and the Tertiary periods, about 65 million years ago, defined by a **mass extinction** that saw the demise of the majority of all life on Earth, including the last of the dinosaurs. The cause of this extinction is now widely believed to be the impact of an asteroid or comet that excavated the **Chicxulub Basin**. Evidence for this event was first found by the American physicist and Nobel laureate Luis Alvarez (1911–1988) and his son, Walter (1940–), a geologist. The father and son team discovered a layer of iridium-enriched clay at the K-T boundary at various places around the world–clear evidence of extraterrestrial debris–and announced the results of their work in 1979.

Kuiper, Gerard Peter (1905–1973)
A Dutch-born American astronomer (his name rhymes with "viper") best known for his pre-space-age observations of the planets. His spectroscopic studies led to the discovery of the atmosphere of **Titan** (1944) and features, afterward known as *Kuiper bands*, in the spectra of Uranus and Neptune, due to **methane**. He discovered Saturn's moon **Miranda** (1948) and Neptune's moon **Nereid** (1949). Born in Harenkarspel, the Netherlands, Kuiper

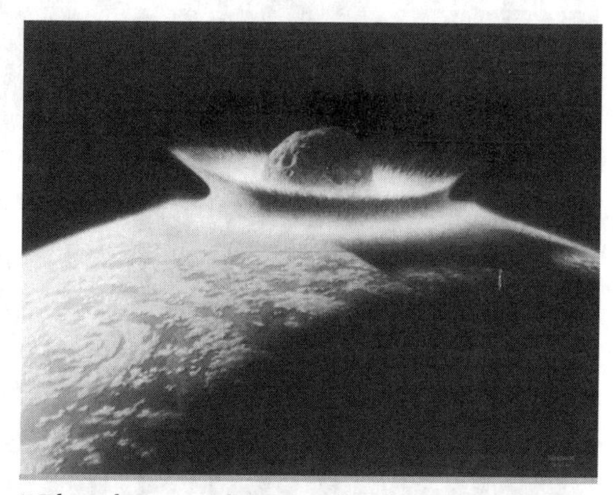

K-T boundary An artist's conception of the event thought to have happened at the K-T boundary–the impact of a giant asteroid. *NASA*

immigrated to the United States in 1933 and joined the staff of the **Yerkes Observatory**. He served as director of Yerkes from 1947 to 1949, and again from 1957 to 1960. From 1960 until his death he held similar positions at the Lunar and Planetary Laboratory at the University of Arizona, and played a vital role in the United States's space probe program during the late 1960s and early 1970s. In recognition of his work, the International Astronomical Union named a ray crater on Mercury after him.

Kuiper Belt

A region of the outer solar system populated by an estimated 10 billion to 1 trillion rock-ice bodies known as **Kuiper Belt objects** (KBOs). It stretches from about 30 AU from the Sun (Neptune's distance) to at least 150 AU and forms an inner, flattened extension of the **Oort Cloud**. The Kuiper Belt is an older structure than the more spherical outer part of the Oort Cloud. KBOs formed pretty much in their present locations–far enough out not to be tossed around by the giant planets–whereas the more distant Oort Cloud objects actually formed closer to the Sun than KBOs and were then slung into their present huge orbits by gravitational interactions with Jupiter and the other gas giants. The Kuiper Belt is thought to be the source of **short-period comets** and of **Centaurs**. It is named after Gerard **Kuiper** who predicted its existence in 1951 but is also sometimes referred to as the *Edgeworth-Kuiper Belt,* in recognition of the amateur astronomer Kenneth Edgeworth (1880–1972). Edgeworth, in his only scientific paper, published in the *Journal of the British Astronomical Association* in 1942, was the first to suggest the existence of a region of cometlike objects beyond the outer planets.

Kuiper Belt object (KBO)

An ice-and-rock body that resides in the **Kuiper Belt**; several hundred have been discovered since the first in 1992, ranging in size up to a thousand kilometers or more.

Kuiper Belt object An artist's impression of the large Kuiper Belt object 2002 LM60, dubbed "Quaoar" by its discoverers. Quaoar is about 1,300 km in diameter (half the size of Pluto) and orbits about 6.5 billion km from the Sun. *NASA/Greg Bacon (STScI)*

KBOs are important because they represent original samples of accreted material from the nebula from which the solar system formed. They fall into two distinct groups: those having eccentric orbits synchronized at a 3:2 ratio with Neptune's orbit, called *plutinos,* and those, in the vast majority, having larger, more circular orbits. According to one estimate, there are at least 35,000 KBOs greater than 100 km in diameter–hundreds of times the number (and mass) of similar sized objects in the main asteroid belt–and billions of smaller ones. The largest known is 2002 LM60, found by researchers at the California Institute of Technology and provisionally named Quaoar (KWAH-o-ar) after the creation force of the Tongva tribe who were the original inhabitants of the Los Angeles basin, where the Caltech campus is located. Quaoar is about 1,300 km in diameter (bigger than the largest asteroid, **Ceres**) and orbits the Sun every 288 years at a mean distance of 6.5 billion km. In second place is 2001 KX76, with an estimated diameter of 1,270 km, followed by 900-km-wide Varuna (2000 WR106). These are all exceeded in size, however, by **Pluto**, which is itself in a 3:2 plutino orbit, and together with its moon Charon, may be a renegade KBO. Occasionally the orbit of a KBO will be disturbed by the interactions of the giant planets so as to cause the object to cross the orbit of Neptune; it will then likely have a close encounter with Neptune that will send it out of the solar system altogether or, alternatively, into an orbit crossing those of the other giant planets. This probably explains the origin of the **Centaurs**, which are very like giant cometary nuclei and move in unstable orbits between Jupiter and Neptune. Occasionally a Centaur may be captured as a moon by one of the giant planets: this could be what has happened to **Triton**, which, if it were a refugee from the Kuiper Belt, would be by far the largest one known.

L

La Silla Observatory

An observatory atop the 2,400-m La Silla mountain that borders the southern extremity of the Atacama desert in Chile, about 160 km north of La Serena at 29° 15′ S, 70° 44′ W. Operated by the **European Southern Observatory** (ESO), its instruments include a 3.6-m reflector opened in 1976; the 3.5-m New Technology Telescope, opened in 1989; a 1.52-m reflector, opened in 1968; a 1-m Schmidt, opened in 1972; and 1-m photometric telescope, opened in 1968. Also at La Silla is the 15-m Swedish-ESO Submillimetre Telescope, the **Leonard Euler Telescope**, and a variety of other instruments owned by individual nations.

La Superba

The star Y Canum Venaticorum; a **semiregular variable** red **supergiant** in **Canes Venatici** with a visual magnitude range of 5.2 to 6.6 and a period of 158 days. It was named by Angelo **Secchi** for its intense red color.

Lacaille, Abbé Nicolas Louis de (1713–1762)

A French astronomer noted for his catalog of nearly 10,000 southern stars, which also included 42 nebulous objects. This catalog, called *Coelum Australe Stelliferum* (Southern sky catalog), was published posthumously in 1763. It introduced 14 new constellations that have since become standard: **Antlia** (Antlia Pneumatica, the Air Pump), **Caelum** (the Chisel), **Circinus** (the Drawing Compasses); **Fornax** (Fornax Chemica, the Chemists' Furnace), **Horologium** (Horologium Oscillatorium, the Pendulum Clock), **Mensa** (Mons Mensa, the Table Mountain), **Microscopium** (the Microscope), **Norma** (Norma et Regula, the Square and Level), **Octans** (Octans Hadleianus, the Octant of Hadley), **Pictor** (Equuleus Pictoris, the Painter's Easel), **Pyxis** (Pyxis Nautica, the Nautical Compass), **Reticulum** (Reticulum Rhomboidalis, the Net), **Sculptor** (Apparatus Sculptoris, the Sculptor's Studio), and **Telescopium** (the Telescope).

Lacerta (abbr. Lac; gen. Lacertae)

The Lizard; a small northern **constellation** that lies on the edge of the Milky Way just south of the midpoint of a line drawn from **Deneb** in Cygnus to Schedir in Cassiopeia. Its brightest star is Alpha Lac (visual magnitude 3.76, absolute magnitude 1.27, spectral type A1V, distance 102 light-years). Its most famous object is BL Lac, the prototype **BL Lacertae object**. (See star chart 1.)

Lagoon Nebula (M8, NGC 6523)

A **diffuse nebula** in **Sagittarius** that is divided by a dark lane; it extends over a patch of sky 3 × 1.3 the apparent diameter of the Moon. Within the brightest part of the Lagoon is a figure-eight feature known as the *Hourglass Nebula,* discovered by John **Herschel** and associated with a number of hot young stars, including Herschel 36 (magnitude 9.5, spectral type O7). Close to this feature is the brightest star associated with the Lagoon Nebula, 9 Sagittarii (magnitude 5.97, spectral type O5), which contributes much of the **ultraviolet** radiation that causes the nebula to glow. One of the most remarkable features of the Lagoon is the presence of dark **globules**, each about 10,000 AU across, that are thought to be collapsing protostellar clouds. Some of the more conspicuous globules are identified in Barnard's catalog of dark nebulae. As so often with diffuse nebulae, a cluster of young stars that formed from the nebula's material was discovered first–in this case the **open cluster** NGC 6530 in the eastern half of M8. This was found by John **Flamsteed** in about 1680, and again seen by Jean de **Cheseaux** in 1746, before Guillaume **Le Gentil** found the nebula in 1747. When Charles **Messier** cataloged this object on May 23, 1764, he also primarily described the cluster and mentioned the nebula separately as surrounding the star 9 Sagittarii; nevertheless, it is the nebula that is now generally regarded as Messier 8.

Visual magnitude:	6.0
Angular size:	90′ × 40′
Actual size:	140 × 60 light-years
Distance:	5,200 light-years
Position:	R.A. 18h 3.8m, Dec. –24° 23′

Lagrange, Joseph Louis de (1736–1813)

A French mathematician and astronomer who made important contributions to **celestial mechanics**. Born in Turin, Italy, and originally named Giuseppe Luigi Lagrangia, he moved to Paris and became a French citizen. He studied the **three-body problem** for Earth, the Sun, and the Moon (1764) and the movement of Jupiter's

La Silla Observatory In this aerial photo of the La Silla Observatory site, the large dome in the background houses the 3.5-m New Technology Telescope. *European Southern Observatory*

satellites (1766), and in 1772 found the special-case solutions to this problem that are now known as **Lagrangian points**.

Lagrangian point

A point in the vicinity of two massive bodies (such as Earth and the Moon) where each others' respective gravities balance. There are five such points, labeled L_1 through L_5. L_1, L_2, and L_3 lie along the centerline between the centers of mass between the two bodies; L_1 is on the inward side of the secondary, L_2 is on the outward side of the secondary, and L_3 is on the outward side of the primary. L_4 and L_5, the so-called **Trojan** points, lie along the orbit of the secondary around the primary, 60° ahead of and behind the secondary, respectively. L_1 through L_3 are points of *unstable* equilibrium; any disturbance will move a test particle there out of the Lagrange point. L_4 and L_5 are points of *stable* equilibrium, provided that the mass of the secondary is less than about 1/25.96 the mass of the primary. These points are stable because centrifugal pseudo-forces work against gravity to cancel it out. Named after Joseph **Lagrange**.

L'Aigle meteorite shower

A **meteorite shower** of more than 3,000 fragments that rained down on the town of L'Aigle in Normandy, France, 70 km west of Paris, in the early afternoon of April 26, 1803. It proved to be a turning point in the understanding of **meteorites** and their origins. Until this time, the idea that rocks came from space seemed fantastic, and even witnessed meteorite falls were treated with skepticism. But, upon hearing of the extraordinary events at L'Aigle, the French Academy of Sciences sent Jean-Baptise **Biot** to investigate. His passionate paper describing how these

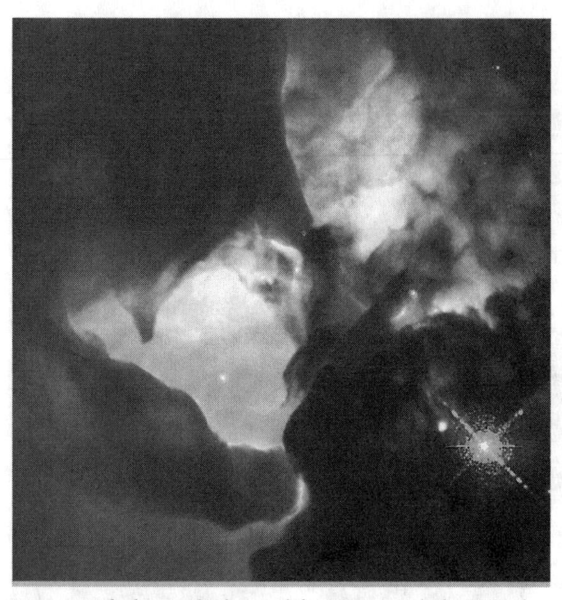

Lagoon Nebula In the heart of the Lagoon Nebula (M8), the Hubble Space Telescope spies a pair of one-half-light-year-long interstellar twisters–funnel-like masses of obscuring matter. *NASA/ESA (A. Caulet)*

stones must undoubtedly be of extraterrestrial origin effectively gave birth to the science of meteoritics.

Laing-Garrington Effect

In a **DRAGN** (a double radio source associated with an **active galactic nucleus**) in which there is a one-sided **jet**, or twin jets in which one is distinctly brighter than the other, the lobe containing the (brighter) jet is almost always less depolarized than the lobe on the other (counter-jet) side. This effect was discovered by the British astronomers Robert Laing and Simon Garrington in 1988 and can be explained neatly by unified schemes of DRAGNs. The asymmetry between the two jets is caused by **relativistic beaming**, so the brighter jet is coming toward us and hence is on the nearer side. The two lobes are embedded in the interstellar medium or halo of the host galaxy, and there is more of this material in front of the farther lobe than the nearer one. As the radio waves from the lobes travel through the intervening matter they are subjected to **Faraday rotation** that causes the depolarization. Hence the farther lobe, on the counter-jet side, is more depolarized.

Lalande, Joseph Jérôme Le Français de (1732–1807)

A Jesuit-trained French astronomer whose *Histoire céleste française* (French celestial history, 1801) contained a catalog of over 47,000 stars, though the observations were reputedly made by his nephew Michel Lalande (1766–1839). His written works include the widely read *Traité d'astronomie* (1764) and *Bibliographie astronomique* (1802). In 1760 Lalande became professor of astronomy in the Collège de France, holding the post for 46 years, and in 1768 was appointed director of the **Paris Observatory**. In the latter capacity he often engaged women astronomers for particular projects, among them the prediction of the exact date of the next return of **Halley's comet**. To this end, he asked a well-known amateur astronomer, Nicole-Reine Lepaute (1723–1788), to work with noted astronomer Alexis Clairaut on the problem. The sheer volume of computational work required that Lepaute work day and night for several months. She and her collaborator released their findings in September 1757, in the nick of time, since by Christmas of that year the first sightings of the comet began. Their work was published in a paper by Clairaut, who, initially, gave full credit to Mme. Lepaute's efforts. Later, sadly, Clairaut retracted his statements and took full credit for himself. Lepaute's efforts, however, were recognized by Lalande who included her in many other projects. She was not his only female collaborator. One of the first large-scale studies of lunar astronomy was undertaken at the Paris Observatory, and the chief investigator was an amateur, Mme. du Piery. Another woman, Lalande's niece by marriage, Marie-Jeanne de Lalande, lectured on astronomy and worked with Lalande in collaboration with her husband. Women soon became well known as "computers," and were eventually employed at observatories around the world, including **Harvard College Observatory**.

Lalande 21185

The sixth closest star to the Sun, a **red dwarf** located in the southeastern corner of **Ursa Major**, northwest of Alula Borealis (Nu UMa); it is about three times too faint to be seen with the naked eye. Moving perpendicular to the galactic plane at a high velocity of 47 km/s, Lalande 21185 appears to belong to the Galaxy's thick disk population and to be considerably older than the Sun.

Visual magnitude:	7.49
Absolute magnitude:	10.46
Spectral type:	M2.1Ve
Luminosity:	0.005 L_{sun}
Radius:	0.35 R_{sun}
Mass:	0.3 M_{sun}
Distance	8.32 light-years

Lambda Boötis star (λ Boo star)

A rare type of **A star** with weak metallic lines (showing a deficiency in heavy elements) and unusually slow rota-

new galaxy interaction Two spiral galaxies, in the constellation Canis Major, are engaged in a close pas de deux. Strong tidal forces from NGC 2207, the larger, more massive galaxy to the left, have distorted IC 2163, the smaller system to the right, flinging out stars and gas into streamers stretching out 100,000 light-years toward the right edge of the image. IC 2163 is swinging past NGC 2207 in a counterclockwise direction, having made its closest approach 40 million years ago. However, IC 2163 doesn't have enough energy to escape from the gravitational pull of NGC 2207, and is destined to fall back and swing past the larger galaxy again in the future. Trapped in their mutual orbit, these two galaxies will continue to distort and disrupt each other until, eventually, billions of years from now, they will merge. *NASA/STScI/AURA*

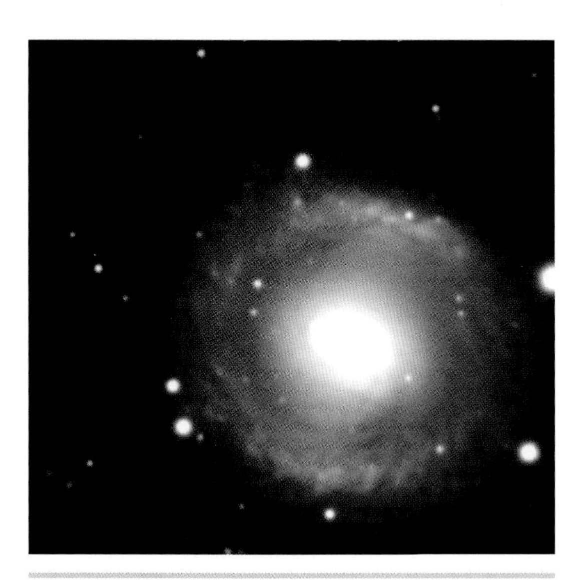

The spiral galaxy ESO 269-57, about 150 million light-years away in the southern constellation Centaurus, as seen by the Very Large Telescope. A tightly bound object of type *Sar,* it shows a bright compact nucleus (overexposed here) and a surrounding ring in which a starburst—a frenzy of new star formation—is taking place. *European Southern Observatory*

Ant Nebula A pair of glowing lobes surrounding a dying star suggest the head and thorax of a common garden ant. However, Hubble Space Telescope's view of this extraordinary planetary nebula, known formally as Menzel 3, reveals a more extended structure that challenges theorists to explain the underlying ejection process. *NASA/ESA/STScI*

Crab Nebula Multiple observations by the Chandra X-ray Observatory and the Hubble Space Telescope are combined in this image to reveal the extraordinary activity around the Crab pulsar, a fast-spinning neutron star the size of Manhattan. Bright wisps move outward at half the speed of light to form an expanding ring that is visible in both X-ray and visible light. These wisps appear to originate with a shock wave that shows up as an inner X-ray ring. This ring consists of about two dozen knots that form, brighten and fade, and occasionally undergo outbursts that give rise to expanding clouds of particles. Another dramatic feature is a turbulent jet that lies perpendicular to the inner and outer rings. *NASA*

Bubble Nebula An expanding shell of glowing gas surrounding a hot, massive star in our Galaxy. The shell is shaped by strong stellar winds of material and radiation produced by the bright star at the left, which is 10 to 20 times more massive than our Sun. *NASA/STScI/AURA*

A color infrared view by the Very Large Telescope of the star-forming region known as RCW38. This stellar nursery lies about 5,000 light-years away and is hard to study at optical wavelengths because of heavy obscuration by clouds of gas and dust. Near-infrared wavelengths, however, penetrate the fog and give a view of the embedded stars and the structure of the interstellar clouds that surround them. The diffuse radiation is a mixture of starlight scattered by the dust and gas in the area, and atomic and molecular hydrogen line emission. *European Southern Observatory*

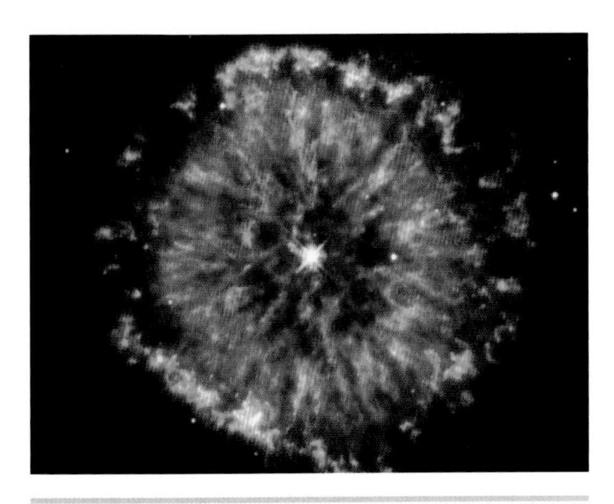

The eye-like planetary nebula NGC 6751 in Aquila, photographed by the Hubble Space Telescope. The nebula shows several remarkable and poorly understood features. Blue regions mark the hottest glowing gas, which forms a roughly circular ring around the central stellar remnant. Orange and red indicate cooler gas, which tends to lie in long streamers pointing away from the central star, and in a surrounding, tattered-looking ring at the outer edge of the nebula. The origin of these cooler clouds within the nebula is uncertain, but the streamers are clear evidence that their shapes are affected by radiation and stellar winds from the hot star at the center. *NASA/P. Harrington and K.J. Borkowski (University of Maryland)*

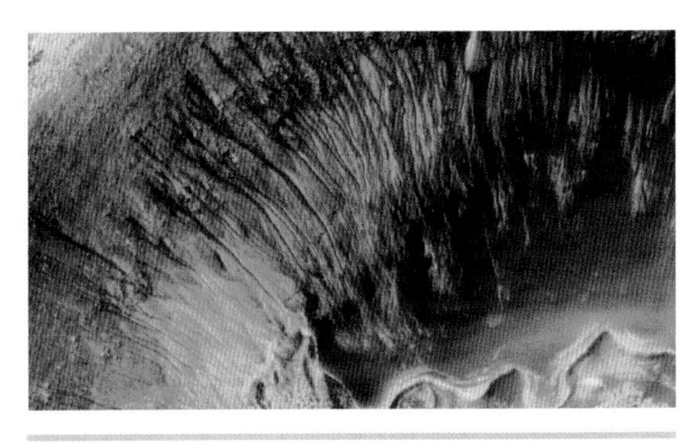

The north wall of a 7-km-wide crater on Mars (near 41.1° S, 159.8° W) showing many narrow gullies thought to have been formed by running water and debris flows. Material carried with the water created lobed and fingerlike deposits at the base of the crater wall where it intersects the floor (bottom center). Many of the fingerlike deposits have small channels indicating that a liquid—most likely water—flowed in these areas. Hundreds of individual water and debris flow events might have occurred to create the scene shown here. The individual deposits at the ends of channels in this mosaic of images from Mars Global Surveyor were used to estimate the minimum amount of water that might be involved in each flow event. For a flow containing only 10% water, these estimates suggest that at least 2.5 million liters of water are involved in each event—enough to fill about 7 community-sized swimming pools. *NASA/JPL/MSSS*

Eagle Nebula Nicknamed "the Pillars of Creation," these eerie, dark, columnlike structures are actually tubes of cool interstellar hydrogen gas and dust that serve as incubators for new stars. The pillars protrude from the interior wall of a dark molecular cloud like stalagmites from the floor of a cavern. *Jeff Hester and Paul Scowen (Arizona State University), and NASA*

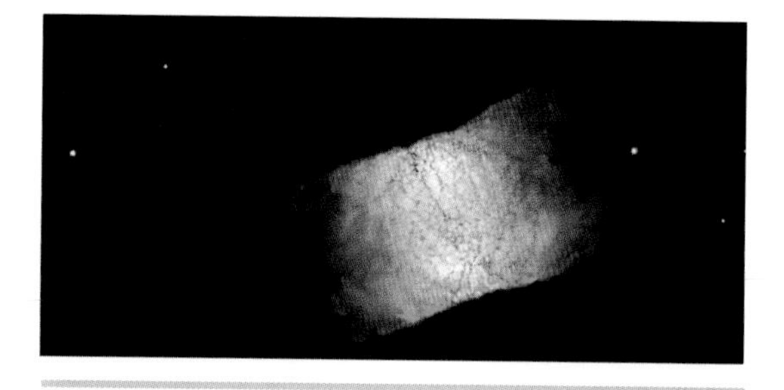

The Retina Nebula, IC 4406: a donut-shaped planetary nebula seen side-on, instead of face-on (as in the case of the famous Ring Nebula) when the central hole would be visible. The donut of material confines the intense radiation coming from the remnant of the dying star. Gas on the inside of the donut is ionized by light from the central star and glows. Light from oxygen atoms appears blue in this image, hydrogen shows as green and nitrogen as red. An interesting feature of IC 4406 is the irregular lattice of dark lanes that crisscross the center. These lanes are about 160 AU wide and lie at the boundary between the hot glowing gas that produces the visual light and the neutral gas seen with radio telescopes. *NASA/STScI/AURA*

galaxy cluster The cluster of galaxies known as 1ES 0657-55 in the southern constellation Carina. Members of the cluster are mostly the yellow fuzzy objects in this image taken by the 3.6-m New Technology Telescope at La Silla. 1ES 0657-55 is a source of strong and very hot X-ray emission and has an asymmetric galaxy distribution, pointing to a large mass and recent formation. Notice the narrow arc to the upper right of the picture, which is the gravitationally distorted image of a much more distant galaxy caught in its formation stage, when the universe was only about 2 billion years old. *European Southern Observatory*

giant molecular cloud A complex of bright and dark nebulae known as RCW108 in the OB association Ara OB1, a star-forming region in the southern constellation Ara. The Ara OB1 association contains many bright, massive stars and lies about 4,000 light-years from the Sun; the part shown in this image covers an area about 40 light-years across and includes most of RCW108. RCW108 is a molecular cloud that is in the process of being destroyed by intense ultraviolet radiation from massive, hot stars in the nearby stellar cluster NGC 6193, seen to the left. Most of this radiation comes from the bright object near the center of the image, which is a binary system composed of two O stars. The small bright patch with several stars near the darkest part of the nebulosity, to the right, is the infrared source IRAS 16362-4845, which marks a site where a small cluster of stars is being formed at present. *European Southern Observatory*

galaxy interaction The spectacular barred spiral galaxy NGC 6872 and its smaller companion, IC 4970, with which it is interacting. (The bright object to the lower right of the galaxies is a star in the Milky Way.) The upper left spiral arm of NGC 6872 is significantly disturbed and is populated by a host of bluish objects—star-forming regions that have been triggered into action by the recent passage of IC 4970 through the arm. This interesting system lies in the southern constellation Pavo at a distance of almost 300 million light-years. From tip to tip it extends across 750,000 light-years, making it one of the largest known barred spirals. *European Southern Observatory*

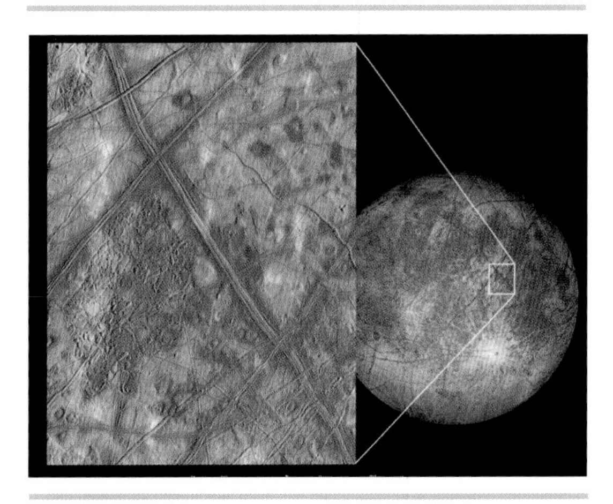

The image on the left shows a region of Europa's crust made up of blocks that are thought to have broken apart and then shifted into new positions. These features are the best geologic evidence to date that Europa had a subsurface ocean at some time in its past. Combined with the geologic data, the presence of a magnetic field encourages the view that an ocean is present today. In this false color image, reddish-brown areas represent non-ice material resulting from geologic activity. White areas are rays of material ejected during the formation of the 25-km-diameter impact crater Pwyll (see global view). Icy plains are shown in blue tones to distinguish possibly coarse-grained ice (dark blue) from fine-grained ice (light blue). Long, dark lines are ridges and fractures in the crust, some of which are more than 3,000 km long. *NASA/JPL*

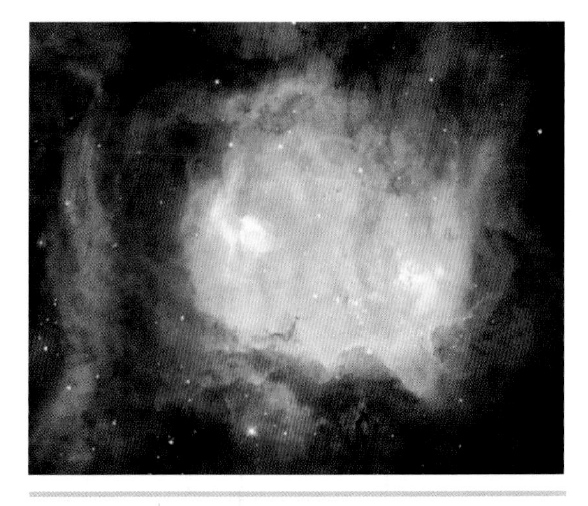

Ghost Head Nebula One of a chain of star-forming regions lying south of the Tarantula Nebula in the Large Magellanic Cloud, the Ghost Nebula has two bright regions (the eyes of the ghost), named A1 (left) and A2 (right), which are very hot, glowing blobs of hydrogen and oxygen. The bubble in A1 is produced by the intense radiation and powerful stellar wind from a single massive star. A2 has a more complex appearance due to the presence of more dust, and it contains several hidden, massive stars. The massive stars in A1 and A2 must have formed within the last 10,000 years since their natal gas shrouds are not yet disrupted by the powerful radiation of the infant suns. *ESA/NASA/Mohammad Heydari-Malayeri (Observatoire de Paris, France)*

globule What appear to be holes in a field of pinkish nebulosity are, in fact, globules—dense, compact, opaque dust clouds—silhouetted against nearby bright stars in the busy star-forming region IC 2944. They are known as Thackeray's Globules after the astronomer A. D. Thackeray who discovered them in 1950. *NASA/STScI/AURA*

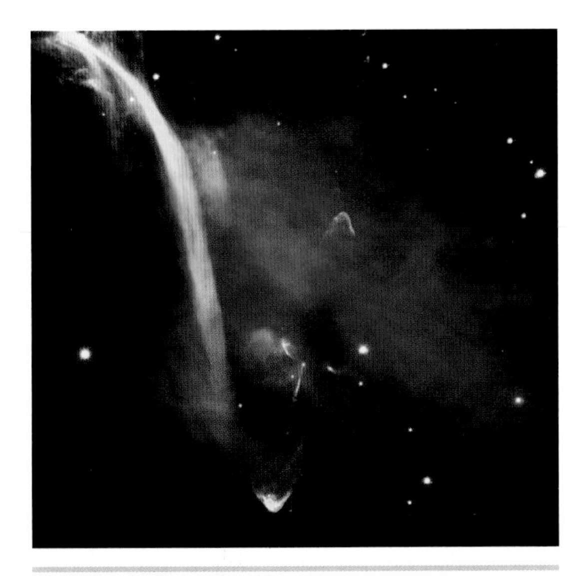

Herbig-Haro Object The young stellar object known as Herbig-Haro 34 (HH-34), which is now in the protostar stage of evolution. Its complex appearance includes two oppositely directed jets, one of which appears as a red line, that ram into the surrounding interstellar matter and produce two arc-shaped shock fronts at equal distances from the central source. These jets suggest that the star experiences episodic outbursts when large chunks of material fall onto it from a surrounding disk. HH-34 lies about 1,500 light-years away, near the Orion Nebula. The spectacular "waterfall" feature to the upper left has yet to be explained. *European Southern Observatory*

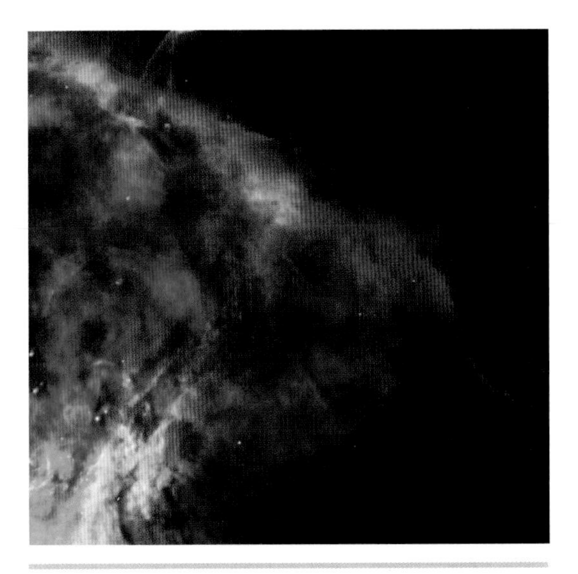

A Hubble Space Telescope image of a region of the Orion Nebula, one of the nearest stellar nurseries to the Sun. The great plume of gas in the lower left in this picture is the result of the ejection of material from a recently formed star. The brightest portions of the image are "hills" on the surface of the nebula, and the long bright bar is where Earth observers look along a long wall on a gaseous surface. The diagonal length of the photo is 1.6 light-years. Red light depicts emission in nitrogen; green is hydrogen and blue is oxygen. *Raghvendra Sahai and John Trauger (JPL)/NASA*

Hubble Space Telescope Space Shuttle astronauts Steven L. Smith, and John M. Grunsfeld carry out a spacewalk to service and upgrade the Hubble Space Telescope in 1999. *NASA*

An active volcanic eruption on Jupiter's moon Io seen on Feb. 22, 2000, by the Galileo spacecraft. It occurred in Tvashtar Catena, a chain of giant volcanic calderas, which was also the scene of a major eruption in November 1999. A dark, L-shaped lava flow to the left of the center marks the location of the November eruption. White and orange areas on the left side of the picture show newly erupted hot lava, visible in this false color image because of infrared emission. The two small bright spots are sites where molten rock is exposed to the surface at the toes of lava flows. The larger orange and yellow ribbon is a cooling lava flow that is more than 60 km long. *NASA/JPL*

black hole An artist's conception of matter swirling around the accretion disk of a supermassive black hole and also captured by a magnetic field. The gravity in such a region, found in the cores of some galaxies, appears to be so intense that the very fabric of space twists around the black hole, dragging magnetic field lines along with it. *NASA*

Tarantula Nebula The most active starburst region in our galactic neighborhood. Near the edge of the Tarantula Nebula, seen in the lower right corner of this image, is a cluster of brilliant, massive stars known as Hodge 301. This cluster is not the brightest, youngest, or most populous star cluster in the Tarantula, however; that honor goes to the spectacular R136 at the center of the nebula. Hodge 301 is almost 10 times older than the young cluster R136 and many of its stars have exploded as supernovae. These explosions have blasted material out into the surrounding region at high speeds. As the ejecta plow into the surrounding Tarantula Nebula, they shock and compress the gas into a multitude of sheets and filaments, seen in the upper left portion of the picture. These features are moving away from Hodge 301 at speeds of more than 300 km per second. Also present near the center of the image are small, dense gas globules and dust columns where new stars are being formed today. *NASA/STScI/AURA*

A close-up view of Saturn's C-ring (and to a lesser extent, the B-ring at top and left) compiled from three separate images taken through ultraviolet, clear, and green filters by Voyager 2. More than 60 bright and dark ringlets are evident. In general, C-ring material is very bland and gray—the color of dirty ice. Color differences between this ring and the B-ring indicate differing surface compositions for the material composing these complex structures. *NASA/ESA/John T. Clarke (University of Michigan)*

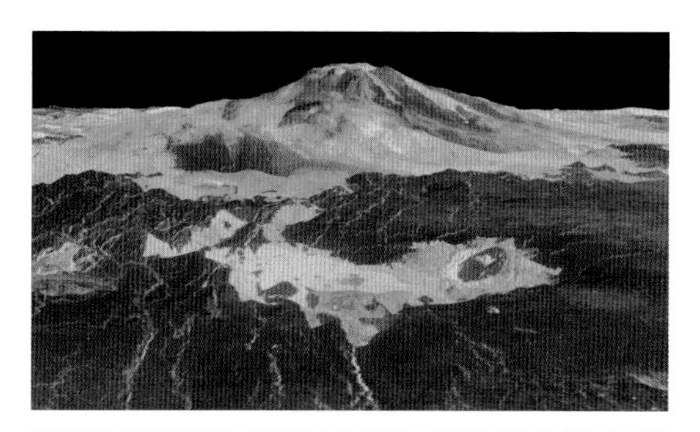

new Venus The volcano Maat Mons on Venus, shown in computer-generated three-dimensional perspective from a viewpoint 634 km north of the mountain and from a height of 3 km above the terrain. Lava flows extend for hundreds of km across the fractured plains in the foreground to the base of the volcano. Maat Mons lies at about 0.9 degrees N, 194.5 degrees E and has a peak that towers 5 km above the surrounding terrain and 8 km above the mean surface of the planet. The vertical scale in this perspective has been exaggerated tenfold. The image details are based on data from the Magellan probe and the simulated hues are derived from color photos recorded by the Soviet Venera 13 and 14 spacecraft. *NASA/JPL*

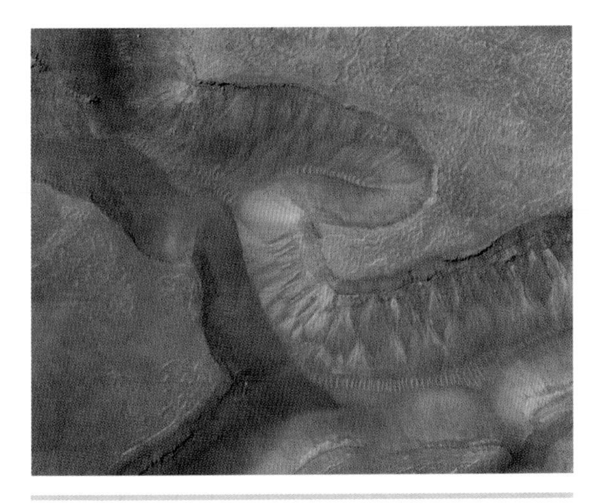

Mars gully Troughs and layered mesas in the Gorgonum Chaos region of the Martian southern hemisphere, imaged by Mars Global Surveyor on January 22, 2000. The area represented is 3 km wide by 2.6 km high. One theory is that the gullies were formed by groundwater seeping from a specific layer near the tops of trough walls, particularly on south-facing slopes (south is toward the bottom of each picture). The presence of so many gullies associated with the same layer in each mesa suggests that this layer is an aquifer. *NASA/JPL/MSSS*

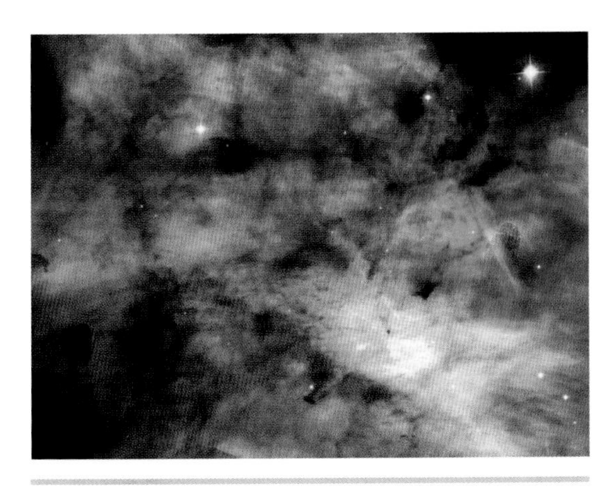

Omega Nebula The center of M17, also known as the Omega Nebula or Swan Nebula—a hotbed of newborn stars wrapped in colorful blankets of glowing gas and cradled in an enormous cold, dark hydrogen cloud. The region of the nebula shown in this picture by the Hubble Space Telescope is about 3,500 times wider than our solar system. Like its famous cousin in Orion, the Omega Nebula is illuminated by ultraviolet radiation from young, massive stars, located just beyond the upper right corner of the image. As the infant stars evaporate the surrounding cloud, they expose dense pockets of gas that may contain developing stars. One isolated pocket is seen at the center of the brightest region of the nebula and is about 10 times larger than our solar system. *NASA/ESA/Holland Ford (JHU)*

Neptune photographed by Voyager 2 in 1989 from a range of 7 million km, 4 days and 20 hours before its closest approach. The picture shows the Great Dark Spot and its companion bright smudge; on the west limb the fast moving bright feature called Scooter and the little dark spot are visible. These clouds were seen to persist for as long as Voyager's cameras could resolve them. To the north, a bright cloud band similar to the south polar streak may be seen. *NASA/JPL*

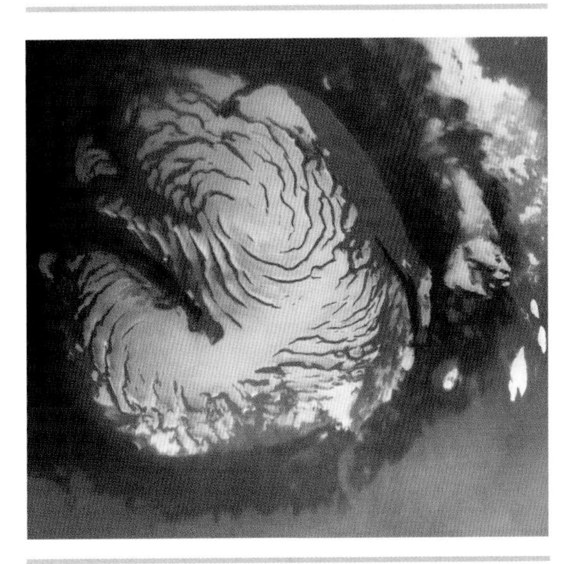

A wide angle view of the Martian north polar cap as it appeared to Mars Global Surveyor on March 13, 1999, in the early northern spring. The light-toned surfaces are residual water-ice that remains through the summer season. The nearly circular band of dark material surrounding the cap consists mainly of sand dunes formed and shaped by wind. The north polar cap is about 1,100 km across. *NASA/JPL*

An area of the sky near the Chameleon I complex of bright nebulae and hot stars in the constellation of the same name, close to the southern celestial pole. This picture was taken by the Very Large Telescope in 1999. *NASA/Bruce Balick (University of Washington), et al.*

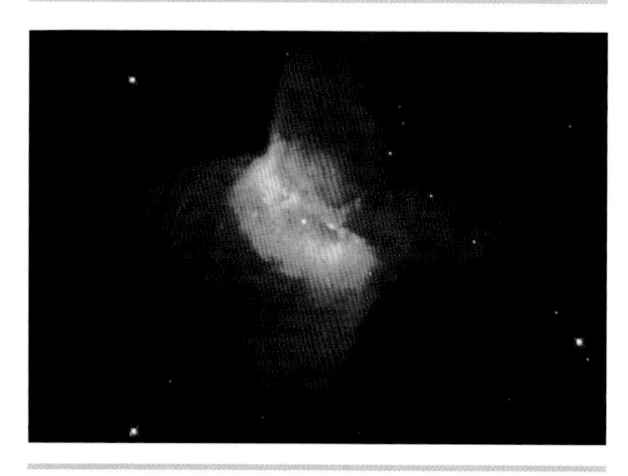

The planetary nebula NGC 2346, remarkable because its central star is known to be actually a close pair of stars, orbiting each other every 16 days. It is thought that the binary was originally more widely separated. However, when one component of the binary evolved and became a red giant, it literally swallowed its companion, which then spiralled down inside the red giant and, in the process, spewed out a ring of gas. Later, when the hot core of the red giant was exposed, it developed a fast stellar wind, which emerged perpendicular to the ring and inflated two huge bubbles. This two-stage process is believed to have resulted in the butterfly-like shape of the nebula seen today. NGC 2346 lies about 2,000 light-years away and is roughly one-third of a light-year across. *NASA*

star cluster The double cluster NGC 1850 in the Large Magellanic Cloud. This cluster is a type of object known as a young, globularlike stellar association, of which there is no counterpart in our own Galaxy. NGC 1850 consists of a main cluster, about 40 million years old, and a second, smaller one (to the right of the main cluster in the photo) that is only 4 million years old and is mostly composed of extremely hot stars. About 1,000 stars in the main cluster are thought to have exploded as supernovae during the past 20 million years. One theory holds that the birth of the younger cluster was caused by the combined effect of these explosions on the residual gas around the main cluster. Much gas still remains. While part of this may be the remnant of the parent gas cloud from which both clusters were born, the presence of filaments and of various sharp shocks supports the theory of supernova-induced star birth in the younger of the two clusters. *European Southern Observatory*

Bright nebulae near the hot binary star BAT99-49 in the Large Magellanic Cloud, captured by the Very Large Telescope in 2003. *NASA/STScI/AURA.*

tion. Lambda Boötis stars may be **pre-main-sequence objects**, or they may be main-sequence stars that formed from gas whose metal atoms had been absorbed by interstellar dust.

Lambda Eridani star (λ Eri star)

A **Be star** that shows very small but extremely regular variations in light output, with a period of 0.4 to 2 days, that may be caused by complex pulsations or by rotation and dark starspots. **Achernar** is an example.

Laplace, Pierre Simon, Marquis de (1749–1827)

A French physicist and mathematician who put the final capstone on mathematical astronomy by summarizing and extending the work of his predecessors in his five-volume *Traité de Mécanique Céleste* (Treaty on Celestial Mechanics), published from 1799 to 1825. This work was important because it translated the geometrical study of mechanics used by Isaac **Newton** to one based on calculus. In *Mécanique Céleste,* Laplace proved the dynamical stability of the solar system (with tidal friction ignored) on short timescales. Over long periods, however, this assertion has proven false because of the effects of chaos. Laplace explained the long-term variations in the orbital speeds of Jupiter and Saturn (1786), and the Moon (1787). His nebular hypothesis of the origin of the solar system (1796) is similar to that of Immanuel Kant, of which he was apparently unaware. After reading *Mécanique Céleste,* Napoleon is said to have questioned Laplace on his neglect to mention God. In contrast to Newton's view on the subject, Laplace replied: "Sir, I have no need of that hypothesis."

Large Binocular Telescope (LBT)

A future telescope that will use twin 8.4-m mirrors on a common mounting, their centers 14.4 m apart, to achieve the light-gathering capability of a single 11.8-m reflector. The LBT is being built at **Mount Graham International Observatory** and is jointly owned by the **Steward Observatory**, Arizona; Arcetri Astrophysical Observatory, Italy; and Research Corporation, a private foundation in Tucson. Delivery of the optics is expected in early 2004.

Large Magellanic Cloud (LMC)

One of the two **Magellanic Clouds** (dwarf irregular galaxies), visible in the Southern Hemisphere, that orbits the Milky Way Galaxy; it spans 8° of the sky in Dorado and Mensa. The LMC is about 30,000 light-years in diameter, has a visible mass of about one-tenth that of our own Galaxy, and, at a distance of some 180,000 light-years, was long considered to be the nearest external galaxy before losing that distinction to the **Sagittarius Dwarf Elliptical Galaxy**. It displays a noticeable bar of stars, some of which, including **S Doradus**, are extremely luminous; from the ends of the bar project a weak spiral structure. The LMC is rich in a variety of diffuse nebulae, including the spectacular **Tarantula Nebula**, planetary nebulae, open clusters, globular clusters, and so-called *blue populous clusters,* which resemble compact, young globulars and are of a type unseen in our own Galaxy. An absence of intermediate-age clusters suggests that the LMC has experienced early and late bursts of star formation. The LMC was also the site of **Supernova 1987A**–the nearest observed supernova since that recorded by Kepler in 1604. See also **Local Group**.

Large Millimeter Telescope (LMT)

A 50-m-diameter millimeter-wave-telescope–the world's largest–designed to operate mainly at wavelengths of 1 to 4 mm. The LMT is a joint project of the University of Massachusetts at Amherst and the Instituto Nacional de Astrofísica, Óptica, y Electrónica (INAOE) in Mexico, and is being built atop Sierra Negra, a volcanic peak in the Mexican state of Puebla. Telescope construction is expected to be complete in 2004.

Large Magellanic Cloud A small patch of the Large Magellanic Cloud (LMC) viewed through the Hubble Space Telescope. Over 10,000 stars can be seen in the photo, covering a region about 130 light-years wide. Also visible are sheets of glowing gas and dark patches of interstellar dust silhouetted against the stars and gas behind them. *NASA/STScI*

large-scale structure

The distribution of galaxies and other forms of matter on a distance scale of about 100 million light-years or greater. The visible large-scale structure of the universe is sponge-like, with galaxy **superclusters** arranged into enormous filaments and sheets that are separated by giant **voids** where very few if any galaxies reside. Among the nearest components of the cosmic large-scale structure are the **Local Supercluster**, the **Great Wall**, the **Great Attractor**, and the **Shapley Concentration**. In addition, observations by the Chandra X-ray Observatory have revealed part of an intergalactic web of hot gas and dark matter that is crucial in defining the cosmic landscape. The hot gas alone, which appears to lie like a fog in channels carved by rivers of gravity, is more massive than all the stars in the universe. Its detection may eventually enable astronomers to map the distribution of dark matter.

Larissa

The fifth moon of **Neptune**, also known as Neptune VII; it was discovered by the American space engineer Harold Reitsema by ground-based stellar occultation observations, although the first (and so far only) images of it were obtained in 1989 by **Voyager** 2. Larissa measures 208 × 178 km, appears to be heavily cratered, and orbits at a mean distance of 73,550 km from (the center of) the planet.

Las Campanas Observatory

An observatory at an altitude of 2,300 m on Cerro Las Campanas, 100 km northeast of La Serena, Chile, which was founded in 1971 and is owned and operated by the Carnegie Institution of Washington, D.C. It is home to the twin 6.5-m **Magellan Telescopes**, the 2.5-m Irénée du Pont Telescope, opened in 1976, and the 1-m Swope Telescope, opened in 1971. The clear, dark skies here, and at the neighboring **La Silla Observatory** and **Cerro Tololo Inter-American Observatory**, offer excellent seeing unsurpassed anywhere on Earth.

Lassell, William (1799–1880)

An English amateur astronomer who discovered Neptune's moon **Triton** (1846), Saturn's moon **Hyperion** (1848), and Uranus's moons **Ariel** and **Umbriel** (1851). Born in Bolton, Lancashire, Lassell was a successful brewer before turning to astronomy. He set up a private observatory at Starfield near Liverpool and developed an interest in techniques for building very large reflectors. In 1845 he completed a 24-in. (0.6-m) reflector, which was the first sizeable reflector on an **equatorial mount**, and used it to discover Neptune's largest moon–just 17 days after Neptune itself was found. Lassell also discovered Hyperion and the crêpe ring of Saturn, independently of William **Bond**. In 1851 he discovered Ariel and Umbriel from observations made in Malta, where the atmosphere was much clearer than in industrial England. In Malta in 1861 he built a 48-in. (122-cm) reflector and used it to observe and catalog hundreds of new nebulae.

late heavy bombardment

The period between about 3.8 and 4.0 billion years ago when the **Moon** and other objects in the solar system were pounded most heavily by wayward **asteroids**. The evidence of this pummeling is clear to see in the maria basins on the Moon and in similar structures elsewhere, such as the Caloris Basin on **Mercury** and the great craters in the southern hemisphere of **Mars**. But, until recently, although the presumption has always been that our planet suffered in a similar way, there had been no direct evidence of this, since very few surviving surface rocks date back to this period, and in those that did, such as the **Acasta Formation**, no signs of asteroid impacts had been found. Now this has changed. In 2002, British and Australian researchers announced they had found traces of an isotope of tungsten in 3.7-billion-year-old rocks from Greenland and Canada that can only be extraterrestrial.

Based on the visible impact history on other worlds, computer estimates suggest that during the 200-million-year period of the late heavy bombardment, Earth received over 22,000 craters larger than 20 km, about 40 **impact basins** larger than 1,000 km, and several continent-sized 5,000-km basins. There would have been an impact that affected global conditions every century or so. Yet amid this cosmic blitzkrieg, life managed to get started on our world. In fact, it may be that impact-generated hydrothermal systems made excellent incubators for prebiotic chemistry and the early evolution of life, an idea that is consistent with other evidence that shows life may have originated in hot water systems around or slightly before 3.85 billion years ago.

late-type galaxy

A spiral or irregular galaxy. Such systems are described as late-type because of their position at the end of the tuning-fork diagram of the Hubble classification of galaxies (see **galaxy classification**). For the same reason, Sc or Sd spirals are referred to as late-type spirals.

late-type star

A star of spectral type K, M, S, or C. The term dates from when astronomers thought that all cooler, redder stars like these were at a later stage of evolution than the hot blue stars of spectral types O and B (called *early-type stars*).

late heavy bombardment Evidence of the period of late heavy bombardment is clearly seen on the side of the Moon that always faces Earth. The dark, lava-filled basins, which help define the features of the "Man in the Moon," were created at the end of this era, 3.9 to 3.8 billion years ago. *NASA/JPL*

latitude

The angular distance north or south from the equator. See **celestial sphere**.

latitude variation

The small semicyclic apparent change in the declinations of stars due to the **Chandler wobble** and fluctuations in atmospheric refraction. Earth's polar wandering causes a maximum variation of 0.36″ (equivalent to a latitude shift of about 10 m on the ground) in a period of 432 days. Seasonal movements of air masses result in an additional variation of 0.18″ (+/–5 m) in a period of 1 year.

lava

Molten rock that erupts onto the surface of a planet or moon, from a volcano or fissure, and is hot enough to flow; also, the rock formed by solidification of this material. Lava flows are described as *pillow lava, pahoehoe,* and *a'a.* Silica-rich lava hardens before flowing far, forming a dense-texture rock of tiny crystals or glass. Basic lava, which contains less than 50% silica, flows farther before it solidifies, giving rise to coarse-grained igneous rock, such as granite or **gabbro**. In many eruptions, lava is ejected with such force that it fragments in the atmosphere, hardens while airborne, and lands to form thick layers of volcanic tuff and related **pyroclastic** rock.

lava fountain

An eruption of molten material through a fissure. Seen on certain parts of Earth, **lava** fountains have also been observed in the Tvashtar region of **Io**, reaching a height of more than 1.5 km and involving lava at a temperature of 1,000 to 1,600 K.

Le Gentil (de la Galaziere), Guillaume Joseph Hyacinthe Jean Baptiste (1725–1792)

A French astronomer who made a valiant effort to observe the **transits** of Venus in 1761 and 1769. The first transit found him stuck in the middle of the Indian Ocean, unable to make any useful observations. After spending four years in Mauritius and Madagascar, and even taking a side trip to the Philippines, Le Gentil arrived in India, built an observatory at Pondicherry, and waited for the next transit, which would occur on June 4, 1769. The weather was clear for the month prior to the transit, but clouded up on transit day, only to clear immediately after the long-awaited event. Le Gentil then contracted dysentery and remained bedridden for nine months. He booked passage home aboard a Spanish warship that was demasted in a hurricane off the Cape of Good Hope and blown off course north of the Azores before finally limping into port at Cadiz. Le Gentil crossed the Pyrenees on foot and returned to France after an absence of more than 11 years, only to learn that he had been declared dead, his estate looted, and its remains divided up among his heirs and creditors. However, Le Gentil did not give up astronomy upon coming home. In fact, he lived at the **Paris Observatory** and the observatory's records contain a complaint that Madame Le Gentil hung out diapers to dry in the observatory gardens.

leading hemisphere

The hemisphere of a moon in **synchronous rotation** that faces forward, into the direction of orbital motion.

leap second

A second sometimes inserted at the end of a calendar year to keep atomic clock time in close agreement with the rotation of Earth, which is gradually slowing.

leap year

(1) In the **Gregorian calendar**, a year lasting 366 days rather than 365, with February 29 (leap day) added as the extra day; this occurs in years whose last two digits are evenly divisible by 4; e.g., 1996. (2) A year with an extra day in any other calendar.

Leavitt, Henrietta Swan (1868–1921)

An American astronomer who discovered the **period-luminosity relation** of **Cepheid variables**. Having graduated from Radcliffe College (1892), she joined **Harvard College Observatory** in 1895 as a volunteer research assistant, receiving a permanent post in 1902. Like her colleague Annie **Cannon**, she was extremely deaf. In 1907 the director of the observatory, Edward **Pickering**, announced plans to redetermine stellar magnitudes photographically, previous estimates having been only visual. Leavitt was made head of the department of photographic **photometry** and while studying photographic plates made at Harvard's field station in Peru, she discovered, in 1912, that Cepheid variables show a simple relationship between period and luminosity. Using Leavitt's work as a springboard, first Ejnar **Hertzsprung**, then Harlow **Shapley**, and finally Walter **Baade**, were able to employ the Cepheids as a cosmic distance indicator. Leavitt also did much work on other variable stars, discovering about 2,400–roughly half of those known in her time.

Leda

The tenth moon of **Jupiter**, also known as Jupiter XIII; it was discovered in 1974 by the American astronomer Charles Kowal (1940–). Leda has a diameter of about 8 km and orbits at a mean distance of 11.17 million km from the planet.

Lemaître, Georges Édouard (1894–1966)

A Belgian cosmologist and priest who was the original proponent of what later became known as the **Big Bang** theory. Trained as a civil engineer, he served as an artillery officer with the Belgian army during World War I, then, in 1923, he entered a seminary, where he was ordained a priest. From 1923 to 1924 he visited the University of Cambridge to study solar physics and there met Arthur **Eddington**; he then spent two years at the Massachusetts Institute of Technology where he was influenced by the ideas of Edwin **Hubble** and Harlow **Shapley** regarding the likelihood of an expanding universe. In 1927 he returned to Belgium and was made professor of astrophysics at the University of Louvain. In 1933 he published his *Discussion on the Evolution of the Universe,* in which he suggested that the universe stemmed from what he called a "primeval atom." This incredibly dense initial form, he argued, contained all the material for the universe in a sphere about 30 times larger than the Sun. Its explosion sent matter flying in all directions and resulted ultimately in the expansion of the galaxies that we see today. The significance of his theory lay not so much in its affirmation of the expansion of the universe as in its presumption of an initial event to start the expansion. In 1946, Lemaître published his *Hypothesis of the Primal Atom*–the same year in which George **Gamow** and his colleagues began to develop the nuclear physics of the Big Bang.

Lemonnier, Pierre Charles (1715–1799)

A French astronomer and physicist at the Collège de France who studied the Moon and the influence of Saturn on the motion of Jupiter, determined the positions of many stars, and conducted extensive research in terrestrial magnetism and atmospheric electricity. His greatest claim to fame, however, is holding the record for the greatest number of sightings of **Uranus**–at least 10 between 1764 and 1771, including six in January 1769 alone–before the planet was officially discovered by William **Herschel**. Lemonnier's failure to recognize its true nature is explained by the fact that, at the beginning of 1769, Uranus was near its **stationary point**, so that its movement among the stars would not have been obvious.

lens

A clear optical element that produces an image by **refraction**. A **convex** lens makes light rays converge at a **focal point**, while a **concave** lens causes rays to diverge. A thinner lens has a longer **focal length** than a thicker one, and is also easier to make and suffers less from **chromatic aberration** and other distortions. In practice, however, combinations of lenses are used to overcome these problems. See **achromatic lens**.

lenticular galaxy

A galaxy with a central bulge and disk but apparently lacking spiral arms and interstellar material. Lenticular galaxies are named because of their lenslike appearance when seen edge-on. In the Hubble classification scheme (see **galaxy classification**), lenticulars are known as S0 galaxies.

Leo (abbr. Leo; gen. Leonis)

The Lion; a large, prominent **constellation**, the fifth sign of the zodiac, which dominates the northern spring sky. It lies between Virgo and Cancer and to the south of Leo Minor and Ursa Major. The Sickle asterism, formed by the stars Alpha (**Regulus**), Eta, Gamma (**Algieba**), Zeta (Adhafera), Mu, and Epsilon Leo, outlines the Lion's head. (See tables, "Leo: Stars Brighter than Magnitude 4.0," and "Leo: Other Objects of Interest." See also star chart 12.)

lenticular galaxy The barred lenticular galaxy NGC 2787, photographed by the Hubble Space Telescope in 1999. Tightly wound, almost concentric arms of dark dust encircle the bright nucleus of NGC 2787; a faint bar, which leads to its classification as a type SB0 system, is not apparent in this image. *NASA/ESA (M. Carollo)*

Leo I

(1) A large, prominent group of mostly spiral galaxies in **Leo**. It consists of two main subgroups: the *M66 Group* and the *M96 Group*. The former lies about 35 million light-years away and is centered on the interacting spirals M65 (NGC 3623), M66 (NGC 3627), and the edge-on NGC 3628, also known as the *Leo Triplet*. Not far from the M66 Group, and almost certainly physically related to it, is the much larger M96 group dominated by M96 itself, about 41 million light-years away, and including M95, M105, and NGC 3384. Fainter members of the M96 Group include NGCs 3299, 3377, 3377A, 3384, 3412, and 3489. The slightly more distant S0 or early Sa galaxy NGC 3593 is probably also a member. (2) A **dwarf spheroidal galaxy** (dE3) that, at a distance of about 820,000 light-years, is the most remote **satellite galaxy** of the Milky Way and hence also a member of the **Local Group**. Discovered in 1950, it has a diameter of somewhat over 3,000 light-years. Because it lies close to **Regulus** in the sky (making study of it difficult in the bright star's light), it is sometimes known as the *Regulus Dwarf*, and also cataloged as DDO 74, UGC 5470, and Harrington-Wilson 1.

Leo II

(1) A large collection of galaxy groups centered about 30 million light-years to one side of the **Virgo Cluster**. (2) A **dwarf**

Leo: Stars Brighter than Magnitude 4.0

Star	Magnitude		Spectral	Distance	Position	
	Visual	Absolute	Type	(light-yr)	R.A. (h m s)	Dec. (° ′ ″)
α Regulus	1.36	−0.52	B7V	77	10 08 22	+11 58 02
γ Algieba	2.01	−0.92	K1IIIbCN+G7IIICN	126	10 19 58	+19 50 30
β Denebola	2.14	1.91	A3V	36	11 49 04	+14 34 19
δ Zosma	2.56	1.32	A4V	58	11 14 06	+20 31 25
ε Asad Australis	2.97	−1.46	G0II	251	09 45 51	+23 46 27
θ Chort	3.33	−0.35	A2V	178	11 14 14	+15 25 46
ζ Adhafera	3.43	1.02	F0III	99	10 16 41	+23 25 02
η	3.48	−5.60	A0Ib	2,130	10 07 20	+16 45 45
ο Subra	3.52	0.43	A5V+F6II	135	09 41 09	+09 53 32
ρ	3.84	−7.38	B1Ib	5,720	10 32 49	+09 18 24
μ Rassalas	3.88	0.83	K0IIIbCNCaBa	133	09 52 46	+26 00 25
ι Tsze Tseang	4.00	2.08	F2IV	79	11 23 55	+10 31 45

spheroidal galaxy (dE0) that, at a distance of about 670,000 light-years, is a remote **satellite galaxy** of the Milky Way (the second farthest, after **Leo I**) and hence also a member of the **Local Group**. Discovered in 1950, it has a diameter of about 3,000 light-years. Also known as *Leo B,* it is cataloged as DDO 93, UGC 6253, and Harrington-Wilson 2.

Leo III

A **dwarf irregular galaxy** that, at a distance of about 2.3 million light-years, is a fairly nearby member of the **Local Group**. Also known as *Leo A,* it is cataloged as DDO 69 and UGC 5364.

Leo Minor (abbr. LMi; gen. Leo Minoris)

The Small Lion; a small, faint northern **constellation**. It lies between the southern border of Ursa Major (at the rear feet) and the northern border of Leo. Leo Minor has no star designated Alpha and only one that carries a Greek letter name at all–Beta–although this ranks second in brightness to 46 LMi, also known as Praecipua (visual

Leo: Other Objects of Interest

Object	Notes
Galaxies	
M65 (NGC 3623)	An edge-on type Sb spiral that forms a conspicuous triple with M66 and NGC 3628. Magnitude 9.3; angular size 10.0′ × 3.3′; distance: 35 million light-years; R.A. 11h 18.9m, Dec. +13° 05′.
M66 (NGC 3627)	A physical partner of M65 and another edge-on Sb spiral. Magnitude 9.0; angular size 8.7′ × 4.4′; R.A 11h 20.2m, Dec. +12° 59′.
M95 (NGC 3351)	A face-on type SBb barred spiral that forms a pair with M96 (see **Leo I**). Magnitude 9.7; angular size 7.4′ × 5.1′; R.A. 10h 44.0m, Dec. +11 42°.
M96 (NGC 3368)	An Sb galaxy in partnership with M95. Like all the above galaxies, it is visible with a small telescope. Magnitude 9.2; angular size 7.1′ × 5.1′; R.A. 10h 46.8, Dec. +11° 49′.
M105 (NGC 3379)	A type E1 elliptical. Magnitude 9.3; angular size 4.5′ × 4.0′; R.A. 10h 47.8m, Dec. +12° 35′.
Leo I, II, III	See separate entries.

Lepus: Stars Brighter than Magnitude 4.0

Star	Magnitude		Spectral	Distance	Position	
	Visual	Absolute	Type	(light-yr)	R.A. (h m s)	Dec. (° ′ ″)
α Arneb	2.58	–5.40	F0Ib	1,280	05 32 44	–17 49 20
β Nihal	2.81	–0.63	G5II	159	05 28 15	–20 45 35
ε	3.19	–1.02	K4IIIv	227	05 05 28	–22 22 16
μ	3.29	–0.47	B9IVpHgMn	184	05 12 56	–16 12 20
ζ	3.55	1.88	A2Vn	70	05 46 57	–14 49 20
γ	3.59	3.82	F7V	29	05 44 28	–22 26 55
η	3.71	2.82	F1V	49	05 56 24	–14 10 04
δ	3.76	1.07	G8IIIwkCN	112	05 51 19	–20 52 45

magnitude 3.79, absolute magnitude 1.41, spectral type K0III, distance 98 light-years). (See star chart 12.)

Leonard Euler Telescope

A 1.2-m Swiss reflector at the **La Silla Observatory**, operated by the University of Geneva, and named in honor of the famous Swiss mathematician. It is used specifically in conjunction with the Coralie spectrograph to conduct high-precision **radial velocity** measurements, mainly to search for large **extrasolar planets** in the Southern Hemisphere. Its first success was the discovery of a planet orbiting Gliese 86.

lepton

An elementary particle, such as an **electron** and **neutrino**, that does not interact via the strong nuclear force.

Lepus (abbr. Lep; gen. Leporis)

The Hare, prey of Orion the Hunter; a small but interesting southern **constellation**, located just south of Orion. Of the few stars making up a figure that looks like a smashed box kite, the brightest is only third magnitude. (See tables, "Lepus: Stars Brighter than Magnitude 4.0," and "Lepus: Other Objects of Interest." See also star chart 3.)

Leverrier, Urbain Jean Joseph (1811–1877)

A French mathematician who, beginning in 1838, studied the causes of **perturbation**s in the solar system. His work led to improved knowledge of the masses of the planets, the scale of the solar system, and the velocity of light. He also predicted the existence of two new planets–one of which was subsequently confirmed. In 1845 he learned from François **Arago** of certain irregularities in the movements of Uranus, which hinted at the existence of an eighth planet. Leverrier's calculated position of this perturbing body enabled Johann **Galle** to confirm it observationally, though John **Adams** had made a similar but unpublished prediction some months earlier. Leverrier, not one to avoid publicity, suggested that Uranus be renamed for Heschel, the finder, and that the new discovery be named after himself. In the event, man lost

Lepus: Other Objects of Interest

Name	Notes
Stars	
Hind's Crimson Star	R Lep. See separate entry.
Open cluster	
NGC 2017	Binoculars or a small telescope reveal five stars, ranging from sixth to tenth magnitude; of these, two are binaries that can be split by a telescope of 15-cm aperture. Magnitude 6.4; diameter 15′; R.A 5h 39.4m, Dec. –17° 51′.
Globular cluster	
M79 (NGC 1904)	A compact, interesting object for small telescopes, lying close to Herschel 3752. Magnitude 9.9; diameter 8.7′; R.A. 5h 24.5m, Dec. –24° 33′.

out to a (Roman) god and the eighth planet was called Neptune. Leverrier's other prediction, first made in 1845, was for a new innermost planet, which became known as **Vulcan**. Searches turned up nothing and we now know that the irregularities in Mercury's orbit are an effect of **general theory of relativity**.

Levy, David H. (1948–)

A Canadian amateur astronomer who, in 1993, along with Gene and Carolyn **Shoemaker**, discovered the most famous comet of recent years. Comet **Shoemaker-Levy 9**, as it became known, attracted the attention of public and professionals alike when it smashed into **Jupiter** in July 1994. Levy discovered his first comet in 1984 and has found a total of 21–8 from his own backyard and 13 in collaboration with the Shoemakers at Palomar Observatory. He is also the author of a number of popular astronomy books.

Lexell, Comet (D/1770 L1)

A **long-period comet** discovered on June 14, 1770, by Charles **Messier**, but named for the Swedish astronomer Anders Johann Lexell (1740–1784) who first calculated its orbit. Several gravitational encounters with Jupiter have changed the comet's orbit so that it no longer comes close enough to Earth to be seen and it may even have been ejected from the solar system altogether.

Libra (abbr. Lib; gen. Librae)

The Scales; a faint southern **constellation** that is the seventh sign of the **zodiac** and the only one that is not named after a living thing. It lies between Virgo to the west and north and Ophiuchus to the east. Appropriate to its name, it once held the autumnal equinox (but no longer does, thanks to **precession**). Its two brightest stars used to represent the outstretched claws of the Scorpion. (See table, "Stars Brighter than Magnitude 4.0," and star chart 15.)

libration

Any of several periodic rocking motions of an orbiting body. The various librations of the **Moon** enable, over time, about 59% of the Moon's surface to be seen from Earth. *Libration in latitude,* a north-south nodding, is caused by the tilt of the Moon's rotation axis relative to its orbital plane. The average up/down latitude wobbling is 5.13°, corresponding to the Moon's orbital inclination with respect to the **ecliptic**, though perturbations by the Sun can add a further +/– 0.9°. *Libration in longitude* arises from the difference between the Moon's varying orbital velocity and its constant rotation rate. Without any other factors it would average 6.29°, but the Sun's contribution pushes the peak oscillation in longitude to 7.75°. *Diurnal libration* is an optical rather than a physical libration, amounting to less than 1°, and stems from the fact that, because of Earth's rotation, we view the Moon from different angles at moonrise and moonset. Libration happens when an orbiting body is locked in a synchronous orbit; **Mercury** shows it, as do some other moons in the solar system.

Lick Observatory

The first mountaintop observatory, established in 1888 at an altitude of 1,283 m on Mount Hamilton, California, with financial backing from James Lick of San Francisco. It houses a 36-in. (91-cm) refractor, the second largest in the world after that at **Yerkes Observatory**. Other instruments include a 120-in. (3-m) reflector and the 36-in. (0.9-m) Crossley reflector. The observatory is run by the University of California at Santa Cruz.

Life Finder

A proposed large successor to the **Terrestrial Planet Finder** that would be capable of detecting the spectroscopic signs of life on nearby **extrasolar planets**. It is still in the planning stage and is unlikely to be launched before 2020.

Libra: Stars Brighter than Magnitude 4.0

	Magnitude		Spectral	Distance	Position	
Star	Visual	Absolute	Type	(light-yr)	R.A. (h m s)	Dec. (° ′ ″)
β Zubenelschamali	2.61	–0.85	B8V	160	15 17 00	–9 22 58
α^2 Zubenelgenubi	2.75	0.88	A3IV	77	14 50 53	–16 02 30
σ Zubenalgubi	3.25	–1.51	M3IIIa	292	15 04 04	–25 16 55
υ	3.60	–0.28	K3III	194	15 37 01	–28 08 06
τ	3.66	–2.02	B2.5V	445	15 38 39	–29 46 40
γ Zubenelhakrabi	3.91	0.56	KOIII	152	15 35 32	–14 47 23

life in the universe

See article, pages 292–293.

light

(1) **Electromagnetic radiation** to which the human eye is sensitive, ranging from red, at a wavelength of about 7,500 Å, to violet, at about 3,800 Å. (2) More generally, electromagnetic radiation from **infrared** to **ultraviolet**.

light curve

A graph showing how the brightness of an object, such as a star, changes with time.

light echo

The reflection of light from a primary source, such as a bright star, off another object such as a dust cloud or planet in its vicinity. Remarkable light echoes were seen from two sheets of dust near **Supernova 1987A**.

light element

(1) **Hydrogen** or **helium**, including any of their **isotopes**; all other elements are regarded as "heavy." (2) Any chemical element produced during the first few minutes after the **Big Bang**, including lithium, beryllium, and boron in addition to hydrogen and helium.

light pollution

Ambient background and direct artificial light that interferes with astronomical observations; it does this by reducing the contrast of celestial objects with the sky, thus limiting the ability to see faint sources. Groups of volunteers around the world are now showing that effective laws and guidelines can be instated at the local and regional levels of government, which mean that proper outdoor night lighting can be a norm so that everybody benefits–auto drivers, sleeping residents, government budgets, and astronomers alike. Laws mandating full-cutoff light fixtures are already in place in states such as Maine and Connecticut and are pending in some others. See also **International Dark-Sky Association**.

light-time

The time taken for light, traveling at 299,792 km/s, to reach Earth from a distant object. A correction for the effect of light-time has to be made when calculating rotation periods of planets, eclipses, and transits. The observed times of maxima and minima of variable stars also need a light-time correction depending upon the position of Earth in its orbit at the time of observation, as periods of variation are stated for the Earth at mean distance from the star.

light-year

A unit of length used in astronomy equal to the distance traveled by light in one year. At the rate of 299,792 km/s, one light-year is equivalent to 9.46053×10^{12} km, 5.88 trillion miles, or 63,240 AU.

limb

The apparent edge of the Sun, the Moon, a planet, or any other body having a detectable disk. The leading limb of an object as it moves due to the diurnal rotation of Earth is known as the *preceding limb,* while the trailing limb is called the *following limb.* The *lower limb* of an object (usually the Sun or the Moon) is the limb that is closer to the observer's horizon. The edges of the visible face of the Moon and of Venus are known as the *limb regions.*

limb correction

A correction that must be made to the distance between the center of mass of the Moon and its **limb**. Such corrections are due to the irregular surface of the Moon and are a function of the **libration**s in longitude and latitude and the **position angle** from the central meridian.

limb darkening

A darkening of the solar **limb** at optical wavelengths, arising because the observer's line of sight passes through shallower, cooler, and more absorbing layers of the **photosphere** as it moves away from the center of the Sun's disk. A similar effect occurs with Jupiter and with the members of eclipsing stars as they move in front of each other, and has been detected in the case of several single stars, including **Hamal**. *Limb brightening* is a brightening of the solar limb at X-ray, extreme ultraviolet, and radio wavelengths, due to the observer's line of sight intercepting more of the hot inner **corona** and less of the cooler photosphere as it moves away from the center of the Sun's disk.

limiting magnitude

The faintest detectable magnitude that can be observed or detected by an instrument or telescope. Limiting magnitude depends not only on the instrument, but also on the observing technique and, crucially, upon the seeing conditions.

Lindblad, Bertil (1895–1965)

A Swedish astronomer who was the first to propose the rotation of the Galaxy. He graduated from Uppsala and spent nearly all his career at the Stockholm Observatory, of which he was director from 1927 to 1965. Having considered Jacobus **Kapteyn**'s work on moving stellar

(continued on page 293)

life in the universe

The prospects of finding life elsewhere in space, both within our own solar system and beyond, have climbed in recent years for a number of reasons. These include the detection of numerous **extrasolar planets**; the realization that liquid water may exist or have existed on nearby worlds, including **Mars** and **Europa**; the finding of increasingly complex **interstellar molecules**; the proof that some rocks found on Earth are **Mars meteorites**, evidence that life began on Earth at a very early stage; and, most important of all, the discovery of a vast menagerie of terrestrial extremophiles–creatures, mostly single-celled, that flourish under physical and chemical extremes. Astonishingly, life has turned up in deep, hot underground rocks; around scalding volcanic vents at the bottom of the ocean; in the desiccated, super-cold Dry Valleys of Antarctica; in places of high acid, alkaline, and salt content; and below many meters of polar ice. The fact that life has adapted to such hostile places on Earth greatly expands the possible range of locales where organisms could be expected to survive elsewhere in the universe. Furthermore, genetic studies suggest some deep-dwelling, heat-loving microbes, are among the oldest species known, hinting that not only can life thrive indefinitely in what appear to us to be totally alien environments, it may actually originate in such places. If so–if the cradles of biogenesis tend to be hot, dark, and subterranean rather than the sun-drenched, moderate surface climes that we humans prefer–then the widespread appearance of life throughout the cosmos is much more likely.

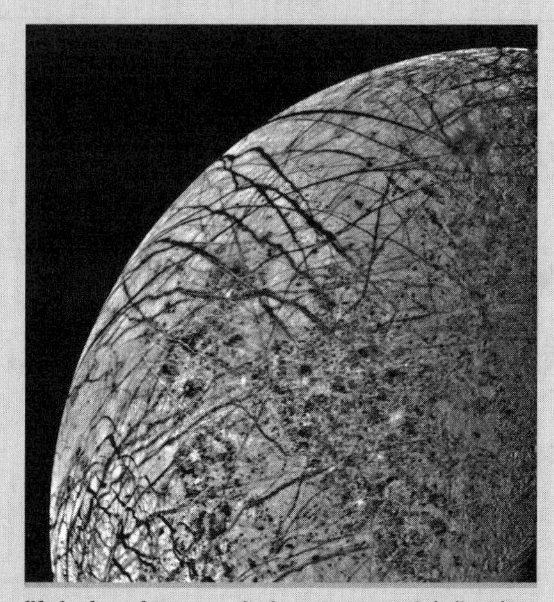

life in the universe Jupiter's moon Europa is believed to have an ocean of liquid water beneath its frozen surface, prompting speculation that it might also harbor primitive forms of life. *NASA/JPL*

Scientific opinion has also shifted toward the view that life is "easy" in the sense that it may be able to assemble itself from simpler components at the slightest opportunity. This view is encouraged by the discovery of possible biological traces in rocks on

life in the universe Mars, whose rocky surface is revealed in this Viking panorama, has long been a favorite target for speculations about extraterrestrial life. Whether Martian life has existed in the past, or perhaps survives today, remains an open question. *NASA/JPL*

Earth that are at least 3.8 billion years old, which arrived during the period of **late heavy bombardment**. An increasingly common claim among astrobiologists and origin-of-life researchers is that life may arise almost inevitably whenever a suitable energy source, a concentrated supply of organic (carbon-based) material, and water occur together. These ingredients are starting to look ubiquitous in space. **Comets**, in particular, are seen as significant vehicles for delivering water and organic cocktails to infant worlds. And with the discovery of meteorites from Mars, the interplanetary transfer of biochemicals or even life itself has become a respectable topic of debate.

Both within and beyond the solar system, the list of potential places where life may have become established is growing fast. Mars, which was once a much warmer and wetter place, may still harbor life underground, similar to the subsurface or hydrothermal communities found on Earth. The Beagle lander, carried aboard the European Mars Express (see **Mars spacecraft**), will be the first probe to put this possibility to the test. With evidence mounting that one or more of the large moons of Jupiter have ice-covered seas, there is much discussion of the potential for life on these outer worlds, too. Even the atmosphere of **Venus** and Saturn's giant moon **Titan** have been eyed with a view to their biological potential.

Beyond the solar system lie the planets of other stars. Future spacecraft such as **Kepler**, **Gaia**, the **Terrestrial Planet Finder**, **Darwin**, and **Planet Imager** will not only search for Earth-like extrasolar worlds but will also look for signs of life in their atmospheric spectra. By 2020 or thereabouts, we should have a much better idea of the frequency with which life has taken hold in our Galaxy.

Lindblad, Bertil (1895–1965)

(continued from page 291)

streams and Howard **Shapley**'s proposal that the center of the Milky Way Galaxy lay tens of thousands of light-years away, he put forward the **differential rotation** theory–namely, that the speed of rotation of stars about the galactic center depends on their distance. Soon after, in 1927, Jan **Oort**'s observations of stellar motion provided support for Lindblad's views, which inspired a number of Swedish astronomers, including Lindblad's son, Per Olaf (1927–), to specialize in the movements of stars. Lindblad also helped demonstrate the spiral nature of the Galaxy and, in the 1940s, put forward the **density wave** theory to explain spiral structure.

Lindsay-Shapley Ring (AM0644-741)

A large southern **ring galaxy**, which, like the **Cartwheel Galaxy**, is host to super starburst clusters that may be analogs of young globular clusters. The collision between two galaxies has resulted in an expanding outer ring, 115,000 light-years across, that essentially contains all of the interstellar matter of the original systems.

line blanketing

The decrease in intensity of a star's spectrum due to many closely spaced, unresolved **absorption lines**. Line blanketing is particularly noticeable in the case of cool stars, whose atoms contain many different types of atoms and molecules that tend to absorb light at higher (bluer) wavelengths and reemit in the red and infrared.

line broadening

A widening of the absorption and emission lines in a spectrum due to any of several factors. These include **Doppler broadening**, caused by movements within the emitting gas, **pressure broadening** or the **Stark effect**, due to collisions between atoms and molecules, and the **Zeeman effect**, due to a strong magnetic field.

line displacement

The change in a **spectral line**'s wavelength due to a **Doppler shift**; approaching sources have spectral lines displaced toward the blue end of the spectrum, receding ones toward the red.

line of nodes

A line that joins the **ascending node** and the **descending node** of an orbit; it marks the intersection of the orbital plane and some reference plane, usually the **ecliptic**.

line profile

A plot of intensity versus wavelength across a **spectral line**.

line ratio

The ratio of the strengths of two particular **absorption lines** or **emission lines** in a spectrum. Because these lines result from different energy transitions in the same element, or transitions involving different elements, they are sensitive to the temperature and/or the density of the source gas.

line spectrum

A pattern of lines, each corresponding to an image of the entrance slit of the **spectrometer**, seen when light is either emitted by or interrupted by a hot rarefied gas. The pattern is characteristic of the gas, and the wavelength at which the features are seen to occur is indicative of the velocity of the object.

line width

The width of an **absorption line** or an **emission line** in a spectrum, usually quoted as *full width at half maximum* (FWHM). Any **line broadening** present may shed light on processes taking place in the absorbing or emitting region.

lineament

A linear topographic feature, such as a fault line, aligned volcanoes, or straight stream course, on the surface of a planet or a moon. It may be positive (such as a range of mountains) or negative (such as a valley).

LINEAR, Comet The destruction of Comet LINEAR is revealed in this image from the Hubble Space Telescope, which shows the comet's nucleus reduced to a shower of glowing mini-comets. The farthest fragment to the left may be the remains of the parent nucleus that cast off the cluster of smaller pieces to the right. The comet broke apart around July 26, 2000, when it made its closest approach to the Sun. This picture was taken on August 5 when the comet was 102 million km from Earth. *NASA/STScI*

LINEAR (Lincoln Near-Earth Asteroid Research) program

A program run by the Massachusetts Institute of Technology Lincoln Laboratory and funded by the U.S. Air Force and NASA, the goal of which is to demonstrate the application of technology originally developed for the surveillance of Earth satellites, to the problem of detecting and cataloging **near-Earth asteroids** that threaten Earth. The LINEAR program uses a pair of telescopes at Lincoln Laboratory's Experimental Test Site on the White Sands Missile Range in Socorro, New Mexico, equipped with **CCD** electro-optical detectors. Observations are then sent to the main Lincoln Laboratory site on Hanscom Air Force Base in Lexington, Massachusetts, where they are linked from night to night, checked, and sent to the **Minor Planet Center** (MPC). The MPC assigns designations to LINEAR's new discoveries of near-Earth asteroids, comets, unusual objects, and main belt asteroids.

LINEAR, Comet (C/1999 S4)

A comet that was observed to break apart as it approached the Sun. C/1999 S4 was discovered by the **LINEAR** program in September 1999. On July 21–22, 2000, a couple of days before perihelion on July 24 (0.75 AU), it suddenly developed a long straight dust tail, suggesting a violent event in the nucleus. On August 5, the Hubble Space Telescope showed a dozen or so tiny comets where the old nucleus had been. The disintegration of C/1999 S4 after perihelion had more to do with solar heating than with the Sun's gravitational force.

LINER (low-ionization nuclear emission-line region)

A type of gaseous region common in the centers of many kinds of galaxies. Some of these have been shown to be low-luminosity **active galactic nuclei**, perhaps an extension of Seyfert activity to the lowest levels, implying that the whole phenomenon of nuclear activity occurs in a significant fraction of bright galaxies. A LINER's spectrum is dominated by emission lines from low-ionization states, such as O II, N II, and S II, with only weak lines from highly ionized states.

Lippershey, Hans (c. 1570–c. 1619)

A German-born Dutch lens maker who appears to have been first to use lenses in combination, though whether he made the first **refracting telescope** is a matter of debate. His compatriots James Metius and Zacharias Jansen were among those who claimed priority. However, it is beyond doubt that Lippershey applied for a patent for a two-lens refractor in 1608, intending it for a military application. **Galileo** heard of the device in the spring of

1609 and the age of telescopic astronomy began in short order.

lithium (Li)

An element with atomic number three; its two stable **isotopes** are the rarer ^{6}Li, with three protons and three neutrons, and the more common ^{7}Li, with three protons and four neutrons. Some lithium was formed in the immediate aftermath of the **Big Bang**, along with huge amounts of hydrogen and helium.

lithium depletion boundary (LDB)

A new age-dating method proposed for **open clusters** based on determination of the **lithium** abundances of cluster members whose masses are near the **hydrogen burning** mass limit. The idea behind this method is that for stars near the substellar mass limit, the age at which stars become hot enough in their cores to burn lithium is a sensitive function of mass. Furthermore, it is argued that the physics required to predict the location of the LDB as a function of age is well understood and not subject to much uncertainty. In the mass range of interest, stars are fully convective, and the core lithium abundance will be directly reflected in the surface lithium abundance, which can be determined from measurements of the 6708 Å Li I doublet.

lithium-rich star

A star with an anomalously high abundance of **lithium** in its upper layers as revealed by a strong **absorption line** of neutral **lithium** (Li I) at 6708 Å. Most lithium stars are **carbon stars**, but lithium enrichment has also been found in some normal late-type giants and **T Tauri stars**. These objects present astronomers with a puzzle because stars destroy most of their lithium soon after formation. Lithium is consumed at nuclear fusion temperatures and not normally remade. For example, the Sun has a lithium abundance about 100 times less than that of the **interstellar medium** (ISM). High amounts of lithium would certainly not be expected in aging stars such as red giants. Yet some of these evolved suns show lithium in quantities far higher than that of the Sun, and, in extreme cases, as high as that of the ISM. One possible explanation is the recent infall of a large planet or **brown dwarf**. Such **planet swallowing** would provide a sudden fresh supply of lithium (which might last 100,000 years or so) and show up as an excess in the star's spectrum. Although this idea, made all the more likely by discoveries of massive planets in tiny orbits around their host stars, works well in most cases, it struggles when faced with the most extreme examples of lithium overabundance found in objects such as the subgiant halo star BD+23 3912. One way a red giant can make fresh lithium is by the transport of ^{7}Be from the core to the envelope where it can be converted to ^{7}Li by electron capture. The only known process for manufacturing ^{6}Li is by **spallation** involving cosmic rays. But these mechanisms, too, are hard pressed to account for the greatest cases of lithium excess. Possibly a combination of factors is at work or there are lithium-generating mechanisms inside stars that are as yet poorly understood.

lithosphere

The **crust** and uppermost **mantle** of a rocky planet or moon. Earth's lithosphere is about 100 km thick, though its thickness is age dependent (older lithosphere is thicker). Below the lithosphere is the **asthenosphere**.

Little Dipper

An asterism formed by the seven brightest stars of **Ursa Minor**, the most conspicuous of which are the North Star (**Polaris**, Alpha Ursae Minoris) and the two front bowl stars, **Kochab** and Pherkad (Beta and Gamma). The Little Dipper looks like a miniature and much fainter version of the well-known **Big Dipper**.

Little Dumbbell Nebula (M76, NGC 650 and 651)

A **planetary nebula** in **Perseus** that resembles, but is smaller and fainter than, the **Dumbbell Nebula**; it is also known as the *Cork Nebula, Butterfly Nebula,* and *Barbell Nebula* and was discovered by Pierre **Méchain** in 1780. The bright bar-shaped main body (measuring 42″ × 87″), is probably a slightly elliptical ring seen edge-on from only a few degrees off its equatorial plane. This ring seems to be expanding at about 42 km/s. Along the axis perpendicular to the ring plane, the gas is moving out more rapidly to form lower surface-brightness wings (157″ × 87″). Finally, there is a faint halo covering a region about 290″ in diameter, consisting of material that was probably ejected in the form of **stellar winds** from the central star when it was still in its red giant phase.

Visual magnitude	
Nebula:	12.0
Central star:	15.9
Distance:	3,400 light-years
Position:	R.A. 1h 42m, Dec. 51° 35′

Little Gem Annular Nebula (NGC 6818)

A small, bright **planetary nebula** in **Sagittarius**, discovered by William **Herschel** in 1787. It shows two distinct layers: a spherical outer region and a brighter, vase-shaped interior bubble.

Visual magnitude:	9.3
Angular diameter:	3′
Distance:	6,000 light-years
Position:	R.A. 19h 44.0m, Dec. –14° 9.2′

lobe

A bright, diffuse region of radio emission on one or both sides of the nucleus of a **radio galaxy** or other active galaxy, and lying outside the visible confines of the galaxy. A lobe is believed to contain matter ejected from the nucleus and carried into intergalactic space along a **jet**.

Local Bubble

A great cavity in the **interstellar medium** (ISM), at least 300 light-years across, within which the Sun and many nearby stars reside. The Local Bubble has a neutral hydrogen density of only about 0.07 atoms/cm^3–at least 10 times lower than the average ISM in the Galaxy, and also contains a thin gruel of million-degree X-ray-emitting plasma. These observations have led astronomers to conclude that the Bubble was formed between a few hundred thousand and a few million years ago by several relatively nearby **supernova** explosions that pushed aside gas and dust in the ISM leaving the current depleted expanse of hot, low-density material. A prime suspect for an object left behind by this supernova activity is **Geminga**. The "Bubble" may be a misnomer since it appears to have an hourglass shape that is narrowest in the **galactic plane** and that widens above and below the plane. In fact, in directions away from the galactic plane the Bubble appears to be opened-ended, bursting into the **galactic halo**, so that "Local Tube" describes it better. Inside it are numerous cloudlets, oriented in sheet-like structures near the Bubble's boundary. The Sun, along with several neighboring stars, is presently embedded in a group of such cloudlets, known as the "Local Fluff Complex" or, more prosaically, as the *local interstellar medium* (LISM), that is passing through the Local Bubble. These floating islands of neutral hydrogen atoms have resulted from the expansion of an even larger bubble, the *Loop I superbubble,* up against our own cavity. The Loop I superbubble was created by supernovae and stellar winds in the **Scorpius-Centaurus Association**, which lies about 500 light-years away. When the Local Bubble and the Loop I collided, the Local Fluff Complex formed at the boundary between the two. This boundary lies 50 to 130 light-years away and through it the cloudlets are invading our Local Bubble. The Sun lies very close to the edge of a cloudlet named the **Local Interstellar Cloud** and is moving roughly perpendicular to it.

Local Group

A collection of more than 40 galaxies, spread across a volume of space some 10 million light-years in diameter, of which our own **Milky Way Galaxy** and the **Andromeda Galaxy** are the dominant and central members. Both of these giant spirals have retinues of **satellite galaxies**, which together account for most of the membership of the Local Group. The Milky Way's satellites include: the **Large Magellanic Cloud**, the **Small Magellanic Cloud**, the **Sagittarius Dwarf Elliptical Galaxy**, the Ursa Minor Dwarf, the **Draco Dwarf**, the **Carina Dwarf Galaxy**, the **Sextans Dwarf**, the **Sculptor Dwarf**, the **Fornax Dwarf**, **Leo I**, and **Leo II**. Among the retinue of the Andromeda Galaxy are: **M32**, **M110**, the fainter and more faraway NGCs 147 and 185, the very faint systems And I, And II, And III, and possibly And IV, And V, And VI (the **Pegasus Dwarf**), and And VII (**Cassiopeia Dwarf**). The third-largest galaxy, the **Triangulum Galaxy** (M33), may or may not be an outlying gravitationally bound companion of M31, but has itself probably the dwarf LGS 3 as a satellite. The other members of the Local Group, including the **Antlia Dwarf Galaxy** and **Wolf-Lundmark-Melotte Galaxy**, fall outside the main subgroups and float alone in the gravitational seas between the giant group members. The substructures of the group are probably not stable. Observations and calculations suggest that the group is highly dynamic and has changed significantly in the past. The galaxies around the large elliptical Maffei 1, for example, were probably once part of our Galaxy group. Indeed, the Local Group is not isolated but is in gravitational interaction and member exchange with the nearest surrounding groups, notably the **Maffei 1 Group**, the **Sculptor Group**, the **M81** Group, and the M83 Group. In the future, interaction between the member galaxies and with the cosmic neighborhood will continue to change the Local Group. Some astronomers speculate that the two large spirals, our Milky Way and the Andromeda Galaxy, may collide and merge to form a giant elliptical. Also, there is evidence that our nearest big cluster of galaxies, the **Virgo Cluster**, will probably stop our cosmological recession away from it, accelerate the Local Group toward itself, and so eventually assimilate the Local Group into its collection. (See table, "The Local Group of Galaxies.")

local horizon

(1) The apparent or visible horizon. (2) The lower boundary of the observed sky or the upper outline of terrestrial objects, including nearby obstructions or irregularities.

local hour angle (LHA)

The **hour angle** of a celestial object measured with respect to the observer's local meridian.

The Local Group of Galaxies

Galaxy	Position R.A.	Position Dec.	Type	Abs. Mag.	Diameter (light-year)	Radial Vel.* (km/s)	Distance (kly)$_2$†
Milky Way Galaxy	17:45.6	−28:56	SBbc I-II	−20.6	90	0	28
Sagittarius Dwarf Elliptical	18:55	−30:30	dSph(E7)	−14.0	10		78
Large Magellanic Cloud	05:19.7	−68:57	Irr III-IV	−18.1	30	+119	179
Small Magellanic Cloud	00:51.7	−73:14	Irr IV-V	−16.2	16	+34	210
Ursa Minor Dwarf	15:08.8	+67:12	dSph	−8.9	2	−47	215
Sculptor Dwarf	01:00.0	−33:42	dSph	−10.7	3	+115	260
Draco Dwarf	17:20.1	+57:55	dSph	−8.6	3	−87	270
Sextans Dwarf	10:13.2	−01:37	dSph	−10.0	4		280
Carina Dwarf	06:14.6	−50:58	dSph	−9.92	2	+13	330
Fornax Dwarf	02:39.9	−34:32	dSph	−13.0	6	−41	450
Leo II	11:13.5	+22:10	dSph	−10.2	3	+36	670
Leo I	10:08.5	+12:18	dE3	−12.0	3	+60	820
Phoenix Dwarf	01:51.1	−44:27	dIrr/dSph	−9.9	2		1450
NGC 6822 (Barnard's Galaxy)	19:44.9	−14:49	Irr IV-V	−16.4	8	+44	1600
Andromeda II	01:16.4	+33:27	dSph	−11.7	2		1700
NGC 185	00:39.0	+84:20	dSph/dE3	−15.3	8	−64	2000
Leo III (Leo A)	09:59.4	+30:45	dIrr	−11.7	4	−19	2250
Andromeda VII	23:27.8	+50:35	dSph	−12.0	2		2250
IC 1613	01:05.1	+02:08	Irr V	−14.9	10	−152	2300
NGC 147	00:33.2	+48:31	dSph/dE5	−14.8	10	+28	2350
Andromeda III	00:35.4	+36:31	dSph	−10.2	3		2500
Cetus Dwarf	00:26.1	−11:02	dSph	−10.1	3		2550
Andromeda VI	23:51.7	+24:36	dSph	−11.3	3		2550
Aquarius Dwarf	20:46.8	−12:51	dIrr/dSph		2	−23	2600
M32	00:42.7	+40:52	dE2	−16.4	8	−28	2600
Andromeda I	00:45.7	+38:00	dSph	−11.7	2		2600
Andromeda V	01:10.3	+47:38	dSph	−9.1			2650
LGS 3 (Pisces Dwarf)	01:03.8	+21:53	dIrr/dSph	−9.7	2	−149	2650
Andromeda Galaxy (M31)	00:42.7	+41:16	Sb I-II 3.4	−21.1	140	−121	2650
M110 (NGC 205)	00:41.3	+41:41	dSph/dE5	−16.3	15	−60	2650
IC 10	00:20.4	+59:18	dIrr	−17.6	8	−146	2700
Triangulum Galaxy (M33)	01:33.9	+30:39	Sc II-III	−18.9	55	−46	2850
Tucana Dwarf	22:41.7	−64:25	dSph	−9.6	2		2850
Wolf-Lundmark-Mellote	00:02.0	−15:28	Irr IV-V	−14.0	10	−61	3000
Pegasus Dwarf	23:28.6	+14:45	dIrr/dSph	−12.7	2	−20	3100
Sagittarius Dwarf Irregular	19:30.1	−17:42	dIrr	−11.0	3	+8	3450
Antlia Dwarf	10:04.1	−27:20	dSph	−10.7	3		4000
NGC 3109	10:03.1	−26:09	Irr IV-V	−15.8	25	+194	4100
UGC-A92	04:27.4	+63:30	dIrr		3	+66	4200
UKS 2323-326	23:26.5	−32:23	dIrr	−13.1	3	+74	4300
Sextans B	10:00.0	+05:20	dIrr	−14.4	8	+168	4400

(continued)

The Local Group of Galaxies *(continued)*

Galaxy	Position R.A.	Position Dec.	Type	Abs. Mag.	Diameter (light-yr)	Radial Vel.* (km/s)	Distance (kly)[†]
Sextans A	10:11.1	–04:43	dIrr	–14.3	10	+164	4700
IC 5152	22:06.1	–51:17	dIrr		8	+80	5200
GR 8	12:58.7	+14:13	dIrr	–12.5	2	+183	5200

*Radial velocity with respect to center of the Milky Way Galaxy.
†Thousands of light-years.

Local Interstellar Cloud (LIC)

A cloud of neutral hydrogen that is flowing away from the **Scorpius-Centaurus Association** and through which the Sun is currently passing. The LIC resides in a low-density hole in the interstellar medium known as the **Local Bubble**. Nearby, high-density **molecular clouds**, including the *Aquila Rift*, surround star-forming regions.

local standard of rest

The motion of the Sun relative to a hypothetical circular orbit about the center of the Galaxy. The local standard of rest is approximately 16.5 km/s in the direction of the **galactic coordinates,** $l = 53°$, $b = 23°$.

Local Group The dwarf irregular galaxy NGC 6822, a member of the Local Group that lies about 2 million light-years away. *European Southern Observatory*

Local Supercluster

The **supercluster** to which our own **Local Group** of galaxies belongs. By supercluster standards it is of very modest size with only one **rich cluster of galaxies**, the **Virgo Cluster**, at its center. The Local Group, which belongs to the Canes Venatici cloud of galaxies, lies in the outskirts. The Local Supercluster appears to be made of two major structures: a flattened disk, which has a thickness of about 5 million light-years and contains about 60% of the bright galaxies, and a roughly spherical halo, which contains the remaining 40% of the bright galaxies in a small number of clouds. The major components (galaxy clusters) of the Local Supercluster are listed in the table "Local Supercluster"; none of the listed galaxy clouds has its center farther from the Virgo Cluster than our own.

Lockman Hole

A patch of the sky, lying roughly between the pointer stars of the Big Dipper (centered at R.A. 10h 45m, Dec. +58° 00′), with an area of 15 square degrees, that is almost free from absorption by neutral hydrogen gas in the Galaxy. Lockman Hole's minimum column density of galactic hydrogen allows very sensitive searches for extragalactic objects and, for this reason, it is one of the best studied sky areas across a very wide range of wavelengths, from radio waves to gamma rays.

Lockyer, (Joseph) Norman (1836–1920)

An English solar astronomer who carried out pioneering studies of the Sun's spectrum independently of William **Huggins** and Thomas Young (1773–1829). With a prism, he also studied solar **prominences**. He concluded that a line reported to him by Jules **Janssen** belonged to a new element, never detected up to that point on Earth, which he named *helium,* the Greek word for Sun. (Helium was later (1897) found in the laboratory by William Ramsay.) Lockyer founded the British journal of science *Nature* in 1869.

Local Supercluster

Name	Distance Mpc	Size Mpc	Volume Mpc^3	N	Density Mpc^{-3}
Virgo Cluster	10.7	2.2	6	62	10
Canes Venatici	9.8	14.6	470	99	0.2
Virgo II (S)	11.6	10.6	200	55	0.3
Leo II	13.2	15.4	510	45	0.1
Virgo III	15.5	11.3	190	40	0.2
Crater (NGC 3672)	14.9	8.1	120	25	0.2
Leo I	7.8	5.4	20	15	0.6
Leo Minor (NGC 2841)	5.1	7.5	60	11	0.2
Draco (NGC 5907)	9.6	8.1	60	6	0.1
Antlia (NGC 2997)	7.5	4.6	40	5	0.1
NGC 5643	10.2	5	2	3	–
Rest			~2,900	8	0.003
Total			~4,500	374	0.07

Notes: "Name" is the common name of the cloud, "distance" is the distance of its center to Earth, "size" is its greatest extent, "volume" is its approximate volume, "N" is the number of bright galaxies in the cloud (galaxies roughly at least as bright as the Milky Way Galaxy), and "density" is the number of bright galaxies per unit volume. All distances are measured in Mpc/*h* (written Mpc in the table). The number of fainter galaxies in these galaxy groups is likely much greater than the number of bright galaxies.

lodranite

A rare type of primitive **achondrite** named after the Lodran meteorite that fell in Pakistan in 1868. Initially, the lodranites were grouped with the **stony-iron meteorites** because they contain components of both nickel-iron metal and stony material, consisting of **olivine**, **orthopyroxene**, and minor **plagioclase feldspar** in nearly equal proportions. However, since the discovery of the closely related **acapulcoite** group, the lodranites have been classified as primitive achondrites. Because both groups have similar mineralogical and oxygen isotopic compositions, it is thought that they came from the same parent body, most likely an S-class asteroid that has not yet been identified. Lodranites have coarser-grained olivines and pyroxenes and experience higher temperatures than acapulcoites, suggesting that they originated within the deeper layers of the acapulcoite/lodranite parent body where they were subjected to a more intense and prolonged thermal processing.

Lomonosov, Mikhail Vasilievich (1711–1765)

A Russian writer, chemist, and astronomer who was the first to obtain direct evidence that Venus has an atmosphere. In 1761, at the University Observatory in St. Petersburg, Lomonosov observed a **transit** of Venus with the intention of finding the planet's diameter by measuring the size of its dark outline. He was able to determine that Venus is similar in diameter to Earth, but the precision of his measurement was hampered by the fact that the edge of the planet's disk, instead of appearing sharp, as he'd expected, was fuzzy and indistinct. From this, Lomonosov concluded that Venus has "an atmosphere equal to, if not greater than, that which envelops our earthly sphere."

longitude

The angular distance measured east or west from the prime meridian. See **celestial coordinates**.

longitude at the epoch (*L*)

The **longitude** that a planet has at a given date and time. It is the angle measured along the ecliptic from the **vernal equinox** eastward to the **ascending node**, and then eastward along the orbital plane to the planet's position at that instant.

longitude of the perihelion (Ω)

The angle measured from the **vernal equinox** eastward along the ecliptic to the **ascending node** of a planet's orbit, and then continued eastward along the orbital plane to the perihelion. It is one of the **orbital elements**.

long-period comet

A **comet** with a period of more than 200 years and as much as several million years. Long-period comets, together with **Halley-type comets**, are believed to come from the **Oort Cloud**, an enormous reservoir of frozen cometary nuclei orbiting the Sun at a distance of tens of thousands of AU. The gravitational influences of passing stars, giant molecular clouds, and the central regions of the Galaxy are thought to be instrumental in occasionally perturbing the orbits of some of the Oort objects and causing them to plunge toward the inner solar system on highly elliptical paths.

long-period variable

Usually taken to be another name for a **Mira star**. In the past, the term was applied more generally to mean any red giant or supergiant **pulsating variable** with a period of 20 to 1,000 days, including, in addition to Mira variables, **semiregular variables** and slow **irregular variables**. *Long-period blue variables* are pulsating **B stars** with periods longer than 1 day and amplitudes generally less than 0.05 magnitude but occasionally up to 0.1 magnitude.

loop prominence

An arch of high-temperature gas that forms above the Sun's limb after the eruption of a large **solar flare**, and in which gas drains down to the solar surface. A *loop prominence system* is a set of such features.

Lost City Meteorite

The second **meteorite** fall to be recorded by a camera network (the first was that of the **Príbram meteorite** in 1957), thus enabling its trajectory and orbit to be determined and facilitating a quick recovery of the fragments. The now-defunct Prairie Meteorite Network, a system of 16 camera stations in seven Midwestern states run by the Smithsonian Astrophysical Observatory, recorded the Lost City fall in 1970. The photos enabled scientists to reconstruct the meteorite's orbit and determine that it originated in the asteroid belt. See also **Innisfree Meteorite**.

Lovell, (Alfred Charles) Bernard (1913–)

An English radio astronomer and physicist, who was the driving force behind the world's first giant **radio telescope**. Before World War II, he studied cosmic rays with Patrick Blackett (1897–1974) and during World War II he did radar research. After the war, with the help of James Hey (1909–) he was able to procure an ex-Army mobile radar unit operating at a wavelength of 4.2 m. After finding that the electric trams at Manchester caused too much interference, he transported the van to the site of the future **Jodrell Bank Observatory** on the Cheshire plain, where he used it to study transient radar echoes. His goal was to study cosmic-ray showers using radar, but instead he found that the echoes came from ionized meteor trails. With J. A. Clegg, he built a 218-foot transit telescope (the "wire bowl"), with which Robert Hanbury Brown (1916–2002) and Cyril Hazard found that the Andromeda Galaxy was a radio source. Pursuing his dream of a large steerable telescope, Lovell supervised construction of the Mark 1 Jodrell Bank telescope (now named after him). Lovell is the author of many popular books, including *In the Center of Immensities,* and the textbook *Radio Astronomy,* which he coauthored with Clegg.

Lowell Observatory

An observatory at Flagstaff, Arizona, founded in 1894 by Percival **Lowell**. The original building, at an altitude of 2,210 m on Mars Hill, houses a 24-inch Alvan Clark refracting telescope and is now a National Historic Landmark. From here Lowell sought evidence of Martian canals and of a ninth planet. The latter was eventually found at the observatory by Clyde **Tombaugh** in 1930. Lowell astronomers were also the first to detect the rings of **Uranus**. In 1961 a dark-sky outstation was opened at Anderson Mesa, 19 km southeast of Flagstaff. Since 1995 this has been the site of the Navy Prototype Optical Interferometer, a joint project of the Naval Research Laboratory, the U.S. Naval Observatory, and Lowell Observatory.

Lowell, Percival (1855–1916)

An American astronomer who established the **Lowell Observatory** in 1894 to search for signs of intelligent life on Mars, notably in the form of markings that he believed to be artificial waterways. Although he searched unsuccessfully for a planet beyond Neptune, this planet (Pluto) was eventually found from the observatory he established.

lower transit

Also known as *lower culmination,* the point at which a celestial object reaches its minimum altitude on the observer's meridian.

low-surface-brightness galaxy

A galaxy that is at most a few percent brighter than the sky background, making it very difficult to see. The first low-surface-brightness (LSB) galaxy to be found, **Malin-1**, was discovered as recently as 1987 and it now appears that up to half the galaxy population is of this type and had been missed in earlier surveys. LSB galaxies have fewer stars per unit volume than normal galaxies, possibly because they are more isolated in space and have not undergone tidal interactions with other galaxies that stimulate bursts of star formation. Recent photographic and **CCD** surveys have uncovered large numbers of new small and medium-sized, moderate-to-low surface brightness spiral galaxies; however, LSB giants such as Malin-1 have remained relatively rare.

low-velocity star

A star whose motion relative to the **local standard of rest** is small, indicating that it moves in an orbit about the galactic center similar to that of the Sun.

luminosity

The total amount of energy radiated by a star, or other luminous object, every second. It is measured in watts (W) or in terms of solar luminosity L_{sun} (3.9×10^{26} W) and is related to **bolometric magnitude** (M_{bol}) by the formula:

$$M_{bol} - 4.72 = 2.5 \log (L/L_{sun})$$

The luminosity of a **blackbody** (which most stars closely approximate) of temperature T and radius R is given by the *Stefan-Boltzmann equation:*

$$L = 4\pi R^2 \sigma T^4$$

where σ is the Stefan-Boltzmann constant (5.67×10^{-8} W/m^2/K^4).

luminosity class

A classification of stellar spectra according to **luminosity** for a given **spectral type**; it was introduced as part of the **Morgan-Keenan classification**. The luminosity class broadly indicates whether a star is a dwarf (that is, a main sequence star), a giant, or a supergiant, since luminosity is directly related to surface area. Luminosity class is expressed as a Roman numeral, from I to V, and appears after the spectral type; for example, Tau Ceti, with spectral type G8, is listed as a G8V object. (See table, "Luminosity Classes.")

luminosity function

(1) The number of stars in the Galaxy with a particular **absolute magnitude**. The luminosity function reveals that luminous stars are rare and intrinsically faint stars are common. (2) The distribution of galaxies by absolute magnitude. Luminous galaxies are rare, while intrinsically faint ones are common.

Luminosity Classes

Class	Description
Ia-0	Hypergiant (extreme supergiant)
Ia	Bright supergiant
Iab	Normal supergiant
Ib	Subluminous supergiant
II	Bright giant
III	Normal giant
IV	Subgiant
V	Main-sequence (dwarf) star

luminous mass

The mass contributed by luminous matter in galaxies; now thought in most cases to be only about 10% of the total mass that is actually present.

Lunar-A

A Japanese lunar probe, scheduled for launch in the third quarter of 2004, that will be the first mission to study the Moon's internal state using penetrators. After entering lunar orbit, the spacecraft will deploy three 13-kg spear-shaped cases, 90 cm long and 13 cm in diameter. These will be individually released over a period of a month and impact the Moon at 250 to 300 m/s, burrowing 1 to 3 m into the surface. The penetrators are equipped with seismometers and devices to measure heat flow, and will transmit their data to the orbiter as it passes over each penetrator every 15 days.

lunar eclipse

An eclipse in which the Moon passes through the shadow cast by Earth. The eclipse may be *total* (the Moon passing completely through Earth's **umbra**), *partial* (the Moon passing partially through Earth's umbra at maximum eclipse), or *penumbral* (the Moon passing only through Earth's **penumbra**). Total lunar eclipses tend to be brighter when the atmosphere is relatively clear of volcanic and other dust: a dirty atmosphere blocks more sunlight and dims the eclipse. The Danjon scale is used to estimate the level of illumination of the Moon during an eclipse. Unlike a total eclipse of the Sun, which lasts a maximum of 7½ minutes, the lunar variety happens at a more leisurely pace. The shadow made by Earth as it blocks the Sun is 16,782 km wide where the Moon crosses through it. The Moon itself is only 3,476 km wide, and is traveling about 3,700 km/h. So roughly 3 hours elapse from the time the Moon first touches the umbra until the last part of the Moon passes out of it. The middle third of the journey is the part of the eclipse that is total. The *magnitude of a lunar eclipse* is the fraction of the lunar diameter obscured by Earth's shadow at the greatest phase of an eclipse, measured along the common diameter.

lunar meteorite

An **achondrite** (a type of **stony meteorite**) that has come from the surface of the Moon. More than two dozen lunar meteorites have been identified since the 1970s with a total mass of about 8.5 kg. Most have been found in Antarctica and in the deserts of northern Africa and Oman, although one 19-g specimen was recovered in 1990 from Calcalong Creek, Australia. These stones are of great importance because, in many cases, they represent specimens of the Moon from regions not visited by the manned Apollo or unmanned Russian sample-return missions. The majority were blasted out of the lunar highlands rather than the low-lying maria, which served as the Apollo landing sites. The lunar meteorites found so far, representing four distinct types of Moon rock, are categorized into groups *LUN A, LUN B, LUN G* (for "gabbro"), and *LUN N* (for "norite"). Among the more interesting specimens is a *LUN B* meteorite, found in Morocco in 2000, that crystallized from lava just 2.8 billion years ago and provides evidence for surprisingly recent lunar volcanism. The only known *LUN N* meteorite, found in three pieces near Dchira in the Western Sahara and named NWA 773, is especially important because it represents a type of rock never sampled by the Luna or Apollo landing missions, but detected from orbit at several sites on the surface. A possible source of NWA 773 is the Aitken Basin, a large impact structure near the lunar south pole that is famous for its noritic composition and secondary impact craters. The large impact that excavated the Aitken Basin removed the upper crust, revealing the lower crustal layers that contain olivine-rich norites and gabbronorites.

lunar parallax

The difference, averaging 3422.5″, in the Moon's apparent celestial position as seen from the center of Earth and when it lies on the horizon for an observer at the surface.

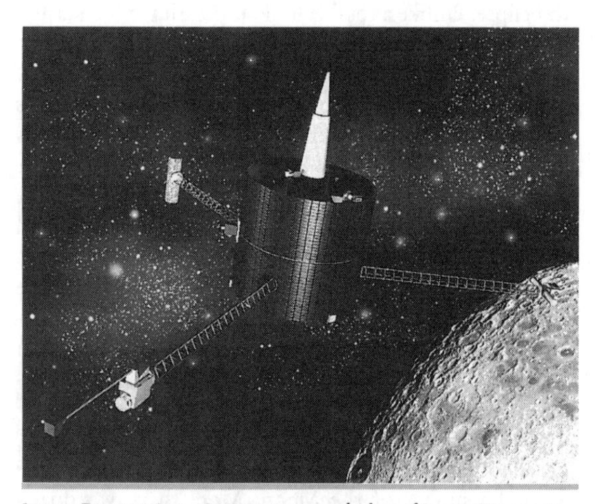

Lunar Prospector A computer rendering of Lunar Prospector in orbit around the Moon. *NASA*

lunar phases

Cyclically recurring appearances of the Moon due to differing amounts of sunlight reflected by the Moon toward Earth during the Moon's orbit. The eight phases are: new moon, waxing crescent, first quarter, waxing gibbous, full moon, waning gibbous, last quarter, and waning crescent. New moon is when the sunlit (daytime) side of the Moon is facing away from us, full moon when it is facing toward us and thus directly opposite the Sun in the sky.

lunar polarization

The partial polarization of sunlight reflected from the lunar surface. The bright highlands show a stronger effect than do the dark maria, both reaching a maximum at the quarter phases.

Lunar Prospector

A NASA spacecraft, launched on January 6, 1998, that circled the Moon in polar orbit for about 18 months, gathering data on the surface chemical composition and lunar magnetic and gravitational fields. The probe's neutron spectrometer confirmed the earlier finding of **Clementine** that large amounts of hydrogen, assumed to be bound up in water-ice molecules, exist in the Moon's polar regions. However, a controlled crash of Lunar Prospector into a south polar crater on July 31, 1999, failed to produce positive results from telescopes that were trained on the impact site to search for spectroscopic evidence of water.

lunisolar

To do with both the Sun and the Moon, as in the case of some **calendars**.

Lupus (abbr. Lup; gen. Lupi)

The Wolf; a southern **constellation** that lies between Scorpius to the east, Centaurus to the west and north, and Circinus to the south. One of its few deep sky objects of interest to the amateur observer is the open star cluster NGC 5822 (visual magnitude 7; apparent diameter 40′; R.A. 15h 5.2m, Dec. –54° 21′). The *Lupus Loop* is a prehistoric **supernova remnant**, about 1,500 light-years away (R.A. 15h 10m, Dec. –40°), in the form of a large broken shell, 4.5° in diameter, that is detectable at radio and X-ray wavelengths. (See table, "Lupus: Stars Brighter than Magnitude 4.0," and star chart 15.)

Luyten, Willem Jacob (1899–1994)

A Dutch astronomer who, from the 1920s to the 1980s, found thousands of nearby stars during the course of several major surveys, including the Bruce Proper Motion Survey (1944) and the Luyten-Palomar Survey (1970s).

Lupus: Stars Brighter than Magnitude 4.0

Star	Magnitude		Spectral	Distance	Position	
	Visual	Absolute	Type	(light-yr)	R.A. (h m s)	Dec. (° ′ ″)
α **Kakkab** (Men)	2.30	–3.83	B1.5III	548	14 41 56	–47 23 17
β KeKouan	2.68	–3.35	B2III	524	14 58 32	–43 08 02
γ	2.80	–3.41	B2IV	567	15 35 08	–41 10 00
δ	3.22	–2.75	B1.5IV	510	15 21 22	–40 38 51
ε	3.37	–2.58	B2IV-V	504	15 22 41	–44 41 21
ζ	3.41	0.65	G8III	116	15 12 17	–52 05 57
η	3.42	–2.48	B2.5IV	493	16 00 07	–38 23 48
ϕ^1	3.57	–1.44	K5III	326	15 21 48	–36 15 41
κ^1	3.88	0.14	B9.5Vne	182	15 11 56	–48 44 16
π	3.91	–2.01	B5V+B5IV	497	15 05 07	–47 03 04
χ	3.97	–0.03	B9III	206	15 50 57	–33 37 38

Among the many proper motion catalogs he compiled were the *Luyten Catalog (L),* the *Luyten-Palomar Catalog (LP),* the *Luyten Two-Tenths Arcsecond Catalog (LTT),* the *Luyten Four-Tenths Arcsecond Catalog (LFT),* and the *Luyten Half-Second Catalog (LHS).* He was educated at the universities of Amsterdam and Leiden, and, after receiving his Ph.D. in 1921, he worked at Lick Observatory, Harvard College Observatory, and, for nearly 60 years, at the University of Minnesota. In the 1920s he used **proper motions** to obtain statistical parallaxes and refine the **Hertzsprung-Russell diagram** of local stars. Using plates taken in the Southern Hemisphere, many of them personally procured at Harvard's Bloemfontein station, he found proper motions of more than 120,000 stars. Later he repeated the National Geographic Society Palomar Sky Survey, built an automated computerized plate scanner and measuring machine, and found 400,000 more– all this despite having lost the sight of one eye in 1925. Luyten determined the density of stars in space as a function of luminosity, discovered the great majority of the white dwarfs known, and also analyzed orbits of many spectroscopic binary stars.

Luyten 726-8 (UV Ceti)

One of the nearest star systems to the Sun at a distance of 8.73 light-years; it lies in **Cetus**, northeast of Deneb Kaitos (Beta Ceti). Luyten 726-8 is a binary in which both components are **red dwarfs** in a highly elliptical orbit, with a period of 26.5 years, that takes the two stars between 2.1 and 8.8 AU apart. Component B is an extreme example of a **flare star** that can surge in brightness by a factor of five in less than a minute, then fall somewhat more slowly back to normal luminosity within two or three minutes before flaring suddenly again after several hours. In 1952, Luyten 726-8 B, better known by its variable-star name UV Ceti, was observed flaring to 75 times its normal brightness in only 20 seconds. It is the prototype of the class known as **UV Ceti stars**. (See table, "The Twin Stars of Luyten 726-8.")

Lyman-alpha

The spectral line at 1216 Å, in the far **ultraviolet**, that corresponds to the transition of an electron between the two lowest energy levels of a hydrogen atom. See also **Lyman series.**

The Twin Stars of Luyten 726-8

Component	Type	Mass (Sun=1)	Radius (Sun=1)	Luminosity (Sun=1)
A	M5.5eV	0.10	0.15	0.000011
B	M6eV	0.10	0.14	0.000004

Lynx: Stars Brighter than Magnitude 4.0

Star	Magnitude		Spectral	Distance	Position	
	Visual	Absolute	Type	(light-yr)	R.A. (h m s)	Dec. (° ′ ″)
α	3.13	−1.03	M0III	222	09 21 03	+34 23 33
38	3.82	0.95	A1V	122	09 18 51	+36 48 09

Lyman-alpha forest

The appearance of many differentially redshifted **Lyman-alpha** absorption lines in a **quasar**'s spectrum, caused by intervening hydrogen clouds along the line of sight to the quasar.

Lyman series

The series of lines in the **hydrogen spectrum** associated with transitions to or from the first energy level or ground state. It lies in the ultraviolet (Lα at 1215.67 Å; Lβ at 1026 Å) with a series limit at 912 Å. The He II Lyman lines have almost exactly one-quarter the wavelength of their hydrogen equivalent (He II Lα, 303.78 Å; He II Lyman limit, 227 Å).

Lynx (abbr. Lyn; gen. Lyncis)

The Lynx; a faint northern **constellation** that lies between Auriga to the west and Ursa Major to the east. Several past changes to the constellation boundaries have left some stars in this region with an identity problem. The star *10 UMa,* for example, is now part of Lynx, whereas *41 Lyn* falls within the bounds of Ursa Major. Lynx is also home to the **Intergalactic Wanderer** and NGC 2419, a type Sc spiral galaxy (magnitude 9.7; diameter 9.3′ × 2.5′; R.A. 8h 52.7m, Dec. +33° 25′). (See table, "Lynx: Stars Brighter than Magnitude 4.0," and star chart 2.)

Lyra (abbr. Lyr; gen. Lyrae)

The Lyre of Orpheus; a small but bright northern **constellation** located between Cygnus to the east, Hercules to the south and west, and Draco to the north. Aside from its several star attractions, Lyra also boasts the most famous of planetary nebulae, the **Ring Nebula**, and M56 (NGC 6779), a **globular cluster** visible with binoculars (magnitude 8.2; diameter 7.1′; distance 45,000 light-years; R.A. 19h 16.6m, Dec. +30° 11′). (See table, "Stars Brighter than Magnitude 4.0," and star chart 7.)

Lysithea

The twelfth moon of **Jupiter**, also known as Jupiter X; it was discovered in 1938 by Seth **Nicholson**. Lysithea has a diameter of about 38 km and orbits at a mean distance of 11.72 million km from the planet.

Lyra: Stars Brighter than Magnitude 4.0

Star	Magnitude		Spectral	Distance	Position	
	Visual	Absolute	Type	(light-yr)	R.A. (h m s)	Dec. (° ′ ″)
α **Vega**	0.03	0.58	A0Va	25	18 36 56	+38 47 01
γ Sulaphat	3.25	−3.20	B9III	635	18 58 56	+32 41 22
β **Beta Lyrae** (Sheliak)	3.52	−3.64	B7Ve+A8p	882	18 50 05	+33 21 46
ε **Epsilon Lyrae**	4.67	1.18		162		
ε^1	4.67		A4V		18 44 20	+39 40 15
ε^2	5.14		A8Vn		18 44 23	+39 36 46

M

M star

A cool red star of **spectral type** M with a surface temperature of less than 3,500 K and a spectrum dominated by molecular bands, especially those of titanium oxide (TiO), and the absorption lines of metals. M-type dwarfs, also known as **red dwarfs**, have masses under 0.5 M_{sun} and luminosities less than 0.08 L_{sun}; examples include **Proxima Centauri** and **Barnard's Star**. M-type giants have masses of 1.2 to 1.3 M_{sun} and luminosities of over 300 L_{sun}. M-type supergiants, such as **Betelgeuse** and **Antares**, have masses of 13 to 25 M_{sun} and luminosities of 40,000–500,000 L_{sun}.

M4 (NGC 6121)

Probably the nearest **globular cluster** to the Sun, just surpassing NGC 6397 in **Ara**, which seems to be about 300 light-years farther away. Discovered by Philippe Loys de Chéseaux (1718–1751) in 1746, M4 lies in **Scorpius**, 1.3° west of Antares, and can just be made out with the naked eye under dark skies. It would be one of the most spectacular globulars on view, if it weren't obscured by dust in the galactic plane, which both dims and reddens its light. Among the most open or loose of its type, it has a remarkable feature in the form of a central bar. In 1987, the first millisecond **pulsar** was discovered in M4: PSR 1821-24, with a period of 3.0 milliseconds.

In July 2003 a planet estimated to be 13 billion years old was detected in a binary system in M4 consisting of another millisecond pulsar, B1620-26, and a white dwarf. This makes it by far the oldest planet known and the first to be found in a globular cluster.

Visual magnitude:	5.6
Angular size:	26′
Linear diameter:	58 light-years
Distance:	7,200 light-years
Position:	R.A. 16hr 23.6m, Dec. –26° 32′

M32 (NGC 221)

A **dwarf elliptical galaxy**, of type E2 in Hubble's classification (see **galaxy classification**), that is a satellite of the **Andromeda Galaxy** (M31). Discovered in 1749 by Guillaume **Le Gentil**, it was the first elliptical to be found. M32 can be seen even in small telescopes as a slightly elongated bright patch due south of M31's central region, overlaid on the outskirts of the spiral arms. Although its diameter is only about 8,000 light-years and its total mass about 3 billion M_{sun}, it has a nucleus comparable to that of M31 itself, with some 100 million solar masses in rapid motion around a central supermassive object. M32 also has a stellar **population** similar to that of larger ellipticals, including a mixture of mostly old, low-mass stars and some intermediate-age stars richer in heavy elements, though it lacks **globular clusters**. These facts suggest that M32 may once have been much larger but then lost its outer stars and globular clusters during one or more close encounters with M31. These stars and clusters would have become part of M31's halo. That M32 has recently undergone a close encounter with its larger neighbor is also indicated by disturbances visible in the big galaxy's spiral pattern. M32 contrasts markedly with the other bright satellite of the Andromeda Galaxy: **M110**. See also **Local Group**.

M80 (NGC 6093)

One of the densest known **globular clusters** in our Galaxy. M80 contains thousands of stars, including a

M32 A swarm of nearly 8,000 hot, young stars resembles a blizzard of snowflakes near the core (lower right) of M32, one of the satellites of the Andromeda Galaxy. This image, by the Hubble Space Telescope, gave astronomers their first glimpse of such stars deep inside an elliptical galaxy. *NASA/STScI*

Visual magnitude:	7.3
Angular size:	8.9′
Distance:	28,000 light-years
Position:	R.A. 16h 17m, Dec. –22° 59′

large population of **blue stragglers** in its core. The cluster was also the site of a **nova** in 1860. (See photo, page 308.)

M81 (NGC 3031)

A giant **spiral galaxy** in **Ursa Major**; it is the brightest and probably dominant member of the *M81 group,* the second nearest galaxy group to the **Local Group**. A few hundred million years ago, a close encounter took place between M81 and its smaller, near-neighbor, **M82**, during which the latter was dramatically deformed. The encounter has also left traces in the spiral pattern of M81. The galaxies are still close together–their centers separated by as little as 150,000 light-years. Discovered by

M82 A revealing view of a cosmic cigar. *Isaac Newton Group of Telescopes/Nik Szymanek*

Johann **Bode** in 1774, and sometimes referred to as "Bode's Nebula," M81 is one of the easiest and most rewarding galaxies for amateur astronomers in the Northern Hemisphere and, under exceptional seeing conditions, has even been glimpsed with the naked eye.

Visual magnitude:	6.9
Angular size:	25.7′
Distance:	11 million light-years
Position:	R.A. 9h 56m, Dec. +69° 4′

M82 (NGC 3034)

Also known as the *Cigar Galaxy,* the prototype disk **irregular galaxy** and a physical partner of **M81**; it lies in **Ursa Major**. The core of M82 has been spectacularly disturbed by a relatively recent encounter with M81, and is now the site of conspicuous dust lanes and a burst of star formation (see **starburst galaxy**). Turbulent explosive gas flows underlie a powerful radio source, known as Ursa Major A or 3C 231. At **infrared** wavelengths, M82 is the brightest galaxy in the sky and much brighter than it is in the visible part of the spectrum. Over 100 young **globular clusters** have been discovered in it with the Hubble Space Telescope, their formation probably another side effect of the close brush with M81. M82 was discovered (together with M81) by Johann **Bode** in 1774.

Visual magnitude:	8.4
Angular size:	9′ × 4′
Distance:	11 million light-years
Position:	R.A. 9h 56m, Dec. +69° 41′

M87 (NGC 4486)

A giant elliptical galaxy in the **Virgo Cluster**, of which it is probably the dominant member. M87 has a diameter of at least 120,000 light-years–greater than that of the Milky Way's disk. As M87 is nearly spherical (type E0 or E1 in Hubble's classification), it fills a far larger volume than does our own galaxy, and thus contains many more stars. Its mass has been estimated at 2 to 3 trillion M_{sun}. A recent study by David Malin of the Anglo-Australian Observatory showed that M87 extends farther than the previously known spherical region into a vast elongated shape more than half a million light-years wide. Its outer layers are noticeably distorted, probably because of gravitational interactions with other Virgo Cluster galaxies and because of material acquired during cannibalistic encounters. M87 contains over 10,000 **globular clusters**–including one of the largest ever seen–compared with the Milky Way's 150 to 200. It also boasts a spectacular **jet** that extends over 5,000 light-years from the center and consists of a string of knots and clouds of gas ejected from the core. A second, less-conspicuous jet points in the opposite direction. Images of the violent active nucleus of M87, taken with the Hubble Space Telescope, have revealed a central object of about 2 to 3 billion solar masses concentrated within a spherical region just 60 light-years across and surrounded by a rapidly rotating gaseous **accretion disk**. Almost certainly at the very heart of this object is a supermassive **black hole**. M87 was identified with the strong radio source Virgo A by Walter **Baade** and Rudolf **Minkowski** in 1954. It is also a powerful source of X rays, and sits near the center of a hot, X-ray emitting cloud that extends over vast reaches of the Virgo Cluster.

Visual magnitude:	8.6
Apparent size:	7′
Distance:	60 million light-years
Position:	R.A. 12h 31m, Dec. +12° 24′

M101 (NGC 5457)

A very large, relatively nearby, face-on spiral galaxy, also known as the *Pinwheel Galaxy.* It was discovered by Pierre **Méchain** in 1781 and was among the first "spiral nebulae" identified by Lord **Rosse**. While appearing superficially symmetric on short exposures that show only the central region, it is actually remarkably unsymmetric, with a core considerably displaced from the center of the disk. Halton **Arp** included M101 as No. 26 in his *Catalogue of Peculiar Galaxies* with the description "Spiral with One Heavy Arm." It is the brightest of a group of at least nine galaxies, among which NGC 5474 (type Sc, magnitude 10.85) to the south-southeast and NGC 5585 (Sa, magnitude 11.49) to the northeast are the other most prominent. The *M101 Group* lies physically close to the larger M51 (NGC 5194) Group, and the two are often included together in lists as one large group.

Visual magnitude:	7.9
Apparent size:	22′
Diameter:	170,000 light-years
Distance:	27 million light-years
Position:	R.A. 14h 3.2m, Dec. +54° 21′

M110 (NGC 205)

One of the two bright satellite galaxies of the **Andromeda Galaxy** (M31)–the other being **M32.** Although discovered by Charles **Messier** in 1773 and depicted in his drawing of the "Great Andromeda Nebula" (1807), it was not included by Messier in his catalog, for unknown reasons. In fact, it became the last object in the extended

M80 One of the densest globular clusters in the Milky Way Galaxy, as imagined by the Hubble Space Telescope. *NASA/STSCI/AURA*

Messier list when added by the Welsh astronomer Kenneth Jones (1915–1995) in 1966. Previously described as a **dwarf elliptical galaxy** of type E5 or E6 in Hubble's classification, M110 is now more often classed as a **dwarf spheroidal galaxy**. However, as it is much brighter than typical dwarf spheroids, the Canadian astrophysicist Sidney van den Bergh recently introduced the term "spheroidal galaxy" to describe it and similar systems, including other **Local Group** members NGC 147 and NGC 185. Despite its small mass of some 10 billion M_{sun}, M110 has a halo containing eight **globular clusters**–the brightest of them, G73, within reach of large amateur telescopes. See also **Local Group**.

MACHO (Massive Compact Halo Object)

A hypothetical compact object, large numbers of which might account for a significant fraction of the **dark matter** in the **galactic halos** of large galaxies, including that of our own. If MACHOs exist they could take the form of **black holes**, **neutron stars**, **red dwarfs**, **brown dwarfs**, large free-floating planets, or some other type of nonluminous body. A signature of these objects would be the occasional amplification of light from more distant sources by **microlensing**. No conclusions have yet been drawn about the existence or nature of MACHOs; however, various microlensing events detected by search efforts such as **Optical Gravitational Lensing Experiment** OGLE and the **MACHO Project** support the idea that invisible compact objects do exist in the Galaxy's halo.

MACHO Project

A collaboration between scientists at the Mount Stromlo and Siding Spring observatories, Australia; the Center for Particle Astrophysics at the Santa Barbara, San Diego, and Berkeley campuses of the University of California; and the Lawrence Livermore National Laboratory to search for **microlensing** events from **MACHOs** (Massive Compact Halo Objects) using a **CCD**-array mounted on the 1.3-m telescope at Mount Stromlo.

macroturbulence

Large-scale motions, at up to 50 km/s, in a stellar atmosphere that smear out a star's **spectral lines** without changing their effective width. The effect of macroturbulence is particularly noticeable in the spectra of giant and supergiant stars. Compare with **microturbulence**.

Mädler, Johann Heinrich von (1794–1874)

A German astronomer who, with Wilhelm **Beer**, published the most complete contemporary map of the Moon–the four-volume *Mappa Selenographica* (1834–1836). It was the first lunar map to be divided into quadrants and it remained the standard until Julius **Schmidt**'s map of 1878. In 1840 Mädler left his Berlin home to become director of the Dorpat Observatory in Estonia. He formed the curious idea that **Alcyone** (Eta Tauri) was the star lying at the center of the Galaxy.

Maestlin, Michael (1550–1631)

A German astronomer and professor of mathematics at Tübingen who taught and corresponded with Johannes **Kepler**. Although he lectured generally using the **Ptolemaic system**, he chose also to teach to a select group of students, including Kepler, about the new **Copernican system** in which Earth and the other planets revolve around the Sun. He thus played an important role in Kepler's adoption of the heliocentric viewpoint.

Maffei 1 Group

The nearest group of galaxies to our own **Local Group**; it was probably once part of the Local Group but was ejected following a violent encounter with the **Andromeda Galaxy**. Its dominant member is the giant **elliptical galaxy**, Maffei 1, which, together with Maffei 2, was discovered on infrared plates in 1968 by the Italian astronomer Paolo Maffei (1926–). Both galaxies lie near the galactic equator in **Cassiopeia** in the so-called **zone of avoidance** and are thus heavily obscured at visible wavelengths by dust and gas. Maffei 2 is a medium-sized, average-brightness **barred spiral galaxy** (type SBc II) about 16 million light-years away, while the distance to Maffei 1 has been put at around 10 million light-years. Other known galaxies in the group are IC 342, **Dwingeloo 1**, Dwingeloo 2, and some smaller systems, including two possible satellites of Maffei 1.

Magellan

A NASA **Venus** radar mapping probe, launched in May 1989, that entered a highly elliptical, near-polar orbit around the second planet in August 1990; it subsequently mapped the entire surface of Venus with an average resolution of better than 300 m. In 1993 its orbit was intentionally lowered and circularized by atmospheric drag to enable the probe to map the planet's gravitational field. Magellan burned up in the atmosphere of Venus in October 1994.

Magellan Telescopes

Twin 6.5-m telescopes that are jointly owned and operated by the Carnegie Institution, the Massachusetts Institute of Technology, and the universities of Arizona, Harvard, and Michigan, and that are located at the **Las Campanas Observatory** in the Chilean Andes. The first telescope was completed in 1998 and achieved **"first light"** in September 2000; the second saw first light in September 2002.

Magellanic Clouds

Two irregular galaxies, known individually as the **Large Magellanic Cloud** and the **Small Magellanic Cloud**, that orbit the **Milky Way Galaxy**. Easily visible as misty patches to naked-eye observers in the Southern Hemisphere, they are named after the Portuguese explorer Ferdinand Magellan (1480–1521), who described them during his voyage around the world.

Magellanic Stream

A filament of neutral hydrogen, spanning roughly 300,000 light-years, that may have been torn out of the **Magellanic Clouds** by the **Milky Way Galaxy** some 200 million years ago. It forms an arc in the southern sky about 150° long, stretching from the region between the Magellanic Clouds and passing close to the south galactic pole.

magma
Molten rock in the interior of a planet or moon. If it reaches the surface it is known as **lava**; upon solidifying it forms **igneous** rock.

magnetar
A **neutron star** with a fantastically strong magnetic field, of the order of 10^{11} tesla–about a thousand trillion times stronger than Earth's, 100 to 1,000 times stronger than the field of a radio **pulsar** (the normal observed form of neutron star), and perhaps stronger than any other magnetic field in the universe. Such an object would be off-limits to humans: because a magnetar acts as a colossal electromagnetic generator, a person in a spacecraft floating nearby would feel 100 trillion volts between his head and feet! A magnetar as close as the Moon would rearrange the molecules in our bodies. Magnetars are the only objects in space powered mainly by magnetism and are now considered to be the most likely explanation for **soft gamma repeaters** (SGRs) and **anomalous X-ray pulsars** (AXPs). Their existence was first theorized in the late 1980s by the American astronomer Robert Duncan and the Canadian astronomer Christopher Thompson, who concluded that, in principal, neutron stars could be formed with magnetic fields that were very much stronger than those seen in familiar pulsars. Such an intense field, they reasoned, would put the star's surface under enormous stress, perhaps triggering **starquakes** and leading to intense bursts of high-energy radiation. In 1992, they used their magnetar theory to explain an enormously powerful **gamma-ray burst** detected on March 5, 1979, and later identified with a young **supernova remnant**, N49, in the **Large Magellanic Cloud** and, later still, recognized as the first-recorded SGR. In 2002, the American astrophysicists Brian Kern and Chris Martin of the California Institute of Technology showed that the magnetar theory is the only reasonable explanation for the behavior of 4U0142+61, a faint object in Cassiopeia. In an ordinary pulsar, the optical pulsations are a diluted byproduct of the X-ray pulsars and are therefore relatively weaker. But in the case of 4U0142+61, a quarter of the pulsed output is visible light compared with only a 3% contribution from X rays–exactly the behavior expected of a magnetar.

magnetic cataclysmic binary
A **cataclysmic binary** in which the **white dwarf** primary has a strong magnetic field that radically affects the **accretion** flow in the system. Magnetic cataclysmic binaries (also known as *magnetic cataclysmic variables*) fall into two main classes based on the strength of their magnetic fields: **AM Herculis stars** (or *polars*) and **DQ Herculis stars** (or *intermediate polars*).

magnetic star
A star with a strong magnetic field. Types include **Ap stars**, **magnetic cataclysmic binaries**, and **magnetars**.

magnetite (Fe_3O_4)
A black, strongly magnetic form of iron oxide. Tiny crystals of magnetite are produced by some terrestrial bacteria and used for orientation. The discovery of similar crystals in some **Mars meteorites**, including ALH84001, has been taken as evidence of past Martian life, although this claim is hotly disputed.

magnetogram
A chart that shows the strength, polarity, and distribution of magnetic fields across the Sun's disk. It is produced by a **magnetograph**, which measures the **Zeeman effect** by taking two narrowband images in a **spectral line** that is sensitive to the magnetic field. Gray areas in a magnetogram indicate that there is no magnetic field along the line of sight, while black and white areas indicate regions where there is a magnetic field.

magnetograph
An instrument attached to a **solar telescope** for the purpose of producing **magnetograms**. The first solar magnetograph was developed by Horace **Babcock** in 1953.

magnetohydrodynamics
The study of how **plasmas** behave in the presence of electric and magnetic fields. It forms an important part of astrophysics since plasma is one of the commonest forms of matter in the universe, occurring in stars, planetary **magnetospheres**, and interplanetary and interstellar space. See also **Alfvén waves**.

magnetopause
The boundary between a planet's **magnetosphere** and the magnetic field of the **solar wind**. It forms at a distance where the solar wind dynamic pressure equals the magnetic pressure of the planet's field. In the case of Earth, this is typically 8 to 11 Earth radii upwind, depending on solar wind conditions. See also **magnetosheath** and **heliosphere**.

magnetosheath
The region between a planet's **magnetopause** and the **bow shock** caused by the **solar wind**. Within the magnetosheath, the magnetic field is turbulent, distorted, and weaker than the magnetospheric field. The particles in the magnetosheath come from the shocked solar wind and have a density that typically decreases from the bow shock to the magnetopause but always remains higher than the magnetospheric plasma density. See also **heliosphere**.

magnetosphere

The region of space in which a planet's magnetic field dominates that of the **solar wind**. It is distorted into a teardrop shape by the solar wind pushing on the dayside and drawing out a long **magnetotail** on the nightside. Earth's magnetosphere normally extends about 10 Earth radii on the dayside, while its tail stretches out several hundred Earth radii in the anti-Sunward direction. It is, however, a highly dynamic structure that responds dramatically to changes in the dynamic pressure of the solar wind and the orientation of the **interplanetary magnetic field**. Its ultimate source of energy is the interaction with the solar wind. Some of the energy extracted from this interaction goes directly into driving various magnetospheric processes, while some is stored in the magnetotail, to be released later in **substorms**. Significant magnetospheres also exist around Mercury and the four gas giants of the solar system: Jupiter, Saturn, Uranus, and Neptune. The Sun's own enormous magnetosphere is called the **heliosphere**.

magnetotail

The part of a planet's **magnetosphere** that is elongated away from the Sun by the **solar wind**. Earth's magnetosphere extends about 65,000 km on the dayside but more than 10 times farther (beyond the Moon's orbit) on the nightside. Jupiter's magnetotail extends beyond the orbit of Saturn.

magnification

The factor by which the **angular diameter** of an object is apparently increased when viewed through a telescope or other optical instrument. It can be calculated by dividing the **focal length** of the telescope by the focal length of the **eyepiece**. The best magnification to use depends on the type of observation and on the

magnetosphere Earth's magnetosphere is shaped and influenced by charged particles flowing out from the Sun. *NASA*

seeing conditions. High magnification may be necessary, for example, for separating close double stars or for resolving fine detail on a planet's surface, but it can also be a disadvantage. It results a smaller **field of view**, a dimmer image with less **contrast**, and the emphasis of any shortcomings in an instrument or atmospheric disturbances. As a rough guide for the amateur, the practical upper limit to a telescope's magnification is twice the instrument's **aperture** in millimeters. A lower limit to magnification is set by the size of the **exit pupil**. When this exceeds the size of the pupil of the eye, light is wasted and the image appears no brighter than if the magnification were increased.

magnitude

The **brightness** of a celestial object, measured on a scale in which lower numbers mean greater brightness. The magnitude system stems from the ancient Greeks who ranked stars from first to sixth magnitude: those of first magnitude being the first to appear after sunset, those of sixth magnitude being at the limit of naked-eye visibility in a dark sky. In the nineteenth century, when it became possible to accurately measure the relative brightness of stars, the system was put on a strict quantitative footing by the English astronomer Norman Pogson (1829–1891). On this new scale (known as the *Pogson scale*), defined so that most of the traditional magnitudes of stars stayed roughly the same, a difference of one magnitude corresponds to a change in brightness by a factor of 2.512, while a jump of 5 magnitudes equals a brightness change of exactly 100-fold. **Apparent magnitude** measures how bright an object looks from Earth. **Absolute magnitude** measures an object's intrinsic brightness and is defined as the apparent magnitude an object would have if viewed from a distance of 10 parsecs (32.6 light-years). **Bolometric magnitude** measures brightness over all wavelengths, not just those of visible light.

magnitude of the tenth brightest member

A way of gauging the approximate distance to a **cluster of galaxies** that involves measuring the **apparent magnitude** of the cluster's tenth brightest member. The method is based on statistical studies that show that clusters of galaxies tend to contain similar distributions of brighter and dimmer members. By assuming that the tenth brightest member of every cluster has the same brightness, and taking a standard value for the intrinsic brightness of this galaxy, the distance can be found.

Maia (20 Tauri)

The fourth brightest star in the **Pleiades** after **Alcyone, Atlas**, and Electra; it is named after one of the seven mythical daughters of Atlas and Pleione. Maia is a **B star** that spins more slowly than any other B star in the Pleiades and, as a result, has an atmosphere in which different kinds of atoms are able to drift down under the pull of gravity, while others are lifted up by radiation. These effects make Maia a **mercury-manganese star**–a star in which these two, and certain other, chemical elements are greatly enhanced. Maia has also been at the center of a debate. In 1955, Otto **Struve** proposed the existence of a group of variables between spectral types B7V-III and A2V-II, with periods of a few hours, of which Maia was the prototype and Pherkad (Gamma Ursae Minoris) another member. However, the reality of the "Maia variables" was disputed and it is now known that Maia itself and some others in the purported class are in fact stable (though others do vary but not for any common reason). Maia illuminates a nearby **reflection nebula** known as the *Maia Nebula* or NGC 1432 (apparent diameter of 30′; R.A. 3h 45.8m, Dec. +24° 22′).

Visual magnitude:	3.87
Absolute magnitude:	–1.35
Spectral type:	B8III
Surface temperature:	12,600 K
Luminosity:	700 L_{sun}
Distance:	360 light-years

main sequence

The curving track on the **Hertzsprung-Russell diagram**, from top left (high temperature, high luminosity) to lower right (low temperature, low luminosity), along which 90% of visible stars lie. A star arrives on the main sequence after it starts **hydrogen burning** in its core and remains there throughout its core-hydrogen-fusion phase. A star's position and length of stay on the main sequence depend critically on mass. The most massive stars–the hot, blue-white **O stars** and **B stars**–occur to the upper left and have main-sequence lifetimes of only a few million or tens of millions of years. The least massive, hydrogen-burning stars, the **red dwarfs**, sit to the lower right on the chart and may remain on the main sequence for hundreds of billions of years.

main-sequence fitting

The process of determining the distance to a star cluster by overlaying its **main sequence** on the standard **zero-age main sequence** and noting the difference between the cluster's **apparent magnitude** and the zero-age main sequence's **absolute magnitude**.

main-sequence turnoff

The point on the **Hertzsprung-Russell diagram** of a star cluster at which stars begin to leave the **main sequence**. The main-sequence turnoff is a measure of age: in general, the older a star cluster, the fainter the main-sequence turnoff.

major axis
The greatest diameter of an **ellipse**; it passes through the two foci.

major planet
Any of the nine **planets**, Mercury through Pluto (though the status of Pluto has been questioned). "Minor planet," by contrast, is another name for an **asteroid**.

Maksutov Telescope
A type of **catadioptric telescope** (an instrument combining mirrors and lenses), developed by the Russian optical specialist Dmitri Maksutov (1896–1964). It uses a spheroidal primary mirror, a deeply curved spheroidal meniscus **corrector plate**, and a small spheroidal secondary mirror fixed to, or simply silvered onto, the back of the corrector plate, to give a triple-folded optical path. The result is a very compact telescope of long **focal length** that gives a wide **field of view** and images of excellent quality. Well-made "Maks" outperform **Schmidt-Cassegrain Telescopes** of equal quality and have became popular with amateur astronomers, especially for **astrophotography** and planetary viewing. However, because their corrector plates are difficult to manufacture for large aperture instruments, their professional application is limited.

Malin-1
The first **low-surface brightness galaxy** to be discovered and the largest **spiral galaxy** known; it lies about 1 billion light-years away in **Coma Berenices**. Despite its diameter of nearly 600,000 light-years and mass of over 2 trillion solar masses, its gas-rich disk is so faint that it was found only by accident in 1987 by the American astronomers Christopher Impey and Gregory Bothun on photographic plates processed by the English-born astronomer and photographer (now based in Australia) David Malin (1941–).

Malmquist effect
A statistical bias that has to be taken into account in surveys of galaxies and other remote objects. It arises because as the distance to a collection of objects increases, so does the tendency to see only the brighter objects present, thus giving a false impression of average brightness with distance. The effect is named after the Swedish astronomer Karl Malmquist (1893–1982).

manganese star
A **B star**, with a surface temperature of 10,000 to 15,000 K whose spectrum shows an unusually high ratio of manganese to iron. This and other chemical anomalies are due to the physical separation of different kinds of ions in the stellar atmosphere. Manganese stars rotate slowly, as **Ap stars** do, but lack their strong magnetic fields. The spectra, temperature, and other properties are intermediate between those of Ap stars and **Am stars**. See also **mercury-manganese star**.

mantle
The main bulk of a rocky planet or a large moon, lying between the crust and the core and differing in composition from both. Earth's mantle extends from about 7 km (beneath the continents) or 30 km (beneath the oceans) to 2,900 km below the surface.

many-body problem
The mathematical problem of finding the positions and velocities of any number of massive bodies that interact with one another gravitationally, at any point in the future or the past, given their present positions, masses, and velocities. The problem can only be solved precisely for all cases of the **two-body problem** and for special cases of the **three-body problem**. High-speed computers enable approximate solutions to be found for general cases of the many-body problem, but because of chaos and rounding errors, these solutions decline in accuracy as the period over which the behavior is calculated increases. Two broad principles govern the overall behavior of a many-body system: the center of mass of the system moves with constant velocity, and the total energy and total angular momentum of the system remain constant.

MAP (Microwave Anisotropy Probe)
A NASA satellite designed to map fluctuations in the **cosmic microwave background** over the whole sky with unprecedented accuracy and to provide insights into the formation of galaxies and the basic parameters of cosmology. MAP was launched on June 30, 2001, on a two-year mission.

Marcy, Geoffrey (1954–)
An astronomer at the University of California, Berkeley, and lead investigator with the California-Carnegie Planet Search. Together with Paul Butler and other colleagues, he has been responsible for the discovery of well over half the presently known **extrasolar planets**. Since 1998, he has served on the NASA working groups concerned with the **Terrestrial Planet Finder**. Marcy obtained a B.A. from the University of California, Los Angeles (1976), and a Ph.D. from the University of California, Santa Cruz (1982).

Maria family
A **Hirayama family** of **asteroids**, with about 70 known members, mostly S-class, that orbits the Sun at a mean

distance of 2.55 AU; its members are unusual for main belt asteroids in that they have a fairly high inclination of about 15°. The prototype and one of the largest of the group is the 40-km-wide (170) Maria, discovered in 1877 by H. Joseph Perrotin (1845–1904); semimajor axis 2.552 AU, perihelion 2.39 AU, aphelion 2.72 AU, inclination 14.4°, period 4.08 years.

Marius, Simon (1573–1624)

A German astronomer (also known as Mayr) who named Jupiter's **Galilean moons**. He and **Galileo** both claimed to have discovered them in 1610 and likely did so independently. Marius was also the first to observe the **Andromeda Galaxy** with a telescope and one of the first to observe **sunspots**.

Markarian Chain

A spectacular arc of galaxies in the central region of the **Virgo Cluster**. The Chain starts with the two bright elliptical galaxies M84 (NGC 4374, at R.A. 12h 25m, Dec. +12° 53′) and M86 (NGC 4406) and curves away to the northeast, taking in NGC 4435, NGC 4438, NGC 4458, NGC 4461, NGC 4473, and NGC 4477. There is no compelling evidence that any of the systems in the chain are interacting. See also **Eyes, the**.

Mars

See article, pages 315–318.

Mars meteorites

More than two dozen meteorites have been found on Earth that have almost certainly come from Mars. Most of them belong to the group known as *SNC meteorites*, which includes the **shergottites**, **nakhlites**, and **chassignites**. All of the SNC meteorites contain minerals that crystallized within the past 1.35 to 0.15 billion years, making them much younger than any other known **achondrites**. Trapped gases within them exactly match the composition of the Martian atmosphere as measured by the **Viking** landers, confirming their place of origin. Oddly enough, the most famous Mars meteorite, ALH 84001, which became the center of a major controversy in 1996 when a group of NASA scientists claimed it

(continued on page 319)

Mars meteorites The meteorite known as EETA 79001, found on the Antarctic ice and believed to have come from Mars. For scale, the cube at the lower right is 1 centimeter per side. The meteorite is partly covered by a black glassy layer—the fusion crust—which formed during the object's fall through Earth's atmosphere. Inside, the meteorite is gray basalt, very similar to terrestrial basalts. EETA 79001 appears to have solidified after having been ejected from a volcano on Mars about 180 million years ago. *Lunar and Planetary Institute*

Mars

The fourth planet from the Sun, named for the Roman god of war and also known as the Red Planet; its ruddy hue is due to iron oxide (rust) in the surface rocks. Mars has only one-tenth the mass and one-quarter the surface area of Earth but, because it lacks oceans, the area of Mars's accessible dry land is roughly equal to that of Earth's dry land. Its atmosphere consists of carbon dioxide (95.3%), nitrogen (2.7%), and argon (1.6%), with small amounts of oxygen (0.15%), water vapor (0.03%), and other gases. The surface pressure averages about 7 millibars (less than 1% of Earth's 1,013 mb) but ranges from almost 9 mb in the deepest basins to about 1 mb at the highest elevations. Though very thin, the atmosphere can still support strong winds, which occasionally whip up vast dust storms that can obscure the planet for months.

One effect of Mars's significantly elliptical orbit is to exaggerate seasonal temperature variations on the surface. While the average temperature on Mars is about –55°C, surface temperatures range from as low as –133°C at the winter pole to almost 27°C on the dayside during summer. Permanent ice caps at both poles are composed almost entirely of water ice with a seasonably variable coating of solid carbon dioxide (dry ice). These caps have a layered structure in which alternating strata contain different amounts of dark dust. What causes the layering isn't certain; one possibility is climate change linked to long-term changes in the tilt of the planet's axis. Seasonal variations in the amount of carbon dioxide at the polar caps alter the global atmospheric pressure by about 25%.

Mars appears to have an iron core, some 1,700 km in radius, that is at least partly molten. Surrounding this is a thick mantle of soft silicate rocks, overlaid in turn by a crust that varies in depth from about 35 km in the northern hemisphere to 80 km in the south.

Mars has a varied and interesting surface, some of the highlights of which are described in the table "Notable Surface Features of Mars." The southern hemisphere is dominated by ancient cratered highlands similar to those of the Moon. In contrast, most of the northern hemisphere consists of plains that are much younger, lower in elevation, and have a more complex history. An abrupt elevation change of several kilometers seems to occur at the boundary. The origin of this global dichotomy and sharp boundary are unknown; one theory invokes a massive impact shortly after Mars formed.

Mars has an important place in human imagination due to long-standing speculation that it harbors, or has harbored, life. This idea was encouraged, in the last quarter of the nineteenth century, by reported observations of linear features on the surface, argued by Percival **Lowell** and some others to be artificial, and of seasonal changes in the brightness of some areas that were thought to be caused by vegetation growth. The linear features are now know to be nonexistent or in some cases, ancient dry water-courses. The color changes have been ascribed to dust storms. However, interest has been rekindled in the possibility of Martian life by several factors. These include controversial remains contained in meteorites that have come from Mars (see **Mars meteorites**), mounting evidence that water has played–and may continue to play–a decisive role in the Martian surface and subsurface environment, and the discovery of hardy microbes on Earth (known as extremophiles)

Equatorial diameter:	6,794 km
Oblateness:	0.0092
Mass (Earth = 1):	0.11
Mean density:	3.95 g/cm^3
Surface gravity (Earth =1):	0.38
Escape velocity:	5.1 km/s
Albedo:	0.16
Surface temperature	
Average:	–25°C
Minimum:	–187°C
Axial inclination:	23° 59′
Rotational period:	24h 37m 23s
Orbit	
Mean distance from Sun:	1.524 AU (227,940,000 km)
Perihelion:	1.381 AU (206,700,000 km)
Aphelion:	1.666 AU (249,100,000 km)
Closest approach to Earth:	55,750,000 km
Eccentricity:	0.0934
Inclination:	1.85°
Period:	687 days

that would probably be capable of surviving on Mars. The **Viking** landers carried out a series of life-seeking experiments but produced either negative or ambiguous results. Further investigations are planned by the Beagle 2 lander carried by Mars Express (see **Mars spacecraft**).

Much of the renewed optimism for life on Mars comes from a wealth of data suggesting that, up to about 3.5 billion years ago, Mars was much more like Earth, with large amounts of surface water–a key ingredient to life as we know it. Some images from MGS, showing what appear to be recently formed gullies and debris flow features, even hint at the continued presence of liquid water, possibly in subsurface aquifers. Especially intriguing are dark stains that have appeared in a period of less than a year in certain regions, such as around Olympus Mons. Measurements by Mars Odyssey strongly indicate that vast amounts of water, most of it probably frozen, lie at shallow depths over much of the planet's surface.

Like Mercury and the Moon, present-day Mars appears to lack active plate tectonics. With no sideways plate motion, hot-spots under the crust are fixed relative to the surface. This, along with the lower surface gravity, may account for the Tharis bulge and its enormous volcanoes. Although there's no evidence of current volcanic activity, observa-

Mars Gullies on the layered north wall of a crater in Newton Basin (41.8°S, 158.0°W) on Mars. The picture was taken by Mars Gobal Surveyor in May 2002 and covers an area of 4.3 km by 2.9 km. Dark sand dunes are visible at the bottom of the image, especially at the lower right. Nearly all the craters in Newton Basin have numerous similar gullies, hinting that Newton is the site of an aquifer. Other theories argue that, instead of being the product of seepage and runoff of groundwater, such gullies may form by melting of ground ice, melting of surface snow (under climatic conditions different than today), or discharge of carbon dioxide that somehow became buried under the Martian surface. *NASA/JPL*

Mars A computer-processed image of the largest volcano on Mars, Olympus Mons, using data from Mars Global Surveyor and the Viking Orbiters. In several places the steep scarp surrounding the volcano has collapsed. Vertical distances have been exaggerated in this picture by a factor of 10 to 1. *NASA/JPL*

Notable Surface Features of Mars

Feature	Description
Argyre Planitia	An 800-km-diameter impact basin in the southern hemisphere that is the youngest such structure on the planet, with an age of about 3.5 billion years.
Chryse Planitia	A relatively smooth, circular plain, possibly an ancient impact basin, in the north region, 1,600 km across and 2.5 km below the mean level of the planet's surface. It seems to have suffered water erosion in the past and was the site of the Viking 1 landing.
Elysium Planitia	The second largest volcanic region on Mars, after Tharsis Montes. It measures 1,700 by 2,400 km and contains three large volcanoes: Elysium Mons, Hecates Tholus (NE of Elysium Mons), and Albor Tholus (SE of Elysium Mons).
Hellas Planitia	Formerly known simply as "Hellas," a near-circular impact basin, some 2,500 km wide and 5 km deep. It is conspicuous by its color, which is frequently lighter than surrounding areas due to overhanging mists and clouds.
Ma'adim Vallis:	One of the largest valley systems on Mars; named after the Hebrew for "Mars," it is about 860 km long, 8 to 15 km wide, up to 2,100 m deep, and located in the highlands of the southern hemisphere. Images sent back by Mars Global Surveyor suggest that Ma'adim Vallis formed some 3.5 billion years ago when a large lake, estimated to have been 1.1 million km^2 in area and 1,100 m deep, overflowed a low point in its perimeter.
Olympus Mons	The highest volcano in the solar system—three times higher than Mount Everest—which rises at a slope of only a few degrees from the surrounding plain, with a steep escarpment that borders the summit. A shield volcano, similar to those in the Hawaiian chain but vastly larger, it measures 624 km in diameter and 25 km in height, and has a **caldera** in the center, 80 km wide with multiple circular, overlapping collapse craters created by different volcanic events. The radial features on the slopes of the volcano were formed by overflowing lava and debris. Olympus Mons is found in the Tharsis Montes region near the Martian equator.
Syrtis Major	A conspicuous dark, roughly triangular marking, about 1,200 km long and 1,000 km wide, centered at about +10° N, 290° W; it was first noted by Christiaan **Huygens** in 1659. Formerly known as the Hourglass Sea or the Kaiser Sea, its present name is Greek for "great sandbank," which is appropriate since it appears to consist of a large area of wind-blown dust.
Tharsis Montes	An extensive upland region, from which rise three giant shield volcanoes, Ascraeus Mons, Pavonis Mons, and Arsia Mons, each to about 27 km above the datum level for the planet. Tharsis Montes, also known as the Tharsis Ridge, extends for 2,100 km and varies in height between about 9 and 11 km above the datum level.
Utopia Planitia	A vast, sparsely cratered, sloping plain, about 3,200 km across, centered at latitude 48° N, longitude 277°. Viking 2 landed in eastern Utopia Planitia, about 200 km west of the crater Mie.
Valles Marineris	The largest system of canyons in the solar system. Just south of the Martian equator, it is about 4,000 km long—as wide as the continental United States. The central individual troughs, generally 50 to 100 km wide, merge into a depression as much as 600 km wide. In places the canyon floor reaches a depth of 10 km—six to seven times deeper than the Grand Canyon. The geologic history of the central canyon system is complex: first the surface collapsed into a few deep depressions that later became filled with layered material, perhaps as lake deposits. Then **graben**-forming faults cut across some of the older troughs thus widening existing troughs, breaching barriers between troughs, and forming additional ones. At that time the interior deposits were locally bent and tilted, and perhaps water, if still present, spilled out and flowed toward the outflow channels. Huge landslides fell into the voids created by the new grabens.

Mars Earth (left) and Mars (right) shown side-by-side and to scale. *NASA*

tions suggest that Mars may have been tectonically active early on, making comparisons with Earth all the more interesting. Lacking plate tectonics today, Mars can't recycle any of the carbon dioxide in its rocks back into its atmosphere and so can't sustain much of a greenhouse effect. The surface of Mars is therefore much colder than Earth would be at that distance from the Sun. Despite a lack of plate tectonics, Mars has clearly been active in other ways. Data collected by the Mars Global Surveyor and Odyssey probes have revealed an extraordinary amount and diversity of layering in the surface rocks, which planetary geologists will now study to uncover secrets of the planet's past.

Large, but not global, weak magnetic fields exist in various regions of Mars, probably remnants of an earlier global field that has since disappeared. Mars has two tiny satellites, **Phobos** and **Deimos**, both believed to be captured asteroids. (See table, "The Moons of Mars.")

The Moons of Mars

		Orbit			
Name	**Diameter (km)**	**Distance* (km)**	**Period (d)**	**Inclination (°)**	**Eccentricity**
Phobos	27×22×19	9,378	0.32	1.1	0.021
Deimos	16×12×10	23,500	1.26	1.6	0.003

*Mean distance from the center of Mars.

(continued from page 314)

contained fossils and other biological remains, is not a member of the SNC group and has a much older crystallization age of 4 billion years. Of the original evidence put forward in favor of past life, the strongest has proved to be chains of magnetite crystals. These crystals, found within ALH 84001, closely resemble similar structures formed by certain types of bacteria on Earth.

Mars spacecraft

See article, pages 320–321.

Mars Trojan

An **asteroid** located at either of the stable **Lagrangian points** (L_4 or L_5) of the orbit of **Mars**. Two are known, (5261) Eureka, discovered in 1990, and the provisionally named 1998 VF31, though it is likely that more are waiting to be found.

Mars-crosser

An **asteroid** whose **perihelion** lies within part or all of the orbit of **Mars**. Mars-crossers are subdivided into *shallow Mars-crossers,* with perihelia between 1.58 and 1.67 AU, and *deep Mars-crossers,* with perihelia between 1.3 and 1.58 AU; by comparison the distance of Mars from the Sun varies between about 1.38 and 1.67 AU. Mars-crossers are believed to have originated as main-belt asteroids that fell into a 3:1 orbital **resonance** (at a heliocentric distance of 2.5 AU) with **Jupiter**. Objects that go around the Sun three times for every orbital period of Jupiter meet up with the giant planet at the same point every third orbit. Jupiter's powerful gravity then perturbs the asteroid's path, increasing its **eccentricity** with each encounter. Over a period of about 100,000 years, as perturbations accumulate, the asteroid becomes a Mars-crosser. Further perturbations from Mars, over a much longer period of several million years, can then transform the asteroid into an **Earth-crossing asteroid**–and a potential hazard.

mascon

A region on the **Moon**'s surface where the gravitational pull is slightly higher than normal due to the presence of heavy rock; "mascon" is short for "mass concentration." Mascons were first detected by spacecraft orbiting the Moon and are associated with basalt-filled basins such as Mare Imbrium.

maser

A **microwave** oscillator; "maser" stands for "microwave amplification by stimulated emission of radiation." Masers (and their optical counterpart, lasers) involve the interaction between an electromagnetic wave of a certain wavelength and an atom or a molecule in a suitable (excited) energetic state. The passage of the wave triggers the atom/molecule to give up energy in the form of more radiation of exactly the same wavelength. This reinforces the passing wave, which can then interact with more excited atoms to build up a well-directed, intense pulse of monochromatic (single-wavelength) radiation. Several types of **interstellar maser** have been identified but, as yet, no optical or infrared astrophysical lasers.

mass

A measure of the total amount of material in a body, defined either by the inertial properties of the body or by its gravitational influence on other bodies.

mass discrepancy

In the study of **clusters of galaxies**, the difference between the mass of a cluster obtained by using the **virial theorem**, M_{VT}, and the mass inferred from the total luminosities of the member galaxies, M_L. Typically $M_{VT} / M_L > 10$, implying the existence of large amounts of **dark matter**.

mass extinction

A relatively sudden, global decrease in the diversity of life-forms on Earth. Some mass extinctions appear to have resulted, at least in part, from extraterrestrial catastrophes involving **asteroid** or **comet** impacts and, possibly, nearby **supernova**e or other stellar outbursts. See also **K-T boundary**. (See table, "The Five Largest Known Mass Extinctions.")

mass function

A numerical relation between the masses of the two components of a **spectroscopic binary** that provides information on the relative masses of the two stars when the **spectral line**s of only one component can be seen. If M_p is the mass of primary (whose spectrum is known), M_s is the mass of secondary, and i is the inclination of the orbit, the mass function is given by

$$(M_s^3 \sin^3 i) / (M_p + M_s)^2.$$

mass loss

The outflow of particles and gas from a star, occurring at varying rates and by a variety of processes, throughout its lifetime. **Bipolar outflow**s are often associated with **protostar**s, while powerful **stellar wind**s blow from **T Tauri star**s, **giant**s, and **supergiant**s. Ejected material builds up as a disk around a **Be star** or as a shell around a **Gamma Cassiopeiae star**; in late evolutionary stars it may give rise to a **planetary nebula**. More violent ejection accompanies a **nova** or **supernova** explosion. Mass loss may

(continued on page 321)

Mars spacecraft

Mars has been more intensively studied by space probes than any other object in the solar system apart from the Moon. More than 30 spacecraft have been launched Marsward since the early 1960s, with varying degrees of success, from failure at liftoff to complete triumph in the case of probes such as Mariner 9 and the **Viking** missions. The most recent successful missions and those scheduled for launch over the next few years are described, in alphabetical order, in the following sections.

Mars Exploration Rovers

Twin NASA missions to Mars, launched in June 2003 and, scheduled to arrive in January 2004, carrying two identical 103-kg rovers. The rovers will land at widely separated sites chosen because they appear to have been associated with liquid water in the past and may therefore have been favorable to life. One of these sites, Gusev Crater, 15° south of the Martian equator, is basically a big hole in the ground with a dry riverbed going right into it and is presumed to have held a lake at some point. The other location, Meridiani Planum, about 2° south of the equator and halfway around the planet from Gusev, is rich in gray hematite–a type of iron oxide characteristic of past water. Although the rovers are not equipped to search for life directly, they will seek to determine the history of climate and water at their sites, which has a direct bearing on the issue of possible Martian biology. Immediately after landing, the rovers will begin a reconnaissance by taking 360° visible and infrared image panoramas. Then they will drive off and begin their exploration. Using images and spectra taken daily from the rovers, mission scientists on Earth will command the vehicles toward rock and soil targets of particular interest and then evaluate their compositions and textures at microscopic scales. The rovers will be able to travel up to 100 m a day–as far as the Mars Pathfinder Sojourner rover went in its lifetime. Each carries a panoramic camera, a rock abrasion tool to expose fresh surfaces, a miniature thermal infrared spectrometer, a Mossbauer spectrometer, and an alpha-proton-X-ray spectrometer. The rovers are expected to function for at least 90 days.

Mars Express

A European Space Agency mission scheduled for launched on June 2, 2003 and due to arrive in late December 2003. It will be the first spacecraft to use radar to penetrate the surface of Mars and map the distribution of possible underground water deposits. It will also release the miniature Beagle 2 lander, which will carry instruments dedicated to looking for traces of life and carrying out geochemical analyses. These include a "mole" that can burrow along or through soil and collect samples for analysis by a mass spectrometer. To expose fresh material from inside rocks, Beagle 2 also will carry a modified dentist's drill. Samples will be subjected to stepped combustion–heated to successively higher temperatures–and then analyzed. An elevated ratio of carbon-12 to carbon-13 would be interesting, because living things preferentially use the lighter form of carbon. Beagle 2 will also sniff the atmosphere for tiny amounts of methane, which would suggest there might be microbes alive today on Mars.

Mars Global Surveyor (MGS)

A NASA orbiter launched in November 1996 and designed to investigate, over the course of a full Martian year, the surface, atmosphere, and magnetic properties of Mars in unprecedented detail. It has been so successful that its operation has been extended until late 2004. MGS carries four main science instruments. The Mars Orbiter Camera provides daily wide-angle images of Mars similar to weather photos of Earth, and narrow-angle images of objects as small as 1.5 m across. The Mars Orbiter Laser Altimeter bounces a laser beam off the surface to measure accurately the height of mountains and the depth of valleys. The Thermal Emission Spectrometer scans emitted heat to study both the atmosphere and the mineral composition of the surface. Finally, the Magnetometer and Electron Reflection experiment provides data on the magnetic state of the crustal rocks, which in turn sheds light on the early magnetic history of the planet.

Mars Odyssey 2001

A Mars orbiter which arrived at its destination in October 2001 after a seven-month cruise. Its main goals are to gather data to help determine whether the environment of Mars was ever conducive to life, characterize the climate and geology of Mars, and study potential radiation hazards to future astronaut missions. It will also act as a communications relay for upcoming missions to Mars over a five-year period.

Mars Pathfinder

A NASA mission that deployed the first Martian rover–a miniature 10-kg vehicle called Sojourner. Pathfinder, which made safe planetfall on July 4, 1997, in the lowland region of Ares Vallis, was the first successful Mars landing since **Viking**. It was aimed at studying the surface meteorology and geology in its immediate vicinity. During almost three months of operation, Pathfinder returned more than 16,000 images from the lander and 550 from the rover, carried out 15 chemical analyses of rocks, and gathered extensive data on winds and other aspects of the weather. Among the highpoint discoveries of the mission were that the rock size distribution at the landing site is consistent with a flood-related deposit, the rock chemistry might be different from that of the SNC meteorites, and dust is the main absorber of solar radiation in the Martian atmosphere.

Mars Reconnaissance Orbiter

A NASA orbiter, to be launched in August 2005, that will make high-resolution measurements of the surface from orbit, including images with resolution better than 1 meter. The main objectives of the mission will be to look for evidence of past or present water and to identify landing sites for future spacecraft. The orbiter will also be used as a telecommunications relay for missions to come.

Mars 2007

A long-range, long-duration NASA rover equipped to perform a variety of surface studies on Mars and to demonstrate the technology for accurate landing and hazard avoidance in difficult-to-reach sites. In the same year, CNES (the French space agency) plans to launch a remote sensing orbiter and four small Net-Landers, and ASI (the Italian space agency) plans to launch a communications orbiter to link to the Net-Landers and future missions.

NetLander

A European mission to Mars, led by CNES (the French space agency), that will focus on investigating the interior of the planet and the large-scale circulation of the atmosphere. NetLander consists of four landers that will be separated from their parent spacecraft and targeted to their locations on the Martian surface several days prior to the main craft's arrival in orbit. Once on the ground, the landers will form a linked network, each deploying a network science payload with instrumentation for studying the interior of Mars, the atmosphere, and the subsurface, as well as the ionospheric structure and geodesy. The mission is scheduled for launch in 2007.

mass loss

(continued from page 319)

also occur through the outer **Lagrangian point** in a contact binary or an overcontact binary (see also **close binary**).

mass-luminosity radius relation

All nondegenerate stars with the same mass and the same chemical composition will have the same radius and the same **luminosity**. See also **Vogt-Russell theorem**.

mass-luminosity ratio

The ratio of the mass of a system, expressed in solar masses, to its visual **luminosity**, expressed in solar luminosities. The **Milky Way Galaxy** has a mass-luminosity ratio in its inner regions of 10, indicating that the typical star is a dwarf of mass about half that of the Sun. A rich **cluster of galaxies** such as the **Coma Cluster** has a mass-luminosity ratio of about 200, indicating the presence of a considerable amount of **dark matter**.

mass-luminosity relation

A relationship between **luminosity** (intrinsic brightness) and mass for stars that are on the **main sequence**. Averaged over the whole main sequence (i.e., for stars of all masses), it is found that $L = M^{3.5}$, where both L and M are measured in solar units. This means, for example, that if the mass is doubled, the luminosity increases more than 10-fold. A more detailed examination shows that the relationship is different in different mass regimes. For

The Five Largest Known Mass Extinctions

Geological Period	Years Ago	Notes
Late Ordovician	438 million	100 families extinct, including more than half of all bryozoan and brachiopod species.
Late Devonian	360 million	30% of animal families extinct.
End of Permian	245 million	Trilobites extinct. 50% of all animal families, 95% of all marine species, and many trees die out.
Late Triassic	208 million	35% of all animal families die out, including most early dinosaur families and most synapsids, except for the mammals.
Cretaceous-Tertiary Boundary	65 million	About half of all life-forms died out, including the dinosaurs, pterosaurs, plesiosaurs, ammonites, and many families of fish, snails, sponges, sea urchins.

stars of less than 0.43 M_{sun} (in which **convection** is the sole energy transport process), $L = 0.23\ M^{2.3}$. For a mass greater than this and up to several solar masses, L varies as the fifth power of M, while for the very massive stars, L varies as M cubed. These relationships are empirical ones based largely on observations of binary stars.

mass-radius relation

The relationship between the radius, R, of a main-sequence star and its mass, M. If R and M are both in solar units, then $R = M^{0.8}$.

mass transfer

The process in which one evolved member of a **close binary** system passes gaseous material to its companion star. The donor star may have grown to fill its **Roche lobe** (in which case the system is a *semidetached binary*) or may be the source of a strong **stellar wind**. The recipient star receives the matter either directly or via an **accretion disk**.

Mathilde (minor planet 253)

A main-belt **asteroid**, discovered by Johann **Palisa**, that was passed at close range, on June 27, 1997, by the **NEAR-Shoemaker** probe on its way to **Eros**. Not only is Mathilde one of the slowest spinning asteroids (only 1220 Clocus and 288 Glauke have longer rotational periods), but it is also one of the blackest objects in the solar system–twice as dark as a chunk of charcoal–reflecting only 3% of the light that strikes it. Its surface has a spectroscopic signature the same as that of **carbonaceous chondrites** (spectral class C), whose typical density is about 2 g/cm^3. The fact that Mathilde has a much lower density than this, around 1.3 g/cm^3, suggests that it is a **rubble-pile asteroid** made of big chunks with large voids in between, or lots of loose, smaller pieces, like gravel. Diameter 66 × 48 × 46 km, rotational period 17.4 days, semimajor axis 2.646 AU, eccentricity 0.266, inclination 6.71°, period 4.31 years.

Mauna Kea Observatories

The cluster of observatories, containing 11 major telescopes, near the summit of Mauna Kea ("White Mountain"), a dormant volcano on the island of Hawaii, the largest and southernmost of the Hawaiian Islands. The highest point in the Pacific Basin, and the highest island-mountain in the world, Mauna Kea rises 9,750 m from the ocean floor to an altitude of 4,205 m above sea level, which places its summit above 40% of Earth's atmosphere. Among the instruments located there are the twin 10-m telescopes of the **Keck Observatory**, the 8.3-m

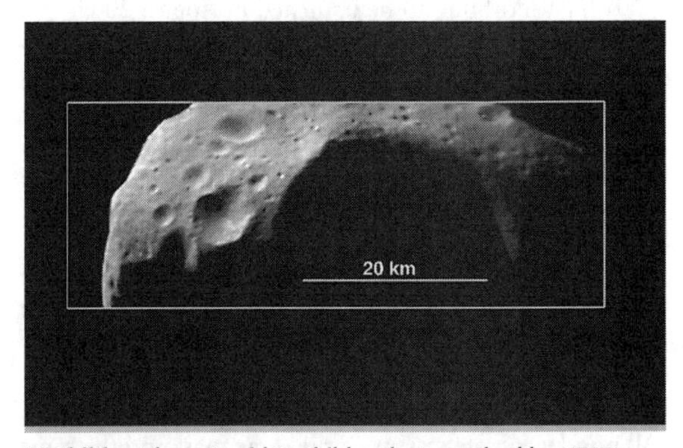

Mathilde The asteroid Mathilde, photographed by NEAR-Shoemaker from a distance of about 1,200 km, shortly after the spacecraft's closest approach. Numerous impact craters can be seen, ranging from over 30 km to less than 0.5 km in diameter. Raised crater rims suggest that some of the material ejected from these craters traveled only short distances before falling back to the surface; straight sections of some crater rims indicate the influence of large faults or fractures on crater formation. *NASA/Johns Hopkins University*

Subaru Telescope, the 8.1-m **Gemini Observatory** (North), the 3.8-m **United Kingdom Infrared Telescope**, the 3.1 **Canada-Hawaii-France Telescope**, the 3-m NASA Infrared Telescope Facility, the 15-m submillimeter-wave **James Clerk Maxwell Telescope**, and the 10.4-m **Caltech Submillimeter Observatory**. The **Submillimeter Array** is currently under construction, while the westernmost antenna of the **Very Long Baseline Array** is situated at a lower altitude 3 km from the summit. The atmosphere above the mountain is extremely dry, which is crucial for **infrared** and submillimeter observations, and cloud-free, so that the proportion of clear nights is among the highest in the world. The exceptional stability of the atmosphere above Mauna Kea permits more detailed studies than are possible elsewhere, while its distance from city lights and a strong island-wide lighting ordinance ensure an extremely dark sky. A tropical inversion cloud layer about 600 m thick, well below the summit, isolates the upper atmosphere from the lower moist maritime air and ensures that the summit skies are pure, dry, and free from atmospheric pollutants.

Maunder, (Edward) Walter (1851–1928)

An English astronomer who, after working briefly in a bank, became photographic and spectroscopic assistant at the **Royal Greenwich Observatory** in 1873. Maunder's appointment allowed Greenwich to branch out from purely positional work, for Maunder began a careful study of the Sun, mainly of **sunspots** and related phenomena. After 1891 he was assisted by his second wife, Annie Scott Dill Maunder, née Russell (1858–1947), a Cambridge-trained mathematician. It had been known since 1843 that the intensity of sunspot activity went through an 11-year cycle. In 1893 Maunder, while checking the history of the cycle found that between 1645 and 1715 there was virtually no sunspot activity at all (a fact known earlier to Gustav **Spörer**). For 32 years not a single sunspot was seen on the Sun, and in the whole period fewer sunspots were observed than have occurred in an average year since. He wrote papers on his discovery in 1894 and 1922 but they aroused no interest. More sophisticated techniques developed in recent years have established that Maunder was correct in his deduction of the so-called *Maunder minimum*. Also, the realization that the period of the minimum corresponds to a prolonged cold spell suggests that Maunder's discovery is no mere statistical freak. It may throw light on the Sun's part in long-term climatic change and on possible variations in the processes within the Sun that produce the sunspots.

Mayall Telescope

The largest optical telescope at **Kitt Peak National Observatory**. The 4-m Nicholas U. Mayall Telescope, named after a former director of the observatory, was completed in 1970.

McDonald Observatory

The observatory of the University of Texas, on the adjacent peaks of Mount Locke (altitude 2,070 m) and Mount Fowlkes (2,070 m), near Fort David in western Texas. It operates six telescopes, including the 9.2-m **Hobby-Eberly Telescope**; the 2.7-m Harlan J. Smith Telescope, opened in 1969, on Mount Locke; and the 2.1-m Otto Struve Telescope, opened in 1939, on Mount Fowlkes. The Observatory was founded in 1932 and is named after its benefactor William McDonald (1844–1926).

McIntosh scheme

A classification scheme for **sunspots** that superceded the older *Zürich scheme* in 1966. A three-letter code describes the class of sunspot group (single, pair, or complex), penumbral development of the largest spot, and compactness of the group.

M-class asteroid

A relatively bright and reflective **asteroid**, made mainly of metallic iron and nickel (the "M" is for "metal"), typically found at the middle of the main **asteroid belt**. M-class asteroids are slightly reddish and have featureless reflectance spectra over the range 0.3 to 1.1 microns. The largest known example is the 248-km-wide (16) Psyche.

McMath-Pierce Solar Telescope

The world's largest **solar telescope**; it opened in 1962 on Kitt Peak, Arizona, and is a facility of the **National Solar Observatory**. It comprises three telescopes in one–the main, east auxiliary, and west auxiliary systems–that can be used concurrently or independently. The main telescope is housed in an inclined shaft, 152 m long, that runs parallel with Earth's axis. A **heliostat** mounted on a 33.5-m-high tower reflects sunlight down this shaft to a 1.6-m mirror, about 50 m below ground level, which reflects the light partway back up the shaft to a flat mirror. From the flat mirror, the light travels vertically down to form an image of the Sun 85 cm wide in a subterranean observation room. The telescope is named after American solar physicists Robert McMath (1891–1962) and A. Keith Pierce (1918–).

Me star

An **M star** with **emission lines** in its spectrum. Some **Mira stars** are of this type, as also are some M-type dwarfs.

mean

The mean value of a set of observations is a straight arithmetic mean or average, i.e., the sum of the individual values divided by the number of values used. If the observations are weighted according to their relative reliability, the individual values are first multiplied by weighting factors before they are summed and the mean taken; this is then called a weighted mean.

mean motion

The constant **angular velocity** that an object, such as a planet, would have if it were moving in a circular orbit (instead of an elliptical one) of radius equal to its mean distance from the Sun, and equal to its actual revolution period. This is a hypothetical concept used in **celestial mechanics**.

mean motion resonance

The dynamical situation where the ratio of the orbital periods of two orbiting objects can be expressed as the ratio of two small integers. For example, **plutinos** are Kuiper Belt objects that are in a 3:2 **resonance** with Neptune, meaning that they orbit twice in the time Neptune completes three orbits. Mean motion resonances may lead to major changes in the orbit of one or both of the bodies or may enhance orbital stability, depending on the precise nature of the resonance.

mean orbital elements

The **orbital elements** of a reference orbit chosen to approximate a real orbit that is perturbed by the presence of another object.

mean parallax

The distance, derived by means of statistical studies of brightnesses and motions, for a group of stars whose individual distances are unmeasurable.

mean place

The position of an object on the **celestial sphere**, as seen from the Sun, referred to the mean equator and **equinox** at the beginning of the year; it is thus the apparent place corrected for annual **parallax, proper motion, precession, aberration of starlight**, and **nutation.** All observations of the apparent place of a star made during a particular year may be reduced to the mean place for the beginning of that year, so that they can be intercompared; in practice, though, all observations made over a period of several years are reduced to a common epoch, e.g., J2000.0.

mean sun

A fictitious body that moves eastward in a circular orbit along the celestial equator (see **celestial sphere**), making a complete circuit with respect to the vernal **equinox** in a tropical **year**. It is the moving point chosen in defining mean **solar time**.

Méchain, Pierre François André (1744–1804)

A French astronomer who, like his friend Charles **Messier**, hunted comets and, in the process, discovered many new deep sky objects. Between 1779 and 1782, he found 30 deep sky objects, 29 of which were original firsts. As he was in close cooperation with Messier at this time, he almost instantly communicated his observations to Messier, who would check their positions and add them to his catalog. Méchain's discoveries include M63, M65, M66, M68, M72, M75, M76, M78, and M102 through M107, the majority of which are galaxies and globular clusters (see **Messier Catalogue**). Méchain also discovered eight comets between 1781 and 1799, including **Encke's comet** (named after the calculator of its orbit).

megamaser

An extremely powerful **maser** associated with an **active galactic nucleus**. Megamasers, the most powerful of which are the OH (hydroxyl) and H_2O megamasers, are up to a million times more powerful than stellar masers found within the Milky Way Galaxy.

Meissa (Lambda Orionis, λ Ori)

A binary star consisting of a hot (35,000 K) and luminous (65,000 L_{sun}) **O star**, of about 25 M_{sun} located 4″ away from a **B star** (25,000 K, 5,500 L_{sun}). Meissa (also known as Heka) is best-known for a huge ring of gas, an astonishing 200 light-years across that surrounds the star and is ionized by it. The *Meissa Ring* lies within an even larger ring of interstellar dust and molecules. These structures may consist of material left over from the formation of Meissa that has been compressed by the action of the O star. Alternatively, they may have resulted from a supernova in the neighborhood of Meissa a few million years ago.

Visual magnitude:	3.39
Absolute magnitude:	–4.16
Distance:	1,060

Mekbuda (Zeta Geminorum, ζ Gem)

One of the few **Cepheid variables** in the sky that can be followed throughout its cycle by the naked eye as it changes from magnitude 3.6 to 4.2 and back every 10.15 days; its Arabic name refers to a "lion's paw." Although classified as a **Delta Cephei star** (a classical Cepheid), it breaks the mold of this type by having a symmetric rather than a "shark fin"–shaped light curve. Observations by optical interferometer in 2000 provided a direct measure of the changing size of the star as it pulsates. Mekbuda (just southwest of brighter Wasat, Delta Geminorum) shines yellow-white and has a temperature similar to that of the Sun.

Visual magnitude:	3.7 to 4.1
Absolute magnitude:	–3.77
Luminosity:	3,000 L_{sun}
Radius:	60 R_{sun}
Distance:	1,170 light-years

meniscus lens

A lens having two spherically curved faces, one **convex** and the other **concave**, so that it has the form of a shell. A *positive meniscus lens* is thicker in the middle than at the edges and serves as a converging lens; a *negative meniscus lens* thickens toward the edges and works as a diverging lens. Very large telescopes may use *meniscus mirrors*, which are lightweight and require an active control system to retain their shape against gravity.

Menkalinan (Beta Aurigae, β Aur)

The third brightest star in **Auriga**; its Arabic name (also written as Menkarlina) means "the shoulder of the rein-holder." Menkalinan lies just 0.5′ away from the *solstitial colure,* the great circle in the sky that passes through both celestial poles and the summer and winter solstices. Menkarlinan is a multiple system in which the dominant members are two almost identical subgiant **A stars** in a tiny orbit that, every 3.96 days, results in a partial eclipse of one star by the other by 0.09 magnitude. The twins, with masses of 2.35 M_{sun} and 2.25 M_{sun}, are separated by only about one-fifth the distance between the Sun and Mercury. They are so close that they distort each other into an ellipsoidal shape through mutual tides. A faint **red dwarf**, of magnitude 14.1, orbits the main pair at a distance of at least 330 AU.

Visual magnitude:	1.90
Absolute magnitude:	–0.11
Spectral type:	A2V
Surface temperature:	9,200 K
Luminosity:	95 L_{sun}
Distance:	82 light-years

Menkent (Theta Centauri, θ Cen)

An orange giant **K star** that is the third brightest star in **Centaurus**; its Arabic name means "the shoulder of the centaur."

Visual magnitude:	2.06
Absolute magnitude:	0.70
Spectral type:	K0IIIb
Surface temperature:	4,500 K
Luminosity:	45 L_{sun}
Radius:	16 R_{sun}
Mass:	4 M_{sun}
Distance:	61 light-years

Menkib (Xi Persei, ξ Per)

One of the few naked-eye **O stars** and one of the hottest and most massive stars visible without the aid of a telescope; the fact that it is only the twelfth brightest star in **Perseus** is due to its great distance and also because about half of its light is absorbed by dust in the plane of the Milky Way. It is probably responsible for illuminating the **California Nebula** (NGC 1499). Menkib's name, meaning "collarbone" refers to a larger Arabic constellation of which Atik is the "shoulder." Menkib is slightly variable, changing between magnitude 2.80 and 2.93, and also blowing a powerful **stellar wind** by which it sheds about 10^{-6} M_{sun} per year. Classified as a supergiant, it has moved off the main sequence and may be in the core helium-burning stage with only a million years or so left before it explodes. It is also one of the sky's few known **runaway stars**, hurtling away from its birthplace in the Perseus OB2 Association (home to its sister star, Atik). Its ejection may have been due to a close encounter with another star or by a nearby supernova explosion. Menkib has a binary companion of magnitude 6.5, lying 12.9″ away, in a 7-day orbit.

Visual magnitude:	3.98
Absolute magnitude:	–4.70
Spectral type:	O7Iae
Surface temperature:	37,000 K
Luminosity:	330,000 L_{sun}
Mass:	40 M_{sun}
Distance:	1,770 light-years

Mensa (abbr. Men; gen. Mensae)

The Table; a small, faint **constellation** near the south celestial pole. It has few objects of interest for the amateur apart from the **Large Magellanic Cloud**, part of which overlaps into Mensa from its northern neighbor, Dorado. (See star chart 4.)

Menzel, Donald Howard (1901–1976)

An American astrophysicist best remembered for his important solar and planetary work. An authority on the

Sun's **chromosphere**, he discovered with J. C. Boyce (1933) that the Sun's **corona** contains oxygen. With W. W. Salisbury, in 1941, he made the first of the calculations that led to radio contact with the Moon in 1946. With Fred Whipple, in 1955, he proposed a maritime model of **Venus**. From 1926 to 1932 Menzel worked at the **Lick Observatory**; in 1932 he joined the faculty at Harvard, where he became professor (1938) of astrophysics and director (1954) of the **Harvard College Observatory**.

Merak (Beta Ursae Majoris, β UMa)

A main sequence **A star** that is the fifth brightest in **Ursa Major** and also one of the "pointers" that, together with **Dubhe**, just to the north, leads the way to Polaris, the North Star. Merak's anatomically correct name comes from an Arabic phrase meaning "the flank of the Greater Bear." Like **Vega**, it radiates excess **infrared** that seems to come from a disklike shroud of heated dust, similar in size to the orbit of Saturn and having a temperature of a few hundred K. Merak is part of the **Ursa Major Moving Cluster**.

Visual magnitude:	2.34
Absolute magnitude:	0.41
Spectral type:	A1V
Surface temperature:	9,000 K
Luminosity:	60 L_{sun}
Mass:	3 $M_{sun.}$
Distance:	79 light-years

Mercury

See article, pages 327–329.

mercury-manganese star

A type of **manganese star** whose spectrum has a prominent line at 3984 Å due to absorption by ionized mercury.

meridian

(1) A great circle on the **celestial sphere** that passes through the celestial poles and the observer's zenith. The *meridian angle* of a celestial object is the angle measured eastward along the celestial equator from an observer's meridian to that of the object. The *meridian altitude* of an object is its altitude when it crosses the observer's meridian. *Meridian transit,* also known as *meridian passage* or **culmination**, is the moment when a celestial object crosses, or transits, an observer's meridian. A *meridian observation* is an observation of the object's **declination** (and, occasionally, brightness) at this time and is made by a **transit circle**.

(2) A line of longitude on Earth, or on some other astronomical body. On Earth, the meridian that passes through Greenwich, London, is known as the *prime meridian.*

Merope (23 Tauri)

A subgiant **B star** that is the fifth brightest star in the **Pleiades**. Like some other members in the cluster, Merope is a **Be star**; its rapid spin (with a rotational velocity of 20Km/s) has caused it to eject a disk of bright, emitting gas (though Merope's disk is thinner than those of some of its neighbors, such as **Pleione**). Merope's greatest claim to fame, however, is not the star itself, but its surroundings: the reflection nebulosity for which the Pleiades is well known is so bright around Merope as to have its own names, the **Merope Nebula** and IC 349.

Visual magnitude:	4.14
Absolute magnitude:	–1.07
Spectral type:	B6IVe
Surface temperature:	14,000 K
Luminosity:	630 L_{sun}
Radius:	4.3 R_{sun}
Mass:	4 M_{sun}
Distance:	389 light-years

Merope Nebula (NGC 1435)

The most conspicuous of the **reflection nebula**e in the **Pleiades**; it was discovered by Wilhelm **Tempel** in 1859 and is sometimes called *Tempel's Nebula.* Within it is a small bright knot, IC 349, discovered by Edwin **Barnard** in 1890, located just 0.06 light-year away from **Merope**–the star that is the cause of its illumination. Long thought to be the remnant of the Pleiades birth, the nebula is instead a chance occurrence, as the cluster is merely passing through an interstellar cloud; in fact, the cluster leaves a wake.

meson

A type of **hadron** composed of just two **quarks**; there are three types: pi-mesons (pions), K-mesons (kaons), and eta-mesons, which can be positively or negatively charged, or electrically neutral. (So-called mu-mesons, or muons, are actually **leptons**.) Mesons are unstable, with half-lives of 10^{-8} to 10^{-16} s, and decay into stable particles. They are a major component of secondary **cosmic rays**.

mesopause

A level in the **ionosphere**, at an altitude of about 85 km, above which temperatures rise as altitude increases instead of falling as they do in the lower ionosphere.

Mercury

The nearest planet to the Sun and the smallest except for Pluto; it is surpassed in size (but not in mass) by both **Ganymede** and **Titan**. In ancient Greece it had two names–Apollo for its appearance as a morning star and Hermes as an evening star–although Greek astronomers knew that a single body was involved. Being named after the fleet-footed god Hermes, or its Roman equivalent Mercury, is especially apt as the planet races around the Sun at an average speed of 48 km/s, completing one circuit in 88 days. Although its mean distance from the Sun is 57,910,000 km (0.387 AU), its orbit is so elongated that its perihelion-aphelion range is 46 million to 70 million km.

Until the 1960s it was thought that Mercury's day (the time it takes to spin on its axis) was the same length as its year (the time it takes to complete one orbit) so that it always kept the same face to the Sun. However, Doppler radar measurements in 1965 showed that Mercury actually rotates three times in two of its years. This fact and Mercury's highly eliptical orbit would lead to strange effects for an astronaut on the surface. At some longitudes the Sun would appear to rise briefly, set, and rise again before traveling westward across the sky. At sunset, the Sun would set, rise again briefly, and then set again. In other places, an observer could watch the Sun come to a standstill and then move backward for a while, before continuing its original motion, having done a complete loop.

Mercury's perihelion precesses around the Sun but at a rate that nineteenth-century astronomers couldn't properly explain. The "advance" of the perihelion was 476" per century–42.6" more than expected. This discrepancy remained a nagging problem for many decades, and it was even suggested that the answer might lie in the existence of an intra-Mercurian planet, sometimes called **Vulcan**. In the end, the puzzle was solved in much more dramatic style, by applying a new theory of gravity: Einstein's **general theory of relativity.**

Despite being fairly close to us, Mercury is one of the least-studied planets in the solar system. It has been visited by only one spacecraft, Mariner 10, which flew by three times in 1974 and 1975, mapping 45% of the surface. Mariner observed a world similar in general respects to the Moon: heavily cratered but with regions of relatively smooth plains, some of which may be the result of ancient volcanic activity, others due to the deposition of ejecta from cratering impacts. Its most distinctive surface feature is the Caloris Basin, a colossal, **multiringed basin** about 1,350 km in diameter, whose inner floor contains mostly smooth plains, known as Caloris Planitia, but also many ridges and fractures, some of them radial and others arranged in two or three concentric rings. Its name, meaning "basin of heat," stems from the fact that it lies near the **subsolar point** when Mercury is at perihelion and thus can experience temperatures as high as 700 K.

There are also great escarpments, up to 1,500 km long and 3 km high, some of which slice through the rings of craters and other features in a way that shows they were formed by compression of the crust when Mercury's interior cooled and shrank. Estimates suggest that the planet's surface area decreased by about 0.1% and its radius by about 1 km.

Mercury is the second densest major body in the solar system, after Earth, pointing to a relatively enormous iron core with a radius of 1,800 to 1,900 km–proportionately much larger than Earth's and making up most of the planet. Above this is a relatively thin silicate mantle and crust, only 500 to 600 km in total from top to bottom. Mercury has a weak magnetic field about 1% as strong as Earth's. It also

Diameter:	4,880 km
Mass (Earth = 1):	0.054
Density:	5.43 g/cm^3
Escape velocity:	4.2 km/s
Albedo:	0.06
Surface temperature:	–180 to 425°C
Axial period:	58.65 days
Axial inclination:	negligible
Orbit	
Distance from the Sun	
Mean:	57.9 million km (0.387 AU)
Aphelion:	69.7 million (0.467 AU)
Perihelion:	45.9 million km (0.306 AU)
Period:	87.97 days
Eccentricity:	0.206
Inclination:	7.00°

Mercury A montage of Mercury made from dozens of separate images sent back by Mariner 10. The north pole is at the top. *NASA/JPL*

Mercury Part of Mercury's Caloris Basin, seen halfway in shadow on the morning terminator. *NASA/JPL*

has an extremely thin atmosphere, about one-trillionth the density of Earth's, composed mainly of atoms of argon, neon, and helium that have been blasted off the surface by the **solar wind**. Radar images taken by astronomers at the Jet Propulsion Laboratory and the California Institute of Technology in 1991 suggest that the polar regions of Mercury may be covered with patches of water-ice. Although this seems impossible due to the planet's sizzling heat, the polar regions receive very little sunlight and may get as cold as –148°C. In 2004, NASA plans to launch a spacecraft, **MESSENGER**, that will orbit Mercury in 2009. Europe and Japan are also developing a robot mission to Mercury, for launch in 2009 or 2010, that will consist of two orbiters and possibly a smart lander.

mesosiderite

One of two main types of **stony-iron meteorite**, the other being **pallasites**. The name comes from the Greek *mesos,* meaning "middle" or "half," and *sideros* for "iron"; hence "half-iron." Mesosiderites are a strange, roughly 50:50 mixture of silicates and iron-nickel. The silicates are heavily brecciated (smashed-up), evolved **igneous** rocks, similar to those in eucrites, diogenites, and other **HED group** members, and clearly come from the crust of an achondritic parent body (see **achondrite**). The metal in mesosiderites, on the other hand, is similar to that in group IIIAB **iron meteorites**, and can only have come from the core of a distinct, differentiated asteroid, genetically unrelated to the precursor of the eucritic and diogenitic portion; there is no mantlelike material at all. One way to explain this odd combination of crustal and core material is in terms of a collision between two differentiated asteroids in which the still-liquid core of one asteroid mixed with the solidified crust of the other. This model assumes the collisional disruption and gravitational reassembly of at least one of the asteroids–the one that later became the parent body of the mesosiderites. It is still much debated whether the HED parent body, **Vesta**, represents one of these asteroids.

mesosphere

The part of Earth's atmosphere immediately above the **stratosphere**, where the temperature drops from about 270 to 180 K. It lies between the **stratopause** and the **mesopause** and overlaps the lower part of the **ionosphere**.

MESSENGER (Mercury Surface, Space Environment, Geochemistry, and Ranging)

A NASA probe to investigate **Mercury** from an orbit around the planet. Scheduled for launch in March 2004, MESSENGER will make two flybys of Venus and two of Mercury during a five-year voyage that will position it for orbital insertion in 2009. During its nominal one-year science mission, the probe will image the surface with a high-resolution camera; analyze the planet's surface composition with X-ray, gamma-ray, infrared, and neutron **spectrometers**; study the magnetic field with a **magnetometer**; and determine the height of features with a laser altimeter. The only previous mission to Mercury was Mariner 10 in 1974.

Messier, Charles Joseph (1730–1817)

A French astronomer famed for his list of more than 100 bright deep sky objects, now known to be a variety of nebulae, star clusters, and galaxies (see **Messier Catalogue**). Principally a comet-hunter–Louis XV called him the "Comet Ferret"–Messier compiled his list of other fuzzy looking objects so that he and others wouldn't keep confusing them for comets. Little is known about him prior to his joining the **Paris Observatory** as a draftsman and astronomical recorder. His interest in comets stemmed from the return of **Halley's comet**, which Edmund **Halley** predicted would take place around the beginning of 1759. Messier sighted its return on January 12, 1759, an experience that inspired him to search for new comets for the rest of his life. (Although he is attributed with being the first person to re-sight Halley's comet on French soil, the German amateur astronomer Palitzch is believed to have been the first in the world to see it, on Christmas Day, 1758.) His final comet tally stood at 15 unique discoveries and six further codiscoveries. The compilation of his famous catalog began in about 1760 and took more than two decades, during which time he used a variety of telescopes, including a 6-inch reflector and a 3½-in. refractor.

Messier Catalogue

A list of 110 nebulous-looking celestial objects, the first 103 of which were compiled by Charles **Messier** in the eighteenth century; it was one of the earliest important astronomical catalogs. His final list, published in 1781, includes discoveries by other observers, notably his colleague Pierre **Méchain**. On star charts and atlases, Messier objects now include the letter "M" before each number. Among the best known examples are M1 (Crab Nebula), M31 (Andromeda Galaxy), M42 (Orion Nebula), and M45 (Pleiades). (See table, "The Messier Objects.")

metal

In astronomy, any element heavier than hydrogen and helium. An object's **metallicity** is its abundance of such elements, or, often, more specifically, the abundance of iron, which is easy to measure. The terms *metal-poor* and *metal-rich* are used to indicate low metallicity and high metallicity, respectively.

metallic hydrogen

A hypothetical form of hydrogen in which the molecules have been forced by extremely high pressures to assume the lattice structure typical of metals. It is estimated that as much as 40% of Jupiter's mass (but not more than 3% of Saturn's) may be in the form of metallic hydrogen.

metallicity

A measure of the proportion of heavy elements or metals (in astronomy, elements heavier than hydrogen or helium) that a star contains. Usually, metallicity is given in terms of the relative amount of iron and hydrogen present, as determined by analyzing absorption lines in a stellar spectrum, compared with the solar value. The ratio of the amount of iron to the amount of hydrogen in the object is divided by the ratio of the amount of iron to the amount of hydrogen in the Sun. This value, denoted as [Fe/H], is plotted on a logarithmic scale. For example, if the metallicity [Fe/H] = –1 then the abundance of heavy elements in the star is one-tenth that found in the Sun; if [Fe/H] = +1, the heavy element abundance is 10 times the solar value.

metal-poor star

A star composed of material that has been little recycled through previous generations of stars. Its proportion of elements heavier than hydrogen or helium is therefore very much less than that of the Sun. Metal-poor stars are **Population II** objects, usually found in the **galactic halo** or in **globular clusters**. Whether they ever have associated planetary systems is an open question.

The Messier Objects

M	NGC	Constellation	Type of Object	Common Name
1	1952	Taurus	Supernova remnant	Crab Nebula
2	7089	Aquarius	Globular cluster	
3	5272	Canes Venatici	Globular cluster	
4	6121	Scorpio	Globular cluster	
5	5904	Serpens	Globular cluster	
6	6405	Scorpio	Open cluster	Butterfly Cluster
7	6475	Scorpio	Open cluster	
8	6523	Sagittarius	Diffuse nebula	Lagoon Nebula
9	6333	Ophiuchus	Globular cluster	
10	6254	Ophiuchus	Globular cluster	
11	6705	Scutum	Open cluster	Wild Duck Cluster
12	6218	Ophiuchus	Globular cluster	
13	6205	Hercules	Globular cluster	Great Hercules Cluster
14	6402	Ophiuchus	Globular cluster	
15	7078	Pegasus	Globular cluster	
16	6611	Serpens	Nebula and open cluster	
17	6618	Sagittarius	Diffuse nebula	Omega Nebula
18		6613	Sagittarius	Open cluster
19	6273	Ophiuchus	Globular cluster	
20	6514	Sagittarius	Diffuse nebula	Trifid Nebula
21	6531	Sagittarius	Open cluster	
22	6656	Sagittarius	Globular cluster	
23	6494	Sagittarius	Open cluster	
24	6603	Sagittarius	Open cluster	
25	IC 4725	Sagittarius	Open cluster	
26	6694	Scutum	Open cluster	
27	6853	Vulpecula	Planetary nebula	Dumbbell Nebula
28	6626	Sagittarius	Globular cluster	
29	6913	Cygnus	Open cluster	
30	7099	Capricornus	Globular cluster	
31	224	Andromeda	Spiral galaxy	Andromeda Galaxy
32	221	Andromeda	Elliptical galaxy	Companion to M31
33	598	Triangulum	Spiral galaxy	Triangulum Spiral
34	1039	Perseus	Open cluster	
35	2168	Gemini	Open cluster	
36	1960	Auriga	Open cluster	
37	2099	Auriga	Open cluster	
38	1912	Auriga	Open cluster	
39	7092	Cygnus	Open cluster	
40	–	–	Missing (comet?)	
41	2287	Canis Major	Open cluster	

(continued)

The Messier Objects *(continued)*

M	NGC	Constellation	Type of Object	Common Name
42	1976	Orion	Diffuse nebula	Great Nebula
43	1982	Orion	Diffuse nebula	Part of Orion Nebula
44	2632	Cancer	Open cluster	Praesepe
45	–	Taurus	Open cluster	Pleiades
46	2437	Puppis	Open cluster	
47	2422	Puppis	Open cluster	
48	2548	Hydra	Open cluster	
49	4472	Virgo	Elliptical galaxy	
50	2323	Monoceros	Open cluster	
51	5194	Canes Venatici	Spiral galaxy	Whirlpool Galaxy
52	7654	Cassiopeia	Open cluster	
53	5024	Coma Berenices	Globular cluster	
54	6715	Sagittarius	Globular cluster	
55	6809	Sagittarius	Globular cluster	
56	6779	Lyra	Globular cluster	
57	6720	Lyra	Planetary nebula	Ring Nebula
58	4579	Virgo	Spiral galaxy	
59	4621	Virgo	Elliptical galaxy	
60	4649	Virgo	Elliptical galaxy	
61	4303	Virgo	Spiral galaxy	
62	6266	Ophiuchus	Globular cluster	
63	5055	Canes Venatici	Spiral galaxy	
64	4826	Coma Berenices	Spiral galaxy	
65	3623	Leo	Spiral galaxy	
66	3627	Leo	Spiral galaxy	
67	2682	Cancer	Open cluster	
68	4590	Hydra	Globular cluster	
69	6637	Sagittarius	Globular cluster	
70	6681	Sagittarius	Globular cluster	
71	6838	Sagittarius	Globular cluster	
72	6981	Aquarius	Globular cluster	
73	6994	Aquarius	Four faint stars	
74	628	Pisces	Spiral galaxy	
75	6864	Sagittarius	Globular cluster	
76	650	Perseus	Planetary nebula	
77	1068	Cetus	Spiral galaxy	
78	2068	Orion	Diffuse nebula	
79	1904	Lepus	Globular cluster	
80	6093	Scorpio	Globular cluster	
81	3031	Ursa Major	Spiral galaxy	
82	3034	Ursa Major	Irregular galaxy	Companion to M31

The Messier Objects

M	NGC	Constellation	Type of Object	Common Name
83	5236	Hydra	Spiral galaxy	
84	4374	Virgo	Spiral galaxy	
85	4382	Coma Berenices	Spiral galaxy	
86	4406	Virgo	Elliptical galaxy	
87	4486	Virgo	Elliptical galaxy (Seyfert)	
88	4501	Coma Berenices	Spiral galaxy	
89	4552	Virgo	Elliptical galaxy	
90	4569	Virgo	Spiral galaxy	
91	–	–	Not identified (comet?)	
92	6341	Hercules	Globular cluster	
93	2447	Puppis	Open cluster	
94	4736	Canes Venatici	Spiral galaxy	
95	3351	Leo	Barred spiral galaxy	
96	3368	Leo	Spiral galaxy	
97	3587	Ursa Major	Planetary nebulae	Owl Nebulae
98	4192	Coma Berenices	Spiral galaxy	
99	4254	Coma Berenices	Spiral galaxy	
100	4321	Coma Berenices	Spiral galaxy	
101	5457	Ursa Major	Spiral galaxy	
102	–	–	Unidentified	
103	581	Cassiopeia	Open cluster	
104	4594	Virgo	Spiral galaxy	Sombrero Galaxy
105	3379	Leo	Elliptical galaxy	
106	4258	Ursa Major	Spiral galaxy	
107	6171	Ophiuchus	Globular cluster	
108	3556	Ursa Major	Spiral galaxy	
109	3992	Ursa Major	Spiral galaxy	
110	205	Andromeda	Elliptical galaxy	Companion to M31

metal-rich star

A star with a relatively high **metal**-to-hydrogen ratio, as indicated by strong lines of magnesium, calcium, iron, and other heavy elements. They are typically members of **Population I**, though the Sun itself is considered a moderately metal-rich star. An unresolved problem is the connection between metal richness of stars and their likelihood of having planets.

metamorphic

Sedimentary or **igneous** rock that has been recrystallized as a result of changes in temperature, pressure, and chemical environment. Schist, for example, is metamorphosed **basalt** or slate.

meteor

(1) The light phenomenon, known popularly as a *shooting star,* that results from the entry into Earth's atmosphere of a solid particle from space. (2) The particle itself during entry.

meteor shower

Meteors seen to fan out from a single point in the sky, known as the **radiant**, in a burst of activity lasting for several hours or days. A meteor shower consists of dusty debris, spread out along part of the orbit of a parent body, usually a comet, which Earth intersects at the same time each year. The main meteor showers are listed in the table "Principal Nighttime Meteor Showers," but there are dozens of others that are more feeble or are detectable only with radar equipment during daylight hours. The Arietids, for example, reach a maximum of 60 per hour on June 7 but take place unseen against the daytime sky.

Some showers vary enormously from year to year. The Leonids are famous for this, normally putting on a modest annual show of up to 15 meteors per hour. But every 33 years or so, when the parent comet **Tempel-Tuttle** is at perihelion and in Earth's neighborhood, the Leonids are capable of staging a **meteor storm**. In 1799, 1833, and 1966, when the Earth passed particularly close to the stream of debris following in the comet's wake, rates of up to 150,000 meteors per hour were reported. In other cases, the variability is more erratic and linked to changes in or a complete breakup of the parent body. For example, the Draconids, also known as the Giacobinids, are usually so weak as to be unrecognizable to the untrained eye, but have been known to produce storms, as happened in 1933 and 1946 when several thousand meteors per hour were seen. The Andromedids, also known as the Bielids, are best known for two sensational displays, on November 27, 1872 and 1885, following the destruction of the parent, **Biela's comet**, in the mid-nineteenth century, when the hourly rate reached 6,000 and 75,000, respectively. Since then, gravitational perturbations have gradually pulled the meteor stream out of Earth's path until today when the Andromedids are very weak.

meteor, sporadic

A random meteor that is not associated with a **meteor shower**. Sporadic meteors can be seen on any clear, dark night, with average rates of 3 to 4 per hour in spring and 8 to 10 per hour in autumn. The average rate also increases during the night hours because of Earth's rotation: a meteor seen in the evening has had to catch up with Earth in its orbit, whereas a meteor seen in the early morning has met Earth head on.

meteor storm

A rare, short-lived event that occurs when Earth encounters a **meteoroid swarm**. Meteor storms involve meteor rates exceeding 1,000 per hour.

Principal Nighttime Meteor Showers

					Radiant		
Shower	**Begins**	**Peak**	**Ends**	**Max. Rate**	**R.A.**	**Dec.**	**Parent Object**
Quadrantids	Jan 1	Jan 3	Jan 6	110	15.5h	+50°	
Alpha Centaurids	Jan 28	Feb 8	Feb 21	6	14.0h	–59°	
Gamma Normids	Feb 25	Mar 22	Mar 13	8	16.6h	–51°	
Lyrids	Apr 16	Apr 22	Apr 25	15	18.1h	+32°	Comet 1861 I (Thatcher)
Eta Aquarids	Apr 19	May 5	May 28	60	22.8h	00°	Halley's comet
June Lyrids	Jun 10	Jun 15	Jun 21	8	18.5h	+35°	
Pisces Australids	Jul 15	Jul 27	Aug 10	8	22.7h	–30°	
Delta Aquarids	Jul 15	Jul 28	Aug 19	20	22.6h	–10°	
Alpha Capricornids	Jul 3	Jul 29	Aug 15	8	20.3h	–12°	
Perseids	Jul 25	Aug 12	Aug 18	100	03.1h	+58°	Comet Swift-Tuttle
Alpha Aurigids	Aug 25	Aug 31	Sep 5	10	05.6h	+42°	
Draconids	Oct 6	Oct 10	Oct 9	var	18.0h	+54°	Comet Giacobini-Zinner
Orionids	Oct 16	Oct 21	Oct 26	30	06.4h	+15°	Halley's comet
Taurids	Oct 20	Nov 4	Nov 25	12	03.7h	+22°	Encke's comet
Cepheids	Nov 7	Nov 9	Nov 11	8	23.5h	+63°	
Leonids	Nov 15	Nov 17	Nov 19	var	10.1h	+22°	Comet Tempel-Tuttle
Puppid-Velids	Dec 1	Dec 7	Dec 15	10	08.2h	–45°	
Geminids	Dec 7	Dec 14	Dec 15	58	07.5h	+32°	3200 Phaethon
Ursids	Dec 17	Dec 22	Dec 24	12	14.5h	+76°	Comet Tuttle

Historic Meteorite Showers

Meteorite	Country	Fell	Total Weight (kg)	Individuals
Pultusk	Poland	1868	200	180,000
Holbrook	United States	1912	220	16,000
Sikhote-Alin	Russia	1947	70,000	15,000
Allende	Mexico	1969	2,000	5,000
L'Aigle	France	1803	37	3,000
Mocs	Romania	1882	300	3,000

meteor train

A trail of ionized dust and gas that remains along the path of a **meteor**. A *persistent train* is one that remains visible for more than a second.

meteorite

See article, page 336.

meteorite shower

Only a few meteorites make their passage through the atmosphere in one piece. Most of them are fragmented on their way due to the high pressure to which they are subjected. Sometimes they burst into only a few fragments, at other times, into thousands of individual pieces creating a meteorite shower. Only a few pieces from such a shower actually arrive on Earth since the smaller fragments usually burn up during their passage. However, sometimes, when the initial pieces are big enough, it literally "rains stones," and thousands of meteorites can be collected from huge **strewn fields**. (See tables, "Historic Meteorite Showers," and "Prominent Prehistoric Strewn Fields.")

meteoroid

A solid object moving in interplanetary space, smaller than an **asteroid** but much larger than an atom or a molecule.

meteoroid stream

A trail of solid particles released from a parent body such as a **comet** or **asteroid**, moving on similar orbits. Various ejection directions and velocities for individual **meteoroids** cause the width of a stream and the gradual distribution of meteoroids over the entire average orbit.

meteoroid swarm

A clump of particles within a **meteor stream**. Such clumps have been released quite recently from the parent body and have not had time to spread out along the orbit. When Earth passes through a swarm, the visible effect is a **meteor storm**.

methane (CH_4)

A gaseous compound, one of the alkanes, in which every carbon atom is surrounded by four hydrogen atoms.

Prominent Prehistoric Strewn Fields

Meteorite	Country	Found	Total Mass (kg)	Individuals
Campo del Cielo	Argentina	1576	70,000	Thousands
Gibeon	Namibia	1836	26,000	Thousands
Toluca	Mexico	1776	2,500	Thousands
Brenham	United States	1882	2,400	Thousands
Imilac	Chile	1822	1,000	>1,000
Plainview	United States	1917	700	>1,000

meteorite

A solid portion of a **meteoroid** that survives its fall to Earth, or some other body. Meteorites are classified as **stony meteorites**, **iron meteorites**, and **stony-iron meteorites**, and further categorized according to their mineralogical content (see table, "Types of Meteorites," and individual entries for each of these types). They range in size from microscopic to many meters across. Of the several tens of tons of cosmic material entering Earth's atmosphere each day, only about 1 ton reaches the ground; an object's survival chances depend on its initial mass, speed of entry, and composition. Incoming meteoroids with masses between 10^{-6} g and 1 kg tend to burn up completely as **meteors**. Smaller objects are dramatically slowed down without being incinerated and fall as a continuous, gentle, invisible rain of **micrometeorites**. Larger objects, up to 1,000 tons, are decelerated to a lesser extent and fall through the lower atmosphere at high speed, causing them to glow brightly as a fireball.

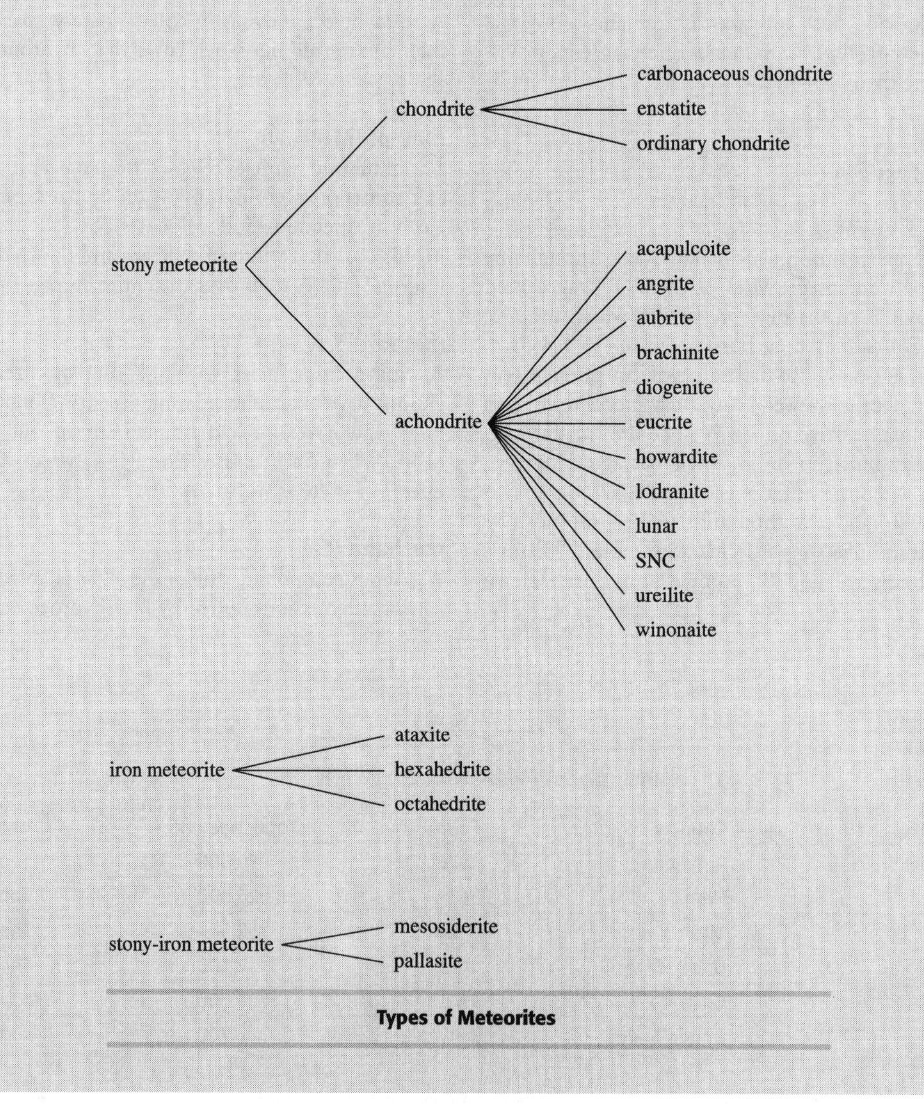

Types of Meteorites

Metis

(1) The innermost moon of **Jupiter**, also known as Jupiter XVI. It is irregular in shape with a mean diameter of about 44 km. Metis orbits just 56,500 km above the Jovian cloud tops and within Jupiter's ring system, the material for which it and **Adrastea** almost certainly supply. It was discovered in 1979 by the American astronomer Steven Synnott from Voyager images. (2) The ninth **asteroid** to be discovered and the only one to be discovered from Ireland, in 1848 by Andrew Graham (diameter 151 km, albedo 0.13, semimajor axis 2.09 AU, perihelion 2.68 AU, orbital period 3.684 yr, eccentricity 0.123, inclination 5.58°).

Metonic cycle

A period of 19 years, after which the phases of the **Moon** recur on the same calendar date and within two hours of the same time. Discovered by Meton of Athens in 432 B.C., it arises from the fact that 235 lunations equal 19 tropical **years** almost exactly–about 6939.5 days. The Metonic cycle formed the basis for the Greek calendar until 46 B.C. when the **Julian calendar** was introduced. The *Callipic cycle* is four 19-year Metonic cycles, or 76 years, and was introduced by the Greek astronomer and mathematician Callipus of Cyzicus (c. 370–c. 300 B.C.) in order to reconcile more closely the lunar month with the solar year.

Miaplacidus (Beta Carinae, β Car)

The second brightest star in **Carina**, a subgiant **A star**. Its name is of uncertain origin, possibly the Arabic *Mi'ah* for "waters" combining with the Latin *placidus* to give "placid waters."

Visual magnitude:	1.67
Absolute magnitude:	–1.00
Spectral type:	A2IV
Luminosity:	210 L_{sun}
Radius:	5.7 R_{sun}
Distance:	111 light-years

Mice, the (NGC 4676 A and B)

Two **spiral galaxies**, about 300 million light-years away in **Coma Berenices**, that are in the process of colliding

Mice, the The most detailed image ever obtained of this famous pair of interacting galaxies. Streams of interchanging material can be seen and, along the straight tail, clumps of young stars separated by faint regions, where less dust and gas seem to exist. These dimmer regions suggest that the clumps of stars have formed from the gravitational collapse of the gas and dust that once occupied these areas. Some of the clumps have luminous masses comparable to dwarf galaxies that orbit in the halo of our own Milky Way Galaxy. *NASA/STScI*

and merging. Their name refers to the long tails produced by tidal action–the relative difference between gravitational pulls on the near and far parts of each galaxy. Each spiral has likely already passed through the other and they will probably collide repeatedly until the two systems coalesce to form a giant elliptical. Because the distances involved are so vast, the whole process seems to take place in slow motion–over hundreds of millions of years. The Mice give us a foretaste of what may happen to our own Milky Way several billion years from now when it collides with the Andromeda Galaxy. They appear to be members of the **Coma Cluster** of galaxies.

microlensing

A small-scale **gravitational lens** effect. In microlensing, the gravitational field of the lensing object is not strong enough to form distinct images of the background source, but instead simply causes an apparent brighten-

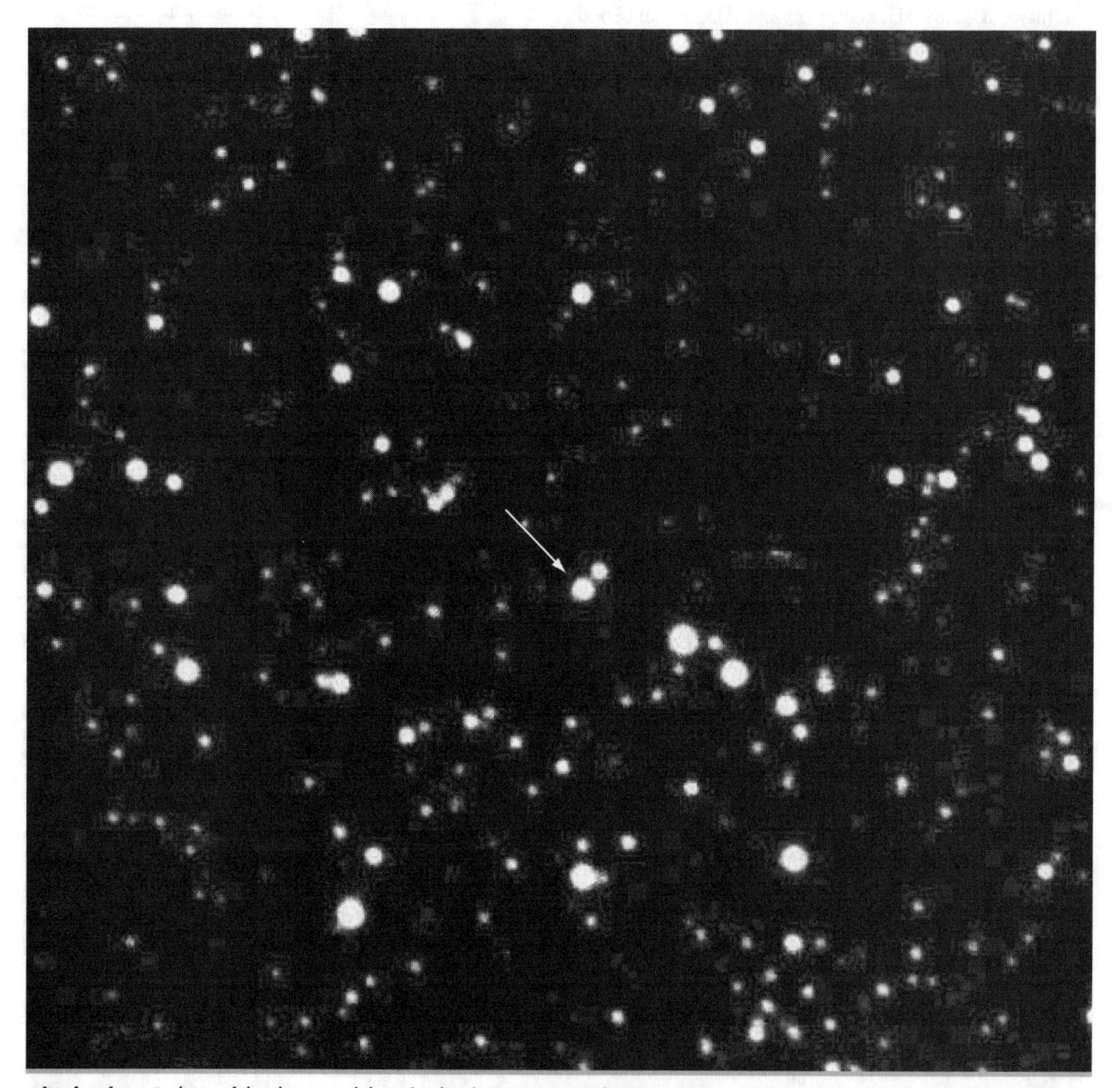

microlensing A photo of the sky around the microlensing event named EROS-BLG-2000-5 taken by the Very Large Telescope. The use of this powerful telescope to study the event allowed spectra to be obtained, as a result of which the star that was microlensed has been identified as a cool giant, some 25,000 light-years away in the galactic bulge. Further investigation of the data collected will allow a detailed analysis of the star's atmosphere–remarkable considering its great distance. *European Southern Observatory*

ing of the source. Stars are expected to vary in brightness in a characteristic way if low-mass stars, **brown dwarfs**, or **extrasolar planets** pass in front of them, and this effect has been detected for stars in the **Large Magellanic Cloud** and in the central bulge of our Galaxy.

micrometeorite

A small extraterrestrial particle that has survived entry into Earth's atmosphere. The actual size is not rigorously constrained but is defined operationally by the collection method. Micrometeorites found on Earth's surface are smaller than 1 mm; those collected in the **stratosphere** are rarely as large as 50 micrometers.

microquasar

An object of stellar mass that displays in miniature some of the properties of **quasars**, including strong emission across a broad range of wavelengths from radio wave to X-ray, rapid variability in X rays, and radio **jets**. A microquasar consists of a binary system in which a normal star orbits around, and loses matter to, a nearby compact object, either a **black hole** or a **neutron star**. The lost matter enters a fast-spinning **accretion disk**, is heated to millions of degrees, and then either falls onto the compact object or is ejected as a **bipolar outflow**. A dozen or so microquasars have been found in the Milky Way Galaxy. One of them, known as GRS 1915+105, lies 40,000 light-years away in **Aquila** and was discovered in 1994 by the GRANAT X-ray satellite. It consists of an ordinary star with about the same mass as the Sun orbiting around the heaviest stellar black hole found to date, with a mass of 14 M_{sun}. In its year of discovery, it was observed to shoot out material with one-third the mass of the Moon in opposite directions at 92% of the speed of light. Another microquasar, LS 5039, is much closer, at a distance of only 9,100 light-years, has twin radio-bright jets, each about 2.6 billion km long, but is surprisingly dim in X rays, suggesting that future searches might reveal many more such X-ray-dim objects. If so, it may be that microquasars are a substantial, if not dominant, source of high-energy particles and radiation in the Galaxy.

Microscopium (abbr. Mic; gen. Microscopii)

The Microscope; a faint southern **constellation** that lies south of Capricornus and east of the southern part of Sagittarius. (See star chart 14.)

microturbulence

Small-scale motions (up to 5 km/s) in a stellar atmosphere that broaden the star's **spectral lines** and may contribute to their effective width. In practice, astronomers may attribute any broadening of a spectral line, not accounted for by known causes, such as motions of the emitting atoms, to microturbulence, whether or not it is actually present. See also **macroturbulence**.

microwaves

Electromagnetic radiation in the **radio wave** region just beyond the **infrared** with a wavelength of about 1 mm to 30 cm (about 10^9 to 10^{11} Hz).

Mie scattering

Scattering of light (without regard to wavelength) by larger particles, such as those of dust or fog in Earth's atmosphere. Compare with **Rayleigh scattering**.

Milankovich cycles

Cyclical changes in the rotation and orbit of our planet that influence the amount of solar radiation striking different parts of Earth at different times of year and have been correlated with climatic effects. There are three cycles: (1) changes in the **eccentricity** of Earth's orbit, with a period of about 100,000 years, that alter the distance between Earth and the Sun at aphelion and perihelion; (2) variations in the tilt of Earth's rotational axis (obliquity of the ecliptic), with a period of about 40,000 years; and (3) a wobble in the angle by which the axis of Earth's rotation is tilted with respect to the orbital plane, altering the seasons at which aphelion and perihelion occur (precession of the equinoxes), with a period of about 25,800 years. They are named after the Serbian mathematician, Milutin Milankovitch (1879–1958), who explained how these orbital cycles cause the advance and retreat of the polar ice caps. Although they are named after Milankovitch, he was not the first to link orbital cycles to climate. Joseph Adhemar (1842) and James Croll (1875) were two of the earliest.

Milky Way

(1) The **Milky Way Galaxy** to which the Sun belongs. (2) A softly glowing band of light across the sky, produced by light from stars in the **galactic disk**.

Milky Way Galaxy

See article, pages 340–341.

millimeter wave astronomy

See **submillimeter wave astronomy**.

Mimas

The seventh moon of **Saturn** and the closest of those known before the Space Age, also called Saturn I; it orbits at a mean distance of 185,520 km from the planet's center. Mimas's low density (1.17 that of water) indicates

(continued on page 341)

Milky Way Galaxy

Our home galaxy, also known simply as *the Galaxy,* a large type Sb or Sbc **spiral galaxy** containing some 200 to 400 billion stars (or possibly many more if brown dwarfs are included). Comprised of a disk, a bulge, and a halo, it has a total mass, including an uncertain but large amount of **dark matter** in the halo, of 750 billion to 1 trillion M_{sun}. The **galactic disk** is home to the various spiral arms of the Galaxy, including the Orion Arm in which the Sun is located (27,700 light-years from the center), the Outer Arm and Perseus Arm (both outside our own arm), the Sagittarius-Carina Arm (immediately inward of the Sun), and the Scutum Arm, Crux Arm, and three-kiloparsec Arm (all even closer to the center). Within these arms are many ordinary, intermediate-age disk stars, such as the Sun, together with the more showy extreme **Population I** objects, in the form of young, hot stars, stellar associations, open clusters, diffuse nebulae, and the bulk of the interstellar matter from which future stars will form. The **galactic bulge** and the much larger **galactic halo** contain **Population II** objects–mostly old stars and roughly 200 globular clusters, of which about 150 are known. These globulars are strongly concentrated toward the **galactic nucleus**.

Milky Way Galaxy The innermost region of the Milky Way, only a few light-years across, seen in infrared light with the Very Large Telescope. The compact objects are stars and there is also diffuse emission from interstellar dust. The two small arrows mark the position of the black hole, SgrA*, believed to exist at the very center of our Galaxy. *European Southern Observatory*

Diameter of disk:	About 100,000 light-years
Thickness of disk:	2,300 to 2,600 light-years
Thickness of bulge:	13,000 light-years
Mass:	At least 100 billion M_{sun}
Proportion of gas/dust:	5 to 10%
Mean density:	0.1 M_{sun} per cubic parsec
Total luminosity:	About 10^{44} ergs/s
Magnetic field:	3 to 5 × 10^{-6} gauss

The nucleus of the Milky Way contains a complex of gas, dust, stars, **supernova remnants**, magnetic filaments, and, almost certainly, a massive **black hole** at the very center; it lies in the direction of **Sagittarius**, around R.A. 17h 46m and Dec. –28° 56′. Lying dead center in the Galaxy is the **Sagittarius A** Complex, which is believed to be associated with a black hole of about 3 million M_{sun}, material in orbit around this object, and a nearby supernova remnant. Surrounding the galactic center are narrow threads known as nonthermal filaments (NTFs), the most prominent of which are called the Arc, the Pelican, and the Snake. These seem to consist of magnetic **flux tube**s filled with relativistic electrons, beaming **synchrotron radiation**, that have been swept up from adjacent molecular clouds and hurled along the field lines at incredible speeds. Another unusual structure in the nucleus is cataloged as 359.1-00.5 and appears to be a **superbubble** with a cluster of as many as 200 newborn stars at its heart.

The Milky Way Galaxy has a number of **satellite** galaxies, as shown in the table "Satellite Galaxies of the Milky Way," (also see individual entries for these galaxies) and is the second largest (after the **Andromeda Galaxy**), but possibly most massive, member of the **Local Group**.

Satellite Galaxies of the Milky Way

Galaxy	Distance (light-years)	Year of Discovery	Absolute Magnitude	Diameter (light-years)
Sagittarius Dwarf Elliptical	78,000	1994	−13.4	>10 000?
Large Magellanic Cloud (LMC)	160,000	–	−18.1	20,000
Small Magellanic Cloud (SMC)	180,000	–	−16.2	15,000
Ursa Minor Dwarf	220,000	1954	−8.9	1,000
Draco Dwarf	270,000	1954	−8.8	500
Sculptor Dwarf	285,000	1938	−11.1	1,000
Sextans Dwarf	290,000	1990	−9.5	3,000
Carina Dwarf	330,000	1977	−9.3	500
Fornax Dwarf	450,000	1938	−13.2	3,000
Leo II	670,000	1950	−9.6	500
Leo I	830,000	1950	−11.9	1,000

Mimas

(continued from page 339)

that it is composed mostly of water-ice with only a small amount of rock. The surface is saturated with impact craters, by far the largest of which is *Herschel* with a diameter of 130 km–⅓ of the moon's 393 km diameter. Herschel's walls are about 5 km high, parts of its floor measure 10 km deep, and its central peak rises 6 km above the crater floor with a base of 30 × 20 km. The impact that made this crater must have nearly split Mimas apart; fractures can be seen on the opposite side of the moon that may have resulted from the same event. Mimas was discovered by William **Herschel** in 1789.

Mimosa (Beta Crucis, β Cru)

The second brightest star in **Crux**; its relatively modern name, which comes from the Latin for "actor" (and is also a word used in botany), is of unknown origin. Mimosa is a hot giant **B star** and also a binary whose components are only about 8 AU apart–too close to resolve–and have an orbital period of 5 years. Mimosa is also a **Beta Cephei star** that varies between magnitudes 1.23 and 1.31 with multiple periods of 5.68, 3.87, and 2.91 hours.

Visual magnitude:	1.25
Absolute magnitude:	−3.92
Spectral type:	B0.5III
Surface temperature:	27,600 K
Luminosity:	34,000 L_{sun}
Mass:	14 M_{sun}
Distance:	353 light-years

Minkowski, Hermann (1864–1909)

A Russian mathematician who developed a four-dimensional geometry of space and time that influenced the formulation of the **general theory of relativity**. He is also well known for the use of geometric methods in the theory of numbers. Minkowski was professor at the University of Königsberg (1894–1896), the Federal Institute of Technology, Zürich (1896–1902), and the University of Göttingen (1902–1909).

Minkowski, Rudolf Leo Bernhard (1895–1976)

A German-American astrophysicist who divided **supernovae** into Types I and II (1941) and, with Walter **Baade**, identified the first optical counterpart to an extragalactic radio source, **Cygnus A**. Minkowski earned his Ph.D. in physics at the University of Breslau, taught at the University of Hamburg (1922–1935), then joined the **Mount Wilson Observatory** staff. After retirement, in 1960, he continued research at the University of California at Berkeley. He studied spectra, distributions, and motions of **planetary nebula**e, more than doubling the number of these objects known, investigated novae and supernovae and their remnants, especially the **Crab Nebula**, and headed the *Palomar Observatory Sky Survey* which photographed the entire northern sky in the 1950s.

Minkowski's Object (Arp 133)

A peculiar galaxy in **Cetus** lying adjacent to and southwest of the elliptical galaxy NGC 541, which is also a **radio galaxy** that has a **jet** of ionized plasma emanating from its nucleus. Evidence now suggests that the star formation in Minkowski's Object was actually triggered by the jet from the nucleus of NGC 541 because the body of the jet can be traced all the way to the region where the new stars are forming. If this turns out to be true, it will be the first known case of a jet from one galaxy influencing a neighboring galaxy and triggering new star birth. The details of just how this could occur are uncertain.

minor axis (*b*)

The axis passing through the center of an **ellipse**, perpendicular to the **major axis**; it is the shortest diameter distance between opposite points of an ellipse.

minor planet

An alternative name for an **asteroid** or an asteroid-like body, such as a **Centaur** or a **Kuiper Belt object**.

Minor Planet Center (MPC)

An organization based at the **Smithsonian Astrophysical Observatory** that collects, computes, checks, and disseminates information about minor planets, including **asteroids**, **comets**, and **Kuiper Belt objects**. The MPC operates under the auspices of Commission 20 of the **International Astronomical Union** and is funded mainly by subscriptions to its various services. The Institute for Theoretical Astronomy in St. Petersburg, Russia, fulfills a complementary function by publishing each year the *Ephemerides of Minor Planets,* which contains the **orbital elements** of all numbered asteroids, together with their **opposition** dates and ephemeredes (see **ephemeris**).

Mintaka (Delta Orionis, δ Ori)

The seventh brightest star in **Orion** and the westernmost and faintest star in **Orion's Belt**; its Arabic name means "the belt of the Central One." Mintaka is a multiple star system, the main components of which are a hot (30,000 K) **B star** and an even hotter **O star,** each with a mass of over 20 M_{sun} and a luminosity of 70,000 L_{sun}. These bright stars form a compact visual binary, with a period of 5.73 days and a maximum separation of 0.3″, and also an **Algol star** system showing a dip of about 0.2 magnitude during mid-partial eclipse. Two much dimmer and remote companions trek around the central O-B couple. A magnitude 6.8 B star orbits at a distance of about 0.25 light-year, while, closer in, circles a magnitude 14 component. Mintaka is famous as a background against which the thin gas of interstellar space was first detected, when the German astronomer Johannes Hartmann (1865–1936) in 1904 discovered absorption in the star's spectrum that could not be produced by the orbiting pair.

Visual magnitude:	2.25
Absolute magnitude:	–4.99
Spectral type:	O9.5II + B2V
Distance:	916 light-years

Mira (Omicron Ceti, ο Cet)

A **red giant** and **pulsating variable** in **Cetus** that ranges in brightness from 2.0 to 10.1 and spectral type from M6e to M9e III over a period of 331.96 days as its surface rises and falls. Measurements by the Hubble Space Telescope show that Mira has an angular diameter of 0.06″, corresponding to a radius of about 700 R_{sun}, and that its shape is far from spherical. Its great size and instability result in a **stellar wind** that will soon blow away the star's outer envelope, exposing a core that will eventually end its days as an Earth-sized white dwarf. Mira is the prototype and brightest long-period variable or **Mira star,** and the second variable of all to be recorded. Its name, meaning "wonderful," was given by **David Fabricius,** who was the first to describe its brightness changes in 1596. Mira has a faint B companion which is itself variable.

Visual magnitude:	2.0 to 10.1
Absolute magnitude:	0.93
Spectral type:	M6e to M9e III
Surface temperature:	just above 2,200 K
Luminosity:	15,000 L_{sun}
Distance:	418 light-years

Mira star

A cool **red giant** of **spectral type** Ke, Me, Se, or Ce (with molecular bands) that pulsates with a period of 80 to 1,000 days and varies in brightness from 2.5 to 11 magnitudes. Mira stars, also known as **long-period variables**, are named for their prototype, **Mira** (Omicron Ceti). Their brightness, large amplitude, and distinctive properties, make them so easy to find that more of them are known than any other type of variable star. Mira stars occupy the high luminosity portion of the **asymptotic giant branch** in the **Hetzsprung-Russell diagram**, along with **semiregular variables**. They have masses similar to that of the Sun but, owing to their much greater size, have a feeble gravitational hold on the material in their outer layers which, as a consequence, escapes into space in the form of a strong **stellar wind** at the rate of about 10^{-7} to 10^{-6} M_{sun}/year. Material thus shed accumulates around the star as an extensive circumstellar shell. The rate of mass loss is such that the Mira stage can only last about a million years before the aging star evolves to become a **white dwarf** surrounded by a **planetary nebula**. The Sun itself is destined to become a Mira star within the next few billion years.

Mirach (Beta Andromedae, β And)

One of the two brightest stars (equal to **Alpheratz**) in **Andromeda**; its name comes from the Arabic for "girdle" or "waist-cloth." A giant **M star**, Mirach measures about 0.8 AU across or roughly the size of Mercury's orbit. It may be at the stage of core helium burning and is destined to end its days as an Earth-sized **white dwarf**. Like many **red giants**, Mirach shows signs of semiregular variability. It also has a low-mass companion, of magnitude 14.4, that is over 60,000 times fainter and separated from it by about 27″, equivalent to at least 1,700 AU.

Visual magnitude:	2.07
Absolute magnitude:	−1.87
Spectral type:	M0IIa
Surface temperature:	3,800 K
Luminosity:	1,900 L_{sun}
Mass:	3 to 4 M_{sun}
Distance:	199 light-years

Mirach's Ghost (NGC 404)

A galaxy in **Andromeda** that, although of high surface-brightness, is difficult to observe because of the glare of **Mirach** (Beta Andromedae), just 7′ away.

Visual magnitude:	10.1
Apparent diameter:	4.4′
Position:	R.A. 1h 9.4 m, Dec. +35° 43′

Miranda

The eleventh closest and fifth largest moon of **Uranus**, also known as Uranus V; it was discovered by Gerald **Kuiper** in 1948 and was, prior to the arrival of **Voyager** 2 in 1986, the innermost known moon. By good fortune, Voyager passed particularly close to Miranda because of the gravity-assist trajectory needed to take the probe on to Neptune. The pictures it sent back revealed one of the most remarkable surfaces in the solar system. Miranda's terrain is an extraordinary mixture of heavily cratered regions, grooves, valleys, and cliffs, including some sheer drops of up to 10 km. It was initially thought that the moon had been completely shattered by giant impacts and reassembled several times in its history, each time burying some parts of the original surface and exposing some of the interior. Now, however, a more mundane explanation is favored that involves the upwelling of partially melted ices. Miranda has a diameter of 472 km and moves in a nearly circular orbit 129,850 km from the center of the planet and is inclined at about 4°.

Miranda A high-resolution image of this bizarre moon of Uranus acquired by Voyager 2 on January 24, 1986, from a range of 36,250 km. Ridges and valleys abound, cut across by numerous faults. The largest fault scarp, or cliff, is seen below and right of center; it shows grooves probably made by the contact of the fault blocks as they rubbed against each other. Movement of the down-dropped block is shown by the offset of the ridges. The fault may be 5 km high—higher than the walls of the Grand Canyon on Earth. *NASA/JPL*

Mirfak (Alpha Persei, α Per)

The brightest star in **Perseus,** also known as Algenib; the name Mirfak comes from an Arabic phrase that means "the elbow of the Pleiades." A supergiant **F star**, Mirfak is of interest to astrophysicists in that it borders on being a **Cepheid variable**, and thus is important in defining the

Visual magnitude:	1.79
Absolute magnitude:	−4.51
Spectral type:	F5Ib
Surface temperature:	7,000 K
Luminosity:	5,400 L_{sun}
Radius:	60 R_{sun}
Distance:	592 light-years

nature of such stars. It is the brightest star of the *Alpha Persei Moving Cluster* (Melotte 20), an **OB association** that contains many of the fainter surrounding stars in Perseus, most of which are hot and massive.

mirror

A reflecting surface that forms an image; it may be either flat (which simply inverts an image side-to-side), concave, or convex. Telescope mirrors are coated on their front surface (unlike everyday mirrors that are coated at the back) by **aluminizing**. Because astronomical images need to be sharp, the surface shape of the mirror must be very precise; typically, it should not alter the reflected wavefront of light by more than one-quarter of a wavelength of light across its entire surface. New techniques, including **segmented mirrors** and **active optics**, have enabled the construction of telescopes with mirrors up to 10 m in diameter. Mirrors up to 10 times larger are envisaged over the next decade (see also **extremely large telescope**).

mirror matter

A hypothetical form of matter, not to be confused with **antimatter**, that would balance out the fact that ordinary matter has a slight left-hand bias in its interactions at the subatomic level; in other words, mirror matter would restore parity to the universe. The idea that every particle in nature has an elusive, unseen mirror partner was first put forward in the 1980s. Then, in 1999, came the suggestion that a small number of **MACHOs**, which had been detected on the outskirts of our Galaxy, might be stars composed of this exotic stuff. Mirror matter would be subject to its own distinct set of physical laws. Although it would feel gravity in the ordinary way and therefore be able to condense into mirror stars and planets, its versions of the three other basic forces–electromagnetism and the strong and weak forces–would be different. One consequence is that mirror stars would be invisible because they would not emit electromagnetic radiation. The only way their presence could become known to us is through their gravitational **microlensing** effects and their subsequent identification as MACHOs. Theoretical considerations suggest that the maximum mass of a stable mirror star would be about 0.5 M_{sun}–just right to explain the dozen or so MACHOs that have been singled out as candidates. Some researchers, notably Robert Foot (1965–) at the University of Melbourne, Australia, have gone further and suggested that the presence of mirror matter in the solar system might explain the **Tunguska event**, possible anomalous movements of the deep space probes **Pioneer 10 and 11**, and even some unusual material at the bottom of crater floors on **Eros**.

Mirzam (Beta Canis Majoris, β CMa)

The fourth brightest star in **Canis Major**; its Arabic name (also spelled "Murzim") suggests that it is an "announcer" of **Sirius**, as Mirzam rises first. Mirzam is a hot giant **B star** that has reached the end of its stay on the main sequence, and is one of the brightest variables of the type known as **Beta Cephei stars** or, sometimes (after itself), as *Beta Canis Majoris stars*. It has a principal period of about 6 hours. It has also been used extensively as a background source for studying the **interstellar medium**, and lies along a corridor in which the gas is especially hot and tenuous.

Visual magnitude:	1.98
Absolute magnitude:	–3.95
Spectral type:	B1III
Surface temperature:	22,000 K
Luminosity:	19,000 L_{sun}
Distance:	500 light-years

missing mass

The unseen matter whose gravitational influence has to be invoked to explain the way galaxies rotate, and also to bind clusters of galaxies together. It is thought to consist, in part, of giant halos of **dark matter** that surround the visible portions of galaxies, and similar material that invisibly occupies the intergalactic voids.

Mizar (Zeta Ursae Majoris, ζ UMa)

The fourth brightest star in **Ursa Major** and the middle star in the handle of the Big Dipper. It forms a well-known naked-eye double, known to the Arabs as the "horse and rider," with fourth magnitude **Alcor** (80 Ursae Majoris), 0.2° to the northeast. Mizar is also famous as the first true binary, in which a pair of stars that orbit around each other, were seen telescopically, by **Giovanni Riccoli** in 1650. The components, Mizar A and B, lie 14″ apart (at least 500 AU) and take some 5,000 years to complete one mutual circuit. Spectroscopic observations reveal that each of these two components is itself a binary. Mizar A consists of a very bright pair, separated by just 0.008″, with an orbital period of 20.5 days; Mizar B's dimmer stars make a round trip of each other every six months. All four stars are main sequence A-type objects (the brighter, both A2 with masses of 2.5 M_{sun} each, the fainter, both about A6 with masses of 1.6 M_{sun}). In 1996, the Navy Prototype Optical Interferometer (NPOI), near Flagstaff, Arizona, was used to compile images of Mizar A that, for the first time, showed it as two separate stars. The NPOI data revealed that the components of Mizar A come as close together as 16 million km and as far apart as 54 million km. The Mizar quartet moves through space at a similar speed and direc-

Statistics for Mizar and Alcor

	Mizar	Alcor
Visual magnitude:	2.23	3.99
Absolute magnitude:	0.33	2.09
Spectral type:	A2V + A2V + A1V	A5V
Distance (light-years):	78	81

tion to Alcor, but measurements by the Hipparcos satellite suggest that Mizar and Alcor are separated by 2–3 light-years–too far for the systems to be physically associated. (See table, "Statistics for Mizar and Alcor.")

molecular cloud

A cold, dense **interstellar cloud** that contains a high fraction of **molecules**, of which well over 100 different types have now been discovered in space (see **interstellar molecules**). It is widely believed that the relatively high density of **dust** particles in these clouds plays an important role in the formation and protection of the molecules. The emission of molecular lines often shows several distinct intensity peaks, each representing individual clumps or clouds of gas and dust in a region that characteristically extends for 50 light-years and is often associated with **T Tauri stars**–young, pre-main-sequence stars–and also hot massive stars and the ionized gas around them. Two distinct types of molecular cloud are known, both associated with star formation: **giant molecular clouds** and **dwarf molecular clouds**.

molecule

The smallest unit of a chemical compound. A molecule is made of two or more atoms, linked by interactions of their **electrons**.

Monoceros (abbr. Mon; gen. Monocerotis)

The Unicorn; a dim **constellation** that lies across the celestial equator. Among its interesting deep sky objects are the **Cone Nebula** and the **Rosette Nebula**, and the open clusters M50 (NGC 2323), near the border with Canis Major (magnitude 5.9; diameter 16′; R.A. 7h 3.2m, Dec. –08° 20′) and NGC 2244, which surrounds the star 12 Mon and is itself surrounded by the Rosette Nebula. The *Monoceros Loop* is a filamentary loop nebula, part of a 300,000-year-old **supernova remnant**, about 3,000 light-years away in the galactic plane northeast of the Rosette. (See table, "Monoceros: Stars Brighter than Magnitude 4.0," and star chart 3.)

Monogem Ring

A **superbubble**, discovered in the soft X-ray energy band by **ROSAT**, that lies about 1,000 light-years away in **Monoceros** and **Gemini**. It is believed to be the highly evolved remains of a **supernova** that took place some 86,000 years ago. If it shone as brightly in visible light, it would illuminate a region of the night sky about 25° across.

month

A period of time connected with the motion of the Moon around Earth. The familiar **calendar** (or civil) month is an artificial unit consisting of a whole number of days. For astronomical purposes there are several other types of month. (1) *Synodic month:* the average interval from one new Moon to the next–29.53059 days; also called a *lunar month*. (2) *Anomalistic month:* the interval in which the Moon passes from perigee to perigee–27.55464 days. (3) *Sidereal month:* the interval in which the Moon passes from a fixed position with respect to the stars back to the same position–27.32166 days. (4) *Draconic month* (or *nodal month*): the interval between successive passages of the Moon through its **ascending node**. (5) *Tropical month:* the interval between successive passages of the Moon through the vernal **equinox**–a mean of 27.21222 days.

Moon

See article, pages 346–348.

Monoceros: Stars Brighter than Magnitude 4.0

	Magnitude				Position	
Star	Visual	Absolute	Spectral Type	Distance (light-yr)	R.A. (h m s)	Dec. (° ′ ″)
β	3.76	–2.87	B3Ve	691	06 28 49	–07 01 58
α	3.94	0.71	KOIII	144	07 41 15	–09 33 04
γ	3.99	–2.49	K3III	645	06 14 51	–06 16 29

Moon

Earth's only known natural satellite (searches for Earth-orbiting moonlets having so far drawn a blank) and the only extraterrestrial body to have been visited by humans–12 in all, aboard six Apollo missions between July 1969 and December 1972. The Moon has been the target of numerous robotic probes including, most recently, **Clementine** (1994) and **Lunar Prospector** (1999). A total of 382 kg of rock samples were returned to Earth by Apollo and the Russian Luna programs; in addition, a number of **lunar meteorites** have been found. Most of these samples have been dated between 3 and 4.6 billion years (the one exception is a lunar meteorite dated at 2.8 billion years), and provide information about the early history of the solar system which is missing on Earth due to a lack of rocks more than about 3.8 billion years old. The lunar rocks have also provided strong support for the impact theory of the Moon's origin, namely, that the Moon formed from material splashed out of Earth's core and mantle by a colliding body as big or bigger than Mars shortly after Earth formed. Proof of this theory could come from the European Space Agency's SMART-1 probe, scheduled for launch in the second half of 2003, which will map the Moon in X rays. These X-ray measurements will show conclusively if the Moon contains less iron than Earth, compared to lighter elements such as magnesium and aluminum–as it should if, as predicted, the Moon came mainly from the light mantle rocks of the young Earth and its impactor.

The Moon has two main types of terrain: bright, densely cratered, ancient highlands, dating back over 4.5 billion years, and dark, relatively smooth, younger maria (Latin for "seas"), aged about 3.8 billion years. The maria, which make up about 17% of the Moon's surface, are huge **impact basins** that were later flooded by molten lava. Most of the surface is covered with a mixture of fine dust and rocky debris, known as regolith, produced by meteor impacts that vary in depth from 3 to 5 m in the maria to 10 to 20 m in the highlands.

The Moon's crust ranges in depth from 60 km on the nearside to 100 km on the farside. This uneveness may account for the fact that the Moon's center of mass is displaced from its geometric center by about 2 km in the direction toward Earth. Crustal uneveness may also explain differences in lunar terrain, such as the striking dominance of maria on the Earth-facing hemisphere. Below the crust is a mantle and probably a small core, some 340 km radius and containing about 2% of the lunar mass. Curiously, the Moon's center of mass is offset from its geometric center by about 2 km in the direction toward the Earth.

The Moon has no atmosphere. But evidence from Clementine suggests that there may be water-ice in some deep craters near the Moon's south pole that are permanently shaded. This was confirmed by

Moon An oblique view of the crater Copernicus on the lunar nearside, photographed from Apollo 17 in lunar orbit. *NASA*

Mean diameter:	3,476 km
Mass (Earth = 1):	0.0123
Mean density:	3.34 g/cm^3
Surface gravity (Earth = 1):	0.165
Escape velocity:	2.38 km/s
Mean albedo:	0.07
Axial tilt:	6° 41′
Orbit	
Mean distance from Earth:	384,404 km (1.28 light-seconds)
Eccentricity:	0.0549
Inclination:	5° 9′
Period:	27d 7h 43m 11s

Moon The sharpest image ever taken of the Moon by a ground-based telescope. The Very Large Telescope photographed this small patch of the Moon's surface, some 6° north of the lunar equator and near the intersection of Mare Tranquillitatis and Mare Foecunditatis, on April 30, 2002, revealing detail as small as 130 meters across. The large crater at the top is named Cameron and has a diameter of about 10 km. The size of the field is about 60 km × 45 km. *European Southern Observatory*

Some Notable Features on the Moon

Feature	Description
Aitken Basin	An **impact basin** in the south polar region. With a diameter of about 2,500 km, a maximum depth of over 12 km, and an average depth of about 10 km, it is the biggest, deepest impact basin in the solar system.
Apennines	A mountain range that rises to 4,572 m at the southeastern edge of Mare Imbrium. The highest escarpment on the Moon, it is higher above the adjacent flatlands than the Himalayan front is above the plains of India and Nepal. The landing site of Apollo 15 was chosen to allow the astronauts to drive from the Lunar Module to the base of the Apennines during two excursions.
Bailly	The largest crater on the nearside of the Moon with a diameter of 295 km and maximum depth of 3.96 km. A highly eroded structure, it is named after the French astronomer Jean Bailly.
Copernicus	A 93-km-wide crater that is one of the most prominent features on the lunar nearside. Made less than 1 billion years ago, and thus one of the Moon's youngest major markings, it has a system of bright rays seen most clearly at full Moon.
Imbrium Basin	The largest and youngest of the giant impact basins on the nearside of the Moon. The asteroid collision that formed it, about 3.9 billion years ago, went close to breaking the Moon apart; in the event, it threw ejecta over much of the lunar surface and created deep fissures. Through these cracks, lava poured out, filling much of the basin and leaving the 1,300-km-wide dark feature known as Mare Imbrium.
Orientale Basin	The youngest and best preserved impact basin on the Moon, visible from Earth only at the extreme western limb as a libration feature. Formed some 3.8 to 3.9 billion years ago, it shows three concentric rings of mountains. Strong radial lineations made by the scouring flow of ejecta are also evident.
Tycho	A spectacular, 85-km-wide crater associated with the brightest and most extensive ray system on the Moon. In some cases, the rays extend for over 1,500 km; their prominence suggests that Tycho was formed relatively recently, perhaps within the past 3 billion years.

Lunar Prospector, which also indicated there may be ice at the north pole. The Moon has no global magnetic field, though some of its surface rocks display remnant magnetism suggesting that there may have been a global magnetic field early in the Moon's history. With no atmosphere and no magnetic field, the Moon's surface is exposed directly to the **solar wind**. Lunar temperatures range from –184°C at night to 214°C during the day, except at the poles where it's a constant –96°C.

The gravitational interaction between Earth and the Moon leads to some important effects, the most obvious of which is the tides. Although the solid rock of Earth and the Moon is distorted a bit by the mutual tugging, the effect is most noticeable on Earth's oceans. These are pulled into an elliptical shape. As Earth rotates under the ocean bulge, it causes high tides to propagate onto beaches. Because there are two bulges, there are two high tides and two low tides every day. In the case of Earth's rocky body, which isn't completely fluid, the bulges raised by the Moon's gravity are carried by Earth's rotation slightly ahead of the point directly beneath the Moon. The Moon's gravity, acting on these out-of-line bulges, then produces a twisting force, or torque, which slows Earth's rate of spin, by about 1.5 milliseconds/century. A similar but much more powerful braking effect by Earth on the Moon, long ago, slowed the Moon's spin rate to match its orbital period. To conserve angular momentum as Earth's spin slows, the size of the Moon's orbit gradually increases, by 3.8 cm/year. At this rate the Moon will look about 15 percent smaller from Earth in about one billion years. See also **lunar eclipse, lunar phases,** and **transient lunar phenomenon**. (See table, "Some Notable Features on the Moon.")

moonquake

A seismic disturbance on the **Moon**. Most moonquakes are small and deep-seated, and take place when the tidal effects of Earth and the Sun are strongest–at times of perigee, apogee, new moon, and full moon.

Mopra Observatory

A 22-m radio antenna that forms part of the **Australia Telescope National Facility** (ATNF), operated by the Commonwealth Scientific and Industrial Research Organisation (CSIRO). It is intended for use in conjunction with other ATNF antennas–the six 22-m dishes at Narrabri, and the 64-m dish at **Parkes Observatory**–to form the Long Baseline Array. Like the Parkes antenna, it is also used for single-dish operation; millimeter-wavelength receivers are to be installed in the future.

Moreton wave

A shock wave on the Sun's **chromosphere** that is produced by a large **solar flare** and expands outward at about 1,000 km/s. It usually appears as a slowly moving diffuse arc of brightening in **H-alpha**, and may travel for several hundred thousand km. Moreton waves are always accompanied by meter-wave radio bursts; they are named after the American solar astronomer Gail Moreton.

Morgan, William Wilson (1906–1994)

An American astronomer who, with Philip Keenan (1908–) and Edith Kellman (1911–), introduced stellar **luminosity class**es and developed the **Morgan-Keenan classification** of stellar spectra in 1943. With Donald Osterbrock (1924–) and Stewart Sharpless (1926–) he demonstrated the existence of **spiral arm**s in the Galaxy using precise distances of O and B stars obtained from spectral classifications. Morgan helped originate the UBV (ultraviolet-blue-violet) system of magnitudes and colors, and, with Nicholas Mayall, developed a spectral classification system for giant galaxies. After three years of undergraduate study at Washington and Lee University, Morgan joined the staff of the **Yerkes Observatory**, where he spent his entire career, including three years (1960–1963) as director. While at Yerkes he earned bachelor's and doctoral degrees at the University of Chicago. Eschewing theory, his research focused on morphology–the classification of objects by their form and structure.

Morgan-Keenan classification

An extension of the Harvard classification of stellar spectra to include **luminosity class**es and a more precise observational description of each type, based on temperature, luminosity, and chemical abundance. It was introduced by the American astronomers William **Morgan**, Philip Keenan (1908–2000), and Edith Kellman (1911–) in 1943 and is also known as the *MKK system*. See also **spectral type**.

mottles

Features that form the **chromospheric network** on the Sun, visible in **H-alpha** and calcium K-line spectrograms.

They cover the whole Sun, quiet and active regions alike. Coarse mottles are bright, up to 20,000 km across, and often elongated; they merge to form **plages**. Fine mottles are narrow (700 km wide and up to 7,000 km long), dark or bright, and are probably **spicules** seen against the disk. Fine mottles form clusters with their bases rooted in the coarse mottles. All coincide with regions of strong magnetic field on the **photosphere**.

Mount Graham International Observatory (MGIO)

A division of **Steward Observatory**, the research arm for the Department of Astronomy at the University of Arizona. MGIO is at a height of 3,260 m on Mt. Graham near Safford, Arizona, 125 km northeast of Tucson. Two telescopes are in operation: the Vatican Observatory/Arizona 1.8-m Lennon telescope (VATT) and the 10-m diameter Heinrich Hertz Submillimeter Telescope (SMT), a joint project of Arizona and the Max Planck–Institut für Radioastronomie, Germany. The **Large Binocular Telescope**, one of the biggest in the world, is under construction.

Mount Hopkins Observatory

Located in Tucson, Arizona, it was renamed the Fred Lawrence Whipple Observatory in 1982. It houses the **Multiple Mirror Telescope**.

Mount Wilson Observatory

An observatory at an altitude of 1,740 m on Mount Wilson in the San Gabriel Mountains near Pasadena, California, about 30 km northwest of Los Angeles; it was founded in 1904 by George **Hale** and is owned by the Carnegie Institution of Washington, D.C. Its main nighttime telescopes are a 60-inch (1.5-m) reflector, built in 1908, and the famous 100-inch (2.5-m) Hooker Telescope, built in 1917. Two solar observatories, the 60-foot tower telescope (operated by the University of Southern California) and the 150-foot tower telescope (operated by the University of California at Los Angeles), maintain long-term exploration of the magnetic activity behavior of the Sun. There are also two **interferometers** on the site: the Infrared Spatial Interferometer (ISI, operated by the University of California at Berkeley) and the NRL Optical Interferometer.

Mount Wilson sunspot classification

A way of classifying **sunspots** in terms of their magnetic properties that was developed at **Mount Wilson Observatory**. *Alpha* denotes a unipolar sunspot group. *Beta* denotes a sunspot group having both positive and negative magnetic polarities, with a simple and distinct division between the polarities. *Beta-Gamma* denotes a sunspot group that is bipolar but in which no continuous line can be drawn separating spots of opposite polarities. *Delta* indicates a complex magnetic configuration of a solar sunspot group consisting of opposite polarity **umbrae** within the same **penumbra**. *Gamma* indicates a complex active region in which the positive and negative polarities are so irregularly distributed as to prevent classification as a bipolar group.

mounting

The structure that holds a telescope and enables it to be pointed at any part of the sky. Since movement in two dimensions is involved, there must be two axes that are perpendicular to each other. The two main types of mounting, the **equatorial mounting** and the **altazimuth mounting**, differ principally in the axes they use.

moving cluster

A group of stars dynamically associated so that they have a common motion with respect to the **local standard of rest**; nearby examples include the **Hyades** and the **Ursa Major Moving Group**. Because of perspective, the **proper motions** appear to converge on a single point in the sky. If the linear velocity of the cluster is known from observations of **radial velocity**, then the distance of each star can be estimated from the total proper motion; this method is known as *moving cluster parallax.*

MS star

A star that has characteristics of both an **M star** and an **S star**; particularly prominent are the molecular **absorption** bands of metal oxides, especially zirconium dioxide (ZrO), titanium dioxide (TiO), and strontium oxide (SrO).

Müller, Johann (1436–1476)

A German astronomer and mathematician, also known as *Regiomontanus.* He studied trigonometry and completed a translation of Ptolemy's ***Almagest*** from the original Greek that had been started by his tutor, Georg von Peurbach (1423–1461). Ironically, this translation, together with some comments and observations added by Müller, helped overthrow the **Ptolemaic system** of the universe. Müller also observed the motion of the Moon, planets, and comets, and erected an observatory at Nürnberg, which included several modified **quadrants** and a few of the first weight-driven clocks.

Multiple Mirror Telescope (MMT)

A 6.5-m telescope on the summit of Mount Hopkins, the second highest peak in the Santa Rita Range of the Coronado National Forest, about 55 km south of Tucson, Arizona, on the grounds of the Smithsonian's Fred Lawrence Whipple Observatory. In 1998, the large single

Musca: Stars Brighter than Magnitude 4.0

Star	Magnitude		Spectral Type	Distance (light-yr)	Position	
	Visual	Absolute			R.A. (h m s)	Dec. (° ′ ″)
α	2.69v	−2.17	B2IV	306	12 37 11	−69 08 07
β	3.04	−1.86	B2.5V	311	12 46 17	−68 06 29
δ	3.61	1.38	K2III	91	13 02 16	−71 32 56
λ	3.63	0.65	A7III	128	11 45 36	−66 43 43
γ	3.84	−1.15	B5V	324	12 32 28	−72 07 58

mirror replaced six identical 1.8-m mirrors on a common mount (with a combined light-collecting power of a 4.5-m mirror), which had been in use since 1979. The MMT is a joint venture of the Smithsonian Institution and the University of Arizona, which operates it.

multiple star

A system of three or more stars orbiting around one another. A three-star system is also known as a *trinary.*

multiplet

A group of spectral lines showing **fine structure** with several components caused by spin-orbit interactions in the atom.

multiringed basin

A large impact feature surrounded by a series of concentric rings. Examples include Mare Orientale on the **Moon**, the Caloris Basin on **Mercury**, and Valhalla and Asgard on **Callisto**. The rings are similar to the ripples caused by throwing a stone into a pond; they are the peaks and troughs of rock made molten by the impact, which then froze in position as it cooled.

Murchison Meteorite

A **carbonaceous chondrite** that exploded into fragments over the town of Murchison, 400 km north of Perth, Australia, on September 28, 1969. About 82 kg of meteorite were recovered. Eyewitnesses arriving at the scene reported smelling something like methanol or pyridine–an early indication that the object might contain organic material. Subsequent analysis by NASA scientists revealed the presence of six **amino acids** commonly found in proteins and 12 that did not occur in terrestrial life. All of these appeared in both right-handed and left-handed molecular forms, suggesting that they were not the result of terrestrial contamination. The meteorite also contained hydrocarbons that appeared abiotic in character and was enriched with a heavy isotope of carbon, confirming the extraterrestrial origin of its organics. Initial studies suggested that the amino acids showed no bias. However, in 1997, further analysis revealed an excess of left-handed versions of four amino acids ranging from 7 to 9%.

Musca (abbr. Mus; gen. Muscae)

The Fly (originally called Apis, the Bee, by **Bayer**); a small **constellation** of the south circumpolar region,

Musca: Other Objects of Interest

Object	Notes
Planetary Nebula	
NGC 5189	A highly irregular planetary. Magnitude 10; diameter 2.6′; R.A. 13h 33.7m, Dec. −65° 58.5′.
Globular clusters	
NGC 4833	A bright cluster, close to Delta Mon, well seen in a small telescope. Magnitude 7.3; diameter 13.5′; R.A. 13h 00m, Dec. −70° 53′.
NGC 4372	Close to Gamma Mon and similar to NGC 4833. Magnitude 7.8; diameter 18.6′; R.A. 12h 25.8m, Dec. −72° 40′.

lying south of Crux (in fact, part of the **Coalsack** extends from Crux into Musca). (See tables, "Musca: Stars Brighter than Magnitude 4.0," and "Musca: Other Objects of Interest." See also star chart 4.)

MUSES-C

The first sample-return mission to an **asteroid**. This Japanese spacecraft, launched on May 9, 2003, will rendezvous with asteroid 1998 SF36 in late 2005. It will then orbit its target for about five months using autonomous navigation and guidance while its optical navigation camera, lidar, laser range finder, and fan-beam sensors gather topographic and range information about the object's surface. After building a three-dimensional model of the asteroid during this mapping phase, the MUSES-C project team will decide on suitable sites for surface exploration. To capture samples, the spacecraft carries a horn that will be brought to the surface as the spacecraft makes a close approach. A small pyrotechnic charge will then fire a bullet into the surface, and fragments of the impact will be captured by the horn and funneled into a sample container. The aim is to carry out several sample extractions, each taking only a second or so, from different locations. MUSES-C will return to the vicinity of Earth and release a small reentry capsule containing the collected samples. Reentering at over 12 km/s directly from its interplanetary return trajectory, the capsule will use an advanced heat-shield as protection and finally deploy a parachute for soft landing.

N

N galaxy
A galaxy with a small, bright, blue nucleus superposed on a much fainter red background. In terms of form, color, spectrum, **redshift**, and optical and radio variability, N galaxies are intermediate between **quasars** and **Seyfert galaxies**. The N designation comes from the Morgan classification (see **galaxy classification**).

N star
An obsolete name for what is now classed as a cool **carbon star** (classes C6 to C9).

nadir
The point on the **celestial sphere**, blocked from view by Earth, that is diametrically opposite the **zenith**.

Naiad
The innermost moon of **Neptune**, also known as Neptune III; it was discovered in 1989 by **Voyager** 2. Naiad has a diameter of 58 km and orbits at a mean distance of 48,230 km from (the center of) the planet.

naked-eye star
A star visible without a telescope or other optical aid, except for spectacles or contact lenses. In principle, the *naked-eye limiting magnitude* is about 5.5. However, what can actually be seen with the naked eye depends upon the individual, the location, and the conditions of the observation.

nakhlite
One of the types of *SNC meteorites* believed to have come from Mars (see **Mars meteorite**); nakhlites probably formed as lava flows with unusual compositions. They are named for the first member of the group to be found, which fell in pieces in El Nakhla, Egypt, in 1911; local legend has it that one of the fragments hit and killed a dog, though this story may be apocryphal. The nakhlites consist mainly of green **augite** crystals with some **olivine** in a very fine-grained blend of **plagioclase feldspar** and other **feldspars**, **pyroxenes**, iron-titanium oxides, sulfides, and phosphates. Most intriguing, there are traces of preterrestrial aqueous alteration products in the form of hydrated minerals, including clay minerals and carbonates. Some researchers think that the presence of these hydrated minerals in the nakhlites, in addition to concentrations of water-soluble ions such as those of chlorine, potassium, sodium, and calcium, suggests that they were once in an environment in which liquid seawater was present for some time, perhaps under an ancient Martian ocean. A problem with this idea is the comparatively young age of the nakhlites: they seem to have crystallized only 1.3 to 1.4 billion years ago and to have been altered by water a mere 700 million years ago, long after Mars supposedly lost its ancient lakes, rivers, and seas.

Naos (Zeta Puppis, ζ Pup)
A blue **supergiant** and the second brightest **O star** in the sky in terms of apparent magnitude. It is embedded in the **Gum Nebula** and used to be the Zeta star of **Argo Navis** (mythical Jason's vessel Argos)–the Greek *naos* meaning "ship"–before becoming Zeta in **Puppis**, the Stern, after the big constellation was broken up. It also belongs to the young Vela R2 association. Typical of its kind, Naos is blowing a fierce **stellar wind**, averaging 2,300 km/s and by which it is losing 10^{-6} M_{sun} per year (10 million times the rate lost by the Sun); this wind has been so well observed at wavelengths from radio wave to X ray that it has become a major proving ground for stellar wind theorists. Naos is a classic O-type **runaway star**, having been ejected at high speed (nearly 100 km/s) from its birthplace in the neighboring constellation **Vela**. Such stars are invariably single and spin much faster than normal O stars: in Naos's case, about 211 km/s at its equator.

Visual magnitude:	2.21
Absolute magnitude:	–5.96
Spectral type:	O5Ib
Surface temperature:	42,000 K
Luminosity:	1 million L_{sun}
Mass:	50 M_{sun}
Radius:	20 R_{sun}
Distance:	1,400 light-years

Nasmyth focus
A focal point to one side of the tube of a telescope with an **altazimuth mount**. It is formed by placing a third mirror (tertiary) so as to direct the beam along the altitude axis and through a hole in the supporting trunnions. Nasmyth foci enable bulky instruments to be

mounted on a permanent platform that needs to rotate only in azimuth; they are commonly used with large modern telescopes, especially for spectrographic work. Named after the Scottish engineer James Nasmyth (1808–1890), best known for his invention of the steam hammer.

National Optical Astronomy Observatories (NOAO)

A group of optical observatories run by the Association of Universities for Research in Astronomy under contract to the U.S. National Science Foundation. Formed in 1984, with headquarters in Tucson, Arizona, NOAO has responsibility for four sites: **Cerro Tololo Inter-American Observatory**, **Gemini Observatory**, **Kitt Peak National Observatory**, and **National Solar Observatory**.

National Radio Astronomy Observatory (NRAO)

The collection of radio astronomy facilities owned by the American government and administered on behalf of the U.S. National Science Foundation by Associated Universities, Inc., a consortium of nine universities. NRAO facilities include **Green Bank**, the **Very Large Array**, the **Very Long Baseline Array**, the **Kitt Peak 12-Meter Telescope** and the future **Atacama Large Millimeter Array**. Its headquarters are in Charlottesville, Virginia.

National Solar Observatory

A branch of the **National Optical Astronomy Observatories**, founded in 1984, that operates solar telescopes at two main sites: Kitt Peak, Arizona, where the **McMath-Pierce Solar Telescope** is located, and **Sacramento Peak Observatory**. NSO also manages the **Global Oscillation Network Group**.

near-Earth asteroid (NEA)

An **asteroid** whose orbit occasionally brings it close to Earth. Most, if not all, such asteroids appear to have come from the **asteroid belt** and have been given new orbits through collisions with other asteroids or the gravitational influence of Jupiter. There are three main groups: the **Aten group** (which stays mainly inside Earth's orbit), the **Apollo group** (which crosses Earth's orbit), and the **Amor group** (which orbits mostly between Earth and Mars). Estimates suggest there are roughly 100 Atens, 700 Apollos, and 1,000 Amors with diameters greater than about 1 km. Because these asteroids travel in orbits that cross or approach Earth's orbit, some of them can, occasionally, pass relatively close to our planet or even threaten to collide with it. **Potentially hazardous asteroids** are coming under increasing scrutiny.

Near-Earth Asteroid Tracking (NEAT)

Two autonomous observing systems at Maui Space Surveillance Site (NEAT/MSSS) and at Palomar Observatory (NEAT/Palomar), that consist of special cameras attached to 1.2-m telescopes to search for near-Earth objects (NEOs)–both **near-Earth asteroids** (NEAs) and comets. NEAT is a Jet Propulsion project funded by NASA.

near-Earth object (NEO)

A **near-Earth asteroid** or an extinct **short-period comet** on an orbit that intercepts, or nearly intercepts, that of Earth.

NEAR-Shoemaker

A NASA spacecraft that became the first probe to orbit and (although this was not originally planned) to land on an asteroid: **Eros**. Known at launch, on February 17, 1996, simply as NEAR (Near-Earth Asteroid Rendezvous), it was later renamed in memory of the American geologist Eugene **Shoemaker**. On June 27, 1997, NEAR-Shoemaker flew by the asteroid (253) **Mathilde** at a distance of 1,200 km. A deep-space maneuver in July 1997 brought the probe back around Earth on January 23, 1998, for a gravity assist that put it on course for its rendezvous with the Manhattan-sized asteroid (433) Eros. NEAR entered an orbit of 323 × 370 km around Eros on February 14, 2000, then moved to gradually smaller orbits over the next year or so, returning a total of 160,000 images. During the final days of its mission, NEAR maneuvered to within 24 km and then, against all the odds, became the first spacecraft actually to land on an asteroid. In the final moments before it touched down without crashing on February 12, 2001, NEAR returned pictures showing surface details as small as a few tens of cm across. (See photo, page 355.)

nebula (pl. nebulae)

A cloud of gas and dust in space. There are three general types: **emission nebulae**, which shine by their own light; **reflection nebulae**, which reflect light from nearby stars; and **dark nebulae**, which absorb light and appear dark against a brighter background. When cloudlike material in space is patchy, or of a form that is difficult to categorize as a particular type of nebula, it is referred to as *nebulosity*.

Nemesis

A hypothetical small stellar companion of the Sun. Once every 30 million years, it was hypothesized, this dark star would pass through the **Oort Cloud**, triggering **comets** that perhaps caused periodic **mass extinctions** on Earth. No evidence for such an object has been found.

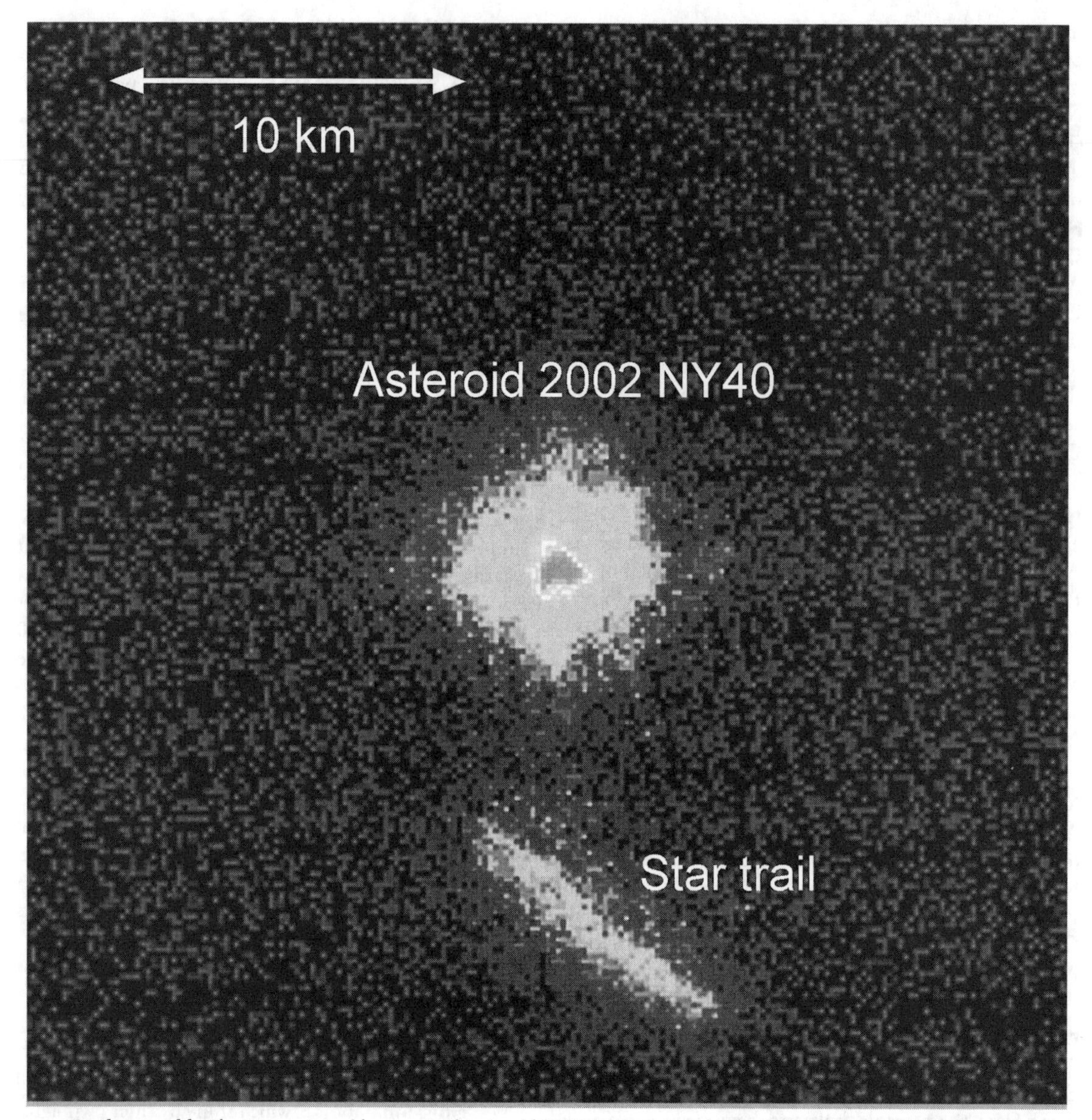

near-Earth asteroid The 800-meter-wide near-Earth asteroid 2002 NY40 as seen by the William Herschel Telescope on La Palma, Canary Islands, on the night of August 17–18, 2002. The asteroid was imagined just before its closest approach to Earth, when it was about 750,000 km away—twice the distance to the Moon—and moving rapidly across the sky (crossing a distance similar to the diameter of the Moon every 6 minutes). *Isaac Newton Group of Telescopes*

Neptune
See article, pages 356–357.

Nereid
The outermost satellite of **Neptune**, discovered in 1949 by Gerard **Kuiper** and named after the Nereids, the 50 daughters of Nereus, a Greek sea god. Nereid is about 340 km in diameter and is so far from Neptune that it takes 360 days to complete one circuit. Its orbit is the most eccentric of any moon in the solar system, taking Nereid from about 1.4 to 9.6 million km from Neptune. **Voyager** 2's best photos, taken from about

NEAR-Shoemaker The planned site of NEAR-Shoemaker's landing on Eros. In this view, taken from an orbital height of 200 km, south is at the top and the terminator (the imaginary line dividing day from night) lies near the equator. The landing site (at the tip of the arrow) is near the boundary of two distinctly different regions. To the south and east (above and to the left) lies older, cratered terrain, while to the north (down) is the saddle-shaped feature Himeros, whose lesser density of superposed craters indicates relatively recent resurfacing by geologic processes. *NASA/Johns Hopkins University*

4.7 million km, give it an **albedo** of 0.14, making it somewhat more reflective than Earth's Moon.

Nereus (minor planet 4660)

A **near-Earth asteroid** that may be an extinct comet. It is the target of Japan's **MUSES-C** that is due to arrive at Nereus in late 2005, collect surface samples, and return to Earth in 2006. Diameter 2 km, period 1.82 years, semimajor axis 1.490 AU, eccentricity 0.360, inclination 1.42°.

Neumann lines

Found in some **iron meteorites**, groups of very fine parallel lines that cross each other at various angles. Irons containing Neumann lines can easily be cleaved in three mutually perpendicular directions. They are named after their discoverer, the German crystallographer and mathematical physicist Franz Ernst Neumann (1798–1895).

neutral point

A point in space at which a particle experiences no net gravitational force. In theory, a unique neutral point would exist between any two static bodies. In practice, when two objects, such as Earth and the Moon are orbiting each other, there are five points, known as **Lagrangian points**, at which gravitational and centrifugal forces are exactly in balance.

neutrino

A ghostly subatomic particle with no charge and very little mass, that interacts only by the weak force and by gravity. It was first postulated by Wolfgang Pauli in 1930 to ensure conservation of energy and angular momentum in beta decay; effectively, it carries away excess energy in nuclear reactions. Three different types of neutrinos exist, known as electron-, mu-, and tau-neutrinos, respectively corresponding to the three massive **leptons**: electrons, muons, and tau mesons. The Sun produces neutrinos from thermonuclear reactions in its core and, since these neutrinos pass entirely through the Sun and then all the way to Earth, they provide a way of glimpsing into the heart of a star. A large flux of neutrinos carries away most of the energy of a **supernova** and neutrinos are one of the candidates for **dark matter**. So, neutrino astronomy offers an important new window on the universe beyond the electromagnetic spectrum. Because neutrinos pass so easily through ordinary matter, they're very hard to detect: large masses of stopping material and indirect detection of the effects of neutrino absorption are needed. Among the most powerful neutrino "telescopes" are the **Sudbury Neutrino Observatory** in Canada and the **Super-Kamiokande** in Japan.

One of the great puzzles of astrophysics in recent decades has been the discrepancy between the number of neutrinos detected coming from the Sun and the number expected from theory. The so-called *solar neutrino problem,* which emerged from measurements by Ray **Davis** and his pioneering neutrino detector in a South Dakota gold mine, suggesting that only one-third the expected number of solar neutrinos were arriving at Earth, has now been effectively cleared up by recent data from the Canadian and Japanese instruments. These data show that some of the electron-neutrinos produced in the Sun's core change into the other types of neutrino while en route to Earth. Earlier experiments, including that of Davis, only registered the electron-neutrinos and therefore suggested a shortfall. The newer experiments, such as that at Sudbury, pick up all the varieties of neutrino and have shown that the total count of solar neutrinos is in line with the rate of electron-neutrino predicted by orthodox theory of nuclear reactions inside the Sun.

neutron

An electrically neutral, massive particle found in the nuclei of atoms. Each neutron has a mass of 939.6 MeV, slightly more than that of the **proton**, and is composed of one up **quark** and two down quarks. Although the

(continued on page 358)

Neptune

The eighth planet from the Sun (except when **Pluto** is around perihelion) and the most distant of the **gas giants**. It was discovered in 1846 by Johann **Galle** and Louis d'**Arrest** following predictions by Urbain **Leverrier**; in fact, Galileo, as his notes reveal, had seen Neptune much earlier but thought it was a star. With an orbital period of 165 years, Neptune will first return to the point in its orbit where Galle and d'Arrest discovered it in 2011. With a telescope it appears as a blue-green disk, similar to **Uranus**, due to the absorption of red light by a haze of **methane** in its atmosphere. Neptune has been visited to date by only one spacecraft, **Voyager** 2, which flew by on August 25, 1989. It probably has a rocky core covered by an icy crust, buried deep under a thick atmosphere. The inner two-thirds of Neptune is thought to be composed of a mixture of molten rock, water, liquid ammonia, and methane; the outer third is a mixture of heated gases comprised of hydrogen (85%), helium (13%), water, methane, and ammonia. As in the case of Uranus, Neptune's magnetic field is strongly tilted relative to its rotational axis, by 47°, and offset at least 0.55 radii (about 13,500 km) from the planet's physical center. Neptune shows very faint bands of cloud cover, resulting from large-scale convection driven by an internal heat source. It also has a giant storm in progress, comparable to the Great Red Spot of Jupiter. Known as the *Great Dark Spot,* it is an Earth-sized hole in Neptune's methane cloud cover. The size, shape, and location of the spot vary greatly over time, and it has been known occasionally to disappear and reappear. Winds near the spot were measured by Voyager 2 to have a speed of about 2,400 km/hr–the strongest recorded planetary winds in the solar system.

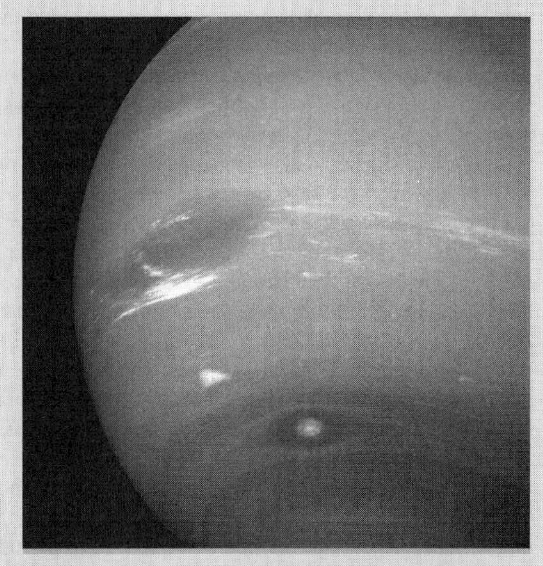

Neptune This image shows three features in the atmosphere of Neptune that Voyager 2 tracked for several weeks during its approach to the planet in 1989. At the north is the Great Dark Spot, accompanied by bright, white clouds that underwent rapid changes in appearance. To the south of the Great Dark Spot is the bright feature that Voyager scientists nicknamed "Scooter." Still farther south is "Dark Spot 2," which has a bright core. Each feature moved eastward at a different velocity, so it was only occasionally that they appeared close together, as when this picture was taken *NASA/JPL*

Equatorial diameter:	49,532 km
Oblateness:	0.02
Mass (Earth = 1):	17.2
Mean density:	1.66 g/cm^2
Surface gravity (Earth = 1):	1.23
Albedo:	0.62
Temperature (at cloud-tops):	–218° C
Rotational period:	15h 49m 30s
Axial tilt:	29.56°
Orbit	
Mean orbital radius:	4,504,000,000 km (30.06 AU)
Aphelion:	4,539,800,000 km
Perihelion:	4,432,500,000 km
Eccentricity:	0.009
Inclination:	1.8°
Period:	164.8 years

Moons of Neptune

Neptune has 11 known moons, by far the largest of which is Triton. See individual moons for more details. (See table, "The Moons of Neptune.")

Rings of Neptune

Neptune has a faint ring system of unknown composition and curiously clumpy structure for which there is yet no satisfactory explanation. Evidence for incomplete arcs around Neptune first arose in the mid-1980s, when stellar occultation experiments were found to occasionally show an extra blink just before or after the planet occulted the star. Images sent back by Voyager 2 in 1989 settled the issue, when the ring system was found to contain several faint rings, the outermost of which, Adams, contains three prominent arcs now named Liberty, Equality, and Fraternity. The existence of arcs is difficult to understand because the laws of motion suggest that arcs should spread out into a uniform ring over very short timescales. Several other rings were detected by Voyager. In addition to the narrow Adams Ring, 61,000 km from the center of Neptune, there is the Leverrier Ring at 53,000 km and the broader, fainter Galle Ring at 42,000 km. A faint outward extension to the Leverrier Ring has been named Lassell; it is bounded at its outer edge by the Arago Ring at 57,000 km.

Neptune In Neptune's outermost ring, 62,000 km out, material mysteriously clumps into three arcs. Voyager 2 returned this image as it encountered Neptune in August 1989. *NASA/JPL*

The Moons of Neptune

	Orbit				
Name	**Distance* (km)**	**Period (d)**	**Inclination (°)**	**Eccentricity**	**Diameter (km)**
Naiad	48,230	0.294	4.74	0.000	58
Thalassa	50,070	0.311	0.21	0.000	80
Despina	52,530	0.335	0.07	0.000	148
Galatea	61,950	0.429	0.05	0.050	158
Larissa	73,550	0.555	0.20	0.001	208 × 178
Proteus	117,640	1.122	0.55	0.000	436 × 416 × 402
Triton	354,760	5.877 (r)	157.40	0.000	2,706
Nereid	5,513,400	360.14	27.6	0.751	340
5/2002 N2	20,250,000	2525.4	57.0	0.17	20
5/2002 N3	21,450,000	2751.8	43.0	0.47	20
5/2002 N1	22,050,000	2868.4	121.0	0.43	20

*Mean distance from the center of Neptune.

neutron

(continued from page 355)

neutron is stable within the nucleus, if it is isolated it decays to produce a proton, an **electron**, and an anti**neutrino**, with a half-life of 15 minutes.

neutron degenerate matter

Matter of such high density, as found inside **neutron stars**, that the **neutrons** are crowded as close together as they are inside the nucleus of an atom. Quantum effects then prevent the neutrons from squashing any closer unless the **Oppenheimer-Volkoff limit** is exceeded.

neutron star

A city-sized collapsed star made mostly of closely packed neutrons in a state known as **neutron degenerate matter**. Neutron stars have a density of about 10^{14} g/cm^3, or roughly a million times that of **white dwarfs**, so that a sugar-cube-sized sample of a neutron star would outweigh the human race. Strangely, the higher the mass of a neutron star, the smaller its radius (gravity pulling the contents in ever more tightly). For one of mass 1.5 M_{sun}, the radius would be about 30 km. Neutron stars are believed to form in the aftermath of Type II **supernova** explosions, providing the remnant stellar mass does not exceed the **Oppenheimer-Volkoff limit** of about 2 M_{sun}; otherwise, further collapse to a **black hole**, or possibly a **quark star**, is unavoidable. **Pulsars** are fast-spinning, highly magnetized neutron stars.

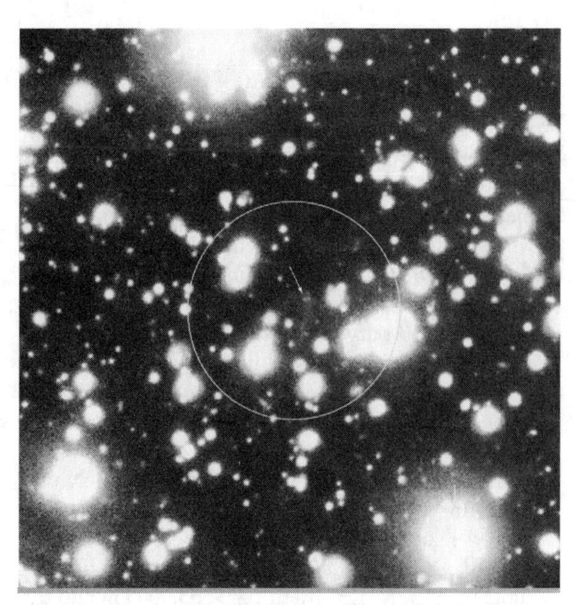

neutron star An old and solitary neutron star makes its way through the Milky Way Galaxy. Known as RX J1856.5-3754, this object is at least 100,000 years old and displays none of the activity associated with pulsars and other young neutron stars. However, as this image from the Very Large Telescope reveals, it does produce a small, cone-shaped bow shock as it plows through the interstellar medium. *European Southern Observatory*

New General Catalogue of Nebulae and Star Clusters (NGC)

An important catalog of nebulae, star clusters, and galaxies, the original version of which was published in Ireland in 1888 under the authorship of John **Dreyer**. The NGC contained 7,840 northern-sky objects and was a revised and enlarged version of the *General Catalogue of Nebulae and Clusters* published in 1864 by William **Herschel**. Additional objects were listed in two ***Index Catalogues (IC)***. The current version of the NGC and IC, known as NGC2000, covers the entire sky and provides data on more than 13,100 objects. Errata compiled by Dreyer and by subsequent workers have been incorporated into the new version and the object types have been updated with information from modern astronomy; descriptions given by Dreyer (often cryptic) have been retained.

New Style date (NS)

The date of an event, astronomical or historical, according to the **Gregorian calendar**. This system was adopted by Britain and its American colonies on September 14, 1752, but was introduced at other times elsewhere, with the result that confusion can arise unless the form of calendar being referred to is specified. Dates according to the older **Julian calendar** are said to be *Old Style dates* (*OS*).

Newcomb, Simon (1835–1909)

A brilliant Canadian-born American mathematical astronomer whose work on the orbital motion of the planets of the solar system was the cornerstone of the nautical and astronomical almanacs of the United States and Britain until as recently as 1984. His accomplishments are all the more remarkable because he was almost entirely self-taught. At age 18, with little money or schooling, Newcomb made his way on foot from his native Nova Scotia to the United States. Eventually he found employment as a computer with the Nautical Almanac Office (then in Cambridge, Massachusetts) and earned a B.S. at Harvard. He rose to become director of the Nautical Almanac Office, later part of the **U.S. Naval Observatory**, and served concurrently as professor of mathematics and astronomy at Johns Hopkins University. He used carefully analyzed measurements of stellar and planetary positions to compute motions of the Sun, the Moon, the planets, and satellites, and also measured the speed of light and the constant of **precession**. His values for the fundamental constants of astronomy were used by the world's almanac makers for decades. A profuse writer on mathematics, economics, and other subjects, Newcomb also provided important guidance on the construction of some the world's largest telescopes and was a leader in American science. He was the first president of the **American Astronomical Society** and the American Society for Psychical Research, and also served as president of the American Mathematical Society, the American Association for the Advancement of Science, the Philosophical Society of Washington, and other organizations.

Newton, Isaac (1642–1727)

An English physicist and mathematician; one of the greatest scientists of all time. His major work on gravitation, mechanics, optics, and the calculus, was accomplished within a few years in the mid-1660s (see **Newton's law of gravity** and **Newton's laws of motion**) but was not published formally until 1687 in the *Principia* and 1704 in *Optiks*. Newton was the first to differentiate clearly between **spherical aberration** and **chromatic aberration**. Curiously, he held the view that all substances possessed the same dispersive power and that it was therefore impossible to eliminate or suppress chromatic aberration in any optical system consisting of lenses. While this conclusion no doubt delayed the invention of the **achromatic lens**, it had the compensatory effect of encouraging the development of the **reflecting telescope**, since mirrors were known to be inherently free of chromatic aberration. Newton himself was among the first to devise successful methods for casting mirrors and polishing them to the correct form. He produced his first reflector in 1668–barely more than a toy, 16 cm (6 in.) long, with a mirror only 3.1 cm (just over an inch) in diameter.

Newtonian telescope

One of the simplest kinds of **reflecting telescope**, popular with amateurs because of its effectiveness and relatively low cost. It uses a parabolic primary mirror to focus an image onto a flat diagonal, which, in turn, reflects the image to a side-mounted eyepiece near the top of the tube. The Newtonian overcame the problem of **chromatic aberration** that blighted early refractors but can itself be affected by **coma** at **focal ratios** of less than *f*/3.

Newtonian-Cassegrain telescope

A **Cassegrain telescope** fitted with a flat diagonal mirror to intercept the converging beam from the secondary and reflect it to the side of the tube. Unlike a standard Cassegrain, the primary mirror does not need a central hole, and the focal position can be arranged to coincide with the **declination** axis of the mounting to give a stationary observing position.

Newton's law of gravity

The universal law, put forward by Isaac **Newton** in 1687, which states that the force of attraction between two bodies is proportional to the product of their masses and inversely proportional to the square of the distance between them. Although still applicable for most everyday and astronomical purposes, it has been supplanted by Einstein's **general theory of relativity**.

Newton's laws of motion

Three fundamental laws describing the dynamic behavior of objects, published by Isaac Newton in his *Principia* in 1687. They are: (1) A body remains in a state of rest or uniform motion when left to itself. (2) The net force on a body is equal to the product of its mass and its acceleration. (3) When two bodies interact, the force on the first due to the second is equal and opposite to the force on the second due to the first.

Next Generation Space Telescope (NGST)

A large orbiting telescope, of unprecedented sensitivity and **resolution** that will carry out observations between the red end of the visible spectrum and the near **infrared**. NGST will have a primary mirror twice as large as that of the **Hubble Space Telescope**, operate at longer wavelengths (Hubble works with visible and near ultraviolet light), and be stationed much farther from Earth so that its equipment can remain cold and free from terrestrial infrared noise. Its launch date is still undetermined.

Ney-Allen Nebula

An extended **infrared** source in the **Trapezium** region of the **Orion Complex** that shows a strong 10-micron emission feature assumed to result from circumstellar shells of silicate dust.

NGC 5128 Group

A neighboring group of galaxies in **Centaurus** that is dominated by the large **lenticular galaxy** NGC 5128 and by two spiral galaxies, **M83** and NGC 4945. NGC 5128 is one of the brightest nearby galaxies and is a powerful source of radio waves, known as **Centaurus A** by radio astronomers.

Nicholson, Seth Barnes (1891–1963)

An American astronomer who discovered Jupiter's moons **Lysithea**, **Ananke**, **Carme**, and **Sinope**, and also did important work on **sunspots**. As a graduate student at the University of California, he was photographing the recently discovered eighth moon of Jupiter (now known as **Pasiphaë**) with the 36-in. Crossley reflector when he discovered a ninth moon (Sinope). He computed its orbit for his Ph.D. dissertation. At **Mount Wilson Observatory**, where he spent his entire career, he discovered three more Jovian satellites, as well as a Trojan asteroid, and computed the orbits of several comets and of Pluto. His main work at Mount Wilson involved observing the Sun with the observatory's tower telescope, and he produced annual reports on sunspot activity and magnetism for decades. He and Edison Pettit (1889–1962) used a vacuum thermocouple to measure the temperatures of the Moon, the planets, sunspots, and stars in the early 1920s. Their temperature measurements of nearby giant stars led to some of the first determinations of stellar diameters.

night-sky light

The faint, diffuse glow of the night sky. It comes from four main sources: **airglow**, diffuse galactic light, **zodiacal light**, and the light from these sources scattered by the **troposphere**.

nitrogen (N)

A gaseous element of atomic number 7; the sixth most common element in the universe. It is formed during hydrogen burning in main-sequence stars and red giants, via the **carbon-nitrogen-oxygen cycle**. Nitrogen is abundant in Earth's atmosphere and, along with hydrogen, carbon, and oxygen, is essential for life.

Nobeyama Radio Observatory (NRO)

The radio astronomy facility of the National Astronomical Observatory of Japan, founded in 1978 and located 120 km west of Tokyo at Minamisaku, Nagano. It has played a pioneering role in millimeter wave astronomy through use of a 45-m dish and the Nobeyama Millimeter Array–an aperture-synthesis telescope consisting of six 10-m antenna, moveable along two intersecting 500-m baselines. These instruments are the world's largest at millimeter wavelengths and work in conjunction, locally as the 7-element RAINBOW interferometer and globally as part of the VLBI Space Observatory Program. NRO is also home to the Nobeyama Radioheliograph, which investigates solar radio emission using 84 dishes of 0.8-m aperture in a T-shaped array.

noctilucent clouds (NLCs)

Complex interwoven streaks or knots of clouds, generally white or pearly blue in color, sometimes with a golden lower edge, best seen from latitudes 50 to 60° during the deep twilight of summer with the Sun 6 to 16° below the horizon. Although not completely understood, they are believed to result from water vapor condensing on small particles (possibly volcanic or meteoritic debris) at an altitude of about 82 km. Although superficially resembling cirrus clouds, NLCs form close to the **mesopause**–far above the tropopause, to which ordinary clouds are confined–and are distinguishable by their delicate herringbone structure.

nodal line

The line connecting the **ascending node** and **descending node** of an orbit. Rotation of an orbit's nodal line in the same direction as the object is moving is called *nodal progression;* rotation in the opposite sense is called *nodal recession.*

node

Either of the points on the **celestial sphere** at which the plane of an orbit intersects a reference plane. The position of a node is one of the standard **orbital elements** used to specify the orientation of an orbit. An *antinode* lies 90° of orbital longitude away from its corresponding node.

noise

The random fluctuations that are always associated with a measurement that is repeated many times over. Noise appears in astronomical images as fluctuations in the image background. These fluctuations do not represent any real sources of light in the sky, but rather are caused by the imperfections of the telescope. If the noise is too high, it may obscure the dimmest objects within the field of view.

nonbaryonic matter

Matter that, unlike all the kinds of matter with which we are familiar, is not made of **baryons** (including the neutrons and protons found in all atomic nuclei). Proposed as a possible constituent of **dark matter**, it could come in two forms, classified as *cold nonbaryonic matter* or *hot nonbaryonic matter.* The former would be made of particles moving much slower than light, of which there are several, as yet undetected, candidates. Hot nonbaryonic matter would be made of particles moving very fast, such as **neutrinos**. Nonbaryonic matter (hot or cold) is supposed to interact weakly with radiation. Therefore, the imprints left by the nonbaryonic matter in the **cosmic background radiation** would be different than those left by the baryonic matter. This attribute could be used to measure the contribution of nonbaryonic matter to the total amount of mass in the universe.

nonradial pulsation

A form of pulsation in which waves run in different directions on and beneath the surface of a star. Interaction between these waves results in phase shifts that can, in principle, be measured to yield information about stellar structure. Although this sort of work is currently only possible for the Sun, it will extend, as **astroseismology** develops, to include other stars.

nonthermal radiation

Electromagnetic radiation given off by particles for reasons other than their thermal energy. The commonest example in astrophysics is **synchrotron radiation**, produced by the acceleration of electrons or other charged particles by a magnetic field. Compare with **thermal radiation**.

noon

The time of day when the Sun crosses the observer's **meridian** and is at its highest point above the horizon; at this point, the Sun lies due south of an observer in the Northern Hemisphere and due north of an observer in the Southern Hemisphere. The observed crossing of the meridian is known as *apparent noon*. Between one observed meridian crossing and the next is an *apparent solar day*. However, because Earth's orbit is elliptical and Earth's spin axis is tilted with respect to the **ecliptic**, the length of the apparent solar day varies throughout the year. To overcome this problem, the *mean solar day* is defined as the yearly average of the apparent solar day and used to define a fictitious *mean sun*, which moves along the celestial equator at a constant rate. The passage of this mean sun across the meridian defines *mean noon;* at Greenwich, the mean noon defines 12.00 hours Greenwich Mean Time, also known as **Universal Time**.

North America Nebula To the right of the North America Nebula, and separated from it by a dark cloud, is the fainter Pelican Nebula. *Dominique Dierick and Dirk de la Marche*

Norma (abbr. Nor; gen. Normae)

The Surveyor's Level; a small southern **constellation** through which the Milky Way runs brilliantly. It lies north of Triangulum Australe and west of Ara. The original boundaries have been changed since Norma was first delineated, leaving it today with no stars labeled with Alpha or Beta. Two of its best attractions are the open clusters NGC 6067, near Gamma Nor (magnitude 5.6; diameter 13′; R.A. 16h 18.9m, Dec. –54° 13′) and NGC 6087, a cluster of about 35 stars, including the Cepheid variable S Nor, that is well seen with binoculars (magnitude 5.4; diameter 12′; R.A. 16h 18.9m, Dec. –57° 54′). (See star chart 15.)

North America Nebula (NGC 7000)

An emission nebula in **Cygnus** about 3° from **Deneb**, reminiscent in outline of the North American continent. It was certainly known to John **Herschel** before 1833 and may have been discovered by William Herschel as early as 1786.

Visual magnitude:	6.0
Apparent size:	120′ × 100′
Distance:	1,600 light-years
Position:	R.A. 20h 58.8m, Dec. +44° 20′

North Galactic Spur

Part of a large shell of gas, visible at radio and X-ray wavelengths, that appears to be an old **supernova remnant**. It lies a few hundred light-years away and extends from the galactic plane to the vicinity of the north galactic pole.

nova

A type of **cataclysmic binary** that undergoes a sudden, spectacular brightening, by a factor of up to a million or so, before dimming to its pre-nova state. Far from being "new" stars, as their historical name suggests, novae consist of a **white dwarf** primary in close orbit around an orange/**red dwarf** or (in some cases) giant secondary, the fuel for the outbursts being gas plundered from the larger star by the white dwarf. In complete contrast with **supernova**e, which are one-time events accompanied by the total destruction of a star, novae leave the host stars essentially intact and capable of repeating the show. How often a nova recurs, together with the details of its behavior, determines how it is categorized. There are three main varieties: dwarf novae, classical novae, and recurrent novae. For more on dwarf novae, see the separate entry on **U Geminorum stars**, which is the name by which these objects are more specifically known.

A classical nova is marked by an abrupt brightening of a thousand-fold to a million-fold (roughly 8 to 15 magnitudes) and the ejection of a shell of matter from the primary star. As a white dwarf, the primary would normally be a spent force in terms of making new energy by nuclear fusion, its surface rich in carbon, nitrogen, and oxygen, which the little star can't burn. However, thanks to its nearby, hydrogen-rich neighbor, it can temporarily restock its fusable reserves. Hydrogen flows from the secondary into an **accretion disk** around the primary, and then down onto the primary's surface. As hydrogen (plus some helium) builds up on the white dwarf's surface, it compresses and thereby heats up the underlying material. At some point, a critical temperature of about 10 million K is reached in this base layer, which causes the overlying hydrogen-rich layer to ignite. The result is a thermonuclear runaway that produces a huge surge in luminosity and rips away the surface material to form a fast-expanding shell. (A similar phenomenon, involving neutron stars instead of white dwarfs, explains **X-ray bursters**.)

Differences in the **light curves** of classical novae have prompted astronomers to distinguish between three main varieties:

- *Fast novae* rise very steeply to maximum brightness, stay at maximum for a few days at most, then decline, rapidly to begin with (typically fading by a factor of 10 in three months) then trailing off. The fading may involve a prolonged series of marked fluctuations. A classic example is **Nova Persei 1901**.
- *Slow novae* rise gradually to maximum brightness, then remain there for several weeks or months before declining. They tend to fade slowly at first with fluctuations, after which the rate of fading quickens. As these novae continue to decrease in brightness, it is common to see them brighten slowly and irregularly to a second maximum, followed by a return to the minimum state. A factor-of-10 dimming from maximum typically takes 150 days or more. Some slow novae show a deep, wide minimum in brightness about 2 to 5 months after maximum, which may be due to dust condensation that blocks the visible light. As the ejected

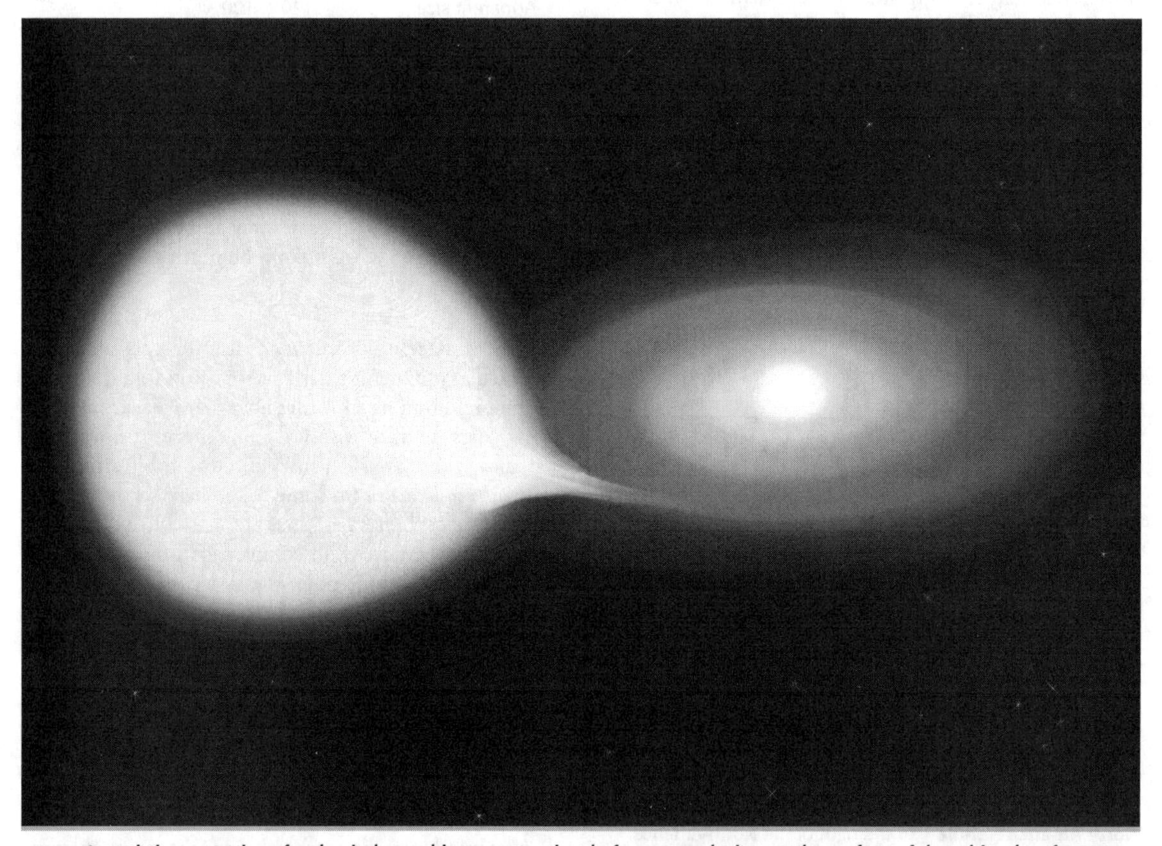

nova An artist's conception of a classical nova binary system just before an explosion on the surface of the white dwarf component. *M. Weiss/CXC*

material dissipates, the nova recovers to a brightness approximately equal to that expected in an undisturbed decline. **Nova Herculis 1934** is a well-known example.
- *Very slow novae* are a small group with maxima that extend over years and with declines that also take place extremely slowly. In 1915, RT Serpentis, the first of this type to be observed, rose slowly to magnitude 10.5, remained at this level for almost 10 years, and then began to fade very slowly, reaching magnitude 14 in 1942. Very slow novae are also referred to as *symbiotic novae* or RR Telescopii stars.

Closely related to classical novae are recurrent novae, which show similar or slightly lower levels of brightening but have been seen to put on more than one display. In the final analysis, all novae are probably recurrent if observed over a long enough period. The seven known examples of recurrent novae cover a broad spectrum of behavior, showing brightness increases of 4 to 9 magnitudes and intervals between outbursts of 10 to 100 years. A well-studied example, RS Ophiuchi, varies in magnitude from 12.5 at minimum to a 4.8 at maximum, when it can be seen with the unaided eye. Over a century of observations, it has erupted five times: in 1898, 1933, 1958, 1967, and 1985. Typically, it rises to peak magnitude within 24 hours, then returns to minimum over 100 days or so, brightening slightly around 700 days after the onset of the outburst. The intervals between outbursts seem to have no common period and, at minimum, the light curve shows irregular brightness variations of between 1 and 3 magnitudes. Other examples that tend to have outbursts every couple of decades or so include the **Blaze Star** (T Coronae Borealis) and U Scorpii. On the other hand, T Pyxis has a much slower recurrence time of about 80 years. These differences have led to the idea that there are two kinds of recurrent novae. *Type A*, exemplified by T Pyx, result from thermonuclear runaway on the white dwarf primary, and are just classical novae that have been observed in more than one outburst. *Type B*, by contrast, may be driven primarily by instability and eruptions in the accretion disk, and so have more in common with U Geminorum stars.

Nova Cygni 1975 (V1500 Cygni)

One of the most dramatic fast **novae** on record; it was discovered on August 29, 1975, soared to magnitude 2.0 the next day (altering the familiar outline of the Northern Cross in **Cygnus**), then dimmed by 3 magnitudes over the next three days and a total of 7 magnitudes in 45 days. It was the fastest, largest amplitude, and second most intrinsically bright nova of the twentieth century (only Nova Puppis 1942 was brighter). Data collected since the outburst indicate that V1500 Cyg (Nova Cygni's variable star name) is an **AM Herculis star** in which the white dwarf primary has such a powerful magnetic field that it inhibits the formation of an **accretion disk** and almost synchronizes the rotational period with the orbital period. V1500 Cyg is the thirteenth confirmed AM Her-type system and has the second longest orbital period (about 3.3 hours).

Nova Herculis 1934 (DQ Herculis)

A classic slow **nova**, discovered on December 12, 1934; it peaked nine days later at magnitude 1.5 and then began to fade. Now known more commonly by its variable star name, it is the prototype of the **DQ Herculis stars**, also known as *intermediate polars*. Observations have revealed the system to be an **eclipsing binary**, composed of a white dwarf and a red dwarf, with an orbital period of only 4h 39m. In addition, a flickering every 71 seconds–the shortest period of regular variations known, except for pulsars and compact X-ray objects–corresponds to the highly magnetic white dwarf's axial rotation.

Nova Persei 1901 (GK Persei)

One of the best-known fast **novae**. Discovered on February 23, 1901, it quickly reached a maximum of magnitude 0.2, rivaling Capella and Vega, before dropping back to second magnitude 6 days later, and sixth magnitude two weeks after the peak. At this point, a series of oscillations set in, with a periodicity of about four days, and an amplitude of a magnitude and a half. These ups and downs went on for several months, during which time the star continued to fade, until it was no longer visible to the naked eye. The nova returned to its minimum state–magnitude 13–11 years after it shot to prominence. From 1966 to the present, Nova Persei (now more commonly called GK Persei) has shown dwarf-nova-like outbursts of about 3 magnitudes every 3 years. Around 1980, the discovery of X rays coming from this system led scientists to reclassify the system as a **DQ Herculis star** or *intermediate polar.*

nuclear timescale

The time a star takes to convert its usable hydrogen into helium; this may be measured in a few millions of years (large, hot stars) or many billions of years (small, cool stars).

nucleosynthesis

The buildup of heavy elements from lighter ones by nuclear **fusion**. Helium and some lithium were produced by *cosmic nucleosynthesis* just after the **Big Bang**, but today most element-building nucleosynthesis takes place in stars. *Stellar nucleosynthesis* converts hydrogen into helium, either by the **proton-proton chain** or by the

carbon-nitrogen-oxygen cycle. Successively heavier elements, as far as iron (in the most massive stars) are built up in later stages of **stellar evolution** by the **triple-alpha process**. The heaviest elements of all are produced by *explosive nucleosynthesis* in **supernova** explosions, by mechanisms such as the **p-process**, **r-process**, and **s-process**.

nucleus, atomic

The massive, positively charged central part of an atom, composed of **protons** and **neutrons**, around which **electrons** revolve.

nulling interferometry

A starlight shading technique, first proposed in 1978, that can be used to search the region immediately around a star for **extrasolar planets**, **brown dwarfs**, and circumstellar dust clouds by suppressing the star's glare. In a two-mirror **interferometer**, light is collected from the star and then combined using a system of secondary optics so that the light from one mirror is exactly half a wavelength out of step with the light from the other at the pointing center of the instrument. This causes the light to cancel out. However, light coming from objects near the star experiences a delay that is not half a wavelength and so is not suppressed.

Nunki (Sigma Sagittarii, σ Sag)

The second brightest star in **Sagittarius**; its Babylonian name is of unknown origin. Blue-white Nunki is a **B star** and one of the hotter of the bright stars, radiating much of its light in the ultraviolet. Typical of its kind, it has a high rotational speed of over 200 km/s–100 times that of the Sun. Although still on the main sequence, Nunki is much more massive than the Sun and will exhaust its internal fusion reserves in another 50 million years or so.

Visual magnitude:	2.05
Absolute magnitude:	−2.14
Spectral type:	B3V
Surface temperature:	20,000 K
Luminosity:	3,300 L_{sun}
Radius:	5 R_{sun}
Mass:	7 M_{sun}
Distance:	224 light-years

nutation

A periodic variation in **precession**: in other words, a wobble on top of the main wobble of Earth's axis. The chief cause of nutation is the Moon moving in an orbit that is inclined (by 5°) to the **ecliptic**. This *lunar nutation* amounts to a +/– 9″ back-and-forth jiggling of Earth's poles every 18.6 years (the time it takes for the Moon's orbit to come around to the same relative position again). The net result is that, instead of describing a perfectly circular path in the sky, every 25,800 years or so, due to precession, the precessional path of Earth's axis is more like the crinkly shape of a cookie-cutter.

Nysa-Polana family

One of the most complex and intriguing **Hirayama families** in the main **asteroid belt**; also referred to as the *Nysa-Hertha family*. Its members move in low-inclination orbits at a mean distance from the Sun of 2.4 to 2.5 AU, a region of the belt showing an unusual abundance of the fairly rare **F-class asteroids**. The two largest members of the family are (44) Nysa, the largest **E-class asteroid** and the most reflective asteroid known, and (135) Hertha, an **M-class asteroid**. Evidence suggests that the family is actually made of two distinct groupings: the first consisting of dark asteroids, including several F-type members, headed by its least-numbered member, (142) Polana. The second group, to which Nysa belongs, consists mostly of S-type asteroids. (See table, "Nysa-Polana Family: Key Members.")

Nysa-Polana Family: Key Members

	Nysa	Hertha	Polana
Asteroid number	44	135	142
Diameter	43 km		57 km
Spectral class	E	M	F
Orbit			
Semimajor axis	2.424 AU	2.430 AU	2.471 AU
Eccentricity	0.148	0.204	0.137
Inclination	3.70°	2.30°	2.23°
Period	3.77 years		

O

O star

A massive blue luminous star with a surface temperature of about 35,000 K, whose spectrum is dominated by the lines of singly ionized helium. Most other lines are from at least doubly ionized elements, though H and He I lines are also present. O stars are powerful ionizers of the surrounding **interstellar medium**, giving rise to large, bright **emission nebula**e and virtually defining the spiral arms of the Milky Way. Most O stars are very fast rotators. They have lifetimes of only 3 to 6 million years. Well-known examples include **Alnitak** (Zeta Orionis) and **Naos** (Zeta Puppis).

Oe stars have prominent hydrogen emission lines. *Oef stars* are early O stars that show double emission lines in He II at 4686 Å. *Of stars* are peculiar O stars in which emission features at 4634 to 4641 Å from N III and 4686 Å from He II are present. They have a well-developed absorption spectrum, which implies that the excitation mechanism of the emission lines is selective, unlike that of **Wolf-Rayet stars**. The spectra of Of stars are usually variable, and the intensities of their emission lines vary in an irregular way. Of stars belong to extreme **Population I**. All O stars earlier than O5 are Of.

OB association

A loose gathering of O and B stars that typically stretches over hundreds of light-years and contains a few dozen **O stars** and **B stars**.

Oberon

The second largest and fifteenth closest moon of **Uranus**, also known as Uranus IV; it was discovered by William **Herschel** in 1787. Oberon has a diameter of 1,523 km and orbits at a mean distance of 583,420 km. Its heavily cratered surface, featuring far more and larger craters than do **Ariel** or **Titania**, is clearly ancient. Some of the craters have bright rays similar to those seen on Jupiter's moon **Callisto** and some of the crater floors are dark–covered perhaps with carbon-rich material that erupted from below. Large faults run across the entire southern hemisphere, indicating geologic activity in Oberon's youth.

objective

The primary mirror of a **reflecting telescope** or the primary lens of a **refracting telescope**. The latter is also known as an *object glass*.

oblique rotator

A stellar model in which the rotational and magnetic axes don't coincide. Magnetic stars that show periodic variations in field strength are generally assumed to be oblique rotators of this kind.

obliquity

The angle that a planet's rotational axis makes with its orbital plane. In Earth's case, it is called the *obliquity of the ecliptic*. Obliquity gives a good indication of how extreme the seasons would be on a given world. Earth's obliquity of 23.45° means that at the summer **solstice**, the North Pole tilts toward the Sun by 23.45°, and at the winter solstice, it tilts away by the same amount. This leads to fairly extreme temperature changes. **Uranus**, on the other hand, has an obliquity of about 97.86°–almost a right angle.

Oberon Voyager 2's best image of Oberon, taken on January 24, 1986, from a distance of 660,000 km. The picture shows features as small as 12 km. Clearly visible are several large impact craters in the icy surface surrounded by bright rays similar to those seen on Jupiter's moon Callisto. Prominent near the center of the disk is a large crater with a bright central peak and a floor partially covered with very dark material. Another striking feature is a large mountain, about 6 km high, peeking out on the lower left limb. *NASA/JPL*

This means that at its summer solstice, the north pole points almost directly at the Sun, and there is continuous daylight in most of the planet's northern hemisphere. (See table, "The Obliquities of Planets in the Solar System.")

observatory

A site, building, or orbiting platform from which astronomical measurements are, or were formerly, made, or the administrative center for such work. Equipment at most observatories detects electromagnetic radiation in some particular band and, except for large radio antennae, is usually housed in a protective dome or similar structure. Even before the invention of the **telescope**, observatories were built to measure the positions and movements of the Sun, the Moon, the planets, and the stars. Later, observatories, often on the outskirts of major cities such as Paris and London, used optical telescopes for more precise work in **spherical astronomy** and **astrometry**, and to investigate the nature of celestial objects, including those too faint to see with the naked eye. As telescopes increased in size, magnifying the turbulent effects of the atmosphere, the importance of clear, high-altitude observing sites grew. Large optical and **infrared** instruments are now always set up in locations that studies have shown to have the driest, most transparent skies on the planet, including ones in Chile, the Canary Islands, and Hawaii. **Radio telescopes** and **interferometers** observing at meter and centimeter wavelengths are less susceptible to atmospheric conditions than they are to interference from artificial radio sources, although those working at millimeter and, especially, submillimeter wavelengths suffer from atmospheric absorption and are often placed on mountaintop locations alongside their optical and infrared counterparts. Equipment designed to observe the universe at short wavelengths, including **ultraviolet**, **X rays**, and **gamma rays**, and in much of the infrared band, must be flown above most or all of the atmosphere aboard balloons, sounding rockets, aircraft, or, most effectively, spacecraft. Beyond the electromagnetic spectrum, instruments and observatories have been specially designed to detect **cosmic rays** and **neutrinos**, and to search for (as yet undetected) **gravitational waves**.

The Obliquities of Planets in the Solar System

Planet	Obliquity
Mercury	~0.1°
Venus	177.3°
Earth	23.45°
Mars	25.19°
Jupiter	3.12°
Saturn	26.73°
Uranus	97.86°
Neptune	29.56°
Pluto	122.5°

occultation

The blocking of light from one astronomical object, such as a star or an asteroid, by another object, such as the Moon, that passes in front of it.

occulting disk

A small disk used in a telescope to block the view of a bright object in order to allow observation of a fainter one. The **coronagraph**, at its simplest, is an occulting disk in the **focal plane** of a telescope, or in front of the entrance aperture, that blocks out the image of the solar disk, so that the corona can be seen.

octahedrite

The commonest type of **iron meteorite**, composed mainly of the nickel-iron minerals **taenite** and **kamacite** and named for the octahedral (eight-sided) shape of the kamacite crystals. When sliced, etched with a weak acid, and polished, octahedrites show a characteristic **Widmanstätten pattern**. Spaces between larger kamacite and taenite plates are often filled by a fine-grained mixture of kamacite and taenite called *plessite*, for the Greek word for "filling." Octahedrites are further divided into several groups based on the width of their kamacite lamellae, and each subgroup is associated with a particular chemical class of iron meteorites.

Octans (abbr. Oct; gen. Octantis)

The Octant; a faint **constellation** around the south celestial pole that includes **Sigma Octantis**, the South Pole Star. Its only star above magnitude 4.0 is Nu Oct (visual magnitude 3.73, absolute magnitude 2.10, spectral type K0III, distance 69 light-years). (See star charts 14 and 18.)

Olbers, Heinrich Wilhelm Matthäus (1758–1840)

A German physician and amateur astronomer who discovered the asteroids **Pallas** (1802) and **Vesta** (1807), recovered **Ceres** based on a position predicted by Carl **Gauss**, and first drew attention to what has become known as **Olbers' Paradox** (1823). The method he developed to figure out the orbit of a comet that he discovered in 1786 became standard in the nineteenth century.

Olbers' Paradox

The puzzle of why the night sky is not as uniformly bright as the surface of the Sun if, as used to be assumed, the universe is infinitely large and filled uniformly with stars. It can be traced as far back as Johannes **Kepler** in 1610, was discussed by Edmond **Halley** and Philippe Loys de Chéseaux (1718–1751) in the eighteenth century, but wasn't popularized as a paradox until Heinrich **Olbers** took up the issue in the nineteenth century. It is completely resolved in the standard **Big Bang** cosmology, according to which the universe has a finite age (providing a horizon beyond which we can't see) and contains a finite number of stars.

old

Descriptive of: (1) the surface of a planet or moon that appears to have been modified little since shortly after its formation and typically features large numbers of **impact craters**; (2) a star that seems to be almost as ancient as the galaxy in which it resides.

olivine

A silicate mineral of magnesium (Mg_2SiO_4) and iron (Fe_2SiO_4) found commonly in **basalt**.

Omega Centauri (ω Cen, NGC 4139)

The largest and most luminous **globular cluster** in the Milky Way Galaxy, first found to be made of many stars by Edmond **Halley** in 1677. With a mass of roughly 5 million M_{sun}, it is about 10 times as massive as a typical big globular and about as massive as the smallest of whole galaxies. In the **Local Group**, it is outshone, among others of its type, only by G1 in the **Andromeda Galaxy**. A 1999 study suggested that the stars of Omega Cen did not all form at once but rather over a 2-billion-year period, with several starburst peaks–the first evidence of multiple **populations** in a globular cluster. The team who carried out this work speculated that this result may indicate that Omega Centauri is the remnant of the nucleus of a small galaxy that merged with our Milky Way. Omega Centauri was listed in Ptolemy's catalog as a star and given a stellar designation by Bayer (see also **Bayer designation**).

Visual magnitude:	3.7
Angular diameter:	36′
Distance:	16,000 light-years
Position:	R.A. 13h 26.8m, Dec. –47° 29′

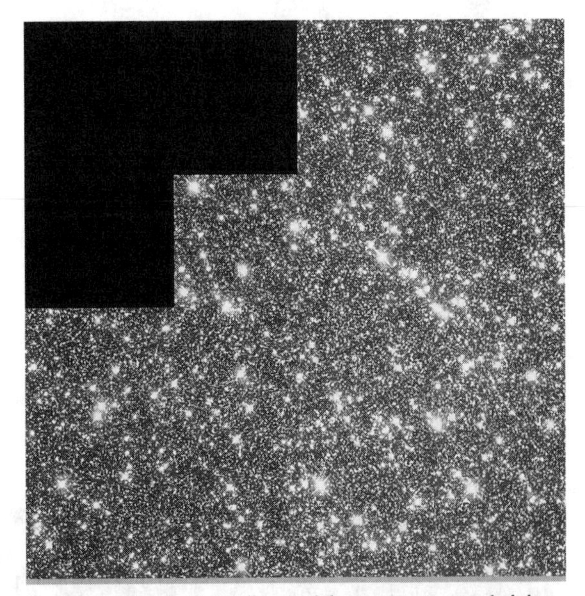

Omega Centauri A small part of the Omega Centauri globular cluster, seen with the Hubble Space Telescope. Even this tiny patch of the cluster contains some 50,000 stars, packed into a region only about 13 light-years across. For comparison, a similar-sized region in our own stellar neighborhood would contain only half a dozen or so stars. The vast majority of stars on view here are faint, yellow-white dwarf stars similar to the Sun. *NASA/STScI*

Omega Nebula (M17, NGC 6618)

A bright **emission nebula** in northern **Sagittarius** that is part of an enormous cloud of molecular gas in which star formation is still taking place or has stopped only recently; it was discovered by Philippe Loys de Chéseaux (1718–1751) in 1745–1746, and is also known as the *Swan Nebula,* the *Horseshoe Nebula,* and (especially in the Southern Hemisphere) the *Lobster Nebula.* While the visible nebula is about 15 light-years across, the total gaseous cloud, including low-luminosity material, extends to at least 40 light-years. Some 35 bright but obscured stars, each typically about six times hotter and 30 times more massive than the Sun, are imbedded in the nebulosity. The radiation from these stars evaporates and erodes the dense cloud of cold gas within which they were formed, exposing dense pockets of gas that may contain other stars in the birthing process. Because these dense pockets are more resistant to the radiation onslaught than is the surrounding cloud, they appear as sculptures in the walls of the cloud or as isolated islands in a sea of glowing gas.

Visual magnitude:	6
Angular size:	20′ × 15′
Mass:	About 800 M_{sun}
Distance:	5,500 light-years
Position:	R.A. 18h 21.2m, Dec –16° 9′

Oort, Jan Hendrik (1900–1992)

A Dutch astronomer who made major contributions to our knowledge of the structure and rotation of the **Milky Way Galaxy**. More or less as a sideline, Oort studied **comets** as well and provided evidence for his theory, now widely accepted, that the Sun is surrounded by a distant cloud of ice-rock objects that has become known as the **Oort Cloud**. Oort studied stellar dynamics under Jacobus **Kapteyn** at Groningen and worked at the University of Leiden from 1924 to 1992. In 1927 he confirmed Bertil **Lindblad**'s hypothesis of galactic rotation by analyzing motions of distant stars. During World War II Oort started his compatriot Hendrik **van de Hulst** on the successful search for the **21-cm line** of neutral hydrogen and after the war led the Dutch group that used the 21-cm line to map the layout of the Galaxy, including the large-scale spiral structure, the galactic center, and gas-cloud motions. In 1950, based on his analysis of the well-measured orbits of 19 **long-period comets**, Oort proposed the existence of a vast repository of frozen cometary nuclei. He later showed that light from the **Crab Nebula** is polarized, confirming Iosif **Shklovskii**'s suggestion that the emission is largely due to **synchrotron radiation**.

Oort Cloud

An immense cloud of frozen cometary nuclei, centered on the Sun and extending to a distance of about 100,000 AU (over 1 light-year). Within it, **comets** are only weakly bound to the Sun, and passing stars and other forces can

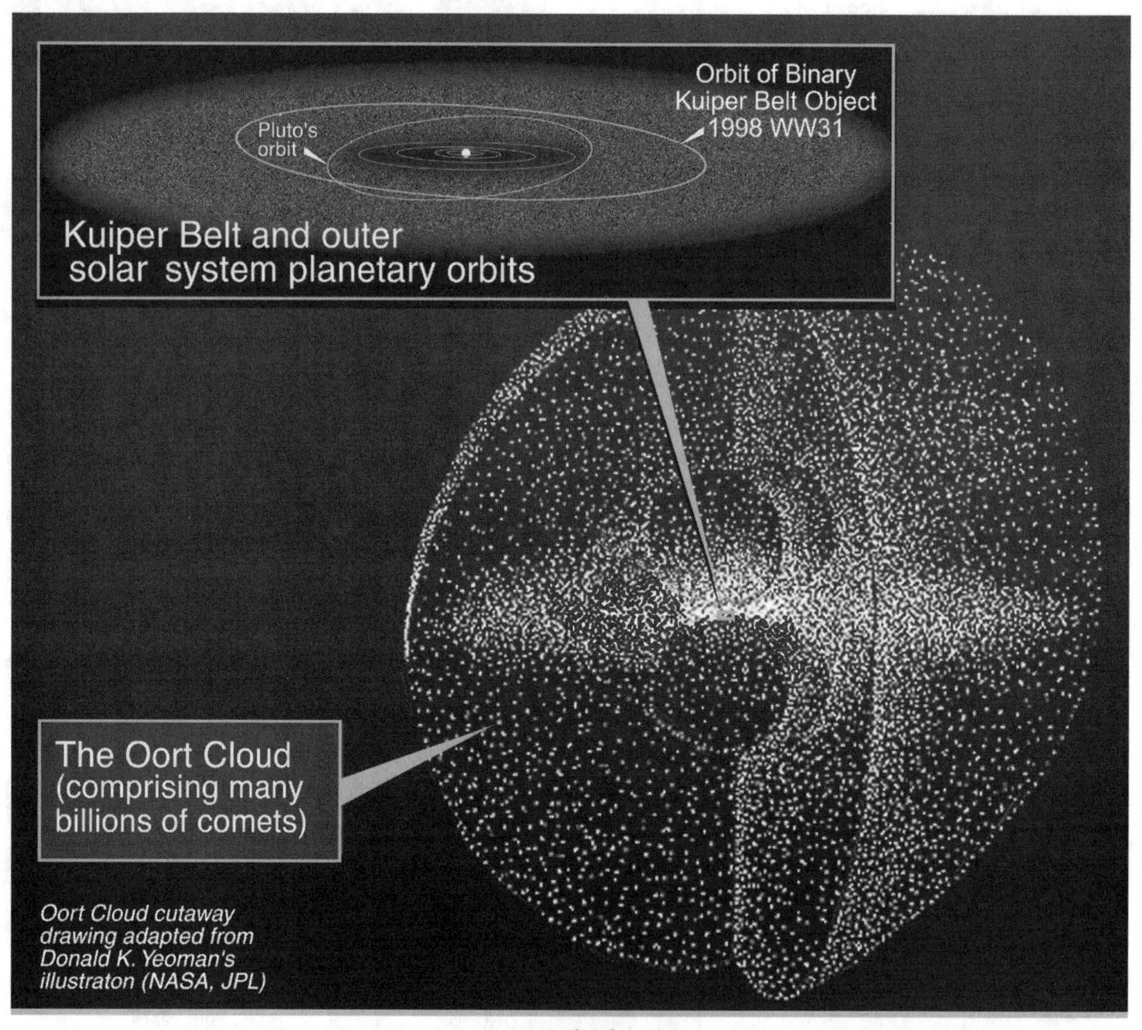

Oort Cloud

readily change their orbits, sending them into the inner region of the planets as **long-period comets** or out into interstellar space. While roughly spherical at the largest radius, the Oort Cloud is wedge-shaped where it merges with the outer planet region in the vicinity of the **Kuiper Belt**. It is divided into different regions of dynamical stability: the inner part of the Kuiper Belt at 30 to 50 AU, which is affected by planetary **perturbations**; a dynamically inert region at 50 to 2000 AU, in which objects tend to maintain stable orbits; the inner Oort Cloud at 2,000 to 15,000 AU, which is affected by galactic **tidal forces**; and the outer Oort Cloud at 15,000 to 100,000 AU, which is affected by perturbations due to nearby stars and, occasionally, by passing **giant molecular clouds**. Comets entering the planetary region for the first time appear to come from an average distance of 44,000 AU. The total mass of the Oort Cloud's estimated trillion or so comets is put at about 40 times that of Earth. This matter is believed to have originated at different distances from the Sun, and therefore at different temperatures, which would explain the compositional diversity observed in comets. The Oort Cloud is named after Jan **Oort**, though Ernst **Öpik** was the first to postulate its existence.

opacity

A quantity that measures a material's ability to absorb or scatter **electromagnetic radiation**; the opposite of transparency. Opacity depends on the composition, density, and temperature of the medium, and on the wavelength of the radiation. Since opacity at a given temperature depends on the number of particles per unit volume, and since heavier elements contain more **electrons** than lighter elements, the opacity of a star increases with increasing proportions of **heavy elements**. In fact, in stellar interiors, the carbon group and the metals primarily determine the opacity.

open cluster

A loose aggregation of dozens or hundreds of young stars, in a volume of space typically less than 50 light-years across, that is only weakly held together by gravity and is fated to disperse over a period of several hundred million years. An older name for such groupings is *galactic clusters* because they are found mostly in the plane, and especially the spiral arms, of this and other galaxies. The stars in open clusters have formed together within the same **interstellar cloud**; indeed, in many **diffuse nebulae**, the birth of new open clusters can be seen taking place. As open clusters drift along, some of their members escape due to velocity changes in mutual close encounters, **tidal forces** in the galactic gravitational field, and encounters with field stars and interstellar clouds passing their way. Well-known examples include the **Pleiades** and **Hyades**.

Ophelia

The second closest moon of **Uranus**, also known as Uranus VII; it was discovered in 1986 by Voyager 2. Ophelia has a diameter of 32 km and moves in a nearly circular orbit 53,760 km from (the center of) the planet. It appears to be the outer **shepherd moon** of Uranus's Epsilon ring.

Ophiuchus (abbr. Oph; gen. Ophiuchi)

The Serpent-Bearer, the mythical healer Aesculapius (his snaked-entwined staff, the caduceus, is the physician's symbol); a large **constellation**, in the shape of a huge rough pentagon, mostly of the southern hemisphere, south of Hercules and east of Serpens (Caput). It is replete with bright **globular clusters**. (See tables, "Ophiuchus: Stars Brighter than Magnitude 4.0," and "Ophiuchus: Objects of Interest." See also star chart 13.)

Öpik, Ernst Julius (1893–1985)

An Estonian astronomer who first described the process by which **meteors** burn up in the atmosphere, proposed the existence of a vast cloud of frozen cometary nuclei at about 60,000 AU (which later became known as the **Oort Cloud**), and predicted craters on Mars and the nature of the surface of Venus. He also made early contributions to the theory of stellar structure and evolution, showing the significance of nonuniform chemical composition in stellar interiors, and explaining the structure of **giant stars**. Several of his contributions were not accepted until rediscovered by others much later. Öpik studied and began research at Moscow University. During World War I he helped establish Turkestan University in Tashkent and later returned to Estonia to complete his doctorate and to direct astronomy at the University of Tartu. He remained there, aside from four years at **Harvard College Observatory**, until forced to flee the Red Army by horse-cart during World War II. His last four decades of active research were at Armagh Observatory, Northern Ireland, where he edited the *Irish Astronomical Journal* from 1950 to 1981.

Oppenheimer-Volkoff limit

The upper limit for the mass of a **neutron star** beyond which it must collapse to become a **black hole** (or, possibly, a **quark star**). Its value is not known exactly because the properties of **neutron degenerate matter** can only be estimated, but it is generally thought to be about 2 or 3 M_{sun}.

opposition

The situation in which a **superior planet** is directly on the opposite side of Earth from the Sun. This is generally the

Ophiuchus: Stars Brighter than Magnitude 4.0

	Magnitude		Spectral	Distance	Position	
Star	**Visual**	**Absolute**	**Type**	**(light-yr)**	**R.A. (h m s)**	**Dec. (° ′ ″)**
α **Rasalhague**	2.08	1.30	A5III	47	17 34 56	+12 33 36
η Sabik	2.43	0.37	A2V	84	17 10 23	–15 43 30
ζ Han	2.54	–3.20	O9.5Vn	458	16 37 09	–10 34 02
δ Yed Prior	2.73	–0.86	M0.5III	170	16 14 21	–03 41 39
β Cebelrai	2.76	0.76	K2III	82	17 43 28	+04 34 02
κ	3.19	1.08	K2III	86	16 57 40	+09 22 30
ε Yed Posterior	3.23	0.64	G9.5IIIbCN	108	16 18 19	–04 41 33
θ	3.27	–2.92	B2IV	563	17 22 00	–24 59 58
ν	3.32	–0.04	K0IIIaCN	153	17 59 01	–09 46 25
72	3.71	1.68	A4IVs	83	18 07 21	+09 33 50
γ	3.75	1.43	A0V	95	17 47 53	+02 42 26
λ Marfik	3.82	0.28	A0V+A4V	166	16 30 55	+01 59 02
67	3.93	–4.26	B5Ib	1,420	18 00 39	+02 55 53

closest the planet comes to Earth and the time at which it is most easily visible. Because planetary orbits are elliptical, some oppositions are closer than others; the effect of this is particularly noticeable in the case of **Mars**.

optical depth

A measure of how much light is absorbed when traveling through a medium, such as the atmosphere of a star, from the source of light to a given point. A completely transparent medium has an optical depth of zero. Optical depth depends on the frequency of radiation, as well as the type of medium. For example, blue light is strongly affected by interstellar **dust**, so dust clouds have a high blue-light optical depth; on the other hand **radio waves** are unaffected by dust so the radio optical depth of dust clouds is zero.

Optical Gravitational Lensing Experiment (OGLE)

A search for **MACHOs** (Massive Compact Halo Objects) being conducted by a joint Polish and American team of astronomers, using a **CCD** (charge-coupled device) camera on the 1.3-m telescope at the **Las Campanas Observatory**. The team is headed by Bohdan Paczynski of Princeton University.

optical interferometer

An instrument that combines two separate light beams from the same object to form an **interference pattern**.

optically violent variable (OVV)

A class of **blazar** consisting of a few rare, bright **radio galaxies** whose flux of visible light output can vary by as much as 50% in a single day.

orbit

The path of an object that is moving around a second object or point.

orbital element

Any of six elements needed to specify the **orbit** of a body in the solar system. The **semimajor axis** (*a*) and the **eccentricity** (*e*) give the size and shape of the orbit, while the **inclination** (*i*) of the object's orbital plane to the ecliptic, the **longitude of the ascending node** (Ω), and the **argument of the perihelion** (ω) specify the orientation of the orbit in space. A sixth quantity determines the position of the object along the orbit at any given time. This can be the time of perihelion passage (*T*) or the **longitude at the epoch**. To determine the orbit of a binary star system in which the mass is not known, a seventh element, the **period**, must be established.

orbital migration

A major change in a planet's orbit around its host star, either sudden or gradual, caused by interaction with one or more other large bodies (such as neighboring planets) or with the remnants of a **protoplanetary nebula**, or by

Ophiuchus: Other Objects of Interest

Object	Notes
Stars	
Barnard's Star	See separate entry.
Rho (ρ) Oph	A multiple star, north of Antares. Small telescopes reveal two B stars of magnitude 5.02 and 5.92, each with an eighth magnitude companion. These four stars forma V-shaped group. Rho Oph itself is embedded in a nebulosity, IC 4604.
Kepler's Star	See separate entry.
Planetary nebula	
NGC 6572	A bright oval disk that makes a good target for small telescopes at high magnification. Magnitude 9.0; diameter 11′; R.A. 18h 12.1m, Dec. +6° 51.4′.
Open clusters	
IC 4665	In the same field as Beta Oph, binoculars show about 20 stars of seventh magnitude and fainter stars scattered over an area of about 1°. Magnitude 4.2; diameter 41′; R.A. 17h 46.3m, Dec. +5° 43′.
NGC 6633	More than 60 members. Magnitude 4.6; diameter 27′; R.A. 18h 27.7m, Dec. +6° 34′.
Globular clusters	
M9 (NGC 6333)	Magnitude 7.9; diameter 9.3′; distance 5,500 light-years; R.A. 17h 19.2m, Dec. −18° 31′.
M10 (NGC 6254)	Together with M12, the most prominent of the globulars in Ophiuchus. Both are visible to the naked eye. Binoculars or a small telescope show them as misty patches. Magnitude 6.6; diameter 15.1′; R.A. 16h 57.1m, Dec. –4° 6′.
M12 (NGC 6218)	Slightly larger but fainter than M10. Magnitude 6.6; diameter 14.5′; R.A. 16h 47.2m, Dec. −1° 57′.
M14 (NGC 6402)	A less condensed center with a slightly elliptical shape. Magnitude 7.6; diameter 11.7′; R.A. 17h 37.6m Dec. –3° 15′.
M19 (NGC 6273)	The most oblate known globular cluster. Magnitude 7.1; distance: 27,000 light-years; R.A. 17h 2.6m, Dec. –26° 16′.
M62 (NGC 6266)	The most irregularly shaped known globular cluster. Magnitude 6.6; diameter 14.1; R.A. 17h 1.2m, Dec. –30° 7′.
M107 (NGC 6171)	Smaller and fainter than M10 or M12 and, unusual for globulars, contains dark regions. Magnitude 8.1; diameter 10.0′; R.A. 16h 32.5m, Dec. +3° 13′.

some other process. *Outward orbital migration* may result in a planet's complete ejection from its orbit to become an unbound **rogue planet**. *Inward orbital migration* almost certainly accounts for the unusually small or highly elliptical orbits of many of the first **extrasolar planet**s to be discovered. Theories of inward migration were hastily put forward following the unexpected discovery of massive planets in extremely small orbits about their central stars. Since it is virtually impossible to understand how such large worlds, which are presumably gas giants like Jupiter, could have formed in their present orbits of radii less than 2 AU, and in some cases less than 0.2 AU, the favored conclusion is that they formed farther out and were then caused to move inward. Several processes have been suggested to account for the drastic orbital shift, including near encounters with other massive worlds, resulting in an effect like the scattering of billiard balls; and viscous drag as the new planet plowed its way through the remnants of the dusty nebula from which it was formed, losing energy as it traveled. Both theory and observation suggest that the migration time to a small orbit for a giant planet that formed at a radius of 5 AU is less than 1 million years.

orbital period

The time taken to go once around a closed **orbit**.

orbital velocity

The velocity of an object at a given point in its **orbit**. If the orbit is perfectly circular, the magnitude of the velocity is constant and given by

$$V_{orb} = \sqrt{GM/r}$$

where G is the gravitational constant, M is the mass of the gravitating body, and r is the radius of the orbit. An object moving faster than circular velocity will enter an elliptical orbit with a velocity at any point determined by **Kepler's laws of planetary motion**. If the object moves faster still, it will travel at **escape velocity** along a **parabolic orbit** or beyond escape velocity in a **hyperbolic orbit**.

ordinary chondrite

A **chondrite** that is composed mostly of **olivine**, **orthopyroxene**, and a certain percentage of more or l ess oxidized nickel-iron. Based on the differing content of metal and differing mineralogical compositions, the ordinary chondrites are divided into three distinct groups.

H Group chondrites have a high content of free nickel-iron (15 to 19%) and are attracted easily to a magnet. Their main minerals are olivine and the orthopyroxene bronzite, earning them their older name of *bronzite chondrites*. Comparison of the reflectance spectra of the H chondrites to the spectra of several main belt **asteroids** has yielded a probable parent body in the asteroid (6) Hebe. However, Hebe might not be the direct source of the H chondrites; most likely, Hebe collided with another asteroid at some point in its history and larger parts of it were dislodged into an elliptical near-Earth orbit, from which fragments eventually came to Earth.

L Group chondrites are low in free nickel-iron (4 to 10%) and also contain olivine and the orthopyroxene **hypersthene**. The asteroid (433) **Eros** is suspected as a parent body, based on reflectance spectra, but most L chondrites show signs of severe shock metamorphism suggesting a violent history of the parent. Possibly the L chondrites came from a relative or a former part of Eros that was entirely broken up when it collided with another asteroid.

LL Group chondrites (low iron/low metal) have only 1 to 3% free metal. Their olivine is more iron-rich than in the other ordinary chondrites, implying that the LL types must have formed under more oxidizing conditions than their H or L cousins. Scientists are still searching for a probable parent body for this group. One small main belt asteroid, (3628) Boznemcová, shows a similar reflectance spectrum, but with a diameter of just 7 km it seems too small to be regarded as the progenitor of the LL members.

Orion (abbr. Ori; gen. Orionis)

The Hunter; a spectacular northern **constellation**, located southeast of Taurus in the equatorial region of the sky. Below **Orion's Belt** is one of the most famous deep sky objects, the **Orion Nebula**. (See tables, "Orion: Stars Brighter than Magnitude 4.0," and "Orion: Other Objects of Interest." See also star chart 16.)

Orion Complex

A giant cloud of interstellar gas and **dust** in the direction of **Orion**, approximately centered on the **Orion Nebula** (M42); it is about 1,500 light-years away and several hundred light-years across. This giant cloud, or complex of clouds, of interstellar matter and young stars contains, besides M42 and the neighboring DeMairan Nebula (M43), and the nebulosity associated with them (NGC 1973-5-7), a number of famous objects, including **Barnard's Loop**, the **Horsehead Nebula** region (also containing NGC 2024, or Orion B), and the reflection nebulae around M78. Within this cloud, stars have formed recently, and are still in the process of formation. These young stars make up the so-called *Orion OB1 Association*. The association can be divided into subgroups: 1a, which includes and surrounds the stars of **Orion's Belt**; 1b, which lies northwest of the Belt stars; and 1c, which contains **Orion's Sword**. The stars of the Orion Nebula and of M43 form a subset of this group, and are sometimes separately counted as subgroup 1d–the very youngest stars of the Orion OB1 association. There are two major **molecular cloud**s in the region. *Orion Molecular Cloud 1* (OMC-1) is centered approximately 1′ northwest of the **Trapezium** and contains the **Becklin-Neugebauer Object** and **Kleinmann-Low Nebula**. *Orion Molecular Cloud 2* (OMC-2) is an infrared and molecular emission complex about 12′ northeast of the Trapezium, centered on a cluster of infrared sources.

Orion Nebula (M42, NGC 1976)

A bright cloud of gas and dust, about 20 light-years across and 1,500 light-years away in **Orion**, which is a birthplace of stars; it is visible to the naked eye as a fuzzy patch known as **Orion's Sword**. Telescopes reveal at its heart the **Trapezium** cluster whose **ultraviolet** light energizes and illuminates the nebula. Numerous **protoplanetary disks** have been observed around the Orion Nebula's newborn stars. (See photo, page 374.)

Orionids

A moderate annual **meteor shower**, consisting of debris from **Halley's comet**, that shows a complex pattern of activity over its range from October 16 to 26 or beyond. The maximum **zenithal hourly rate** of about 30 meteors per hour occurs for a couple of days around October 21.

Orion: Stars Brighter than Magnitude 4.0

Star	Magnitude		Spectral	Distance	Position	
	Visual	Absolute	Type	(light-yr)	R.A. (h m s)	Dec. (° ′ ″)
β Rigel	0.12v	–6.69	B8Iae	773	05 14 32	–08 12 06
α Betelgeuse	0.5v	–5.14v	M2Iab	427	05 55 10	+07 24 26
γ Bellatrix	1.64	–2.73	B2III	243	05 25 08	+06 20 59
ε Alnilam	1.69	–6.39	B0Iae	1,340	05 36 13	–01 12 07
ζ Alnitak	1.74	–5.26	O9.5Ibe	817	05 40 46	–01 56 34
κ Saiph	2.07	–4.66	B0.5Iav	721	05 47 45	–09 40 11
δ Mintaka	2.25	–4.99	B0II+O9V	916	05 32 00	–00 17 57
ι Na'ir al Saif	2.75	–5.30	O9III	1,330	05 35 26	–05 54 36
π^3	3.19	3.66	F6V	26	04 49 50	+06 57 41
η Algjebbah	3.35v	–3.86	B1V+B2e	901	05 24 29	–02 23 50
λ Meissa	3.39	–4.16	O8e+B0.5V	1,060	05 35 08	+09 56 02
τ	3.59	–2.56	B5III	555	05 17 36	–06 50 40
π^4	3.68	–4.25	B2III+B2IV	1,260	04 51 12	+05 36 18
π^5	3.71v	–4.37	B3III+B0V	1,340	04 54 15	+02 26 26
σ	3.7	–3.96	O9.5V	1,150	05 38 45	–02 36 00

Radiant: R.A. 6h 24m, Dec. +15° (between *Betelgeuse* and Gamma Geminorum).

Orion's Belt

A line of three brilliant white stars, from left to right (east to west), **Alnitak**, **Alnilam**, and **Mintaka**. The names of all three refer to the set: the outer two are named after the "belt" of the Arabs' "Central One" (a mysterious female figure), while Alnilam comes from an Arabic word that aptly means "the string of pearls." The proximity of the trio in the sky is an illusion. In fact, the stars at either end of the Belt, Alnitak and Mintaka, are the closest together in space, Alnitak being a little over 800 light-years away, and Mintaka 100 light-years farther off. The central star, Alnilam, is much more distant than either of these, lying on the edge of the *Orion Molecular Cloud,* more than 1,300 light-years from the Sun. Alnilam is also easily the most massive and luminous of the three stars, so that despite its greater distance it still shines more brightly than its two companions. Like much of its parent constellation, Orion's Belt is a highly nebulous region of the sky. This is especially true in the direction of Alnitak, where numerous nebular structures are found. The most famous of these is **Horsehead Nebula**, a dark cloud

Orion: Other Objects of Interest

Object	Notes
Star	
Trapezium	Theta[1] Orionis. See separate entry.
Open cluster	
NGC 1981	10 stars, including the binary Struve 750, with sixth and eighth magnitude components. Magnitude 4.6; diameter 25′; R.A. 5h 35.2m, Dec. –4° 26′.
Emission nebulae	
Orion Nebula	M42. See separate entry.
DeMairan Nebula	M43. Part of the Orion Nebula.
Dark Nebula	
Horsehead Nebula	IC 434. See separate entry.

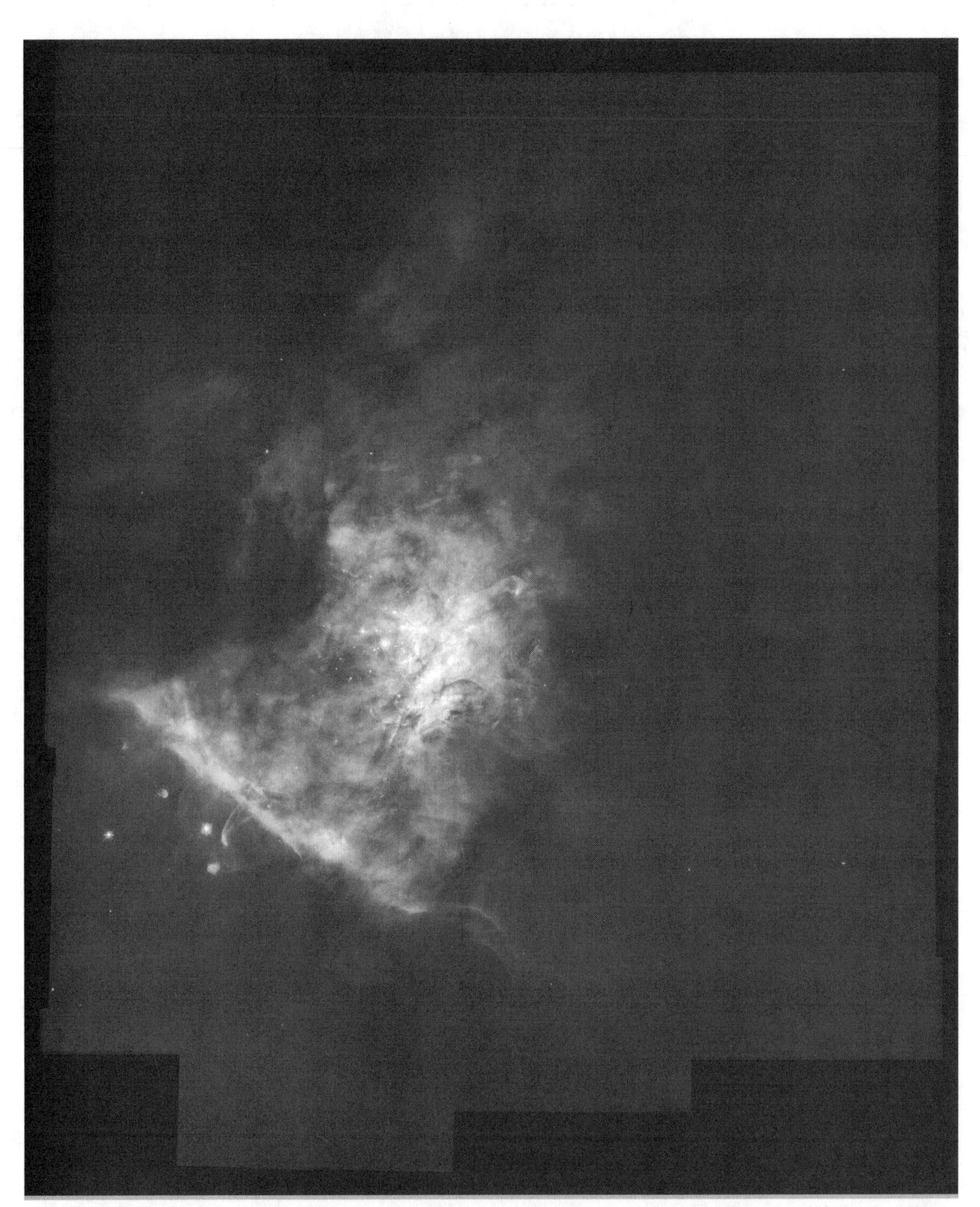

Orion Nebula The center of the Orion Nebula is revealed in one of the largest pictures ever assembled from individual images taken with the Hubble Space Telescope. This panorama, a seamless mosaic of 15 separate fields, covers an area of sky about one-twentieth the size of the Moon. *NASA/STScI*

blotting out the light from the red-pink streak of IC 424. Also occupying this region are the less well-known **Flame Nebula** (NGC 2024) and the Lump Star (NGC 2023).

Orion's Sword

A region of sky below **Orion's Belt** that includes the **Orion Nebula** (plus the **Trapezium** stars within it), the diffuse nebula NGC 1977, the open cluster NGC 1981, and **Iota Orionis**.

Orion-type star

Any kind of **eruptive variable** associated with nebulosity, including **T Tauri stars** and **FU Orionis stars**. Orion-type stars are young objects that undergo irregular outbursts of up to several magnitudes but that may also show more regular changes linked to their rotatation.

orrery

A working model of the solar system showing the planets, possibly with some of their moons, in orbit around the Sun. It is named after Charles Boyle (1676–1731), Fourth Earl of Cork and Orrery, who commissioned one from John Rowley in 1712.

orthopyroxene

A group of **pyroxene** minerals that have crystallized in the orthorhombic system. Examples include **enstatite** and **hypersthene**.

orthopyroxenite

A type of **Mars meteorite** that is currently represented by only one known specimen, ALH 84001, which was found in the Allan Hills region of Antarctica in 1984. It is a cumulate rock consisting of 97% coarse-grained, magnesium-rich **orthopyroxene**, with small amounts of **plagioclase**, chromite, and carbonate. It was initially classified as a member of the HED group, specifically a diogenite, because these achondrites are also made mostly of orthopyroxene. However, the presence of oxidized iron in the chromite of ALH 84001 led to its reclassification as a Martian meteorite; a fact that has been subsequently confirmed by its oxygen isotope composition. With a crystallization age of about 4.4 billion years, it is by far the oldest meteorite from the fourth planet yet discovered. It probably represents a sample of the early Martian crust, providing evidence for the earliest geologic history of Mars. However, public and much scientific attention has been focused on a minor aspect of this unique rock during the past years–the presence of small orange-colored carbonate spherules that probably formed 3.9 billion years ago. These "orangettes" are barely visible to the naked eye, ranging up to 200 microns in size, and they seem to have formed in the presence of liquid water within fractures inside ALH 84001. In 1996, a team led by NASA's David McKay (1936–) published a paper announcing the discovery of traces of fossil Martian life within the orangettes, consisting of organic molecules, several biominerals, and microfossils that resemble terrestrial nanobacteria. Ever since then, his discovery has been vigorously debated, splitting the scientific community into advocates and prosecutors of the existence of (former) primitive life on Mars. Future missions to Mars will confirm whether we are alone in a hostile universe, or if there are unsuspected neighbors in our own solar system, thus implying a biophile nature for our universe where life is not the exception but the rule.

osculating orbit

The path that a planet or other orbiting body would follow if it were subject only to the inverse-square attraction of the Sun or some other central body. In practice, secondary bodies, such as Jupiter, produce **perturbations**.

overcontact binary

A **close binary** in which both stars have exceeded their **Roche lobes**, so that the system consists of a common, dumbbell-shaped envelope surrounding two stellar cores. Overcontact binaries are also often eclipsing binaries of the **W Ursae Majoris star** type.

Owens Valley Radio Observatory (OVRO)

The radio astronomy observatory of the California Institute of Technology, located near Bishop, California, about 400 km north of Los Angeles on the east side of the Sierra Nevada, at an altitude of 1,220 m. Its main instruments are: the *Millimeter-Wavelength Array*, a new **interferometer** array of six high-accuracy radio telescopes, each 10.4 m in diameter and moveable along a T-shaped railroad track to give the equivalent resolution of a single 300-m dish; the *40-Meter Telescope*, a single dish built in 1965; and the *Solar Array*, consisting of two 27-m telescopes, dating from 1960, that is now used as an interferometer for studying magnetic field variations on the Sun.

Owl Nebula (M97, NGC 3587)

A **planetary nebula** in **Ursa Major**; it was the third to be discovered, by Pierre **Méchain** in 1781, and was named by Lord **Rosse** in 1848. The Owl is one of the more complex planetary nebulae. Its appearance has been interpreted as that of a cylindrical torus shell (or globe without poles), viewed obliquely, so that the projected matter-poor ends of the cylinder correspond to the Owl's eyes. This shell is enveloped by a fainter nebula of lower ionization. The mass of the nebula has been estimated at 0.15 M_{sun} and that of the magnitude 16 central star at 0.7 M_{sun}.

Visual magnitude:	9.8
Angular diameter:	3.4′ × 3.3′
Age:	6,000 years
Distance:	1,300 light-years
Position:	R.A. 11h 14.8m, Dec. +55° 1′

oxygen (O)

The most abundant **heavy element** in the universe, and the third most abundant element overall, after **hydrogen** and **helium**. Oxygen has the **atomic number** eight and is produced by massive stars–those born with over 8 M_{sun}– which eject the element into interstellar space when they explode. *Oxygen burning* is the fusion of oxygen into the even heavier elements silicon and sulfur. Oxygen is the most common element in the crusts and mantles of the inner planets and of rocky moons, making up all silicate minerals.

ozone layer

A layer in the lower part of Earth's **stratosphere** (about 20 to 60 km above sea level) where the greatest concentration of ozone (O_3) appears. This is the layer responsible for the absorption of most ultraviolet radiation.

P

P Cygni profile

A combination of features in a star's spectrum that points to an outflow of material in the form of either an expanding shell of gas or a powerful **stellar wind**. The P Cygni profile is characterized by strong **emission lines** with corresponding **blueshifted absorption lines**. The latter are produced by material moving away from the star and toward Earth, whereas the emission comes from other parts of the expanding shell. P Cygni itself (also known as *34 Cygni*) is a **Be star**, one of the most luminous stars known and only the third **variable star** to be discovered; it lies about 7,000 light-years away. It was first noted in 1600 in a place where no star had been recorded before, as a third magnitude star. The first well-documented observations were made by Willem Blaeu (1571–1638). On a globe made by Blaeu, now in a Prague museum, is written: "The new star in Cygnus that I first observed on August 8, 1600, was initially of third magnitude. I determined its position . . . by measuring its distance from Vega and Albireo. It remains in this position but now is no brighter than 5th magnitude." Over the next few years, the star faded below naked-eye visibility, but returned to magnitude 3.5 in 1655, where it remained until 1659. It faded again below sixth magnitude and rose once more in 1665. After some fluctuations, it became steady at about magnitude 5 around 1715 and, during the last 200 years, has oscillated in brightness around this value. The brightness bursts are thought to be due to the star throwing off shells of gas. Satellite observations have shown, that below 2000 Å, many O and B stars display these kind of spectral lines. It is therefore ironic that the object that gave its name to the subclass is no longer considered a member of it! P Cygni is now grouped with an even more elite class of stellar celebrities–the **S Doradus stars**. The recent detection of a P Cygni profile in the X-ray emission from **Circinus X-1** means that this powerful tool for probing stellar winds and other stellar outflows is now available in an entirely new region of the spectrum.

palimpsest

An ancient, relatively bright, circular feature on the surface of a dark icy moon, such as **Ganymede** or **Callisto**. Palimpsests lack the relief associated with normal craters and are thought to be **impact craters** of which the topographic relief has been eliminated by viscous relaxation (creep) of the icy surface, probably during the impact itself. Typical is Ganymede's 340-km-wide *Memphis Facula.* Such structures hold important clues to the early thermal history and composition of the bodies on which they are found. The original meaning, "scraped again," refers to a manuscript on a waxen tablet or other writing material from which an early text was removed and then written over. The value of palimpsests, in both planetary astronomy and literature, is that, they preserve a record of the past in the form of something that is partly hidden from view.

Palisa, Johann (1848–1925)

An Austrian astronomer at the Vienna Observatory who discovered 120 **asteroids** and published catalogs with the position of nearly 5,000 stars. He was born in Troppau, Silesia (now Czech Republic).

Pallas (minor planet 2)

The second largest **asteroid** and the second to be discovered, by Heinrich **Olbers** in 1802. It is slightly irregular in shape, measuring 570 × 525 × 482 km, has a density of

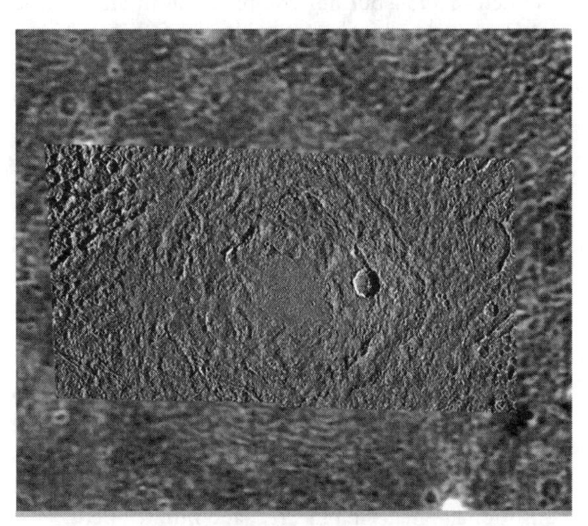

palimpsest Buto Facula, a palimpsest located in Marius Regio on Jupiter's largest satellite Ganymede. The higher resolution image in the center, showing features as small as 360 m across, was acquired by the Galileo spacecraft in May 1997 from a distance of about 18,600 km, while the lower resolution background came from Voyager in 1979. *NASA/JPL*

3.4 g/cm^3 and an albedo of 0.05, and appears to have the same composition as meteorites classed as low-grade **carbonaceous chondrites** or **enstatite achondrites**. With a maximum visual magnitude of 8.0, it is brighter than all other asteroids except **Vesta** and **Ceres**. Semimajor axis 2.773 AU, perihelion 2.136 AU, aphelion 3.409 AU, eccentricity 0.230, inclination 34.85°, period 4.62 years.

pallasite

One of two main classes of **stony-iron meteorite**, the other being **mesosiderites**. Pallasites are **igneous** in nature and characterized by crystals of **olivine**, often *peridots* (clear, green olivine crystals of gem quality), embedded in a matrix of iron-nickel. The type specimen, weighing 680 kg, was found in the mountains near Krasnojarsk, Siberia, and first documented by the German naturalist Peter Pallas in 1772, although it was some decades later before the extraterrestrial nature of what became known as the *Pallas Iron* was realized. Pallasites are believed to have come from the core/mantle boundary of differentiated **asteroids** that were broken apart by impact. In most cases, they have chemical, elemental, and isotopic features that link them to specific chemical groups of **iron meteorites**, suggesting that they come from the same parent bodies as these irons.

Palomar Observatory

An observatory at an altitude of 1,700 m on Mount Palomar, 80 km northeast of San Diego, California. Founded in 1934 but not completed until after World War II, it is owned and operated by the California Institute of Technology, and is home to the 5-m **Hale Telescope** and the 1.2-m Oschin Schmidt Telescope, both opened in 1948. The latter was used for the ***Palomar Observatory Sky Survey***.

Palomar Observatory Sky Survey (POSS)

A photographic survey of the sky from the north celestial pole to Dec. –30°, carried out using the 1.2-m Schmidt telescope at **Palomar Observatory** and financed by the National Geographic Society. The original survey, completed in 1958, involved 936 pairs of plates 0.355-m square with a field of view of 6.5°, each consisting of one plate sensitive to red light and one to blue light. A second survey was started in 1985 from the north celestial pole to the equator, using modern plates; when complete, it will consist of 894 fields in blue, red, and infrared light.

Pan

The innermost of moon of **Saturn**, also known as Saturn XVIII; Pan lies within the Encke Division of Saturn's A-ring at a distance of 133,583 km from the planet's center and is responsible for keeping the Encke Division open. It has a diameter of about 20 km. Prior to its discovery, an analysis of the patterns at the edge of the A-ring predicted the size and location of a small moon. Pan was found by the American astronomer Mark Showalter in 1990 in **Voyager** 2 photos, taken nine years earlier, at the predicted spot.

Pandora

The fourth moon of **Saturn**, also known as Saturn XVII; it orbits at a mean distance of 141,700 km from the planet's center and, like **Prometheus**, serves as a **shepherd moon** for the outer edge of Saturn's F-ring. Pandora measures about 114 × 84 × 62 km and is more heavily cratered than Prometheus, with at least two craters 30 km in diameter. It was discovered in 1980 by S. Collins and others from photographs taken by **Voyager** 1. Pandora and Prometheus both follow chaotic orbits as a result of their gravitational interaction. Each time Pandora passes inside Prometheus, which happens about every 28 days, the two moons give each other a gravitational kick. Because neither moon's orbit is quite circular, the distance between them on those occasions–hence the strength of the kick–varies. The perturbations lead to changes in motion that are not periodic or predictable. The Pandora/Prometheus combination is the first observation ever of chaotic orbital motions in the solar system. A larger moon of Saturn, **Hyperion**, had earlier been found to have chaotic rotation around its axis.

parabola

A curve with an **eccentricity** equal to 1 that is obtained by slicing a cone with a plane parallel to one side of the cone. A parabola can be considered to be an **ellipse** with an infinite **major axis**.

parabolic orbit

An open (escape) orbit with an eccentricity of 1. *Parabolic velocity* is the minimum velocity needed to escape from a given closed orbit.

parallactic displacement

The apparent change in a star's position caused by the movement of Earth in its orbit; also known as *parallactic motion*. Over the course of a year, a star appears to follow a small path in the sky called a *parallactic ellipse*, or a *parallactic orbit*, the shape of which ranges from a circle for a star at the ecliptic pole to a line for a star on the **ecliptic**.

parallax

The apparent change in position of a relatively close object compared to a more distant background as the location of the observer changes. *Annual parallax* is the change in the apparent position of a star resulting from Earth's annual motion around the Sun; it is defined as the angle subtended at the object by the **semimajor axis** of Earth's orbit. The largest parallax known for an object outside the solar system is that of the nearest star, **Proxima Centauri**, 0.762″. Knowing the parallax of a star and the size of Earth's orbit, the star's distance can be worked out by simple trigonometry. There are other ways to measure stellar distances, which, though not involving trigonometric parallax, are still referred to as varieties of parallax. *Spectroscopic parallax* is the most widely used technique for determining the distances of stars that are too distant for their trigonometric parallaxes to be measured. From an analysis of a star's spectrum, its position is determined on the **Hertzsprung-Russell diagram**, which gives the **absolute magnitude**. By comparing the star's absolute magnitude with its apparent brightness, its distance follows directly. *Dynamical parallax* is a method for determining the distance to a visual **binary star**. The angular diameter of the orbit of the stars around each other and their apparent brightness are observed. The distance comes from an application of **Kepler's laws of planetary motion** and the **mass-luminosity relation**. See also **lunar parallax**.

Paranal Observatory

The site of the European Southern Observatory's **Very Large Telescope**; it is located on the top of Cerro Paranal in the Atacama Desert in northern Chile in what is believed to be the driest area on Earth. Cerro Paranal is a 2,635-m-high mountain, about 120 km south of the town of Antofagasta and 12 km inland from the Pacific Coast. There is minimal light pollution, atmospheric conditions are unusually stable, humidity is virtually zero, and there are around 330 clear nights a year. The geographical coordinates are 24° 40′ S, 70° 25′ W.

Paranal An aerial view of Paranal Observatory, which sits atop the highest peak to the right, and its immediate surroundings in the Atacama Desert. The Pacific Ocean is visible in the background. *European Southern Observatory*

Parkes Observatory

An Australian radio astronomy observatory located 20 km north of Parkes, New South Wales. It is home to a 64-m radio dish that can be used as an independent instrument or networked with other Australian and international radio telescopes for **very long baseline interferometry**. Parkes is part of the **Australia Telescope National Facility**.

parsec (pc)

A standard unit of stellar and galactic distance; parsecs, rather than light-years, are generally used in professional astronomy. One parsec is the distance at which the mean Earth-Sun distance (one astronomical unit) subtends an angle of one second of arc. 1 parsec = 3.2616 light-years. 1 kiloparsec (kpc) = 1,000 pc; 1 megaparsec (Mpc) = 1,000,000 pc.

partial eclipse

(1) A **solar eclipse** in which the Moon doesn't completely cover the Sun's disk. (2) A **lunar eclipse** in which the Moon doesn't pass completely into the darkest part (the umbra) of Earth's shadow.

Paschen-Back effect

The splitting of **spectral lines** into a number of components by a very strong magnetic field. The effect is named after the German physicists Friedrich Paschen (1865–1947) and Ernst Back (1881–1959).

Pasiphaë

One of the outer moons of **Jupiter**, also called Jupiter VIII; it was discovered in 1908 by Philbert Melotte (1880–1961). Pasiphaë has a diameter of 58 km and moves in a **retrograde** orbit at a mean distance of 23.6 million km from the planet.

Paul Wild Observatory

A radio astronomy observatory, near Narrabri, that forms part of the **Australia Telescope National Facility** (ATNF). The Narrabri site contains the *Australia Telescope Compact Array*, which consists of five antennas located along a 3-km rail-track, and a sixth antenna 3 km farther to the west.

Pauli exclusion principle

A fundamental rule in subatomic physics which says that no two **electrons**, or other **fermions**, can occupy the same quantum state; it was formulated by the German physicist Wolfgang Pauli in 1925. The exclusion principle prevents **white dwarfs** that are below the **Chandrasekhar limit** from collapsing: the electrons are squashed together to form **electron degenerate matter**, which resists any further attempts at compression.

Pavo (abbr. Pav; gen. Pavonis)

The Peacock; a southern **constellation** lying between Telescopium to the north and Octans to the south. Its most noteworthy deep sky object is the globular cluster NGC 6752 (magnitude 5.4; diameter 20.4′; R.A. 19h 10.9m, Dec. –59° 59′), large enough to be well seen with binoculars or a small telescope. (See table, "Pavo: Stars Brighter than Magnitude 4.0," and star chart 15.)

Payne-Gaposchkin, Cecilia Helena (1900–1979)

A British-American astronomer (née Payne) who became the first female full professor at Harvard and one of the founders of modern astrophysics. Her Ph.D. dissertation, entitled "Stellar Atmospheres: A Contribution to the Observational Study of High Temperature in the Reversing Layers of Stars" (1925), was later acclaimed as the best in twentieth-century astronomy. In it, she argued that the great variation in stellar **absorption lines** was due to differing amounts of ionization (related to differing temperatures), and not, as was generally supposed, to significant

Pavo: Stars Brighter than Magnitude 4.0

Star	Magnitude		Spectral	Distance	Position	
	Visual	Absolute	Type	(light-yr)	R.A. (h m s)	Dec. (° ′ ″)
α Peacock	1.94	–1.81	B2IV	183	20 25 39	–56 44 06
β	3.42	0.29	A5IV	138	20 44 57	–66 12 12
δ	3.55	4.62	G5IV	20	20 08 43	–66 10 56
η	3.61	–1.67	K1III	371	17 45 44	–64 43 25
κ	3.9max	–1.71	F5Ib	543	18 56 57	–67 14 01
ε	3.97	1.41	A0V	106	20 00 35	–72 54 38

differences in chemical composition. She correctly posited that silicon, carbon, and other common **heavy elements** seen in the Sun occurred in about the same relative amounts as on Earth but that helium and particularly hydrogen were vastly more abundant (by about a factor of one million in the case of hydrogen). When she sent a draft of her paper to Henry **Russell**, he replied that such a result was "clearly impossible." Russell had earlier written a paper in which he argued that if the Earth's crust were heated to the temperature of the Sun, its spectrum would look the same. Deferring to Russell's stature as an astronomer, Payne added the comment that her results were "almost certainly not real." Within a few years, however, her claim had been fully vindicated.

P-class asteroid

A dark **asteroid**, spectrally indistinguishable from those of classes M or E but of lower **albedo** (0.02 to 0.06). P-class asteroids show featureless reflectance spectra across the range 0.3 to 1.1 microns and are found especially in the outer part of the main **asteroid belt**. The largest example is the 282-km-wide (87) Sylvia.

Pearce, Joseph Algernon (1893–1988)

A Canadian astronomer who became the first director of the **Dominion Astrophysical Observatory** in Victoria along with John **Plaskett**. In the 1930s, he published the first detailed analysis of the rotation of the **Milky Way Galaxy**, demonstrating that the Sun is two-thirds from the galactic center and revolves around it once in 220 million years. He also studied **radial velocities** of O and B stars, cataloged the observable B stars (finding that 40% of them are binary systems), and estimated the temperatures and dimensions of representative giant eclipsing double stars.

Pease, Francis Gladheim (1881–1938)

An American astronomer and designer of optical instruments, born in Cambridge, Massachusetts. He served as an observer and optician at **Yerkes Observatory** (1901–1904), and instrument-maker at the **Mount Wilson Observatory** (1908–1913), where he designed the 100-in. (2.5-m) telescope, as well as the 50-foot (15-m) interferometer telescope by means of which he made direct measurements of star diameters. He was also involved in the design of the **Hale Telescope**. In 1928 he found the first **planetary nebula** in a **globular cluster** (M15), now cataloged as Pease 1; it had been previously cataloged as a star by Friedrich Kustner in 1921.

peculiar galaxy

A galaxy that has some abnormality of shape, size, or content that sets it apart from the normal ellipticals, spirals, and irregulars in the Hubble classification (see **galaxy classification**). Peculiar galaxies often result from **galaxy interactions** or **galaxy mergers**, or may show some other distinctive feature such as **jets** emerging from the nucleus, unusual amounts of **dust**, or low surface brightness. They are denoted by the addition of "*p*" or "*pec*" to their main classification type. Halton **Arp**'s catalog (1966) is devoted to peculiar galaxies.

peculiar motion

(1) The real or true motion of a star with respect to the **local standard of rest**; also known as *peculiar velocity.* (2) The motion of a cosmological object other than the apparent recession caused by the expansion of the universe.

peculiar star

A star whose spectrum has unusual features compared with the majority of stars of its **spectral type**; "*p*" is appended to the spectral type to show this. The term is most often used to describe stars between spectral types B5 and F5, known as **Bp stars**, **Ap stars**, and **Fp stars**, whose spectra provide evidence of chemical anomalies.

Peekskill meteorite

A 12.4-kg **meteorite**, classified as an H6 **chondrite**, that fell in Peekskill, New York, on October 9, 1992, penetrating the trunk of a 1980 Chevy Malibu that was sitting in a driveway. The descent of the space rock was witnessed by thousands in the eastern United States as a brilliant **fireball** and was caught on at least 14 amateur videotapes. First seen over Kentucky, it traveled in a near-grazing trajectory, of only 3.4° for more than 700 km in a north-northeasterly direction for at least 40 seconds, before landing. A sonic boom was heard as it broke into over 70 pieces, one of which impacted Michelle Knapp's red Malibu coupe outside her home. When police arrived on the scene, they filed a report for criminal mischief. The smell of gas from the punctured gas tank finally prompted the fire department to investigate, at which time they found the culprit. The stone was impounded by the police and subsequently cracked open to see what was inside before being returned to its rightful owner. Analysis of its recorded descent enabled a pre-atmospheric velocity of 14.72 km/s to be calculated, together with orbital parameters (perihelion 0.886 AU, aphelion 2.1 AU). Only a few other meteorite falls have corresponding photographic references enabling their orbits to be worked out, including the **Lost City**, **Príbram, Innisfree**, Tagish Lake, and Morávka meteorites. Peekskill was the first to be recovered following videotaped atmospheric passage. Through analysis of radionuclides and cosmic-ray tracks, an **exposure age** of 32 million years was obtained for the Peekskill rock that experienced a two-stage exposure history. During the first stage the sample was located 20 to 25 cm deep on a meteoroid having a radius of 40 to 60 cm; during the second stage, lasting less than 200,000 years, the depth was greater than 7 cm within an overall radius of 25 to 40 cm. The meteorite and Malibu have been taken on tour to several countries including Germany, France, Japan, and Switzerland, as well as throughout the United States.

Pegasus (abbr. Peg; gen. Pegasi)

The Winged Horse; a large northern **constellation.** Alpha, Beta, and Gamma Peg, together with Alpha Andromedae (which was originally Delta Peg), form the *Square of Pegasus.* The most prominent deep sky object is the **globular cluster** M15 (NGC 7078), one of the densest of its type and notable for its many variable stars and pulsars (magnitude 6.3; diameter 12.3′; R.A. 21h 30.0m, Dec. +12° 10′). Binoculars and small telescopes show it as a misty patch; telescopes of at least 15-cm aperture resolve individual stars. (See table, "Stars Brighter than Magnitude 4.0," and star chart 1.)

Pegasus Dwarf (Peg dSph, DDO 216)

A **dwarf spheroidal galaxy**, discovered in 1998, that is a small, newly recognized member of the **Local Group.** Also known as Andromeda VI, because it is probably a satellite companion of the **Andromeda Galaxy** and the sixth such to be found, the Pegasus Dwarf is almost hidden in the glare of relatively bright foreground stars in our own Milky Way. The absence of young and intermediate-age stars in Peg dSph, and another recently found satellite of the Andromeda Galaxy, the **Cassiopaeia Dwarf,** suggests they have been stripped of star-forming gas by Andromeda's gravitational field. This is surprising in view of their considerable distance from the great spiral and may indicate that Andromeda is more massive than previously thought.

Visual magnitude:	13.2
Angular size:	5.0′ × 2.7′
Linear diameter:	2,000 light-years
Position:	R.A. 23h 28.6m, Dec. +14° 45′

Pelican Nebula (IC 5070)

A large area of **emission nebula** in **Cygnus,** close to Deneb, and divided from its brighter, larger neighbor, the **North America Nebula,** by a **molecular cloud** filled with dark **dust.** The Pelican receives much study, however, because it has a particularly active mix of star formation and evolving gas clouds. The light from young energetic stars is slowly transforming cold gas to hot and causing an **ionization front** gradually to advance outward. Particularly dense filaments of cold gas are seen to still remain. Millions of years from now this nebula might no longer be known as the Pelican, as the balance and placement of stars and gas will leave something that appears completely different.

Visual magnitude:	8.0
Size:	60′ × 50′
Position:	R.A. 20h 51m, Dec. +44° 00′

penumbra

(1) The lighter, outer shadow of a double-shadow effect produced by a large, unfocused source of light shining on an interfering opaque object. A *penumbral eclipse* is one in which the eclipsed object passes through the penumbral shadow. (2) The lighter area, marked by radial filaments, that may surround the darker, central **umbra** of a **sunspot.** *Penumbral waves* are waves observed in the light of **H-alpha** that move out from the center of a sunspot across the penumbra.

Pegasus: Stars Brighter than Magnitude 4.0

	Magnitude		Spectral	Distance	Position	
Star	Visual	Absolute	Type	(light-yr)	R.A. (h m s)	Dec. (° ′ ″)
ε **Enif**	2.38	−4.19	K2Ib	673	21 44 11	+09 52 30
β Scheat	2.44v	−1.49	M2.5II	199	23 03 46	+28 04 58
α Markab	2.49	−0.67	B9III	140	23 04 46	+15 12 19
γ Algenib	2.83v	−2.22	B2IV	333	00 13 14	+15 11 01
η Matar	2.93	−1.17	G2II+F0V	215	22 43 00	+30 13 17
ζ Homan	3.41	−0.62	B8V	209	22 41 28	+10 49 53
μ Sadalbari	3.51	0.74	M2III	117	22 50 00	+24 36 06
θ Biham	3.52	1.16	A2V	97	22 10 12	+06 11 52
ι	3.77	3.42	F5V	38	22 07 01	+25 20 42

periapsis
The point in an **orbit** when two objects are closest together. Special names are given to this point for commonly used systems, including **periastron**, **perihelion**, and **perigee**. Compare with **apoapsis**.

periastron
The point of closest approach between two stars in a binary system.

peridotite
A course-grained **igneous** rock formed mainly of the minerals **olivine** and **pyroxene**.

perigalacticon
The point in a star's **orbit** around the Milky Way Galaxy when the star lies closest to the galactic center. The Sun is near perigalacticon now.

perigee
The point in its **orbit** where the Moon (or an artificial satellite) is closest to Earth.

perihelion
The point where an object in solar **orbit** is closest to the Sun. The *perihelion distance* is the distance between the object and the Sun at this point. The *time of perihelion passage* is the time at which an object is at perihelion.

period-age relation
A relationship between the inferred evolutionary age of a **Cepheid variable** and its period. Because high-mass stars evolve more rapidly, they are younger when they enter the **instability strip**. At the same time, their high mass is reflected in a longer pulsation period.

periodic comet
Another name for a **short-period comet**.

period-luminosity relation
A correlation between the periods and mean luminosities of **Cepheid variable**s; it was discovered by Henrietta **Leavitt** in 1912. The longer a Cepheid's pulsation period, the more luminous the star. Since measuring a Cepheid's period is easy, the period-luminosity relation allows astronomers to determine the Cepheid's intrinsic brightness and hence its distance. If the Cepheid is in another galaxy, the Cepheid's distance gives the distance to the entire galaxy.

period-mass relation
A correlation between the mass of a **pulsating variable** and its period. Roughly speaking, the greater the mass of the star, the larger its radius, the lower its mean density, and therefore the longer its pulsation period.

period-spectral relation
A correlation between the periods of **Cepheid variable**s, and similar pulsating stars, and their spectra. Stars that appear redder have longer periods, while stars that are bluish (having higher temperatures) have shorter periods.

Perseids
One of the three most prominent annual **meteor showers,** with activity that ranges from July 25 to August 18 and reaches a peak **zenithal hourly rate** of 80 to 140 meteors per hour on August 12. The Perseids move swiftly (about 60 km/s) and consist of debris from Comet **Swift-Tuttle.** Radiant: R.A. 3h 04m, Dec. +58° (near the Double Cluster).

Perseus (abbr. Per; gen. Persei)
Perseus the Hero, mythical rescuer of Andromeda from the jaws of Cetus; a large northern **constellation**, bordering on Aries to the north and Taurus to the south. The figure of Perseus is thought to hold the head of Medusa in his hand (the star Beta Per representing her evil eye). As well as a number of interesting stars, most notably **Algol**, Perseus boasts the **Little Dumbbell Nebula**, the **Double Cluster**, and another large, bright open cluster, M34 (NGC 1039), which can be seen with the naked eye on a dark night and can be resolved by a small telescope even at low power (magnitude 5.5; diameter 35′; distance 1,400 light-years; R.A. 2h 42.0m, Dec. +42° 47′). (See table, "Stars Brighter than Magnitude 4.0," and star charts 2 and 16.)

Perseus Cluster (Abell 426)
A diffuse, irregular **cluster of galaxies**, containing about 500 members and located about 250 million light-years away, that is dominated by and centered on the **Seyfert galaxy** NGC 1275 associated with the radio source Perseus A. The mass of the cluster has been estimated at 200 trillion M_{sun}.

Perseus-Pisces Supercluster
One of the largest known structures in the universe. Even at a distance of 250 million light-years, this chain of galaxy clusters extends more than 40° across the northern winter sky.

persistent train
The remaining glow, due to ionization in the upper atmosphere, after the passage of a **meteoroid**. The intensity and duration depend on the meteoroid's atmo-spheric entry velocity, its size, and its composition. Bright **fireball**s occasionally give rise to trains visible for several minutes.

Perseus: Stars Brighter than Magnitude 4.0

Star	Magnitude		Spectral	Distance	Position	
	Visual	Absolute	Type	(light-yr)	R.A. (h m s)	Dec. (° ′ ″)
α Mirfak	1.79	−4.51	F5Ib	592	03 24 19	+49 51 40
β Algol	2.09v	−0.18	B8V	93	03 08 10	+40 57 21
ζ	2.84	−4.56	B1Ib	982	03 54 08	+31 53 01
ε	2.90	−3.10	B0.5V+A2V	538	03 57 51	+40 00 37
γ	2.91	−1.57	G8III+A2V	256	03 04 48	+53 30 23
δ	3.01	−3.04	B5III	528	03 42 55	+47 47 15
ρ Gorgonea Terti	3.32v	−1.68	M3III	325	03 05 11	+38 50 25
η Miram	3.77	−4.29	K3Ib	1,330	02 50 42	+55 53 44
ν	3.77	−2.39	F5II	557	03 45 12	+42 34 43
κ Misam	3.79	1.10	K0III	112	03 09 30	+44 51 27
ο Atik	3.84	−4.44	B1III	1,480	03 44 19	+32 17 18
τ Kerb	3.93	−0.48	G4III+A4V	248	02 54 15	+52 45 45
ξ Menkib	3.98	−4.70	O7Iae	1,770	03 58 58	+35 47 28

perturbation

A change in the **orbit** of a body, usually as a result of the gravitational effect of another, typically much larger, body. The planets mutually perturb one another; the orbits of satellites are perturbed by their mutual interactions and also by the Sun. Most dramatically, the giant outer planets, especially **Jupiter**, can radically alter the orbits of **asteroids** and **comets** and send them careening on new paths into the inner solar system.

PG 1159 stars

Very hot stars with strong O VI and C IV lines, that are X-ray emitters. They are probably the central stars of **planetary nebulae** that have dissipated. Also known as *pre-degenerates.*

Phaethon (minor planet 3200)

The parent body of the Geminids **meteor shower** and, since the parent bodies of all other meteor streams identified to date are **comets**, probably a defunct comet. Discovered by the Infrared Astronomy Satellite in 1983, Phaethon has one of the most extreme orbits of any minor planet, approaching the Sun much closer than Mercury ever does at perihelion and retreating into the main asteroid belt at aphelion. Diameter 5 km, semimajor axis 1.271 AU, perihelion 0.140 AU, aphelion 2.403 AU, eccentricity 0.890, inclination 22.12°.

phase

The ratio of the lit to dark surface of the Moon or that of an **inferior planet**. Changes of phase are caused by changes in the relative positions of Earth, the Sun, and the illuminated body. Conventionally, 0° phase occurs when the hemisphere facing Earth is fully sunlit. See **lunar phases**.

phase angle

For an object in the solar system, the angle "Sun-object-Earth"; that is, the angle between the Sun and the observer as seen from the given object. It is 0° when the object is fully illuminated, 90° when the object is half-illuminated (like the Moon at first quarter and last quarter), and 180° when the object is between Earth and the Sun (like the Moon at new moon). The Moon, Mercury, and Venus can have phase angles covering the full range 0 to 180°. Mars, on the other hand, has a maximum phase angle of about 45°, meaning that it is always almost fully illuminated.

Phase function is the change in the brightness of an object as a function of the phase angle. In general, an object gets brighter as the phase angle approaches 180° or 0°. The function is usually fairly smooth except for at small phase angles where there may be a "spike" of increased brightness. The phase function is usually described as the change in magnitude (brightness) per degree of phase angle. *Phase defect* is the angular extent of the illuminated portion of the disk of the Moon or a planet compared to its full disk.

Phobos

The inner and larger, potato-shaped moon of **Mars**; it was discovered by Asaph **Hall** in 1877. Temperature data and close-up images have shown its surface to be composed of powdery material at least 1 m thick, caused by millions of years of meteoroid impacts. The largest crater on Phobos, *Stickney,* measures 5 km in diameter and is named after Angeline Stickney (1830–1892), the maiden name of Asaph Hall's wife, known for her persistent encouragement of her husband as he strove to track down the Martian satellites. Since Phobos orbits around Mars faster than the planet itself rotates, **tidal forces** are slowly but steadily decreasing its orbital radius. At some point in the future Phobos will impact on Mars's surface or be broken apart, possibly to form a ring, when it falls within the planet's **Roche limit**. **Deimos**, on the other hand, is far enough away that its orbit is being slowly boosted instead. (See **Mars** table, "The Moons of Mars.")

Diameter:	27 × 21.6 × 18.8 km
Density:	2.0 g/cm^3
Visual albedo:	0.06
Rotational period:	7h 39m 14s
Orbit around Mars	
Semimajor axis:	9,378 km (measured from the center of Mars)
Period:	same as rotational period
Eccentricity:	0.021
Inclination:	1.1°

Phobos The larger of the two Martian moons, snapped by Mars Global Surveyor on August 19, 1998. The 10-km-wide crater Stickney yawns wide at the top of the image. Individual boulders, visible on its near rim and presumed to be ejecta blocks from the impact that formed the great bowl, measure more than 50 m across. Crossing at and near the rim of Stickney are grooves, which may be fractures also associated with its formation. *NASA/JPL*

Phocaea group

A cluster of **asteroids** in the main **asteroid belt** that were perturbed into high inclination (21 to 25°) orbits by the gravitational effect of the major planets, notably Jupiter, during the earliest period of the solar system. The group is named after its first discovered member, the 72-km-wide, S-class (25) Phocaea, found in 1853 by the French astronomer Jean Chacornac (1823–1873); semimajor axis 2.402 AU, perihelion 1.79 AU, aphelion 3.01 AU, inclination 21.6°, period 3.72 years. Its largest member is the C-class asteroid (105) Artemis, with a diameter of 126 km. Most of the Phocaea group is a ragbag assemblage of objects without a common ancestry; however, certain members, including (323) Brucia, (852) Wladilena, and (1568) Aisleen, may form a true **Hirayama family**.

Phoebe

The twentieth moon of **Saturn**, also known as Saturn IX; it is 240 km in diameter and was discovered in 1898 by Edward **Pickering**. Phoebe's highly inclined, elongated, and retrograde orbit–averaging almost 13 million km from the planet–together with the moon's low **albedo** (0.05) suggest that it is a captured **asteroid** or **Kuiper Belt object**. Material ejected from Phoebe's surface by meteor impacts may be responsible for the dark surfaces of **Hyperion** and the leading hemisphere of **Iapetus**.

Phoenix (abbr. Phe; gen. Phoenicis)

The Firebird; a southern **constellation** lying near **Achernar** between Hydrus and Sculptor. Among its interesting objects are SX Phoenicis, a **Delta Scuti star** (a kind of pulsating variable) with the remarkably short period of 79 minutes (magnitude range 6.8 to 7.5; R.A. 23h 46.5m, Dec. –41° 35′) and the **Phoenix Dwarf Irregular** galaxy. (See table, "Stars Brighter than Magnitude 4.0," and star chart 6.)

Phoenix: Stars Brighter than Magnitude 4.0

Star	Magnitude		Spectral	Distance	Position	
	Visual	Absolute	Type	(light-yr)	R.A. (h m s)	Dec. (° ′ ″)
α Ankaa	2.40	0.52	K0III	77	00 26 12	–42 18 22
β	3.32	–0.60	G8IIIv	198	01 06 05	–46 43 07
γ	3.41	–0.87	K5II	234	01 28 22	–43 19 06
ζ	3.94v	–0.73	B6V+B9V	280	01 08 23	–55 14 45
ε	3.88	0.71	K0III	140	00 09 25	–45 44 51
κ	3.93	2.07	A7V	77	00 26 12	–43 40 48
δ	3.93	0.66	K0III	147	01 31 15	–49 04 22

Phoenix Dwarf
A **dwarf irregular galaxy**, or possibly a **dwarf spheroidal galaxy**, that is a member of the **Local Group**. Discovered in 1976, it lies at a distance of 1.45 million light-years and was at first suspected of being a **globular cluster**.

Pholus (minor planet 5145)
The second **Centaur** to be found, in 1992 by the American astronomer David Rabinowitz (1960–). It has a diameter of about 190 km and an orbit similar to that of the first discovered Centaur, **Chiron**. Unlike Chiron, however, which is pretty much gray all over, Pholus is very red–in fact, the reddest known object in the solar system; also unlike Chiron, it doesn't appear to develop a comet-like **coma** when near perihelion. Its infrared spectrum has a feature in the 2-micron region that had never been seen before and is probably related to whatever makes Pholus so red. The most likely explanation is that Pholus is covered with organic molecules that started off as simple ices but were converted into more complex, stable molecules by millions of years of bombardment by **cosmic rays**. According to one theory, it may be a potential giant cometary nucleus that has never been activated. Semimajor axis 20.35 AU, perihelion 8.68 AU (just inside Saturn's orbit), aphelion 31.91 (just outside Neptune's orbit), eccentricity 0.574, inclination 24.7°, period 91.8 years.

photodissociation
The breakup of molecules through exposure to light. It occurs, for example, at the edges of a cloud of cool molecular gas when illuminated by ultraviolet radiation from nearby hot stars.

photographic magnitude
The **magnitude** of a celestial object as measured on a photographic emulsion–a method of measuring apparent brightness that is now obsolete. Because emulsions are especially sensitive in the blue region of the spectrum, a blue star records as being brighter than a red star of the same apparent visual magnitude.

photoionization
The **ionization** of an atom or molecule by the absorption of a high-energy photon. It is an important source of **opacity** in stars.

photometer
In astronomy, an instrument used to measure the brightness of stars or other celestial objects. A *photoelectric photometer* records the electric current generated when light from a source falls onto a light-sensitive surface, kicking on electrons, which are then amplified into a measurable current by a photomultiplier. An *area photometer,* or *imaging photometer,* on the other hand, measures simultaneously the light intensity at each point of an extended object such as a nebula or galaxy using **CCD** detectors or an infrared array.

photometry
In optics: (1) the measurement of the intensity, distribution, color, absorption factor, and spectral distribution of visible light, and occasionally of near-ultraviolet and near-infrared light; (2) the science of making such measurements. In astronomy: the use of photometric instruments and techniques to make a precise measurement of the amount of electromagnetic energy that is received from a celestial object.

photon
The smallest possible quantity of light or other electromagnetic energy: the quantum of the electromagnetic field and the exchange particle of the electromagnetic force. Photons have zero mass and no electric charge.

photosphere

The visible surface of the Sun or some other star; it lies just below the **chromosphere** and just above the convective zone and has a temperature of about 6,000 K. The photosphere ends (and the chromosphere begins) at about the place where the density of negative hydrogen ions falls too low to result in appreciable **opacity**. Almost all the features of the Sun's visible-light spectrum originate in the photosphere, including the dark **Fraunhofer lines**. It has a texture known as **granulation**, caused by rising convection cells of hot gas, and is the location of **sunspots**, **faculae**, and **filigrees**–all associated with strong magnetic fields.

Piazzi, Guiseppe (1746–1826)

An Italian astronomer, born in Ponte di Valtellina, who became a Theatine monk, professor of theology in Rome (1779), and professor of mathematics at the Academy of Palermo (1780). He set up an observatory at Palermo in 1789, published a star catalog (in 1803, revised in 1814) showing the positions of 7,646 stars, and discovered and named the first (and largest) asteroid, **Ceres**. Although he was able to make only three observations of Ceres, Carl **Gauss** had recently developed a mathematical technique that allowed the orbit to be calculated. Piazzi also discovered that the star 61 Cygni has a large proper motion, which led Friedrich Bessel (1784–1846) to choose it as the object of his **parallax** studies. The thousandth asteroid to be found was named after him.

Pickering, Edward Charles (1846–1919)

A prominent American astronomer who pioneered the development of the **color index** method for cataloging stars and who encouraged many young scientists in astronomy, including (unusual for the time) many women. Pickering graduated from Harvard and then taught physics for 10 years at the Massachusetts Institute of Technology, where he built the first instructional physics laboratory in the United States. Appointed at age 30 as director of the **Harvard College Observatory**, he served in this post for 42 years. He and his staff made visual photometric studies of 45,000 stars. With funds provided by Henry Draper's widow, Anna Palmer Draper, he hired a number of women, including Williamina **Fleming**, Annie Jump **Cannon**, Antonia Maury, and Henrietta **Leavitt**, and produced the ***Henry Draper Catalogue***, with objective prism spectra of hundreds of thousands of stars classified according to Cannon's "Harvard sequence." He established a station in Peru to make the southern photographs and published the first all-sky photographic map. He and Hermann Vogel (1842–1907) independently discovered the first **spectroscopic binary** stars. He also discovered a new series of spectral lines, now known as the *Pickering series,* that turned out to be due to ionized helium. Pickering encouraged amateur astronomers and was a founder of the American Association of Variable Star Observers.

Pickering, William Henry (1858–1938)

An American astronomer, younger brother of Edward **Pickering**, who helped set up the **Lowell Observatory**. He later lost favor with Percival **Lowell** when his photographs of **Mars**, among the earliest obtained, undermined belief in Lowell's canals. From **Harvard College Observatory**'s southern station in Peru, Pickering discovered **Phoebe** in 1898. His photographic atlas of the Moon was published in 1903.

Pictor (abbr. Pic; gen. Pictoris)

The Painter's Easel; a faint southern **constellation** that lies between Dorado and Carina. (See table, "Pictor: Stars Brighter than Magnitude 4.0," and star chart 8.)

Pigott, Edward (1753?–1825)

A British amateur astronomer and star-gazing colleague of John **Goodricke** who, in 1784, discovered the variability of **Eta Aquilae**, later recognized as the first known **Cepheid variable**. In 1795, he also found R Coronae Borealis, the prototype of another important class of variable (see **R Coronae Borealis star**), and **R Scuti** (the brightest **RV Tauri star**). He was the son of Nathaniel Piggott (1725?–1804), also an astronomer, who built an observatory at his estate in Glamorgan, Wales.

Pinwheel Galaxy

A name variously ascribed to the face-on spiral galaxies **M101** (most frequently), the **Triangulum Galaxy** (M33), and M99 (NGC 4254, the Virgo Pinwheel).

Pictor: Stars Brighter than Magnitude 4.0

Star	Magnitude		Spectral	Distance	Position	
	Visual	Absolute	Type	(light-yr)	R.A. (h m s)	Dec. (° ′ ″)
α	3.24	0.83	A7IV	99	06 48 11	–61 56 29
β Beta Pictoris	3.85	2.42	A3V	63	05 47 17	–51 03 59

Pioneer 10 and 11

Twin probes, launched in December 1973 and December 1974, that became the first to cross the **asteroid belt** and to fly past **Jupiter**. Pioneer 11 used a Jupiter gravity-assist to redirect it to an encounter with **Saturn**. Both spacecraft, along with **Voyager** 1 and 2, are now leaving the solar system. Of this quartet, only Pioneer 10 is heading in the opposite direction to the Sun's motion through the Galaxy. It was tracked until 2003 in an effort to learn more about the interaction between the **heliosphere** and the local **interstellar medium**. Pioneer 10's course is taking it generally toward **Aldebaran** (65 light-years away) in Taurus and a remote encounter about 2 million years from now. It was superceded as the most distant human-made object by Voyager 1 in mid-1998. The last communication from Pioneer 11 was received on November 30, 1995. With its power source exhausted, it can no longer operate any of its experiments or point its antenna toward Earth, but it continues its trek in the direction of **Aquila**. Pioneer 10's final signal may have been picked up on January 22, 2003.

Pipe Nebula

A rift in the star fields of **Ophiuchus**, shaped like a smoker's pipe, that extends from the region of **Rho Ophiuchi** and the **Lagoon Nebula**. It includes a number of dark nebulae, including Barnard 59 (at the western tip of the Pipe, just south of NGC 6293), 65, 66, 67, 77, 244, and 256.

Pisces (abbr. Psc; gen. Piscium)

The Fish; a large but faint north equatorial **constellation** and the twelfth sign of the **zodiac**. It lies just north of Aquarius and Cetus and is surrounded by Pegasus, Andromeda, Triangulum, and Aries. The vernal **equinox** lies within Pisces. Directly south of the Square of Pegasus lies another asterism known as the Circlet, or the *Western Fish,* formed by the stars Gamma, b, Theta, Iota, 19, Lambda, and Kappa Psc. The main deep sky object of note is the classic type Sc galaxy M74. Very good observing conditions are needed to see more than the bright core; under such circumstances a scope with an aperture of about 10 cm reveals a bit of the spiral arms. (See table, "Pisces: Stars Brighter than Magnitude 4.0," and star chart 1.)

Pisces Cloud (NGC 379, 380, 382-5)

A small group of galaxies that is part of the elongated **Perseus-Pisces Supercluster.** Its brightest members are the elliptical NGC 382 and the larger lenticular NGC 383, which are so close as to be interacting.

Pisces Dwarf (LGS 3)

A **dwarf irregular galaxy** that is one of the smallest members of the **Local Group** and is probably a satellite of the **Triangulum Galaxy** (M33). It was discovered in 1979 by Trinh Thuan and G. E. Martin.

Visual magnitude:	15.4
Angular size:	2′
Linear diameter:	about 2,000 light-years
Distance:	2.65 million light-years
Position:	R.A. 1h 3.8 m, Dec. +21° 53′

Piscis Austrinus (abbr. PsA; gen. Piscis Austrini)

The Southern Fish (also known as *Pisces Australis*); a small **constellation** of the Southern Hemisphere that lies west of Sculptor and south of Aquarius. By far its brightest star (the only one brighter than magnitude 4.0) is Alpha PsA, or **Fomalhaut**.

Pistol Star

The most luminous and most massive star known; it is about 10 million times brighter than the Sun and 100 times more massive. The Pistol Star lies at the center of the *Pistol Nebula,* which it created by expelling up to 10 M_{sun} of gas in giant outbursts some 4,000 to 6,000 years ago. Both star and nebula are in the **Quintuplet Cluster**, close to the center of our Galaxy, about 25,000 light-years away; yet despite this immense distance, the Pistol Star would be visible to the naked eye, as a fourth magnitude star, if it were not obscured by **dust** clouds in the plane of the Milky Way. The most powerful telescopes cannot see the Pistol Star at visible wavelengths; however, 10% of its **infrared** light reaches Earth, putting it within reach of infrared telescopes. Factoring in all the material the Pistol Star has shed in its brief lifetime, this extraordinary object may have started out

Pisces: Stars Brighter than Magnitude 4.0

	Magnitude		Spectral	Distance	Position	
Star	Visual	Absolute	Type	(light-yr)	R.A. (h m s)	Dec. (° ′ ″)
η Alpherg	3.62	−1.16	G8IIIa	294	01 31 29	+15 20 45
γ	3.70	0.68	G7III	131	23 17 10	+03 16 56
α Alrescha	3.79	0.96	A3m+A0pSiSr	139	02 02 03	+02 45 49

Pistol Star An infrared view by the Hubble Space Telescope of the Pistol Star and its surrounding nebula. *NASA/D. Figer (UCLA)*

with a mass of up to 200 M_{sun}–larger than the maximum stellar mass anticipated by contemporary theories of **stellar evolution.**

pitch angle

(1) The angle between a tangent to a **spiral arm** in a spiral galaxy and the perpendicular to the direction of the galactic center; it gives a measure of how tightly the spiral arms of a galaxy are wound. (2) The angle between the direction of a magnetic field and a particle's spiral trajectory in this field; in a nonuniform magnetic field, it changes as the ratio between the perpendicular and parallel components of the particle velocity changes. In the case of the **geomagnetic field**, pitch angle is important because it is a key factor in determining whether a charged particle will be lost to Earth's atmosphere or not.

plage

A bright spot in the solar **chromosphere**, usually found in or near an active region such as a **sunspot**; the name is French (with a short "a") and means "beach." Plages were formerly known as flocculi. Not to be confused with the photospheric **faculae**, they are normally only visible in the monochromatic light of spectral lines, such as H-alpha or Ca II. They can last for several days and mark areas of nearly vertical emerging or reconnecting magnetic field lines. A *plage corridor* is a space in a plage lacking plage intensity and coinciding with a polarity inversion.

plagioclase feldspar

A group of common **feldspar** minerals, ranging from $NaAlSi_3O_8$ to $CaAl_2Si_2O_8$, that form pale, glassy crystals and are important ingredients in some **igneous** and **metamorphic** rocks.

Plancius, Petrus (1552–1622)

A Dutch astronomer, cartographer, and theologian who, on the occasion of the first Dutch expedition to the East Indies, in 1595, asked Pieter Keyser (?–1596), the chief pilot on the *Hollandia,* to make observations to fill in the blank area around the south celestial pole on European maps of the southern sky. Keyser died in Java the following year, but his catalog of 135 stars arranged in 12 new constellations (some perhaps acquired from the myths of local peoples), prepared with the help of explorer-colleague Frederick de Houtman, was delivered to Plancius, who inscribed the new constellations on a globe he prepared in 1598. Keyser and de Houtman's constellations are **Apus** the Bird of Paradise, **Chamaeleon, Dorado** the Goldfish (or Swordfish), **Grus** the Crane, **Hydrus** the Sea Monster, **Indus** the (American) Indian, **Musca** the Fly, **Pavo** the Peacock, **Phoenix, Triangulum Australe** the Southern Triangle, **Tucana** the Toucan, and **Volans** the Flying Fish. These constellations, together with three other new southern patterns added by Plancius himself, **Camelopardalis, Columba**, and **Monoceros**, were then incorporated by Johannes **Bayer** in his sky atlas, the *Uranometria,* in 1603.

Planck

A European Space Agency spacecraft designed to search for tiny irregularities in the **cosmic microwave background** over the whole sky with unprecedented sensitivity and angular resolution. Named after the quantum theorist Max Planck (1858–1947), it is expected to provide a wealth of information relevant to several key cosmological and astrophysical issues, such as theories of the early universe and the origin of cosmic structure. Planck is scheduled for launch in 2007.

planet

An object that orbits a star and is intermediate in size between a large **asteroid** and a small **brown dwarf** (the least massive kind of star). In some cases it is difficult to decide if an object satisfies this criterion. For example, **Pluto** is not much larger than the largest known asteroid and is about the same size as the largest known **Kuiper Belt objects**. At the other end of the scale, some **extra-solar planets** have been discovered that are near or at the lower mass limit for brown dwarfs. To add to the confusion, binary systems such as Earth-Moon and Pluto-Charon are sometimes referred to as *double planets,* while asteroids are commonly known as *minor planets.* The nine bodies in the solar system conventionally referred to as planets are often further classified in a variety of ways. By composition they are divided into two groups. *Terrestrial,* or *rocky,* planets (Mercury, Venus, Earth, and Mars), are made primarily of rock and metal and have relatively high densities, slow rotation, and solid surfaces. *Jovian* planets, or *gas giants* (Jupiter, Saturn, Uranus, and Neptune), are composed mostly of hydrogen and helium and generally have low densities, rapid rotation, and deep atmospheres. Pluto falls outside this scheme. By size, the planets Mercury, Venus, Earth, Mars, and Pluto, all with diameters less than 13,000 km, are described as *small,* while Jupiter, Saturn, Uranus, and Neptune, all with diameters greater than 48,000 km, are said to be *giant.* By position relative to the Sun, there are the *inner* planets (Mercury, Venus, Earth, and Mars) and the *outer* planets (Jupiter, Saturn, Uranus, Neptune, and Pluto), with the **asteroid belt** marking the boundary between the inner and outer solar system. By position relative to Earth, Mercury and Venus are described as *inferior* planets, whereas the planets Mars through Pluto are said to be *superior.*

planet formation

Planets form from the gas and **dust** in the rotating disk of debris that surrounds a star in its infancy; however, the details of the formation process are not well understood. Dust grains chemically stick together to make larger grains, which, in turn, aggregate into even bigger clumps. At some point, the gravitational pull of the accumulating masses overtakes chemical stickiness as the prime factor in further growth. Localized mass concentrations begin to acquire eddy-like **accretion disks** that form within the large accretion disk of the stellar nebula. Boulder-size masses grow to become mountain-size chunks, which then combine to make **planetesimals,** which form the building blocks of planets. A planet the size of Earth might take shape by this means in about one million years. A **gas giant** such as Jupiter, however, might take 10 times as long, because of the need to grow a sizeable core first, containing 5 to 10 Earth masses, which has sufficient gravity to then draw in a huge atmosphere of light gases. The trouble is that the accretion disks around newborn stars probably don't survive this long. Observations suggest that protoplanetary disks last no more than about 7 million years, while studies of the environment in which stars form indicate that many disks may evaporate in much less time. Yet gas giants appear to be very common and account for virtually all the **extrasolar planets** discovered to date. A rival model of planet formation allows Jupiter-like worlds to accumulate much faster–perhaps in mere centuries instead of millions of years. This alternative scheme of gas giant formation draws upon the surprising fact that computer modeling of a protoplanetary disk shows that the disk, after just a few orbits about its parent star, suddenly fragments into clumps as big as a modest-size planet. These clumps would be so massive that they'd continue pulling in more and more material at a rapid rate. Much additional work needs to be done to test this promising hypothesis.

Planet Imager

A possible future NASA mission that would succeed the **Terrestrial Planet Finder** (TPF) and produce images of Earthlike planets. To obtain a 25 × 25 pixel image of an **extrasolar planet**, an array of five TPF-class optical **interferometers** flying in formation would be needed. Each interferometer would consist of four 8-m telescopes to collect light and to null it before passing it to a single 8-m telescope, which would relay the light to a combining spacecraft. The five interferometers would be arranged in a parabola, creating a very long baseline of 6,000 km with the combining spacecraft at the focal point of the array.

planet swallowing

The process by which planets are diverted on to collision courses with their central stars and end up being destroyed by them. Planet swallowing is likely to be fairly common during the early stages of a planetary system, when interplanetary collisions and close encounters are rife, young planets may end up being diverted onto a collision course with their parent stars. Planet swallowing may also happen much later when the central star swells to become a **red giant** and engulfs any worlds that happen to be in close orbits. The devouring of a planetary midget, such as Mercury, would hardly alter a star at all. But with the discovery of many giant **extrasolar planets** in very small orbits, astronomers are beginning to realize that, in some cases, planetary swallowing might have a marked effect on stellar evolution. It could explain the unusual properties of many giant stars and some of the features of **planetary nebulae**.

Amazingly, a large planet could continue to orbit *inside* a red giant for thousands of years, only gradually being vaporized as it spiraled down toward the core. Gravitational energy given up by the sinking planet could heat the star, causing it to puff off its outer layers as expanding shells of warm gas and dust glowing at infrared wavelengths. Although red giants normally rotate slowly, a rapidly orbiting devoured planet would boost the stellar spin rate and inject the star with a dose of heavy elements.

If a **brown dwarf** were swallowed, it would have an even greater effect on a star and might even merge with the stellar core.

Some of the effects of the ingestion of a planet or brown dwarf, such as a high spin rate and lithium abundance, could last for hundreds of thousands of years. Lithium doesn't normally survive long in a star, so its observed presence points to a recent ingestion. Many red giants have all of these telltale traits: fast spin, excess infrared emission, and unusually high lithium abundance. Between 4 and 8% of red giants show evidence of planet swallowing, in agreement with empirical estimates of how common planets are. The planet-swallowing hypothesis seems to be the best explanation for the origin of these lithium-rich giants.

It may also explain why many **planetary nebulae** are bipolar, with material ejected along a preferred axis. According to an older theory, a bipolar shape may arise when the nebula has another star as a close companion. The companion becomes enveloped in the main star's outer layers and spirals inward, dumping enough energy to eject these layers at about 20 km/s. This material stays in the plane of the companion's orbit to form a torus-shaped cloud around the main star. Later, when the star's core is exposed, its radiation drives the ejecta in the **stellar wind** at velocities up to 1,000 km/s. The wind blows in all directions, but because it is impeded by the slow-moving gas in the doughnut, it emerges perpendicularly as a bipolar outflow. The trouble with this theory is that only a few planetary nebulae seem to have both a white dwarf and a spiraled-in companion at their centers. However, this isn't a problem if the companion is a massive planet or brown dwarf because it could have completely dissipated or merged with the white dwarf.

Planet X

A hypothetical tenth planet that moves around the Sun beyond the orbit of **Pluto**. There have been many theories about and searches for such an object, but none has been found and there is currently no compelling evidence for the existence of Planet X. Many **Kuiper Belt objects**, however, do circle the Sun beyond Pluto and may reach diameters of more than 1,000 km. See also **Vulcan**.

planetary features

Planetary nomenclature is handled by the **International Astronomical Union** (IAU) according to a set procedure. When images are first obtained of the surface of a planet or moon, a theme for naming features is chosen and a few important features are labeled, usually by members of the appropriate IAU task group. Later, as higher resolution images and maps become available, suggested names for additional features go to the task group and, if successfully reviewed, are submitted to the IAU Working Group for Planetary System Nomenclature (WGPSN). Upon successful review by the members of the WGPSN, names are considered provisionally approved and can be used on maps and in publications as long as the provisional status is clearly stated. Provisional names become official only if

Some Types of Planetary Features

Feature	Description	Example
Catena	A **crater chain** or line of (usually overlapping) craters	Gipul Catena, Callisto
Chasma	A steep-walled trough or large canyon	Diana Chasma, Venus
Dorsum	A meandering, elongated elevation or ridge, also known as a *wrinkle ridge*	Antoniadi Dorsum, Mercury
Farrum	A pancake-like feature when seen from above. Applied especially to flat-topped volcanoes on Venus	Aegina Farrum, Venus
Fluctus	Terrain produced by the flow of molten material	Tung Yo Fluctus, Io
Fossa	A long, straight, narrow depression (literally a "ditch")	Isbanir Fossa, Enceladus
Labes	A landslide or structure caused by a landslide	Candor Labes, Mars
Labyrinthus	An intersecting valley complex	Noctis Labyrinthus, Mars
Lacus	A small plain (literally, "lake")	Lacus Somniorum, Moon
Macula	A dark round or irregular spot	Tyre Macula, Europa
Mare (pl. maria)	A lava-flooded impact basin on the Moon	Mare Tranquilitatis
Mensa	A small flat-topped prominence with cliff-like edges	Cydonia Mensa, Mars
Mons (pl. montes)	A mountain or volcano	Maxwell Montes, Venus
Palus	A small plain (literally "swamp")	Palus Somnii, Moon
Patera	A shallow crater with a scalloped or complex edge	Orcus Patera, Mars
Planitia	A broad low-lying plain on another planet or moon	Borealis Planitia, Mercury
Planum	A large plateau or smooth high plain	Lakshmi Planum, Venus
Regio	A large area marked by reflectivity or color distinctions from adjacent areas, or a broad geographic region	Beta Regio, Venus
Rupes	A scarp or line of cliffs produced by faulting or erosion	Rupes Recta, Moon
Scopulus	An irregular, degraded scarp	Eridana Scopulus, Mars
Sinus	An indentation at the edge of high ground (literally "bay")	Sinus Iridum, Moon
Sulcus	An intricate network of linear furrows and ridges	Kishar Sulcus, Ganymede
Terra	A large highland region	Roncevaux Terra, Iapetus
Tessera	Polygonally patterned, tile-like terrain	Fortuna Tessera, Venus
Tholus	A small dome-shaped mountain or hill	Apis Tholus, Io
Vallis (pl. valles)	A sinuous channel, often with tributaries	Valles Marineris, Mars
Vastitas	A widespread lowland plain	Vastitas Borealis, Mars

they are approved by the IAU's General Assembly, which meets triennially. Names for most planetary features include a term that describes the type of feature; for example, Ithaca Chasma (on Saturn's moon Tethys). (See table, "Some Types of Planetary Features.") See also **basin, caldera, graben, grooved terrain, impact feature, lineament, massif, mesa, multiringed basin, palimpsest, plateau, pseudocrater, ray crater, rift valley, rille, shatter cone, shield volcano**, and **volcanism**.

planetary nebula

A luminous shell of gas, often of complex structure, cast off and caused to fluoresce by an evolved star of less than about 4 M_{sun}. Planetary nebulae have nothing to do with planets other than their name, which is a historical one stemming from their occasional likeness to the greenish disk of Uranus in small telescopes. About 1,600 planetary nebulae have been discovered in the Milky Way Galaxy out of a total galactic population of about 10,000. This relatively small number reflects the short life span of planetary nebulae–no more than about 50,000 years.

The formation of a planetary nebula begins when a star has evolved to become a **Mira star**, a pulsating red giant that sheds matter in the form of a strong **stellar wind**. At this stage, the star has an inactive carbon core that is surrounded by a helium-burning shell. Instabilities gradually build up in the outer layers of the star until they break free, expanding at a speed of about 20 km/s and

planetary nebula NGC 6369, known to amateur astronomers as the Little Ghost Nebula because it appears as a small, ghostly cloud surrounding the faint, dying central star. This planetary nebula, captured by the Hubble Space Telescope, lies in the direction of the constellation Ophiuchus, at a distance between 2,000 and 5,000 light-years. *NASA/STScI*

leaving behind a hot, dead stellar core. High-energy ultraviolet radiation pouring from this exposed core, whose surface temperature is around 100,000 K, is absorbed by the nebular material and reemitted, mostly in certain unusual spectral lines, known as **forbidden lines**, the brightest of which is the green forbidden line of doubly ionized oxygen at 6007 Å. The ejected gas shell becomes visible as a glowing disk, ring, or more elaborate shape, typically about 0.5 light-year across. Infant planetary nebulae sometimes show evidence of a bipolar flow, as in the case of the **Butterfly Nebula** or **Ant Nebula**. In deep exposures, the matter ejected at the precursor Mira-variable stage can sometimes also be detected as an extended halo. Planetary nebulae are classified by appearance in the **Vorontsov-Velyaminov scheme**.

The first planetary nebulae to be discovered, and the only ones to be recorded in the **Messier Catalogue** are the **Dumbbell Nebula** (M27) in 1764, the **Ring Nebula** (M57) in 1779, the **Little Dumbbell Nebula** (M76) in 1780, and the **Owl Nebula** (M97) in 1781. A handful of planetaries have been found in **globular clusters**, including Pease 1 in M15, and a number in nearby galaxies of the **Local Group**, such as the **Magellanic Clouds** and the **Andromeda Galaxy**, and beyond. It is a stage that our Sun is destined to pass through some 5 billion years from now.

planetary system

All the celestial bodies that revolve around a star, including planets, moons, asteroids, comets, and dust.

planetesimal

A rocky and/or icy body, a few km to several tens of km in size, that was produced in the **solar nebula**. According to the *planetesimal theory,* the solar system formed from the collisional aggregation of a great many of these objects.

planetocentric and planetographic coordinates

Two systems in which the positions of features on the surface of a near-spherical body such as a planet or satellite may be recorded. Specific terms are used when referring to familiar bodies: heliocentric and heliographic coordinates for the Sun; selenocentric and selenographic for the Moon; geocentric and geographic for Earth; areocentric and areographic for Mars; zenocentric and zenographic for Jupiter; saturnicentric and saturnigraphic for Saturn; and so forth.

Planetocentric coordinates refer to the equatorial plane of the body concerned and are much used in the calculations of **celestial mechanics**. Planetocentric longitude is measured around the equator of the body from a **prime meridian** defined and adopted by international agreement. (The prime meridian may be referred to a visible feature in the case of a solid-surfaced body such as Mars, but in the case of a gaseous planet such as Jupiter it is a purely hypothetical concept.) Planetocentric latitude is measured in an arc above or below the equator of the object in the same way that latitude is measured on Earth.

Planetographic coordinates are used for observations of the surface features of those planets whose figures are not truly spherical, but oblate. They refer to the mean surface of the planet, and the coordinates are actually determined by observation. They can readily be converted to planetocentric coordinates if required. As the oblate planets are symmetrical about their axes of rotation, there is little difference in practice between planetocentric and planetographic longitudes. However, the differences between planetocentric and planetographic latitudes are quite significant for very oblate bodies such as Jupiter and Saturn.

planisphere

(1) A circular star map drawn for a given latitude, having a rotary overlay that shows which constellations are visible at any time and date in the year. (2) A map in which the **celestial sphere** is projected onto a plane surface, especially in a polar projection, and to which a grid system is added to help locate celestial objects.

plano-concave lens

A negative (diverging) spherical **lens** with one convex surface and one flat (plano) surface.

plano-convex lens

A positive (converging) spherical **lens** with one convex surface and one flat (plano) surface.

Plaskett, John Stanley (1865–1941)

A Canadian mechanical engineer and astronomer who, in 1918, became director of the newly established **Dominion Astrophysical Observatory** (DAO) in British Columbia, for which he had organized the design, construction, and installation of a new 72-in. (1.8-m) reflecting telescope. Plaskett's field of research was **spectroscopy**, in particular the measurement of radial velocities of celestial bodies, i.e., their velocities along the line of sight, from the shift in their spectral lines. Using the 72-in. (1.8-m) reflector and a highly sensitive spectrograph which he had also designed, many spectroscopic binary systems were discovered. In 1922, Plaskett identified an extremely massive star as a binary, now known as **Plaskett's Star**. In 1927 he provided confirmatory evidence for the theory of galactic rotation put forward by Bertil **Lindblad** and Jan **Oort**. By 1928 Plaskett, in collaboration with Joseph **Pearce**, had obtained evidence for the hypothesis formulated by Arthur **Eddington** in 1926 that interstellar matter was widely distributed throughout the Galaxy. Their discovery that interstellar **absorption lines**, mainly of calcium, took part in the galactic rotation, showed that interstellar matter was not confined to separate star clusters. Although this result was first announced by Otto **Struve** in 1929, Plaskett felt he had priority and was convinced that Struve had obtained his results from him. He retired from the DAO in 1935.

Plaskett's Star (V640 Mon)

A **spectroscopic binary**, of visual magnitude 6.05, lying about 5,000 light-years away in **Monoceros**. It is named after John **Plaskett** who, in 1922, discovered the system's great claim to fame: it is the most massive binary star system known. According to a recent estimate, the two components, both blue **supergiants**, are respectively 51 and 43 times more massive than the Sun. They orbit around their common center of gravity with a period of 14.4 days. Plaskett's Star is found just east of 13 Monocerotis and is very probably a member of NGC 2244, which in turn is part of the **Rosette Nebula**.

plasma

A low-density gas in which the some of the individual atoms or molecules are ionized (and therefore charged), even though the total number of positive and negative charges is equal, maintaining an overall electrical neutrality; considered to be a fourth state of matter. Strictly speaking, almost all gas in space is a plasma, though only a tiny fraction of the atoms are ionized when the temperature is below about 1,000 K. The very low densities in space allow the **electrons** to travel without much obstruction, so paradoxically space is an almost perfect electrical conductor. Although charges can move around freely, averaged over even small volumes (say a million km across) cosmic plasmas are always neutral. Plasmas in space are permeated by magnetic fields. A good way to think about cosmic magnetic fields is in terms of field lines. These behave like rubber bands embedded in the plasma, so that as the plasma flows the field lines are pulled and stretched along with it. When they are stretched enough they can pull back on the plasma. Individual electrons and **ions** in the plasma feel a magnetic force, which makes them travel in a helical path around the field lines so that they can travel long distances only in the direction along the field. This binds the plasma together so that it behaves like a continuous medium, even when the individual electrons and ions almost never collide.

plasma frequency

The maximum frequency of internal oscillation of a **plasma**. The plasma frequency is proportional to the square root of the electron density.

plasmasphere

A donut-shaped region of relatively cool (2,000 K), dense (tens to thousands of particles per cm^3) **plasma** within Earth's **magnetosphere** that extends out to a distance of about 4 Earth radii (25,000 km) and is bounded by the inner **Van Allen Belt**. The outer limit of the plasmasphere is called the *plasmapause*.

plate tectonics

The now widely accepted theory, formulated in the late 1960s, that Earth's crust and upper mantle (**lithosphere**) consist of moving pieces (plates) lying above a weaker semiplastic **asthenosphere**. Plate tectonics provides a convincing explanation for such phenomena as continental drift, earthquake activity, and the formation of mountain ranges and of chains of volcanoes. It does not appear to have played a significant role in the development of any other planet in the solar system, although the Tharsis Region on **Mars** shows evidence of embryonic but stalled plate activity.

Platonic year

The time required for one complete cycle of the **precession** of the equinoxes–about 25,800 years. Also known as a *great year*.

Pleiades (M45, NGC 1432)

A nearby, fairly loose **open cluster** of about 500 stars in **Taurus**. It is also known as the Seven Sisters, although most people can see only six stars, all of them bright blue-white B or Be stars: **Alcyone**, Atlas, Electra, **Maia**, **Merope**, and Taygeta (**Pleione** is the seventh brightest). The Pleiades are currently moving through a great dusty cloud of interstellar matter, the **dust** grains in the nebula reflecting the light of the blue stars. This cloud is *not* a remainder of the nebula from which the cluster formed, as can be seen from the fact that the nebula and cluster have different **radial velocities**, crossing each other with a relative velocity of 11 km/s. Though not visible to the eye, long photographic exposures show this reflection nebula enmeshing the whole crowd. The nebula is particularly bright around Merope (see **Merope Nebula**).

Pleiades Wispy tendrils stretch out from a dark interstellar cloud being destroyed by the passage of one of the brightest stars in the Pleiades star cluster. This Hubble Space Telescope image shows dust being drawn off the Merope Nebula by the star after which it is named, just out of picture to the upper right. *Hubble/STScl*

Visual magnitude:	1.6
Apparent size:	110′
Distance:	380 light-years
Age:	about 100 million years
Position:	R.A. 3h 47m, Dec. +24° 7′

Pleione (28 Tauri)

The seventh brightest star in the **Pleiades** star cluster; in mythology, Pleione was the mother, by Atlas, of the seven daughters known as the Pleiades. In contrast to most of its bright neighbors, which are subgiant or giant **A stars,** Pleione (like Sterope) has not yet left the main sequence. Along with **Gamma Cassiopeiae,** it is one of the sky's best known **Be stars,** spinning rapidly (with an angular velocity at the equator of about 330 km/s) and occasionally casting off a ring of hot gas to become a *shell star.* Since 1888, Pleione has ejected three gaseous shells. On each such occasion, it has brightened at first and then faded by several tenths of a magnitude below its normal level, to about magnitude 5.5, because of absorption of starlight by material in the ring. These brightness fluctuations have led to Pleione being given a variable star designation: BU Tauri. A shell ejection in 1936 was followed by a light decrease between 1938 and 1952, while another ejection in 1972 led to a dimming that lasted until 1987. Phase changes in the shell spectrum every 17 or 34 years have led to the suggestion that Pleione may have a binary companion in an eccentric orbit with a roughly 34-hour period.

Visual magnitude:	5.05
Absolute magnitude:	–0.33
Spectral type:	B7V
Distance:	387 light-years

plume

(1) A rising column of gas over a maintained source of heat. (2) A buoyant mass of hot, partially molten mantle material that rises to the base of the **lithosphere**.

plutino

A **Kuiper Belt object** whose motion, like that of **Pluto,** is controlled by the 3:2 resonance with **Neptune**. Plutinos are found in the inner part of the **Kuiper Belt**.

Pluto

See article, pages 397–398.

Pluto-Kuiper Belt mission

Also known as New Horizons, the first probe to the outermost planet, **Pluto**, and its only moon, Charon. By mid-2002, the mission, repeatedly threatened by budget cutbacks, had successfully completed its first major product review. New Horizons is working toward a 2006 launch, arrival at Pluto and Charon in 2015, and exploration of various objects in the **Kuiper Belt** up to 2026. Mission planners are anxious to intercept the ninth planet while it is still in the near-perihelion part of its orbit; at its greater distances from the Sun, Pluto's atmosphere may completely freeze and any surface activity, such as ice geysers, become less frequent.

Poincaré, (Jules) Henri (1854–1912)

A French mathematician (professor of mathematics at the University of Paris from 1881 to 1912) who revolutionized **celestial mechanics**, inaugurating a rigorous treatment and initiating studies of the stability of the solar system. One of the greatest mathematicians of all time, Poincaré made significant advances in function theory, algebraic functions, algebraic geometry, algebra, number theory, algebraic topology, and above all, differential equations. He determined the shape of a

(continued on page 398)

Pluto

The smallest planet (with only one-fifth the mass of Earth's Moon and two-thirds of its diameter) and, usually, the most distant planet from the Sun. It was discovered by Clyde **Tombaugh** in 1930 and is named after the Roman god of the underworld. Pluto's elongated orbit carries it closer to the Sun than **Neptune** for a decade or so on either side of perihelion, as happened at the end of the twentieth century (January 21, 1979, to February 11, 1999, with perihelion on September 5, 1989). Pluto's orbit is also unusual in being highly inclined and in having a 3:2 resonance with that of Neptune, so that Pluto completes three orbits for every two of Neptune. There is no danger, however, that these two worlds will collide. As Pluto approaches perihelion, it also reaches its maximum distance from the ecliptic due to its 17° orbital tilt, so that it is always far above or below the plane of Neptune's orbit. In fact, Pluto and Neptune never come closer than 18 AU apart. Unlike most planets, but similar to Uranus, Pluto spins on its side–that is, with its poles almost in its orbital plane. This extreme axial tilt, together with the high orbital tilt, makes for pronounced seasonal climatic changes (though, of course, it is always very cold.) The old idea that Pluto is an escaped moon of Neptune has given way to the modern view that it is a displaced **Kuiper Belt object**.

Pluto has one satellite, **Charon**, which is so large in comparison with the primary–more than half its diameter–that the Pluto-Charon system is often referred to as a **double planet**. The average separation

Pluto Images of Pluto taken by the Hubble Space Telescope in late June and early July 1994, which show the first hint of surface detail on the planet ever seen. Opposite hemispheres are shown to the left and right. The variations in brightness may be due to topographic features and/or surface composition, frost layers, and interactions with Pluto's nitrogen-methane atmosphere. *NASA/STScI*

distance is 19,640 km (roughly eight Pluto diameters). Uniquely, among planets in the solar system, Pluto has a rotation period exactly equal to that of its moon. Although it is common for a satellite to travel in a synchronous orbit with its planet, Pluto is the only planet to rotate synchronously with the orbit of its satellite. Thus tidally locked, Pluto and Charon keep the same faces toward each other as they travel through space. From 1985 to 1990, Earth was aligned with the orbit of Charon around Pluto enabling daily observations of the mutual eclipses of the pair. The data collected showed that Pluto has a highly reflective south polar cap, a dimmer north polar cap, and both bright and dark features in the equatorial region. Its geometric **albedo** ranges from 0.49 to 0.66, compared with Charon's (darker) 0.36 to 0.39. This difference, together with Pluto's higher density (between 1.8 and 2.1 g/cm^3 compared with Charon's 1.2 to 1.3 g/cm^3), suggests that the two bodies formed separately rather than as the result of a single collision, though this is by no means certain. Pluto's higher density indicates it is made of 50 to 75% rock mixed with ice, whereas Charon seems to be mostly ice with very little rock.

Pluto has a very thin (surface pressure 100,000 times less than that on Earth) atmosphere composed of 98% nitrogen and small amounts of methane and carbon monoxide, much like that of Neptune's moon Triton. This atmosphere gradually freezes and collects on the surface as the planet moves away from the Sun. Yet, interestingly, observations have shown that between 1989, when Pluto was at perihelion, and 2002, the atmospheric pressure increased threefold. The explanation probably has to do with the fact that materials take time to warm up and cool off, which is why the hottest part of the day on Earth, for example, is usually around 2 or 3 P.M. rather than local noon, when sunlight is the most intense. The fact that Pluto's atmosphere is still building up rather than freezing out, as many scientists expected, is good news from the standpoint of what a flyby probe might learn of the ninth world. Pluto is the only planet unvisited by spacecraft. However, this may change if NASA's long-anticipated **Pluto-Kuiper Belt mission** is developed and launched in the near future.

Diameter:	2,274 km (0.178 of Earth)
Mass:	0.0021 M_{earth}
Mean density:	1.8 to 2.1 g/cm^3
Axial tilt:	122.5°
Escape velocity:	1.22 km/s
Albedo range:	0.49 to 0.66
Mean surface temperature:	37 K
Mean distance from the Sun:	39.53 AU (5,913,520,000 km)
Perihelion:	29.65 AU (4,434,990,000 km)
Aphelion:	48.83 AU (7,304,330,000 km)
Rotation period:	6.387 days
Orbit	
Period:	248.54 years
Eccentricity:	0.248
Inclination:	17

Poincaré, (Jules) Henri (1854–1912)

(continued from page 62)

rotating fluid subject only to gravity. He also wrote extensively on the philosophy of science, worked in mathematical physics, and developed many of the ideas and equations of the special theory of relativity independently of Albert Einstein (1879–1955) and Hendrik Lorentz (1853–1928).

point source

A source whose angular extent is so small that it cannot be measured with existing equipment. Compare with **extended source**.

polar distance

The angular distance of an object from a celestial pole. It is equal to 90° minus the object's **declination** and is sometimes used as an alternative coordinate to declination since it avoids negative values.

polar ring galaxy (PRG)

A galaxy with a ring of gas, dust, and stars nearly at right angles to the disk of the galaxy. Polar rings occur in almost 5% of **lenticular galaxies** and in a few **spiral galaxies**. Understanding their kinematics and evolution is important especially because it can help determine the three-dimensional structure of **dark matter** halos. There are two main types of PRGs: bulge-dominated lenticulars and disk-dominated later-type galaxies. Lenticulars have narrow rings, while disk-dominated galaxies have wide extended rings. The ring may be a tenuous disk of gas and stars where the central region depleted its quantity of gas and dust into the disk of the host galaxy. Estimated ages of ring/disk systems are on the order of a few billion years. Long exposures capture a small glow coming from the center from stars that formed in between the ring and the disk of the target galaxy. Three main theories have been put forward to explain PRG formation, all involving an influx of mass from outside. The first theory calls for tidal stripping from a host galaxy and can be tested simply by looking at galaxies within a distance that would be reasonable given the age of the given PRG. The other two theories are harder to confirm. The second theory suggests that mass can be acquired from a gas-rich environment or intercluster medium (ICM). However, if the gas has since been removed from the system, it's a difficult mechanism to confirm. The third theory proposes a formation process that involves the total absorption of another galaxy–**galaxy cannibalism,** for which there are observational tests. The topic of dark matter halos comes into play in connection with the rotation curves of PRGs. Halos are used to explain the extended kinematics, such as rotation curves, out to tens of thousands of light-years. Early halo models suggested a spherical structure, but more recent observations reveal better fits to the rotational curves using an elliptical representation–a dark matter halo resembling an E5 to E7 ellipsoid seeming to offer the best fit. One of the nearest PRGs is the **Helix Galaxy** (NGC 2685).

polar wandering

The prehistoric wandering of Earth's magnetic poles, revealed by the different magnetic alignment of particles in rocks formed at different times.

polarimeter

A device that measures the **polarization** of any form of electromagnetic radiation, but particularly light.

Polaris (Alpha Ursae Minoris, α UMi)

The current **pole star,** often referred to as the Pole Star, the brightest star in **Ursa Minor**, and the brightest **Cepheid variable** in the sky. Polaris is actually about 1° away from the north celestial pole and travels each day in a tiny circle around it. Because of a 26,000-year wobble in Earth's axis, the pole will move slightly closer to Polaris until about the year 2100, then it will start to drift away. Polaris is an aging yellow supergiant **F star**, passing through a phase of instability in which it pulsates over a period of about 3.97 days,

polar ring galaxy NGC 4650A, a polar ring galaxy located about 130 million light-years from Earth. The vertical ring, seen edge-on in this Hubble Space Telescope image, is thought to consist of material stripped from a smaller galaxy when it collided with the large elliptical galaxy in the center. *NASA/STScI*

varying in brightness by 0.15 magnitude. Polaris is notable because the pulsations have almost come to a halt. Just as a plucked string sounds a fundamental note that gives its pitch, it also vibrates in overtones at higher frequencies. Comparison with other Cepheids shows that Polaris is pulsating not with its natural fundamental period, but in its first overtone. In fact, it may be in the process of moving to its fundamental period of 5.7 days to become a conventional Cepheid. It has two companions: a type F3 main sequence star in a 2,000-AU-wide orbit with a period of many thousands of years and a close partner at an average distance of just 5 AU.

Visual magnitude:	1.97
Absolute magnitude:	–3.64
Spectral type:	F7Ib-II
Luminosity:	2,200 L_{sun}
Distance:	430 light-years

polarization

A condition in which electromagnetic waves are constrained to vibrate in a certain plane or planes. Electromagnetic waves consist of vibrating electric and magnetic fields, with the direction of vibration perpendicular to the direction of motion of the wave. If the direction of vibration remains steady with time, the wave is said to be 100% *linearly polarized* in that direction. If the direction of vibration rotates at the same frequency as the wave, the wave is said to be 100% *circularly polarized.* Most naturally occurring electromagnetic waves have a direction of vibration that jiggles around at random: these are said to be *unpolarized.* Intermediate states, where there is some jiggling around an average direction, are said to be *partially polarized;* the amount of order is specified by the *degree of polarization,* which ranges from 0 to 100%. Plane polarization is usually caused by scattering, and circular polarization by strong magnetic fields.

pole

(1) Either of two such points on Earth's surface; the North Pole or the South Pole. (2) Either of two similar points in the heavens about which the stars seem to revolve.

pole star

Although **Polaris**, the North Star, sits within half a degree of the north celestial pole, this was not always so. Earth's rotational axis undergoes a slow, 26,000-year wobble around the perpendicular to its orbit around the Sun, as a result of which the position of the sky's rotational pole, around which all the stars seem to go, constantly changes. Around the time of the Greek poet Homer, **Kochab** was the North Pole star. Among the best ever, however, was **Thuban**, which was almost exactly at the pole in 2700 B.C. It remained better than Kochab up to around 1900 B.C., and was therefore the pole star during the time of the ancient Egyptians. Other bright stars, including **Alderamin**, have served as pole star, and will do so again in the remote future. The star currently closest to the south celestial pole is Sigma Octantis, which is barely visible to the naked eye and lies 1° 3′ from the pole (though it was as close as 45′ just a century ago).

Pollux (Beta Geminorum, β Gem)

The brightest star in **Gemini**, despite its Beta designation, and the seventeenth brightest in the sky. Pollux is a giant orange **K star** that makes an interesting color contrast with its white "twin," Castor. Evidence has been found for a hot, outer, magnetically supported **corona** around Pollux, and the star is also known to be an X-ray emitter.

Visual magnitude:	1.16
Absolute magnitude:	1.08
Spectral type:	K0IIIb
Surface temperature:	4,500 K
Luminosity:	32 L_{sun}
Radius:	10 R_{sun}
Distance:	34 light-years

polycyclic aromatic hydrocarbons (PAHs)

A group of hydrocarbons made up from multiple interconnected 6-carbon (benzene) rings, like pieces of chicken wire, the simplest of which is napthalene (the chemical in mothballs). More complex PAHs may contain elements other than carbon and hydrogen, including nitrogen and sulfur. PAHs can form in a variety of ways, as a result of both biological and nonbiological processes. They occur in **carbonaceous chondrites**, and in 1998, scientists from the NASA Ames Research Center and the Astrophysikalisches Institut in Germany announced they had found the spectroscopic signatures of PAHs in interstellar space. Ames researchers simulated conditions in space using extreme cold, a near vacuum, and artificial starlight. Then they measured the spectra of large, carbon-bearing molecules in the ultraviolet and visible bands of the spectrum and compared their results to astronomical data from various observatories. The German scientists carried out a similar study using other forms of carbon molecules. Out of the joint work came the conclusion that previously unidentified absorption features in the spectra of the interstellar medium, known as **diffuse interstellar bands**, are due to PAHs in space that originated in the atmospheres of **red giants**. This adds to the growing impression that complex carbon chemistry is a universal phenomenon.

Pons, Jean Louis (1761–1831)

A French astronomer whose intimate knowledge of the night sky helped him to discover or codiscover 37 comets–still a record. At the age of 28 he became a porter and doorkeeper at the Marseille Observatory. Noting his interest in astronomy, the directors of the observatory gave him instruction, and he turned out to be good at practical observation. Pons was made assistant astronomer in 1813 and assistant director in 1818. In 1819 he became director of a new observatory at Lucca, in northern Italy, before moving to the Florence Observatory in 1822. Pons would be better known if the comet with the shortest period of all, which he discovered, had been named after him. But, in fact, it is called **Encke's Comet** after Johann **Encke** who first calculated its orbit and predicted its return. Actually, Encke wanted the comet to be named after Pons, but glory was thrust upon him anyway.

population

A way of classifying stars on the basis of several properties, including location in the host galaxy, type of orbit, and heavy element content or **metallicity**; it was introduced by Walter **Baade** in 1943. In Baade's scheme there are two main population types: **Population I** and **Population II**. A more refined system, based on modern knowledge, sees our Galaxy and others made of four stellar populations: *thin disk, thick disk, stellar halo,* and *bulge.* The thin disk population is confined to within about 1,000 light-years of the **galactic plane** and includes the Sun and 96% of its neighbors. Thin disk stars are metal rich, vary in age from newborn to 10 billion years, and revolve around the Galaxy fast in fairly circular orbits. The thick disk population, which probably includes **Arcturus** and 4% of the Sun's neighbors, is generally older than the thin disk and extends several thousand light-years above the galactic plane. Its members move in elliptical orbits and have metallicities around one-quarter that of the Sun. The stellar halo population is a roughly spherical system of very metal-poor stars (1 to 10% of solar metallicity), mostly **subdwarfs**, that move in highly elliptical orbits that may reach up to 100,000 light-years from the galactic center at apogalacticon and as little as a few thousand light-years at perigalacticon. The nearest example of a halo star to the Sun is **Kapteyn's Star**. The bulge population occupies the central few thousand light-years of the Galaxy and consists of old, metal-rich stars. Since no such objects are in the solar neighborhood, this is the least explored stellar population in the Milky Way.

Population I

Stars and star clusters that lie in the **galactic disk** and follow roughly circular orbits around the center. They are younger than **Population II** objects, have a relatively high heavy element content, and have probably been formed continuously throughout the lifetime of the disk. *Extreme Population I* are found in spiral arms and consist of young objects, such as **T Tauri stars**, **O stars**, **B stars**, and stars newly arrived on the **zero-age main sequence**.

Population II

Stars and **globular clusters** that occur in the **galactic halo** and the **galactic bulge**. These objects follow highly elliptical orbits around the galactic center and, with the exception of bulge components, have a very low heavy element content, indicating that they formed early in their galaxy's history, before the interstellar medium became enriched with elements heavier than helium by the debris from supernovae and red giants. See also **Population I**.

pore

A short-lived dark spot in the Sun's **photosphere** resembling a small **sunspot** without a penumbra. Pores last less than an hour and range up to 2,500 km in diameter. They occasionally form where several granulation channels meet and sometimes precede the development of sunspots.

Portia

The seventh moon of **Uranus**, also known as Uranus XII; it was discovered in 1986 by **Voyager** 2. Portia has a diameter of 110 km and moves in a nearly circular orbit 66,970 km from (the center of) the planet.

position angle (P.A.)

The direction of an imaginary arrow in the sky, measured in degrees from north through east. Applied to a **binary system**, it is the angle between the primary and secondary components. Applied to a galaxy, it measures the object's tilt–that is, the angle between the galaxy's long axis and a line from its center headed north.

potentially hazardous asteroid

A **near-Earth asteroid** (NEA) whose orbit can bring it so close to Earth that there is a risk (typically very small) of a future collision. Past asteroid and comet impacts have had dramatic effects both on a local scale (as in the case of the **Tunguska event**) and globally (for example, the mass extinction at the **K-T boundary**). The growing realization that Earth will inevitably be hit hard again at some point, with potentially catastrophic results for human and other life, has led to the setting up of several programs to detect and monitor NEAs (or, more generally,

NEOs–near-Earth objects–which also include short-period comets) and near-Earth threatening objects. These include the **Spacewatch Program, LINEAR**, the Spaceguard Foundation, and the Jet Propulsion Laboratory's NEO Program. Among recent close encounters was one on June 14, 2002, when an asteroid the size of a football pitch, cataloged as 2002 MN, passed Earth at a distance of around 120,000 km–less than a third the distance to the Moon–traveling at over 10 km/s (23,000 miles per hour). This was bettered by 1994XM1, which came to within 105,000 km of Earth in December 1994.

Poynting-Robertson effect

An effect of **radiation pressure** on a small particle orbiting a star that causes it to spiral slowly into the star. The radiation falls preferentially on the leading edge of the orbiting particle and acts as a drag force. For example, a dust grain one micron wide located at the position of Earth would spiral into the Sun in a period of about 3,000 years. Also called *Poynting-Robertson drag*, the effect is named after the English physicist John Poynting (1852–1914) and the American mathematician and cosmologist Howard Robertson (1903–1961).

p-process

A type of nuclear reaction thought to occur in the early stages of a **supernova** explosion that results in the formation of certain neutron-deficient isotopes of elements from krypton to zirconium. It involves the capture of protons by nuclei previously formed by the **r-process** and the **s-process**.

Praesepe (M 44, NGC 2632)

A large, bright, relatively nearby **open cluster** in **Cancer**, easily visible to the naked eye; also known as the *Beehive cluster*, its Latin name means "manger." **Ptolemy** includes it as one of seven "nebulae" in his ***Almagest*** and Galileo first resolved it into stars. Large telescopes have revealed that more than 200 of the 350 stars in the region of sky covered by the cluster (about 1.5° across) are actual members of Praesepe; they include the eclipsing binary TX Cancri, the metal line star Epsilon Cancri, several **Delta Scuti stars** of magnitudes 7 to 8 in an early post-main-sequence state, and one peculiar blue star. Interestingly, the age and the direction of **proper motion** of M44 are very similar to those of the **Hyades**, another famous naked-eye cluster. Probably these two clusters, though now separated by hundreds of light-years, have a common origin in some great diffuse gaseous nebula which existed 400 million years ago. This would also explain the similarity of the stellar populations–both clusters containing red giants (Praesepe claiming at least five of them) and some white dwarfs.

Visual magnitude:	3.7
Angular diameter:	95′
Distance:	577 light-years
Position:	R.A. 8h 40.1m, Dec. +19° 59′

preceding

In astronomy, the side of an object that comes into view first, or that leads in motion across the sky or across the face of a rotating body. Examples include the *preceding limb* of a planet and the *preceding spot*, or *p-spot*, which is the more western of a pair of **sunspots**.

precession

A slow, periodic conical motion of the rotation axis of a spinning body, most familiar in the wobbling of a toy top or gyroscope. Earth's axis precesses, once around every 25,800 years, as a result of it not being perpendicular to the ecliptic (the plane of Earth's orbit around the Sun). Being tipped over at about 23.5°, it is affected by gravitational **perturbation**s from other bodies in the solar system, including the Sun, Moon, and planets. These bodies pull harder on the part of Earth's equatorial bulge nearest them than on the part farthest away, resulting in a torque that is the underlying cause of precession. The component of precession caused by the Sun and the Moon is called *lunisolar precession;* that caused by the action of the planets is called *planetary precession.* The sum of lunisolar and planetary precession, known as *general precession,* results in a westward movement of the equinoxes (the nodes of the ecliptic) known as the *precession of the equinoxes.* This movement amounts to an average of about 50.3″ per year and is called the *precession constant.* Precession means that a number of different stars in the circumpolar regions of the celestial sphere serve as **pole stars** on a cyclical basis.

pre-main-sequence object

An infant star that has exhausted its core supply of **deuterium** and is undergoing further gravitational contraction, along the **Hayashi track** or, at a slightly later stage, along the **Henyey track**, before arriving on the **main sequence**. For a star of about the mass of the Sun, this phase lasts several million years. A familiar type of pre-main-sequence object is the **T Tauri star**.

pressure broadening

The broadening of **spectral lines**, particularly in **white dwarfs**, caused by the pressure of the stellar atmosphere, which in turn is caused by the high surface gravity of the star.

pressure ionization

A state found in **white dwarfs** and other degenerate matter in which the atoms are packed so tightly that the **electron** orbits encroach on each other to the point where an electron can no longer be regarded as belonging to any particular nucleus and must be considered free.

Príbram Meteorite

The first meteorite to have its fall to Earth recorded by a camera network, enabling its inbound trajectory and orbit to be determined and the meteorite to be recovered. Cameras operated by the Ondrejov Observatory in the Czech Republic recorded a brilliant fireball on April 7, 1959. Subsequently, 19 fragments of an H6 **ordinary chondrite**, with a total mass of 9.5 kg and a largest individual mass of 4.3 kg, were found outside the town of Príbram, near Prague. The calculated aphelion of the object was in the outer part of the main asteroid belt.

primary

(1) The most massive body in a system of celestial bodies that revolve about a local common center of gravitational forces. (2) The star that appears most luminous in a multiple or binary system.

primary mirror

The first mirror encountered by incoming light in a telescope system.

prime focus

The point in a telescope at which the reflected rays from the **primary mirror** or the transmitted rays from the objective lens come to a focus.

prime meridian

The meridian adopted as the zero of longitude measurement on the surface of a planet. In the case of Earth, the prime meridian adopted by international agreement in 1894 is that of the Airy Transit Circle at the Old **Royal Greenwich Observatory**.

Probing Lensing Anomalies Network (PLANET)

An international collaboration of astronomers, with access to a variety of telescopes in the Southern Hemisphere for detailed monitoring of **microlensing** events. Whereas other projects, such as **Optical Gravitational Lensing Experiment** (OGLE), sample microlensing events (which typically last 15 to 90 days) every 10 to 100 days, PLANET is set up to collect data on a daily or hourly basis. One of the fruits of this approach was to observe atmospheric details in a giant **K star** in the **galactic bulge** that was the source of a microlensing event in 2000.

Procyon (Alpha Canis Minoris, α CMi)

The brightest star in **Canis Minor**, the eighth brightest star in the sky, and the easternmost star of the *Winter Triangle*. Its Greek name means "before the dog," since in northern latitudes it rises before **Sirius**, the Dog Star, and its constellation, Canis Major. Procyon is a nearby yellow-white subgiant **F star**, about to shut down the fusion of hydrogen to helium in its core. Its tiny companion, Procyon B, is a **white dwarf**, first detected visually in 1896 by John Schaeberle with the 36-inch (91-cm) refractor at Lick Observatory, though its existence and an orbital period of 40 years had been deduced in 1861 by Arthur von Auswers (1838–1915) based on wobbles he measured in the brighter star. Procyon B is about half the size of Earth, has a surface temperature of 8,700 K, and is separated from Procyon A by an average of only 14.9 AU–roughly the distance between Uranus and the Sun. See also **stars, nearest**.

DATA FOR PROCYON A	
Visual magnitude:	0.40
Absolute magnitude:	2.68
Surface temperature:	6,500 K
Spectral type:	F5IV
Luminosity:	7 L_{sun}
Distance:	11.41 light-years

prograde

See **direct motion**.

Prometheus

(1) The third moon of **Saturn**, also known as Saturn XVI; it and **Pandora** act as **shepherd moons** for the inner edge of Saturn's F-ring. Prometheus is extremely elongated, measuring about 145 × 85 × 61 km, and has a number of ridges and valleys and several craters about 20 km in diameter. It was discovered in 1980 by S. Collins and others from photographs taken by Voyager 1 during its encounter with Saturn. (2) A volcano on **Io** known as the Jovian moon's "Old Faithful." Prometheus has been active during every observation of it since it was first seen by Voyager 1, although its plume, which is 80 km tall, has migrated about 85 km to the west.

prominence

A strand of relatively cool (about 10,000 K) gas in the solar **corona** seen as a bright structure above the Sun's limb against the blackness of space. When the same feature is seen dark in projection against the solar disk it is called a **filament**. Prominences come in two broad classes: *active* and *quiescent* (including quiet region

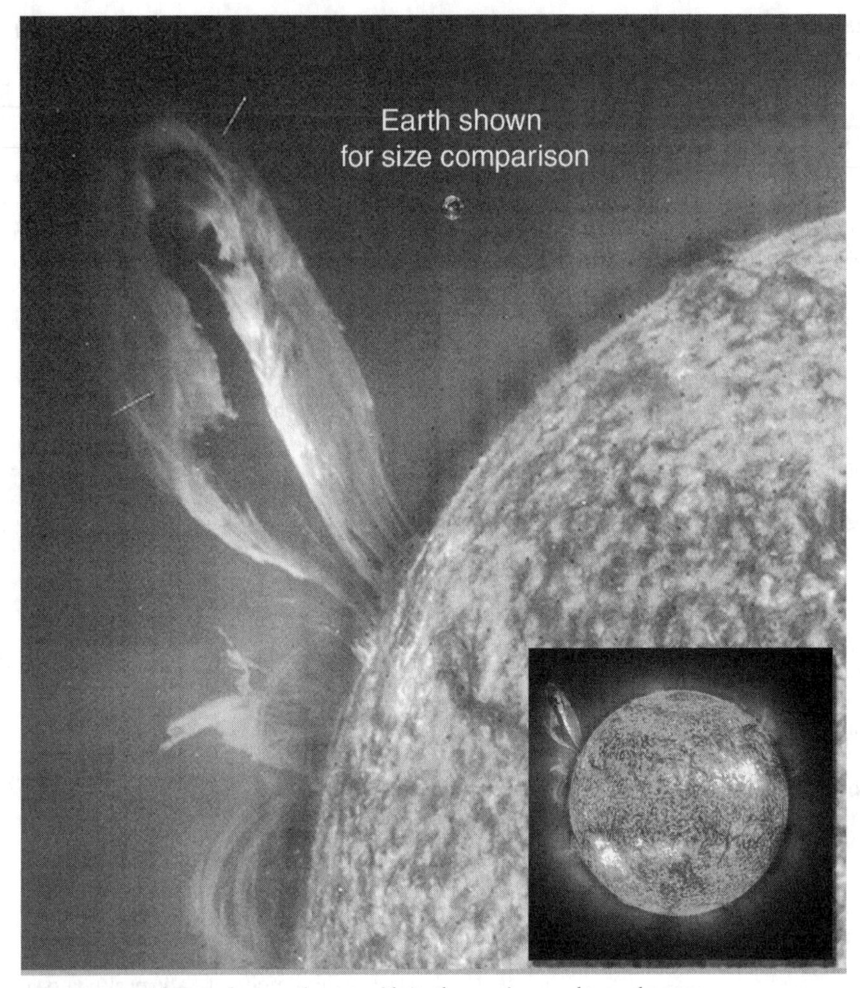

prominence A large solar prominence with Earth superimposed to scale. *NASA*

filaments and active region filaments). An active prominence changes in appearance over a few minutes and involves high-speed material motion of up to 2,000 km/s. Active prominences are associated with **sunspots** and **flares**, can last up to several hours, and take the form of **loop prominences**, sprays, and **surges**. Cooler material flowing back from prominences may be seen as **coronal rain**.

proper motion

The apparent angular rate of motion of a star or other object across the line of sight on the **celestial sphere**. If the star's distance is also known, its **tangential velocity** can be calculated. **Barnard's Star** has the largest proper motion of any star, 10.3″ per year.

Prospero

The nineteenth moon of **Uranus**, also known as Uranus XVIII; together with **Setebos** and **Stephano**, it was discovered in 1999 by Brett Gladman, Matthew Holman, and J. J. Kavelaars using the Canada-France-Hawaii Telescope. Prospero has a diameter of about 20 km and is an **irregular moon** with a highly inclined, retrograde orbit at a mean distance of some 16.1 million km from the center of the planet.

Proteus

The sixth moon of **Neptune**, also known as Neptune VIII; it was discovered in 1989 by **Voyager** 2. Although larger than **Nereid**, it was not discovered earlier because it is very dark and so close to Neptune that it is difficult to see in the glare of the much brighter planet. Proteus measures 436 × 416 × 402 km and is probably about as big as an irregular body can be before gravity pulls it into a more spherical shape. It orbits at a mean distance of 117,640 km from the planet.

proton

A subatomic particle with a positive charge found in the nuclei of all atoms. The number of protons determines the element; for example, all atoms with one proton are hydrogen, all atoms with two protons are helium, and so on. Protons belong to the **hadron** family.

proton flare

A solar **flare** producing significant fluxes of greater than 10 MeV **protons** in the vicinity of Earth.

proton-proton chain

The most important energy-producing nuclear process taking place inside relatively low mass stars, such as the Sun. It involves a chain of **fusion** reactions in which four **protons** combine to form a nucleus of **helium**; the process was first described in 1938 by Hans **Bethe** and the American physicist Charles Critchfield (1910–1944). The proton-proton chain begins with two protons (^{1}H) colliding to form a nucleus of **deuterium** (^{2}H). Any time such fusion takes place, a tiny amount of mass is turned into a comparatively huge amount of energy. When the two protons fuse to make the deuterium, one of the protons turns into a **neutron** and releases energy in the form of a **positron** (e^+) and a **neutrino** (ν); the positron annihilates with an **electron**, creating two **gamma rays** (γ). The deuterium then combines with another proton, releasing a gamma ray and giving a nucleus of helium-3 (^{3}He). Finally, the helium-3 nucleus fuses with another helium-3 to form normal helium (^{4}He). This last step sets free two protons to start the whole process again. The gamma rays produced in the proton-proton reaction take 1 to 10 million years to work their way out from the star's core, being scattered numerous times and losing energy as they go, until they emerge from the surface as rays of light and heat. Inside the Sun, about 655 million tons of hydrogen are converted into 650 million tons of helium every second. In stars heavier than about 2 M_{sun}, in which the core temperature is more than about 18 million K, the dominant process in which energy is produced by the fusion of hydrogen into helium is a different reaction chain known as the **carbon-nitrogen-oxygen cycle**.

protoplanet

A planet in the process of accretion from a **protoplanetary nebula**. Protoplanets are believed to form **planetesimals** that draw together and aggregate through gravity.

protoplanetary disk

Another name for a **protoplanetary nebula**, sometimes abbreviated to *proplyd.*

protoplanetary nebula

(1) A cloud of gas and dust from which planets are formed around a newborn star, as in the **solar nebula**. (2) An early stage in the development of a **planetary nebula** in which the central star has shed its red giant envelope, exposing its hot core. Examples include the **Calabash Nebula** and the **Egg Nebula**. (See photo, page 406.)

protostar

A contracting gas cloud in the embryonic stage of **star formation** before the onset of thermonuclear reactions in its interior. A protostar is surrounded by a dense cocoon of gas and dust that blocks visible light, but allows through large amounts of far **infrared** and **microwave** radiation.

protostellar core

The smallest opaque clump into which a collapsing interstellar gas cloud fragments. The characteristic mass of a protostellar core is about 0.01 M_{sun}. It subsequently grows by accretion as surrounding matter falls toward it, attaining stellar mass, and becoming a **protostar**, within 100,000 years after its initial formation.

Proxima Centauri (Gliese 551)

The nearest star to the Sun at a distance of 4.22 light-years (268,000 AU). Proxima Centauri is a dim **red dwarf** of spectral type M5.5Ve, 7,200 times fainter than the Sun (visual magnitude 11.01, absolute magnitude 15.45), discovered in 1915 by Robert **Innes**. It may be gravitationally bound to **Alpha Centauri** A and B, 0.29 light-years (13,000 AU) away, with an orbital period of 0.5 to 2 million years. Alternatively, it may be on an open (hyperbolic) trajectory past Alpha Centauri that will eventually take it away from the system. According to this idea, Proxima is an independent member of a moving group that includes α Cen A and B and a number of other nearby stars. Because of its nearness, both to the Sun and its α Cen neighbors, Proxima has well-determined physical properties, including a mass of 0.12 M_{sun}, a radius of 0.15 M_{sun}, and an age of about 5 billion years. Despite its considerable age, Proxima has an active **chromosphere** and is also a **flare star** (variable-star designation V645 Centauri), capable of brightening a magnitude or more in minutes. Observations of its chromosphere at ultraviolet wavelengths suggest a rotation period of about 31 days. Claims made in the mid-1990s, based on data from the Hubble Space Telescope, that Proxima may be orbited by a large planet or a brown dwarf, have not been substantiated. See **stars, nearest**.

protoplanetary nebula Gomez's Hamburger (named after its discoverer Arturo Gomez) is a nebula in the early stages of evolving to become a fully fledged planetary nebula. The dark band across the middle is the shadow of a thick disk of dust around the central star, which is seen edge-on from Earth. This celestial food-item lies about 6,500 light-years away in the constellation Sagittarius. *NASA/STScI*

Przybylski's Star (HD 101065)

One of the most chemically peculiar stars known; it is named informally after its discoverer, the Polish-Australian astronomer, Antoni Przybylski (pronounced "jebilski"). Apart from the usual lines of hydrogen and the calcium H and K lines, the strongest lines in HD 101065 are due to singly ionized lanthanides, presenting a spectrum similar to that of an **S star**, a highly evolved object whose atmosphere is enriched with recently synthesized material from deep within its interior. Yet, in other respects, HD 101065 appears to be a main sequence star or subgiant. One possibility is that it is a cool, extreme **Ap star**, a theory supported by the discovery of a several kilogauss magnetic field in HD 101065 similar to that of many other Ap stars. The outstanding difficulty with HD 101065 as an Ap star was that its spectrum didn't *look* like the others. Lines of neutral and first ionized iron are prominent in the spectra of Ap stars, and in some cases, the iron is clearly overabundant.

pseudocrater

A generally circular crater produced by a phreatic eruption resulting from emplacement of a lava flow over wet ground.

Ptolemaic system

A theory, developed by **Ptolemy of Alexandria** about A.D. 150, in which a motionless Earth is said to be at the center of the universe with the Sun, the Moon, and the planets revolving around it in eccentric circles and **epicycles**; the fixed stars are attached to an outer sphere concentric with Earth. The Ptolemaic system gave the positions of the planets accurately enough for naked-eye observations, although it also made some ridiculous predictions, such as that the distance to the Moon should vary by a factor of two over its orbit. It held sway in Europe, supported by the Church of Rome, until the rise of the **Copernican system**.

Ptolemy of Alexandria (A.D. 87–150)

An Alexandrian astronomer, mathematician, and geographer, also known as Claudius Ptolemaeus, who based his astronomy on the belief that all heavenly bodies revolve around Earth (see **Ptolemaic system**). He authored a book called *Mathematical Syntaxis,* widely known as the ***Almagest***, which includes a star catalog containing 48 **constellations**, using the names we still use today. Although no longer in serious use, the catalog lists 1,022 stars visible from Alexandria and was used as a standard in the Western and Arab worlds for over a thousand years. Ptolemy's catalog was based to some extent on an earlier one by **Hipparchus**. An even earlier star catalog was that of Timocharis of Alexandria, written about 300 B.C. and later used by Hipparchus. See also **Greek astronomy**.

Ptolemy's Cluster (M7, NGC 6475)

A large, brilliant **open cluster**, easily seen with the naked eye, projected on a background of numerous faint and distant Milky Way stars. It was known to Ptolemy who, in A.D. 130, described it as the "nebula following the sting of Scorpius." (The description may also include M6 but this is uncertain.) M7 was observed by **Hodierna** before 1654 who counted 30 stars. In fact there are about 80 stars brighter than magnitude 10 within a region some 25 light-years across. The brightest star is a yellow giant (magnitude 5.6, spectral type G8), the hottest main sequence star is of spectral type B6 (magnitude 5.9). M7's age has been estimated at 220 million years.

Visual magnitude:	4.1
Apparent diameter:	80′
Distance:	1,000 light-years
Position:	R.A. 17h 53.9m, Dec. –34° 49′

Puck

The tenth moon of **Uranus**, also known as Uranus XV. It was one of 10 new moons discovered in 1986 by **Voyager** 2 and the only one found soon enough to allow images of it to be captured by the spacecraft. These reveal a roughly spherical shape pockmarked by several large craters. Puck has a diameter of 154 km and moves in a nearly circular orbit 86,010 km from the planet.

pulsar

See article, page 408.

pulsar planet

A planet that orbits a **pulsar**. The planets around the millisecond pulsar PSR 1257+12, about 1,000 light-years away, became the first known **extrasolar planets** when they were discovered in 1991. Subsequently, a planet was found in association with a second millisecond pulsar, PSR B1620-26, at the much greater distance of 12,000 light-years. The nature and origin of these strange worlds is a matter of debate. One possibility is that the pulsar planets existed before their host star exploded as a **supernova**. However, it is hard to see how this could be so. The problem is not that the supernova would destroy any nearby planets but that it would effectively loosen the gravitational glue holding the planetary system together. A planet orbiting a star that suddenly lost a large fraction of its mass would fly off into space. The alternative and more likely scenario is that pulsar planets formed after the pulsar came into existence. Millisecond pulsars are believed to spin so fast because they have acquired material from a companion star. Planets could condense from some of this material as it entered an **accretion disk** in orbit around the pulsar. (See table, "Known Pulsar Planets," page 409.)

pulsar wind nebula

A characteristic form, consisting of a bull's-eye ring pattern and escaping **jets**, that is created by radiation and particles flowing away from some rapidly spinning neutron stars at the heart of **supernova remnants** (SNRs). The magnetic fields around the pulsar accelerate the particles,

pulsar

A fast-spinning, highly magnetized **neutron star**, formed (in most cases) following a **supernova** explosion, that sends out regular directional pulses of radiation as it rotates, in the manner of a lighthouse beam; the pulsar effect is seen if the beam happens to sweep in our direction. Pulsars were found originally at radio wavelengths but have since been observed at optical, X-ray, and gamma-ray energies; the first was discovered in 1967 by Jocelyn **Bell Burnell**. The emission comes from the acceleration of electrons to near-light speed above the pulsar's magnetic poles, which are displaced from the object's geographic poles. Pulses last on the order of microseconds, while the interval between pulses, known as the *pulse period,* is typically 0.25 to 2 seconds, but can be as short as 1 to 10 milliseconds in the case of *millisecond pulsars.* The pulse period gradually lengthens as the neutron star loses rotational energy, though some young pulsars are prone to **glitch**es, probably due to **starquake**s, when their period abruptly changes. Precise timing of pulses has revealed the existence of **binary pulsars** and of one pulsar, PSR 1257+12, that has a set of planet-sized companions (see **pulsar planet**).

The discovery of millisecond pulsars, with pulse periods of less than 10 ms (the record holder is PSR 1937+21, also the first to be found, with a period of 1.56 ms) was initially puzzling. The reason for this is that the youngest known pulsar is the one in the **Crab Nebula** with a period of 33 ms. Since pulsars slow down with age, how could older pulsars have shorter periods? The answer seems to be that millisecond pulsars have been "spun up," thereby becoming what are called *recycled pulsars,* by the transfer of matter from a companion. Almost all of the 90 or so known millisecond pulsars have been found to be part of binary systems in which the partner is a white dwarf, the presumption being that the pulsars were rejuvenated by matter transfer while their companions were still in the red dwarf phase.

Although most pulsars are thought to form during supernovae, mounting evidence suggests that some originate as white dwarfs. Matter transfer is again the underlying phenomenon at work. If a white dwarf acquires enough mass from a close companion and doesn't somehow manage to get rid of it, as in a nova outburst, for example, it will eventually collapse to become a neutron star and may be seen as a pulsar.

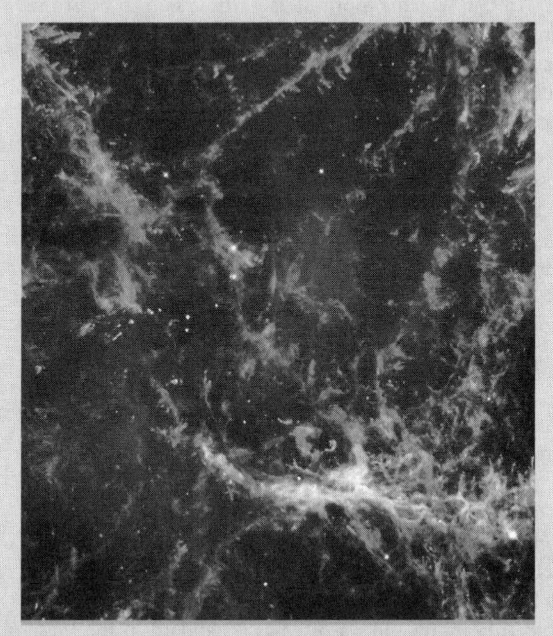

pulsar The pulsar at the heart of the Crab Nebula, visible in this Hubble image as the lower of the two moderately bright stars to the upper left of center. The Crab pulsar spins on its axis 30 times a second, and heats its surroundings, creating the diffuse glowing gas cloud in its vicinity, including an arc just to its right. *NASA/STScI*

forming jets in the vicinity of the poles of the collapsed star and a ring flowing away from the pulsar's equatorial region. Pulsar wind nebulae have been observed in the X-ray region by the Chandra X-ray Observatory around several young SNRs, including the **Crab Nebula**, the Vela supernova remnant, and SNR G54.01+0.3.

pulsating variable

A type of **variable star** whose light output, surface temperature, and spectrum change because of a periodic expansion and contraction of the star's upper layers. If the pulsation is radial, it occurs symmetrically over the whole surface so that the star remains spherical. This is

Known Pulsar Planets

Star	Planet	Mass	Mean Orbital Distance (AU)	Period	Eccentricity
PSR 1257+12	a	0.015 M_{Earth}	0.19	25.3 days	0.0
	b	3.4 M_{Earth}	0.36	66.5 days	0.02
	c	2.8 M_{Earth}	0.47	98.2 days	0.03
	d	~100 $M_{Jupiter}$	~40	170 years	?
PSR B1620-26		1.2-6.7 $M_{Jupiter}$	10-64	62-389 years	0.0–0.5

the case with most giant and supergiant pulsating variables, including **Cepheid variables**, **RR Lyrae stars**, **RV Tauri stars**, and **Mira stars**. If the pulsation is nonradial, as for example with **ZZ Ceti stars**, waves run in all directions over the star's surface giving rise to multiple periods and a complex pattern of nodes and antinodes.

Puppis (abbr. Pup; gen. Puppis)

The Stern, of mythical Jason's ship (the Argos); a southern **constellation** that spans a region of the Milky Way rich in interesting star fields. (See tables, "Puppis: Stars Brighter than Magnitude 4.0," and "Puppis: Other Objects of Interest." See also star chart 3.)

Puppis A

A **supernova remnant**, one of the brightest radio and X-ray sources in the sky, about 6,000 light-years away; also known by the X-ray designation name 2U 0821-42. A **neutron star** has been found in association with this remnant that is believed to be the remains of the progenitor star that exploded some 4,000 years ago. Observational evidence suggests that the pre-explosion mass of the star was about 25 M_{sun}–the heaviest progenitor star known with which a neutron star has been linked. In order for the neutron star to have reached its present location, it has traveled at a speed of at least 1,000 km/s. While neutron stars associated with several other supernova remnants seem to be traveling much faster, the one in Puppis A has a unique aspect. It is traveling across the sky in the *opposite direction* from the knots of ejecta, providing strong evidence of an asymmetric supernova explosion. The star's outer layers blew off preferentially in one direction, sending the central neutron star in the other; there is currently no satisfactory theory that explains how this could happen.

Purkinje effect

The tendency of the peak sensitivity of the human eye to shift toward the blue end of the spectrum at low illumination levels. It can affect visual estimates of variable

Puppis: Stars Brighter than Magnitude 4.0

Star	Magnitude		Spectral	Distance	Position	
	Visual	Absolute	Type	(light-yr)	R.A. (h m s)	Dec. (° ′ ″)
ζ Naos	2.21	–5.96	O5Ibf	1,400	08 03 35	–40 00 12
π	2.71	–4.92	K3Ib	1,090	07 17 09	–37 05 51
ρ Turais	2.83v	1.41	F2IIp	63	08 07 33	–24 18 15
τ	2.94	–0.81	KOIII	183	06 49 56	–50 36 53
ν	3.17	–2.40	B8III	423	06 37 46	–43 11 45
σ	3.25	–0.51	K5III	184	07 29 14	–43 18 05
ξ Asmidiske	3.34	–4.74	G6I1	1,350	07 49 18	–24 51 35
c	3.62	–4.53	K4III	1,390	07 45 15	–37 58 07
a	3.71	–1.41	G5III	345	07 52 13	–40 34 33
3	3.94	–8.05	A2Iabe	8,150	07 43 48	–28 57 18

Puppis: Other Objects of Interest

Name	Notes
Planetary nebulae	
NGC 2438	Located near the edge of M46 (see below) but as a foreground object. Magnitude 10.8; diameter 1.1′; distance 2,900 light-years; R.A. 7h 41.8m, Dec. –14° 44′.
NGC 2440	A young, butterfly-shaped planetary containing one of the hottest white dwarfs known. Magnitude 10.8; diameter 16″; distance 3,600 light-years; R.A. 7h 42m, Dec. –18° 13′.
Supernova remnant	
Puppis A	See separate entry.
Open clusters	
M46 (NGC 2437)	A very rich cluster with more than 500 stars. Magnitude 6.1; diameter 27′; R.A. 7h 41.8m, Dec. –14° 49′.
M47 (NGC 2422)	A cluster of about 50 stars, marginally visible with naked eye and well seen with binoculars. Magnitude 4.4; diameter 30′; R.A. 7h 36.6m, Dec. –14° 30′.
M93 (NGC 2447)	A small, bright cluster of about 80 stars in a triangular arrangement. Magnitude 6.2; diameter 22′; R.A. 7h 44.6m, Dec. –23° 52′.
NGC 2451	A large, scattered group of bright stars, near the center of which is the orange supergiant *c Pup*. Magnitude 2.8; diameter 45′; R.A. 7h 45.4m, Dec. –37° 58′.
NGC 2477	Among the most beautiful clusters in this area, containing about 300 stars. Magnitude 5.8; diameter 27′; distance 3,300 light-years; R.A. 7h 52.3m, Dec. –38° 33′.

stars when using comparison stars of different colors, especially if one of the stars is red. See **dark adaptation**.

PV Telescopii star

A blue supergiant **pulsating variable** of spectral type Bp with weak hydrogen lines and strong lines of helium and carbon, sometimes called *strong helium stars.* They pulsate with amplitudes of about 0.1 magnitude and with periods as short as 2 hours and as long as about a year. Only about a dozen examples are known and they have been relatively poorly studied.

pyroclastic

Of clastic (broken and fragmented) rock material formed by volcanic explosion or aerial expulsion from a volcanic vent. A *pyroclastic eruption* is an explosive eruption of lava producing and ejecting hot fragments of rock and lava.

pyroxene

Any of a group of dark, dense, rock-forming silicate minerals rich in calcium, iron, and magnesium and commonly found in basalt. The group includes **augite, hypersthene**, and diopside, with composition varying as a mixture of $FeSiO_3$, $MgSiO_3$, and $CaSiO_3$. *Pyroxenite* is an igneous rock composed largely of pyroxene.

Pythagoras of Samos (c. 580–500 B.C.)

A Greek philosopher, mathematician, and astronomer who founded a philosophical and religious school, the

Pyxis: Stars Brighter than Magnitude 4.0

Star	Magnitude		Spectral	Distance	Position	
	Visual	Absolute	Type	(light-yr)	R.A. (h m s)	Dec. (° ′ ″)
α	3.68	–3.39	B1.5III	845	08 43 36	–33 11 11
β	3.97	–1.41	G5II	388	08 40 06	–35 18 30

Pythagorean school in Croton, Italy. Pythagoras believed that Earth was a sphere at the center of the universe, and correctly realized that the morning star and the evening star were the same object: Venus. Pythagoras (or the Pythagoreans) made a number of fundamental mathematical discoveries: that for a right triangle, the sum of the squares of the two shorter sides is equal to the square of the hypotenuse (known as the Pythagorean theorem); that the sum of the angles of a triangle is equal to two right angles; and that irrational numbers exist.

Pyxis (abbr. Pyx; gen. Pyxidis)

The Mariner's Compass (originally called *Pyxis Nautica*); a faint southern **constellation** that lies north of Vela and west of Antlia. (See table, "Pyxis: Stars Brighter than Magnitude 4.0," and star chart 3.)

Q

Q-class asteroid

A rare class of **asteroid** of which only **Apollo** and a few other **near-Earth asteroids** are known to belong. Q types are fairly bright and have reflectance spectra with a strong absorption feature at wavelengths less than 0.7 micron and a mild one near 1 micron–spectra that are similar to those of ordinary **carbonaceous chondrites**.

quadrant

An old instrument, based on a quarter of a circle and designed to measure the altitude above the horizon of astronomical bodies. As it was originally used, the plane of the quadrant was adjusted to lie in the plane of the meridian. Vertical alignment was indicated by a plumb-bob suspended from the quadrant's center. Pivoted from this center was one end of a movable rod approximately equal in length to the radius of the quadrant. Sights mounted on the rod enabled observations to be made of stars and planets as they crossed the observer's meridian, and an angular scale inscribed on the periphery of the quadrant gave their meridian altitudes. It isn't certain whether **Ptolemy** actually constructed such an instrument or not. The Arabians, however, subsequently adopted the idea of the quadrant and greatly improved upon its design: in particular, quadrants were developed that could rotate about a vertical axis. They culminated in an enormous masonry device, 55 m high, erected in the fifteenth century by Ulugh Beg at Samarkand.

Quadrantids

One of the three most prominent annual **meteor showers**, with activity that ranges from January 1 to 6 but is mostly concentrated in a 12-hour period on January 3 to 4 with a maximum **zenithal hourly rate** as high as 110 meteors per hour. The parent body is unknown. Radiant: R.A. 15h 28m, Dec. +50°.

quadrature

The configuration when the Moon or a planet lies at an angular distance of 90° east or west of the Sun as seen from Earth.

quantum

A discrete quantity of energy $h\nu$ associated with a wave of frequency ν. It is the smallest amount of energy that can be absorbed or radiated by matter at that frequency.

quantum cosmology

The study of the earliest moments of the universe after the **Big Bang**, a stage known as the *Planck era.* Because this field demands an understanding of how spacetime behaves at the quantum level, and there is not yet an adequate quantum theory of gravity, it is highly speculative.

quark

One of the two families of fundamental particles–the other is **leptons**–from which all known matter is made. Quarks, which are the building blocks of **hadrons**, come in three pairs: *up* and *down, strange* and *charm,* and *top* and *bottom.* A proton is made of two up quarks (each with an electric charge of +2/3) and one down quark (with a charge of –1/3), while a neutron is made of two downs and one up. Although quarks existed in a free state in the first fraction of a second after the **Big Bang**, it is an open question whether they can still do so anywhere in the universe today; one possibility is inside **quark stars**.

quark star

A hypothetical star composed of free **quarks** with a density intermediate between that of a **neutron star** and a **black hole**. First theorized in the 1980s, it had been seriously doubted whether these objects really existed in nature. However, in April 2002, observations by the **Chandra X-ray Observatory** of an object known as RX J185635-375, about 450 light-years away, seemed to fit the bill. RX J185635-375 was previously thought to be a neutron star but Chandra's measurements suggest that, at just over 11 km across, it is too small to be composed of solid nucleons (neutrons and protons). Instead, at the density it appears to have, nucleons would burst apart releasing their constituent quarks. A second candidate quark star is 3C58, previously believed to be an ordinary neutron star resulting from a supernova that oc-

quasar

The highly energetic core of a remote **active galaxy**; quasars are the most luminous objects in the universe, capable of radiating over a trillion times as much energy as the Sun from a region little larger than the solar system. The first quasars were found because of their radio emission and were called *quasistellar radio sources*. But these represent only about 1% of the quasar population. When others turned up that were radio quiet, the name was changed to *quasistellar object* (QSO). In either case, quasar is a contraction of "quasistellar," in reference to the fact that the visual appearance is starlike. Following Maarten **Schmidt**'s discovery that quasars have very high **redshifts**, it became clear that they lie at remote cosmological distances and, therefore, to appear as bright as they do, must be fantastically luminous. Moreover, some quasars show marked variability over a period of just a few days, pointing to an incredibly compact source. Their radio structures often include **jets** and lobes similar to what we see from **radio galaxies**. Indeed, quasars and radio galaxies are simply different aspects of the same **active galactic nucleus** phenomenon, all powered ultimately by the same engines–supermassive **black holes**. The nearest quasar is 3C273, at a distance of about 2.5 billion light-years in **Virgo** (R.A. 12h 29.1m, Dec. +20° 3.1′), the first of its type to be identified (1963) and the brightest quasar in apparent magnitude (12.8); it is visible with a good 25-cm telescope under dark skies and therefore qualifies as the most remote object normally accessible to serious amateurs. 3C273 is unusually luminous for being, in quasar terms, relatively nearby: most of its brilliant brethren are found populating the early universe at distances of around 10 billion light-years. 3C273 also has by far the brightest optical **jet** known among quasars.

quasar Quasar HE 1013-2136 (center) and its surroundings, as seen by the Very Large Telescope. A spectacular arc-like tidal tail stretches from the quasar toward the southeast (lower left) over a distance of more than 150,000 light-years. Almost lost in the brilliance of the tail is a very close companion galaxy, just 20,000 light-years from the quasar, which appears to be the cause of the disruption to its brighter neighbor. Quasar activity is believed to be triggered by such dramatic events but this is the first time that the process has been seen in action. *European Southern Observatory*

Several quasars have been discovered with **redshifts** greater than 6, which places them at distances of around 13 billion light-years and a time of less than 1 billion years after the **Big Bang**. Observations of these remote objects thus shed valuable light on conditions in the early universe.

curred in the year 1181. However, with a surface temperature of only about 1 million °C, it is much cooler than it should be if made purely of nucleons; theory suggests that some of 3C58's material has degenerated into a quark broth.

quartz

A hard mineral made of silica (silicon dioxide) that is a constituent of sand, sandstone, and many other rocks. It is the second most abundant mineral on Earth after **feldspar**.

quasar
See article, page 413.

quiescent prominence
A long, sheetlike **prominence** of relatively cool solar material that hangs nearly vertically to the Sun's surface for days or months.

quiet Sun
The Sun when the 11-year **solar cycle** of activity is at a minimum; typically during such times there is less than one chromospheric event (a mild **solar flare**) per day.

Quintuplet Cluster
A bright cluster of stars located within 100 light-years of the center of the **Milky Way Galaxy**, that formed about 4 million years ago and is now slowly dispersing. It is one of the most massive **open clusters** yet discovered, weighing about 10,000 M_{sun}, and is home to what is believed to be the brightest star yet cataloged: the **Pistol Star**. Because of obscuring dust between us and the galactic center, the Quintuplet is only visible at **infrared** wavelengths. It is more dispersed than the nearby **Arches Cluster**.

Quintuplet Cluster A neighbor, in the galactic core, of the more compact Arches Cluster. The Quintuplet harbors stars on the verge of blowing up as supernovae, one of which is the most luminous known star in the Galaxy. This photo was taken by the Hubble Space Telescope. *NASA/STScI*

R

R association

A group of young stars, typically less than a million years old, embedded in **reflection nebula**e that are the remains of the material from which the stars formed.

R Canis Majoris star (R CaM star)

A semidetached **eclipsing binary**, in which the less massive component almost fills its **Roche lobe**. The prototype varies in brightness from magnitude 6.7 to 5.9 (roughly doubling) in a period of 27 hours.

R Coronae Borealis star (R CrB star)

A luminous, **eruptive variable** supergiant, rich in carbon and poor in hydrogen, of spectral type F or G, unusual in that it goes into outburst not by brightening, but by fading. Each minimum is thought to be caused by carbon-rich material puffed off the star during a pulsation cycle. As the cloud moves out, it cools and eventually condenses into carbon **dust** particles, which absorb much of the light coming from the star's photosphere. Only when the dust has been blown away by **radiation pressure**, does the star return to its normal brightness. While at maximum light, R CrB stars are observed to undergo small-scale, Cepheid-like variations with fluctuations of several tenths of a magnitude and with periods of 20 to 100 days.

Typical is the behavior of R Coronae Borealis (R CrB), the prototype of the class. R CrB is an F8 or G0 Ib supergiant whose variability was discovered nearly 200 years ago by Edward **Pigott**. Most of the time, R CrB is at its maximum of around magnitude 6. But at irregular intervals it goes, over a period of several weeks, into a decline of up to 8 magnitudes. The star may remain faint for many months or have several recoveries and declines in succession. Often the final rise back to maximum light is slow, taking several months to a year.

The R CrB phase is probably fairly short-lived, maybe on the order of 1,000 years, as evidenced by the fact that less than 50 of these stars are known. Their evolutionary status is uncertain, though there are two main theories: (1) the *Double Degenerate* (DD) model or (2) the *Final Helium Shell Flash* (FF) model. Both involve the expansion of **white dwarfs** to the supergiant phase. The DD model invokes the merging of two white dwarf stars, while the FF model assumes that a single white dwarf star expands to the supergiant stage by means of a final helium flash (see **final helium flash object**).

R Leonis (R Leo)

A **Mira star**, 390 light-years away, and one of the brightest and easiest to observe variable stars in the sky. With a mean visual magnitude range of 5.8 to 10.0, it can be followed throughout its average 312-day-cycle with a small telescope, and for much of that time with binoculars. Discovered by J. A. Koch of Danzig in 1782, R Leo was the fifth variable star, and fourth long-period variable to be found after Mira, Chi Cygni, and R Hydrae. It lies about 5° west of Regulus in the direction of Xi Leo. As a result of data collected by the Hubble Space Telescope, it is one of a handful of Mira stars known to be somewhat egg-shaped, with an apparent diameter of 70 × 78 milliarcseconds. This asymmetry is thought to involve the star's extended atmosphere, but its cause is not properly understood. It may be a result of **non-radial pulsations** (the star not pulsating equally in all directions) or it may be an optical illusion as a result of large dark spots on the star, perhaps caused by giant convection cells. Similar indications of an oblong shape have been found in the case of Mira, R Cassiopeia, R Leo, and W Hydrae.

R Scuti (R Sct)

The brightest of the class of variables known as **RV Tauri stars**; R Sct was discovered in 1795 by Edward **Pigott** when only a handful of variables of any type were known. It lies about 1° northwest of the **Wild Duck Cluster** and 1° north of Beta Sct, on the northern edge of the *Scutum Star Cloud*. At maximum, it shines at a magnitude of about 4.5, making it visible to the naked eye. Even as the star fades to its deepest minimum of 8.8, it can still be seen using binoculars or a small telescope. It has a primary period of about 144 days. According to one suggestion, the spectral behavior of R Sct at minimum resembles that of an **R Coronae Borealis star** (R CrB stars). Both classes of star erupt by fading to deep minima. Both classes fade by 3 to 7 magnitudes in about a month and can stay at minimum for weeks (RV Tauri stars) to years (R CrB stars). Also, both the RV Tauri and R CrB stars are known to be enshrouded in a shell of circumstellar dust. An evolutionary link between the two types is a possibility.

R star
An obsolete name for what is now classed as a **carbon star** toward the warm end of the carbon star range.

radar astronomy
The use of radar to measure the topography, rotation rate, and distance of objects in the solar system, including the Moon, the planets, and asteroids, and to detect meteors. Ground-based radar astronomy, with the exception of meteor work, demands the use of large radio telescopes. Some space probes, notably Magellan at Venus and Mars Express, have been equipped with radar to study the surface and subsurface of planets from orbit.

radar meteor
A **meteor** detected by bouncing a radar signal off the train of ionized air left in its wake.

radial pulsation
A pulsation in which a star oscillates around an equilibrium state by changing its radius while maintaining a spherical shape. The radial pulsation is just a special case of **non-radial pulsation,** in which some parts of the stellar surface move inward, while others move outward at the same time.

radial velocity
The speed at which an object is moving away from or toward an observer along the line of sight. It can be calculated from the displacement of **spectral lines** from their normal position: an object moving toward us has a **blueshift**ed spectrum, while an object moving away has a **redshift**ed spectrum. The larger the blueshift or redshift, the larger the radial velocity. The present radial-velocity champion for a star in the Milky Way Galaxy is Giclas 233-27 in **Lacerta,** which is approaching us at 583 km/s. External galaxies and, especially, **quasars,** may show very large recessive radial velocities due to the overall expansion of the universe.

radiant
The point where the backward projection of a **meteor** trajectory intersects the **celestial sphere**. More generally, the point in the sky from which meteors of a specific shower seem to come. This point moves against the background of the stars as Earth passes through the meteor stream, a movement known as *radiant drift.*

radiation
Energy emitted in the form of waves or particles that radiates out from a source. Examples include **electromagnetic radiation**, **cosmic rays**, and streams of **alpha particle**s or **beta particles**.

radiation belt
A region of charged particles in a **magnetosphere**.

radiation pressure
Pressure exerted on an object by **electromagnetic radiation**, best thought of in this context as a stream of **photons**. Radiation pressure on **dust** grains in space can dominate over gravity and explains why the tail of a **comet** always points away from the Sun.

radiation temperature
The surface temperature of an object in space, calculated assuming that it behaves as a **blackbody**. The radiation temperature is the same as the **effective temperature** but is usually measured over a narrow region of the **electromagnetic spectrum**, such as the visible range.

radiation-bounded nebula
An **emission nebula** whose central star or stars is not hot enough to ionize the entire cloud of gas. The visible extent of the nebula is thus limited by how much ionizing radiation is available, not by the overall size of the cloud or by the amount of material it contains. Compare with a **gas-bounded nebula**.

radio astronomy
The branch of **astronomy** devoted to making observations of the universe in **radio waves** using **radio telescopes**, radio **interferometers**, and other types of antennae; it began through the pioneering work of Karl **Jansky** and Grote **Reber**. Among the processes by which astronomical bodies emit radio waves are: (1) **thermal radiation** from solid bodies such as the planets; (2) thermal, or **"bremsstrahlung,"** radiation from hot gas in the interstellar medium; (3) **synchrotron radiation** from relativistic electrons in weak magnetic fields; (4) spectral line radiation from atomic and molecular transitions that occur in the interstellar medium or in the gaseous envelopes around stars; and (5) pulsed radiation from fast-spinning neutron stars with powerful magnetic fields. Among the important **radio source**s that have been probed by radio astronomers are **quasars**, **radio galaxies**, the center of the **Milky Way Galaxy**, **pulsars**, **maser**s and other emission by **interstellar molecules** including the **21-centimeter line** of neutral hydrogen, **solar flares**, **sunspots**, and the **cosmic microwave background**.

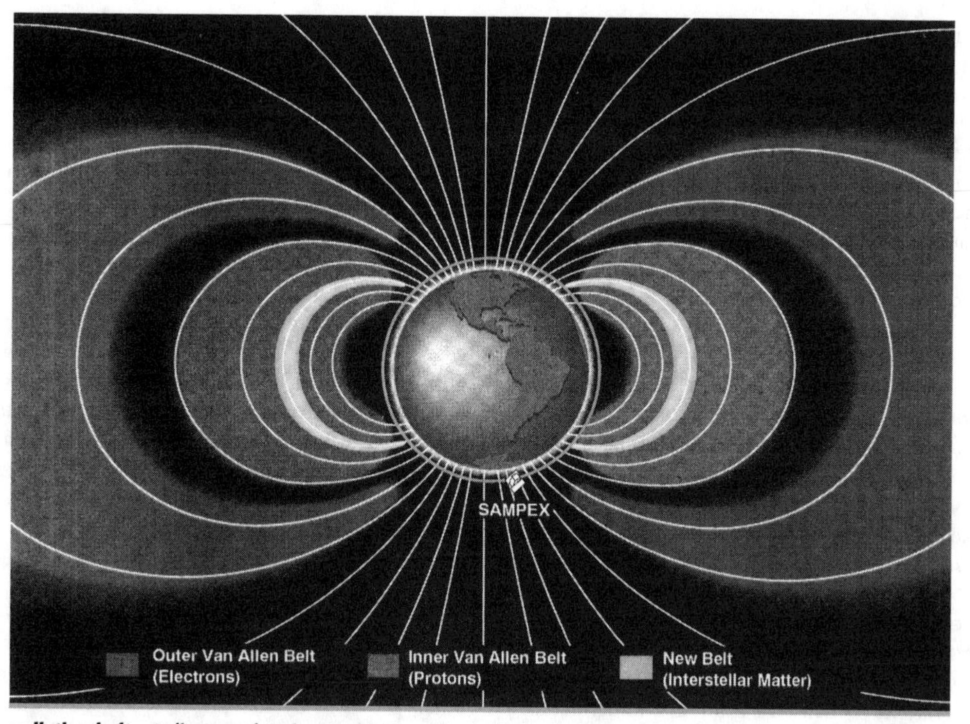

radiation belt A diagram showing Earth's Van Allen radiation belts together with a more recently discovered belt that contains trapped interstellar matter. *NASA*

radio galaxy

A **galaxy** that is extremely luminous at radio wavelengths. It is usually a giant elliptical–the largest galaxy in a cluster. The radiation emission may come only from the nucleus, but more typically also comes from a pair of more or less symmetric lobes stretching as far as a million light-years either side of the nucleus. Many radio galaxies also show emission from **jets** connecting the nucleus to these lobes. Whatever produces the radio emission must have a long "memory," preserving its direction over millions of years. The only reasonable explanation for the central energy source, as in the case of all **active galactic nuclei**, is a supermassive **black hole**, that squirts matter out along its spin axis and so delivers a steady supply of energy to the lobes. Among the brightest radio galaxies are **Centaurus A**, **Cygnus A**, and Virgo A (see **M87**).

radio source

A source of extraterrestrial radio radiation. The strongest known is **Cassiopeia A**, followed by **Cygnus A** and the **Crab Nebula** (Taurus A). (The capital letters following the name of a constellation refer to the radio sources of the constellation, A being the strongest source, B the second strongest, and so on). Radio sources are divided into two main categories: Class I, which are those associated with our Galaxy, and Class II, which are extragalactic. Most radio sources are galaxies, **supernova** remnants, or **H II regions**.

radio telescope

One or more directional receiving aerials connected to very sensitive amplifiers. Since radio wavelengths range from millimeters to about 10 m, the actual form of the aerials may vary greatly. Many are parabolic dishes with precise surfaces, while others look like a forest of large television antennae. The ability to see fine detail in sources depends on the ratio of the size of the telescope to the radio wavelength. Several approaches have been taken to make this ratio as large as possible: large single instruments (such as the fixed dish at the **Arecibo Observatory**, or the steerable **Green Bank Telescope** and the Lovell Telescope at **Jodrell Bank Observatory**), **aperture synthesis**, and radio **interferometers**.

radio waves

Electromagnetic radiation of wavelengths longer than about 1 mm (30 GHz). The shortest radio waves, from about 1 mm to 30 cm, are known as *microwaves;* the longest detected are about 30 m long (10 MHz) but, in theory, could have a wavelength of 1 km or more.

radioactivity

The spontaneous breaking apart, or decay, of unstable atomic nuclei. The three types of radioactive emission are **alpha particles** (nuclei of helium), **beta particles** (fast electrons) and **gamma rays** (high-energy photons). Radioactive nuclei are formed in **supernova** explosions.

radius

(1) The distance from the center to the circumference of a circle or sphere. (2) An old instrument for measuring the angular distance between two celestial objects.

radius vector

An imaginary line connecting the center of an orbiting body to the center of the body (or point) that it is orbiting.

Raman effect

In **spectroscopy**, the change in the wavelength of light scattered by molecules. It arises from radiation exciting (or de-exciting) atoms or molecules from their initial states.

Ramsden disk

The small disk of light visible in the back focal plane of an **eyepiece**. See also **exit pupil**.

rapid irregular variable

A type of **irregular variable** that changes in brightness by about 0.5 to 1 magnitude for a few hours or days. Unlike **Orion-type stars**, which they resemble, they are not embedded in nebulae and must therefore vary due to a different process.

Rasalhague (Alpha Ophiuchi, α Oph)

The brightest star in **Ophiuchus**; its name, which refers to the whole constellation, comes from an Arabic phrase meaning "the head of the serpent charmer." Rasalhague is a giant **A star** that has probably recently exhausted its core hydrogen reserves. It has a faint, very close companion only 0.065″ away, equivalent to 5 AU in real terms, that orbits with a period of 8.5 years.

Visual magnitude:	2.08
Absolute magnitude:	1.30
Spectral type:	A5III
Surface temperature:	8,500 K
Luminosity:	25 L_{sun}
Mass:	2 to 4 M_{sun}
Distance:	47 light-years

Ra-Shalom (minor planet 2100)

The largest known of the **Aten group** of Earth-crossing asteroids with a diameter of 2.4 km. Discovered in 1978 by E. Helin, it has a shorter year than that of Earth. Spectral class C, rotational period 19.79 hours, semimajor axis 0.832 AU, perihelion 0.469 AU, aphelion 1.195 AU, eccentricity 0.436, inclination 15.7°.

ray crater

A relatively young **impact crater** from which radiate streaks of material thrown out during the collision. Bright ray craters are found on the Moon, Mercury, Ganymede, Callisto, and Oberon. The lunar craters Tycho and Copernicus are particularly conspicuous examples with bright rays extending for hundreds of kilometers. Dark ray craters are rare and occur mostly on Ganymede, e.g., the 30-km-wide Kittu. Whether a ray crater is bright or dark depends on the relative albedo of the underlying material that is splashed out and the surrounding surface material.

Rayleigh criterion

A rule for how finely a set of optics may be able to distinguish the location of objects that are near each other, proposed by the English physicist Lord Rayleigh (1842–1919). The criterion for resolution is that the central ring in the **diffraction pattern** of one image should fall on the first dark interval between the **Airy disk** of the other and its first diffraction ring. For an objective lens of diameter d employing light with a wavelength λ (usually taken to be 5600 Å), the **resolution** is approximately $1.22 \times \lambda/d$.

Rayleigh scattering

The scattering of light by particles that are small in relation to the wavelength of the light. Rayleigh scattering of short wavelengths in sunlight by gas molecules in the atmosphere is the reason the sky appears blue. The phenomenon is named after the English physicist John William Strutt, Lord Rayleigh (1842–1919).

Rayleigh-Taylor instability

A type of fluid instability that occurs any time a dense, heavy fluid is being accelerated by light fluid. Rayleigh-Taylor instabilities are common in **supernova remnants**, such as the **Crab Nebula**, in which hot gas from the explosion is ramming into the surrounding interstellar medium, and they give rise to the familiar clumpy appearance of material in these objects.

R-class asteroid

An extremely red **asteroid** with a fairly high **albedo** and strong absorption features at wavelengths shorter than 0.7 micron and also near 1 micron. The only known example is **Dembowska**.

Reber, Grote (1911–2002)
An American radio engineer who was the first to follow up Karl **Jansky**'s 1933 announcement of the discovery of **radio waves** from space. Reber built a 9-m dish antenna in his backyard and equipped it with three different detectors before finding signals at a wavelength of 1.9 m. His 1940 and 1944 publications of articles titled "Cosmic Static" in the *Astrophysical Journal* marked the beginning of intentional radio astronomy. He was the first to express received radio signals in terms of **flux density** and **brightness**, the first to find evidence that galactic radiation is nonthermal, and the first to produce radio maps of the sky.

recombination
The capture of an **electron** by a positive **ion**. It is the opposite of **ionization**. As the electron drops through the energy levels of the atom with which it has recombined, it gives rise to **emission lines** that are known as *recombination lines* at specific wavelengths.

recovery
The first time a **periodic comet** is observed on its inbound journey toward **perihelion**.

rectangular coordinates
A set of coordinates that gives the position of an object with respect to two or three mutually perpendicular axes (x, y, and z) that cross at some specified origin. In astronomy, heliocentric or geocentric rectangular coordinates are occasionally used to give positions of bodies in the solar system. See **celestial coordinates**.

red dwarf
A main-sequence **M star**. Red dwarfs are much fainter, cooler, and smaller than the Sun but are the most common type of star in the Galaxy, accounting for 70% of all stars (excluding brown dwarfs). The nearest red dwarf, **Proxima Centauri**, lies just 4.22 light-years away, but neither it nor any other of its kind is visible to the naked eye.

red giant
A large, luminous, and relatively cool **M star** or **K star**, that has exhausted its core supply of hydrogen and is now fusing hydrogen to helium in a shell outside the core, fusing helium in the core, or doing both. Red giants are typically at least 25 times as big as the Sun and hundreds of times more luminous. Their bloated outer parts are only weakly held together by gravity.

red giant tip
The upper tip of the red-giant branch in the **Hertzsprung-Russell diagram**; it represents the helium or carbon flash point where density and temperature of the core have become so high that the "ash" in the core is ignited and serves as the fuel for a new series of nuclear reactions.

Red Rectangle
An X-shaped nebula around the star HD 44179, detected in 1975 during a survey of **infrared** sources. The emission, which is intrinsic to the nebula and not scattered or reflected light, is in a broad band in the red part of the spectrum, peaking at about 6400 Å. It is the strongest known source of this red emission, which may be a result of luminescence from hydrogenated carbon **dust**. At the nebula's center is a young binary star system that probably created the nebula. This type of nebula shows a **bipolar outflow** that carries away a significant amount of mass from the central stars. It has been speculated that the central stars create a pair of **jets** that **precess** like a spinning top. These jets might throw gas into a thick disk, which we see edge-on from Earth, so that it appears as a rectangle. The nebula emission is also unusual in that some of the infrared light it emits might be associated with unusual carbon-containing molecules.

Red Spider Nebula (NGC 6537)
A bipolar **planetary nebula** in **Sagittarius** centered on one of the hottest **white dwarfs** ever observed, probably a member of a binary system. **Stellar winds** have been measured blowing from the central stars at over 300 km/s. These hot winds expand the nebula, flow along the nebula's walls, and cause gas and dust to collide. See also **Bug Nebula**.

Visual magnitude:	13
Apparent diameter:	0.2′
Distance:	1,900 light-years
Position:	R.A. 18h 5.2m, Dec. –19° 51′

redshift (z)
The shift of **spectral lines** toward longer wavelengths in the spectrum of a receding source of radiation. More precisely, the ratio of the observed change in wavelength of light emitted by a moving object to the original wavelength emitted by an object at rest. This ratio is related to the velocity of the moving object. In general, if $\Delta\lambda$ is the observed change in the wavelength, λ is the original wavelength, v is the velocity of the moving object, c is the speed of light, then z is given by

$$z = \Delta\lambda/\lambda = (\sqrt{(1 + v/c)}/\sqrt{(1 - v/c)}) - 1$$

If the velocity of the object is small compared to the speed of light, this reduces to the nonrelativistic form:

Red Spider Nebula The complex patterns in the Red Spider are due to supersonic shocks formed when the local interstellar gas is compressed and heated ahead of the rapidly expanding lobes of material thrown out by the central star. Atoms caught in the shocks radiate the visible light seen in this Hubble Space Telescope photo. The process has been under way long enough to make the edges of the lobe walls look as if they have started to fracture into wave crests. *NASA/ESA/Garrelt Mellema (Leiden University, the Netherlands)*

$$z = \Delta\lambda/\lambda = v/c$$

Objects at the farthest reaches of the known universe have values of $z = 5$ or more. See **Doppler effect**.

redshift-distance relation

The correlation between **redshift** in the spectra of galaxies and their distances. See **Hubble law**.

reflecting binary

A type of **variable star** in which light from the hotter star is reflected from the cooler star, so that we see increased brightness when the hotter star is closer to us; the amplitude is typically 0.5 to 1 magnitude. Since the stars have to be close together for this effect to be really noticeable, the system may be an **eclipsing binary** as well. An example of this class is KV Velorum.

reflecting telescope

A telescope that uses mirrors to magnify and focus an image onto an **eyepiece**. Its main advantage is that, unlike a **refracting telescope**, it doesn't suffer from **chromatic aberration**. The main designs of reflectors in use today are the **Newtonian, Cassegrain**, and **Ritchey-Chrétien telescopes**.

reflection nebula

A cloud of interstellar gas and **dust** that shines because of reflected or scattered starlight. The nebula's spectrum contains **absorption lines** characteristic of the spectrum of nearby illuminating stars. The emission component of its spectrum is due to gas; the reflection component, to dust. Reflection nebulae often appear bluish because blue light is more efficiently reflected by dust particles than red light. A well-known example is the reflection nebula that surrounds the **Pleiades**.

refracting telescope

A telescope that uses lenses to magnify and focus an image onto an **eyepiece**. The biggest drawbacks of such a system are **chromatic aberration**, which degrades the sharpness and accuracy of the image by effects such as *fringing* and *false color*, and the difficulty and expense of making large lenses. An *apochromatic refractor* avoids the problem of chromatic aberration to a large extent by bringing all primary colors to focus at the same point. Refractors are no longer built for professional use. The two largest ones in existence are the 40-in. (1-m) telescope at the **Yerkes Observatory**, and the 36-in. (91-cm) instrument at the **Lick Observatory**. Both instruments

reflection nebula NGC 1999, a reflection nebula that lies close to the famous Orion Nebula. It is illuminated by a hot, bright, recently formed star, V380 Orionis, visible in this Hubble Space Telescope photo just to the left of center, and consists of material left over from this star's formation. The remarkable jet-black shape, resembling a letter T tilted on its side, is a globule—a very compact, dark cloud. *NASA/STScI*

date back to roughly 1890. The 40-inch glass lens at Yerkes represents the practical limit in size of refractors. Although glass blanks larger than 40 in. in diameter have been cast during the past 50 years, they are seldom, if ever, sufficiently free of internal defects to make satisfactory lenses. Even if an acceptable blank were obtained, the resulting lens, supported only by its edge, would distort so badly from its own weight it would be optically useless.

refraction

The bending of light as it passes from one transparent material to another with a different **refractive index**. The effect is used in the design of **lens**es and **prism**s, and of combinations of lenses such as **eyepiece**s and **refracting telescopes**. In nature, the effect of refraction is to make objects appear higher in the sky than they actually are. Objects more than halfway from the horizon to the zenith (i.e., with an altitude greater than 45°) are almost totally unaffected; however, objects near the horizon can be shifted by a degree or so.

refractive index

The ratio of the speed of light in a vacuum versus the speed of light through another medium. It is always greater than one, unless in a perfect vacuum, as the presence of matter retards the speed of light. When passing through a medium, long wavelengths refract less than short wavelengths, so the refractive index of a medium depends on the wavelength of light passed through it.

regmaglypt

Any small, well-defined indentation or pit on the surface of a **meteorite**. Regmaglypts are caused by the **ablation** of certain minerals as the meteorite passes through Earth's atmosphere.

regolith

A layer of loose material, including soil, subsoil, and broken rock, that covers bedrock. On the Moon and many other bodies in the solar system, it consists mostly of debris produced by **meteorite** impacts and blankets most of the surface. If the layer is very deep (perhaps a kilometer or more) it is known as *megaregolith*.

regression of nodes

The slow westward motion of the **node**s of the Moon's orbit due to **perturbation**s of Earth and Sun. It amounts to 19.35° per year, or a full circle in 18.6 years.

Regulus (Alpha Leonis, α Leo)

The brightest star in **Leo**; its Latin name means "the little king," the reference to a royal star going back to ancient times. Regulus marks the bottom of the *Sickle* that outlines the Lion's head. It is a trinary system, the chief component of which is a young, fast-spinning **B star**. This is orbited by a lighter-weight companion about 4,200 AU away with a period of at least 130,000 years. The companion is itself a double separated by at least 100 AU and with an orbital period of about 2,000 years. Both stars are less massive and dimmer than the Sun. The brighter star is an orange dwarf, while the fainter is a red (class M) dwarf.

Visual magnitude:	1.36
Absolute magnitude:	–0.52
Surface temperature:	12,000 K
Luminosity:	240 L_{sun}
Radius:	3.5 R_{sun}
Mass:	3.5 M_{sun}
Distance:	78 light-years

relative sunspot number

A number devised by Rudolf **Wolf** of Zurich in 1849 to describe **sunspot** activity. Sometimes called the *Wolf number*, it is defined as:

$$R = k\,(10g + f)$$

where R is the sunspot number, k is a constant depending on the instrument used, g is the number of disturbed regions, and f is the total number of sunspots.

relativistic beaming

An effect in which charged particles moving at a significant fraction of the speed of light will emit **electromagnetic radiation** in a narrow beam in the direction of motion–the faster the particles, the narrower the beam. This explains why a **jet** from an **active galactic nucleus** (AGN) may appear anomalously bright if it is pointing directly at us, or quite faint if pointed away from us. Since the relativistic beaming effect increases with jet speed, the brightest AGN jets represent the fastest known outflows in the universe.

relaxation

The process of gravitational interaction in a cluster of stars or galaxies, whereby a random distribution of motions (an equipartition of energy) is eventually established. The system is said to relax to a state of thermal equilibrium in a period known as the *relaxation time.* For a cluster of N stars this is about $0.1N$ times the time it takes a typical star to cross the cluster.

relief

The maximum regional difference in elevation on the surface of a planet or a moon.

resolution

Also known as *resolving power,* the ability of a telescope to differentiate between two objects in the sky that have a small **angular separation**. The closer two objects can be while still allowing the telescope to see them as two distinct objects, the greater the resolution of the telescope. Two standards for testing resolution are the **Rayleigh criterion** and the **Dawes Limit**. Both were developed before modern advances in lens coating, glass formulation, and improved optical precision figuring and design, not to mention such innovations as **active optics** and **interferometry**. A related concept is *spectral* (or *frequency*) *resolution,* which is the ability of a telescope to differentiate two light signals that differ in frequency by a small amount. The closer the two signals are in frequency while still allowing the telescope to separate them as two distinct components, the greater the spectral resolution of the instrument.

resolving power

See **resolution**.

resonance, orbital

A state in which the orbital period of one body is related to that of another by a simple integer fraction, such as 1/2, 2/3, or 3/5.

resurfacing

The creation of a new surface on a planet or moon by volcanic or **tectonic** processes.

Reticulum (abbr. Ret; gen. Reticuli)

The Reticule, or Crosshair (of a telescope's eyepiece); a small, dim, southern **constellation** lying between Dorado to the northeast, Horologium to the northwest, and Hydrus. One of its few objects of interest to the amateur observer is the wide double, Zeta Ret, consisting of two **G stars**, similar to the Sun, of magnitudes 5.2 and 5.5 (just visible to the naked eye), lying about 40 light-years away. (See table, "Reticulum: Stars Brighter than Magnitude 4.0," and star chart 8.)

retrograde motion

Rotation or orbital motion in a clockwise direction when viewed from above the north pole of the primary (i.e., in the opposite sense to most satellites); the opposite of *direct motion.* The north pole is the one on the same side of the **ecliptic** as Earth's North Pole.

revolution

(1) The orbiting of one celestial body around another. (2) A single complete course of such an orbit.

Rhea

The second largest moon of **Saturn**, with a diameter of 1,530 km; Rhea was discovered in 1672 by Giovanni **Cassini**. Its density of 1.33 gm/cm^3 suggests a rocky core taking up less than one-third of the moon's mass, with the rest composed of water-ice. Rhea is similar to **Dione** in composition, average albedo (0.57), and variety of terrain. Its surface can be divided into two geologically different areas based on crater density: the first has craters larger than 40 km in diameter, the second, in parts of the

Reticulum: Stars Brighter than Magnitude 4.0

Star	Magnitude		Spectral Type	Distance (light-yr)	Position	
	Visual	Absolute			R.A. (h m s)	Dec. (° ′ ″)
α	3.33	–0.17	G7III	163	04 14 26	–62 28 26
β	3.84	1.41	KOIV	100	03 44 12	–64 48 26

Rhea The icy, cratered surface of Saturn's moon Rhea seen by Voyager 1 on November 12, 1980, at a range of 85,000 km as the spacecraft passed over the satellite's north pole. The largest craters on view are 50 to 100 km across. *NASA/JPL*

polar and equatorial regions, has craters under 40 km in diameter. This suggests that a major resurfacing event occurred following the earliest era of bombardment. As with Dione, the trailing hemisphere is darkish, with wispy features, while the leading hemisphere is brighter.

RHESSI (Reuven Ramaty High Energy Solar Spectroscopic Imager)

A NASA satellite, launched in February 2002, to explore the basic physics of particle acceleration and energy release in **solar flares**. It carries out simultaneous, high resolution imaging and spectroscopy of flares from 3 keV X rays to 20 MeV gamma rays with high time resolution.

Rheticus (1514–1574)

An Austrian mystic and mathematician who studied under, and was the most outspoken advocate of, **Copernicus**, and who arranged for the posthumous publication of Copernicus's *De Revolutionibus Orbium Coelestium* (1543). Rheticus, also known by his nonLatinized name Georg Joachim von Lauchen, was the son of an alchemist, astrologer, and magician beheaded for sorcery. He studied mathematics at Zürich, where he met Paracelsus, became professor of mathematics and astronomy at Wittenberg from 1536, and, in 1540, published *Narratio prima de libris revolutionum Copernici,* the first written account of Copernican theory. Having convinced Copernicus shortly before his death to allow publication of *De Revolutionibus,* Rheticus unfortunately left the manuscript in the hands of a Lutheran minister, Andreas Osiander, who didn't believe in the **Copernican system** as a physical model. Consequently, Osiander added an unauthorized preface stating that the contents were merely a device to simplify calculations.

Rho Cassiopeiae (ρ Cas)

An F-type **hypergiant** and one of the most luminous known stars in the Galaxy. Despite its tremendous distance and the absorption of much of its light by interstellar matter, Rho Cassiopeiae is still easilly visible to the naked eye. Put in place of the Sun, its surface would reach out beyond the main asteroid belt. It is one of only seven known yellow hypergiants in the Milky Way Galaxy. These stars have an unusual combination of temperature and luminosity that places them in or near a region of the **Hertzsprung-Russell diagram** known as the *yellow evolutionary void.* Unlike luminous blue variables, such as the **Pistol Star**, which are sufficiently massive to keep their photospheres stable, yellow hypergiants are in a highly unstable state, pulsating wildly and occasionally ejecting large amounts of matter. Theory indicates that stars of about 10 to 60 M_{sun} evolve from main sequence **O stars** to become blue **supergiants** and then red supergiants. In the 40 to 60 M_{sun} range, however, it seems that they loop back to become hotter, smaller, blue supergiants. The evidence suggests that ρ Cas is in this loop-back stage, exhibiting violent and largely unpredictable behavior as it makes the transition. Multiple periods of variability have been detected, each measured in hundreds of days, but these change and can be interrupted by sudden outbursts and fadings. In mid-1946, the star dropped from fourth to sixth magnitude and, more surprisingly, altered its spectral type to that of an **M star**, having apparently pumped a huge amount of gas into an expanding atmosphere. A year later, it had reverted to its previous state. In 2000, it erupted again, blasting three Jupiter masses of material into space, and since then has been pulsating in a strange manner. Its outer layer appears to be collapsing again, as happened prior to the last outburst, so that an even larger eruption may be imminent. The remaing lifetime of ρ Cas, before it explodes as a supernova, is probably no more than a few tens of thousands of years.

Visual magnitude:	4.51
Absolute magnitude:	–7.48
Spectral type:	F8Ia
Surface temperature:	7,500 K
Luminosity:	550,000 L_{sun}
Radius:	450 M_{sun}
Mass:	40 M_{sun}
Distance:	8,150 light-years

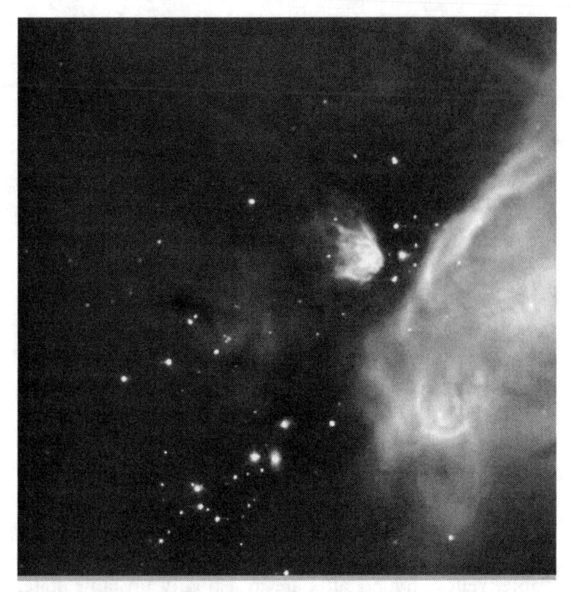

Rho Ophiuchi Nebula Three of the most massive young stars in this stellar nursery are easy to find in this image from the Infrared Space Observatory. One is in the center of the right border, a second is in the middle of a comet-shaped nebula in the lower right of the image, and a third is in the middle of the small nebula close to the center-right. Other young stars and protostars are visible as pointlike sources. The dust surrounding these newborn suns is rich in carbon grains, which are heated by the stellar radiation and made to glow at infrared wavelengths (seen here as extended halos). *European Space Agency*

Rho Ophiuchi Nebula

A complex of nebulae in **Ophiuchus** that is one of the nearest star-forming regions to the Sun at a distance of some 540 light-years. Colorful **emission nebula**e and **reflection nebula**e together with dark clouds make up the complex, which takes its name from one of the bright stars within the nebulosity, Rho Ophiuchi.

Riccioli, Giovanni Battista (1598–1671)

An Italian astronomer best known for his *Almagestum novum* (New Almagest), published in 1651, in which he tried to remedy many defects of traditional astronomy while, at the same time, rejecting the new ideas of the **Copernican system**. His two-volume work includes a lunar atlas in which major craters are named after supporters of the **Ptolemaic system**, including Hipparchus, Tycho, and Ptolemy himself, and smaller craters after the revolutionaries, Copernicus and Aristarchus. The only contemporary map of the Moon of comparable quality was created by Johannes **Hevelius**, with whom Riccioli profoundly disagreed over the question of lunar water and life. Above his map, Riccioli declared, "No Man Dwell on the Moon."

rich cluster of galaxies

A **cluster of galaxies** with hundreds or thousands of large members (and many more smaller ones) within a reasonably compact region of space. The nearest rich cluster is the **Virgo Cluster**. Most of the richest clusters in the nearby universe are known as **Abell clusters**.

richest field telescope (RFT)

A telescope with a fairly large aperture and low power that gives a wide **field of view**, up to several degrees. RFTs are especially useful when searching for unusual events such as comets, novae, and meteor showers. Good **binoculars** offer a similar capability.

rift valley

A long, steep-sided, flat-bottomed depression formed when a block of crust slips down between two approximately parallel faults or groups of faults.

Rigel (Beta Orionis, β Ori)

A blue **supergiant**, normally the brightest star in **Orion** (although variable **Betelgeuse** can occasionally overtake it), and the seventh brightest in the sky. Its name comes from the Arabic phrase *Rijl Jauzah al Yusra'* for "the Left Leg of the Central One," from which "Betelgeuse" also derives. Rigel is easily the most luminous star in our local part of the Milky Way Galaxy; the next nearest star that clearly outshines it is **Deneb**, at more than four times the distance. Because of its brilliance and the fact that it is moving through a region of nebulosity, Rigel illuminates several dust clouds in its general neighborhood. Of these, the most notable is the **Witch Head Nebula**. Rigel is commonly considered to be part of the *Taurus-Orion R1 Association* and is sometime classified as an outlying member of the *Orion OB1 Association.* It is also a multiple star system. The primary, Rigel A, is orbited at a distance of some 2,000 AU (9″ angular diameter) by a pair of similar B stars, Rigel B and C, each of magnitude 7.6, which orbit one another closely at 28 AU.

Visual magnitude:	0.18
Absolute magnitude:	–6.69
Spectral type:	B8Iae
Surface temperature:	11,000 K
Luminosity:	66,000 L_{sun}
Radius:	70 R_{sun}
Mass:	17 M_{sun}
Distance:	773 light-years

right ascension (RA)

One of the **equatorial coordinates** that, along with **declination**, may be used to locate any position in the sky.

Right ascension is analogous to longitude for locating positions on Earth.

rille

A long, narrow furrow, usually with steep sides and sometimes meandering (known as a *sinuous rille*), formed either as an open channel in a lava flow, or as an underground tube carrying hot lava that collapsed as the lava flowed out. Rilles up to several hundred kilometers long and 1 to 2 km wide are common on the Moon and on the surface of several other moons and planets in the solar system.

ring galaxy

A galaxy with a ringlike appearance. The ring contains luminous blue stars; relatively little luminous matter is present in the central regions. It's believed that such a system was an ordinary galaxy that recently suffered a head-on collision with another galaxy.

Ring Nebula (M57, NGC 6720)

One of the best known **planetary nebula**e and the second to be discovered, by the French astronomer Antoine Darquier (1718–1802) in 1779 (15 years after the first one, the **Dumbbell Nebula**); it is located in **Lyra**. Although appearing as a flattened ring, the nebula is actually toroidal or possibly even cylindrical in shape; we simply happen to be looking more or less down its pole. Viewed along its equatorial plane, the Ring would more closely resemble a filled-in planetary, such as the Dumbbell. Sensitive observations have also shown that the toroid or cylinder has lobe-shaped extensions in polar directions, and also a halo of material extending to over 3.5′ (2.4 light-years)–remnants of the central star's earlier **stellar winds** before the nebula itself was ejected. The central star, which has a visual magnitude of 14.7 and a surface temperature of over 100,000 K, was discovered by the German astronomer Friedrich von Hahn (1742–1805) in 1800. M57 is easy to find, between Beta and Gamma Lyrae, at about one-third the distance from Beta to Gamma. Binoculars show it as an almost stellar object. Small amateur telescopes above 100× magnification reveal the elliptical ring, with a major axis at a **position angle** of about 60°. Under good seeing conditions, large instruments may show the central star, together with several very faint stars that lie in the foreground or background of the nebula's extension. A large aperture may also capture the nebula's predominantly greenish tinge–a result of the bulk of its light being emitted in a few green spectral lines.

Apparent size:	1.4′
Linear diameter:	0.9 light-year
Distance:	2,300 light-years
Position:	R.A. 18h 53.6m, Dec. +33° 2′

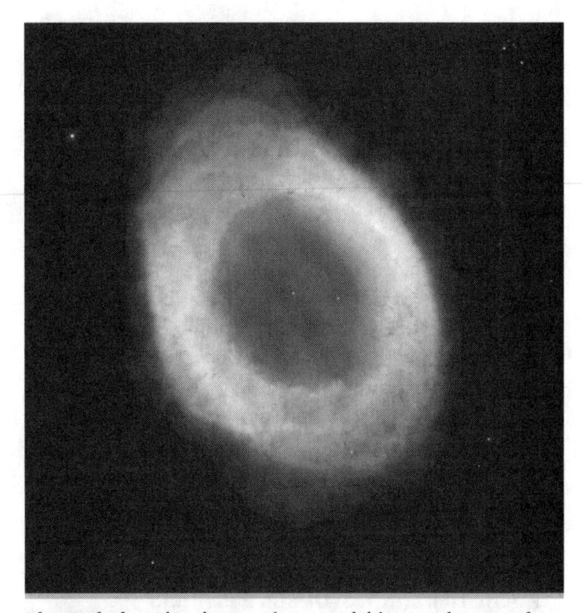

Ring Nebula The clearest view yet of this most famous of planetary nebulae, obtained by the Hubble Space Telescope. It reveals elongated dark clumps of material embedded in the gas at the edge of the glowing ring and the central star surrounded by a gossamer haze. *NASA/STScI*

ring, planetary

An annulus of material, in the form of countless numbers of small objects, in orbit around a planet. Ring matter–whether it is **dust** particles or huge boulders–will orbit a planet without coalescing into a single body if it remains within a certain distance, known as the **Roche limit**. Any bodies that orbit outside this limit won't be adversely affected by the gravity of the parent planet and may therefore accrete into a larger body. All four giant planets in the solar system have ring systems, though only Saturn's is bright enough to be seen at visible wavelengths from Earth. The various rings have different proportions of ice and rock, and different distribution of particle sizes.

Ritchey-Chrétien telescope

A variation on the **Cassegrain telescope** in which the primary mirror is a hyperboloid–slightly more strongly figured than the Cassegrainian paraboloid. It is named after its coinventors, the American astronomer George Ritchey (1864–1945) and the French optician Henri Chrétien (1879–1956). Because Ritchey-Chrétiens are corrected for **coma** as well as for **spherical aberration,** they can give relatively sharp images across a wider **field of view** than do Cassegrains. Good optical performance combined with a short tube length have made this the design of choice for many of the world's largest reflectors, including the **Very Large Telescope** and the twin 10-m telescopes at the **Keck Observatory**.

Roche limit

The closest a fluid body can orbit to its primary without being pulled apart by **tidal forces**. A solid body may survive within the Roche limit if the tidal forces don't exceed its structural strength. For a satellite of negligible mass, zero tensile strength, and the same mean density as its primary, in a circular orbit around its primary, this critical distance is 2.44 times the radius of the primary. (For the Moon, whose density is lower than that of Earth, the Roche limit would be 2.9 Earth radii.) It was first described by the French mathematician Édouard Roche (1820–1883).

ROCHE LIMIT (IN KM) OF SOME PLANETS	
Earth:	18,470
Jupiter:	175,000
Saturn:	147,000
Uranus:	62,000
Neptune:	59,000

Roche lobe

The volume around a star in a binary system in which, if you were to release a particle, it would fall back onto the surface of that star. A particle released above the Roche lobe of either star will, in general, occupy the *circumbinary* region that surrounds both stars. The point at which the Roche lobes of the two stars touch is called the inner **Lagrangian point**. If a star in a **close binary** system evolves to the point at which it fills its Roche lobe, calculations predict that material from this star will overflow both onto the companion star (via the L1 point) and into the environment around the binary system.

rogue planet

Also known as a *free-floating planet*, an **extrasolar planet** that is not gravitationally bound to a star but is instead moving on its own through interstellar space. Rogue planets may have been ejected from their original planetary systems by powerful gravitational interactions with other planets (see **orbital migration**) or interactions between stars in a binary or multiple star system. Alternatively, they may have come about through a different process of **planet formation** such as that which gave rise, for example, to the planets in the solar system. In this alternative scenario, rogue planets may have been formed in the same way as stars and **brown dwarfs**, directly as a result of the collapse of interstellar clouds. A number of rogue planets, no more than a few million years old, have been identified in the **Orion Nebula**.

Ronchi test

An efficient way of checking the shape of a telescope mirror. Conceived by the Italian optician Vasco Ronchi (1897–1988) in 1923, it is based on the **Foucault knife-edge test** but instead of a straight edge it uses a small **diffraction grating** with 20 to 40 lines per millimeter. The grating is put in front of the eye after removing the eyepiece and pointing at a star. Any curvature of the lines indicates deviations from a spherical mirror.

Roque de los Muchachos Observatory (ORM)

An observatory located on La Palma in the Canary Islands, at an altitude of 2,400 m. Founded in 1979, it is owned and operated by the Instituto de Astrofiscia de Canarias. ORM is the site of a number of telescopes belonging to various nations, including the 4.2-m **William Herschel Telescope**.

Rosalind

The eighth moon of **Uranus**, also known as Uranus XIII; it was discovered in 1986 by **Voyager** 2. Rosalind has a diameter of 54 km and moves in a nearly circular orbit 69,930 km from (the center of) the planet.

ROSAT (Röntgen Satellite)

A German-British-American X-ray and ultraviolet astronomy satellite, named for Wilhelm Röntgen (1845–1923) who discovered X rays; it operated successfully for almost nine years following its launch on June 1, 1990. The opening six months of the mission were dedicated to the first imaging survey of the sky in X rays using a German 0.8-m telescope, and the first extreme ultraviolet survey of the whole sky using a British wide-field camera. Subsequently, the satellite made detailed observations of individual sources.

Rosetta

A European Space Agency probe, originally scheduled for launch in January 2003, to rendezvous with Comet **Wirtanen** in 2011. However, a problem with its launch vehicle forced a postponement. Revised plans call for a retargeting of the probe to Comet Churyumev-Gerasimenko and a launch in February 2004. Rosetta will encounter its new target in November 2004 and then start remote sensing investigations. Following the selection of a landing site, the ROLAND lander, carrying the surface science package, will be released. From the comet's nucleus, the lander will transmit data to the main spacecraft, which will relay it to Earth.

Rosetta An artist's impression of the Rosetta lander on the surface of a comet's nucleus while the main probe flies overhead. *European Space Agency*

Rosette Nebula (M16, NGC 2237-2239, 2246)

A large **emission nebula** in **Monoceros** surrounding a cluster of about six hot, young **O stars** known as the *Rosette Cluster* (NGC 2244) whose radiation energizes the nebula and whose **stellar winds** have swept out a hollow at the center of the Rosette. The brightest parts of the nebula have their own NGC catalog numbers.

Visual magnitude:	4.8
Angular size:	80′ × 60′
Distance:	3,600 light-years
Position:	R.A. 6h 31.7m, Dec. +5° 4′

Ross, Frank Elmore (1874–1960)

An American astronomer best remembered for his photographic search for high **proper motion** stars between 1925 and 1939, which yielded the *Ross Catalogue* of 869 nearby stars. He also built the first photographic zenith tube (1911), developed emulsions and lenses for astrophotography (1915–1924), and took the first good infrared and ultraviolet photographs of Venus (1923).

Rosse, Third Earl of (William Parsons) (1800–1867)

An Irish astronomer who, in 1845, built by far the largest telescope in the world at the time, a 72-in. (1.8-m) reflector, on the grounds of Birr Castle, Parsonstown, Ireland.

Its mirror was made of **speculum** and its wood and metal tube, 17 m long, was supported between two parallel masonry walls. The so-called Leviathan of Parsonstown was used mainly for the study of nebulae and (what we know today to be) galaxies, its most notable discovery being the spiral nature of M51 (the **Whirlpool Galaxy**). Although a great milestone in the development of large telescopes, the Leviathan was hampered by its cumbersome mounting, the limited area of sky to which it had access, and the southern Irish weather.

rotating variable

A star that changes in brightness as it spins around, due to variations in brightness on its surface or to a nonspherical shape (see **ellipsoidal variable**). Irregular surface brightness may be due to the presence of large starspots (see **BY Draconis star**) or magnetic effects, as in **magnetic variables** (see **Alpha^2 Canum Venaticorum star** and **SX Arietis star**) or **oblique rotators**. See also **FK Comae Berenices star**.

rotation

The spinning of an object, such as a planet, a star, or a galaxy, about a central axis.

rotation curve

For a **galaxy**, a graph showing how the measured orbital velocity of stars and interstellar matter varies with distance from the center of the galaxy. A **Keplerian rotation curve** would indicate that most of the mass was concentrated toward the center. However, in many cases, the rotation curve is "flat"; in other words, the rotational velocity of the galaxy remains constant with increasing radius. This suggests that large amounts of **dark matter** exist beyond the visible region.

Royal Astronomical Society

A British organization of professional astronomers and geophysicists with headquarters in London. It was founded in 1820 by a group of prominent astronomers and scientists, including John **Herschel** as the Astronomical Society of London and received its Royal Charter in 1831. It publishes several learned journals including the *Monthly Notices* (issued twice a month).

Royal Greenwich Observatory (RGO)

Until its closure in 1998, the oldest scientific institution in Britain. It was founded in 1675 by decree of Charles II. The first **Astronomer Royal**, John **Flamsteed**, was charged "to apply himself with the most exact care and diligence to the rectifying of the tables of the motions of the heavens, and the places of the fixed stars, so as to find out the so much desired longitude of places for the perfecting of the art of navigation." The RGO was first sited at Greenwich, southeast London, but in 1958 moved to the castle at Herstmonceux, Sussex. From the outset, directorship of the Observatory entailed appointment as Astronomer Royal. In its 323-year history, the RGO maintained a tradition of high-precision **astrometry** of planets and satellites and was always at the forefront of technology and observational techniques. In the 1980s the Observatory lost its primary national status with the completion of the Northern Hemisphere Observatory in Las Palmas, the Canary Islands. In 1990 the RGO moved again, to Cambridge, before being closed on October 31, 1998, by the UK Particle Physics and Astronomy Research Council.

r-process

The creation of elements heavier than zinc through the intense bombardment of other elements by **neutrons**. The r-process ("r" for "rapid") occurs on a timescale so short compared to the time needed for nuclear decay via the emission of a **beta particle** that it enables chains of reactions involving highly unstable intermediate nuclei to take place. It is believed to occur in **supernovae**, which unleash a fierce flux of neutrons, and to be responsible for the manufacture of most of the elements in the periodic table more massive than iron.

RR Lyrae stars

Short-period, yellow or white giant **pulsating variables** that belong to **Population II** and are often found in **globular clusters** (hence one of their older names of *cluster variables*) or elsewhere in the **galactic halo**. They have periods of 0.2 to 2 days, amplitudes of 0.3 to 2 magnitudes, and **spectral types** of A2 to F6. Some of them have similar **light curves** to those of **Cepheid variables** (earning them the other obsolete name of *cluster Cepheids* or *short-period Cepheids*) and, like Cepheids, obey a **period-luminosity relation** that enables them to serve as reliable distance indicators. RR Lyrae variables, however, are older, less massive, and fainter (with luminosities typical around 45 L_{sun}) than Cepheids. Two subgroups are recognized: RRAB, which pulsate in the fundamental mode, and RRC, which pulsate in the first overtone. Type RRAB shows an asymmetric light curve, with a steep rise and a more gradual decline, while type RRC varies roughly sinusoidally and with an amplitude of less than 0.8 magnitude. The presence of scatter and large amplitude variations in the light curves are often the signature of a **double-mode variable** and the **Blazhko effect** in some of these variables.

RS Canum Venaticorum star (R5 CVn star)

A type of **eruptive variable** that is also a **close binary** system and shows modest light variations of up to 0.2 magnitude with a period close to that of the orbital period. Superimposed on these short-period changes are longer-term cycles of chromospheric activity, similar to the **solar cycle**, that last 1 to 4 years.

rubble-pile asteroid

An **asteroid** that consists of a weak aggregate of large (boulder size) and/or small (gravel size) components held together by gravity rather than material strength. (45) **Eugenia** and (253) Mathilde are both suspected of having this kind of structure. If they do, it tells us something of the severity of the collisions that have taken place in the **asteroid belt**.

rumurutiite (R group chondrite)

A member of a rare group of **chondrites**, formerly known as the Carlisle Lakes group, for a meteorite that was found in Australia in 1977. It is now named for the type specimen Rumuruti that fell in Kenya, Africa, in 1934. Rumuruti is the only witnessed fall of this group and only one small individual has been preserved in the collection of the Humboldt Museum Berlin, Germany, since 1938. It was thought to be an anomalous chondrite until it was reclassified in 1993 and the R group was formed. The R chondrites are quite different from ordinary chondrites and are the opposite of E chondrites when it comes to mineralogy and their state of oxidation. The members of this group are highly oxidized, containing high amounts of iron-rich **olivine**. There is practically no free metal inside the R chondrites since most of the iron is either oxidized or found in the form of iron sulfides. The iron-rich olivines, along with the oxidized nature of the iron, give most R chondrites a typical red appearance. The meteorites of this group contain fewer **chondrules** than do ordinary chondrites or enstatite chondrites, but they often contain xenolithic inclusions that indicate a **regolith** origin, representing samples of the surface of an asteroid. Another indicator for a regolith origin is the fact that most members of the R group contain high amounts of noble gases implanted into the rock by the **solar wind**. The parent body of the R chondrites has yet to be found, but it must have been subjected to many impact events during its history to result in the high degree of brecciation that most R group members show.

runaway star

A young star, usually of spectral type O or early B, with an unusually high **space velocity**. The best known are **Naos** (Zeta Puppis), and the trio AE Aurigae, 53 Arietis, and Mu Columbae, all of which are racing away on diverging paths from a comparatively small region in **Orion**. Runaway stars are thought to be produced when there is a **supernova** explosion in a **close binary** system–an idea proposed in 1961 by the Dutch astronomer Adriaan **Blaauw**.

Russell, Henry Norris (1877–1957)

An American astronomer who, independently of Ejnar **Hertzsprung**, realized the relationship between a star's temperature (color) and its brightness, and designed a diagram illustrating this relationship in 1913, later called the **Hertzsprung-Russell diagram**. Russell spent six decades at Princeton University–as student, professor, observatory director, and active professor emeritus. From 1921 on he also made lengthy annual visits to **Mount Wilson Observatory**. He measured **parallax**es in Cambridge, England, with Arthur Hinks (1873–1945) and found the correlation between spectral types and absolute magnitudes of stars that is shown in the diagram partly named after him. He popularized the distinction between giant stars and dwarfs while developing an early theory of **stellar evolution**. With his student, Harlow **Shapley**, he analyzed light from **eclipsing binary** stars to determine stellar masses. Later he and his assistant, Charlotte Moore Sitterly, determined masses of thousands of binary stars using statistical methods. With Walter **Adams**, Russell applied Meghnad Saha's theory of ionization (see **Saha equation**) to stellar atmospheres and determined elemental abundances, confirming Celia **Payne-Gaposchkin**'s discovery that the stars are composed mostly of hydrogen. Known as the "Dean of American astronomers," Russell was a dominant force in the community as a teacher, writer, and adviser.

RV Tauri star (RV Tau star)

A luminous yellow supergiant **pulsating variable**, the **light curve** of which shows alternating deep and shallow minima with a period (measured between one deep minimum and the next) of 30 to 150 days and a brightness range of up to 4 magnitudes. The **spectral type** is typically F to G at minimum and G to K at maximum. RV Tau stars seem to be intermediate between the **Cepheid variable**s and **Mira variable**s. They probably represent the low mass, and, at least in some cases, the low **metallicity** portion of stars that are in transition from the **asymptotic giant branch** (AGB) to **white dwarf**s. Because of their previously high mass-loss rates, many will probably become **planetary nebula**e. Others, however, may evolve so slowly that the envelopes may dissipate before becoming photoionized. Since the transition from the AGB to the white dwarf stage of stellar evolution isn't well understood, RV Tau stars stand as a potential bridge across this evolutionary gap. This post-AGB phase of stellar evolution is short, astronomically speaking, lasting only a few thousand years.

There are two main varieties of RV Tau stars: the RV*a* types, of which **R Scuti** is an example, maintain a roughly constant mean brightness; RV*b* types, which include RV Tau itself, have long-term (600- to 1,500-day) periodicity. Infrared studies suggest that RV Tau stars have dusty circumstellar shells, which may be initiated by pulsation via a **shock wave**. Based on the seemingly smooth transition between the RVa and RVb stars, the two groups may not be physically distinct. The RVb stars may be in an active phase in which the dust shell is replenished by dust formation close to the star. The dust may be swept out with this gaseous outflow, and in the absence of fresh dust production the star will become an RVa, with a much less dense shell. The RVa-types may have thinner dust shells or have concentration of dense dust located at large radii.

RXTE (Rossi X-ray Timing Explorer)

A NASA X-ray astronomy satellite, launched in December 1995, designed to study variability in the energy output of X-ray sources with moderate spectral resolution. It was named for the X-ray astronomy pioneer Bruno Rossi (1905–1993). Changes in X-ray brightness lasting from microseconds to months are monitored across a broad spectral range of 2 to 250 keV.

Ryle, Martin (1918–1984)

An English radio astronomer (the first professor of the subject in Britain) who shared the 1974 Nobel Prize for physics with Anthony **Hewish** for the discovery of **pulsars**. He graduated from Oxford and helped develop radar for British defense during World War II before moving to the University of Cambridge, where he developed the technique of **aperture synthesis** and began a series of surveys of radio sources that were published as the Cambridge Catalogues, the best known of which is the third (objects in it are preceded by the designation 3C). Ryle's counts of radio sources versus brightness supported evolving universe cosmologies, and he became a leading opponent of the **steady state hypothesis**. He served as the **Astronomer Royal** from 1972 to 1982.

S

S Andromedae (S And)

Also known as *Supernova 1885,* the only **supernova** seen to date in the **Andromeda Galaxy** and the first supernova observed beyond our own Galaxy. It was recorded on August 20, 1885, by Ernst Hartwig (1851–1923) at Dorpat Observatory in Estonia; others discovered it independently, but Hartwig was the first to realize its significance. S Andromedae reached magnitude 6 between August 17 and 20, and faded to magnitude 16 by February 1890.

S Doradus (S Dor)

The brightest star in the **Large Magellanic Cloud** (LMC), one of the brightest stars known, and the prototype for the class of variable known as **S Doradus stars**. It lies in the young **open cluster** NGC 1910 on the northern rim of LMC's central bar. S Doradus has a spectrum very similar to that of **P Cygni**, another variable of the same type. These stars are extremely massive (at least 60 M_{sun}) and luminous, using up their nuclear fuel so fast that their lifetimes can't exceed a few million years; they will then explode as **supernova**e. The very high luminosity also results in an enormous **radiation pressure** at the star's surface, which tends to blow away significant portions of the stellar mass by way of an intense **stellar wind** and, occasionally, in the form of an ejected gaseous shell. The **light curve** of S Doradus also points to a long-period (40-year) eclipsing variable behavior.

Visual magnitude:	8.6 to 11.7
Spectral type:	A0eq
Distance:	180,000 light-years
Position:	R.A. 5h 18.2m, Dec. –69° 15′

S Doradus star (S Dor star)

A massive, blue, **supergiant** variable star, also known as a *Hubble-Sandage variable* or a *luminous blue variable.* S Doradus stars are the most luminous stars in the Galaxy and are easily identified in other nearby galaxies, including the **Andromeda Galaxy** and the **Triangulum Galaxy**. They are named after the prototype, **S Doradus**, in the Large Magellanic Cloud.

S star

A red giant of **spectral type** S, similar to an **M star** except that the dominant oxides in its spectrum are those of the metals of the fifth period of the periodic table (zirconium, yttrium, and so forth) instead of the third (titanium, scandium, and vanadium). S stars also have strong cyanogen (CN) molecular bands and contain spectral lines of lithium and technetium. Pure S stars, also called *zirconium stars,* are those in which zirconium oxide bands are very strong and titanium oxide bands are either absent or only barely detectable. Almost all S stars are **long-period variable**s. So-called *SC stars* appear to be intermediate in type between S stars and C stars and have a carbon/oxygen abundance ratio near unity.

S Vulpeculae star (S Vul star)

A type of **long-period variable** consisting of a yellow giant or supergiant that behaves like a **Cepheid variable** most of the time but that occasionally goes through phases when it becomes erratic and unpredictable. S Vulpeculae itself has a period of 68.5 days.

Sacramento Peak Observatory

A facility of the **National Solar Observatory** (NSO), located 2,800 m up in the Lincoln National Forest of the Sacramento Mountains, near Sunspot (65 km southeast of Alamogordo), New Mexico. Its largest instrument is the Richard B. Dunn Solar Telescope, a vacuum tower telescope, 108.5 m high–only the top 41.5 m of which rises above ground level–with a 0.76-m entrance window and a 1.6-m main mirror. Other instruments include the John W. Evans Solar Facility, equipped with a 0.4-m **coronagraph** and a 0.3-m **coelostat**, and the Hilltop Dome Facility, which operates a number of solar patrol cameras and test cameras. The Observatory was founded in 1949 by the U.S. Air Force Cambridge Research Laboratories and became part of the NSO when that body was founded in 1984.

Sadr (Gamma Cygni, γ Cyg)

The third brightest star in **Cygnus** and the center star of the Northern Cross; it lies at the northern end of the **Great Rift** with a spectacular portion of the Milky Way as a backdrop; its name comes from an Arabic phrase meaning "the hen's breast." Unlike most **supergiants**, which are either fairly hot or quite cool and reddish, Sadr is yellow-white and of moderate range temperature. It has left the main sequence and is now near a region of

Sagitta: Stars Brighter than Magnitude 4.0

	Magnitude				Position	
Star	Visual	Absolute	Spectral Type	Distance (light-yr)	R.A. (h m s)	Dec. (° ′ ″)
γ	3.51	–1.12	K5III	274	19 58 45	+19 29 32
δ	3.68	–2.01	M2II+A0V	448	19 47 23	+18 32 03

temperature and luminosity in which stars become unstable and pulsate.

Visual magnitude:	2.23
Absolute magnitude:	–6.12
Spectral type:	F8Ib
Surface temperature:	6,500 K
Luminosity:	65,000 L_{sun}
Mass:	12 M_{sun}
Distance:	1,520 light-years

Sagan, Carl Edward (1934–1996)

An American astronomer best known as a science popularizer through his many books and PBS series *Cosmos.* His research focused on topics such as the greenhouse effect on Venus; windblown dust as an explanation for the seasonal changes on Mars; organic aerosols on Titan, Saturn's moon; the long-term environmental consequences of nuclear war; and the origin of life on Earth. A pioneer in the field of astrobiology, he was David Duncan Professor of Astronomy and Space Sciences and director of the Laboratory for Planetary Studies at Cornell University at the time of his death from a rare bone-marrow disease.

Sagitta (abbr. Sge; gen. Sagittae)

The Arrow; a very small northern **constellation,** located south of Vulpecula and north of Aquila, that shows clearly the shape of an arrow flying toward Cygnus. It contains the prototype **WZ Sagittae star** and M71 (NGC 6838), formerly thought to be an open cluster but now considered to be a globular cluster of low condensation. Binoculars show M71 as a misty, round patch (magnitude 8.3; diameter 7.2′; R.A. 19h 53.8 m, Dec. +18° 47′). (See table, "Sagitta: Stars Brighter than Magnitude 4.0," and star chart 7.)

Sagittarius: Stars Brighter than Magnitude 4.0

	Magnitude				Position	
Star	Visual	Absolute	Spectral Type	Distance (light-yr)	R.A. (h m s)	Dec. (° ′ ″)
ε **Kaus Australis**	1.79	–1.45	B9.5III	145	18 24 10	–34 23 05
σ **Nunki**	2.05	–2.14	B2.5V	224	18 55 16	–26 17 48
ζ Ascella	2.60	0.41	A2III+A4IV	89	19 02 37	–29 52 49
δ Kaus Media	2.72	–2.14	K3IIIa	306	18 21 00	–29 49 42
λ Kaus Borealis	2.82	0.94	K1IIIb	77	18 27 58	–25 25 18
π Albaldah	2.82	–2.77	F2II	440	19 09 46	–21 01 25
γ Alnasl	2.98	0.63	K0III	96	18 05 48	–30 25 26
η	3.10	–0.20	M2III	149	18 17 38	–36 45 42
φ	3.17	–1.08	B8III	231	18 45 39	–26 59 27
τ	3.32	0.48	K1III	120	19 06 56	–27 40 13
ξ^2	3.52	–1.77	G8II	372	18 57 44	–21 06 24
o	3.76	0.61	K0III	139	19 04 41	–21 44 30
μ Polis	3.84v	–8.15	B1ape	8,150	18 13 46	–21 03 32
ρ^1	3.92	1.06	F0III	122	19 21 40	–17 50 50
β^1 Arkab Prior	3.96	–1.37	B9V	378	19 22 38	–44 27 32
α Rukbat	3.96	0.37	B8V	170	19 23 53	–40 36 58

Sagittarius: Other Objects of Interest

Object	Notes
Star	
W Sgr	A **Cepheid variable**. Magnitude range 4.4 to 5.0; period 7d 14h; R.A. 18h 5.0m, Dec. –29° 35′.
Planetary nebula	
NGC 6818	Irregular oval shape. Magnitude 10; diameter 0.3′; R.A. 19° 44′.
Diffuse nebulae	
Lagoon Nebula	M8. See separate entry.
Omega Nebula	M17. See separate entry.
Trifid Nebula	M20. See separate entry.
Open clusters	
M18 (NGC 6613)	A loose scattering of about 20 stars over 0.2° best viewed with binoculars or a small telescope. Magnitude 6.9; diameter 9′; R.A. 18h 19.0m, Dec. –17° 8′.
M21 (NGC 6531)	Easy to resolve with a tight central grouping. Magnitude 6.9; diameter 9′; R.A. 18h 19.0m, Dec. –17° 8′.
M23 (NGC 6494)	A large cluster spanning nearly 0.5°, excellent with binoculars and resolvable with a small telescope into about 100 stars. Magnitude 5.5; diameter 27′; R.A. 17h 56.8m, Dec. –19° 1′.
M25 (IC 14725)	About 50 stars, the brightest of sixth magnitude; good binocular object. Magnitude 4.6; diameter 31′; R.A. 18h 31.6m, Dec. –19° 15′.
NGC 6530	About 25 stars near the Lagoon Nebula, the brightest of seventh magnitude; another good binocular cluster. Magnitude 4.6; diameter 14′; R.A. 18h 4.8m, Dec. –24° 20′.
Star field	
M24	The Sagittarius Star Cloud, part of Milky Way. To the naked eye, it appears as a foggy patch; large telescopes reveal a rich field of stars. Magnitude 4.6; diameter 90′; R.A. 18h 16.9m, Dec. –18° 29′.
Globular clusters	
M28 (NGC 6626)	Magnitude 6.9; diameter 11.0′; R.A. 18h 24.5m, Dec. –24° 52′.
M54 (NGC 6715)	Close to Zeta Sgr and lying within the **Sagittarius Dwarf Elliptical Galaxy**, it is so dense that it can be mistaken through binoculars for a star. Magnitude 7.7; diameter 9.1′; R.A. 18h 55.1m, Dec. –30° 29′.
M55 (NGC 6809)	A large but loose globular that can be viewed with binoculars. Magnitude 6.9; diameter 19.0′; R.A. 19h 19.0m, Dec. –17° 8′.
M69 (NGC 6637)	Magnitude 7.7; diameter 7.1′; R.A. 18h 31.4m, Dec. –32° 21′.
M70 (NGC 6681)	Magnitude 8.1; diameter 7.8′; R.A. 18h 43.2m, Dec. –32° 18′.
M75 (NGC 6864)	Magnitude 8.6; diameter 6.0′; R.A. 20h 06.1m, Dec. –21° 55′.
Galaxies	
Sagittarius Dwarf Elliptical Galaxy	See separate entry.
Sagittarius A	See separate entry.

Sagittarius (abbr. Sgr; gen. Sagittarii)

The Archer (actually, a centaur with a bow); a large constellation of the Southern Hemisphere and ninth sign of the zodiac. It lies north of Corona Australis, south of Aquila, west of Capricornus, and east of Ophiuchus. In the center of the constellation lies the conspicuous asterism known as the Milk Dipper or the Teapot, with a spout (formed by the stars Gamma, Delta, and Epsilon Sgr), a handle (Zeta, Sigma, Tau, and Phi Sgr), and a lid (Lambda Sgr). Rich in star fields, star clusters, and nebulae, Sagittarius straddles the plane of the Milky Way. At its very edge, at the border with Ophiuchus and Scorpius, lies the center of the Milky Way Galaxy, about 30,000 light-years distant. (See tables, "Sagittarius: Stars Brighter than Magnitude 4.0," and "Sagittarius: Other Objects of Interest," see also star chart 14.)

Sagittarius A (Sgr A)

A strong source of **radio waves** at the heart of the **Milky Way Galaxy** made up of several components. *Sagittarius A West* is a thermal radio source associated with three small arms of ionized gas and dust that are spinning around the center, and resembles a miniature spiral galaxy. *Sgr A East* is a nonthermal source, about 39 light-years across, that appears to be a **supernova remnant**. The very center of the Milky Way is marked by an intense radio source, called *Sagittarius A**, which is probably a massive **black hole**.

Sagittarius B2 (Sgr B2)

A massive, dense **H II region** and **molecular cloud** complex that lies about 26,000 light-years away, near the galactic center. With a mass of some 3 million M_{sun} and a density up to 10^8 particles/cm^3, it is the richest known interstellar concentration of molecules in the Milky Way Galaxy.

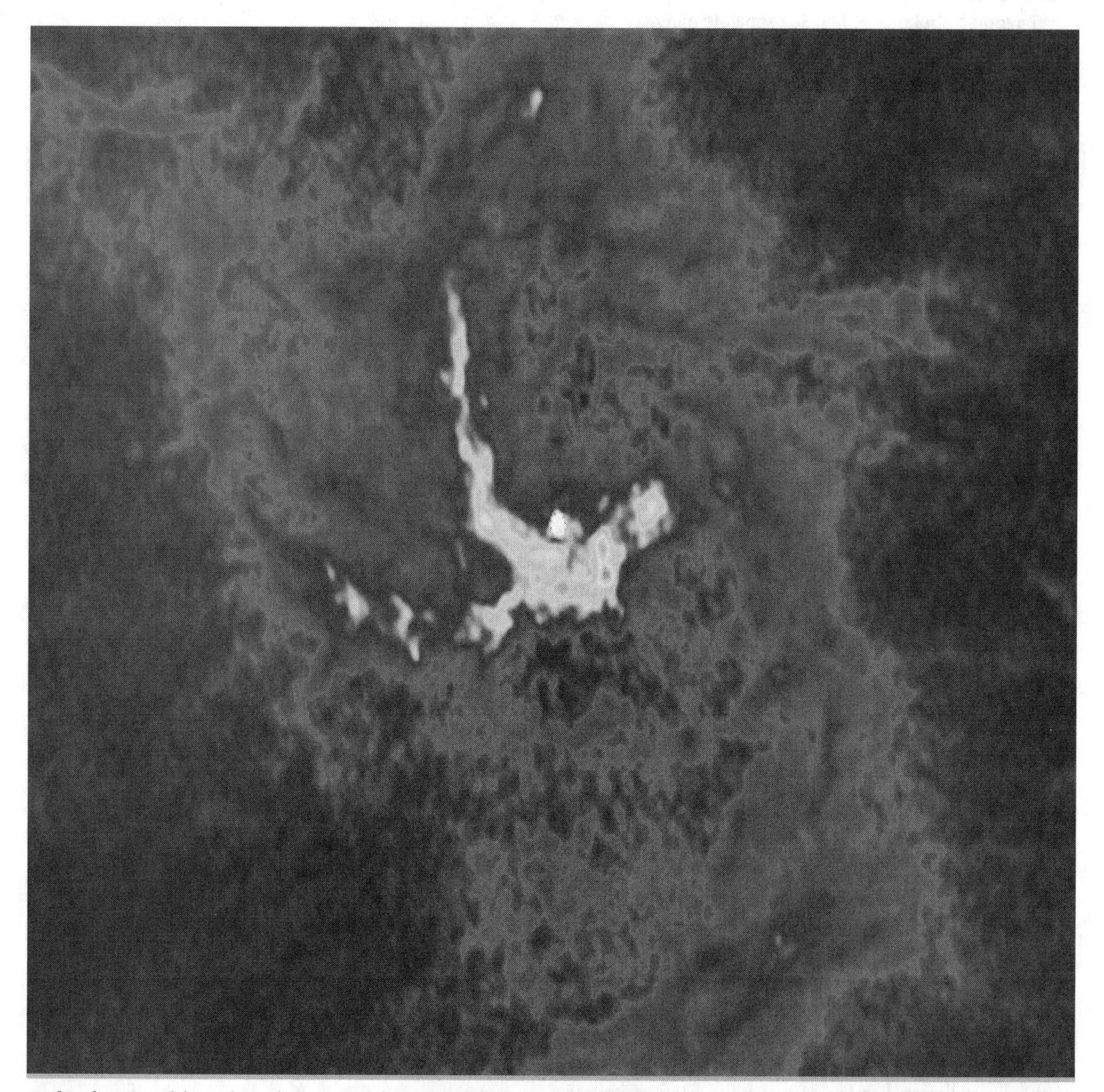

Sagittarius A A false-color radio image of the H II region known as Sagittarius A West. The spiral pattern is formed by ionized gas falling toward, or orbiting, the center of our Galaxy. The bright white dot at the center of the image is the supermassive black-hole candidate Sagittarius A*. *NSF Very Large Array and F. Yusef-Zadeh (NWU)*

Sagittarius Dwarf Elliptical Galaxy (SagDEG, Sag dSph)

A **satellite galaxy** of the Milky Way and the closest external galaxy at a distance of only about 80,000 light-years; it is populated, as usual for a **dwarf elliptical galaxy**, by old yellowish stars. Obscured by large amounts of dust in the galactic plane, SagDEG was discovered as recently as 1994 at R.A. 18h 55m, Dec. –30° 29′. It has four known **globular clusters**–M54, Arp 2, Terzan 7, and Terzan 8–of which M54 is easily the brightest and was the first extragalactic globular ever found, by Charles **Messier** in 1778. SagDEG orbits our Galaxy in less than 1 billion years and must therefore have passed through the dense central region of the Milky Way at least 10 times during our Galaxy's lifetime. The fact that it has stayed intact suggests that SagDEG may contain a significant amount of **dark matter** that helps to bind it together. It is, however, apparently now in the process of being disrupted by **tidal forces** of its massive neighbor. This may lead to its globular clusters and many of its other stars finding a new home in the Milky Way's halo, while its remaining stars escape to become solitary intergalactic travelers. SagDEG should not be confused with another member of the **Local Group**–the **Sagittarius Dwarf Irregular Galaxy** (SagDIG).

Sagittarius Dwarf Irregular Galaxy (SagDIG, Sag dIrr)

A **dwarf irregular galaxy**, discovered in 1977, that is a member of the **Local Group** of galaxies. It has a diameter of about 3,000 light-years and lies about 3.5 million light-years away at celestial coordinates R.A. 19h 30m, Dec. –17° 42′. See also **Sagittarius Dwarf Elliptical Galaxy**.

Saha equation

A formula, derived by the Indian astrophysicist Meghnad Saha (1894–1956) in 1920, that gives the fraction of atoms in each different excited state in a gas in thermal equilibrium at some specified temperature and total density. This in turn predicts the strength of **spectral lines**.

Saiph (Kappa Orionis, κ Ori)

A blue **supergiant** that is the sixth brightest star in **Orion**, lying to the lower left in the Hunter's main seven-star figure. Its name comes from a longer Arabic phrase for "sword of the giant," however, it lies outside the region known today as **Orion's Sword**. Although at about the same distance as Rigel, Saiph looks fainter because its much higher temperature causes it to radiate much of its light in the ultraviolet. Confusingly, its luminosity and temperature place it close to the region of hydrogen-fusion stability, as if it were just in the process of developing into a supergiant. Although it has a slightly variable spectrum, it makes a good background source of light that we use to study the interstellar medium.

Visual magnitude:	2.07
Absolute magnitude:	–4.66
Spectral type:	B0.5Iav
Surface temperature:	26,000 K
Luminosity:	57,500 L_{sun}
Radius:	11 R_{sun}
Mass:	15 to 17 M_{sun}
Distance:	722 light-years

Sakurai's Object (V4334 Sgr)

One of only three observed **final helium flash objects**; it lies in **Sagittarius** and was discovered in February 1996 by the Japanese amateur astronomer Yukio Sakurai.

Salpeter function

A mathematical description of the way that newly formed stars are distributed by mass. Also referred to as the *initial mass function,* the Salpeter function (the number of stars formed per unit mass range) is proportional to $m^{-2.35}$, where m is the mass of a star. It is named after the Australian American astrophysicist Edwin Salpeter (1924–).

Sargas (Theta Scorpii, θ Scorpii)

The third brightest star in **Scorpius**; its Sumerian name is of unknown meaning. Sargas is the most southerly bright star in the Scorpion and cannot be seen north of latitude 50° N. Its southerly position has allowed northern observers to use as a test of the night-sky brightness near the horizon. A yellow-white giant **F star**, Sargas has left the main sequence, probably has a dormant, helium-filled core, and is evolving toward the **Cepheid variable** stage before becoming a **red giant** several million years from now. It has a companion, 6.5″ away, at a projected distance of 540 AU. Just 1.7° to the south of Sargas lies the bright globular cluster NGC 6388.

Visual magnitude:	1.86
Absolute magnitude:	–2.75
Spectral type:	F1II
Surface temperature:	7,200 K
Luminosity:	960 L_{sun}
Radius:	20 R_{sun}
Mass:	3.7 M_{sun}
Distance:	272 light-years

Saros

A period of 18 years 11.32 days (6585.32 days), or 223 lunations, discovered by the Babylonians, after which the Sun, the Moon, and the nodes of the Moon's orbit return to virtually the same relative positions. This begins a new cycle of lunar and solar eclipses, the difference of a fraction of a day causing each eclipse to fall about 120° west of its counterpart in the previous cycle.

satellite

A body that revolves around a larger body, normally either a planet or an asteroid; also known as a *moon.* For example, Europa is a satellite of Jupiter, and Dactyl is a satellite of the asteroid Ida. There are exceptions, for example, several moons are larger than Pluto and two (Ganymede and Titan) are larger than Mercury. Since 1957 the term has also been applied to human-made objects, so that a distinction is now drawn between natural and artificial satellites.

satellite galaxy

A small galaxy–typically, an irregular or dwarf system–that orbits a larger, more massive galaxy. Both the **Milky Way Galaxy** and the **Andromeda Galaxy** have substantial families of satellite galaxies.

Saturn

See article, pages 437–439.

Saturn Nebula (NGC 7009)

A **planetary nebula** in **Aquarius,** discovered by William **Herschel** in September 1782, and nicknamed by Lord **Rosse** in the 1840s, because of its vague resemblance to the sixth planet. Like the **Blinking Nebula,** it has a bright central star at the center of a dark cavity bounded by a football-shaped rim of luminous gas. The cavity and its rim are trapped inside smoothly distributed material in the shape of a barrel and comprised of the star's former outer layers. At larger distances, and lying along the long axis of the nebula, is a pair of ansae, or "handles," each joined to the tips of the cavity by a long **jet** of material.

Visual magnitude	
Nebula:	8.0
Central star:	11.5
Angular diameter	
Bright portion:	36″
Extended halo:	100″
Distance:	2,400 light-years

scatter ellipse

The pattern of distribution on the ground formed by the fragments from a meteorite that broke up in the atmosphere; also known as a *dispersion ellipse* or *strewn field.* The long axis of the scatter ellipse coincides with the inbound flight path. The terminus of the route the meteorite was traveling contains the larger fragments, since, being heavier, they required more braking, and traveled farther before arcing down under gravity. The specifics of a scatter ellipse are determined by the height at which the meteorite broke up and the momentum of the individual fragments.

Scattered Disk

A disk of ice-rock objects in high eccentricity orbits in the ecliptic plane beyond Neptune. *Scattered Disk Objects (SDOs)* may be escapees from the **Kuiper Belt** and/or scattered Uranus-Neptune planetesimals.

scattering

The process whereby light is absorbed and reemitted in all directions, with essentially no change in wavelength (or energy). Different types of scattering involve different types of particles. Photons scatter off electrons by **Thomson scattering** or, if the photon has a lot of energy (as in the case of X rays and gamma rays), by **Compton effect.** If the scattering particle is small compared with the wavelength, the photon experiences **Rayleigh scattering**; if it's large, the process is known as **Mie scattering.**

Schiaparelli, Giovanni Virginio (1835–1910)

An Italian astronomer who explained annual **meteor showers** as the result of the dissolution of comets and proved it for the Perseids. He studied at the University of Turin and the observatories at Berlin and Pulkovo, and worked at the Brera Observatory in Milan for 40 years, all but two as director. Although a major observer of double stars, he is best known for his observations of solar system objects. He made extensive studies, both observational and theoretical, of comets, determining from the shapes of their tails that there was a repulsive effect (now known to be caused by the **solar wind**) from the Sun. He showed that Venus and Mercury rotate very slowly, and observed Mars over the course of seven oppositions, naming its "seas" and "continents." His observations of the *canali,* mistranslated as *canals,* stimulated Percival **Lowell** to found his Arizona observatory and search for life on Mars. Expert in both ancient and modern languages, Schiaparelli took up the history of ancient astronomy after retiring from observing.

Schmidt, (Johann Friedrich) Julius (1825–1884)

A German astronomer best known as a lunar cartographer. He reportedly became interested in the Moon when, at the

(continued on page 440)

Saturn

The sixth planet from the Sun and the second largest; its most spectacular feature is a large, bright system of rings. Saturn is a gas giant, with a diameter more than nine times that of Earth and the most flattened shape of any of the major planets. Its mean density is so low that if placed in a sufficiently large tub of water, it would easily float. Saturn's thick atmosphere consists of about 94% hydrogen and 6% helium, with traces of methane, ammonia, ethane, ethylene, and phosphine. Because Saturn is colder than Jupiter, the more colorful chemicals occur lower in its atmosphere and are not seen; this results in markings that are much less dramatic than those on Jupiter, though bands and some smallish spots are still in evidence. At the planet's center is believed to lie a core of rocky material about the size of Earth, but more dense. Around this is a metallic hydrogen shell some 30,000 km deep, surmounted, in turn, by a region composed of liquid hydrogen and helium with a gaseous atmosphere some 1,000 km deep. Saturn has a modest magnetic field similar in strength to Earth's. In 1997 the Hubble Space Telescope observed **auroral** displays caused by the interaction between the **solar wind** and the planetary magnetic field.

Saturn A Voyager 1 view of the planet and its ring system taken on November 16, 1980, four days after its closest approach, from a distance of 5.3 million km. This viewing geometry, which shows Saturn as a crescent, is never achieved from Earth. The translucent nature of the rings is apparent where Saturn can be seen through parts of the rings. Other parts of the rings are so dense with orbiting ice particles that almost no sunlight shines through them and a shadow is cast onto the yellowish cloud tops of Saturn, which in turn, casts a shadow across the rings at right. The black strip within the rings is the Cassini Division, which contains much less orbiting ring material than elsewhere in the rings. *NASA/JPL*

Equatorial diameter:	116,340 km
Oblateness:	0.0978
Mass (Earth = 1):	95.2
Mean density:	0.7 g/cm^3
Surface gravity:	11 m/s^2
Escape velocity:	33.1 km/s
Rotation period	
Equator:	10h 14m
Poles:	10h 38m
Obliquity:	26° 44′
Temperature (cloud tops):	160 K
Albedo:	0.50
Orbit	
Mean distance from Sun:	9.540 AU
Eccentricity:	0.056
Inclination:	2° 30′
Period:	29.458 years

Moons of Saturn

Saturn has 31 known **satellites**–18 that are named and 13 that were discovered so recently that they still have only temporary designations. The largest of Saturn's moons, Titan, is the only satellite in the solar system to have a substantial atmosphere. Hyperion is unique in that it rotates chaotically and, along with Phoebe, is unusual in that its rotation is not in lock-step with its orbital motion. Three pairs of moons interact gravitationally so as to produce orbital **resonances:** Mimas and Tethys (1:2), Enceladus and Dione (1:2), and Titan and Hyperion (3:4). See individual entries for more details. (See table, "Saturn's Moons.")

Rings of Saturn

Saturn has the largest and most spectacular ring system in the solar system; composed of swarms of ice-rock particles that range from the size of a dust grain to that of an iceberg, they are much brighter (with an albedo of up to 0.6) than any other known planetary rings. There are seven main rings (A–G), which are divided into hundreds of discrete ringlets. Though some 170,000 km wide (more if the tenuous outer portion of the E-ring is included), the rings are only about 1.5 km or so thick. They have a total mass of about

Saturn's Moons

Name	Orbit				Diameter (km)
	Distance* (km)	Period (d)	Inclination (°)	Eccentricity	
Pan	133,583	0.575	0.0	0.000	20
Atlas	137,670	0.602	0.3	0.000	40 × 20
Prometheus	139,350	0.613	0.0	0.003	145 × 85 × 61
Pandora	141,700	0.629	0.0	0.004	114 × 84 × 62
Epimetheus	151,422	0.694	0.34	0.009	144 × 108 × 98
Janus	151,472	0.695	0.14	0.007	196 × 192 × 150
Mimas	185,520	0.942	1.53	0.020	392
Enceladus	238,020	1.370	0.00	0.005	498
Tethys	294,660	1.888	1.86	0.000	1,060
Telesto	294,660	1.888	0	0.000	34 × 36 × 28
Calypso	294,660	1.888	0.00	0.000	34 × 22 × 22
Dione	377,400	2.737	0.02	0.002	1,120
Helene	377,400	2.737	0.0	0.005	36 × 32 × 30
Rhea	527,040	4.518	0.35	0.001	1,530
Titan	1,221,830	15.95	0.33	0.029	5,150
Hyperion	1,481,100	21.28	0.43	0.104	410 × 260 × 220
Iapetus	3,561,300	79.33	14.72	0.028	1,460
S/2000 S5	11,319,000	454.4	48.4	0.160	17
S/2000 S6	11,345,000	458.1	49.4	0.367	14
Phoebe	12,874,000	550.5 (r)	173	0.175	240
S/2000 S2	15,086,000	698.3	46.1	0.458	25
S/2000 S8	15,554,000	731.5	148.6	0.215	8
S/2000 S11	16,358,000	786.9	37.2	0.472	30
S/2000 S3	17,790,000	894.1	48.6	0.361	45
S/2000 S10	17,746,000	894.1	33.7	0.612	10
S/2000 S4	17,847,000	901.5	34.9	0.634	16
S/2003 S1	18,719,000	956.2	134.6	0.352	7
S/2000 S9	18,819,000	975.4	169.7	0.229	7
S/2000 S12	19,696,000	1,046	174.7	0.109	7
S/2000 S7	20,051,000	1,075	174.9	0.557	7
S/2000 S1	22,846,000	1,304	172.8	0.363	20

*Mean distance from the center of Saturn.

0.01 that of the Moon and, if compacted, would make a single body only 100 km or so across. When Earth is occasionally at the same angle as the rings, which are slanted at 27°, they are almost impossible to see. The rest of the time, only the outer A-ring, the brighter B-ring, and the bluish inner C-ring are clearly visible through Earth-based telescopes, together with several dark gaps, including the Cassini Division and the Encke Division, in which ring material is much sparser. Four additional faint rings together with a wealth of intricate and puzzling features were discovered or confirmed by the Voyager probes. Among these features are intriguing dark radial markings called *spokes,* first reported by amateur astronomers, that may be an effect caused by Saturn's magnetic field. The F-ring, a narrow, wavy structure just outside the A-ring, is confined by two small satellites and consists of several strands with embedded knots that may be clumps of ring material or minimoons. Voyager 1 images also showed the F-ring to have an odd braided structure. Complex tidal resonances exist between a number of Saturn's moons and the ring system. The so-called **shepherd moons**–Atlas, Pandora, and Prometheus–are important in preventing the rings from spreading out. Mimas seems to be responsible for keeping the Cassini division relatively clear; while Pandora orbits inside the Encke Division. (See table, "Saturn's Rings and Divisions.")

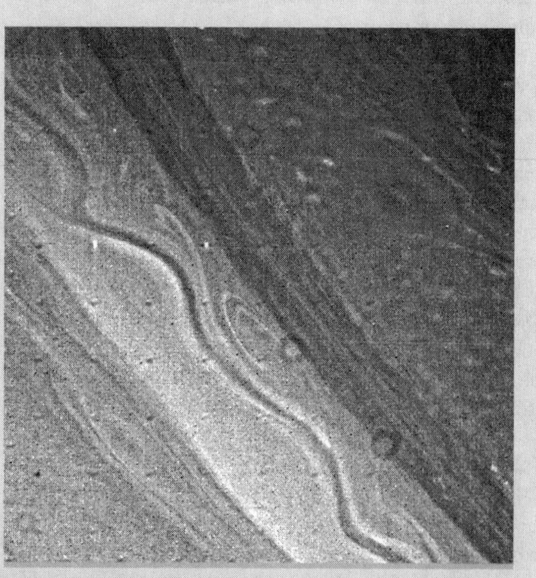

Saturn A ribbonlike structure in Saturn's atmosphere captured by Voyager 2 in this green-filter photograph on August 23, 1977, from a range of 2.5 million km. Some scientists have interpreted the ribbon to be a large-scale atmospheric wave; it is believed to lie in a rapid eastward-moving jet stream. The presence of vortices adjacent to the ribbon itself is a clue to the relationship between such structures and the strong jet streams present in Saturn's atmosphere. The smallest features visible in this photograph measure about 50 km across. *NASA/JPL*

Saturn's Rings and Divisions

Name	Radius (km)		Width (km)	Notes
	Inner	Outer		
D-ring	60,000	72,600	12,600	Very tenuous. Discovered in 1969.
Guerin Division	72,600	73,800	1,200	Also found in 1969 by Pierre Guerin.
C-ring (Crepe Ring)	73,800	91,800	18,000	Gauzy appearance. Discovered by George Bond in 1850.
Lyot (Maxwell) Division	91,800	92,300	500	Named after Bernard Lyot.
B-ring	92,300	115,800	23,500	The brightest ring, seen by Cassini.
Cassini Division	115,800	120,600	4,800	Discovered by Cassini in 1675.
Huygens Gap	117,200	117,600	250–400	Named after Christiaan Huygens.
A-ring	120,600	135,200	14,600	Seen by Cassini.
Encke Minima	126,430	129,940	3,500	First seen by Encke in 1837.
Encke Division	133,580	133,910	330	Named after Johann Encke.
F-ring	140,210	variable	30–500	Discovered by Pioneer II, 1979.
G-ring	165,800	173,800	8,000	Discovered by Voyager, 1980.
E-ring	180,000	480,000	300,000	Discovered by Walter Feibelman, 1967.

Schmidt, (Johann Friedrich) Julius (1825–1884)

(continued from page 436)

age of 14, he came across a copy of *Selenotopographical Fragments* by Johann Schröter (1745–1816). Schmidt devoted the rest of his life to observing, measuring, and drawing the Moon, amassing in the process an incredible amount of selenographic information. He began his career in Germany, spent some time in Moravia, and in 1858 became director of the Athens Observatory in Greece. His observations were made with a variety of telescopes, most notably a 15.8-cm refractor by Plössl. *Chaptre der Gebirge des Mondes* (The topographical chart of the Moon) published in Berlin in 1878, is his piece de resistance, representing the visible surface of the Moon in an area two meters in diameter, and showing about 30,000 craters.

Schmidt, Maarten (1929–)

A Dutch-American astronomer who, in 1963, identified the first **quasar**, showing that these starlike objects exhibit ordinary hydrogen lines, but with extreme cosmological **redshifts** that place them at distances of billions of light-years away. A native of Gröningen, Schmidt earned his Ph.D. under Jan **Oort** at the University of Leiden in 1956. Three years later he went to the California Institute of Technology, where at first he continued working on the mass distribution and dynamics of the Galaxy. When Rudolf **Minkowski** retired, Schmidt took over his project of determining the spectra of objects that had been found to be radio emitters. Thus he came to examine the puzzling spectral lines of 3C273 and to realize that they were none other than the familiar pattern of hydrogen lines but vastly redshifted. Returning home that night, he commented to his wife, "Something really incredible happened to me today." He went on to investigate the evolution and distribution of quasars, discovering that they were more abundant when the universe was younger. He has long sought to find the redshift above which there are no quasars, and in recent years he has joined teams studying X-ray and gamma-ray sources.

Schmidt camera

A type of photographic telescope that uses a spheroidal primary mirror with a **corrector plate** at its center of curvature to eliminate **spherical aberration**; it is named after its inventor, the Estonian-born instrument-maker Bernhard Schmidt (1879–1935). Schmidt cameras address a weakness of reflecting telescopes that are based on parabolic mirrors, namely, that the accurate focusing only occurs for light falling on the center of the paraboloid. Light falling at some distance from the center is not correctly focused owing to the optical distortion in the image, known as coma (see **coma, optical**). This limits the use of parabolic reflectors to a narrow field of view and thus precludes them from accurate survey work and the construction of star maps. The Schmidt camera, by contrast, has extremely fast focal ratios that provide very high-quality images, free from coma and **astigmatism**. The focal plane of the simplest design is not flat, but strongly curved, and is furthermore positioned well inside the body of the instrument, where it is difficult to reach with an eyepiece. Hence the system is almost always used as a camera. Because the focal surface is curved, the photographic plates and films must also be curved. A Schmidt camera has a very wide **field of view** free from astigmatism, **distortion**, and coma. A Baker-Schmidt telescope is a modified type of Schmidt camera designed, by James Baker (1914–) of Harvard in 1940, to produce an image free of astigmatism and coma. It uses a secondary convex mirror and a photographic plate that faces outward from the main mirror (rather than toward it, as in a conventional Schmidt).

Schmidt-Cassegrain telescope

A variation of the **Schmidt camera** to enable visual observation. It uses a convex spheroidal secondary mirror to reflect the converging beam from the primary back through a central hole in the primary, allowing an eyepiece to be mounted at the back end of the tube. While a Schmidt-Cassegrain is compact and gives excellent results when not pushed near its design limits, the relatively large secondary obstruction reduces image contrast. This, and other design compromises, mean that this telescope cannot match the optical performance of **Maksutov telescopes** (Maks) or **Newtonian telescopes** of equal manufacturing quality.

Schönberg-Chandrasekhar limit

The maximum mass of a star's helium-filled core that can support the overlying layers against gravitational collapse, once the core hydrogen is exhausted; it is believed to be 10 to 15% of the total stellar mass. If this limit is exceeded, as can only happen in massive stars, the core collapses, releasing energy that causes the outer layers of the star to expand to become a red giant. It is named after Subrahmanyan **Chandrasekhar** and the Brazilian astrophysicist Mario Schönberg (1916–).

Schwabe, (Samuel) Heinrich (1789–1875)

A German pharmacist and amateur astronomer who discovered that **sunspots** follow a roughly decade-long cycle. His work went largely unnoticed until publicized in 1851 by the German naturalist Friedrich von Humboldt (1769–1859) and used, together with other data, by J. R. **Wolf** in 1857 to deduce a cycle of about 11 years. Interestingly, Schwabe's purpose in keeping daily records of sunspots was to help him find an intra-Mercurian planet (see **Vulcan**) in transit across the Sun's disk.

Schwarzschild, Karl (1873–1916)

A German mathematician and astronomer who predicted the existence of **black holes**. Schwarzschild, who became director of the Potsdam Observatory in 1909, did important practical work, establishing a method for determining stars' brightnesses from photographs by comparing their visual and photographic magnitudes to obtain their color index. However, his lasting contributions are theoretical and were made during the last year of his short life. In 1916, while serving on the Russian front, he wrote two papers on the recently published **general theory of relativity**, one of which develops the idea that the gravitational field of a collapsing star could become so intense as to prevent the escape of light. The boundary of such an object is defined by the **Schwarzschild radius**.

Schwarzschild, Martin (1912–1997)

A German-born American astronomer and son of Karl **Schwarzschild** best remembered for his work on the evolution of stars and galaxies. He earned his Ph.D. at Göttingen (1935), left Germany in 1936, researched and taught at Oslo, Harvard, and Colombia, and, after serving in the U.S. Army in World War II, joined the faculty of Princeton University in 1947. His work on stellar structure and evolution led to improved understanding of pulsating stars, differential solar rotation, post-main sequence evolutionary tracks on the **Hertzsprung-Russell diagram** (including how stars become red giants), hydrogen shell sources, the helium flash, and the ages of star clusters. His 1958 book, *Structure and Evolution of the Stars,* taught a generation of astrophysicists how to apply computers to testing stellar models. In the 1950s and '60s he headed the Stratoscope projects, which took instrumented balloons to unprecedented heights, to make solar, planetary, and deep-space observations. In his later years he made significant contributions toward understanding the dynamics of **elliptical galaxies**.

Schwarzschild radius

The radius, according to the **general theory of relativity**, at which a body would become a **black hole**; alternatively, the radius of the event horizon of a spherical black hole, from within which the strength of gravity is so strong that light cannot escape. Named after Karl **Schwarzschild**, it is given by

$$R_s = 2GM/c^2$$

where G is the universal constant of gravitation, M is the mass of the body, and c is the speed of light. For the Sun, $R_s = 2.5$ km; for Earth $R_s = 0.9$ cm.

Schwassmann-Wachmann 1, Comet 29P/

A periodic **comet** with an unusual, nearly circular orbit between Jupiter and Saturn, which is prone to occasional, unexplained outbursts in brightness of about 5 magnitudes (a factor of 100) over a few days. It was discovered in 1927 by the German astronomers (Friedrich Carl) Arnold Schwassmann (1870–1964) and Arthur Arno Wachmann (1902–1990). Perihelion 5.8 AU, eccentricity 0.04, inclination 9.4°, period 15 years.

scintillation

(1) Variation in the brightness, wavelength, and mean position of stars–twinkling–caused by turbulence high in Earth's atmosphere. (2) Rapid changes in the detected intensity of radiation from compact cosmic radio sources due to disturbances in ionized gas at some point between the source and Earth's surface (usually in Earth's upper atmosphere).

S-class asteroid

A moderately bright, slightly reddish **asteroid** believed to be composed largely of silicate minerals such as **olivine** and **pyroxene**. S-class asteroids are quite numerous in the inner part of the main asteroid belt, their proportion decreasing at greater distances from the Sun. They have an albedo of 0.10 to 0.28 and a reflectance spectrum that is flat at wavelengths longer than 0.7 micron. Examples include Eunomia (see **Eunomia family**) and the 219-km-diameter (29) Amphitrite.

18 Scorpii (18 Sco)

The star nearest to Earth that is virtually identical to the Sun. It closely matches the Sun in mass, luminosity (just 5% brighter), temperature, color, surface gravity, speed of rotation, surface activity, abundance of iron, and the fact that it is solitary.

Visual magnitude:	5.49
Absolute magnitude:	4.75
Distance:	46 light-years

Scorpius (abbr. Sco; gen. Scorpii)

The Scorpion; a prominent **constellation** of the Southern Hemisphere and the eighth sign of the zodiac. It is one of the few constellations to actually look like what it represents. It lies south of Ophiuchus and west of the northern part of Centaurus. (See tables, "Scorpius: Stars Brighter than Magnitude 4.0," and "Scorpius: Other Objects of Interest." See also star chart 15.)

Scorpius X-1 (3U 1617-15, Sco X-1)

A compact eclipsing X-ray source that lies 800 to 1,600 light-years away; discovered in 1962, it is the brightest

Scorpius: Stars Brighter than Magnitude 4.0

	Magnitude				Position	
Star	**Visual**	**Absolute**	**Spectral Type**	**Distance (light-yr)**	**R.A. (h m s)**	**Dec. (° ′ ″)**
α **Antares**	1.06v	–5.28	M1.5Iab+B4V	604	16 29 24	–26 25 55
λ **Shaula**	1.62	–5.05	B2IV+B	703	17 33 36	–37 06 14
θ **Sargas**	1.86	–2.75	F1II	272	17 37 19	–42 59 52
ε Wei	2.29	0.78	K2.5III	65	16 50 10	–34 17 36
δ **Dschubba**	2.29	–3.16	B0.3IV	402	16 00 20	–22 37 18
κ Girtab	2.39	–3.38	B1.5III	464	17 42 29	–39 01 48
β Graffias	2.56	–3.50	B1V+B2V	530	16 05 26	–19 48 19
υ Lesath	2.70	–3.31	B2IV	519	17 30 46	–37 17 45
τ	2.82	–2.78	B0V	430	16 35 53	–28 12 58
σ Alniyat	2.90v	–3.87	B2III+O9.5V	735	16 21 11	–25 35 34
π	2.89	–2.86	B1V+B2V	459	15 58 51	–26 06 50
ι^1	2.99	–5.71	F3Iae	1,790	17 47 35	–40 07 37
μ^1	3.00	–4.01	B1.5V+B6.5V	822	16 51 52	–38 02 51
η	3.32	1.61	F3Vp	72	17 12 09	–43 14 21
μ^2	3.56	–2.44	B2IV	517	16 52 20	–38 01 03
ζ^2	3.62	0.30	K4III	151	16 54 35	–42 21 41
ρ	3.87	–1.63	B2IV	409	15 56 53	–29 12 50
ω^1 Jabhat al Akrab	3.93	–1.64	B1V	424	16 06 48	–20 40 09

X-ray source in the sky (apart from the Sun). Sco X-1 shows day-to-day variations (with a possible period of about 0.78 days) of as much as 1 magnitude; it also has optical and radio counterparts but no correlation has been found among the flares observed at the three different wavelengths. A thermal X-ray source, it is probably associated with a rotating collapsed star surrounded by an extensive **envelope.** Tentative optical identification has been made with the thirteenth-magnitude blue variable V818 Scorpii. (See table, "Scorpius: Other Objects of Interest," page 443.)

Scorpius-Centaurus Association

The nearest **OB association** to the Sun; it is centered about 470 light-years away in **Gould's Belt** and contains several hundred stars, mostly of type B, including **Shaula,** Lesath, and **Antares.** Though born roughly at the same time, the association's stars are not gravitationally bound, and the association is rapidly expanding. Subcomponents of the Scorpius-Centaurus Association include the Upper Scorpius (the youngest), Upper Centaurus-Lupus (the oldest), and Lower Centaurus-Crux associations. The star formation process in Scorpius-Centaurus started in the Upper Centaurus-Lupus Association some 15 million years ago. About 12 million years ago the most massive star in Upper Centaurus-Lupus went **supernova,** creating a large **shock wave.** This shock wave passed through the Upper Scorpius cloud about 5 million years ago, and triggered the star formation process there. Shortly after, the strong winds of the numerous massive stars in Upper Scorpius started to disperse the molecular cloud and halted the star formation process. About 1.5 million years ago the most massive star in Upper Scorpius exploded as a supernova. This shock wave fully dispersed the Upper Scorpius molecular cloud and now passes through the **Rho Ophiuchi Nebula,** where it might well have started the star formation process about 1 million years ago. The **superbubble** known as Loop I, generated by supernovae around the Association is impinging on our own **Local Bubble** and directly affecting the environment in the solar neighborhood.

Sculptor (abbr. Scl; gen. Sculptoris)

The Sculptor's Workshop; a faint southern **constellation** whose brightest star is Alpha Scl (visual magnitude 4.30, absolute magnitude –2.28, spectral type B7III, distance 673 light-years). It contains the **Sculptor Dwarf** and the **Sculptor Group** of galaxies. (See star chart 6.)

Scorpius: Other Objects of Interest

Object	Notes
Open clusters	
Butterfly Cluster	M6. See separate entry.
M7 (NGC6475)	One of the finest open clusters in the Northern Hemisphere, best seen with binoculars. It encompasses many bright stars loosely concentrated at the center. Telescopes show, at the western edge, but still within the cluster's boundaries, the faint globular cluster NGC 6453. Magnitude 3.3; diameter 80′; R.A. 17h 53.9m, Dec. –22° 59′.
NGC 6231	Over 100 stars in a compact 15′ area; this cluster lies on an inner spiral arm of our Galaxy.
Globular clusters	
M4 (NGC 6121)	A large, loosely concentrated cluster, 1.5° W of Antares. Some of its brightest members appear to form a bar through the center. Magnitude 5.9; diameter 26.3′; R.A. 16h 23.6m, Dec. –26° 32′.
M80 (NGC 6093)	A small, tightly concentrated object, difficult to resolve into stars, and then only around the edges; a fuzzy ball seen in binoculars. Magnitude 7.2; diameter 8.9′; R.A. 16h 17.0m, Dec. –22° 59′.
Other	
Scorpius X-1	See separate entry.

Sculptor Dwarf

A **dwarf elliptical galaxy** that is a **satellite galaxy** of the Milky Way. Discovered in 1937 by Harlow **Shapley**, it lies about 260,000 light-years away and has a diameter of about 3,000 light-years. See also **Local Group**.

Sculptor Group

The nearest group of galaxies to the **Local Group**, at a distance of 4 to 10 million light-years. It is dominated by five galaxies, four spiral–NGC 247, 253, 300, and 7793–and one irregular, NGC 55, which is the nearest and lies on the border between the Sculptor Group and our own. The brightest of the five is NGC 253, also known as the *Silver Coin Galaxy* or the *Sculptor Galaxy*, a beautiful edge-on spiral (visual magnitude 7.1; apparent size 25′ × 7′; R.A. 0h 47.6m, Dec. –25° 17′). The Sculptor Group lies around the south galactic pole and is sometimes called the *South Polar Group*.

Sculptor Group The spiral galaxy NGC 300, a Milky Way–like member of the Sculptor Group, located some 7 million light-years away. This picture was taken with the MPG/ESO 2.2-m telescope. *European Southern Observatory*

Scutum (abbr. Sct; gen. Scuti)

The Shield; a small southern **constellation**, one of seven added by **Hevelius**, which honors the Polish king John Sobieski for his defense of Vienna in 1683. It lies southeast of Ophiuchus and southwest of Aquila, 10° south of the celestial equator. Its brightest star is Alpha Sct (visual magnitude 3.85; absolute magnitude 0.21; spectral type K2III; distance 174 light-years). It also contains the prototype **Delta Scuti star** and the yellow giant variable R Sct, an **RV Tauri star** that varies semiregularly between fifth and eighth magnitude. Although faint and the fifth smallest constellation, Scutum occupies an interesting part of the sky, rich in star fields of the Milky Way. It is home to two particularly attractive open clusters: the **Wild Duck Cluster** (M11) and M26 (NGC 6694), a compact grouping (magnitude 8.0; diameter 15′; R.A. 18h 45.2m, Dec. –9° 24′). (See star chart 13.)

seasons

Any of the four climatically different periods of the year that arise because of the tilt of Earth's rotational axis to the orbital plane. The fixed tilt means that, at any given location, the altitude of the Sun and the length of daylight hours, changes on an annual cycle. The effects of these differing amounts of solar heating are most pronounced at high latitudes. The four seasons are astronomically defined by the different positions of the Sun with respect to the equator. In the Northern Hemisphere, spring is measured from the vernal **equinox** (March 21) to the summer **solstice** (June 22), and summer begins with the summer solstice and ends with the autumnal equinox (September 23). Autumn starts at the autumnal equinox and ends at the winter solstice (December 22), and winter goes from the winter solstice to the vernal equinox. In the Southern Hemisphere, the seasons are reversed.

Secchi, Father (Pietro) Angelo (1818–1878)

A prolific Italian astronomer and Jesuit priest, director of the Roman College Observatory (the forerunner of today's Vatican Observatory), who pioneered the recording, measurement, and classification of stellar spectra. Between 1863 and 1867, before the introduction of photography, Secchi carried out a remarkable study of the spectra of some 4,000 stars, using a visual spectroscope on the College's telescope. He noted the absorption-line (dark-line) spectra of the stars, and developed a four-category classification scheme that was later refined into the currently used Draper classification (see ***Henry Draper Catalogue***).

secondary crater

A small crater formed by ejecta thrown out of a larger **impact crater.** Secondary craters tend to cluster in a ring around the main crater, the greatest number lying slightly more than one crater diameter away on the Moon, but closer on Mercury because of the higher surface gravity. Ejecta falling near to the main crater tends to be moving too slowly to form craters but instead piles up as an ejecta blanket.

secondary minimum

The shallower of the two minima in the light curve of an **eclipsing binary**, corresponding to the eclipse of the secondary star by the brighter primary.

secondary mirror

(1) The second reflecting surface encountered by the light in a telescope. (2) A mirror other than the primary mirror of any type of **reflecting telescope**.

secular

Gradual or taking place over a long period. *Secular change* is a continuous, nonperiodic change in one of the attributes of the states of a system. *Secular perturbation* is a slow, continuous change in one of the elements of an orbit. *Secular stability* is a condition in which the equilibrium configuration of a system is stable over a long period of time. A *secular variable* is a star whose brightness changes over centuries.

secular acceleration

The apparent acceleration of the Moon and the Sun across the sky, caused by an extremely gradual reduction in Earth's rotation rate (one fifty-millionth of a second per day).

secular parallax

The angle subtended at a star by a baseline that is the distance the Sun moves in a given interval of time with respect to the **local standard of rest** (4.09 AU per year).

sedimentary

Rock formed at or near the surface of a planetary body from compacted sediments. Such sediments consist of solid rock or mineral fragments transported and deposited by wind, water, gravity, or ice; precipitated by chemical reactions; secreted by organisms; or accumulated as layers in loose, unconsolidated form. Sedimentary rocks are often deposited in layers (strata) and hardened by natural cements, and include shale, sandstone, and limestone.

seeing

The blurring of a stellar (pointlike) image due to turbulence in Earth's atmosphere. Seeing estimates are often given in terms of the full-width in arcseconds of the image at the points where the intensity has fallen to half its peak value. The typical value at a good site is a little better than 1″. Amateur astronomers often used the five-point *Antoniadi Scale,* which rates seeing as I (perfect), II (good), III (moderate), IV (poor), and V (appalling).

segmented mirror

A large mirror construction technique in which many smaller elements are built and then actively controlled to conform to the shape of the required large mirror.

seleno-

To do with the Moon, as in *selenocentric* (with reference to the center of the Moon) and *selenography* (the study of the Moon's surface and description of its topography).

semimajor axis (abbr. *a*)

Half the length of the major axis of an ellipse–a standard element used to describe an elliptical orbit. The semimajor axis is also the average distance of an orbiting object from its primary. The periapsis and apoapsis distances, r_p

and r_a, can be calculated from the semimajor axis, a, and the eccentricity, e, by

$$r_p = a\,(1 - e) \text{ and } r_a = a\,(1 + e)$$

semiregular variable

A late- or intermediate-type giant or supergiant pulsating variable that changes in brightness in a more or less predictable way. Periods range from 20 to more than 2,000 days, accompanied by light variations typically of one or two magnitudes. Four types are distinguished. SRa stars, also known as Z Aquarii stars, have periods longer than 35 days, amplitudes less than 2.5 magnitudes, and fairly regular variability, making them similar to **Mira stars** but with a smaller amplitude. SRb stars, of which Z Ursae Majoris is a good example, have periods greater than 20 days and amplitudes of less than 2.5 magnitudes. Unlike their SRa cousins, however, their periodicity is sometimes broken by lacunae of slow, irregular variations or constant brightness. SRc stars, of which **Betelgeuse** is a famous example, are supergiants characterized by low amplitudes and occasional standstills. Members of subgroups SRa to SRc are all of spectral types M, C, or S, sometimes with emission lines. SRd stars, on the other hand, are a mixed bag of yellow giants and supergiants (spectral types F to K), much hotter than the other semiregulars, covering a wide magnitude range (0.1–4 magnitudes), and prone to occasional irregularities.

Serpens (abbr. Ser; gen. Serpentis)

The Serpent, wrapped around Ophiuchus, the Serpent Bearer; a faint **constellation** of the Northern Hemisphere, it is unique in being the only constellation split into two parts, neither of which touches the other. Entwined around and separated by Ophiuchus, Serpens Caput (the Serpent's Head) is the western portion and Serpens Cauda (the Tail) the eastern portion. The Greek letter names of the constellation's main stars are distributed between the two portions: Alpha through Epsilon in Serpens Caput, Zeta and onward in Serpens Cauda. Serpens contains M16 (NGC 6611), which is a nebula and a sixth magnitude open cluster embedded in the **Eagle Nebula** (diameter 35′ × 28′; R.A. 18h 18.8m, Dec. –13° 47′). Among its other interesting deep sky objects are the globular clusters known as M5 (NGC 5904) (magnitude 5.8; diameter 17.4′; R.A. 15h 18.6m, Dec. +2° 5′) and the **Serpens Dwarf**. (See table, "Serpens: Stars Brighter than Magnitude 4.0," and star charts 11 and 13.)

Serpens Dwarf (Palomar 4, UGC 9792)

Not, as its name might suggest, a dwarf galaxy, but the second most remote **globular cluster** of the Milky Way. It lies in the remote halo in the direction of Ursa Major (R.A. 11h 29m 16.8s, Dec +28° 58′ 25″).

Setebos

The outermost known moon of **Uranus**, also known as Uranus XIX; together with **Stephano** and **Prospero**, it was discovered in 1999 by Brett Gladman, Matthew Holman, and J. J. Kavelaars using the **Canada-France-Hawaii Telescope**. Setebos has a diameter of about 20 km and is an **irregular moon** with a highly inclined, retrograde orbit at a mean distance of some 21.7 million km from the planet.

Serpens: Stars Brighter than Magnitude 4.0

	Magnitude				Position	
Star	Visual	Absolute	Spectral Type	Distance (light-yr)	R.A. (h m s)	Dec. (° ′ ″)
α Unukalhai	2.63	0.87	K2IIIbCNFe	73	15 44 16	+06 25 32
η Alava	3.23	1.84	K0III	62	18 21 18	–02 53 56
Θ Alya	3.4			132	18 56 14	+04 12 10
$Θ^1$	4.62	1.58	A5V			
$Θ^2$	4.98	1.94	A5Vn			
μ	3.54	0.14	A0V	156	15 49 37	–03 25 49
ξ	3.54	0.99	F0III	105	17 37 35	–15 23 55
β Chow	3.65	0.29	A3V	153	15 46 11	+15 25 18
ε	3.71	2.04	A2Vm	70	15 50 49	+04 28 40
δ Tsin	3.80	–0.25	F0IV	210	15 34 48	+10 32 21
γ	3.85	3.62	F6V	36	15 56 27	+15 39 42

setting circles

Graduated scales on the polar and **declination** axes of an amateur equatorial telescope mount used to point the instrument accurately at a particular point in the sky. The declination axis setting circle is usually set and locked in place while the polar axis circle may be moved to match the **right ascension** of a celestial body.

Sextans (abbr. Sex; gen. Sextantis)

The Sextant; a faint **constellation** on the celestial equator. Its brightest star is Alpha Sextantis (visual magnitude 4.48, absolute magnitude –0.25, spectral type A0III, distance 287 light-years). Among its deep sky objects are the **Sextans Dwarf** and the Spindle Galaxy. (See star chart 12.)

Sextans Dwarf

A low-luminosity **dwarf elliptical galaxy** (dE3) that is a satellite of the Milky Way. Discovered in 1990, it lies about 280,000 light-years from the galactic center. See also **Local Group.**

Seyfert galaxy

A galaxy with a small, intensely bright nucleus which shows strong, broad emission lines. Seyfert galaxies, named after Carl Seyfert (1911–1960), who first defined them as a group of similar objects in 1943, are nearly all **spiral galaxies** or barred spirals. Two distinct classes were identified by Edward Khachikian and Daniel Weedman on the basis of their emission lines. Class 1 Seyferts, such as NGC 5548, show broadening only of their hydrogen lines–those that can be associated with the highest gas densities. Class 2 Seyferts, such as NGC 1068 (the brightest known Seyfert), show broadening of all their lines. In both cases, the spectrum points to the presence of large amounts of very hot, fast-moving gas close to the galactic center that is being prevented from escaping by a large central mass. Variations in brightness over periods of a few months exhibited by some Seyferts, for example NGC 4151, show that the bulk of the radiation comes from regions at most a few light-months across. Although the visual luminosity of Seyferts is not unusual for spirals, their total luminosity, including radio wave, X-ray, and, most significantly, infrared emission, is roughly 100 times the norm. Taken as a whole, Seyferts display many of the properties of **quasars** and almost certainly produce the bulk of their energy output in the same way–through the gravitational influence of a supermassive **black hole** at the heart of an **active galactic nucleus.**

Seyfert Galaxy A photo of the nearest Seyfert galaxy, NGC 4395, taken with the Palomar 200-in. telescope. NGC 4395 is the least luminous and nearest Seyfert galaxy known, located 8 million light-years away in the direction of the constellation Canes Venatici. *Allan Sandage/Carnegie Institution*

Seyfert's Sextet (Hickson 79)

A compact group of six galaxies on the border of Hercules and Serpens, discovered in 1951 by Carl Seyfert (1911–1960), which provides an interesting example of **discordant redshifts**. Five of the galaxies have similar redshifts, indicating velocities of recession ranging from 4,017 to 4,482 km/s. However, the sixth, a spiral, is receding from us very much faster–at 19,813 km/s. If, as is generally accepted, all of these redshifts are due to the general expansion of the universe, then Seyfert's Sextet is really a quintet of interacting galaxies that just happens to lie along the same line of sight as a much more remote galaxy. The dissenting view is that the sixth galaxy is physically related to the other five and that the redshift discrepancy is noncosmological.

shadow transit

The passage of the shadow of a moon, cast by the Sun, across the face of a planet. The easiest to see are those of the Galilean satellites on Jupiter and of Titan on Saturn. Occasionally, shadow transits of several of the other larger moons of Saturn can be seen through very large telescopes.

Shapley, Harlow (1885–1972)

An American astronomer who calibrated Henrietta **Leavitt's period-luminosity relation** for **Cepheid variables** and used it to determine the distances to **globular clusters**. He boldly and correctly proposed that the globulars outline the Galaxy, that the Galaxy was far larger than had been generally believed, and that it was centered thousands of light-years away in the direction of Sagittarius. In 1920 he held what was later dubbed Astronomy's Great Debate with Heber **Curtis** on the scale of the universe. Shapley graduated from the University of Missouri in his home state, then wrote an important doctoral dissertation on eclipsing binary stars under Henry **Russell** at Princeton. From 1914 to 1921 he was at Mount Wilson Observatory, and from 1921 to 1952, he was director of the Harvard College Observatory, where he did the seminal work on Cepheids, studied the Magellanic Clouds and cataloged galaxies. He wrote many books, was an important popularizer of science, built an outstanding graduate school, and played a major role in national and international affairs.

Shapley Concentration

The largest **cluster of galaxies** known; it is made up of 25 separate clusters grouped together in one giant supercluster some 700 million light-years away and with a total mass of about 10^{16} (10,000 trillion) M_{sun}. Its gravitational attraction contributes perhaps 25% of the motion of the **Local Group**. The central and most massive cluster of the Concentration is A3558, which is dominated by a cD galaxy.

Shapley-Ames Catalogue

A catalog of 1,246 bright galaxies, down to magnitude 13.2, published by Harlow **Shapley** and Adelaide Ames in 1932. It became the standard listing of bright galaxies and played a major role in studies of galaxies in the local region. A revised version was published by Allan Sandage (1926–) and Gustav Tammann in 1981.

shatter cone

A conical fragment of rock with regular thin grooves (striae) that radiate from the top (apex) of the cone. Shatter cones range in size from less than one centimeter to more than one meter across and are formed as a result of the high pressure, high velocity **shock wave** produced by an impacting meteorite. The exact mechanism of formation, however, is not well understood, and only nuclear tests generate enough heat and pressure to even roughly mimic the process; the first manmade shatter cone was produced in 1959 during an underground nuclear explosion. Years of hunting for and mapping of natural shatter cones and their orientations have revealed that the apex of the cones, from all sides of an impact site, point to the exact center of impact. Consequently, in the early 1960s shatter cones and impact minerals were both considered adequate criteria for suggesting an impact site.

Shaula (Lambda Scorpii, λ Sco)

The second brightest star in **Scorpius**; its name comes from the Arabic *Al Shaulah,* which means "the Stinger," and accurately reflects its position at the tip of the scorpion's barbed tail. Shaula is actually a close binary system in which two similar hot subgiant **B stars**, with a total luminosity of about 35,000 L_{sun}, orbit each other with a period of 5.9 days. One of the pair is a **Beta Cephei star**, varying in brightness by less than 0.1 magnitude with a primary period of 0.21 day and a secondary period of 0.11 day. The observation of an X-ray flare in the vicinity of Shaula on June 1, 1975, together with the general low-level X-ray activity of the star, suggests that there may be a third, **white dwarf** component. Along with Lesath and several other stars in southern Scorpius, **Shaula-Centaurus** belongs to the huge, relatively nearby **Scorpius Association.**

Visual magnitude:	1.62
Absolute magnitude:	–5.05
Spectral type:	B1IV
Distance:	703 light-years

Shedar (Alpha Cassiopeiae, α Cas)

An orange giant **K star** that is normally the brightest star in **Cassiopeia**, but it is occasionally well outshone by the unusual variable **Gamma Cassiopeiae**. Its name (also spelled Schedar, Shadar, or Shedir) may come from the Arabic *Al Sadr* for "the breast," in reference to its position in the ancient Queen or it may derive from the Persian *Shuter* for "constellation." Shedar is one of the first two stars beyond the Sun (the other is **Hamal**) whose limb darkening has been directly observed. It was also among the first few stars to be classified as a variable, some nineteenth-century astronomers claiming that it could dip to mid-third magnitude. However, no variability has been detected in modern times and it is now suspected that Shedar may be stable.

Visual magnitude:	2.24
Absolute magnitude:	–1.99
Surface temperature:	4,530 K
Luminosity:	855 L_{sun}
Radius:	42 R_{sun}
Mass:	4 to 5 M_{sun}
Distance:	229 light-years

shell burning

The fusion of hydrogen, helium, and other elements in regions of a star's interior outside the stellar core. As the particular fuel is progressively exhausted, the shell moves outward until it comes to regions too cool to sustain the reactions.

shell galaxy

An **elliptical galaxy** that is surrounded by faint arcs or shells of stars, often more blue than the galaxy as a whole.

Shell systems have a variety of morphologies; some galaxies have shells transverse to the major axis and interleave on opposite sides of the center of the galaxy, while other galaxies have shells distributed at all position angles. The fraction of field ellipticals with shell-like features is at least 17% and may be as much as 44%. They may be the result of **galaxy cannibalism**.

shell star

A hot main-sequence star, usually of spectral class B, A, or F, whose spectrum shows bright emission lines presumed to be due to a gaseous ring or shell surrounding the star. Variable shell stars, in which the ejection of a shell is accompanied by a temporary fade, are known as **Gamma Cassiopeiae stars**.

shepherd moon

A moon that constrains the extent of a planetary ring through its gravitational influence. Examples include Saturn's Prometheus and Pandora, which shepherd the narrow outer F-ring. Peter Goldreich and Scott Tremaine first proposed the idea of shepherd moons in 1979 to explain why the rings of Uranus were so narrow.

shergottite

The most abundant type of the *SNC meteorites* believed to have come from Mars (see **Mars meteorites**), with 17 known examples by mid-2002; the type member is the Shergotty meteorite, which fell in India in 1865. Shergottites are igneous rocks of volcanic or plutonic (formed deep under the surface) origin, and they resemble terrestrial rocks more closely than do any other **achondrite** group. They all have exceptionally young crystallization ages of 150 to 200 million years, and usually show signs of severe shock metamorphism. Typically, the **plagioclase feldspar** in shergottites has been converted to maskelynite, a glass that is produced when plagioclase is subjected to high shock pressures. The maskelynite was probably formed by the impact forces that blasted the shergottites away from the Martian surface.

shield volcano

A broad volcanic cone with gentle slopes constructed of successive nonviscous, mostly basaltic, lava flows.

Shklovskii, Iosif Samuilovich (1916–1985)

A Ukrainian astrophysicist, best known professionally for explaining that the continuum radiation of the Crab Nebula is due to **synchrotron radiation** and best known generally for his book *Intelligent Life in the Universe,* which was expanded and introduced to the West by Carl **Sagan**. Shklovskii investigated the solar corona, showing that its temperature is around 1 million K and that it is confined by magnetic fields; proposed a new distance scale to planetary nebulae, and made theoretical and radio studies of **supernovae**. He also suggested that cosmic rays from occasional nearby supernovae may be the cause of mass extinctions on Earth, and successfully explained some X-ray stars as binary systems containing neutron stars. After the 1957 launch of Sputnik, he played a major role in Soviet space science. Shklovskii attended Vladivostok University and Moscow State University before earning his doctorate at Sternberg Astronomical Institute. He worked as a professor at Moscow State University, as founding head of the radio astronomy department at the Sternberg Institute, and from 1972 to his death in 1985, as chief of the astrophysics department of the Institute of Space Research in Moscow.

shock wave

A sharp change in the pressure, temperature, and density of a fluid, which develops when the velocity of the fluid begins to exceed the local velocity of sound.

Shoemaker, Eugene Merle (1928–1997) and Caroline, neé Spellmann (1929–)

Husband and wife astrogeologists who made numerous contributions to the study of impact craters on Earth, lunar science, asteroids, and comets. Eugene Shoemaker had hoped to travel to the Moon as an Apollo astronaut/geologist. When this ambition was sidelined by a health problem, he helped train other astronauts in geological methods for use on the lunar surface. In 1993, Eugene and Carolyn Shoemaker and David H. **Levy** discovered the short-period comet **Shoemaker-Levy 9**. Following his death in a car accident, some of Eugene Shoemaker's ashes were placed aboard **Lunar Prospector**, a spacecraft that was later intentionally crashed into the Moon; his are the first human remains resting on another world.

Shoemaker-Levy 9, Comet (D/1993 F2)

The only comet ever seen to collide with a planet. It was discovered by Carolyn and Eugene **Shoemaker** and David **Levy** on March 25, 1993, while it was in orbit around Jupiter. After coming too close to the giant in 1992, it had broken into a chain of 21 fragments, the largest some 2 to 3 km in diameter. These fragments rammed into Jupiter's atmosphere from July 16 to 22, 1994, creating fireballs and producing a series of black mushroom clouds, some larger than Asia, that rose several thousand kilometers above the Jovian clouds. The vast black dark markings remained visible for months. Among the many space-borne instruments trained on

Shoemaker-Levy 9, Comet Jupiter, showing the temporary scars left by the impact, in July 1994, of two fragments of Comet Shoemaker-Levy 9. The feature on the left is surrounded by an outer ring with a diameter of 12,000 km—about the size of Earth. *NASA/STScI*

this spectacular event were those of Voyager 2, Galileo, the International Ultraviolet Explorer, Ulysses, and the Hubble Space Telescope.

short-period comet

A **comet** with an orbital period of less than 200 years. There are two major families: the **Jupiter-family comets** with periods of less than 20 years, and the **Halley-type comets** with periods between 20 and 200 years. The vast majority of short-period comets have orbits inclined to the ecliptic by an angle of less than 35°. Short-period comets are thought to originate in the **Kuiper Belt**, between 35 and 1,000 AU from the Sun. It is also believed that the orbits of some **long-period comets** may become gravitationally perturbed so as to make them indistinguishable from their short-period relatives.

sidereal time

Time based on Earth's rotation with respect to the stars. *Local sidereal time* is given by the **right ascension** of the observer's **meridian.** *Greenwich sidereal time* is the sidereal time on the **prime meridian.** A *sidereal day* is Earth's rotation period with respect to the stars (23h 56m 4s). A *sidereal month* is the Moon's orbital period around Earth with respect to the stars (27.32166 days). A *sidereal year* is Earth's orbital period around the Sun with respect to the stars. A *sidereal period* is the orbital period of a planet around the Sun, or of a moon around its primary, with respect to the stars.

Sigma Orionis star cluster

A young **open cluster**, 1 to 5 million years old, that is part of the Orion OB association, adjacent to the **Horsehead Nebula**; the star Sigma Orionis lies at its center. Within this cluster have been identified a number of free-floating **brown dwarfs** and large planet candidates.

sign of the zodiac

(1) Any of the 12 irregularly sized **constellations** that lie along the ecliptic: Aquarius, Aries, Cancer, Capricornus, Gemini, Leo, Libra, Pisces, Sagittarius, Scorpius, Taurus, and Virgo. (2) Any of 12 divisions of the ecliptic, each spanning 30°.

Sikhote-Alin meteorite

A huge **iron meteorite** that fell as a shower near the village of Paseka in the western part of the Sikhote-Alin mountain range in southeast Siberia at 10:38 A.M. local time on February 12, 1947. Witnesses reported a fireball that was brighter than the Sun. Coming out of the north and descending at an angle of 40°, it left a trail of smoke and dust 30 km long that lingered for several hours. Light and sound from the fall were reported in a radius of some 300 km around the point of impact. Having entered Earth's atmosphere at about 14.5 km/s (31,000 mph), the great iron mass began to break into fragments, which fell together over an elliptical area of about a square kilometer, the biggest making craters and pits, up to 26 m across and 6 m deep. The original mass of the meteorite has been put at more than 70 tons; the largest fragment, a 1,745 kg specimen, is now on display in Moscow. The Sikhote-Alin is classified structurally as a coarsest **octahedrite** with a **Widmanstätten pattern** of nearly a centimeter. Chemically it is a Group IIB, with 5.9% nickel, 0.42% cobalt, 0.46% phosphorus, about 0.28% sulfur and other trace elements, and all the rest iron.

silicate

The most plentiful group of rock-forming minerals, usually consisting of silicon and oxygen combined with a metal. Silicates include **feldspars** and **quartz**.

silicon star

A kind of **Ap star** that shows a particular enhancement of the 4200 Å strontium line. Silicon stars typically lie toward the middle of the Ap star temperature range.

Silverpit Crater

The first impact structure to be found in Britain; oil geologists stumbled across the ancient crater under the North Sea, 130 km east of England's Yorkshire coast. The 3-km-wide chasm, with a central peaked cone, was punched into the crust 60 to 65 million years ago, and seems very well preserved 1,000 m beneath the seabed. It was probably made by an asteroid or comet weighing around 2 million tons and measuring 120 m across that crashed through the shallow sea that covered Britain back then and would have sent disastrous tsunami tidal waves surging across the nearby Cretaceous land masses. The crater has been uniquely preserved by sediments settling into the depression, and is unlike any other **impact crater** so far found on Earth. Other craters we know about were created in hard rocks, whereas Silverpit would have been formed in soft underwater sediments, creating a very different shape of crater. A tall conical central peak is buried inside the crater that is itself surrounded by a series of concentric rings that extend out a further 8 km in each direction. Unparalleled three-dimensional mapping of these concentric features down to a resolution of tens of meters shows that the outer ripples are caused by concentric faults in chalk on the sea floor around the central crater that were probably triggered by the impact. Its shape and size set Silverpit apart from other craters in the inner solar system and its closest relative appears to be the crater Valhalla on Jupiter's moon Callisto.

singularity

A place where the known laws of physics no longer apply, where, as Caltech physicist Kip Thorne (1940–) puts it, gravity "unglues" space and time. Singularities may be points, one-dimensional lines, or even two-dimensional sheets. The **general theory of relativity** predicts that singularities form inside black holes but are concealed from the rest of the universe behind **event horizons**. A proper formulation of quantum gravity may well avoid the infinities associated with classical singularities.

Sinope

One of the outer moons of **Jupiter**, also known as Jupiter IX; it was discovered in 1914 by Seth **Nicholson**. Sinope has a diameter of 38 km and moves in a **retrograde** orbit at a mean distance of 23.9 million km from the planet.

Sirius (Alpha Canis Majoris, α CMa)

The brightest star in the sky, also known as the *Dog Star* after its constellation of **Canis Major**; its proper name comes from the Greek for "searing" or "scorching." Sirius is one of the points in the Winter Triangle together with Betelgeuse in Orion and Procyon in the smaller dog, Canis Minor. A main sequence **A star**, its brilliance comes from a combination of fairly high intrinsic luminosity and proximity (only double the distance of the nearest star to the Sun). Its most remarkable feature, though, is its companion, Sirius B, the "Pup," which is a **white dwarf** (visual magnitude 8.44; absolute magnitude 11.33), almost as massive as the Sun (0.96 M_{sun}) but not much bigger than Earth (0.03 R_{sun}), that completes an orbit every 49.9 years. In the past the Pup must have been more massive than the Dog but managed to shed more than a Sun's-worth of matter in its old age. See also **stars, nearest**.

Visual magnitude:	−1.44
Absolute magnitude:	1.45
Spectral type:	A1V
Surface temperature:	9,400 K
Mass:	2 M_{sun}
Distance:	8.60 light-years

slew

The relatively rapid motion of a telescope under computer control as it moves to point at a new position in the sky. Once at the new position the motion of the telescope returns to that required to cancel the effect of Earth's rotation relative to the stars (the sidereal rate).

Slipher, Vesto Melvin (1875–1969)

An American observational astronomer best known for providing some of the earliest evidence that the universe is expanding. Educated at the University of Indiana, Slipher spent his entire career at the **Lowell Observatory** (1902–1952), serving as its director from 1916 on. His visible and infrared spectroscopic studies of planets led to the determination of rotation periods and the identification of molecules in planetary atmospheres. In 1912 he found that the spectrum of nebulosity in the **Pleiades** closely matched that of the surrounding stars and concluded that the nebula's brightness was due to reflected stellar light. He thus discovered **reflection nebula**e and proved there is **dust** in interstellar space as well as gas. He also supervised the search for a ninth planet, which culminated in Clyde **Tombaugh**'s discovery of Pluto in 1930. But his most significant breakthrough centered on the objects then known as spiral nebulae. In 1912 Slipher obtained a set of spectrographs that indicated the Andromeda spiral was approaching the Sun with a velocity of 300 km/s. Most of the other spirals he examined, however, seemed to be flying away from Earth. His 1925 catalog, which included the radial velocities of almost all of the 44 known spirals, provided powerful evidence that these systems lay outside our Galaxy and paved the way for Edwin **Hubble**'s discovery of the expanding universe.

Small Magellanic Cloud (SMC)

The lesser of the two **Magellanic Clouds**–irregular galaxies that are satellites of our own Galaxy. The SMC is about 16,000 light-years in diameter, has a visible mass of about one-fiftieth that of the Milky Way, and, at a distance of some 210,000 light-years, is the third-nearest external galaxy after the **Sagittarius Dwarf Elliptical Galaxy** and the **Large Magellanic Cloud** (LMC). It contains fewer clusters and nebulae than does the LMC, and relatively more gas and dust. However, like the LMC, it seems to have experienced one burst of star formation early in its history followed by a second much more recently. The SMC is important historically as the location in which Henrietta **Leavitt** discovered the period-luminosity relation of **Cepheid variables**. See also **Local Group**.

SMC X-1 (2U 0115-73)

The most luminous known **X-ray pulsar**. Located in the **Small Magellanic Cloud**, it consists of a neutron star and the blue supergiant Sanduleak 160 (type B0I) in a close, 3.89-day-period orbit.

Smithsonian Astrophysical Observatory (SAO)

A research arm of the Smithsonian Institution, headquartered in Cambridge, Massachusetts, where it is joined with the **Harvard College Observatory** (HCO) to form the **Harvard-Smithsonian Center for Astrophysics**. The SAO owns and operates the Fred Lawrence Whipple Observatory on Mount Hopkins, Arizona, jointly owns and operates the **Multiple Mirror Telescope** at the Whipple Observatory, operates the Oak Ridge Observatory (owned by HCO) in Harvard, Massachusetts, and is a partner in the **Submillimeter Array**. It also runs the Central Bureau for Astronomical Telegrams and the **Minor Planet Center**.

Smithsonian Astrophysical Observatory Catalog (SAO Catalog)

A catalog of 258,997 stars, mostly down to magnitude 8.5 plus a few dimmer stars, published by the **Smithsonian Astronomical Observatory** in 1966. It identifies objects in the form: SAO ####; for example, Vega is listed as SAO 67174.

SMM (Solar Maximum Mission)

A NASA spacecraft equipped to study **solar flares** and other high-energy solar phenomena. Launched in 1980, during a peak of solar activity, SMM observed more than 12,000 flares and over 1,200 coronal mass ejections during its 10-year lifetime. It provided measurements of total solar radiative output, transition region magnetic field strengths, storage and release of flare energy, particle accelerations, and the formation of hot plasma. Observations from SMM were coordinated with in situ measurements of flare particle emissions made by ISSE-3 (International Sun-Earth Explorer 3).

Snake Nebula (Barnard 72)

Also known as *Barnard's Dark S Nebula*, a distinctively shaped dark nebula in **Ophiuchus**, 1.5° north-northeast of Theta Ophiuchi (R.A. 17h 23m, Dec. −23° 38′); it was first cataloged by Edward **Barnard**. To its right is B68, an extremely dense dark nebula, and below it are B69, B70, and B74.

SNC meteorites

The collective name for a group of meteorites believed to have come from Mars (see **Mars meteorites**). They are named for the initial letters of the type examples **shergottite**, **nakhlite**, and **chassignite**.

SOFIA (Stratospheric Observatory for Infrared Astronomy)

A Boeing 747-SP aircraft modified to carry a 2.5-m optical/infrared/submillimeter telescope. SOFIA, a joint project of NASA and DLR, the German center for aerospace research, is the largest airborne telescope in the world, capable of observations that are impossible for even the largest and highest of ground-based instruments.

soft gamma repeater (SGR)

Probably a **magnetar** (a highly magnetic neutron star) that emits bursts of soft (low-energy) **gamma rays**, each typically lasting less than a second, at irregular intervals. The bursts may be caused by **starquake**s on the surface crust of the neutron star. SGR bursts, discovered in 1979, are different from **gamma-ray burst**s, which are one-time events.

SOHO (Solar and Heliospheric Observatory)

A joint European Space Agency and NASA mission, launched in December 1995, to investigate the dynamics of the Sun. SOHO was the first spacecraft to be placed in an orbit around the first **Lagrangian point**, 1.5 million km ahead of Earth, a location in which it avoids terrestrial magnetospheric interference and eclipses. The data it collects is helping astronomers understand how the solar **corona** is heated and how it expands into the **solar wind**.

solar apex

The point on the celestial sphere, in **Hercules**, toward which the Sun is moving with respect to the **local standard of rest** at a rate of about 19.4 km/s (about 4.09 AU/year). The point directly opposite this is called the *solar antapex* and lies in **Columba**.

solar constant

The rate at which solar energy, at all wavelengths, is received per unit area at the top level of Earth's atmosphere. The solar constant actually varies by about 0.3%

over the 11-year solar cycle but averages about 1,368 W/m^2. Each planet has its own *planetary solar constant*, ranging from Mercury (9936.9 W/m^2) to Pluto (0.87 W/m^2).

solar cycle
The approximately 11-year quasiperiodic variation in frequency or number of solar active events.

solar eclipse
An **eclipse** in which Earth passes through the shadow cast by the Moon. Solar eclipses only happen when the Moon is new *and* when the Moon lies close to the node of its orbit (i.e., when it's roughly in the same plane as Earth's orbit). To see a *total solar eclipse* the observer has to fall within the Moon's umbra, the darkest part of the lunar shadow, as it races across our planet. The *path of totality* is never any wider than 270 km and, since it is swept out at some 3,200 km/hr, the length of totality at a given location is never more than 7m 31s, and usually no longer than 3 or 4 minutes. A *partial eclipse* is seen by observers in the Moon's penumbra, the partial shadow, on either side of the path of totality. An *annular eclipse* happens when the Moon is near apogee so that its apparent size is less than that of the Sun and a ring, or annulus, of the Sun's disk is still visible even when the Moon and the Sun are seen directly in line. The *magnitude of a solar eclipse* is the fraction of the solar diameter obscured by the Moon at the greatest phase of an eclipse, measured along the common diameter.

solar flare
A sudden and dramatic release of energy through a break in the Sun's **chromosphere** in the region of a **sunspot**, which may last from a few minutes to a few hours. Following an intense solar flare (electron density 10^{11} compared with 10^8 in solar quiet times) the ionization in Earth's atmosphere may increase by several orders of magnitude leading to effects such as bright aurorae, magnetic storms, and radio interference. Solar flares are classified on a scale of importance ranging from 3^+ (largest area) to 1^- (smallest area). The largest solar flares eject a mass of about 10 billion tons at a speed of roughly 1,500 km/s.

solar longitude
The angular distance along Earth's orbit measured from the intersection of the ecliptic and the celestial equator where the Sun moves from south to north; it gives the position of Earth in its orbit.

solar mass (M_{sun})
A unit of mass equivalent to the mass of the Sun–about 2×10^{30} kg.

solar maximum (or minimum)
The month(s) during the solar cycle when the 12-month mean of monthly average **sunspot** numbers reaches a maximum (or minimum).

solar nebula
The cloud of gas and dust that began to collapse about 5 billion years ago to form the solar system.

solar parallax
The angle subtended by Earth's equatorial radius at the center of the Sun at the mean distance between Earth and the Sun (1 AU); equal to 8.794″.

solar radio emission
Emissions of the Sun at radio wavelengths from centimeters to decameters, under both quiet and disturbed conditions, of which several types are recognized. Type I is a noise storm composed of many short, narrowband bursts in the metric range (50 to 300 MHz). Type II is a narrowband emission that begins in the meter range (300 MHz) and sweeps slowly (over tens of minutes) toward decameter wavelengths (10 MHz). Type II emissions occur in loose association with major flares and are indicative of a **shock wave** moving through the solar atmosphere. Type III are narrowband bursts that sweep rapidly (in a matter of seconds) from decimeter to decameter wavelengths (0.5 to 500 MHz). This type often occurs in groups and is an occasional feature of complex solar active regions. Type IV is a smooth continuum of broadband bursts primarily in the meter range (30 to 300 MHz). These bursts are associated with some major flare events beginning 10 to 20 minutes after the flare maximum, and can last for hours.

solar rotation number
A parameter used by solar observers to describe their data. The system starts, for historical reasons, with rotation #1 on November 9, 1854, and a new rotation begins when the solar central meridian is equal to zero (that is, the Sun has made one apparent rotation). This takes, on average, about 27.275 days.

solar system
The Sun and all objects gravitationally bound to it. Its outer boundary is roughly a sphere with a radius of some 100,000 AU, centered on the Sun, that marks the farthest extent of the **Oort Cloud**. The Sun is overwhelmingly the dominant object. Planets, satellites, and all interplanetary material together comprise only about 1/750 of the total mass. Geochemical dating shows that the solar system chemically isolated itself from the rest of the Galaxy some 4.7 billion years ago.

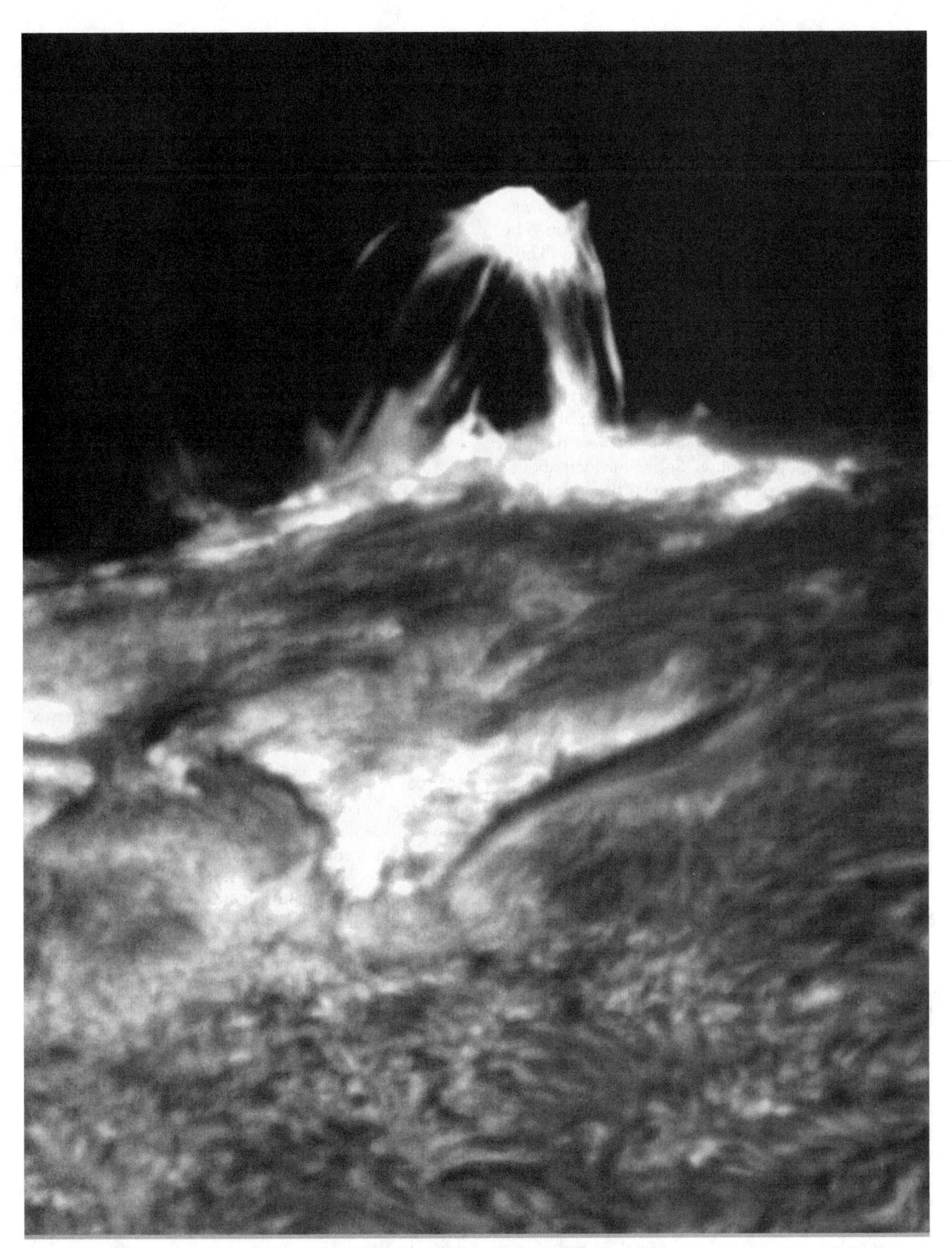

solar flare A spectacular solar flare caught by the Advanced Composition Explorer satellite. *NASA*

solar telescope

A telescope designed specifically for observing the Sun. Major ones, including the **McMath-Pierce Solar Telescope** and the **Big Bear Solar Observatory**, are of the **solar tower** variety.

solar time

Time measured with respect to the Sun. *Apparent solar time* is based on the daily motion of *true Sun,* in other words, the actual position of the Sun in the sky. This is the version of solar time that a sundial shows, but it runs unevenly for the two main reasons: Earth's axis is tilted with respect to its orbital plane and Earth's orbit is elliptical rather than circular. A better version of solar time is *mean solar time,* which is used for accurate timekeeping, including all civil timekeeping. It is based on how the Sun would move in the sky if Earth had no axial tilt and had a perfectly circular orbit and there were no other irregularities in Earth's motion. Solar time loses about 4 minutes a day compared with **sidereal time** because of Earth's orbital movement.

solar tower

A **solar telescope** that is mounted on a tall tower in order to place the optics above the image-distorting layer of Sun-heated air near the ground.

solar wind

A tenuous, radial flow of gas and energetic charged particles, mostly protons and electrons, from the Sun's **corona**. The expanding solar wind also drags the solar magnetic field outward, forming the **interplanetary magnetic field** (IMF). The region of space in which this solar magnetic field is dominant is called the **heliosphere**. Although the solar wind moves out almost radially from the Sun, the rotation of the Sun gives the magnetic field a spiral form (like that of the water jet from a two-way rotating garden sprinkler). The solar wind produces **aurorae**, causes the tails of comets to stream back from the Sun, and distorts the symmetry of planetary **magnetospheres**. Its speed varies widely, from a low-speed wind of (at Earth's distance) about 300 km/s, to a high-speed wind, flowing from **coronal holes**, of 750 km/s.

solstice

Either of two points on the ecliptic, midway between the **equinoxes**, at which the Sun reaches its greatest northern **declination** (summer solstice in the Northern Hemisphere, winter solstice in the Southern Hemisphere, on June 21) and greatest southern declination (vice versa, on December 22). At these points occur the longest day and shortest night in one hemisphere of Earth, and vice versa in the other. The *solstitial points* are the Sun's positions when it is momentarily stationary on these occasions. The *solstitial colure* is the **hour angle** that passes through the solstices on the celestial sphere.

Sombrero Galaxy (M104, NGC 4594)

A type Sa **spiral galaxy** in **Virgo** whose unusually large central bulge, richly populated with several hundred globular clusters, and dark prominent dust lanes give it the appearance of a Mexican hat; we see it from about 6° south of its equatorial plane. Discovered by Pierre **Méchain** in 1781, it became the first galaxy for which a large **redshift** (equivalent to a recession velocity of about 1,000 km/s) was found, by Vesto **Slipher** in 1912. It is also the first galaxy in which rotation was detected, again by Slipher. Modern studies have shown the Sombrero to have both a very extended faint halo and a mildly active nucleus indicative of the presence a central supermassive **black hole**.

Visual magnitude:	8.0
Apparent size:	9′ × 4′
Distance:	50 million light-years
Position:	R.A. 12h 40.0m; Dec. −11° 37′

Sothic cycle

A 1,460-year calendric cycle used in ancient Egypt; Sothis was the Greek name for Sirius, whose heliacal rising determined the Egyptian's fixed year of 365 days. Because of the lack of a leap day, the calendar regressed every Sothic cycle by one year.

source

Anything that emits electromagnetic radiation. A distinction is made between **point sources**, **compact sources**, and **extended sources**.

South African Astronomical Observatory (SAAO)

The national facility in South Africa for optical/infrared astronomy. Located near Sutherland, Northern Cape, at an altitude of 1,760 m, it is owned and operated by the government's Foundation for Research Development. The SAAO was founded in 1972 by combining the resources of the Royal Observatory at the Cape of Good Hope (founded in 1820) and those of the Republic Observatory, Johannesburg (founded in 1905). It operates several reflecting telescopes ranging in aperture from 0.5 m to 1.88 m. However, its capabilities are about to be vastly enhanced by the addition of the **South African Large Telescope**.

South African Large Telescope (SALT)

An 11-m-aperture telescope–the largest single telescope in the Southern Hemisphere–due for completion in late 2004 at the Sutherland observing station of the **South**

Sombrero Galaxy Details as small as 170 light-years are resolved in this picture of the Sombrero Galaxy (M104, NGC 4594) by the Very Large Telescope. *European Southern Observatory*

African Astronomical Observatory. Built using a hexagonal mirror array, SALT will be similar to the **Hobby-Eberly Telescope** in Texas. Partners in the project, in addition to South Africa, include institutions in Germany, Poland, the United States, and New Zealand.

Southern Pleiades (IC 2602)

A cluster of more than 50 stars centered on the blue-white star Theta Carinae (see **Carina**) and forming a triangle in the sky with Beta and Iota Car. At a distance of less than 500 light-years, the Southern Pleiades is relatively nearby compared with many star clusters, and can be seen with the naked eye.

Magnitude:	2
Diameter:	50′
Position:	R.A. 10h 43.2m, Dec. –64° 24′

Southern Reference Stars (SRS)

A list of 20,488 stars in the Southern Hemisphere intended to supplement the AGK3R (see ***Astronomische Gesellschaft Katalog***) in the north, to extend coverage of reference stars across the whole sky for calibrating photographic surveys. Observations of stars, mostly of magnitudes 7.5 to 9.5, were made at a dozen observatories between 1961 and 1973, and coordinated at the U.S. Naval Observatory and Pulkovo Observatory.

Southern Sky Survey

A photographic survey of the sky south of declination –17°, carried out jointly by the 1.2-m **United Kingdom Schmidt Telescope** at Siding Spring and the European Southern Observatory's 1-m Schmidt camera. Both these instruments have the same focal length as the 1.2-m Oschin Schmidt, which produced the Northern Hemisphere ***Palomar Observatory Sky Survey***, so the scale of the plates in each survey is identical.

Space Infrared Telescope Facility (SIRTF)

The fourth and final element in NASA's family of **Great Observatories**. SIRTF consists of a 0.85-m telescope and three cryogenically-cooled science instruments capable of performing imaging and spectroscopy in the 3- to 180-micron wavelength range. It was scheduled for launch in August 2003.

Space Interferometry Mission (SIM)

The first spacecraft to carry an optical **interferometer** as its main instrument; SIM is designed specifically for the precise measurement of star positions. One of its prime goals will be to search for **extrasolar planets** as small as Earth in orbit around nearby stars. It will combine the light from two sets of four 30-cm-diameter telescopes arrayed across a 10-m boom to achieve a resolution approaching that of a 10-m-diameter mirror. This will allow it to perform extremely sensitive astrometry so that it will be able to detect very small wobbles in the movement of a star due to unseen companions. Objects of Earth mass could be inferred around a star up to 30 light-years away. SIM's seven-year mission is scheduled to begin in 2005.

space velocity

The velocity of a star, with respect to the **local standard of rest**, expressed as components along three axes. The components are usually given as U (radially outward from the galactic center), V (in the direction of galactic rotation), and W (in the direction of the galactic north pole, i.e., perpendicular to the galactic plane), with positive signs if the velocity is directed as indicated or negative signs if oppositely directed. The resultant space velocity is given by Pythagoras as $\sqrt{(U^2 + V^2 + W^2)}$. In the case of the Sun, U = –9 km/s, V = +12 km/s, and W = +7 km/s, giving a space velocity of 17 km/s.

space-time

The union of space and **time** into a four-dimensional whole.

Spacewatch Project

A project the University of Arizona's Lunar and Planetary Laboratory founded in 1980 to explore the various populations of small objects in the solar system, especially Centaurs, Trojans, main belt asteroids, trans-Neptunian objects, and near-Earth objects, in order to better understand their dynamical evolution. Spacewatch also finds potential targets for interplanetary spacecraft missions, provides follow-up astrometry of such targets, and finds objects that might present a hazard to Earth. **CCD**-scanning observations are carried out using the Steward Observatory's original 0.9-m Spacewatch telescope and the new 1.8-m Spacewatch telescope, both on **Kitt Peak**. The program has discovered more than 300 new near-Earth asteroids and many thousands of new objects in the main belt and beyond.

spallation

A nuclear reaction in which several particles result from a collision. In astronomy, spallation typically involves the collision of high-energy cosmic rays with atoms, either in interstellar space or on the surface of objects such as meteorites, asteroids, or the Moon, that are unprotected by an atmosphere. Spallation reactions between cosmic rays and interstellar matter result in the production of fast-moving nuclei of lithium, beryllium, and boron, so that these elements are found to be a million times more common in cosmic rays than in stars.

Special Astrophysical Observatory

Russia's main center for ground-based space research. Operated by the Russian Academy of Sciences, it is located at an altitude of 2,100 m on Mount Pastukhov in the Caucasus mountains of Karachaevo-Cherkesia, in the south of the country. Its main instruments are the 6-m Bolshoi Azimuthal Telescope and the 600-m RATAN-600 radio telescope.

speckle interferometry

A method of extracting detail in celestial objects that is normally lost because of atmospheric turbulence. It involves taking hundreds of short snapshots, with exposure times of 0.001 to 0.1 second, each fast enough to freeze the blurring effect of the atmosphere, and computer-processing the results.

spectral line

Emission or absorption at a discrete wavelength (or frequency) caused by a specific electron transition within an atom, molecule, or ion. The dark lines visible in an **absorption spectrum,** or bright lines that make up an **emission spectrum** are caused by the transfer of an electron in an atom from one energy level to another. The essential difference between optical line spectra and X-ray spectra is that the former correspond to energy changes in the *outer* electrons in an atom, and the latter to energy changes in the *inner* electron orbitals. Gamma rays usually correspond to energy changes in the nucleus. Infrared radiation is produced by high-*n* transitions of atoms or by the vibration or rotation of molecules. Thermal radio emission is usually produced by still higher-*n* transitions.

spectral series

The set of **spectral lines** of a given atom produced by electron jumps from different initial energy levels to the same final one; for example, the **Balmer series** of hydrogen.

spectral type

The classification of a star according to the details of its spectrum. Since these details are largely determined by the star's surface temperature, spectral type is also a good indicator of color. There are seven main spectral types. From hot and blue to cool and red, they are O, B, A, F, G, K, and M. For further precision, each spectral type is broken down into 10 subclassifications. For example, from warmest to coolest, spectral type G is G0, G1, G2, G3, and so on to G9. The Sun is spectral type G2. (The exception is the "O" classification. It is subclassified as O4, O5, O6, . . . , O10.) (See table, "The Classification of Stars by Spectral Type.")

spectrograph

An instrument that disperses (breaks apart) light into a spectrum so that the intensity at different wavelengths can be measured by a detector. It consists of a slit for isolating the light from a particular object in the telescope's field of view, a collimator for directing this light into a parallel beam, a diffraction grating or prism for creating the spectrum, and a detector for recording the spectrum. In the past the detector would have been a photographic plate but today **CCDs**, which are far more sensitive, are typically used instead.

spectroheliograph

An instrument used to photograph the Sun at the wavelength of a strong **Fraunhofer line**, usually the hydrogen-alpha or calcium K line. The image of the Sun formed by a telescope is focused onto the slit of a high-dispersion **spectrograph**. A second slit, immediately below the first and lying in the spectrograph's focal plane, isolates the chosen narrow band of wavelengths from the solar continuum. By moving the primary slit across the Sun's image and moving the secondary slit to keep its spectral location, a monochromatic photograph of the Sun is obtained. A *spectrohelioscope* works in the same way, but visually rather than photographically. By placing the eye instead of a photographic plate behind the second slit, and then vibrating both slits in unison through a small amplitude, a monochromatic image of part of the solar disk can be seen. The image appears steady because of the persistence of vision.

spectrometer

An instrument connected to a telescope that separates the light signals into different frequencies, producing a **spectrum.** A *dispersive spectrometer* is like a prism: it scatters light of different energies to different places. A *nondispersive spectrometer* measures the energy directly.

spectrophotometer

A combination of **spectrograph** and **photometer** used to analyze a spectrum that has been recorded on a photographic plate. By passing a narrow beam of light through the recorded image, the original intensity of light from

The Classification of Stars by Spectral Type

Spectral Type	Characteristic Spectra	Color	Temperature Range (K)	Examples
O	He II; He I	blue	28,000–50,000	Alnilam
B	He I; H	blue-white	9,900–28,000	Rigel, Spica
A	H	white	7,400–9,900	Vega, Sirius
F	metals; H	yellow-white	6,000–7,400	Procyon
G	Ca II; metals	yellow	4,900–6,000	Sun, Alpha Centauri A
K	Ca I, Ca II; other mols	orange	3,500–4,900	Arcturus
M	TiO; Ca I; other mols	orange-red	2,000–3,500	Betelgeuse

the source at various wavelengths can be recovered. This information can then be displayed as a graph or stored as a digitized spectrum for further processing.

spectroscope

An instrument that produces a **spectrum** for visual observation. If a camera or **CCD** detector is used to record the spectrum (as is always the case in modern observational work), the device is known as a **spectrograph**.

spectroscopic binary

A **binary star** whose dual nature is only apparent through measurements made with a **spectroscope**. All **close binaries** and most **eclipsing binaries** fall into this category. The existence of two stars is revealed by periodically varying **Doppler shifts** of the spectral lines. Double-lined spectroscopic binaries have two sets of spectral features, oscillating with opposite phases. Single-lined spectroscopic binaries have only one set of oscillating spectral lines, owing to the dimness of the secondary component.

spectroscopy

In astrophysics, the study of the spectra (see **spectrum**) of celestial bodies with a view to learning about their chemical composition, physical conditions, and state of motion. The primary instrument used is a **spectrograph**.

spectrum (pl. spectra)

A plot of the intensity of **electromagnetic radiation** at different **wavelengths**; in the case of visible light, this is the familiar rainbow of colors. An object that emits radiation in a continuous range of colors is said to have a **continuous spectrum**. An object that emits radiation only at certain wavelengths is said to have **emission lines**; objects that absorb radiation only at certain wavelengths are said to have **absorption lines**. A good spectrum of a star, for example, reveals, among other things, its **spectral type**, **radial velocity**, and **metallicity**.

speculum

A brittle white alloy of two parts copper to one part tin that takes a very high polish and was used for telescope mirrors until the late nineteenth century. The manufacture of speculum mirrors reached its apex with William Parsons, third Earl of **Rosse**, who managed to produce several 36-in. (91-cm) **Newtonian telescopes** and later, in 1845, a colossus 72 in. (1.8 m) in diameter. However, speculum, which is difficult to work with and quick to tarnish (forcing a repolishing of the optical surface itself) was made obsolete in astronomy with the development of silvering and **aluminizing**.

Spencer Jones, Harold (1890–1960)

An English astronomer who became the tenth **Astronomer Royal** and led a decade-long worldwide effort to determine the distance to the Sun by triangulating the distance of the asteroid Eros when it passed near Earth in 1930–1931: he photographed Eros more than 1,200 times and reduced the data from other observers in one of the most impressive computational feats of the precomputer era. Educated at Cambridge, Spencer Jones was a successively astronomical assistant at the Royal Observatory, Greenwich; His Majesty's Astronomer at the Cape of Good Hope; and, from 1933 to 1955, director of the Royal Observatory and Astronomer Royal. His work was devoted to fundamental positional astronomy. At the Cape he worked on proper motions and parallaxes. After his monumental effort in solar astrometry, he showed that small residuals in the apparent motions of the planets are due to the irregular rotation of Earth. After World War II he supervised the move of the Royal Observatory to Herstmonceux, where it was renamed the Royal Greenwich Observatory. (Some sources list his surname as Jones; he preferred Spencer Jones.)

spherical aberration

A defect of spherical mirrors and some lenses and eyepieces that causes parallel light rays striking the element at different distances from the optical axis to be brought to a focus at different points along the axis. See also **aberration, optical**.

spherical astronomy

One of the oldest branches of astronomy, concerned with determining the positions and motions of objects on the **celestial sphere**.

spherical coordinates

A system for specifying positions in terms of angles on a sphere, such as the **celestial sphere** or the surface of a planet or large moon.

Spica (Alpha Virginis, α Vir)

The brightest star in **Virgo** and the fourteenth brightest in the sky; its name comes from the Latin for "an ear of wheat" and may stem from the fact that the Sun passes Spica in autumn, thus rendering the star a harvest symbol. It lies about 10° south of the celestial equator and almost on the ecliptic, so that it is regularly occulted by the Moon. Spica is a binary system that consists of a giant or subgiant **B star** and a main sequence B star separated by some 18 million km, in an orbit with a 4-day period. The orbit is inclined at 24° to our line of sight; however, the stars still partially eclipse one another and give rise to variations of about 0.07 magnitude. The bright, giant component has a mass of 10.9 M_{sun} and is a **Beta Cephei star** with a 4.2-hour period, while its partner, which accounts for about 20% of the total light output, has a mass of 6.8 M_{sun}. Strong X-ray emission has been recorded, at least some of which seems to be produced when the winds that flow from the brilliant pair violently collide. Additional evidence suggests that Spica may in fact be a multiple star system, containing two other fainter components.

Visual magnitude:	0.98
Absolute magnitude:	–3.55
Spectral type:	B1III–IV + B2V
Combined luminosity:	2,100 L_{sun}
Distance:	262 light-years

spicule
A short-lived (about 5 minutes), narrow jet of gas, under 10,000 km long, in the upper **chromosphere** of the Sun. Visible in the light of H-alpha, spicules tend to cluster at the edges of **supergranulation cell**s and are usually seen as a mass of tiny brighter spikelike features at the limb or as tiny darker spikes coming out of network elements. Macrospicules are about 10 times larger and are best seen in the He II ultraviolet line at 304 Å.

spin temperature
A way of measuring the temperature of neutral **hydrogen** (HI) in space in terms of the relative populations of its two possible spin states–parallel (electron and proton spins aligned) and antiparallel (spins opposed). The spin temperature of an **H I region** is the temperature at which its observed ratio of parallel to antiparallel spins would occur if the gas were in thermal equilibrium. It differs from the **kinetic temperature** at low densities, where the spin temperature is raised by the tendency of resonant scattering of Lyman-alpha photons to equalize the occupation probabilities in the two spin states.

spin-down
The gradual lengthening of the period of a **pulsar** as it loses energy with age. Observed spin-down rates range from 10^{-12} s per second for the youngest pulsars to 10^{-19} s per second for recycled pulsars; the Crab pulsar is slowing at a rate of 3.5×10^{-13} s per second. Occasionally a *glitch* occurs in the spin-down in which there is a sudden increase in rotation rate. This is thought to be the equivalent of an earthquake where the crust of the neutron star settles a little bit, decreasing the radius and moment of inertia.

spiral arm
A curved feature, containing young stars, open clusters, H II regions, and dust, that winds outward from the nucleus of a **spiral galaxy** into the disk. Usually two arms are present that wrap around in a well-defined pattern; however, some spirals have four or, in rare cases, three arms, in an arrangement that may be complex and fragmentary. The arms are visible because of bright, massive stars, the formation of which is triggered by the movement through the disk of a **density wave**.

spiral galaxy
A **galaxy** with a nuclear bulge from which emanate luminous **spiral arms**. There are two main types, **barred spirals** (type *SB*) and ordinary spirals (type *S*), each of which are divided into subtypes *a* (tightly wound arms, large bulge), *b*, *c*, and (sometimes) *d* (loosely wound arms, small bulge). The trend from *a* to *d* is toward decreasing brightness and mass. In diameter, spirals range from about 10,000 to over 300,000 light-years, and in mass from about a billion to 500 billion M_{sun}. Spiral structure can apparently exist only in disk galaxies above a certain size and, although spirals represent 80% of the bright galaxies in regions outside of rich clusters, there are no spirals with masses as low as those of many irregulars and dwarf ellipticals.

Spitzer, Lyman Jr. (1914–1997)
An American astrophysicist whose most important work was on the physics of the **interstellar medium**. He showed that there must be at least two phases–high temperature clouds around hot stars and cooler intercloud regions–and pioneered studies of interstellar dust grains and magnetic fields. Spitzer studied at Yale and Cambridge Universities and earned his Ph.D. under Henry **Russell** at Princeton. Following research at Harvard, teaching at Yale, and war work at Columbia, he succeeded Russell as professor and observatory director at Princeton in 1947. He promptly hired Martin **Schwarzschild**, and the two built a major research department. An early leader in attempts to harness controlled thermonuclear fusion, Spitzer was the founder and first director of the Princeton Plasma Physics Laboratory (originally called Project Matterhorn). He was the first to propose a large telescope in space (1946) and he led the development and operation of the ultraviolet astronomy satellite Copernicus. He was analyzing data from the Hubble Space Telescope the day he died.

Spörer, Gustav Friedrich Wilhem (1822–1895)
A German astronomer who, independently of Richard **Carrington**, used observations of **sunspot**s to determine the position of the Sun's equator (and hence the tilt of its axis) and to establish that it rotates at different rates at different latitudes. He was the first to draw attention to

spiral galaxy NGC 4414, a flocculent spiral galaxy that lies about 62 million light-years away in the constellation Coma Berenices. *NASA/STScI/AURA*

the lack of sunspots between 1645 and 1715, which became known as the Maunder minimum after Walter **Maunder**, but didn't discover an earlier period of low solar activity, which *is* named after him (see **Spörer minimum**). To add to the misnomenclature, **Spörer's law** was first noticed by Carrington.

Spörer minimum

A period of low **sunspot** activity that lasted from about A.D. 1420 to 1570, based on evidence such as carbon-14 in tree rings and records of auroral activity. The Spörer minimum and the similar Maunder minimum coincide with a period of low global temperatures known as the *Little Ice Age*. It was discovered by the American solar physicist John Allen Eddy (1931–) who named it after Gustav **Spörer**.

Spörer's law

The observational rule that the solar latitude at which new **sunspots** appear gradually decreases, from 30 to 40° north or south of the solar equator, at the beginning of a solar cycle, to 5 to 10°, at the end of the cycle. This tendency is revealed very clearly on a **butterfly diagram**. Although named after Gustav **Spörer**, the "law" was first discovered by Richard **Carrington**.

s-process

A type of nuclear reaction, believed to take place inside massive stars, in which elements heavier than copper are formed by the slow ("s") absorption of neutrons by atomic nuclei. The capture of neutrons occurs on timescales that are sufficiently long to enable unstable nuclei to decay via the emission of a beta particle before absorbing another neutron. This contrasts with the **r-process** that is thought to operate in supernovae. Prominent s-process elements include barium, zirconium, yttrium, and lanthanum.

SS 433

An **X-ray binary** and the first discovered member of a group of very exclusive objects that have come to be known as **microquasars**; it lies about 17,000 light-years away in Aquila within a 40,000-year-old supernova remnant designated W50. SS 433 consists of a normal O or B star, with a mass of 10 to 20 M_{sun}, in a 13.1-day orbit around a compact object that is either a neutron star or a black hole. Material transfers from the normal star into an **accretion disk** surrounding the compact object, blasting out two **jets** of ionized gas in opposite directions at about a quarter of the speed of light. In addition the jets precess (wobble like a top) with a period of about 164 days. Radio studies show that the jets extend to 0.16 light-year from the central "engine," while X-ray studies reveal emission from about 100 light-years on either side, where the jets interact with the surrounding supernova remnant. SS 433 gets its name because it is object number 433 in the Stephenson-Sanduleak catalog of stars with strong emission lines compiled by two astronomers at Case Western Reserve University, Bruce Stephenson and Nicholas Sanduleak, in 1977.

SS Cygni (SS Cyg)

The brightest **U Geminorum star** (dwarf nova) and the second to be discovered (after U Gem itself in 1855), by Louisa D. Wells of the Harvard College Observatory in 1896. It is the type object for the **SS Cygni star** subclass and one of the most observed variable stars in the sky. SS Cyg lies about 90 light-years away and consists of an orange-yellow dwarf of mass 0.60 M_{sun} and a white dwarf of mass 0.40 M_{sun} orbiting around each other in 6 hr 38 min at distance (surface to surface) of 160,000 km or less. Three-quarters of the time, SS Cyg is quiescent with a visual magnitude of 12.2; then it begins to brighten without warning and reaches peak brightness of 8.3 in a day or so. Outbursts last 1 to 2 weeks and are repeated every 4 to 10 weeks with a mean time between eruptions of 54 days. A closer look at SS Cyg's **light curve** reveals ever-changing intervals of long (L) and short (S) outbursts, of about 18- and 8-day durations, respectively. In addition, there are occasional anomalous outbursts, wide and symmetrical in shape, which occur with a slow rise. Although the star has typically displayed this changing outburst characteristic since its discovery, it went through a period from 1907 to 1908 when it abandoned its normal outburst behavior and underwent only minor fluctuations. A statistical study of SS Cygni's light curve revealed that the most common sequence of outbursts is LS (with 134 occurrences), followed by LLS (69), LSSS (14), and LLSS (8). Together these strings represented 89% of the outbursts studied. According to one suggestion, what determines whether an outburst will be long or short is the amount of mass in the disk at the start of a thermal instability: a short outburst corresponding to moderate mass transfer, while a long outburst results from a major mass transfer.

SS Cygni star (SS Cyg star)

One of three subclasses of **U Geminorum stars** (dwarf novae), the others being **SU Ursae Majoris stars** and **Z Camelopardalis stars**. Named after the first to be discovered, **SS Cygni**, these objects undergo well-defined outbursts with an amplitude of 2 to 6 magnitudes, a rise time of 1 to 2 days, and a somewhat longer decline. The mean interval between outbursts ranges from about 10 days to several years. Although most SS Cygni stars display distinct short and long outbursts, the latter are quite different from the supermaxima of SU Ursae Majoris types.

standard candle

An object–usually a type of star or a galaxy–of known intrinsic brightness that can be used in setting up a distance scale. Measurement of the apparent brightness of a standard candle, such as a **Cepeheid variable**, yields its distance.

standard star

A star used to calibrate observations of previously unstudied stars, particularly for photometry. In **spectrophotometry**, standard stars are compared with a blackbody source close to the telescope so that the amount of radiation emitted at each wavelength is known. In conventional photometry, standard stars have accurately known magnitude and colors against which stars under study are compared.

star

A large ball of gas that produces and emits its own radiation; stars form from gas and dust in space. At least at some stage of their evolution, they release energy through nuclear fusion reactions in their core (see **stellar evolution**). However, dead stellar cores, including **white dwarfs** and **neutron stars**, continue to shine because of their stored thermal energy. Stars are classified primarily by their **spectral type**. Many are **variable stars** that change noticeably in their light output, over periods ranging from seconds to years. The mass of stars ranges from about 120 M_{sun}, in the case of the largest **hypergiants**, to 0.0013 M_{sun}, in the case of the lightest **brown dwarfs**. Star luminosities range from 500,000 L_{sun} to less than 0.001 L_{sun}. Although the most prominent stars in the night sky are all brighter than the Sun, most stars (largely **red dwarfs** and brown dwarfs) are much dimmer than the **Sun**. All shine because of nuclear **fusion** reactions, of which the most important are those which convert hydrogen into helium.

star cluster

A gravitationally bound collection of stars that formed from the same gas cloud. There are two main types. **Globular clusters** are found in the halos of galaxies and contain from tens of thousands to millions of ancient stars crowded into a more or less spherical volume of space. **Open clusters** inhabit the disks of galaxies and contain up to a few hundred relatively young stars in a looser arrangement from which stars may occasionally escape. Looser still are **stellar associations**.

star diagonal

A gadget resembling a plumbing elbow fitting, containing a mirror or prism, attached to a downward-pointing focuser to redirect the beam of light upward for more convenient viewing. That is, a star diagonal fits into a focuser, has an eyepiece fitted into it, and bends the light path through a right angle.

star formation

See **stellar evolution**.

star formation A region of very active star formation in our Galaxy known as NGC 3603, imaged in the near-infrared by the Very Large Telescope. The central cluster is the densest concentration of massive stars known in the Milky Way and contains more than 50 hot O stars. *European Southern Observatory*

starburst galaxy The starburst galaxy NGC 3310, whose spiral arms are peppered with the bright points of newly formed star clusters. Each of the clusters represents the formation of up to a million stars, a process that takes less than 100,000 years. In addition, hundreds of individual young, luminous stars can be seen throughout the Galaxy. Measurements of the wide range of cluster colors suggest that the starburst began over 100 million years ago, triggered perhaps by the collision of a companion galaxy. NGC 3310 lies in Ursa Major at a distance of about 59 million light-years. *NASA/STScI*

star streaming

The statistical tendency of stars in the Sun's neighborhood to move in two preferred directions, one roughly away from the center of the Galaxy and the other toward it. The effect is due to the noncircular orbits of stars around the galactic center. Discovered in 1902 by Jacobus **Kapteyn**, star streaming provided an early clue that the Galaxy is rotating, though it wasn't recognized as such at the time.

starburst galaxy

Any galaxy in which star formation is taking place on an unusually large or rapid scale; specifically, a galaxy that is making stars so fast that it would convert all of its unconsolidated material into stars in a timescale, known as the *exhaustion timescale,* that is much less than the age of the universe. In some cases, starburst galaxies have star-forming rates of hundreds of M_{sun} per year, corresponding to an exhaustion timescale of the order of 100 million years (about 1% the age of the universe). It follows that observed starbursts must have started in the relatively recent past. The burst may be galaxy-wide or confined to a small region (less than 1,000 light-years across) about the nucleus. Typically, much of the star formation occurs in very luminous, compact star clusters, 10 to 20 light-years across, with luminosities up to 100 million L_{sun}. These clusters are the most dense and intense star-forming environments known, and may be analogues of typical objects in the early epochs of galaxy formation. If they remain gravitationally bound after the mass loss from massive members is complete, they may eventually come to look a great deal like **globular clusters**.

Observationally, the overwhelming signature of starburst galaxies is intense emission in the far-infrared, caused by the ultraviolet emitted by numerous hot, young stars being absorbed by dust and reemitted at longer wavelengths. At these wavelengths, and to a lesser degree in the radio region, starburst galaxies rank second only to **active galactic nuclei** (AGN). The two types can readily be distinguished, however: a compact, flat-spectrum radio source indicates an AGN, while more diffuse radio emission suggests a star-forming nucleus. Furthermore, AGN are more radio-loud and show high-ionization species, whereas the lack of these species combined with the presence of diffuse interstellar bands, due to molecules that are destroyed by the intense hard radiation in AGN, point to a starburst. The nearest example of a starburst galaxy is **M82** (the Cigar Galaxy).

Stardust

A NASA space probe, launched in February 1999 and designed to collect samples of **dust** of both interplanetary and interstellar origin, during a seven-year mission that will take it to within 160 km of Comet Wild-2 (pronounced "vihlt") in December 2004. Stardust will send back pictures, take counts of the number of dust particles striking it, and analyze in real-time the composition of substances it has collected before returning its samples to Earth in January 2006. Analysis of the captured material, which will include pre-solar grains and condensates left over from the formation of the solar system, together with dust that has blown in from interstellar space, is expected to yield important insights into the evolution of the Sun and planets and possibly the origin of life itself.

Stark effect

The broadening or splitting of a **spectral line** that results when an electric field slightly changes the energy levels of a radiating atom or ion. Stark broadening is proportional to the ion and electron density in a plasma and is therefore a good indicator of pressure in a stellar atmosphere and hence of the star's luminosity. The effect is named after the German physicist Johannes Stark (1874–1957).

starquake

The equivalent of an earthquake in a neutron star: the favored explanation for **glitch**es, or sudden changes in the period of a **pulsar**. As a pulsar gradually slows down in its rate of rotation (see **spin-down**), its slightly oblate shape will tend to become spherical. The readjustments in its solid crust needed for this to happen will be accompanied by a fracturing and resettling of its crust–a starquake.

stars, brightest

The brightest stars in the night sky stand out because they are intrinsically luminous or relatively close to the Sun, or a combination of both. Every one of the twenty-five stars that appear brightest to the unaided eye would outshine the Sun if put in place of the Sun. However, they represent a huge luminosity range, from **Alpha Centauri** (the nearest star system, the brightest member of which has only one and a half times the solar luminosity) to **Deneb,** which would outshine 200,000 Suns. The most intrinsically luminous stars known are **hypergiant**s such as **Eta Carinae**, **Rho Cassiopeiae**, and the **Pistol Star**. (See table, "The Brightest Stars in the Sky.")

stars, nearest

Of the 29 stars that lie within 12 light-years of the Sun, 18 (or 62%) are **red dwarf**s, 5 are **K stars** less luminous than the Sun, 2 are **white dwarf**s, 2 (Sirius A and Procyon A) are stars significantly bigger and brighter than the Sun, and 2 (Alpha Centauri A and Tau Ceti) are **G star**s of similar spectral type to the Sun. See the separate entries for many of these stars. A new nearby star, a red dwarf cataloged as SO025300.5+165258, was discovered in 2003, though its distance hasn't yet been precisely determined. It is thought to lie somewhere between 6.5 (third nearest star system) and 16.3 light-years (forty-sixth nearest star system) away. (See table, "Stars within 12 Light-years of the Sun," page 465.)

stationary point

(1) A point in space at which a body can remain essentially motionless with respect to other bodies; the best known examples are **Lagrangian point**s. (2) A position on the celestial sphere at which an outer planet ceases to move in right ascension before switching from direct motion to retrograde motion, or vice versa.

statistical parallax

The mean **parallax** (from which the distance can be derived) of a group of stars obtained by statistical analysis of their proper motions with the effects of the Sun's motion removed.

The Brightest Stars in the Sky

Star		Magnitude		Distance	Position	
		App.	Abs.	(light-years)	R.A. (h m)	Dec.(° ′)
Sun		–26.72	4.83	–	–	–
Sirius	Alpha Canis Majoris	–1.44	1.46	8.6	06 45	–16 43
Canopus	Alpha Carinae	–0.62	–5.53	312	06 24	–52 42
Rigil Kentaurus	Alpha Centauri	–0.27	4.43	4.22	14 40	–60 50
Arcturus	Alpha Bootis	–0.05	–0.30	37	14 16	+19 11
Vega	Alpha Lyrae	0.03	0.58	25	18 37	+38 47
Capella	Alpha Aurigae	0.08	–0.48	42	05 17	+46 00
Rigel	Beta Orionis	0.18	–6.69	773	05 15	–08 12
Procyon	Alpha Canis Minoris	0.40	2.68	11.4	07 39	+05 13
Achernar	Alpha Eridanus	0.45	–2.77	144	01 38	–57 14
Betelgeuse	Alpha Orionis	0.45v	–5.14	427	05 55	+07 24
Hadar	Beta Centauri	0.61v	–5.43	525	14 04	–60 20
Altair	Alpha Aquilae	0.76	2.20	16	19 51	+08 52
Acrux	Alpha Crucis	0.77	–4.19	321	12 27	–63 06
Aldebaran	Alpha Tauri	0.87v	–0.64	65	04 36	+16 31
Spica	Alpha Virginis	0.98v	–3.55	262	13 25	–11 10
Antares	Alpha Scorpii	1.06v	–5.28	604	16 30	–26 26
Pollux	Beta Geminorum	1.16	1.08	34	07 45	+28 01
Formalhaut	Alpha Piscis Australis	1.17	1.74	25	22 58	–29 37
Deneb	Alpha Cygni	1.25	–8.73	3,230	20 41	+45 17
Mimosa	Beta Crucis	1.25v	–3.92	353	12 48	–59 41
Regulus	Alpha Leonis	1.36	–0.52	77	10 08	+11 58
Adhara	Alpha Canis Majoris	1.50	–4.11	430	06 59	–28 58
Castor	Alpha Geminorum	1.58	0.58	52	05 55	+07 24
Gacrux	Gamma Crucis	1.59v	–0.56	88	12 31	–57 07
Shaula	Lambda Scorpii	1.62v	–5.05	703	17 34	–37 06

steady-state theory

The theory that the universe looks pretty much the same now as it has always looked and that, to maintain a constant density of matter in the face of cosmic expansion, matter is continuously created out of empty space (at a rate of 2.8×10^{-46} g/cm^3/s, or roughly one nucleon per cubic km per year). It rose to prominence in the 1960s under its three greatest champions, Fred **Hoyle**, Thomas **Gold**, and the Austrian-born British cosmologist Hermann Bondi (1919–). However, the discovery of the microwave background persuaded most astronomers to reject the steady-state model in favor of its rival, the **Big Bang** theory.

stellar association

A loose grouping of young, hot stars, born more or less at the same time, with a total mass of 100 to 10,000 M_{sun}; it lacks enough collective gravity to hold itself together against the individual motions of its stars and is doomed to disperse in about 10 million years. Stellar associations are concentrated along the spiral arms of our Galaxy and come in several types. **OB associations** contain predominately high-mass stars of spectral types O to B2. **R associations** are characterized by medium-mass stars. **T associations** are made up of numerous lower-mass **T Tauri stars**. All three types may be found together. The internationally approved designation for associations is the name of the constella-

Stars within 12 Light-years of the Sun

Star	Distance (light-years)	Magnitude		Class
		Visual	Absolute	
Proxima Centauri	4.22	11.01	15.45	M5V
Alpha Centauri A	4.40	–0.01	4.34	G2V
Alpha Centauri B	4.40	1.35	5.70	K0V
Barnard's Star	5.94	9.54	13.24	M5V
Wolf 359	7.78	13.45	16.56	M6V
Lalande 21185	8.32	7.49	10.46	M2V
Sirius A	8.60	–1.44	1.45	A1V
Sirius B	8.60	8.44	11.33	dA2
Luyten 726-8 A	8.73	12.41	15.27	M5.5V
Luyten 726-8 B	8.73	13.25	16.11	M5.5V
Ross 154	9.69	10.37	13.00	M4.5V
Ross 248	10.32	12.29	14.79	M6V
Epsilon Eridani	10.50	3.72	6.18	K2V
Lacaille 9352	10.73	7.35	9.76	M2V
Ross 128	10.89	11.12	13.50	M4.5V
Luyten 789-6 A	11.08	13.3	15.6	M5.5V
Luyten 789-6 B	11.08	13.3	15.6	M5V
Luyten 789-6 C	11.08	14.0	16.3	M7V
Procyon A	11.41	0.40	2.68	F5IV
Procyon B	11.41	10.7	13.0	dA
61 Cygni A	11.43	5.20	7.49	K5V
61 Cygni B	11.43	6.05	8.33	K7
Struve 2398 A	11.64	8.94	11.18	M4V
Struve 2398 B	11.64	9.70	11.97	M5
Groombridge 34 A	11.64	8.09	10.33	M2
Groombridge 34 B	11.64	11.06	13.30	M6
G51-15	11.83	14.81	17.01	M6.5V
Epsilon Indi	11.83	4.69	6.89	K5V
Tau Ceti	11.90	3.49	5.68	G8V

tion followed by an Arabic numeral; for example, Perseus OB2.

stellar atmosphere

The outer envelope of tenuous gas and plasma that surrounds a star and is transparent to light except at certain wavelengths. The wavelengths at which the atmosphere absorbs light, giving rise to dark **Fraunhofer lines**, provide information on the atmospheric composition and physical conditions of stars.

stellar evolution

See article, pages 466–467.

stellar wind

The continuous flow of gas from the surface of a star into space. It is most intense toward the beginning and the end of a star's life, as manifested by **T Tauri stars**, on the one hand, and red giants and supergiants, on the other. The most extreme stellar winds, resulting in a loss of 10^{-6}

(continued on page 467)

stellar evolution

The life history of a star, from birth to death, the timescale and details of which depend crucially on the star's initial mass and, to a far lesser extent, on its composition. A star's lifetime is roughly proportional to $1/M^{2.5}$, where M is the stellar mass, so that, for example, doubling the mass cuts the star's lifetime by a factor of more than 5, while increasing the mass 10-fold drops the life expectancy by a factor of about 300. Stellar evolution is shown succinctly by the **Hertzsprung-Russell diagram** (HR diagram).

Stars form in **giant molecular cloud**s within localized regions that have undergone gravitational collapse. Dense concentrations of gas arise in which gravitational potential energy is converted into light and heat. A young **protostar** becomes hot enough (2,000 to 3,000 K) to glow red, but is obscured by a thick cocoon of gas and dust that allows only infrared and microwaves to pass through. At some point, the central temperature of the protostar climbs high enough for **deuterium**, a heavier isotope of hydrogen, to undergo fusion, which temporarily halts the object's gravitational contraction. When the deuterium is used up, the embryonic star continues its gravitational infall as a **pre-main-sequence object**, until its core temperature reaches the point at which normal **hydrogen burning** can take place and the star joins the **main sequence**. Here it remains, in a stable condition, for about 90% of its lifetime. When the star exhausts its central hydrogen supply, the core, deprived of outward radiation pressure, begins to shrink; meanwhile fusion starts in a shell around the core. This phase is relatively rapid and sees the star's outer layers swell and cool, through the **subgiant** stage, to yield a **red giant**, or, in the case of a very massive star, a red **supergiant**.

A star's subsequent career continues to hinge on its mass. For stars heavier than 1 M_{sun}, the hydrogen-burning shell eats its way outward leaving more and more helium behind. As the helium piles up, the core gains mass and so contracts, becoming denser and hotter. Eventually, the core becomes so dense that the electrons inside it enter the state of **electron degenerate matter**–a condition in which they resist further contraction or expansion. As the hydrogen shell continues to burn, the degenerate core grows hotter, until, at a temperature of about 100 million K, helium begins to fuse into carbon by the **triple-alpha process**. The energy released by this process raises the core's temperature still further. Under normal circumstances, this would cause the core to expand and cool. But, as the core is degenerate, there is no expansion and the temperature rise goes unchecked. Higher temperature means a faster triple-alpha rate, which produces more energy, which raises the temperature further, and so it goes on. When the temperature of the core reaches 300 million K, a nearly explosive consumption of the helium takes place in the **helium flash**. In this remarkable event, which lasts only a few minutes, the star generates energy at 100 times the rate of the rest of the Galaxy put together. However, this huge surge of energy never reaches the surface but instead goes into removing the degeneracy of the electrons and expanding the core. For stars more than 2 M_{sun}, the triple-alpha process starts *before* the electrons become degenerate and so there is no helium flash–just a gradual shift to a core-helium-burning.

After helium-burning begins, either explosively or gradually, the star has two sources of energy: hydro-

stellar evolution Various stages of the life cycle of stars are visible in this single image of the giant galactic nebula NGC 3603 taken by the Hubble Space Telescope. *NASA/STScI*

gen fusion in a shell around the core and helium fusion in the core itself. Helium fuses to give carbon, and some of this carbon combines with helium to yield oxygen. At the same time that the core of the star becomes rich in carbon and oxygen nuclei, the surface temperature climbs until the star reaches a part of the HR diagram known as the horizontal branch. Stars with masses greater than or equal to 1 M_{sun} become smaller and hotter at a constant luminosity. They evolve to the horizontal branch by moving across the HR diagram at constant brightness–low mass stars at about 10 L_{sun}, high mass stars (10 M_{sun}) at about 200 L_{sun}. Notice that as they evolve, HB stars cross the instability strip. For a relatively short time, high mass stars will be **Cepheid variable**s and low mass stars will be **RR Lyrae stars.**

After existing as horizontal branch stars for a few million years, the helium in the core of the star is exhausted (now being mostly carbon and oxygen nuclei) and a helium-burning shell develops below the hydrogen-burning shell. The electrons and nuclei in the core again become degenerate and the star expands and cools to become an **asymptotic giant branch** (AGB) star. Most of the energy comes from the hydrogen-burning shell since the helium-burning shell is small at this time. However, the hydrogen shell is dumping helium ash onto the helium shell. After some time, enough helium is built up so that the helium shell undergoes an explosive event called a *thermal pulse*. This pulse is barely noticed at the surface of the star, but it serves to increase the mass of the carbon/oxygen core so that the size and luminosity of the star gradually increases with time. As the star climbs the asymptotic giant branch, a wind develops in the star's **envelope** that blows the outer layers into space. In this wind, dust particles (important for interstellar clouds and proto-solar systems) are formed from carbon material dredged up from the core by convective currents. During this time, a thick dust shell blocks the visible light from the star such that even though it is 10,000 times brighter than the Sun, it is only seen in the infrared. The stellar wind causes mass loss for AGB stars. This loss is around 10^{-4} M_{sun} per year, which means that in 10,000 years the typical star will dissolve, leaving the central, hot core (the central star in a planetary nebula). If the star is larger than 8 M_{sun}, then the core continues to heat. Carbon and oxygen fuse to form neon, then magnesium, then silicon, all forming into burning shells surrounding an iron ash core. Iron is unusual in that it is extremely stable and resistant to fusion. The temperature of an iron core can reach 3 billion degrees. When the iron core reaches a critical mass, it collapses, violently, into a **supernova** explosion.

As an AGB star becomes larger and more luminous, the rate at which it loses mass also increases. For stars less than 8 M_{sun}, a strong **stellar wind** develops and the outer layers of the star are removed to expose the hot degenerate core. As the gas is expelled and the core is visible, the color of the star becomes much bluer and moves to the left in the HR diagram at constant luminosity. Only a few thousand years are needed for the temperature of a star to grow to 30,000 K. At this temperature, the star begins to emit large quantities of ultraviolet (UV) radiation. This UV radiation is capable of ionizing the hydrogen shell of matter that escaped from the star during the AGB phase. This shell of ionized hydrogen glows deep red as a **planetary nebula**. In the center of the planetary nebula is the remnant core.

stellar wind

(continued from page 465)

M_{sun} or more, occur in X-ray binaries, in which O and B stars are being stripped by a compact companion (either a neutron star or a black hole). During their time on the main sequence, most stars blow a very modest stellar wind; in the Sun's case this amounts to a mass loss of only about 10^{-14} M_{sun} per year (see also **solar wind**).

Stephano

The seventeenth moon of **Uranus**, also known as Uranus XX; together with **Prospero** and **Setebos**, it was discovered in 1999 by Brett Gladman, Matthew Holman, and J. J. Kavelaars using the **Canada-France-Hawaii Telescope**. Stephano has a diameter of about 20 km and is an **irregular moon** with a highly inclined, retrograde orbit at a mean distance of some 7.9 million km from the center of the planet.

Stephan's Quintet

A highly disturbed cluster of five peculiar galaxies–NGC 7317, NGC 7318A, NGC 7318B, NGC 7319, and NGC 7320–lying about 250 million light-years away in Pegasus, that shows signs of gaseous connecting bridges and

Stephan's Quintet More than 100 star clusters and several dwarf galaxies can be seen amid the debris that surrounds the top two galaxies in this famous quintet, testament to the fact that collisions between galaxies can be at least as creative as they are destructive. *NASA/STScI*

(in three of the objects) of strong tidal distortions due to gravitational interaction. Bright new star clusters and dwarf galaxies have been born as a result of the mêlée. Four of the galaxies have large **redshifts**, corresponding to recession velocities of 5,700 to 6,700 km/s; however, the fifth member (NGC 7320) has a much smaller redshift equivalent to a line-of-sight recession of only 800 km/s. This anomaly has led a few astronomers to question the validity of the cosmological explanation. However, the lower redshift system is almost certainly a foreground object that is unconnected with what should properly be called Stephen's *Quartet.* The cluster is named after the French astronomer Éouard Stephan (1837–1923), who discovered it in 1877.

Steward Observatory

The observatory of the University of Arizona, which, together with the Department of Astronomy, form one of the largest academic centers of astronomical studies in the world. Established in 1916 by its first director, Andrew Ellicott Douglass and a bequest made by Lavinia Steward in memory of her late husband, the Steward Observatory now owns and operates the **Multiple Mirror Telescope** and **Mount Graham International Observatory**, and also operates a number of other major optical and submillimeter telescopes at several sites in the state (see table, "The Largest Instruments Operated by the Steward Observatory"). The Steward Observatory Mirror Laboratory has pioneered new techniques of large mirror production, including spin-casting and honeycomb mirrors, while the Center for Astronomical Adaptive Optics is at the forefront of developments in **adaptive optics**. Recent and current construction projects within the Observatory have included making the mirrors for the **Magellan Telescopes** and the **Large Binocular Telescope**.

The Largest Instruments Operated by the Steward Observatory

Telescope	Aperture (m)	Wavelength Band	Location
Kitt Peak 12-Meter Telescope	12	Millimeter	Kitt Peak
Heinrich Hertz Submillimeter Telescope	10	Submillimeter	Mount Graham
Multiple Mirror Telescope	6.5	Optical	Mount Hopkins
Bok Telescope	2.3	Optical	Kitt Peak
Lennon Telescope	1.8	Optical	Mount Graham
Spacewatch Telescope	1.8	Optical	Kitt Peak
Kuiper Telescope	1.6	Optical	Catalina
NASA Telescope	1.5	Optical	Mount Lemmon
1.0-Meter Telescope	1.0	Optical	Mount Lemmon
Spacewatch Telescope	0.9	Optical	Kitt Peak

Largest Known Stony Meteorites

Meteorite	Country	Fell	Class	Group	Weight (kg)
Jilin	China	1976	Chondrite	H5	1,770
Norton County	United States	1984	Achondrite	Aubrite	1,073
Long Island	United States	1948	Chondrite	L6	564 (broken)
Paragould	United States	1930	Chondrite	LL5	371
Bjurbole	Finland	1899	Chondrite	L/LL4	330 (broken)

stishovite

A dense, high-pressure phase of **quartz** that has so far been identified only in shock-metamorphosed, quartz-bearing rocks from meteorite impact craters.

stony meteorite

A **meteorite** composed of silicate (stony) minerals, but that may have up to 25% metal by weight. Stony meteorites are extremely heterogeneous as a group, ranging from samples of primordial matter that have remained more or less unchanged for the last 4.5 billion years to highly evolved rocks from differentiated worlds, such as the Moon or Mars. They are divided into two main classes: **chondrites** and **achondrites**. (See table, "Largest Known Stony Meteorites.")

stony-iron meteorite

A **meteorite** composed of roughly equal amounts by weight of stony minerals and nickel-iron. Modern meteoritics considers the stony-irons to consist of just two groups, the **mesosiderites** and the **pallasites**; however, there is a gradual shading into metal-rich **stony meteorites** such as the **lodranites** (once considered to be stony-irons) and silicate-rich **iron meteorites**. Stony-iron meteorites are less abundant than their stony and iron cousins, comprising a total known mass of some 10 tons–about 1.8% of the entire mass of all known meteorites. (See table, "Largest Known Stony-Iron Meteorites.")

stratosphere

The region of Earth's atmosphere immediately above the **troposphere** that extends roughly from 15 to 50 km above the planet's surface. Its temperature increases with altitude from about 240 to 270 K (roughly the freezing point of water). It includes the ozone layer. The stratosphere is separated from the **mesosphere** by the *stratopause.*

strewn field

A large area over which fragments from a **meteorite shower**, or impact melt spherules or **tektites** are found. The largest strewn field covers the whole of southern Australia and Tasmania and yields the variety of tektite known as *australite.*

Strömgren, Bengt Georg Daniel (1908–1987)

A Swedish astronomer who did important research in stellar structure in the 1930s but is best known for his work on ionized gas clouds–**H II regions**–around hot stars. He surveyed H II regions and found relations between the gas density, the luminosity of the star, and the size of the "Strömgren sphere" of ionized hydrogen around it. Son of the director of the Copenhagen University Observatory, Svante Strömgren (1870–1947), he had published many research papers, mostly in celestial mechanics, and coauthored a book by the time he received his Ph.D. from Copenhagen at age 21. After working there and at the Yerkes Observatory he succeeded his father at Copenhagen

Largest Known Stony-Iron Meteorites

Meteorite	Country	Found	Class	Mass (kg)
Huckitta	Australia	1937	Pallasite	1,400
Krasnojarsk	Russia	1749	Pallasite	680
Brenham	United States	1947	Pallasite	450

in 1940. He was director of Yerkes and McDonald observatories from 1951 to 1957 and spent 10 years at the Institute for Advanced Study in Princeton before returning to Copenhagen.

Strömgren sphere

The region of fully ionized gas around a hot star, also known as an **H II region**; named after Bengt **Strömgren** who first derived the relationship between the density of the nebular gas, the temperature of the central star, and the radius, known as the *Strömgren radius,* within which hydrogen is almost completely ionized.

Struve, Friedrich Georg Wilhelm von (1793–1864)

A German astronomer who was an expert on **double stars** and one of the first astronomers to measure stellar parallax; he was also the patriarch of a dynasty of famous astronomers that spanned four generations. Born in Altona, Schleswig-Holstein, he fled to Dorpat (now Tartu) in Estonia in 1808, to avoid conscription into the German army. In 1810 he graduated from the University of Dorpat and from 1817 on he served as director of the Dorpat Observatory. In 1822 he published the first of many double-star catalogs, the identifying numbers of which are still used today. Struve's stars, however, are now often named in his honor (for example, Struve 2398), whereas the original catalog prefix was the Greek letter sigma (Σ). In 1833 he moved to Russia to set up the Pulkovo Observatory near St. Petersburg, of which he was director until his retirement in 1862, when his son Otto Wilhelm Struve took over the post. In total, Friedrich Struve produced 272 astronomical works and 18 children; his great-grandson Otto, by contrast, produced 907 works but zero children.

Struve, Otto (1897–1963)

A Russian-born German-American astronomer, grandson of Otto Wilhelm **Struve**, who made detailed spectroscopic investigations of stars, especially close binaries, the interstellar medium (where he discovered **H II regions**), and gaseous nebulae. His education at the University of Kharkov was interrupted by World War I and the Russian Civil War, which left him a refugee in Turkey. From there Edwin Frost (1866–1935) brought him to Yerkes Observatory, where he completed his doctorate at the University of Chicago and promptly joined the faculty. He directed four observatories: Yerkes, McDonald (which he founded and where a telescope is named for him), Leuschner, and the National Radio Astronomy Observatory (where he was the first director and encouraged the first search for extraterrestrial intelligence).

Struve, Otto Wilhelm (1819–1905)

A Russian-born German astronomer, son of Friedrich G. W. **Struve**, who accurately determined the rate of change of **precession** and estimated the velocity of the Sun from the proper motions of nearby stars. He worked at Pulkovo Observatory, where he collaborated with his father in the search for double stars.

Struve's Lost Nebula (NGC 1554)

Nebulosity reported by Otto **Struve** in 1868, and verified by Heinrich **d'Arrest**, in the vicinity of **Hind's Variable Nebula** (NGC 1555) in Taurus. Although given a separate catalog number (NGC 1554) by John **Dreyer**, it was not seen by observers until a decade after Struve's discovery, nor has it been seen since. Modern sources, such as *Sky Catalogue* 2000.0, group NGC 1554 and NGC 1555 together as a single object, although 1554 is not visible on the Palomar Sky Survey plates. At the reported position of the Lost Nebula is a magnitude 14 star, 4′ west-southwest of **T Tauri**. It appears that the Lost Nebula may have been a transient portion of the reflection nebula complex in this part of the sky.

SU Ursae Majoris (SU UMa)

The prototype **SU Ursae Majoris star**, located near the tip of the nose of **Ursa Major**, about 3° northwest of Omicron UMa. Its unusual variability was first noted in 1908 by the Russian astronomer L. Ceraski. SU UMa shows normal **U Geminorum**-type outbursts every 11 to 17 days and *super-outbursts* every 153 to 260 days; its range is from magnitude 15 at minimum to a peak magnitude of 10.8 at super-outburst. Since the defining features of SU UMa stars are a narrow outburst, a **super-outburst**, and a **super-hump**, it is strange that for a period of nearly three years in the early 1980s the prototype itself did not exhibit such behavior. Thus, it was questioned whether this variable even belonged to the category named after it! Another absence of super-outburst activity happened between April 1990 and July 1991.

SU Ursae Majoris star (SU UMa star)

One of three subcategories of **U Geminorum stars** (dwarf novae), the others being **SS Cygni stars** and **Z Camelopardalis stars**. In addition to normal U Gem-type outbursts (which consist of a rise from quiescence of 2 to 6 magnitudes and 1- to 3-day durations), SU UMa stars are characterized by **super-outbursts**. These occur once for every 3 to 10 normal outbursts, last for 10 to 18 days, and add one to two magnitudes (up to a factor of 5) in brightness over the normal maxima. Super-outbursts are both brighter and longer in duration than normal outbursts. The rise to super-outburst is indistinguishable from the rise to a normal outburst and, while in super-outburst, a small periodic fluctuation of several tenths of a magnitude known as a **super-hump** is observed at maximum. Super-humps are unique in that the period of fluctuation is 2 to 3% longer than the orbital period of the system. Therefore, by observing the super-humps, the orbital period can be obtained. In almost all cases, SU UMa stars have been found to have orbital periods of less than 2 hours. While the normal outbursts observed in SU UMa are thought to be the same in nature as U Gem-type outbursts, several theories exist to explain super-outbursts. The most favored one, known as the *thermal-tidal instability model*, supposes that both the narrow outbursts and the super-outbursts are governed by **accretion disk** instability. But whereas outbursts are explained in terms of thermal instability, super-outbursts are held to be due to an additional tidal instability. Specifically, physical processes cause the disk to expand until it reaches a critical radius at which a 3:1 resonance is achieved and tidal instabilities produce the super-outburst, bringing the disk back to its normal size. Super-hump activity is always associated with super-outbursts, and never with normal outbursts, so that super-humps and super-outbursts seem to be inherently related. Super-humps appear a day or so after the start of a super-outburst and decrease in amplitude as the super-outburst comes to an end. The super-hump can contribute up to 30% of the total light output. According to the thermal-tidal instability model, super-humps result from a precessing eccentric disk.

The interval from one super-outburst to the next is called the *super-cycle*. The super-cycle lengths of most known SU UMa stars are around a few hundred days, but a few systems have much shorter or much longer super-cycles. The short super-cycle systems have become known as **ER Ursae Majoris stars**, while the long super-cycle systems are called **WZ Sagittae stars**.

Subaru Telescope

An 8.3-m optical-infrared telescope at the summit of Mauna Kea, Hawaii, operated by the National Astronomical Observatory of Japan (NAOJ).

subdwarf (sd)

A star whose luminosity is 1.5 to 2 magnitudes lower than that of a **main sequence** (dwarf) star of the same spectral type. Subdwarfs belong primarily to **Population II** and lie just below the main sequence in the **Hertzsprung-Russell diagram**. Subdwarfs are not really too faint for their temperatures, but too hot for their luminosities–a result of the low metal abundance in their atmospheres.

subgiant

A star that has stopped fusing hydrogen in its core and has begun to evolve off the **main sequence**, increasing in size and luminosity, on its way to becoming a **red giant**. Subgiants produce energy in a hydrogen-burning shell that surrounds the core. The nearest example is Beta Hydrii, 24.4 light-years away, with an estimated mass of 1.1 M_{sun}, a luminosity of 3.5 L_{sun}, and a radius of 1.5 R_{sun} (see also **Hydrus** for more details).

Submillimeter Array

An array of (initially) eight moveable reflectors, each 6 m in diameter, that is being set up near the summit of Mauna Kea, Hawaii. It will see cosmic sources at high resolution at submillimeter wavelengths (see **submillimeter wave astronomy**). The array is a being constructed by the Smithsonian Astrophysical Observatory and the Institute of Astronomy and Astrophysics of Taiwan's Academia Sinica and is expected to be completed by the end of 2008.

Submillimeter Telescope Observatory (SMTO)

An observatory at an altitude of 3,180 m on Emerald Peak of Mount Graham, about 120 km northeast of Tucson, Arizona. It houses the Heinrich Hertz Telescope, a 10-m dish for use at submillimeter wavelengths that was opened in 1994. SMTO is jointly owned and operated by the Max Planck-Institut für Radioastronomie (Bonn) and the University of Arizona's **Steward Observatory**.

submillimeter wave astronomy

The study of the universe in the last waveband of the electromagnetic spectrum to be explored from Earth: the submillimeter window, whose wavelengths range from about 0.3 to 1 millimeter. This bandwidth is ideally suited for studies of the structure and motions of the matter that forms stars; of the spiral structure of galaxies, as outlined by their giant molecular clouds; and of quasars and active galactic nuclei. The submillimeter window lies between the longest **infrared** wavelengths and the short **radio waves** in the millimeter band.

subsolar point

The point on a body at which the Sun is directly overhead at a given point.

substorm

A brief (2- to 3-hour) disturbance in Earth's **magnetosphere** that happens when the interplanetary magnetic field turns southward, allowing interplanetary and terrestrial magnetic field lines to merge at the dayside **magnetopause** and energy to be transferred from the **solar wind** to the magnetosphere. The storage of some of this energy in Earth's **magnetotail** marks the growth phase of a substorm. During the expansion phase, the stored energy is released when the field lines in the inner magnetosphere snap back from their stretched, tail-like configuration. This process results in the energization of charged particles in the plasma sheet and their injection deeper into the inner magnetosphere. The recovery phase is when the magnetosphere returns to its quiet state. The storage and release of energy in the magnetosphere during a substorm leads to characteristic changes in the appearance of aurorae and of emission intensity and to the enhancement of currents flowing in the polar ionosphere and associated disturbances in the strength of the high-latitude surface magnetic field. Substorms occur, on the average, six times a day, and happen more often and are more intense during **geomagnetic storms**.

Sudbury Crater

The eroded remains of a giant 1.8-billion-year-old **impact crater** in Ontario, Canada. The Sudbury Crater, at almost 200 km across, is roughly the size of the much younger **Chicxulub Basin**. But whereas the latter is believed to have been caused by an asteroid that killed the dinosaurs, along with many other life forms at the **K-T boundary**, it appears that the Sudbury Crater was formed by a **comet**. Scientists have reached this conclusion by comparing the mass of the **impact melt** at the two sites: 18,000 cubic km at Chicxulub and 31,000 cubic km at Sudbury.

Sudbury Neutrino Observatory (SNO)

An astronomical **neutrino** observatory the size of a 10-story building, 2 km underground, in the deepest section of the Creighton Mine near Sudbury, Ontario; SNO is an international collaboration of scientists from Canada, the United States, and the United Kingdom. The SNO detector consists of 1,000 tons of ultrapure heavy water (water in which the hydrogen atom in the water molecule has an extra neutron) enclosed in a 12-m-diameter acrylic-plastic vessel, which in turn is surrounded by ultrapure ordinary water in a giant 22-m-diameter by 34-m-high cavity. Outside the acrylic vessel is a 17-m-diameter geodesic sphere containing 9,600 photomultiplier tubes, which detect tiny flashes of light emitted as neutrinos that are stopped or scattered in the heavy water. At a detection rate of about one neutrino per hour, many days of operation are required to provide sufficient data for a complete analysis. Because SNO uses heavy water it is able to detect not only electron-neutrinos through one type of reaction but also all three known neutrino types through a different reaction.

Summer Triangle

An **asterism**, formed by the bright stars **Vega** in Lyra, **Deneb** in Cygnus, and **Altair** in Aquila, which lies high overhead on summer and early autumn evenings at mid-northern latitudes. All three of these white stars have similar surface temperatures; Vega, at 9,500 K, is the warmest. Though Vega and Altair are really quite luminous, they are first magnitude primarily because they are close to us, averaging only 25 light-years away, while Deneb is an extraordinarily luminous star at a distance of some 3,200 light-years.

Sun

See article, page 473.

sundial

A device for showing apparent solar time by the position of a shadow cast by an indicator called a *gnomon*.

Sunflower Galaxy (M63, NGC 5055)

A spiral galaxy (type Sb or Sc) in **Canes Venatici** that appears to lie in the M51 Group dominated by the **Whirlpool Galaxy** (M51); it was discovered in 1779 by Pierre **Méchain**. The Sunflower's spiral arms show up as a grainy background, which brightens slowly at first, going in from the periphery, but then rapidly increases in luminosity near the nuclear region. It hosted a Type I supernova (1971I), first seen on May 25, 1971, that reached magnitude 11.8.

Visual magnitude:	8.6
Apparent size:	10′ × 6′
Distance:	31,000 light-years
Position:	R.A. 13h 15.8m, Dec. +42° 2′

sun-grazing comet

A comet that comes within about 50,000 km of the Sun, so that it passes through the solar atmosphere and may actually fall into the Sun. Records of sun-grazing comets go back many centuries. In the late 1880s and early 1890s, the German astronomer Heinrich Kreutz (1854–1907) studied the possible sun-grazing comets that had been observed until then and determined which were true sungrazers. He also found that the genuine ones all followed the same orbit, with a period of about 800 years, indicating that they were fragments of a single comet that had broken up. To this day, all comets seen to graze the Sun have been members of the *Kreutz sungrazers* group. The parent may have been a bright comet seen by the Greek astronomer Ephorus in 372 B.C. that came close to the Sun and then broke in two. Several hundred sungrazers have been observed by the **SOHO** satellite out of a total population of perhaps 200,000, the smallest of which may be less than 10 m across.

sunspot

An area of the Sun's **photosphere**, typically 2,500 to 50,000 km across, that appears dark because it is cooler than its surroundings. Sunspots are concentrations of magnetic flux and usually occur in pairs of opposite polarity that move in unison across the face of the Sun as

(continued on page 474)

Sun

The star around which Earth revolves and the gravitational center of the solar system. It is a G-type **dwarf**, about 4.6 billion years old, and about 270,000 times closer to us than the next nearest star. The Sun consists largely of hydrogen (71% by mass); plus some helium (27%), and heavier elements. It puts out 400 trillion trillion watts of energy, produced by the fusion of hydrogen to helium by the **carbon-nitrogen-oxygen cycle** in its core, where the temperature reaches 15.6 million K. The Sun is as wide as 109 Earths. It spins on its axis with a period that varies from 25 days at the equator to 33.5 days near the poles. The visible surface, or **photosphere**, is surrounded by the **chromosphere** and, beyond this, the **corona**. Important features of the chromosphere include the **chromospheric network**, **plages**, **prominences**, **filaments**, and **spicules**. Important features of the corona include **coronal holes**, **coronal loops**, **coronal mass ejections**, **helmet streamers**, polar **plumes**, and **solar flares**.

Visual magnitude:	–26.74
Spectral type:	G2 V
Surface temperature:	5,785 K
Diameter:	1,392,000 km
Mean density:	1.41 g/cm^3
Escape velocity:	617 km/s
Rotational period (equator):	24 days 6 hours
Axial inclination:	7°15′

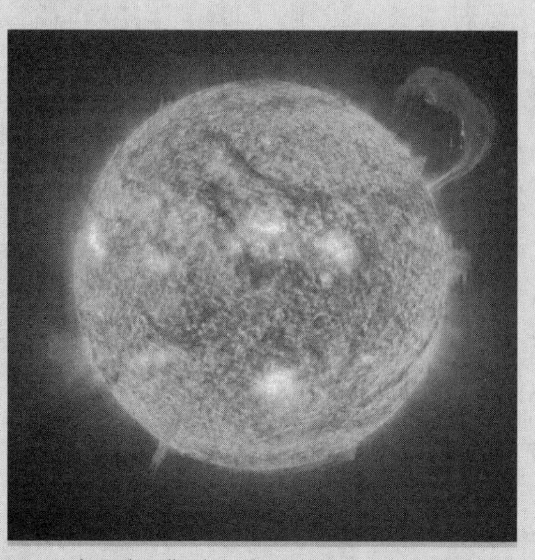

Sun A huge handle-shaped prominence erupts from the Sun in this image taken by the SOHO spacecraft on September 14, 1999. Every feature visible in the image traces the underlying magnetic field structure. The hottest areas show up white, while the cooler areas appear in darker shades. *ESA/NASA*

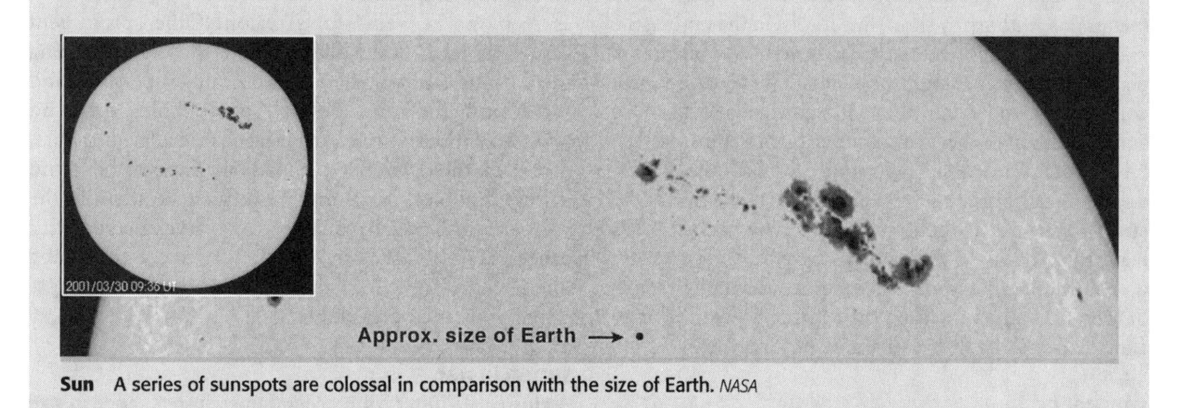

Sun A series of sunspots are colossal in comparison with the size of Earth. *NASA*

sun-grazing comet Comet SOHO-6, recorded in this image of the solar corona by the SOHO spacecraft on December 23, 1996, just before the comet plunged into the Sun. The round shape at the center is part of the coronagraph used to block light from the solar disk in order to see the extremely faint emission from the region around the Sun. *ESA/NASA*

sunspot

(continued from page 472)

it rotates; these pairs are linked by loops of magnetic field that arch through the Sun's **corona**. The magnetic flux of a spot (normally measuring about 0.4 tesla but occasionally reaching 1.0 tesla or more), inhibits the rising of convective heat from below and so keeps the spot at a lower temperature–1,500 to 2,500 K cooler than the rest of the photosphere. Moderate to large spots usually have a darker central region (umbra) surrounded by a lighter halo (penumbra) with many short, fine fibrils. In the umbra, the magnetic field lines tend to be nearly vertical, while in the penumbra, they are almost horizontal. Sunspots are most commonly found within about 30° either side of the Sun's equator, although they can occur at higher latitudes. The lifetime of individual sunspots varies, from as little as a few days, or even hours, to several months (as in the case of the largest spots). As a whole, sunspot activity rises and falls regularly on an 11-year **sunspot cycle**. Sunspots are described using the **McIntosh scheme**, classified in terms of their field structure using the **Mount Wilson Scheme** and counted using the **relative sunspot number**.

sunspot cycle

The roughly 11-year cycle during which the frequency of sunspots varies from a maximum to a minimum level and back again. During a given sunspot cycle, the leading sunspots in groups in the northern solar hemisphere all tend to have the same polarity, while the same is true of sunspots in the southern solar hemisphere, except that here the common polarity is reversed. During the next sunspot cycle, the leading spots in each hemisphere flip to the opposite polarity. From this it's clear that the fundamental period governing solar activity is actually the 22-year magnetic cycle.

Sunyaev-Zel'dovich effect

An apparent temperature dip in the **cosmic microwave background** (CMB) radiation in certain directions, caused by inverse Compton scattering (see **Compton effect**) of CMB photons by electrons in the hot, thin plasma in clusters of galaxies. Measurement of the effect, which amounts to less than 0.001 K at centimeter and millimeter wavelengths but is greater at submillimeter wavelengths, is an important way of determining absolute distances and hence the **Hubble Constant**. It is named after the Uzbekistani astrophysicist Rashid Sunyaev (1943–) and Yakov **Zel'dovich**.

superbubble

A large cavity in the **interstellar medium** created by the explosion of several **supernovae** in the same region of space, typically belonging to the same **OB association**. Because the lifetimes of massive O and B stars are measured in only tens of millions of years, after one supernova has swept clear a bubble around itself, there isn't enough time for the interstellar medium to back-fill the cavity before other stars explode in the same region. Each subsequent supernova will rejuvenate the cavity left by the previous ones, causing the formation of a superbubble with a diameter of 150 to 300 light-years or more. The superbubble interior may be quite irregular, containing high-velocity filaments moving chaotically, as seen in the Vela-Puppis region. Other prominent superbubbles in the Milky Way are those surrounding the Cygnus OB1 association, the Aquila supershell, and the **Monogem Ring**. Several superbubbles surround OB associations in the Large Magellanic Cloud. In fact, most supernova remnants (SNRs) are likely to be found in superbubbles, but it may be difficult to identify the old ones individually because they have merged with other old SNRs. We can only be sure to see the young ones, which are still interacting with circumstellar gas expelled by their progenitors.

supercluster

A cluster of clusters of galaxies. Superclusters range in size from 100 million to at least 300 million light-years across and typically contain tens of thousands of galaxies. They are usually embedded in large sheets and walls of galaxies surrounding immense voids in which very few galaxies exist. Superclusters are believed to have formed in the early universe when matter clumped together under the influence of gravity. Our Galaxy lies within the **Local Supercluster**. The largest known supercluster is the **Shapley Concentration**. According to one recent survey, there are 130 superclusters out to a distance corresponding to a **redshift** of z = 0.1 (about 300 Mpc/*h*). The ones that are closer to us than 100 Mpc/*h*, those that contain at least 10 galaxy clusters, and those that have a proper name in the survey are listed below. (See table, "Nearby Superclusters.")

supergalactic plane

A sheetlike structure that contains the **Local Supercluster**, the **Coma Supercluster**, the Pisces-Cetus Supercluster, and the **Shapley Concentration**, and that separates two giant voids–the Northern and the Southern Local Supervoids. The supergalactic plane is the reference plane for the system of *supergalactic coordinates*, denoted by *L* (longitude) and *B* (latitude). The zero point of supergalactic latitude and longitude is set at R.A. 2h 49m 14s, Dec. +59° 31′ 42″, and the supergalactic north pole is at R.A. 18h 55m .01s, Dec. +15° 42′ 32″ (epoch 2000 coordinates).

supergiant

With the exception of **hypergiants**, the brightest, largest kind of star. Supergiants have luminosities of 10,000 to 100,000 L_{sun} and radii of 20 to several hundred R_{sun} (about the size of Jupiter's orbit). The two commonest types are red supergiants, exemplified by **Betelgeuse** and **Antares**, and blue supergiants, exemplified by **Rigel**. When a star of

Nearby Superclusters

Id	Name	Distance Mpc/*h*	Members
E1	Pegasus-Pisces A	261	4
E6	Pisces-Cetus	179	17
E8		82	5
E18	Cetus A	274	11
E20	Perseus	54	3
E27	Horologium-Reticulum	176	32
E41	Lepus	115	8
E53	Sextans	259	3
E56	Leo	94	8
E66	Ursa Major	215	3
E71		64	2
E74	Hydra-Centaurus	41	6
E80	Shapley Concentration	129	25
E83	Boötes	202	12
E90	Corona Borealis	212	10
E92	Hercules	104	10
E100		55	3
E108	Aquarius-Cetus	167	8
E109	Grus-Indus	218	8
E111	Aquarius-Capricornus	243	5
E122	Aquarius A	229	6
E123	Perseus-Pegasus A	120	4
E125	Pegasus-Pisces B	188	6
E126	Aquarius	260	12
E127	Perseus-Pegasus B	90	2
E128		86	2

"Id" is the identification in the survey, "Name" is the proper name of the supercluster (usually made of the one or two constellations that the supercluster appears in, but the Shapley concentration is named for its discoverer), "Distance" is measured from the center of the supercluster to Earth, in Mpc/*h*, and "Members" shows the number of rich galaxy clusters that are members of the supercluster.

at least 15 M_{sun} exhausts the hydrogen in its core, it first swells to become a **red giant**. But when it reignites through the **triple-alpha process** it expands to an even larger volume. This much brighter but still reddened star is a red supergiant. Through a vigorous **stellar wind**, red supergiants steadily lose their extended atmospheres and turn into smaller but much hotter blue supergiants. A blue supergiant may then develop a fresh distended envelope and revert to the red supergiant phase. Both types, red and blue, can explode as **supernovae**. This came as something of a surprise to astronomers, since stellar evolution theory had long taught that supernovae always come from the red variety. However, the great **Supernova 1987A** was found to have had a blue supergiant precursor.

supergranulation cell

A convective cell, typically 15,000 to 30,000 km in diameter, in the solar **photosphere** that may last as long as a day. Supergranulation cells are distributed fairly uniformly over the solar disk. Their intersections are where most of the magnetic flux through the photosphere is concentrated and where new **sunspots** develop.

super-hump

An additional modulation of the light curve of an **SU Ursae Majoris star** caused by **precession** of the accretion disk. Super-humps show up in the light curve of a **superoutburst** as a modulation with a period a few percent longer than the orbital period. They continue until the star returns to quiescence, although their period usually drifts to slightly shorter periods and smaller amplitudes over time. Nicholas Vogt was the first to propose that superhumps were caused by the disk becoming elliptical during super-outburst. He suggested that such a disk would precess, meaning that the direction in which the disk was elongated would gradually rotate, on a timescale much longer than the orbit (in the same way, the axis of a spinning top precesses, but more slowly than it spins). The long precessional period of the disk would then interact with the orbital cycle to create a new periodicity–the super-hump.

superior planet

Any of the planets Mars, Jupiter, Saturn, Uranus, Neptune, or Pluto; so-called because their orbits are farther from the Sun than Earth's orbit. Mercury and Venus are known as *inferior planets.*

Super-Kamiokande

The world's largest underground **neutrino** observatory; a joint Japanese-American facility, it is located in the Kamioka Mine, about 200 km north of Tokyo. It consists of a tank of ultra-pure water, 40 meters in diameter by 40 meters tall, that is monitored by thousands of sensitive phototubes. Super-Kamiokande was badly damaged in November 2001 when a large number of the phototubes imploded in a chain reaction. However, repairs were carried out following an inquiry into the accident and the observatory resumed collecting data in January 2003.

superluminal radio source

A radio source in which internal motions (e.g., the increasing separation between the core and a knot of material in a **jet**) appear faster than the speed of light in our frame of reference. The data are consistent with this being a transformation effect from seeing jets moving almost directly toward us, so that the emitting material almost catches up with its own radiation. This has the effect of compressing the scale of time that we measure for it, and so increasing the observed speed.

supernova

See article, pages 477–478.

Supernova 1987A

A Type II **supernova** in the **Large Magellanic Cloud** (LMC) discovered on February 24, 1987; it was the first naked-eye supernova since 1604 and reached a peak brilliance, on May 20, 1987, of magnitude 2.9. Contrary to expectations that Type II supernovae always involve red **supergiants**, the progenitor star of 1987A was found to be a much smaller 18-M_{sun} *blue* supergiant, known as Sanduleak –69° 202. One explanation for the unexpectedly small size of the progenitor is connected with the LMC's lower abundance of heavy elements, notably oxygen. If Sanduleak –69° 202 had been oxygen-poor then the star's envelope would have been more transparent to radiation thus making it more likely to contract to a smaller size. The star could have been a red supergiant that experienced contraction until it became a blue supergiant of smaller size and then exploded. (See photo, page 478.)

supernova remnant (SNR)

An expanding **diffuse nebula** that consists of material ejected at a speed of about 10,000 km/s by a **supernova** explosion together with swept-up interstellar matter. Supernova remnants are generally powerful radio and X-ray sources, and may or may not be visible at optical wavelengths. There are several different types. *Shell remnants,* of which **Cassiopeia A** and the **Cygnus Loop** are well-known examples, radiate mainly from the shell itself. A **shock wave** travels out ahead of the ejected material, plows into the surrounding **interstellar medium** (ISM), heats it to several million degrees, and causes it to emit thermal X rays. Electrons accelerated by the shock emit **synchrotron radiation** at radio wavelengths. *Filled-center*

(continued on page 478)

supernova

The death explosion of certain types of stars, resulting in a sudden, vast increase in brightness followed by a gradual fading. At peak light output, a supernova can outshine an entire galaxy. Most of the star is blown into interstellar space at speeds of several percent of the speed of light, eventually forming a **supernova remnant**.

Supernovae come in two main types: type I, which lack hydrogen lines in their spectra, and type II, which have hydrogen lines. These types are further subdivided. Types Ia, Ib, and Ic are distinguished by other details of their spectra. Types II-L and II-P are distinguished on the basis of their **light curves**: type II-Ls showing a linear ("L") decline, while type II-Ps remain on a plateau ("P") for a few weeks before falling off. Type Ia supernovae are the brightest of all, reaching a maximum absolute magnitude of –19. Types Ib and Ic tend to be a couple of magnitudes fainter, while Type II supernovae vary widely in their peak output. Type Ia supernovae occur in older **populations** of stars, such as those making up **elliptical galaxies** and the halos of spiral galaxies. Types Ib, Ic, and II are largely confined to the younger populations of stars found in the disks of spirals.

In terms of the underlying mechanism of the explosion, types Ib, Ic, and II have more in common with each other than they do with type Ia. The latter is thought to occur exclusively in binary systems in which a **white dwarf** has acquired a large amount of material from its companion. When the white dwarf's mass exceeds the **Chandrasekhar limit** of about 1.4 M_{sun}, runaway carbon burning is triggered, which results in the incineration of the dwarf and the violent disruption of its contents. The other varieties of supernova are all thought to be due to the core collapse of massive stars. Types Ia and Ib involve stars that have lost their hydrogen **envelopes**, either through a powerful **stellar wind** or the transfer of material to a companion. **Wolf-Rayet stars** are considered likely progenitors of these kinds of supernovae, the slight chemical differences between types Ia and Ib arising from different degrees of stripping of the Wolf-Rayet's outer layers prior to the explosion.

After a massive star has left the main sequence (see **stellar evolution**), it progresses to a stage where it burns helium in its core, and then carbon. Each burning stage provides less total energy to the system and consequently lasts a shorter period of time. Carbon burning, which produces neon and magnesium, lasts around 100,000 years. When the carbon is gone, the core resumes its contraction and heats until the oxygen residue starts to burn, producing silicon and sulfur–a stage that takes less than 20 years. Then *in a week* the silicon turns to iron. The supergiant is now layered like an onion as each stage of nuclear burning moves outward in a shell around an iron core of nearly 1.4 M_{sun}. Iron, the most tightly bound of all atomic nuclei, marks the end of the road for a star since no energy can be gained from its fusion. Now the core is briefly supported by degenerate electrons. The density of the iron nuclei is so high that the electrons start to combine with them to form manganese, and the heat is so strong that extremely energetic gamma rays penetrate them and begin to break them back down into helium nuclei. Now that the electron degeneracy support and gamma-ray energy are gone from the interior, the core contracts faster and faster, then goes into a catastrophic collapse. The iron core flies inward at a quarter the speed of light. When the center of the incipient **neutron star** exceeds the density of an atomic nucleus, the inner 40% of the core rebounds as a unit. The outer core, still plunging inward, smashes into the rebounding inner core and rebounds in turn, generating a **shock wave**. In about a hundredth of a second, it races out through the infalling matter to the edge of the core. Modelers of supernova had then hoped that the shock wave would continue outward through all the layers of the star blowing it apart. However, calculations done by a number of theorists in 1989 suggest that in **Supernova 1987A** the shock didn't make it out of the core on its own. **Neutrino** emission may have provided the power that revived the shock. The core needed to contract even more before it could become a true neutron star and it did so by vast neutrino losses. The neutrinos were produced by the annihilation of electron-positron pairs made by the energetic gamma rays that pervade material at such high temperatures. The total energy emitted in the 10-second neutrino burst was enormous, about 250 times the energy of the material explosion. It is believed that a small fraction of these neutrinos revived the stalled shock and powered the great explosion of the star.

Supernovae are the primary suppliers of heavy elements for the universe. Elements necessary for life, such as carbon and oxygen, as well as heavier elements like iron, are produced by nucleosynthesis within the star. In the explosive death of the star, these elements are thrown out so that they may be recycled by other stars and gases. The amount of heat and pressure released from a supernova explosion may create new regions of star birth by compressing the surrounding interstellar medium. In addition, supernovae are used as light beacons to measure cosmological distances. Important as they are, few supernovae have been observed nearby. The last one in our Galaxy exploded in 1604 and was observed by Johannes **Kepler** before the invention of the telescope. The rate of supernova discoveries went up immensely with the onset of automatic searches that probe faint magnitudes.

supernova remnant (SNR)

(continued from page 476)

remnants or *plerions,* of which the **Crab Nebula** is the prime example, emit the bulk of their radiation from within the expanding shell because of the presence of a **pulsar**. The pulsar continuously supplies high-speed electrons which give off intense synchrotron radiation in the inner part of the SNR. *Composite remnants* are a cross between the shell remnants and plerions. They may appear shell-like or filled or both depending in which part of the electromagnetic spectrum they are being observed.

SNR tends to involve three main phases. During the first, known as *free expansion,* the front of the expansion is formed from the shock wave interacting with the ambient ISM. This phase is characterized by constant temperature within the SNR and constant expansion velocity of the shell. In the second phase, known as the *Sedov* or *adiabatic* phase, the SNR material slowly begins to decelerate and cool. The main shell of the SNR experiences **Rayleigh-Taylor instability**, which causes the SNR's ejecta to become mixed with the gas that was just shocked by the initial shock wave. This mixing also enhances the magnetic field inside the SNR shell. The third phase, known as the *snowplow* or *radiative* phase, begins after the shell has cooled to about 10^6 K, so the shell can more efficiently radiate energy. This, in turn, cools the shell faster, making it shrink and become more dense, which cools it faster still. Because of the snowplow effect, the SNR quickly develops a thin shell and radiates away most of its energy as optical light. Outward expansion stops, the SNR starts to collapse under its own gravity and, after millions of years, the remnant is absorbed into the ISM.

Supernova 1987A Three glowing rings encircling the site of Supernova 1987A appearing in this Hubble Space Telescope picture taken in 1997. The two larger rings are believed to be caused by twin jets of high-energy particles created by the interaction of the supernova debris and the remnant star—either a neutron star or black hole. *NASA/ESA/STScI*

super-outburst

A type of outburst characteristic of **SU Ursae Majoris stars**, which lasts 5 to 10 times longer and is slightly brighter than the usual dwarf nova outburst. Most super-outbursting dwarf novae have short orbital periods (less than 2 hours).

super-rotation

The rotation of a planet's atmosphere independent of the surface rotation. It is very marked in the case of Venus, where the top layer of the atmosphere rotates in 4 days compared with the 243 days of the surface. The cause of super-rotation remains unclear but is probably linked to the fact that most of the Sun's heat is absorbed by the top of the atmosphere.

surface brightness

The measure of the amount of light that an object, especially a galaxy, emits per area of the sky. Even a galaxy that has a high total **luminosity** can be hard to see if it has a low surface brightness. See also **low-surface-brightness galaxy**.

surface gravity (*g*)

The rate at which a small object in freefall near the surface of a body is accelerated by the gravitational force of the body. It is given by

$$g = GM/R^2$$

where G is the universal constant of gravitation, M is the mass, and R is the radius of the gravitating body. On Earth g is approximately 9.8 m/s^2.

surge
A short-lived collimated jet of material, produced by a **solar flare** or a very active region, that reaches coronal heights and then either fades or returns to the chromosphere along the trajectory of ascent.

Swan bands
Spectral bands of the carbon radical C_2 first investigated in 1856 by the Scottish physicist William Swan (1818–1894). They are a characteristic of the spectra of **carbon stars** and of comets.

SWAS (Submillimeter Wave Astronomy Satellite)
A NASA satellite, launched in December 1988, that was equipped with a 0.6-m telescope for making observations in the 490–550 GHz submillimeter range, and also with an acousto-optical spectrometer. It enabled studies of the cooling of molecular cloud cores–the sites of star formation in the Galaxy–by measuring spectral lines of molecular oxygen and water.

Swift Gamma-Ray Burst Explorer
A NASA probe designed to detect and study the position, brightness, and physical properties of **gamma-ray** bursts–the most powerful energy blasts in the universe. Because gamma-ray bursts are fleeting and unpredictable, Swift has been designed to detect them, collect images and measurements, and send data back to Earth all within about a minute. During its three-year mission, scheduled to begin in December 2003, Swift is expected to record more than 1,000 gamma-ray bursts.

Swift-Tuttle, Comet 109P/
A periodic comet and the parent body of the Perseids **meteor shower**. It was discovered on July 16, 1862, by Lewis Swift (from Marathon, New York), and, independently, on July 19, 1862, by Horace Parnell Tuttle (from Harvard University); however, it is now known to be the same as Comet Kegler of 1737 and appearances have been identified as far back as 69 B.C. Last seen in 1992, Swift-Tuttle will be in our vicinity again in 2126.

Eccentricity:	0.964
Inclination:	113.4°
Period:	134 years

Sc Arietis star (Sx Ari star)
A type of **rotating variable** of spectral type B0p to B9p with an uneven surface distribution of helium and a strong magnetic field. As SX Ari stars rotate, the amount of helium seen in the spectrum and the intensity of the magnetic field appears to vary, while the brightness changes by about 0.1 magnitude. These stars, also called *helium variables,* are high-temperature versions of the **Alpha² Canum Venaticorum stars.**

SX Phoenicis star (SX Phe star)
A pulsating subdwarf star, smaller than the Sun, with a spectral type A2 to F5, typically found in **globular clusters**. The period of pulsation is in the range 0.04 to 0.8 day with an amplitude of about 0.7 magnitude. SX Phe stars are close cousins of **Delta Scuti stars.**

Sycorax
The eighteenth moon of **Uranus**, also known as Uranus XVII; together with **Caliban**, it was discovered in 1997 by Brett Gladman, Phil Nicholson, Joseph Burns, and J. J. Kavelaars using the 200-inch Hale telescope. Sycorax has a diameter of about 160 km and is an **irregular moon** with a highly inclined, retrograde orbit at a mean distance of some 12.2 million km from the planet. Like Caliban it is unusually red in color, suggesting it came originally from the **Kuiper Belt.**

synchronous rotation
The situation when a satellite spins on its axis in the same time it takes to orbit a planet; also known as *captured rotation.* Synchronous rotation means that the satellite always keeps the same face toward its primary, as the Moon does toward Earth.

synchrotron radiation
Electromagnetic radiation given off by a charged particle, such as an electron, that is spiraling along a magnetic field at a relativistic speed (a speed that is a significant fraction of the speed of light).

synodic
Relating to a conjunction between two celestial objects.

synodic period
The period of apparent revolution of one body about another with respect to Earth. For planets, the mean interval of time between successive conjunctions of a pair of planets, as observed from the Sun; for satellites, the mean interval between successive conjunctions of the satellite with the Sun, as observed from the satellite's primary.

syzygy
The condition when three astronomical bodies are in a straight line. The Moon at the moment of new moon or full moon, or a planet at the time of opposition or solar conjunction, are in syzygy.

T

T association

A grouping of **T Tauri stars**, up to several thousand strong, typically embedded in and partly obscured by the nebula from which it formed; it is one of the main types of **stellar association**. Examples include the Taurus-Auriga T Association, the nearby TW Hydrae Association, and Vela T1 and T2. Additionally, many T Tauri stars are found in other groupings not named as T associations; for example, at least 2,000 T Tauris are believed to exist in the **Scorpius-Centaurus Association** and a handful in the young open cluster Collinder 197. T associations gradually break up after 10 million years or so as the component stars go their separate ways. The recently found and nearby Horologium Association, lying about 200 light-years away and believed to be about 10 million years old, contains at least 10 young stars of which half a dozen are characterized as Post-T Tauri.

T Tauri

The type **T Tauri star**, discovered in October 1852 by John **Hind**. Like all of its kind, T Tauri is an **irregular variable**. It has been known to shine as brightly as magnitude 9.3 and as dimly as magnitude 14, though for the past century or so it has tended toward the brighter end of this range. It can vary by a few tenths of a magnitude on nearly a daily basis without any discernable pattern. Not far from T Tauri, Hind found a reflection nebula, now called **Hind's Variable Nebula,** illuminated at the whim of its unreliable neighbor. Then in 1890, nearly 40 years after the discovery of the strange star, Shelburn **Burnham** found that T Tauri itself is nestled within a very small nebula, now known popularly as **Burnham's Nebula**. Much more recently, 30″ west of the brightest point in Hind's nebula a **Herbig-Haro object**–a jet of the type commonly associated with young, mass-ejecting stars has been uncovered. As if all this were not enough, T Tauri turns out to have a binary companion, detected by its infrared glow, 0.5 to 0.7″ to the south (hence known as T Tauri S, while T Tauri becomes T Tauri N), and there is even data to show that T Tauri is a triple-star system.

T Tauri star

A very young, lightweight star, less than 10 million years old and under 3 M_{sun}, that it still undergoing gravitational contraction; it represents an intermediate stage between a **protostar** and a low-mass main sequence star like the Sun. T Tauri stars are found only in nebulae or very young clusters, and have low-temperature (G to M type) spectra with strong **emission lines** and broad **absorption lines**. They are more luminous than main sequence stars of similar spectral types, and they have a high lithium abundance, which is a pointer to their extreme youth, as lithium is rapidly destroyed in stellar interiors. T Tauri stars often have large **accretion disks** left over from stellar formation. Their erratic brightness changes may be due to instabilities in the disk, violent activity in the stellar atmosphere, or nearby clouds of gas and dust that sometimes obscure the starlight. Two broad T Tauri types are recognized based on spectroscopic characteristics that arise from their disk properties: *classical T Tauri* and *weak-lined T Tauri* stars. Classical T Tauri stars have extensive disks that result in strong emission lines. Weak-lined T Tauri stars are surrounded by a disk that is very weak or no longer in existence. The weak T Tauri stars are of particular interest since they provide astronomers with a look at early stages of stellar evolu-

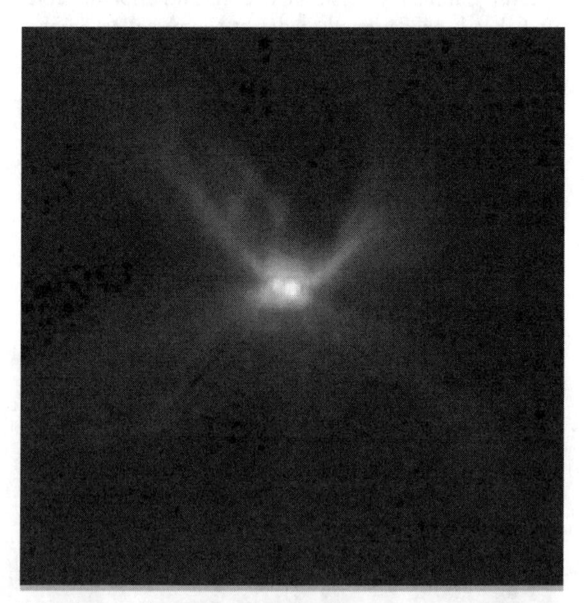

T Tauri star The extremely young binary star system known as CoKu Tau/1, which is in its T Tauri stage of development. *Keck Observatory*

tion unencumbered by nebulous material. Some of the absent disk matter may have gone into making planetesimals, from which planets might eventually form. Infant star relatives of T Tauris include **FU Orionis** stars.

Tadpole Galaxy

A spiral galaxy that has been grossly distorted by a small interloper–a nearby, blue compact galaxy. The Tadpole lies about 420 million light-years away in **Draco**. Strong gravitational forces from the interaction have created an incredibly long tail of stars and gas stretching out more than 280,000 light-years. Numerous young blue stars and star clusters, spawned by the galaxy collision, are seen in the spiral arms, as well as in the long **tidal tail** of stars. Each of these clusters contains up to a million stars and will redden with age to become globular clusters similar to those found in essentially all halos of large galaxies, including our own Milky Way. Two prominent clumps of young bright blue stars are visible in the tidal tail and are separated by a gap. These will likely become dwarf galaxies that orbit in the Tadpole's halo.

taenite

The less common of the two nickel-iron alloys found in **iron meteorites**; the other is **kamacite**. Taenite contains 27 to 65% nickel, and forms small crystals that appear as

Tadpole Galaxy The wildy distorted shape of this object is the result of gravitational interaction with the small bright galaxy visible in the upper left corner of the more massive Tadpole. *NASA/STScI*

highly reflecting thin ribbons on the etched surface of a meteorite; its name comes from the Greek word for "ribbon."

Tarantula Nebula (NGC 2070, 30 Doradus)

The largest and brightest **emission nebula** in the **Large Magellanic Cloud** (LMC) and one of the largest emission nebulae known; it lies at the eastern end of the LMC's stellar bar. It was first cataloged as a star, 30 Doradus, then discovered to be a nebula by Nicolas **Lacaille** in 1751–1752. Also known as the *Looped Nebula* (a name that goes back to John **Herschel**), it is roughly 100 times larger than the famous **Orion Nebula** but is illuminated in the same way: by the ultraviolet radiation from a collection of hot, young, massive stars embedded within it. Several **OB associations** have been observed

Tarantula Nebula One of the largest known star-forming regions in the Local Group of galaxies—the Tarantula Nebula, in the Large Magellanic Cloud—sprawls across this image taken by the Very Large Telescope. *European Southern Observatory*

inside the Tarantula, including the extremely luminous and compact cluster R136 near its center; it is a hotbed of **Wolf-Rayet stars. Supernova 1987A** occurred in an outlying part of the Tarantula–a harbinger of what lies in store for many of the nebula's stars.

Visual magnitude:	8
Apparent size:	40′ × 25′
Diameter:	900 light-years (with faint extensions out to 6,000 light-years)
Mass:	about 500,000 M_{sun}
Position:	R.A. 5h 38.7m, Dec. −69° 6′

Tau Ceti (τ Cet)

A nearby star that is similar, though slightly smaller than the Sun. It has been the target of many SETI (Search for Extraterrestrial Intelligence) programs including the first, Project Ozma, which also looked at **Epsilon Eridani**. Dust, like that in the solar system, has been found around Tau Ceti but, so far, no planets have been detected. Tau Ceti's low **metallicity** makes it questionable whether the nebula from which it condensed had sufficient heavy element content to allow planets to form. See also **stars, nearest**.

Visual magnitude:	3.49
Absolute magnitude:	5.68
Spectral type:	G8V
Luminosity:	0.62 L_{sun}
Mass:	0.85 M_{sun}
Distance:	11.9 light-years

Taurus (abbr. Tau; gen. Tauri)

The Bull; a large northern **constellation** and the second **sign of the zodiac**, lying northwest of Orion and south of Auriga and Perseus. It contains the **Crab Nebula**, the **Hyades**, and the **Pleiades**. (See table, "Taurus Stars Brighter than Magnitude 4.0," and star chart 16.)

Taurus Moving Cluster

A large group of several hundred stars spread over much of **Taurus**. It shares the same motion through space as the **Hyades**, which lies more or less at its center.

Taurus: Stars Brighter than Magnitude 4.0

	Magnitude		Spectral	Distance	Position	
Star	Visual	Absolute	Type	(light-yr)	R.A. (h m s)	Dec. (° ′ ″)
α **Aldebaran**	0.87	−0.64	K5III	65	04 35 55	+16 30 33
β **Al Nath**	1.65	−1.37	B7III	131	05 26 18	+28 36 27
η **Alcyone** (Pleiades)	2.85	−2.41	B7III	368	03 47 29	+24 06 18
ζ Alheka	2.97	−2.57	B4III	417	05 37 39	+21 08 33
λ	3.41v	−1.87	B3V	370	04 00 41	+12 29 25
θ^2 (Hyades)	3.40	−0.03	A7III	158	04 28 40	+15 52 15
ε Ain (Hyades)	3.53	0.14	K0III	155	04 28 37	+19 10 49
o	3.61	−0.45	G8III	212	03 24 49	+09 01 44
27 Atlas (Pleiades)	3.62	−1.72	B8III	381	03 49 10	+24 03 12
γ Hyadum Primus (Hy)	3.65	0.28	G8III	154	04 19 48	+15 37 39
17 Electra (Pleiades)	3.72	−1.56	B6III	371	03 44 53	+24 06 48
ξ	3.73	−0.44	B9Vp	222	03 27 10	+09 43 58
δ^1 (Hyades)	3.77	0.41	G8III	153	04 22 56	+17 32 33
θ^1 (Hyades)	3.84	0.41	G7III	158	04 28 34	+15 57 44
20 **Maia** (Pleiades)	3.87	−1.35	B8III	360	03 45 50	+24 22 04
ν	3.91	0.92	A1V	129	04 03 09	+05 59 21
Other Pleiades:						
Taygeta	4.30	−0.99	B6V	373	03 45 12	+24 28 02
Merope	4.14	−1.07	B6IV	359	03 46 20	+23 56 54
Pleione	5.05v	−0.33	B7Vp	387	03 49 11	+24 08 12

T-class asteroid
A rare type of asteroid with a fairly low albedo (0.04 to 0.11) and a moderate absorption feature at wavelengths shorter than 0.85 micron. Examples include (114) Kassandra and (233) Asterope.

Tebbutt, Comet (C/1861 J1)
A **long-period comet** discovered on May 13, 1861, by Australian amateur astronomer John Tebbutt (1834–1916) and also known as the *Great comet of 1861*. It reached perihelion on June 12 and was 0.13 AU from Earth on June 30 when our planet passed through its tail. At its brightest, Comet Tebbutt was said to have outshone every star and planet except for Venus and to have had a tail 100° long stretching right across the north polar region of the sky. Perihelion 0.82 AU, eccentricity 0.985, inclination 85.4°, period 409 years.

technetium star
A star whose spectrum reveals the presence of the element technetium. The first such stars were discovered in 1952 and provided the first direct evidence for stellar **nucleosynthesis**–the manufacture of heavier elements from lighter ones in stellar interiors. Because the most stable isotopes of technetium have half-lives of only a few million years, the only way this element could be present inside stars is if it had been made there in the relatively recent past. Technetium is observed in some **M stars**, **MS stars**, MC stars, **S stars**, and C stars.

tectonic
To do with the deformation of the crust of a planet or large moon, the forces that cause such deformation, and the resulting structures. Landforms that arise from tectonic processes include mountain ranges, rift valleys, faults, fractured rock, and folded rock masses.

tektite
A small, dark, glassy sphere or aerodynamically shaped buttons formed from molten rock ejected probably as a result of an asteroid or comet impact with silicon-rich rocks on the surface. Tektites are found scattered on Earth's surface in several major locations, known as **strewn fields**. The largest of these covers the whole of south Australia and Tasmania and has yielded *australites* dated at 600,000 to 750,000 years old. A similar age has been measured for the *billitonites* found on Billiton Island, Indonesia. The *moldavites,* dating back 14.7 million years, occur in the Moldau River valley in the Czech Republic and are thought to have been produced in the impact that created the 24-km-diameter Ries Crater, about 500 km away near Nordlingen, Germany. Translucent and dark green, moldavites have been used for jewelry, religious articles, and decorative objects since prehistoric times. The oldest known tektites, with ages of 33 to 35 million years old, are the *bediasites,* found in Texas and named after the local Bedias tribe, and georgiaites, named after the state where they occur.

telescope
An instrument to collect light, or other forms of electromagnetic radiation, from distant objects, magnify the image, and allow the object to be viewed. **Reflecting telescopes** gather light by means of a mirror, **refracting telescopes** by means of a lens, and **catadioptric telescopes** by a mirror-lens combination. **Radio telescopes** gather radio energy typically by using a metallic dish antenna. Telescopes have also been built that can gather X rays, gamma rays, and other forms of energy.

Telescopium (abbr. Tel; gen. Telescopii)
The Telescope; a faint southern **constellation** lying south of Corona Australe and east of Ara. Its brightest star is Alpha Tel (visual magnitude 3.49, absolute magnitude –0.93, spectral type B3IV, distance 249 light-years) and it has few objects of interest for amateur astronomers. (See star chart 14.)

Telesto
The tenth moon of **Saturn**, also known as Saturn XIII; it was discovered in 1980 by Smith, Reitsema, Larson, and Fountain from ground-based observations. Telesto measures 34 × 36 × 28 km and is co-orbital with **Tethys** and **Calypso**, lying at Tethys's leading **Lagrangian point** (i.e., 60° of orbital longitude ahead of the larger moon).

telluric lines
Lines or bands in the spectrum of a celestial object that are due to absorption by gases such as oxygen, water vapor, or carbon dioxide in Earth's atmosphere. The most prominent are the A and B bands (see **Fraunhofer lines**).

Tempel, (Ernst) Wilhelm (Leberecht) (1821–1889)
A German-born astronomer who discovered numerous deep sky objects and comets. After leaving his hometown in Saxony in 1837, he trained and worked as a lithographer in Copenhagen, Christiania (now Oslo), and Venice. In 1860 he went to work at Marseilles Observatory, in 1871 he joined Giovanni **Schiaparelli** at Brera Observatory in Milan, and in 1874 he went to Arcetri Observatory. Tempel discovered the **Merope Nebula** (NGC 1435) in the Pleiades in October 1859. Later he discovered many more nebulous objects: 156 *NGC* (***New General Catalogue***) entries were credited to him, of which at least 123 belong to real deep sky objects. He is also

credited with 13 original comet discoveries and 5 independent codiscoveries, as well as 8 first rediscoveries of periodic comets; his original discoveries include 4 short-periodic comets (9P/Tempel 1, 10P/Tempel 2, 11D/Tempel-Swift, and the Leonid comet 55P/Tempel-Tuttle).

Tempel-Tuttle, Comet 55P/

A periodic comet that is the parent object of the Leonids **meteor shower**. It was discovered independently in 1865 by the German astronomer Wilhelm Tempel (1821–1889) and the American astronomer Horace Tuttle (1837–1923), though past records of appearances have been traced back to A.D. 1366. Perihelion 0.982 AU, eccentricity 0.904, inclination 162.7°, period 33 years.

temperature

A physical parameter characterizing the thermal state of a body. Measured in units of degrees Celsius (°C), Fahrenheit (°F), or Kelvin (K).

temperature minimum

A level in the Sun's atmosphere, about 550 km above the base of the **photosphere**, where the temperature reaches a lowest point of 4,400 K. Above this level, to a height of about 1,000 km, the temperature rises steadily to about 6,000 K, then more rapidly at greater heights.

terminator

The boundary between the lit and the unlit part of a planet's or moon's disk.

terracing

Stepped terrain around the inside slopes of large craters, which has resulted from slumping and sliding of blocks of land. Although usually associated with **impact craters** with diameters of over 20 km, terracing can occur within volcanic craters, especially collapse craters and calderas.

terrestrial age

The time that has elapsed since a **meteorite** fell to Earth. It can be calculated from the abundances of some relatively short-lived radioactive **isotope**s that were formed, while the meteorite was in space, as a result of bombardment by **cosmic rays**. After the meteorite lands, no more of these radioactive isotopes are made and the ones already in the rock continue to decay at a steady (known) rate, thus serving as a kind of clock. Terrestrial ages are usually determined from the isotopes carbon-14, beryllium-10, and chlorine-36. Most meteorites weather quickly in the oxidizing environment of the Earth. However, other meteorites fell at more fortuitous locations and were preserved, e.g., in the ice fields of Antarctica and in the hot deserts of Africa, some for 40,000 years or more. The oldest of all are *fossil meteorites* held in sediments or in other geologic strata conducive to preservation, of which the record-holder is the meteorite of Osterplana, Sweden, that was found in 1987 imbedded in some limestone. This limestone, which dated from Ordovician times, showed that the embedded meteorite had fallen 480 million years ago. This just surpasses the Brunflo meteorite, also found in Swedish limestone in 1980, which has a terrestrial age of 450 million years. As with true fossils, most of the original meteoritic minerals in fossil meteorites have been replaced by terrestrial minerals leaving only the outer structure of the meteorite unchanged. Although both meteorites have been classified as **chondrites**, only their chondritic structure remains as evidence of their extraterrestrial origins. The oldest intact meteorite is the Lake Murray iron. A single mass with a thick iron-shale was found in a gully in Oklahoma in 1933. The meteorite was imbedded in some Antler Sandstone dating from the Lower Cretaceous, suggesting that the Lake Murray meteorite landed in a near-shore, shallow sea, while these beds were being deposited about 110 million years ago. In addition, although the exterior of this meteorite has been heavily corroded, the inner nickel-iron core has remained unaltered, establishing the Lake Murray iron as the oldest known meteorite on Earth.

terrestrial planet

The planets of the inner solar system–Mercury, Venus, Earth, and Mars–composed of rocky materials and iron/nickel, with a density between 4.0 and 5.5 g/cm^3. More generally, any planet made primarily of rock and metal.

Terrestrial Planet Finder (TPF)

A mission, currently under study, that would form an important part of NASA's Origins Program. TPF would search for small, terrestrial-type planets around nearby stars and analyze the spectra of these worlds for the chemical signatures of life. Two different approaches are being considered to achieve the same goal, namely, to block the light from the parent star in order to see its much dimmer planets–a feat likened to finding a firefly near the beam of a faraway searchlight. One of the technologies under study involves an infrared **interferometer**, possibly consisting of four 8-m telescopes, with a total surface area of 1,000 m^2. The interferometer would use a technique called *nulling* to reduce the starlight by a factor of one million (see **nulling interferometry**), thus enabling the detection of the very dim infrared emission from small planets. The other kind of instrument under review is a visible light **coronagraph**, which would consist of a large optical telescope, with a mirror three to four times bigger and at least 10 times more precise than the Hubble Space Telescope. The telescope would have special optics to reduce the starlight by a factor of one billion, thus enabling astronomers to detect the faint planets. A final choice of strategy is expected by 2005 or 2006 and a launch in about 2012.

Tethys

The ninth closest and fifth largest moon of **Saturn**, also known as Saturn III; it was discovered by Giovanni **Cassini** in 1684. Tethys has a diameter of 1,060 km and moves at a mean distance of 294,660 km from the center of Saturn, sharing its orbit with **Telesto** and **Calypso** (which lie at its **Lagrangian points**). Its density is so low–just 1.1 times that of water–that it must consist largely of water-ice. Its surface is thickly cratered and has two extraordinary features: the 400-km-wide crater Odysseus and the 2,000-km-long, 80-km-wide valley Ithaca Chasma, which extends three-quarters of the way around the moon.

Thalassa

The second moon of **Neptune**, also known as Neptune IV; it was discovered in 1989 by Voyager 2. Thalassa has a diameter of 80 km and orbits at a mean distance of 50,070 km from (the center of) the planet.

Tethys Features as small as 5 km across can be seen in this photo of Tethys taken by Voyager 2 from a range of 282,000 km. A boundary between heavily cratered regions (top right) and more lightly cratered areas (bottom right) points to a period of internal activity early in Tethys' history that partially resurfaced the older terrain. The large crater in the upper right lies almost on the huge trench system, several km deep, that girdles nearly three-fourths of the circumference of the satellite. *NASA/JPL*

Thales of Miletus (624–547 B.C.)

Traditionally regarded as the founder of Greek philosophy and therefore of Western rational speculation about the nature of the universe. What we know of him and his works comes through the commentaries of others, since none of Thales's original writings has survived. He asked: What is the raw material from which all the universe is made? He thought the answer was "water." (He also believed that Earth floated on water and was disk-shaped.) But the crucial point is that he broke new ground by suggesting that Earth and everything beyond it had a common physical basis and was subject to natural, rather than supernatural, laws. Thales is said to have proposed that the stars were other worlds–an important departure from the view that they were simply lights suspended from a celestial vault. He also supposedly predicted a solar eclipse in 585 B.C. See also **Greek astronomy**.

Thebe

The fourth moon of **Jupiter**, also known as Jupiter XIV; it was discovered in 1979 by Stephen Synnott from Voyager 1 images. Thebe is about 110 km across at its widest and orbits 222,000 km from the center of the planet.

Themis

(1) Minor planet 24. A C-class (carbonaceous) asteroid with a diameter of 220 km, discovered in 1853 by Annibale de Gasparis (1819–1892). It orbits in the main body of the asteroid family that bears its name (see **Themis family**); semimajor axis 3.13 AU, perihelion 2.71 AU, aphelion 3.55 AU, inclination 0.8°. (2) A satellite of Saturn reported by William **Pickering** in 1898 to be in orbit between **Titan** and **Hyperion**, but subsequently not seen again.

Themis family

A **Hirayama family** of asteroids in the outer part of the main asteroid belt at a mean distance of 3.13 AU from the Sun. It is one of the most populous and well-defined families, consisting of a core of large objects surrounded by a cloud of mostly smaller ones. The core includes (24) **Themis**, (62) Erato, (90) Antiope, (468) Lina, (526) Jena, and (846) Lipperta. Most family members are **C-class asteroids** with low albedos. Themis itself is of spectral class C and has a diameter of 228 km; semimajor axis 3.130 AU, perihelion 2.71 AU, aphelion 3.55 AU, inclination 0.8°.

Themisto

A small, **irregular moon** of **Jupiter**, also known as Jupiter XVIII and, previously, as S/2000 J1, that doesn't belong to either of the two main outer families of irregular Jovian moons. Instead, it moves in a **prograde** orbit with an inclination of 43° and a mean distance from Jupiter of 7.5 million km. Its discovery was announced in 2000 by Scott Sheppard, David Jewitt, Yan Fernandez, and Eugene Magnier.

thermal radiation

Electromagnetic radiation resulting from interactions between electrons and atoms or molecules in a hot dense medium.

thermocouple

A sensitive instrument used to measure the heat radiated from a celestial body. It uses the junction of small pieces of dissimilar metals, such as platinum and bismuth, which are connected to a galvanometer. The thermocouple is placed at the focus of a large reflector, and the heat from a star or other object causes a small electric current to be produced, the strength of which is proportional to the intensity of heat from the celestial source.

thermosphere

An upper layer of Earth's atmosphere, extending from an altitude of about 85 km (above the **mesopause**) to the base of the **exosphere** at about 500 km, within which the temperature increases with height to a maximum of 1,500°C. The thermosphere includes the **ionosphere**.

thick disk

See **population**.

thin disk

See **population**.

tholin

A hard, red-brownish substance made of complex organic compounds. Tholins don't exist naturally on Earth, because our present oxidizing atmosphere blocks their synthesis. However, tholins can be made in the lab by subjecting mixtures of methane, ammonia, and water vapor to simulated lightning discharges. Conditions like this probably exist in many places in the universe, including the icy moons of the outer solar system. The presence of tholins may help explain the orange-red hue of **Titan**'s atmosphere and the reddish surface of some **Centaurs** and outer asteroids.

Thomson scattering

The scattering of **photons** by free electrons, such as occurs in stellar atmospheres; it is independent of wavelength and equal numbers of photons are scattered forward and backward. Thomson scattering is named after the English physicist Joseph John Thomson (1856–1940).

three-body problem

The mathematical problem of finding the positions and velocities of three massive bodies, which orbit each other gravitationally, at any point in the future or the past, given their present positions, masses, and velocities. An example would be to completely solve the behavior of the Sun-Jupiter-Saturn system, or that of three mutually orbiting stars. It is a vastly more difficult exercise than the **two-body problem**. In fact, as Henri **Poincaré** and others showed, the three-body problem is impossible to solve in the general case; that is, given three bodies in a random configuration, the resulting motion nearly always turns out to be chaotic: no one can predict precisely what paths those bodies would follow. However, the problem becomes tractable in certain special cases.

In the *restricted three-body problem,* one of the masses is taken to be negligibly small so that the problem simplifies to finding the behavior of the massless body in the combined gravitational field of the other two. In the *circular restricted three-body problem* and the *elliptical restricted three-body problem,* the two masses pursue circular and elliptical orbits, respectively, about their common center of mass. In the *coplanar restricted three-body problem* the massless body moves entirely in the plane of the massive bodies' orbits; in the *three-dimensional three-body problem,* it is free to move in all three dimensions. These restricted cases cover systems such as Sun-planet-asteroid, Sun-planet-comet, or binary star-planet.

For three interacting bodies, mathematicians have found a small number of special cases in which the orbits of the three masses are periodic. In 1765, Leonhard **Euler** discovered an example in which three masses start in a line and rotate so that they stay lined-up; such a set of orbits is unstable, however, and it would be unlikely to occur in nature. Then, in 1772, Joseph **Lagrange** identified a stable periodic orbit in which three masses, one of which is negligible, are at the corners of an equilateral triangle (see **Lagrangian points**). Each mass moves in an ellipse in such a way that the triangle formed by the three masses always remains equilateral. A **Trojan** asteroid, which forms a triangle with Jupiter and the Sun, moves according to such a scheme. In 2001, mathematicians Richard Montgomery of the University of California, Santa Cruz, and Alain Chenciner of the University of Paris added another exact solution to the equations of motion for three gravitationally interacting bodies. The three equal masses chase each other around the same figure-eight curve in the plane. Computer simulations by Carles Simó of the University of Barcelona demonstrated that the figure-eight orbit is stable: the orbit persists even when the three masses aren't precisely the same, and it can survive a tiny disturbance without serious disruption. This means there's a chance that the figure-eight orbit might actually be seen in some stellar system. However, it's a pretty small chance–somewhere between one per galaxy and one per universe! The existence of the three-body, figure-eight orbit has prompted mathematicians to look for similar orbits involving four or more masses. In 2000, Joseph Gerver of Rutgers University, for instance, found one set in which four bodies stay at the corners of a parallelogram at every instant, while each body follows a curve that looks like a figure-eight with an extra twist. Using computers, Simó has found hundreds of exact solutions for the case of *N* equals masses traveling a fixed planar curve. However, they are not stable and thus of no practical significance.

Thuban (Alpha Draconis, α Dra)

The fourth brightest star in **Draco**; it used to be the **pole star**, from some time prior to 3000 B.C. to 1900 B.C. (when it was superceded by **Kochab**). In 2700 B.C. it lay just 1/180 the width of the full Moon from the pole. Although it lies in the Dragon's tail, its name comes from an Arabic phrase that means "the Serpent's head." Thuban belongs to the fairly rare class of giant **A stars**. It has a faint unseen companion in a 51.4-day orbit that may be a red dwarf or a low-mass white dwarf.

Visual magnitude:	3.67
Absolute magnitude:	−1.21
Spectral type:	A0III
Surface temperature:	9,500 K
Luminosity:	265 L_{sun}
Distance:	309 light-years

Thule (minor planet 279)

A 135-km-wide, D-class **asteroid** whose nearly circular orbit, in 3:4 resonance with Jupiter, is considered to mark the outer boundary of the main asteroid belt. Thule was discovered in 1888 by Johann **Palisa**. Semimajor axis 4.271 AU, perihelion 4.22 AU, aphelion 4.32 AU, inclination 2.3°, period 8.83 years.

tidal dwarf galaxy

A self-gravitating entity of dwarf-galaxy mass, built from tidal material expelled during interactions between larger galaxies.

tidal force

The differential gravitational pull exerted on any extended body in the gravitational field of another body; the result is to raise **tides**. When the tidal forces of a planet and several moons are focused on certain moons, particularly if the orbits of the various objects bring them into alignment on a repeated basis, the tidal forces can generate a tremendous amount of energy within the moon. The intense volcanic activity of **Io** is the result of the interaction of such tidal forces.

tidal heating

Frictional heating of a moon's interior due to flexure caused by the gravitational pull of its parent planet and possibly neighboring satellites.

tidal radius

The distance from the center of a planet to the point at which its gravitational pull equals that of the Sun.

tidal tail

A tail-like structure of stars that has been flung out behind a galaxy when it merges with another system. Tidal tails are produced by the galaxies falling into each other's potential wells.

tide

An effect that happens when a large object is moving in an orbit in a gravitational field. The object behaves, as far as the field is concerned, as if it were concentrated at a single point, the center of mass. So the center of mass moves in exactly the right orbit. But every part of the object that isn't at the center of mass is in a "wrong" orbit. If the object is a planet orbiting the Sun, for example, bits of the planet that are farther out from the Sun than the center of mass is, are being dragged around faster than their correct orbital speed for their distance from the Sun, and bits of the planet that are closer to the Sun than the center of mass is, are being held back from their correct orbital speeds. The result is tidal forces, which stretch the planet in both directions outward from the center of mass along a line joining the center of mass to the Sun. Tidal forces produce tides even in the solid surface of a world. But they have a much bigger effect on the oceans. Consequently, there are two bulges of water in the oceans, one on the side of Earth facing the Sun and one on the opposite side. The Moon has double the tide-raising effect of the Sun because it is so much closer. These effects combine at New Moon and at Full Moon giving rise to the spring tides.

time

(1) One of the fundamental dimensions of the universe, intimately connected with space to form a four-dimensional *space-time continuum.* (2) A quantity, the accurate measurement of which is crucial in astronomy. Time can be measured astronomically with respect to the Sun, when it is known as **solar time**, or with respect to the stars, when it is known as **sidereal time**. **Universal Time** (or Greenwich Mean Time) is a world timekeeping

standard used in astronomical observations. When calculating orbital motions with the solar system, however, astronomers switch to a system known as **dynamical time**.

TiO bands
Absorption bands due to the molecule titanium dioxide (TiO). They are prominent in the spectra of K and M stars, which are cool enough for TiO to survive.

Titan
The largest moon of **Saturn** and the second largest moon in the solar system (after Jupiter's **Ganymede**), with a diameter of 5,150 km–more than that of Mercury. Also known as Saturn VI, it was discovered by Christian **Huygens** in 1655. Titan orbits Saturn at a mean distance of 1,221,800 km and, as with most of Saturn's other moons, is gravitationally locked so that its rotational period is the same as its orbital period. Uniquely among satellites in the solar system, it has a substantial atmosphere, with a surface pressure of about 1.5 bar–50% greater than Earth's and comparable with the pressure at the bottom of a 3-m-deep swimming pool. About 92% of this atmosphere is nitrogen, most of the rest is methane, and there are smaller amounts of ethane, ethylene, hydrogen cyanide, and other gases. A dense orange haze, rich in a variety of complex hydrocarbons, forms high in Titan's stratosphere as a result of the breakdown of methane molecules by solar ultraviolet and the subsequent reaction of the molecule fragments with other substances in the atmosphere. Organic droplets and particulates also rain down onto the moon's surface. The latter is believed to be partly covered by ethane or ethane/methane seas that are interspersed with frozen masses of these same materials. However, a recent analysis of Titan's spectrum suggests that much of the surface may consist of exposed icy bedrock. If this is the case, it remains to be understood how this bedrock has come to be cleared of the organic sediment that continually drizzles from the sky. Another puzzle is why Titan has such a dense atmospheric shroud, when other big, cold moons, such as Ganymede, have none. Much discussion has centered around the idea that Titan, despite its low surface temperature of about –180°C, may reveal much about the early steps leading to life. Beginning in 2004, **Cassini/** Huygens will shed more light on how far Titan's organic chemistry has progressed in the direction of true biology.

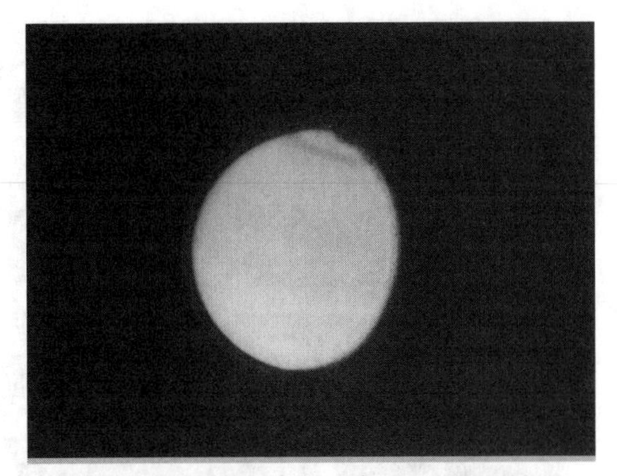

Titan Voyager 2's image of Titan, taken on August 23, 1981, from a range of 2.3 million km, shows some detail in the cloud systems. The southern hemisphere appears lighter in contrast, a well-defined band is seen near the equator, and a dark collar is evident at the north pole. All these bands are associated with cloud circulation in Titan's atmosphere.
NASA/JPL

Titania
The largest and fourteenth closest moon of **Uranus**, also known as Uranus III; it was discovered by William **Herschel** in 1787. Titania orbits at a mean distance of 436,270 km from the planet and has a diameter of 1,578 km. Its surface contains a mixture of cratered terrain (including some craters that are partly submerged) and systems of fault valleys up to 1,500 km long. Deposits of a highly reflective material, which may be frost, were photographed by Voyager 2 along the sun-facing valley walls. According to one theory, Titania was once hot enough to be liquid; the surface cooled and hardened, then the interior froze and expanded, cracking the surface and resulting in the interconnected valleys seen today. (See photo, page 491.)

Titius-Bode Law
A prescription for calculating planetary distances, made famous by Johann **Bode** but first pointed out in 1766 by the German mathematician Johann Titius (1729–1796). It states that the distance to the nth planet is $0.4 + (0.3)^n$ AU. The Titius-Bode law works surprisingly well out to Uranus but then breaks down.

TLP
See **transient lunar phenomenon**.

Toby Jug Nebula (IC 2220)
A **reflection nebula** in **Carina** with a bipolar structure that gives it the shape of an old English drinking vessel. It surrounds and is illuminated by the red giant HD 65670, which is also an irregular variable (ranging between sixth and seventh magnitude) known by the variable star name V341 Carinae. The dust and gas in the Toby Jug have come from mass lost by this star while in its giant stage.

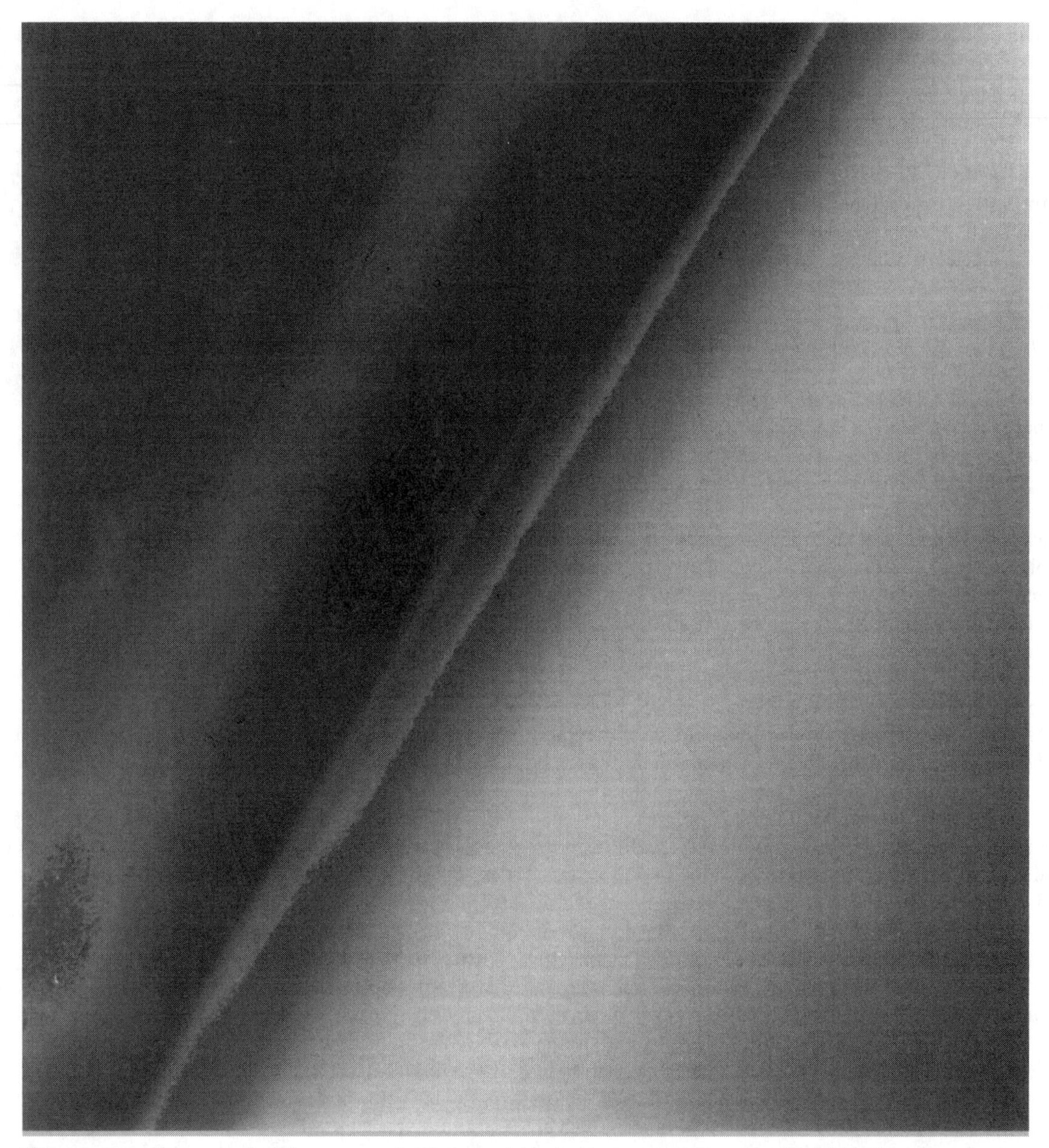

Titan Layers of haze covering Titan are seen in this image taken by Voyager 1 on November 12, 1980, at a range of 22,000 km. The divisions in the haze occur at altitudes of 200, 375, and 500 km above the limb of the moon. *NASA/JPL*

Tombaugh, Clyde William (1906–1997)

An American astronomer who discovered **Pluto** in 1930. Tombaugh began his search for the trans-Neptunian planet, which Percival **Lowell** had predicted, in 1929 when he became an assistant at **Lowell Observatory**. Using a blink **comparator**, he compared pairs of photographs taken a week apart. He was rewarded on February 18, 1930, when he spotted an object that showed movement between two plates exposed the previous month. Tombaugh later searched, without success, for a tenth planet and for small moonlets of Earth, though he did find in the process new star clusters and galaxy clusters and almost 800 asteroids.

topocentric coordinates

A system of **celestial coordinates** with its origin at a specific point on Earth's surface. Usually, the difference in the position of an object in the sky measured using topocen-

Titania Details as small as 9 km across show up in this view of Titania compiled from several Voyager 2 images taken from a range of about 500,000 km on January 24, 1986. As well as impact scars, Titania displays evidence of other geologic activity at some point in its history. The large, trenchlike feature near the terminator at middle right suggests at least one episode of tectonic activity. Another, basinlike structure near the upper right signals an ancient period of heavy impact activity. *NASA/JPL*

tric and geocentric (Earth-centered) coordinates is very small because most celestial objects are so far away. The Moon's position, however, can vary as much as 2° depending on where on Earth it is measured from, and, within the past few years, an observatory in Australia missed spotting an asteroid that passed close to Earth because it used geocentric rather than topocentric positions.

topography

The configuration (shapes, positions, and arrangements) and relief (elevations and slopes) of the land surface of a planet, a moon, or an asteroid.

Torino Scale

The official scale for quantifying the Earth impact hazard of near-Earth asteroids and comets. It was introduced at an international conference on **near-Earth objects** held in June 1999 in Torino, Italy, as a revised version of the "Near-Earth Object Hazard Index." It is a two-parameter scale that uses a number from 0 to 10 to indicate the chance of a collision, and a color to give information about the danger of the event (from white, nondangerous, to red, catastrophic). An object that will make several different close approaches to Earth will have a different Torino Scale value for each approach; normally, only the highest of these values is considered to identify an object. The Torino Scale value will change with time as an object's orbit becomes better known.

Toro (minor planet 1685)

A member of the **Apollo group** of Earth-crossing asteroids, discovered by Carl Wirtanen (1910–1990) in 1948 and rediscovered in 1964, whose closest approach takes it within 0.13 AU of Earth. Radar observations indicate a rocky surface, thinly covered with dust. It has a diameter of 5 × 3 km, an albedo of 0.15, and a rotational period of 10.18 hours. Semimajor axis 1.37 AU, perihelion 0.77 AU, aphelion 1.96 AU, eccentricity 0.44, period 584.2 days.

total eclipse

A condition when one celestial body completely obscures or shades another. See **solar eclipse**.

totality

The period of total obscuration of a celestial body such as the Sun during a **solar eclipse** or the Moon during a **lunar eclipse**.

Toutatis (minor planet 4179)

One of the largest **near-Earth asteroids**, measuring 4.5 × 2.4 × 1.9 km, and potentially one of the most dangerous. A member of the **Apollo group**, it was discovered in 1989 by French astronomers and named (somewhat inappropriately) after a Celtic god that was the protector of the tribe in ancient Gaul. Its eccentric, four-year orbit extends from just inside Earth's orbit to the main asteroid belt; the danger comes from the fact that the plane of Toutatis's orbit is closer to the plane of Earth's orbit than any known Earth-orbit-crossing asteroid. In December 1992, Toutatis came within about 4 million km of Earth enabling radar images to be acquired using the Goldstone Deep Space Communications Complex in California's Mojave desert. These images revealed two irregularly shaped, cratered objects about 4 and 2.5 km in average diameter, which are probably in contact with each other. Such contact binaries may be fairly common since another one, **Castalia**, was observed in 1989. Numerous surface features on Toutatis, including a pair of large craters, side by side, and a series of three prominent ridges–a type of asteroid mountain range–are presumed to result from a complex history of impacts.

Toutatis shows an extraordinarily complex rotation. Whereas the vast majority of asteroids, and all the planets, spin about a single axis, Toutatis tumbles around two axes with different periods, of 5.47 and 7.35 Earth days, that combine in such a way that Toutatis's orientation with respect to the solar system never repeats.

On September 29, 2004, Toutatis will pass by Earth at a range of four times the distance between Earth and the Moon. One consequence of the asteroid's frequent close approaches to Earth is that its trajectory more than several centuries from now cannot be predicted accurately. In fact, of all the Earth-crossing asteroids, the orbit of Toutatis is thought to be one of the most chaotic. Density 2.1 g/cm^3, spectral class S, semimajor axis 2.516 AU, perihelion 0.92 AU, aphelion 4.11 AU, inclination 0.5°, period 4 years.

trailing hemisphere

The hemisphere of a moon in **synchronous rotation** that faces backward, away from the direction of orbital motion.

transfer lens

A lens used to transfer a light beam to another place within an optical system without changing the image; also known as a *relay lens*. In a **Maksutov telescope**, for example, a transfer lens is used to make the **focal point** accessible by bringing it outside the instrument.

transient lunar phenomenon (TLP)

Any kind of short-lived unusual activity or event seen on the Moon. TLPs reported over the years have included red glows, flashes, obscurations, abnormal albedo, and shadow effects. Unfortunately there is no *unambiguous* photographic or photometric record of such an event, and some authorities are inclined to dismiss TLPs as either aberrational effects in Earth's atmosphere, changes in lunar lighting conditions, or outright illusions.

Accounts of TLPs go back as far A.D. 557. Most are visual reports of bright spots, flashes, hazes, and curious temporary colorations of the lunar soil. Reputable observers such as William **Herschel**, Wilhelm **Struve**, and Edward **Barnard** have seen them. About 200 of some 30,000 lunar features visible in telescopes have been recorded as TLP sources. Half have shown activity only once. Of the remainder, a mere dozen features contribute three-fourths of all reports. One area, the Aristarchus-Herodotus-Schroters Valley, is responsible for one-third of the total number sighted.

Among the more famous sightings, on the night of November 2–3, 1958, Soviet astronomer Nikolai Kozyrev (1908–1983) saw something strange while making spectrograms of the crater Alphonsus with the Crimean Astrophysical Observatory's 50-inch reflector. As he watched through the telescope's guiding eyepiece, he saw the crater's central peak blur and turn an unusual reddish color. He reported that spectrograms confirmed his impression of an anomalous event, showing the emission spectrum of carbon vapor.

On July 19, 1969, Apollo 11 had just achieved lunar orbit when the Mission Control Center received word that amateur astronomers reported transient phenomena in the vicinity of the crater Aristarchus. Asked to check out the situation, Neil Armstrong looked out his window and observed an "area that is considerably more illuminated than the surrounding area. It just has . . . seems to have a slight amount of fluorescence to it." Although he wasn't sure, Armstrong believed the region was Aristarchus.

Before the Apollo landings, TLPs were often explained in terms of minor volcanic activity. But it now seems certain that the Moon has been geologically dead for over 3 billion years (aside from the odd minor moonquake caused by tidal forces). Other possibilities, however, remain. Shortly after a large flare erupted on the Sun in 1963, Zdenek Kopal (1914–1993) and Thomas Rackham (1919–2001) at Pic du Midi Observatory in southern France photographed a local brightening around the craters Copernicus, Kepler, and Aristarchus. Kopal proposed that energetic particles from the flare caused lunar rocks to fluorescence. Such activity might be expected especially at full phase when the Moon passes through Earth's magnetosphere, where solar wind particles become trapped.

transit

(1) The passage of a smaller astronomical object across the face of a larger one, as in a transit of Venus or Mercury across the Sun, or the transit of a moon across its primary. In the latter case, if the transit involves the moon's shadow rather than the moon itself, it is known as a *shadow transit*. Transits of Venus and Mercury can only take place when the planets are close to the **nodes** of their orbits at inferior **conjunction**: in early December or early June for Venus, and in early November or early May for Mercury. Only five transits of Venus have been observed, those of 1639, 1761, 1769, 1874, and 1882; the next is due in 2004. (2) The passage of a **meridian** or surface marking across the **central meridian** of the disk of a rotating planet. (3) The instant at which a star or other celestial object crosses the observer's meridian; also known as **culmination** or *meridian passage*.

transit circle

A telescope mounted so that it can only move along the **meridian** from the **zenith** to the horizon, that is, about a fixed horizontal axis in a north-south vertical plane. Transit circles, also known as *meridian circles*, are used to measure accurately the altitudes of stars and to time their passage across the observer's meridian.

transition region
A thin and very irregular layer of the Sun's atmosphere, a few hundred km deep, that lies between the **chromosphere** and the **corona**. The temperature climbs steeply within the transition region from about 20,000 K in the chromosphere to over 1 million K at the base of the corona.

transition variable
A variable star that is in the process of evolving from one type into another. Transition variables are either just starting or ceasing to vary, changing type, or simply have not been studied well enough to tell what type they really are.

trans-Neptunian object (TNO)
A class of objects that includes **Kuiper Belt objects** and **Scattered Disk** objects.

trans-Plutonian planet
See **Planet X**.

transverse velocity
The component of the velocity of an object, such as a star, that is at right-angles to the observer's line of sight; also known as *tangential velocity*. To calculate a star's transverse velocity, the star's distance and proper motion must be known. See also **radial velocity**.

Trapezium (Theta[1] Orionis, θ[1] Ori)
A star cluster at the heart of the **Orion Nebula** consisting of approximately 1,000 young, hot O and B stars crowded into a space about 4 light-years in diameter (roughly the distance between the Sun and the next nearest star). With an average stellar age of only 1 million years, it is the youngest cluster known. Most of its member stars are hidden by dust or the glare of the nebula but are visible at infrared wavelengths. The four brightest stars (A, B, C, and D), which form the vertices of a trapezium and give the cluster its name, can be seen easily with a small telescope. Two fainter stars of eleventh magnitude, E and F, show up in moderately sized amateur instruments under good seeing conditions. A further two, of sixteenth magnitude, G and H, are only visible in very large amateur scopes. θ^1 Ori A, also known as V1016, is an eclipsing binary with a period of 65.432 days and a magnitude range of 6.72 to 7.65. θ^1 B, also known as BM Ori, is another eclipsing binary, spectral type B2-B3, with a period of 6.471 days and a magnitude range of 7.90 to 8.65. An infrared companion has also been discovered in the θ^1 A system, making it a triple star, and θ^1 B is now known to be a quadruple system, with three components detected separately in the near-infrared. Infrared observations have shown that θ^1 C, which is actually the brightest star in the cluster, is a close binary as well.

The Trapezium was first drawn as a triple star (A, B, and C) by Giovanni **Hodierna** before 1654 and described by Christen **Huygens** in 1656. Star D was independently discovered by Jean **Picard** and Huygens in 1684, while E and F were discovered by William **Struve** and John **Herschel** in 1826 and 1830, respectively. Star G was found in 1888 by Alvan **Clark** while testing the 36-inch refractor that he made for Lick Observatory, and Edwin **Barnard** discovered H later in the same year with the same instrument.

Triangulum (abbr. Tri; gen. Trianguli)
The Triangle; a small northern **constellation** with a distinctive shape located at the southeast edge of Andromeda. Its prime deep sky object is M33, the **Triangulum Galaxy**. (See table, "Triangulum: Stars Brighter than Magnitude 4.0," and star chart 1.)

Triangulum Australe (abbr. TrA; gen. Trianguli Australis)
The Southern Triangle; a small **constellation** close to the southern circumpolar region and touched by the southern edge of the Milky Way. Its three leading stars are bright enough to serve as pointers to other constellations nearby, including Apus in the south, Norma in the north, Circinus in the west, and Ara and Pavo in the

Trapezium These four brilliant stars give the Trapezium cluster its name and provide most of the illumination for the entire Orion Nebula. *NASA/STScI*

Triangulum: Stars Brighter than Magnitude 4.0

Star	Magnitude		Spectral	Distance	Position	
	Visual	Absolute	Type	(light-yr)	R.A. (h m s)	Dec. (° ′ ″)
β	3.00	0.09	A5III	124	02 09 33	+34 59 14
α Mothallah	3.42	1.95	F6IV	64	01 53 05	+29 34 44

east. The open cluster NGC 6025 at the boundary with Norma, contains about 30 stars of seventh magnitude and fainter and is a good object for binoculars (distance 2,000 light-years; magnitude 5.1, diameter 12′; R.A. 16h 3.7m, Dec. −60° 20′. (See table, "Triangulum Australe: Stars Brighter than Magnitude 4.0," and star chart 5.)

Triangulum Galaxy (M33, NGC 598)

A nearby type Sc **spiral galaxy** that is a prominent member of the **Local Group**. With about half the diameter of the Local Group's two dominant systems, the Andromeda Galaxy and our own Milky Way, M33 is an average-sized spiral, with a mass of between 10 and 40 billion M_{sun}. It was probably first found by **Hodierna** before 1654 and independently rediscovered by Charles **Messier** in 1764. Despite its modest size, M33 is home to one of the largest H II regions known: NGC 604, with a diameter of nearly 1,500 light-years and at least 200 newly formed hot massive stars. It also has globular clusters and, possibly, its own satellite galaxy, the **Pisces Dwarf** or LGS 3. (M33, in turn, may be a remote but gravitationally bound companion of the Andromeda Galaxy.) For the amateur observer, the Triangulum Galaxy can be seen with the naked eye under exceptionally good conditions, making it, for those with keen eyesight, the most distant object visible without optical aid. It is outstanding in good binoculars, but as its considerable total brightness is distributed quite evenly over an area of nearly four times that covered by the Full Moon, its surface brightness is extremely low.

Visual magnitude:	5.7
Angular diameter:	73′ (about 2.5 times the Full Moon)
Diameter:	About 55,000 light-years
Distance:	2.85 million light-years

Trifid Nebula (M20, NGC 6514)

A large emission and reflection nebula in **Sagittarius**, famous for its three-lobed appearance. It was discovered by Charles **Messier** in 1764 and named the "Trifid" by John **Herschel**. The dark nebula, which is the reason for the Trifid's appearance, was cataloged by Edward **Barnard** as Barnard 85 (B 85). The red emission portion of the nebula, illuminated by a young star cluster near its center (the brightest member of which is a very hot O star), is surrounded by a blue reflection nebula that is particularly conspicuous toward the northern end.

Visual magnitude:	9.0
Apparent size:	28′
Distance:	2,700 light-years
Position:	R.A. 18h 2.6m, Dec. −23° 2′

triple-alpha process

A chain of nuclear fusion reactions, also known as the *Salpeter process* after the Austrian-American astrophysicist Edwin Salpeter (1924–) who first identified it, by which helium is converted into carbon. It is the main source of energy production in **red giants** and red **supergiants** in which the core temperature has reached at least 100 mil-

Triangulum Australe: Stars Brighter than Magnitude 4.0

Star	Magnitude		Spectral	Distance	Position	
	Visual	Absolute	Type	(light-yr)	R.A. (h m s)	Dec. (° ′ ″)
α Atria	1.91	−3.62	K2IIb	415	16 48 40	−69 01 39
β	2.83	2.38	F2III	40	15 55 08	−63 25 50
γ	2.87	−0.88	A1V	183	15 18 55	−68 40 46
δ	3.86	−2.54	G5IIa	621	16 15 26	−63 41 08

lion K. Two nuclei of helium (alpha particles) collide, fuse, and form a nucleus of beryllium. During a third collision between beryllium and helium, carbon is formed. Successive collisions build oxygen and neon.

triplet

Three simple lenses used in combination, placed close together or in contact.

tritium

The heaviest isotope of **hydrogen**, in which each nucleus contains one proton and two neutrons, instead of only one proton as in normal hydrogen or one proton and one neutron as in **deuterium**. Tritium is radioactive, with a half-life of 12 years.

Triton

The seventh nearest and by far the largest moon of **Neptune**. It was discovered on October 10, 1846, by William **Lassell**, only 17 days after the discovery of Neptume itself. Most of what we know about Triton has come from observations by Voyager 2, which flew past it on August 25, 1989, at a distance of about 40,000 km. Triton is unique among big moons in that it orbits its planet in the opposite direction to which its planet spins. This kind of *retrograde* motion, which is also shown by many of the small outer moons of Jupiter and Saturn, is a sure sign that a body has been captured at some point after the solar system formed. Most likely, Triton, like Pluto, is an escaped **Kuiper Belt object** (if so, it would be the largest one known), that strayed too close to Neptune and became ensnared by its gravitational pull. A capture scenario also provides an explanation for the extremely eccentric orbit of Neptune's outermost moon, **Nereid** (which would have been displaced by the massive newcomer's arrival), and the fact that Triton's interior appears to be differentiated (layered into a core, a mantle, and a crust): tidal heating from an eccentric postcapture orbit being circularized could have kept Triton liquid for a billion years, during which time its heavier materials would have sunk toward the center. Triton's retrograde motion, however, spells its doom. The already close orbit is slowly decaying further, as a result of tidal interactions, so that, in about 250 million years, Triton will fall within Neptune's **Roche limit** and be torn apart to form a ring system even more spectacular than that of Saturn.

The stretching and flexing of Triton in the past, following its capture, led to global resurfacing and the complete erasure of densely cratered areas and large impact basins. Three types of terrain now dominate the surface: a large area of fractured plains that looks like the skin of a cantaloupe and shows a complex network or dimples and ridges; a region of smoother, lava-flooded plains; and the bright polar ice caps, composed of frozen methane, frozen nitrogen, water ice, carbon monoxide ice, and carbon dioxide ice. The large polar caps are largely responsible for the moon's exceptionally high **albedo**. Because Triton reflects so much solar energy back into space, it is unusually cold, with a surface temperature similar to that of **Pluto**. Also like Pluto, it has a very thin atmosphere, made up of nitrogen and a dash of methane, with a surface pressure of about 15 microbars (0.000015 the surface pressure on Earth). In such an atmosphere, organic compounds could form by chemical reactions between the methane and the nitrogen, as they certainly do on **Titan**. A faint haze and tenuous clouds lie at an altitude of about 13 km. Observations by the Hubble Space Telescope in 1998 showed that Triton had warmed up by a couple of degrees since the Voyager rendezvous almost a decade earlier. This is due to additional nitrogen entering the atmosphere as midsummer approaches in the southern hemisphere and giving rise to an extra (although still small) **greenhouse effect**.

Perhaps the most unexpected discovery by Voyager 2 at Triton was that of active ice volcanoes, which occur particularly around the margins of the south polar cap. One of the images returned by the probe shows a plume rising 8 km abouve the surface and extending 140 km downrange. Triton thus joins an elite group of volcanically active bodies in the solar system, the only other known members of which are Earth, Venus (probably), and **Io**. Different mechanisms are involved in the four bodies. Eruptions on Earth and Venus (and also, in the past, on Mars) are driven by internal heat and consist of molten rock. Io's massive volcanism is powered by tidal energy and involves sulfur and its compounds. Triton's eruptions, by contrast, are of volatile compounds, including nitrogen and methane and probably some organic material (which would account for the red-brown colorations on the surface), and are driven by seasonal heating from the Sun. This heating comes about because Triton's axis is tilted 157° with respect to Neptune's axis (which in turn, leans 30° away from the plane of the planet's orbit) with the result that Triton's polar and equatorial regions alternately face sunward for long periods.

Diameter:	2,706 km
Density:	2.0 g/cm^3
Surface temperature:	34.5 K (–235°C)
Albedo:	0.7 to 0.8
Rotational period:	5.877 days
Orbit	
Mean distance from Neptune:	354,760 km
Eccentricity:	0.00
Inclination:	157°
Period:	5.877 days

Triton Voyager 2 returned this image of Triton, showing detail of the "cantaloupe terrain" as small as 2.5 km across, from a range of about 130,000 km on August 25, 1989. The long linear feature extending vertically across the image is probably a graben (a narrow down-dropped fault block) about 35 km across. The ridge in the center of the graben probably is ice that has welled up by plastic flow in the floor of the graben. The surrounding terrain is a relatively young icy surface with few impact craters. *NASA/JPL*

troctolite

A type of igneous rock, found in the lunar highlands, composed of **plagioclase feldspar** and **olivine**.

troilite

A nonmagnetic form of iron sulfide (FeS) found in a wide variety of meteorites.

Trojan

An object located at either of the stable **Lagrangian points** (L_4 or L_5) of a larger object's orbit. The term comes from the theme of the names given to the first asteroids to be found at Jupiter's Lagrangian points, **Achilles**, Patroclus, and **Hektor**–characters in Homer's epic poem about the Trojan War, the *Iliad.* Jupiter's Trojans occupy two banana-shaped regions, centered 60° ahead and behind the planet, which can be considered the "Greek" and "Trojan" nodes, respectively; only Hektor in the Greek node and Patroclus in the Trojan node are, so to speak, in the camps of the enemy. Perturbations by other planets, principally Saturn, cause the Trojans to oscillate about the Lagrangian points in an arc about 45 to 80° from Jupiter with a period of 150 to 200 years. Several hundred Jupiter Trojans are known out of a total population that includes an estimated 2,300 objects bigger than 15 km across and many more of smaller size; most do not move in the plane of the planet's orbit but rather in orbits inclined by up to 40°. Unless qualified, "Trojan" is assumed to mean "Jupiter Trojan." There are three known **Mars Trojans**, but searches have so far failed to uncover any similar objects in the orbits of Venus, Earth, Saturn, or Neptune. Saturn's satellites **Helene**, **Calypso**, and **Telesto** are also sometimes called Trojans because they lie at the leading Lagrangian point of **Dione**, and trailing and leading Lagrangian points of **Tethys**, respectively.

tropic

Either of two latitudes on Earth at which the Sun appears directly overhead at mid-summer and mid-winter. The tropic of Cancer is the parallel of latitude 23° 27′ north of the equator where the Sun lies in the zenith at the summer solstice, around June 22. The tropic of Capricorn is the parallel of latitude 23° 27′ south of the equator where the Sun lies in the zenith at the winter solstice, around December 22.

tropopause

The boundary in Earth's atmosphere between the **troposphere** and the **stratosphere**. Its height is about 15 to 17 km over the tropics and 10 km nearer the poles, but it varies seasonally and with changes in weather.

troposphere

The lowest level of Earth's atmosphere, extending from the surface up to a height of 10 to 16 km. Its temperature decreases steadily with altitude from about 290 to 240 K at a rate of about 6.5 K per km. Almost all the atmospheric water vapor and weather systems occur here.

true pole

The direction toward which Earth's axis points at a given time. The *true equator* is the great circle on the celestial sphere drawn perpendicular to this direction. The *true equinox* is the intersection of the ecliptic (the plane of Earth's orbit) with the true equator.

Trümpler classification

A classification scheme for **open clusters** devised by the American astronomer Robert Trümpler (1886–1956). It uses three criteria: degree of central concentration (from I, most, to IV, least concentrated); range of brightness of member stars (from 1, narrow, to 3, wide); and total number of stars in the cluster (p for poor, less than 50; m for moderately rich, 50 to 100; r for rich, more than 100). The addition of "n" indicates that the cluster has some nebulosity; for example, the Trümpler classification of the Pleiades is 13rn.

Tucana (abbr. Tuc; gen. Tucanae)

The Toucan; a small **constellation** of the southern circumpolar region. (See tables, "Tucana: Stars Brighter than Magnitude 4.0," and "Tucana: Other Objects of Interest," see also star chart 8.)

47 Tucanae (NGC 104)

The second brightest globular cluster in the sky, after **Omega Centauri**. To the unaided eye it looks like a misty star, while binoculars clearly show an increase in brightness toward the center and a telescope with an aperture of at least 10 cm resolves some of the roughly 100,000 member stars. Although a conspicuous naked-eye sight, it lies so far south that it wasn't discovered by astronomers until 1751 when Nicholas de **Lacaille** cataloged it in his list of southern nebulous objects.

Tucana: Stars Brighter than Magnitude 4.0

Star	Magnitude		Spectral	Distance	Position	
	Visual	Absolute	Type	(light-yr)	R.A. (h m s)	Dec. (° ′ ″)
α	2.87	–1.06	K3III	199	22 18 30	–60 15 35
β	3.7	–0.20		140		
β^1	4.36	1.20	B9V		00 32 33	–62 57 30
β^2	4.53	1.37	A2V+A7V		00 32 34	–62 57 57
γ	3.99	2.27	F1III	72	23 17 26	–58 14 08

Tucana: Other Objects of Interest

Object	Notes
Globular clusters	
47 Tucanae (NGC 104)	See separate entry.
NGC 362	A foreground object at the northern edge of the Small Magellanic Cloud, visible through binoculars. Distance 40,000 light-years; magnitude 6.6; diameter 12.9′; R.A. 0.1h 03.2m, Dec. –70° 51′.
Galaxy	
Small Magellanic Cloud	See separate entry.

Visual magnitude:	4.0
Apparent size:	31′ (about the size of the full Moon)
Diameter:	120 light-years
Position:	R.A. 0h 24.1m, Dec. –72° 5′
Distance:	13,400 light-years

Tully-Fisher relation

An observed relationship between the luminosity of **spiral galaxies** and their maximum rotation velocity; it is used as a way of estimating distances to spirals. The form is a linear relation between the **absolute magnitude** of a galaxy and the logarithm of the velocity at the flat part of the rotation curve, although the slopes and intercepts of these relations are different for Sa, Sb, and Sc type galaxies. Approximations made in deriving the relation are that the mass to light ratios are constant for all galaxies and that the average surface brightness of all galaxies is also equal.

Tunguska event

A violent explosion that took place over the valley of the Stony Tunguska river, in central Siberia, at 7:17 A.M. on June 30, 1908, and is now generally attributed to the detonation of icy material from a comet in the atmosphere. The blast, which stripped or felled trees out to a radius of 40 km, burned reindeer to death, and sent the tents of nomads flying through the air, was preceded by the observed passage through the air of a dazzling blue **bolide** trailing a column of dust. Soviet mineralogist Leonid Kulik (1883–1942), who first investigated the event, initially assumed that it had been caused by the fall of a giant iron meteorite but the absence of any crater quickly led to abandonment of this idea. The cometary theory was first put forward in the 1930s by the Soviet astronomer Igor Astapovich (1908–1976) and the English meteorologist Francis Whipple (1876–1943).

Turner, Herbert Hall (1861–1930)

An English astronomer generally credited with coining the term **"parsec"** and being partly responsible for naming the ninth planet. Having served as chief assistant at the Royal Greenwich Observatory for nine years, he spent most of his career as Savilian professor of astronomy at Oxford. A leader in the worldwide effort to produce an astrographic chart of the sky, he developed improved methods for obtaining both positions and magnitudes from photographic plates, and was adept at training unskilled workers to take part in this project. Most of his later work was in seismology; he compiled and published worldwide earthquake data starting in 1918, and discovered the existence of deep-focus earthquakes in 1922. He was responsible for forwarding the suggestion of Venetia Phair that the newly found world, discovered at the Lowell Observatory, be christened "Pluto."

turnoff point

The point on the **Hertzsprung-Russell diagram** at which stars in a cluster depart from the main sequence. The brighter the turnoff point, the younger the cluster age. The *turnoff mass* is the least mass of a star (in a group) that has had time to begin evolving away from the main sequence.

21-centimeter line

An emission line in the radio region of the spectrum, at a wavelength of 21.1 cm (corresponding to a frequency of 1420 MHz) that is due to the "spin-flip" transition in neutral hydrogen atoms. The proton and electron making up a hydrogen atom both spin. When their spins are parallel, the atom has slightly more energy than when their spins are antiparallel. As the atom flips from the parallel to the antiparallel state, it emits a radio wave in the 21-cm line. Most of what is known about the distribution of cold gas in the Galaxy, including the mapping of the spiral arms, has come from detailed studies of the variation of 21-cm emissions across the sky.

twilight

The interval that follows sunset and comes before sunrise, during which the sky is partially illuminated by light from the Sun. Twilight lengthens with the distance of the observer from the equator, and is shortest as seen from anywhere on Earth at the equinoxes. *Astronomical twilight* is defined to begin in the morning and to end in the evening, when the center of the Sun is geometrically 18° below the horizon. Before the beginning of astronomical twilight in the morning and after the end of astronomical twilight in the evening, the Sun does not contribute at all to sky illumination. This corresponds to the older definition, going back to ancient times, that hinged on the ability to discern a sixth magnitude star (at the limit of naked-eye visibility) at the zenith. *Civil twilight* and *nautical twilight* are defined in exactly the same way but with 6° and 12°, respectively, in place of the 18° for the astronomical variety.

two-body problem

The problem of finding the positions and velocities of two massive bodies that attract each other gravitationally, given their masses, positions, and velocities at some initial time. It was first solved by Isaac **Newton**, who showed mathematically that the orbit of one body about another was either an ellipse, a parabola, or a hyperbola, and that the center of mass of the system moved with constant velocity.

Tycho Brahe

See **Brahe, Tyco**.

Tycho Catalogue

Data collected as part of the overall mission of the **Hipparcos** satellite; it represents a catalog of position, parallax, proper motion, and magnitude data collected for over 1 million stars. In most cases, its precision is much greater than all earlier catalogs. About the only case in which the Tycho data would be ignored would be if Hipparcos data is available instead. The Tycho data are a survey of all stars that were bright enough to be measured by the detector and are essentially complete to about magnitude 10.5, with somewhat incomplete coverage to magnitude 11 or 11.5.

Tycho's Star

A Type I **supernova** in Cassiopeia (also known by its variable star name B Cassiopeiae) that was observed and described by Tycho **Brahe** in November 1572. At its peak it was as bright as Venus and was visible in the daytime, reaching a magnitude of about –4. The present supernova remnant, which lies about 8,000 light-years away, is associated with some faint optical nebulosity, the intense radio source 3C 10, and the weak X-ray source 2U 0022+63.

U

U Geminorum star (U Gem star)

Also known as a *dwarf nova,* a type of **cataclysmic binary** that brightens abruptly and unpredictably by a factor of 5 to 250 (2 to 6 magnitudes). The rise to maximum takes less than a day and is followed by a decline to quiescence of several days or weeks. U Geminorum stars, of which several hundred are known, consist of a white dwarf (or hot blue subdwarf) primary and an orange or red (K- or M-type) subgiant secondary that has grown to fill its **Roche lobe**. Each star has a mass of 0.5 to 1 M_{sun} and the orbital period of the system is 3 to 15 hours. Matter from the larger star enters an **accretion disk** around the dwarf, where it forms a *hot spot,* and from the inner edge of the disk spirals down onto the dwarf's surface. The light from a U Gem star, therefore, comes from four sources: the white dwarf primary, the cooler secondary, the accretion disk, and the hot spot. Most of the time, a U Gem star shows small, sometimes rapid variations in light. Occasionally, however, the system will brighten rapidly by several magnitudes and then, over a period of days to months, return to normal. Such outbursts are assumed to be caused by the explosive nuclear fusion of hydrogen-rich material on the surface of the white dwarf. Two competing theories have been proposed to explain the outbursts. In the *mass-transfer burst model,* an outburst stems from a sudden increase in mass transfer from the secondary to the white dwarf primary. Such an increase may be triggered by an instability in the atmosphere of the cooler star. The sudden mass transfer may then cause the disk to collapse, dumping a load of fusionable matter on the white dwarf. The rival *disk-instability model* postulates that the mass transfer from the secondary to the primary is relatively constant and that the relocated gas builds up steadily in the cool outer region of the disk, far from the white dwarf. When a critical surface density is achieved, thermal instabilities within the disk cause the matter to be accreted onto the white dwarf, giving rise to an outburst. The latter theory is currently favored because the instability offers a definite mechanism for the cause of the outburst, and simulations of outbursts based on the model have successfully replicated many of the characteristics of U Gem eruptions. The details of outbursts vary from system to system, but any given U Gem star will blaze up at semiregular intervals: the longer the interval, the greater the increase in brightness. The prototype, U Gem itself, remains at magnitude 13.5 for periods varying between 40 and 130 days, and rises to 9.5 or brighter in the course of a day or two, before reverting to its usual state over the next couple of weeks. It also experiences a minor minimum every 4h 11m as its larger, secondary component passes in front of the white dwarf primary. U Gem subtypes include **SS Cygni stars**, **SU Ursae Majoris stars**, and **Z Camelopardalis stars**. See **nova**.

UBV system

A system of stellar magnitudes devised by Harold Johnson (1921–1980) and William **Morgan** at Yerkes Observatory, which consists of measuring an object's apparent magnitude through three color filters: the ultraviolet (U) at 3,600 Å; the blue (B) at 4,200 Å; and the visual (V) in the green-yellow spectral region at 5,400 Å. It is defined so that for A0 stars, $B - V = U - B = 0$; it is negative for hotter stars and positive for cooler stars. The Stebbins-Whitford-Kron six-color system (U, V, B, G, R, I) is defined so that $B + G + R = 0$. The difference, $B - V$, is a good measure of a star's actual color. For example, Betelgeuse has $B - V = 1.85$, indicating that it is quite red. On the other hand, Rigel has $B - V = -0.03$, indicating that it is bluish. Most stars fall between these extremes, except for a few redder-than-red stars (mostly carbon stars) and a few bluer-than-blue stars (mostly young, high-mass stars).

ultra-cool star

A dwarf star of spectral type M7 or later–the warmest kind of **brown dwarf**. Studies of ultra-cool stars in the Trapezium cluster have revealed them to have a dusty atmosphere.

ultra-luminous cluster

A young **globular cluster** that has formed in the past few tens or hundreds of millions of years, and that is composed mainly of bright blue stars. Ultra-luminous clusters are several hundred times brighter than the ancient variety found in the Milky Way and are characteristic of galaxies that have recently undergone collisions or mergers, such as the **Atoms for Peace Galaxy** and the **Antennae**, and are also believed to have been common in galaxies in the early universe.

ultra-luminous infrared galaxy (ULIRG)

A galaxy that emits most of its total energy output in the far (long-wavelength) infrared–a sure sign that it contains a huge amount of dust. Such large dust concentrations, in turn, suggest that ULIRGs are the scene of prodigious star formation. Although ULIRGs are still found in the universe today, recent surveys suggest they were much more common in the remote past and contribute significantly to the cosmic background in the far-infrared and submillimeter bands.

ultraviolet

Electromagnetic radiation beyond the violet with wavelengths in the approximate range 100 Å to 4,000 Å. The *near ultraviolet* (the long wavelength end of the range) extends from just beyond the violet end of the visible spectrum to about 2,000 Å. The *far ultraviolet* stretches from 2,000 Å to the Lyman limit at 912 Å. Beyond this, from the Lyman limit to the start of the X-ray regime at about 100 Å, lies the *extreme ultraviolet.*

Ulysses

A joint NASA/European Space Agency probe to study the Sun, the interplanetary medium, and the makeup of the **solar wind**. Launched in 1990, it provided the first opportunity to take measurements over the Sun's poles.

umbra (pl. umbrae)

(1) The portion of a shadow cone in which none of the light from an extended light source (ignoring refraction) can be observed. (2) The dark core or cores in a **sunspot** with a penumbra, or all of a sunspot that lacks a penumbra.

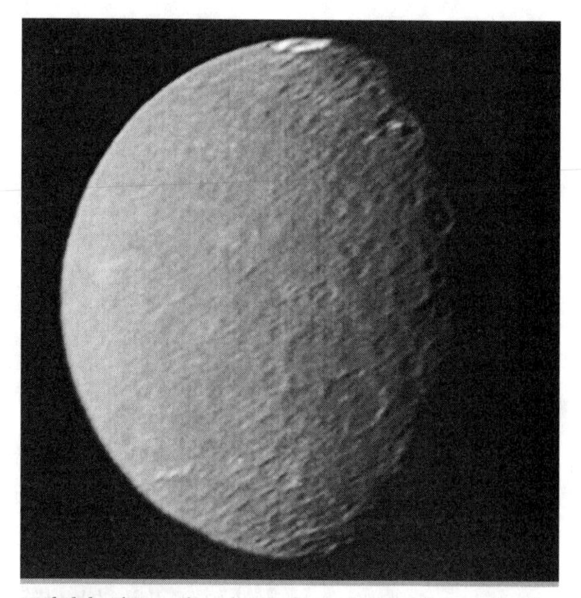

Umbriel The southern hemisphere of Umbriel shows heavy cratering in this Voyager 2 image, taken on January 24, 1986, from a distance of 557,000 km. The prominent crater on the terminator (upper right) is about 110 km across and has a bright central peak. The strangest feature (at top) is a bright ring, nicknamed the "fluorescent cheerio," about 140 km in diameter. Its nature is unknown, although it might be a frost deposit, perhaps associated with an impact crater. *NASA/JPL*

Umbriel

The thirteenth closest and third largest moon of **Uranus**, also known as Uranus II; it was discovered by William **Lassell** in 1851. Umbriel orbits at a mean distance of 265,980 km from the center of the planet and has a diameter of 1,172 km. Its surface is heavily cratered, obviously very ancient, and dark, reflecting only about half as much light as its inner neighbor, Ariel.

United Kingdom Infrared Telescope (UKIRT)

A 3.8-m telescope that is the world's largest dedicated solely to infrared astronomy. It opened in 1979, is sited near the summit of Mauna Kea, Hawaii, at an altitude of 4,194 m, and operates at wavelengths between 1 and 30 microns. In 2002, it was equipped with a revolutionary new spectrometer, the UKIRT Imaging Spectrometer (UIST), that can effectively slice any celestial object into sections and so, for example, produce a three-dimensional view of the conditions throughout an entire galaxy in a single observation. UKIRT is owned by the U.K. Particle Physics and Astronomy Research Council and operated along with the James Clerk Maxwell Telescope by the staff of the **Joint Astronomy Centre**.

United Kingdom Schmidt Telescope (UKST)

A 1.2-m **Schmidt camera** at Siding Spring, New South Wales, opened in 1973. Its initial task was to construct a photographic survey of the entire southern sky. UKST is jointly owned by Britain and Australia and is part of the **Anglo-Australian Observatory**.

Universal Time (UT)

A worldwide standard that serves as the basis of civil and most astronomical timekeeping, also known as *Greenwich Mean Time* (GMT). UT is defined as the **hour angle** of the **mean sun** as seen by an observer on the **prime meridian** plus 12 hours (so that 0000 UT corresponds to midnight rather than noon). In practice, it is determined from a formula that links it to **sidereal time** so that it actually comes from observations of the daily motion of stars or extraterrestrial radio sources. The version of UT thus produced, known as UT0, depends slightly on the place of observation. When UT0 is corrected for the shift in longitude of the observing station caused by polar motion, UT1 is obtained. If a further correction is made for seasonal changes in Earth's rotation rate, UT2 is obtained. *Coordinated Universal Time* (UTC) is used for broadcast time signals (available via shortwave radio, for example). It differs from **International Atomic Time** by an integral number of seconds and is kept within +/–0.9s of UT1 by introducing leap-seconds when necessary. See **time**.

upper transit

Also known as *upper culmination,* the meridian passage from east to west of an object above the pole, that is, between the North Pole and the south point on the horizon, for an observer in the Northern Hemisphere. See **lower transit**.

Uranometria

The first star atlas to cover the entire sky, published by Johann **Bayer** in 1603 and based on positions taken from Tycho **Brahe**'s catalog. It contained 51 charts, one for each of **Ptolemy**'s 48 constellations, one for the southernmost skies, which were unknown to Ptolemy (introducing 12 new southern constellations defined by the Dutch navigators Pieter Keyser and Frederick de Houtmann), and two planispheres. *Uranometria* also introduced the **Bayer designations**, which are still used today.

Uranus

See article, pages 503–505.

Urca process

A cycle of nuclear reactions in which a nucleus loses energy by first absorbing an electron and then reemitting a beta particle (a high-speed electron) plus a neutrino-antineutrino pair. It is an important process in pre-**supernova** stars because, by cooling and decreasing the pressure in the star's core, it renders the core unable to support the weight of the overlying layers. These layers collapse onto the core and subsequently rebound and escape in the supernova explosion. "Urca" is not an acronym but the name of a casino in Rio de Janeiro at which George **Gamow** commented to the Brazilian astrophysicist and art critic Mario Schenberg (1914–1990) (who first pointed out the importance of neutrino emission in supernovae to Gamow): "the energy disappears in the nucleus of the supernova as quickly as the money disappeared at that roulette table." The process was dramatically confirmed by **Supernova 1987A** whose observation coincided with a burst of 11 neutrinos, detected by the **Kamiokande Observatory** in Japan, and a further eight registered independently by a detector in Ohio.

ureilites

A subgroup of primitive **achondrites** named for Novo Urei, a village in the Mordova Republic, Russia, where several meteorites fell in late 1886. It has been reported that one stone was soon recovered by local peasants, but not to preserve it for science; on the contrary, the crumbly stone was immediately broken apart and eaten! The report doesn't reveal why this happened–perhaps the freshly fallen meteorite smelled good, or perhaps because it was shaped like a loaf of bread, which some ureilite are. However, not all of the stones were eaten, and Novo Urei became the type specimen of one of the best-represented achondrite groups. There are two subgroups. *Main group ureilites* are composed primarily of coarse-grained **olivine** and minor **pyroxene**, mostly in the form of calcium-poor pigeonite, set in a dark carbonaceous matrix of graphite, diamond, nickel-iron metal, and troilite. *Polymict ureilites* consist of a mixture of different lithologies. Besides clasts from main group ureilites, they contain magmatic inclusions, dark carbonaceous clasts, chondritic fragments of different origins, and various other inclusions. This suggests a surface or regolith origin for the polymict ureilites, an assumption supported by the values for noble gases that have been implanted into the regolith by the **solar wind**. However, both the origin and the formation history of the ureilites remain enigmatic. Their mineral and oxygen isotopic compositions suggest that they formed as residues from partial melting, and therefore represent primitive achondrites that probably formed on several parent bodies. On the other hand, rare-element patterns and other chemical characteristics indicate that ureilites are highly fractionated igneous rocks that formed in different regions of the same parent body; probably a moderately differentiated C-class asteroid that was disrupted by an impact event and then rapidly cooled. An impact history would also explain the occurrence of high-pressure minerals such as diamond and londsdaleite that are formed by intense shock metamorphism. Even this theory isn't without its problems though. Recently, a ureilite from the Libyan Sahara, named DaG 868, was found to contain diamonds, but paradoxically, appears to be nearly unshocked.

Ursa Major (abbr. UMa; gen. Ursae Majoris)

The Great Bear; a very large and prominent northern **constellation**, located east of Lynx and north of Leo and Leo Minor. Ursa Major is a good starting point to find other stars and constellations in the sky. A line through the Pointers, Alpha UMa (Dubhe) and Beta UMa (Merak), leads north to the pole star, Polaris, and south toward **Regulus**. Capella in Auriga can be found by following a line from Delta UMa to Alpha UMa. Likewise, a line from Delta UMa to Beta UMa and beyond leads to Alpha Gem. Ursa Major is most famous for the Big Dipper (or Plough) asterism, made from the stars Alpha,

(continued on page 505)

Uranus

The seventh planet from the Sun and the third largest of the **gas giants** (narrowly surpassing Neptune) with a diameter about four times that of Earth. It was discovered by William **Herschel** in 1781. Others had seen it earlier (at its brightest it is just visible to the naked eye) but had taken it to be a star because it doesn't move perceptibly from one night to the next. The earliest recorded sighting was in 1690 when John **Flamsteed** cataloged it as 34 Taurus. William Herschel originally called it Georgium Sidus (George's Star) in honor of King George III, while French astronomers began calling it Herschel. It was Johann **Bode** who proposed the name Uranus, after the Greek god of the heavens, but this didn't come into common usage until around 1850.

Uranus's most extraordinary feature is the tilt of its axis–almost 98° (or 82° if it's taken to be retrograde), so that the planet effectively spins around on its side. As a result, for part of its orbit, one pole continually faces the Sun while the other is in total darkness. Half an orbit later, the roles (and poles) are reversed. In between, the Sun rises and sets around the equator normally. For Uranus to be in such a position, it was presumably struck a formidable blow by another massive object, perhaps even before it was properly formed.

The atmosphere of Uranus is composed of hydrogen (83%), helium (15%), methane (2%), and trace amounts of acetylene and other hydrocarbons. Its bluish hue stems from an upper methane haze that absorbs strongly at red wavelengths–leaving a featureless blue planet in our telescopes. During Voyager 2's flyby in 1986, Uranus's banded cloud patterns were extremely bland and faint. More recent Hubble Space Telescope observations, however, have shown a more strongly banded appearance now that the Sun is getting closer to being directly overhead at Uranus' equator, a position it will assume in 2007.

Uranus's magnetic field is strange in being offset from the center of the planet and in being tilted 59° from the axis of rotation. As a result, the field twists like a corkscrew as Uranus spins around. (Neptune has a similarly displaced magnetic field, suggesting that this is not necessarily a result of Uranus's axial tilt.) The magnetosphere is twisted by the planet's rotation into a long corkscrew shape behind the planet. Uranus probably has a solid, rocky core, surrounded by a deep layer of ammonia and water that has condensed into an icy slush. The outer layer is made up primarily of liquid hydrogen and helium.

Diameter (equatorial):	51,118 km
Oblateness:	0.07
Mass (Earth = 1):	15
Density:	1.29 g/cm^3
Surface gravity:	7.77 m/s^2
Escape velocity:	22 km/s
Mean cloud temperature:	−193°C
Rotational period:	17.9 hours (retrograde)
Axial tilt:	97.86°
Albedo:	0.66
Visual magnitude (max.):	+5.7
Orbit	
Mean radius:	2,870,990,000 km (19.1914 AU)
Aphelion:	3,004,000,000 km
Perihelion:	2,735,000,000 km
Eccentricity:	0.0461
Inclination:	0.77°
Period:	84.1 years

Moons of Uranus

Uranus has 21 known moons, the first two of which were discovered by William **Herschel** in 1787 and named, by his son, after characters from Shakespeare's *A Midsummer Night's Dream,* Titania and Oberon. Two more moons were found by William **Lassell** in 1851 and named Ariel and Umbriel; Gerard **Kuiper** discovered Miranda in 1948. All moons of Uranus are named after characters from Shakespeare or Alexander Pope. Voyager 2's flyby in January 1986 led to the discovery of another 10. Seven additional moons have since been discovered by telescope. In an unusual move, in late 2001, the International Astronomical Union (IAU) stripped the title of "moon" from a small body seen in images taken by Voyager and previously reported to be orbiting Uranus. The IAU concluded that there wasn't yet enough data to confirm that the object, designated S/1986 U 10, is a satellite. It appeared to be in nearly the same orbit as another

Moons of Uranus

	Orbit				
Name	Distance* (km)	Period (d)	Inclination (°)	Eccentricity	Diameter (km)
Cordelia	49,752	0.335	0.08	0.000	26
Ophelia	53,764	0.376	0.10	0.010	32
Bianca	59,165	0.435	0.19	0.001	44
Cressida	61,767	0.464	0.04	0.000	66
Desdemona	62,659	0.474	0.01	0.000	58
Juliet	64,358	0.493	0.07	0.001	84
Portia	66,970	0.513	0.06	0.000	110
Rosalind	69,927	0.558	0.28	0.000	54
Belinda	75,255	0.624	0.03	0.000	68
Puck	86,006	0.762	0.32	0.000	154
Miranda	129,850	1.413	4.2	0.003	472
Ariel	190,930	2.520	0.3	0.003	1,158
Umbriel	265,980	4.144	0.36	0.005	1,172
Titania	436,270	8.706	0.14	0.002	1,578
Oberon	583,420	13.46	0.10	0.001	1,523
Caliban	7,165,000	579.4	139.7	0.082	80
Stephano	7,900,000	676	152.6	0.512	20
Sycorax	12,214,000	1,289	152.7	0.509	120
Prospero	16,100,000	1,947	146.3	0.324	20
Setebos	21,650,000	2,235	146.3	0.562	20
S/2001U1	n/a				

*Mean distance from the center of Uranus.

moon, Belinda, about 75,000 km from Uranus, and, based on its brightness in the Voyager images, to have a diameter of about 40 km. S/1986 U 10 has not been seen since the Voyager 2 discovery, despite observing efforts that have discovered several other small moons orbiting the planet. (See table, "Moons of Uranus.")

Uranus The nine slender rings of Uranus are all visible in this Voyager 2 image taken on January 22, 1986. The most prominent and outermost of the nine, called epsilon, is seen at top. The next three in toward Uranus, called delta, gamma, and eta, are much fainter and narrower. Then come the beta and alpha rings and finally the innermost grouping, known simply as the 4, 5, and 6 rings. *NASA/JPL*

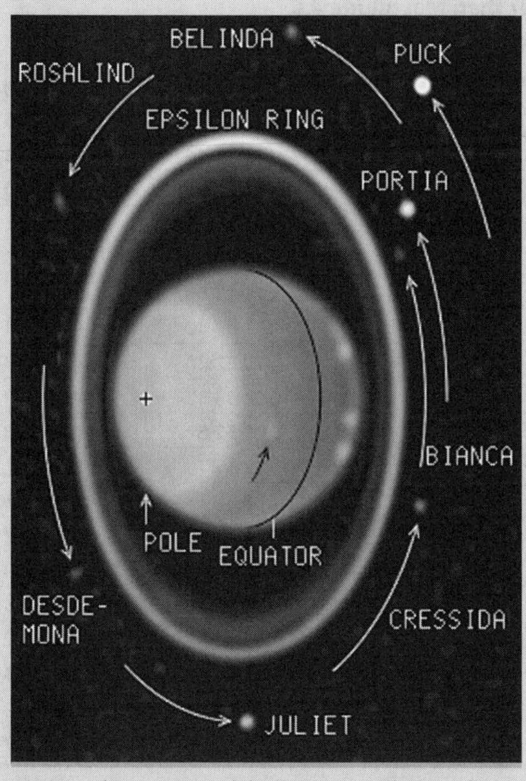

Uranus The equator and the ring system of Uranus lie in the same plane as the orbits of its inner moons. This plane is roughly perpendicular to the planet's orbital plane. Note that the moons are not shown to scale in this picture.
NASA/STScI

Rings of Uranus

The seventh planet has a faint system of 11 narrow, elliptical-shaped rings, composed of dark dust and rocks up to about 10 m in diameter. It was discovered in 1977, during an occultation of a star by the planet. As the star first approached and then receded behind the planet it appeared to blink off and on–a blinking found to be due to a system of rings. (See table, "Rings of Uranus.")

Rings of Uranus

Ring	Distance* (km)	Width of Ring (km)
Epsilon	51,140	20 to 96
1986 U1R	50,020	1 to 2
Delta	48,290	3 to 7
Gamma	47,630	1 to 4
Eta	47,190	<1 to 2
Beta	45,670	7 to 12
Alpha	44,720	7 to 12
Ring 4	42,580	2 to 3
Ring 5	42,230	2 to 3
Ring 6	41,840	1 to 3
1986 U2R	38,000	

*Mean distance from the center of Uranus.

Ursa Major (abbr. UMa; gen. Ursae Majoris)

(continued from page 502)
Beta, Gamma, Delta, Epsilon, Zeta, and Eta UMa. (See tables, "Ursa Major: Stars Brighter than Magnitude 4.0," and "Ursa Major: Other Objects of Interest." See also star chart 17.)

Ursa Major groups

A large spur of spiral galaxies on one side of the **Virgo Cluster** stretching across 20 million light-years. There are probably two main groups: Ursa Major North (around NGC 3631, 3953, and M109) and Ursa Major South (around NGC 3726, 3938, and 4051).

Ursa Major Moving Cluster

A loose group of stars, widely scattered about the sky, with similar space velocities (about 14 km/s in the direction of eastern Sagittarius) that includes five of the seven major members of the Big Dipper–**Merak** (Beta), Phad (Gamma), Megrez (Delta), **Alioth** (Epsilon), and **Mizar** (Zeta)–as well as **Alphecca** (Alpha Coronae Borealis) and **Sirius** (Alpha Canis Majoris). With its center located about 75 light-years from us, about half the distance of the **Hyades**, it is the nearest star cluster to the Sun. Its proximity and its size (some 30 light-years long by 18 light-years wide) mean that it covers an enormous portion of the sky.

Ursa Minor (abbr. UMi; gen. Ursae Minoris)

The Little Bear, also often referred to as the Little Dipper; a **constellation** near the north celestial pole that is almost completely surrounded by Draco. The stars Beta UMi (Kochab) and Gamma UMi (Pherkad) are often called the Guardians of the Pole. (See table, "Ursa

Ursa Major: Stars Brighter than Magnitude 4.0

Star	Magnitude Visual	Magnitude Absolute	Spectral Type	Distance (light-yr)	Position R.A. (h m s)	Position Dec. (° ′ ″)
ε **Alioth**	1.76	–0.22	A0VpCr	81	12 54 02	+55 57 35
α **Dubhe**	1.81	–1.09	F7Va	124	11 03 44	+61 45 03
η **Alkaid**	1.85	–0.60	B3V	101	13 47 32	+49 18 48
ζ **Mizar**	2.23	0.33	A1VpSrSi+A1m	78	13 23 06	+54 55 31
β **Merak**	2.34	0.41	A1V	79	11 01 50	+56 22 56
γ Phekda	2.41	0.36	A0Ve	84	11 53 50	+53 41 41
ψ	3.001	–0.27	K1III	147	11 09 40	+44 29 54
μ Tania Australis	3.06	–1.35	M0III	249	10 22 20	+41 29 58
ι Talita	3.12	2.29	A7IV	48	08 59 12	+48 02 29
θ	3.17	2.52	F6IV	44	09 32 51	+51 40 38
δ Megrez	3.32	1.33	A3V	81	12 15 26	+57 01 57
ο Muscida	3.35	–0.41	G4II	184	08 30 16	+60 43 05
λ Tania Borealis	3.45	0.37	A2IV	134	10 17 06	+42 54 52
ν Alula Borealis	3.49	–2.07	K3IIIBa0.3	421	11 18 29	+33 05 39
κ Al Kaprah	3.57	–2.00	A1Vn	423	09 03 38	+47 09 23
h	3.65	1.82	F0IV	76	09 31 32	+63 03 42
χ Alkafzah	3.69	–0.21	K0III	196	11 46 03	+47 46 45
υ	3.78v	1.04	F0IV	115	09 50 59	+59 02 19
10	3.96	2.88	F5V	54	09 00 38	+41 46 57

Ursa Major: Other Objects of Interest

Object	Notes
Planetary nebula	
Owl Nebula M97 (NGC 3587)	See separate entry.
Galaxies	
M81 (NGC 3031)	See separate entry.
M82 (NGC 3034)	See separate entry.
M101 (NGC 5457)	See separate entry.
M108 (NGC 3556)	Nearly edge-on type Sc spiral. Distance 45 million light-years; magnitude 10.0; diameter 8′ × 1′; R.A. 11h 11.5m, Dec. +55° 40′.
M109 (NGC 3992)	A type SBc barred spiral that looks like a "theta" in the sky. Distance 55 million light-years; magnitude 9.8; diameter 7′ × 4′; R.A. 11h 57.6m, Dec. +53° 23′.
Ambartsumian's Knot	See separate entry.

Ursa Minor: Stars Brighter than Magnitude 4.0

Star	Magnitude		Spectral	Distance	Position	
	Visual	Absolute	Type	(light-yr)	R.A. (h m s)	Dec. (° ′ ″)
α Polaris	1.97v	–3.64	F7Ibv	431	02 31 50	+89 15 51
β Kochab	2.07	–0.88	K4IIIBa0.3	126	14 50 42	+74 09 19
γ Pherkad	3.00	–2.84	A3II	480	15 20 44	+71 50 02

Minor: Stars Brighter than Magnitude 4.0," and star chart 17.)

Ursa Minor Dwarf

A **dwarf spheroidal galaxy**, consisting mostly of old stars, that is one of the small satellite galaxies of the Milky Way. It was discovered by Albert Wilson at Lowell Observatory in 1954 and it is notable for being one of the most **dark matter**–dominated galaxies known. See also **Local Group**.

Apparent size:	41′ × 26′
Position:	R.A. 15h 8.8m, Dec. +67° 12′
Distance:	240,000 light-years

U.S. Naval Observatory

A government observatory, founded in 1844 and originally based at the aptly named Foggy Bottom on the banks of the Potomac River in Washington, D.C. It was from here, using the observatory's 26-inch Clarke refractor–the largest refracting telescope in the world at the time–that Asaph **Hall** discovered the two Martian moons, Phobos and Deimos. The atrocious seeing conditions at the original site led to the observatory being relocated to northwest of the capital in 1893, at which time it also absorbed the U.S. Navy's office of the *Nautical Almanac*. Since 1955, the U.S. Naval Observatory has operated an observing station near Flagstaff, Arizona, at an altitude of 2,315 m.

UU Herculis star (UU Her star)

A **semiregular variable**, somewhat Cepheid-like in its behavior, but that pulsates with two different periods (see **double-mode variable**) and has occasional standstills. One theory suggests that UU Her stars are relatively low-mass post-asymptotic giant branch stars in a stage of evolution in which their spectra mimic those of more massive supergiants.

UV Ceti star (UV Cet star)

A red or orange emission-line dwarf (spectral-type dMe or dKe) that is also an **eruptive variable**; also known as a **flare star**. Within seconds the star may brighten by up to 6 magnitudes, then decline, rapidly at first, then more gradually to a preflare level after several minutes or hours. An increase in radio and X-ray emission accompanies the optical outburst. The flare activity is similar to that found on the Sun but appears comparatively more dramatic because of the normal faintness of red/orange dwarfs. UV Ceti itself is a faint M6eV dwarf that is part of a red dwarf binary system (the B component Luyten 726-8). The two members are of nearly equal brightness (visual magnitude 15.3 and 15.8) and orbit one another every 26.5 years.

UX Orionis (UX Ori)

The prototype for a class of pre-main sequence stars (UXORs) that may eventually evolve into systems like that of **Beta Pictoris** or **Vega**. Photometric, spectroscopic, and polarimetric studies indicate that UXORs have circumstellar disks seen edge-on, which may be in the protoplanetary stage.

UX Ursae Majoris star (UX UMa star)

A novalike variable that may be a **Z Camelopardalis star** with a very long standstill.

V

V404 Cygni

The best case for a stellar **black hole**. Observations at X-ray and optical wavelengths have shown that this is a binary system in which a late G or early K star revolves, every 6.47 days, around a compact companion with a probable mass of 8 to 15 M_{sun}, well above the mass limit at which a collapsed star must become a black hole. Most of the time the X-ray emission is quiescent, indicating only a slow trickle of material from the bright star onto the black hole's **accretion disk**.

V magnitude

The **apparent magnitude** of a star, determined by **photometry**, in a standard wavelength band in the yellow-green region chosen to correspond with that to which the human eye is most sensitive. In the widely used Johnson photometry system, the filter used to measure V magnitude has a central wavelength of 5450 Å and a bandwidth of 880 Å.

Van Allen, James Alfred (1914–)

An American space scientist responsible for experiments aboard early probes in the Explorer series with which Earth's radiation belts, now known as the **Van Allen Belts**, were discovered. Van Allen's space research began with instruments placed aboard captured German V-2 missiles and later included work on the **solar wind**, solar X-ray missions, and the magnetospheres of other planets (especially that of Saturn). Among the missions he contributed to were those of **Pioneer 10 and 11**.

Van Allen Belts

Two doughnut-shaped regions of high-energy charged particles trapped in Earth's magnetic field; they were discovered in 1985 by James **Van Allen** based on measurements made by Explorer 1, the first U.S. satellite. The inner Van Allen Belt lies about 9,400 km (1.5 Earth radii) above the equator, and contains protons and electrons from the **solar wind** and Earth's **ionosphere**. The outer belt is about three times farther away and contains mainly electrons from the solar wind.

van de Hulst, Hendrik ("Henk") Christoffel (1918–2000)

A Dutch astronomer, born and educated in Utrecht, who made his most important contribution while still a student: his 1944 prediction that clouds of cold hydrogen gas in space should emit and absorb a spectral line at a wavelength of 21 cm. Following the detection of this radiation in 1951 by American physicists Edward Purcell (1912–1997) and Harold Ewen (1922–1997), the Dutch team of van de Hulst, Jan **Oort**, and C. Alex Muller began mapping the distribution of neutral hydrogen clouds in the Milky Way and delineating its spiral structure. Subsequently, van de Hulst made studies of interstellar grains and their interaction with electromagnetic radiation, investigated the solar corona, and wrote important books on light scattering and radio astronomy. Based at the University of Leiden from 1948 until his retirement in 1984, he was a leader in several international organizations and in the development of Dutch and European space research.

van der Kamp, Peter (1901–1995)

A Dutch-American astronomer, director of Sproul Observatory, and pioneer in the search for **extrasolar planets**. In 1937, using a technique brought to Sproul by Kaj A. Strand (1907–2000), he began a search for unseen companions of 54 stars known to lie within 16 light-years of the Sun. Over the next two decades, the Sproul group, which also included Sarah Lee Lippincott, reported evidence for planetary bodies around several of the target stars, including 61 Cygni, Ross 614, and Lalande 21185. In 1963, van der Kamp claimed that a giant planet, about 11 times the mass of Jupiter, was in a 24-year orbit around **Barnard's Star**, in which he had a special interest. Over the next few years, he published further evidence for unseen companions around other nearby stars. However, doubts began to grow among fellow astronomers about the validity of his data. In particular, Robert Harrington pointed out that all of the "wobbles" supposedly caused by van der Kamp's planets were of an identical form, suggesting a systematic instrumentation effect. Undeterred, van der Kamp further analyzed his data on Barnard's Star and, in 1976, claimed he now had evidence for two companions–a 0.7 Jupiter-mass planet with a period of 12 years and a 1.2 Jupiter-mass planet with a period of 26 years, orbiting at distances of 3 and 5 times that of the Earth from the Sun, respectively. Three years earlier, he suggested that the nearby Sun-like star, Epsilon Eridani, had a planetary companion six times as massive as Jupiter. Though the planetary discoveries claimed by van der Kamp and his contemporaries at Sproul were all subsequently rejected, van der Kamp's optimism in general has been vindicated. In a 1963 article he argued that there was no reason to doubt the existence of "large numbers" of extrasolar bodies larger than Jupiter and smaller than stars. Recent confirmation of the existence of giant planets and brown dwarfs beyond the solar system have exonerated that view.

van Maanen's Star

The nearest **white dwarf** to the Sun that isn't part of a binary star system. It lies in Pisces, west of Delta Piscium and east of Omega Piscium, and was discovered in 1917 by the Dutch-American astronomer Adriaan van Maanen (1884–1946) who, in comparing photographs made between 1914 and 1917, noticed the star's high annual proper motion of 2.98″. Its relative coolness suggests that it is a very old star, perhaps near to or older than 10 billion years old. Van Maanen's Star has a computed diameter of about 20,200 km (slightly smaller than Earth), which results in a density of about 1.2 tons/cm^3, or about 10 times the computed density of **Sirius** B, the closest known white dwarf.

Luminosity:	0.0002 L_{sun}.
Radius:	0.013R_{sun}
Mass:	0.7 M_{sun}
Distance:	14.4 light-years

variable star

See article, pages 510–511.

variable star naming

The International Astronomical Union appoints a committee that determines the names given to **variable stars**. The assignments are made in the order in which the variable stars were discovered in a constellation. If one of the stars already having a Greek (Bayer) letter name is found to be variable, the star will be referred to by the Greek name. Otherwise, the first variable found in a constellation would be given the letter R, the next S, and so on to the letter Z (a method of naming first adopted by Friedrich **Argelander**). The next star is named RR, then RS, and so on to RZ, SS to SZ, and so on to ZZ. Then the naming starts over at the beginning of the alphabet: AA, AB, and continuing to QZ. This system (the letter J is always omitted) can accommodate 334 names. There are so many variables in some constellations, however, that an additional nomenclature is needed. After QZ, variables are named V335 (since 334 variables have already been named), V336, and so on. The letters representing stars are then combined with the genitive Latin form of the constellation name the same way that the Greek alphabet is used for complete identification of the variable star. Examples are SS Cygni (SS Cyg), AZ Ursae Majoris (AZ Uma), and V338 Cephei (V338 Cep).

V-class asteroid

A rare class of **asteroid**, of which **Vesta** is the preeminent member, characterized by fairly high **albedo** and a reflectance spectrum with strong absorption at wavelengths shorter than 0.7 micron and a separate dip near 0.95 micron. The latter suggests the presence of the silicate mineral **pyroxene**.

Vega (Alpha Lyrae, α Lyr)

The brightest star in **Lyra**, the brightest star in the northern summer sky (forming the northwestern apex of the **Summer Triangle**), and the fifth brightest star in the whole sky; its name comes from an Arabic phrase that means "the swooping eagle." Vega was the first star to be photographed (in 1850), the first to have its spectrum photographed (in 1872), and probably the first to have its **parallax** measured (by Friedrich **Struve** in 1837). Precession of Earth's axis will make it the **pole star** in about A.D. 14000. It is a main sequence **A star** and also a short-period pulsating variable of the type known as a **Delta Scuti star**, with a period of 4.6 h. In 1983, observations by the Infrared Astronomy Satellite (IRAS), revealed a large luminous infrared-radiating halo around Vega that suggests a circumstellar cloud of warm dust. Since Vega seems to be rotating with its pole directed toward Earth, the dust cloud probably represents a face-on disk not unlike the disk surrounding the Sun and that contains the planets. In 2002, astronomers announced that two prominent peaks of dust emission around Vega, one offset 60 AU to the southwest of the star, and the other offset 75 AU to the northeast, could be best explained by the dynamical influence of an unseen planet in an eccentric orbit. Since quite a number of known massive extrasolar planets follow highly eccentric orbits, asymmetric dust concentrations may be common features of extrasolar planetary systems.

Visual magnitude:	0.03
Absolute magnitude:	0.58
Spectral type:	A0Va
Surface temperature:	9,500 K
Luminosity:	50 L_{sun}
Mass:	2.5 M_{sun}
Distance:	25 light-years

variable star

A star that changes in brightness. Variable stars fall into two general categories: *intrinsic variables,* in which physical changes, such as pulsations or eruptions, are involved, and *extrinsic variables,* in which the light output fluctuates due to eclipses or stellar rotation. Further classification is complex; originally it was based on a star's light curve, amplitude, and periodicity (or lack of it), but is now more closely linked to the physical processes underlying the variability. The standard reference for classification is the *General Catalogue of Variable Stars.* Specific types of variable star are often named after a prototype (see **variable star naming**). Additionally, a star may belong to more than one category of variable. See individual entries listed in the chart below for further details.

CLASSIFICATION OF VARIABLE STARS

- **Intrinsic variable**
  - **Pulsating variable** (size and/or shape of star changes)
    - **Cepheid variable** (period of 1–70 days; strict period-luminosity relationship)
    - **RR Lyrae star** (period of 0.05–1.2 days; light variations of 0.3–2 mag.)
    - **RV Tauri star** (alternating deep and shallow minima)
    - **Long-period variable**
      - **Mira** variable (giant red variable; well-defined period of 80–1,000 days)
      - **Semiregular variable** (giant star; periodic with intervals of irregular variation)
    - Nonradial pulsating variable
      - **ZZ Ceti star**
  - **Eruptive variable** (flares or shell ejection)
    - **UV Ceti star** (red/orange dwarf prone to sudden flares)
    - **FU Orionis star** (slow rise to long-lasting maximum; development of emission spectrum)
    - **Gamma Orionis star** (fast-spinning blue giant; occasional ejection of ring of matter)
    - **R Coronae Borealis star** (sudden fades followed by slow recovery)
    - **RS Canum Venaticorum star** (close binary with active chromosphere)
    - **S Doradus star** (massive, luminous blue star often with expanding envelope)
    - **Wolf-Rayet star** (luminous)
    - Orion variable (associated with nebulosity; often irregular)
      - **T Tauri star**
      - **YY Orionis star**
      - **Irregular variable**
    - **Cataclysmic variable** (outbursts on star's surface or accretion disk, or stellar explosion)
      - **Supernova** (sudden, dramatic, final magnitude increase as result of stellar explosion)
      - **Nova** (fusion explosion increases in brightness and then fades)
      - Recurrent nova (system that has undergone 2 or more recorded novalike eruptions)
      - **U Geminorum star** (close binary system with Sun-like star and white dwarf)
        - **Z Camelopardalis star** (no well-defined quiescence; "standstills" of brightness)
        - **SU Ursae Majoris star** (orbital period <2 hours; two distinct outbursts)
      - Polar (close binary system with highly magnetic white dwarf)
        - **AM Herculis star** (strong circular polarization; dwarf in synchronized rotation)
        - **DQ Herculis star** (similar to AM Her but no synchronized rotation)
      - Symbiotic star (semiperiodic novalike outbursts of up to 3 mag.)
        - **Z Andromedae star**
      - X-ray variables (X-ray emissions from binary star, including compact component)

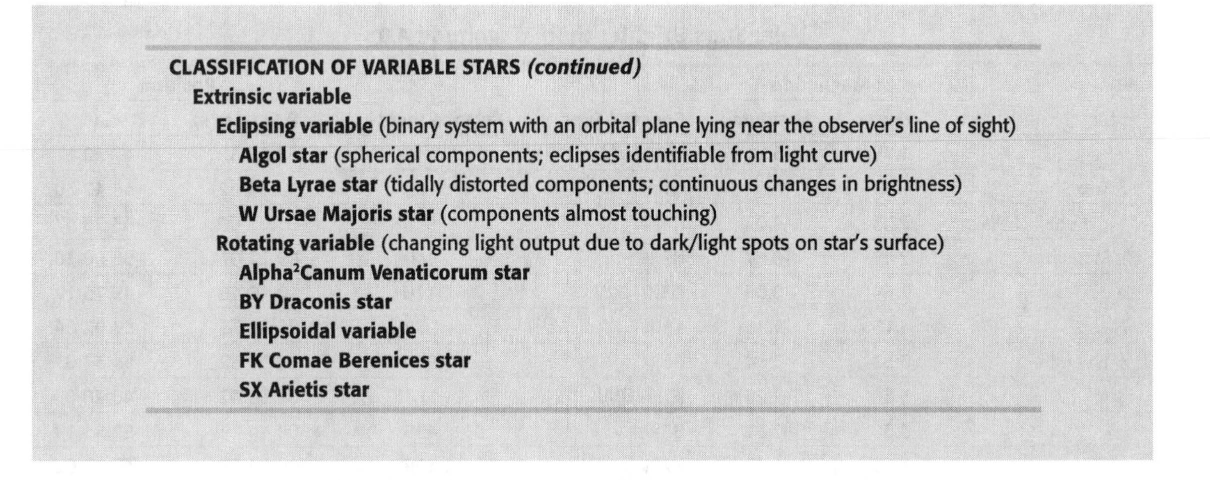

CLASSIFICATION OF VARIABLE STARS *(continued)*

Extrinsic variable

Eclipsing variable (binary system with an orbital plane lying near the observer's line of sight)

Algol star (spherical components; eclipses identifiable from light curve)

Beta Lyrae star (tidally distorted components; continuous changes in brightness)

W Ursae Majoris star (components almost touching)

Rotating variable (changing light output due to dark/light spots on star's surface)

Alpha^{2}Canum Venaticorum star

BY Draconis star

Ellipsoidal variable

FK Comae Berenices star

SX Arietis star

Veil Nebula (NGC 6960, NGC 6979, NGC 6992/5)

The most spectacular part, at visible wavelengths, of the **Cygnus Loop**, an old supernova remnant. The Veil is so large–six times the diameter of the Moon–that earlier observers treated its brighter portions as distinct diffuse nebulae and gave them different NGC numbers: 6960 (close to line of sight of the star 52 Cyg, and also known as the *Filamentary Nebula*), 6979, and 6992/5 (the brighter eastern half of the nebula). Fainter extensions received other catalog identities: NGC 6974 for an appendix to the northern part NGC 6979, and IC 1340 for an extension to the southwest of NGC 6995. Other faint luminous material can be found throughout the Veil's expanse. For example, a patch of nebulosity above and to the left of NGC 6960 is known as Pickering's Triangular Wisp or, less prosaically, as Simeis 3-188. Despite its overall brightness of about magnitude 5, the Veil is only visible to the naked eye under exceptionally good viewing conditions because its light is distributed over such a large area.

Vela (abbr. Vel; gen. Velorum)

The Sails; a **constellation** of the Southern Hemisphere that together with Carina, Puppis, and Pyxis, form **Argo Navis**. It lies west of Centaurus and north and east of Carina, and occupies a region of sky extremely rich in star fields. Its brightest star is the extraordinary **Gamma Velorum** (Regor). (See tables, "Vela: Stars Brighter than Magni-

Veil Nebula Resembling a delicate silk scarf, a tiny portion of the Veil Nebula is revealed by the Hubble Space Telescope. *NASA/STScI*

Vela: Stars Brighter than Magnitude 4.0

	Magnitude				Position	
Star	**Visual**	**Absolute**	**Spectral Type**	**Distance (light-yr)**	**R.A. (h m s)**	**Dec. (° ′ ″)**
γ Regor	1.75	–5.31	WC8+07.5e	841	08 09 32	–47 20 12
δ Koo She	1.93	–0.01	A1V	80	08 44 42	–54 42 30
λ Al Suhail al Wazn	2.23v	–4.00	K4Ib	573	09 08 00	–43 25 57
κ Markab	2.49	–0.67	B9III	140	09 22 07	–55 00 38
μ	2.69	–0.06	G5III+G2V	116	10 46 46	–49 25 12
N	3.16	–1.16	K5III	238	09 31 13	–57 02 04
ϕ Tseen Ke	3.52	–5.34	B5Ib	1,930	09 56 52	–54 34 03
ψ	3.60	2.25	F3IV+F0IV	61	09 30 42	–40 28 00
o	3.60	–2.31	B3IV	495	08 40 18	–52 55 19
c	3.75	–1.13	K2III	309	09 04 09	–47 05 02
p	3.84	1.72	A3V	86	10 37 18	–48 13 32
q	3.85	1.35	A2V	103	10 14 44	–42 07 19
a	3.87	–4.52	A1III	1,550	08 46 02	–46 02 30

tude 4.0," and "Vela: Other Objects of Interest." See also star chart 4.)

Vela Pulsar (PSR 0833-45)

A pulsar associated with the **Vela Supernova Remnant** and lying at a distance of about 1,500 light-years. It is one of the youngest pulsars known and one of the few that has been detected at visible wavelengths. The lengthening of its short period of 89 milliseconds by 10.7 nanoseconds per day gives a maximum age (see **spin-down**) of some 11,000 years. In 1977, nine years after its discovery, it was seen to be flashing at visible wavelengths and thus became the second optical pulsar on record, after the Crab pulsar (see **Crab Nebula**). It is also one of the strongest radio pulsars, the strongest gamma-ray source in the sky, and a powerful X-ray source associated with Vela X-2.

Vela Supernova Remnant

The expanding cloud of gaseous debris thrown out by the supernova in Vela that left behind the **Vela Pulsar**. It extends over about 5° and lies within the more extensive and much older **Gum Nebula.**

Vela: Other Objects of Interest

Object	Notes
Star	
Vela Pulsar	See separate entry.
Open clusters	
IC 2391	A bright, close group around Omicron Vel, visible to the naked eye but best seen with binoculars. Magnitude 2.5; diameter 50′; R.A. 8h 40.2m, Dec. –53° 04′.
IC 2395	Magnitude 4.6; diameter 8′; R.A. 8h 41.1m, Dec. –48° 12′.
NGC 2547	About 50 stars of seventh magnitude and fainter, visible with binoculars. Magnitude 4.7; diameter 20′; R.A. 8h 10.7m, Dec. –49° 16′.
Globular cluster	
NGC 3201	Magnitude 6.7; diameter 18′; R.A. 10h 17.6m, Dec. –46° 25′.
Planetary Nebula	
Eight-burst Planetary	NGC 3132. See separate entry.

Vela X-1
An eclipsing X-ray binary pulsar with an orbital period of 8.96 days in **Vela**; it is a particularly intense emitter of hard X rays.

velocity curve
A plot of the changes in the radial (line-of-sight) velocity seen in the spectrum of a **spectroscopic binary** star or a **pulsating star**. In the case of a spectroscopic binary, it is sometimes possible to obtain a separate velocity curve for each star.

velocity dispersion
The spread of velocities of stars or galaxies in a cluster; it can be estimated by measuring the radial velocities of selected members. Once the velocity distribution is known, the cluster's mass can be calculated using the **virial theorem**.

Venus
See article, pages 514–515.

Very Large Array (VLA)
A **radio interferometer** consisting of 27 linked antennas, each 25 m in diameter, in a Y-shaped configuration. Operated by the U.S. National Radio Astronomy Observatory, and located on the Plains of San Augustin about 100 km west of Socorro, New Mexico, the VLA has the **resolution** of a single antenna 36 km across and the sensitivity of a dish 130 m across.

Very Large Telescope (VLT)
Four identical 8.2-m telescopes, known as Antu ("Sun" in the language of Chile's indigenous Mapuche people), Kueyen (Moon), Melipal (Southern Cross), and Yepen (Venus), at the European Southern Observatory's **Paranal Observatory** in Chile that can work independently or

(continued on page 515)

Very Large Telescope The four giant instruments of the Very Large Telescope prepare for a night of observation. *European Southern Observatory*

Venus

The second closest planet to the Sun and almost a twin of Earth in size. It is the brightest object in the sky after the Sun and the Moon, and is popularly known as the Evening Star or the Morning Star depending on when it is on view. Venus has a slow retrograde (east-to-west) spin, opposite in direction to that of every other planet in the solar system and presumably the result of a massive ancient collision. Its dense atmosphere, of carbon dioxide (96%) and nitrogen (4%), with minor amounts of water vapor, oxygen, and sulfur compounds, has a surface pressure 94 times greater than Earth's. Walking on Venus would be as difficult as walking a half-mile under the ocean. Even more uncomfortable would be the average surface temperature of 460°C, hot enough to melt lead and the result not so much of Venus's proximity to the Sun as of the powerful greenhouse effect caused by the dense carbon dioxide atmosphere. Due to the thermal inertia and convection of its atmosphere, Venus' surface temperature alters very little between the night and day sides of the planet despite its extremely slow rotation period of 243 days.

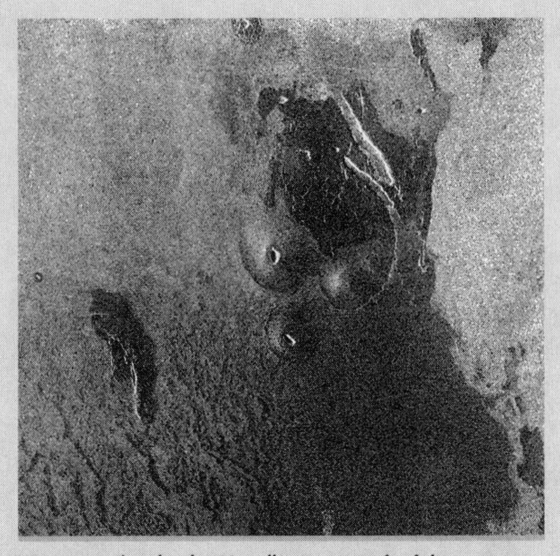

Venus A view by the Magellan spacecraft of the eastern Ovda region of Venus, covering an area of 90 km by 80 km, which shows small volcanic domes on the flank of the volcano Maat. The bright flows to the east are most likely rough lava flows, while the darker flows to the west are probably smoother flows. Small domes are common on Venus, indicating that there has been much volcanic activity on the surface. *NASA/JPL*

Venus A simulated three-dimensional view of 8-km-high Maat Mons on Venus, seen from 560 km north of the volcano at an elevation of 1.7 km above the terrain. Lava flows extend for hundreds of km across the fractured plains, shown in the foreground, to the base of the mountain. To generate this image, Magellan synthetic aperture radar data were combined with radar altimetry and ray-tracing carried out by computer. The vertical scale has been exaggerated 22.5 times. *NASA/JPL*

Measurements of Venus's atmosphere and its cloud patterns show nearly constant high-speed zonal winds, about 100 m/s at the equator. The winds decrease toward the poles so that the atmosphere at cloud-top level rotates almost like a solid body. The wind speeds at the equator correspond to Venus's rotation period of four to five days at most latitudes. Layers of haze in the atmosphere contain small aerosol particles, possibly droplets of sulfuric acid. A concentration of sulfur dioxide above the cloud tops has been seen to decrease since 1978. The source of sulfur dioxide at this altitude is unknown, though one possibility is injection by volcanic explosions.

The issue of present-day volcanic activity of Venus is unresolved. Radar images from the **Magellan** probe provided compelling evidence that the planet has been dominated by volcanism on a global scale in the past.

The photos also showed that the second highest mountain, Maat Mons, appears to be covered with fresh lava. Venus's surface seems to consist largely of recently solidified basalt, with very few meteor craters, which suggests some recent global resurfacing event. According to one theory, although Venus lacks mobile plate tectonics of the kind found on Earth, it undergoes massive volcanic upwellings at regular intervals that inundate its surface with fresh lava.

Venus has two major continent-like plateaus: the northern plateau of Ishtar Terra, which has the planet's highest mountains, the Maxwell Montes, and the larger, Africa-sized southern plateau of Aphrodite Terra. These "continents" make up only 8% of the planet's surface compared to 25% for Earth's. Venus also has an extreme lowland basin, Atalanta Planitia, about the size of Earth's North Atlantic Ocean basin, with a smooth surface that resembles the mare basins of the Moon. The lowest point on the planet is in the rift valley Diana Chasma in Aphrodite Terra, which has much in common with Valles Marineris on **Mars**; it lies up to 2.9 km below the planet's mean surface level and is deeper than any comparable feature on Earth. Another channel, known as Baltis Vallis, is 6,800 km long (slightly longer than the Nile), averages about 1.8 km wide, and presents a riddle. High-temperature lava is unlikely to have caused such a long-distance flow, and there are no known substances that could remain liquid long enough under the planet's atmospheric pressure and temperature to have carved out this snakelike feature.

Diameter:	12,104 km
Mass (Earth = 1):	0.82
Mean density:	5.24 g/cm^3
Escape velocity:	10.36 km/s
Rotation period:	243.1 days (retrograde)
Inclination of axis to orbit:	117.4°
Orbit	
Mean distance from Sun:	108.2 million km
Eccentricity:	0.007
Inclination:	3.4°
Period:	224.7 days

Very Large Telescope (VLT)

(continued from page 513)

together as the world's most powerful optical telescope purposely built as an **interferometer.** The Very Large Telescope Interferometer (VLTI) combines the light from the four big unit telescopes and from several moveable 1.8-m Auxiliary Telescopes, spaced across baselines of up to 200 m, by way of the Interferometric Tunnel. Inside this 130-m-long underground cavern, the light beams gathered by the telescopes are passed through delay lines to compensate for the slightly different path-lengths they have taken in reaching the instruments. The delay lines help to synchronize the beams, before redirecting them to a central laboratory. The interference fringes produced when the beams are finally recombined provide the information needed to reconstruct the original image in unprecedented detail, giving a picture as sharp as if it had come from a single telescope 200 m across. If there were cars on the Moon, then the Very Large Telescope would be able to read their number plates.

Very Long Baseline Array (VLBA)

A dedicated array of ten 25-m radio telescopes for **very long baseline interferometry** (VLBI) completed in 1993 and operated by the National Radio Astronomy Observatory. It shares headquarters with **Very Large Array**, in Socorro, New Mexico. Eight of the VLBA instruments are strung out across the continental United States, while the other two are in Hawaii and the Virgin Islands, giving a maximum baseline of about 8,000 km and a resolution better than 0.001″ at its shortest wavelength.

very long baseline interferometry (VLBI)

A form of **aperture synthesis** in which the individual telescopes aren't directly connected together, but instead make their observations and record their data separately. At a later time, the data is sent to a central correlator and the signals combined. The advantage of this method is that the telescopes in the array can be arbitrarily far apart (often on different continents), and so the technique provides the highest resolution images in astronomy, with beamwidths in the range 0.1 to 10 milliarcsecond (depending on the radio frequency).

Vesta (minor planet 4)

The third largest **asteroid**, with a diameter of 525 km, and fourth brightest; it was discovered in 1807 by

Heinrich **Olbers** and named for the ancient Roman goddess of the hearth. Vesta has a basaltic surface composition and an average density of 3.3 g/cm^3–not much less than that of Mars. Evidently lava once flowed here indicating that the interior was at one time molten. A deep impact crater 456 km wide (on a world not much over 500 km across!), visible to the Hubble Space Telescope, has exposed the mantle beneath Vesta's outer crust, showing that Vesta has been differentiated into layers, like the terrestrial planets, and so must have had an internal heat source in addition to the heat released by long-lived radio isotopes.

Not only can we peer (albeit dimly, for the moment) at Vesta's interior, but we have, scientists believe, quite a few samples of Vesta here on Earth. These consist of the so-called **HED group** of meteorites–the howardites, eucrites, and diogenites–which have spectral and geological fingerprints linking them to Vesta and to each other. But it's very unlikely that the HED stones came from Vesta directly. Vesta is located in a part of the main asteroid belt that makes it almost impossible for it to send meteorites to us. So there are probably intermediate asteroids, which were once part of Vesta, located in more favorable orbits that provided delivery. This theory has been bolstered by the discovery that the asteroid (1929) Kollaa, based on its reflectance spectrum, was once a part of Vesta, and, moreover, that it moves in an orbit from which meteorites could much more easily be launched Earthward. Spectral class V, rotational period 5.34 hours, semimajor axis 2.361 AU, perihelion 2.149 AU, aphelion 2.574 AU, eccentricity 0.090, inclination 7.13°, period 3.63 years.

Vesta A meteorite believed to have come originally from the crust of the asteroid Vesta. The rock is unique in being made almost entirely of the mineral pyroxene, a substance common in lava flows. *NASA/R. Kempton (New England Meteoritical Services)*

Viking

Twin spacecraft, launched in 1975, that studied Mars both from orbit and on the surface, and carried out the first in situ experiments to look for life on another world. The Viking 1 lander touched down in the western part of Chryse Planitia followed by its sister craft 7,420 km to the northeast. The results of the biological experiments were intriguing but inconclusive. Activity such as would be expected of microbes was found in some of the tests, but an instrument known as a gas chromatograph mass spectrometer used to search for organic material found none.

Virgo (abbr. Vir; gen. Virginis)

The Virgin; a very large **constellation** of the celestial equator and the sixth constellation of the zodiac. It lies south of Coma Berenices and west of Libra, and is home to the **Virgo cluster** of galaxies. (See tables, "Virgo: Stars Brighter than Magnitude 4.0," and "Virgo: Other Objects of Interest." See also star chart 18.)

Viking Sand dunes and large rocks are revealed in this panoramic picture of Mars, the first photograph taken by Viking I's Camera 1 on July 23, 1976. The horizon is about 3 km away and the late afternoon Sun is high in the sky to the left. In the middle third of the picture, the rocky surface is covered by thick deposits of wind-blown material, forming numerous dunes. A cloud layer is visible halfway between the horizon and the top of the picture. *NASA/JPL*

Virgo: Stars Brighter than Magnitude 4.0

	Magnitude				Position	
Star	**Visual**	**Absolute**	**Spectral Type**	**Distance (light-yr)**	**R.A. (h m s)**	**Dec. (° ′ ″)**
α **Spica**	0.98v	–3.55	B1III-IV+B2V	262	13 25 12	–11 09 41
γ Porrima	2.74	2.37	F0V+F0V	39	12 41 40	–01 26 57
ε Vindemiatrix	2.85	0.37	G8IIIab	102	13 02 11	+10 57 33
ζ Heze	3.38	1.62	A3V	73	13 34 42	–00 35 46
δ Minelauva	3.39	–0.58	M3III	202	12 55 36	+03 23 51
β Zavijava	3.59	3.40	F8V	36	11 50 42	+01 45 53
109	3.73	0.75	A0V	129	14 46 15	+01 53 34
μ Rijlal Awwa	3.87	2.51	F2III	61	14 43 04	–05 39 30
η Zaniah	3.89	–0.53	A2IV	250	12 19 54	–00 40 00

Virgo Cluster

The nearest large **cluster of galaxies** and the gravitational heart of the **Local Supercluster.** It lies in Virgo and Coma Berenices, is centered about 50 million light-years away, and contains around 2,500 galaxies, about 150 of which are large. The Virgo Cluster completely dominates our small corner of the universe, and our entire **Local Group** of galaxies is being gravitationally drawn toward this huge concentration of matter (an effect known as the *Virgo infall*). At its core lie three supergiant ellipticals, M84, M86, and M87, which probably formed from the merger of many smaller galaxies. Recent observations have shown that the Cluster's principal axis is aligned with an immense filament that is part of the **large-scale structure** of the universe. The major axis orientations of Virgo's brightest elliptical galaxies, as well as M87's **jet**, also appear to fall in line with this filament, which can be traced to even larger distances, where it eventually intersects with the rich cluster Abell 1367 some 160 million light-years away. Abell 1367 forms one node of a well-known supercluster with the **Coma Cluster**, raising the intriguing possibility that the Virgo, Abell 1367, and Coma clusters may all be members of a colossal filamentary network.

Virgo: Other Objects of Interest

Object	**Notes**
Star	
S Vir	A long-period variable of a vivid red color. Magnitude range sixth to thirteenth; period 377 days.
Galaxies	
M49 (NGC 4472)	A type E3 elliptical. Magnitude 8.4; diameter 8.9′ × 7.4′; R.A. 12h 29.8m, Dec. +08° 00′.
M58 (NGC 4579)	A type SB spiral. Magnitude 9.8; diameter 5.4′ × 4.4′; R.A. 12h 37.7m, Dec. +11° 49′.
M59 (NGC 4621)	A type E3 elliptical. Magnitude 9.8; diameter 5.1′ × 3.4′; R.A. 12h 42.0m, Dec. +11° 39′.
M60 (NGC 4649)	A type E1 elliptical. Magnitude 8.8; diameter 7.2′ × 6.2′; R.A. 12h 43.7m, Dec. +11° 33′.
M61 (NGC 4303)	A type Sc spiral. Magnitude 9.7; diameter 6.0′ × 5.5′; R.A. 12h 21.9m, Dec. +04° 28′.
M84 (NGC 4374)	A type E1 elliptical. Magnitude 9.3; diameter 5.0′ × 4.4′; R.A. 12h 25.1m, Dec. +12° 53′.
M86 (NGC 4406)	A type E3 elliptical. Magnitude 9.2; diameter 7.4′ × 5.5′; R.A. 12h 26.2m, Dec. +12° 57′.
M87 (NGC 4486)	See separate entry.
M89 (NGC 4552)	A type E0 elliptical. Magnitude 9.8; diameter 9.5′ × 4.7′; R.A. 12h 35.7m, Dec. +12° 33′.
M90 (NGC 4569)	A type Sb spiral. Magnitude 9.5; diameter 9.5′ × 4.7′; R.A. 12h 36.8m, Dec. +13° 10′.
Sombrero Galaxy	M104 (NGC 4594). See separate entry.

virial theorem

For gravitationally bound systems in equilibrium, the total energy is equal to one-half the time-averaged potential energy. The virial theorem provides a way of estimating the mass of star clusters, galaxies, or clusters of galaxies from observations of the movement of individual members.

viscous relaxation

The process whereby topographic features become subdued over time due to the flow of the surrounding geologic material.

Visible and Infrared Telescope for Astronomy (VISTA)

A 4-m wide-field survey telescope for the Southern Hemisphere, being built at Cerro Paranal, close to the **Very Large Telescope**, by a consortium of 18 British universities. It is expected to be complete by 2006.

visible horizon

A projection of the local horizon outward to intersect with the celestial sphere. Also known as the *apparent horizon.*

visible light

Electromagnetic radiation at wavelengths that the human eye can see, extending from about 4,000 Å (violet) to 6,800 Å (red). This is the region in which optical astronomy is done.

visual magnitude

The **apparent magnitude** of a celestial body in a wavelength range approximating that of the human eye. This is usually different from **photographic magnitude** because photographic emulsions do not exactly match the way the human eye's sensitivity varies with wavelength. Visual magnitude is now essentially synonymous with **V magnitude**, which is determined photometrically.

Vogt-Russell theorem

If the values of temperature, density, and chemical composition govern the pressure, opacity, and rate of energy generation in a star, then mass and chemical composition are sufficient to describe the structure of the star. The **mass-radius relation** and **mass-luminosity relation** follow from this theorem. Named after the German astronomer Heinrich Vogt (1890–1968) and Henry **Russell.**

void

A large region of the universe containing few or no galaxies. The first of these voids to be discovered lies in the direction of Boötes and is about 300 million light-years across. Voids take up about 98% of the volume of the universe, with galaxies concentrated in the thin walls that surround them. But this doesn't necessarily mean that the voids are empty. On the contrary, it is possible that they contain most of the mass of the universe in the form of **dark matter**.

Volans (abbr. Vol; gen. Volantis)

A small, faint **constellation** in the south polar region of the sky to the southwest of Carina. (See table, "Volans: Stars Brighter than Magnitude 4.0," and star chart 4.)

volatiles

Substances that vaporize (boil) at relatively low temperatures. Examples include water (H_2O), carbon dioxide (CO_2), carbon monoxide (CO), methane (CH_4), and ammonia (NH_3).

volcanism

The eruption of molten material or gas-driven solid fragments onto the surface of a planetary body. On icy moons the phenomenon is sometimes called cryovolcanism to distinguish it from silicate or sulfur volcanism.

Volans: Stars Brighter than Magnitude 4.0

	Magnitude				Position	
Star	Visual	Absolute	Spectral Type	Distance (light-yr)	R.A. (h m s)	Dec. (° ′ ″)
γ	3.6			147		
γ^2	3.78	0.60	G8III		07 08 42	–70 29 50
γ^1	5.68	2.41	F2V		07 08 45	–70 29 57
β	3.77	1.17	K2III	108	08 25 44	–66 08 13
ζ	3.93	0.86	K0III	134	07 41 49	–72 36 22
δ	3.97	–2.56	F6II	660	07 16 50	–67 57 27

Vorontsov-Velyaminov scheme

A system for classifying **planetary nebula**e according to their visual appearance. 1: Stellar image. 2: Smooth disk (*a,* brighter toward center; *b,* uniform brightness; *c,* traces of a ring structure). 3: Irregular disk (*a,* very irregular brightness distribution; *b,* traces of ring structure). 4: Ring structure. 5: Irregular form, similar to a diffuse nebula. 6: Anomalous form. More complex structures are characterized by combinations such as "4+2" (ring and disk), or "4+4" (two rings). The scheme was devised by the Russian astrophysicist Boris Vorontsov-Velyaminov (1904–).

Voyager

Two identical spacecraft launched in 1977 to Jupiter and Saturn to build on the knowledge acquired by **Pioneer 10 and 11**. Voyager 2 went on to fly past Uranus and Neptune. Both Voyagers, like the twin Pioneers, are now leaving the solar system.

Vulcan

A hypothetical intra-Mercurian planet, the existence of which was proposed by Urbain **Leverrier** to explain anomalies in Mercury's orbit (now accounted for by the **general theory of relativity**). His claim was supported by the observation in 1859 of a black speck that traveled across the face of the Sun, described in a letter to Leverrier by Edmond Lescarbault, a physician and amateur astronomer. But despite intense efforts by other astronomers, especially during a total eclipse of the Sun in 1860, to see this new world that Leverrer called "Vulcan," nothing was found. In 1868, Richard Covington of Washington Territory reported another possible sighting of Vulcan, whereupon Leverrier published yet another theory stating that the putative tenth planet had an orbital period of least 9 years, and was probably made of some optically transparent material such as quartz or diamond, which was why it was so hard to spot. Neither the sighting nor Leverrier's new theory convinced the scientific establishment, and Leverrier died early in 1877 a broken man only months before Vulcan's next predicted appearance. Nine and a half years later, Vulcan was spotted again–by the young George **Hale**–who published his results in mid-1897 and then privately published a monograph in 1899 in which he claimed that there was a 90% probability that Vulcan was composed of some strange silica diamond mix and that it probably supported life of a totally alien kind. Vulcan has now entered the realms of astronomical myth. However, a number of extraterrestrial Vulcans have been found in recent years–large **extrasolar planets** in orbits much smaller than that of Mercury.

vulcanoid

A hypothetical asteroid with an aphelion of less than 0.4 AU–within the orbit of Mercury; the name is unofficial. Some searches have been conducted for vulcanoids but without success.

Vulpecula (abbr. Vul; gen. Vulpeculae)

The Fox; a faint northern **constellation** that lies north of Sagitta and Delphinus and south of Cygnus. Although its chief star, Alpha Vul, is no brighter than magnitude 4.4, Vulpecula is home to some interesting deep sky objects including the **Dumbbell Nebula** and the **Coathanger** open cluster. (See star chart 7.)

W

W Ursae Majoris star (W UMa star)

A type of **eclipsing binary** with a very short period of a few hours up to a day. In W UMa systems, the component stars, which are of spectral type F or G and lie on or near the main sequence, form a **contact binary** and have pulled each other into teardrop shapes. The primary and secondary minima are virtually the same (some 0.6 to 0.8 magnitude deep), and there's a continuous light variation throughout the orbit, with no clear start or end to an eclipse. W Ursae Majoris itself lies about 200 light-years away and consists of two stars not unlike the Sun, the primary with a mass of 0.99 M_{sun}, a radius of 1.14 R_{sun}, and a luminosity of 1.45 L_{sun}, and the secondary with a mass of 0.62 M_{sun}, a radius of 0.83 R_{sun}, and a solar-equivalent luminosity.

W Virginis star (W Vir star)

Also known as a Type II **Cepheid variable**, a Population II **pulsating variable** with a period of 1 to 35 days. Although W Vir stars show a **period-luminosity relation**, which is characteristic of Cepheids, the relation is different than that of Type I Cepheids, or **Delta Cephei stars**. W Vir stars are typically 1.5 magnitudes fainter than their Type I cousins and have a mass of less than 1 M_{sun}, so that they are clearly at a different evolutionary stage. They also have a distinctive **light curve** with a variation of 0.3 to 1.2 magnitude and a bump on the decline side. Two subgroups are recognized: long-period W Vir stars, exemplified by W Vir itself, with periods of greater than 8 days, and short-period W Vir stars, exemplified by BL Herculis, with periods of less than 8 days.

wavelength

The length of one complete cycle of a wave: say, between two successive crests. It is determined by the speed of the wave divided by its frequency.

Weizsäcker, Carl Friedrich von (1912–)

A German physicist and philosopher at the University of Strasbourg who, in 1944, published an important new version of the nebular hypothesis that explained the puzzling discrepancy between the comparatively low angular momentum of the Sun and the high angular momentum of the outer planets. Previously, in 1938, he had put forward, independently of Hans **Bethe**, the **carbon-nitrogen-oxygen cycle** as an important energy-making process in stars.

West, Comet (C/1975 V1)

One of the brightest **long-period comets** of recent decades. Discovered in November 1975 by the Danish astronomer Richard Martin West (1941–), it brightened irregularly making it difficult to predict. It reached perihelion on February 25, 1976, and a few days later, its nucleus was seen to have split into four pieces. At its nearest point to Earth (0.80 AU) in early March, its magnitude was –1, its head was yellow, and its fan-shaped dust tail, up to 35° long, appeared dull red to the unaided eye. Perihelion 0.20 AU, eccentricity 0.99997, inclination 43.1°, period 500,000 years.

Westerbork Synthesis Radio Telescope (WSRT)

A 3-km east-west array of fourteen 25-m dish antennae near Hooghalen in the Netherlands. The signals from the array are combined, in a technique called **aperture synthesis**, to provide the resolution of a single large instrument. A dozen of the dishes were completed in 1970: 10 fixed in a 1.2-km line and two moveable on a 300-m length of track at the eastern end. In 1980, two more mobile dishes were set up on a 200-m length of track 1.4 km to the east. The WSRT is operated by the Netherlands Foundation for Research in Astronomy.

Wezen (Delta Canis Majoris, δ CMa)

The second brightest star in **Canis Major**; its name comes from the Arabic *al wazn*, for "weight." Wezen is a yellow **supergiant** F star, as wide as Earth's orbit. Although only about 10 million years old, it has already

Visual magnitude:	1.83
Absolute magnitude:	–6.87
Spectral type:	F8Ia
Surface temperature:	6,200 K
Luminosity:	50,000 L_{sun}
Radius:	200 R_{sun}
Mass:	17 M_{sun}
Distance:	1,790 light-years

left the main sequence and is evolving to the stage at which it will begin core helium burning and become a bloated red supergiant.

Whipple, Fred Lawrence (1906–)

An American astronomer who, in 1950, proposed the "dirty snowball" model for comet nuclei. According to this idea, confirmed in 1986 when the **Giotto** probe snapped close-up pictures of **Halley's comet,** comets have icy cores inside thin insulating layers of dirt; jets of material ejected as a result of solar heating give rise to orbital changes. Whipple graduated from the University of California at Los Angeles, and then helped compute the first orbit of newly discovered Pluto while a graduate student under Armin Leuschner (1868–1953) at the University of California at Berkeley. Whipple worked at Harvard University from 1931 to 1977, and directed the Smithsonian Astrophysical Observatory from 1955 to 1973. In the 1930s, using a new, two-station method of photography, he determined meteor trajectories and found that nearly all visible meteors are made up of fragile material from comets. He also found that the Taurids meteor stream moves in the same orbit as Encke's comet.

Whirlpool Galaxy (M51, NGC 5194)

A large, bright spiral galaxy that is interacting with a much smaller neighbor, NGC 5195. The main galaxy was discovered by Charles **Messier** in 1773, and the companion by Pierre **Méchain** in 1781. M51 is the first galaxy in which spiral structure was discovered, by Lord **Rosse** in 1845. This structure, however, has clearly been influenced by the gravitational effect of the companion, which has triggered a huge burst of star formation in the Whirlpool and has also distorted its arms. M51 is the dominant member of a small group of galaxies that also includes the **Sunflower Galaxy** (M63) and half a dozen other smaller systems. (See photo, page 522.)

Visual magnitude:	8.4
Apparent size:	11′ × 7′
Distance:	31,000 light-years
Position:	R.A. 13h 29.9m, Dec. +47° 12′

white dwarf

A dense, planet-sized star that marks the evolutionary endpoint for all but the most massive stars. White dwarfs form from the collapse of stellar cores in which nuclear fusion has stopped, and are exposed to space following the loss of the old star's bloated outer envelope, typically as a **planetary nebula**. Even a fairly large white dwarf, with a mass similar to that of the Sun, is only about as big as Earth. A piece of it the size of a sugar cube would weigh about as much as a hippopotamus. White dwarfs consist of **electron degenerate matter,** which provides the pressure needed to prevent further collapse, providing that the mass of the dwarf doesn't exceed the **Chandrasekhar limit** of about 1.4 M_{sun}. Lighter mass stars that never get around to burning carbon in their cores give rise to carbon-oxygen white dwarfs, while stars that start out with a mass of at least 4 M_{sun} may give rise to neon-oxygen dwarfs. In addition, white dwarfs differ in terms of their spectra, which are dictated by the elements that dominate their surfaces. Three varieties–dA, dB, and dO (where "d" stands for "degenerate")–have nearly pure surfaces of hydrogen or helium lying atop their cores, whereas **PG 1159 stars** appear to be partially exposed cores. White dwarfs may also have a mixture of elements on their surfaces, and are named accordingly. For example, dAB stars contain hydrogen and neutral helium, whereas dAO stars have hydrogen and ionized helium. (See table, "Spectral Classification of White Dwarfs.")

Spectral Classification of White Dwarfs

Type	Characteristics
dA	Only Balmer lines: no He I or metals present.
dB	He I lines: no H or metals present.
dC	Continuous spectrum, no lines deeper than 5 spectrum.
dO	He II strong: He I or H present.
dZ	Metal lines only: no H or He lines.
dQ	Carbon features, either atomic or molecular in any part of the spectrum.

Whirlpool Galaxy One of the most famous galaxies in the sky, the Whirlpool is seen almost face-on with its smaller companion galaxy, NGC 5195, appearing to hang from the end of one of its spiral arms. *Isaac Newton Group of Telescopes/Nik Szymanek*

Widmanstätten pattern

A characteristic cross-hatched pattern that becomes visible on the surface of **octahedrites** (the most common type of **iron meteorites**) and **pallasites** after polishing and etching with nital (nitric acid in solution with ethanol). It is due to an intergrowth of crystals of nickel-rich **taenite** and nickel-poor **kamacite**. The pattern is named after its discoverer, the Austrian mineralogist Alois von Beckh Widmanstätten (1754–1849).

Wild Duck Cluster (M11, NGC 6705)

One of the richest and most compact known **open clusters**, with an estimated 2,900 stars, about 500 of which are brighter than magnitude 14. A planet at the center of M11 would have an astonishing night sky filled with several hundred first magnitude stars. The Wild Duck Cluster was discovered in 1681 by the German astronomer Gottfried Kirch (1639–1710) of the Berlin Observatory and is thought to be about 250 million years old.

Visual magnitude:	6.3
Apparent diameter:	14′
Distance:	6,000 light-years
Position:	R.A. 18h 51.1m, Dec. −6° 16′

Willamette meteorite

A 15.5-ton **iron meteorite** (type IIIA) that is the largest meteorite ever found in the United States. It was discovered in 1902 near Oregon City, Oregon. Aside from its size, its most striking features are a well-defined nose-cone shape and a deeply pitted rear surface, the pits and grooves having been produced by weathering of the exposed surface of the meteorite in the wet Oregon climate. Its name comes from the fact that it was purchased in 1906 for the American Museum of Natural History in New York by Mrs. William E. Dodge.

William Herschel Telescope

A 4.2-m reflector at the **Roque de los Muchachos Observatory** in the Canary Islands. Opened in 1987, it is jointly owned by Britain and the Netherlands.

winonaite

A class of primitive **achondrite**, named for a most unusual find. The Winona meteorite was found in a stone cist in the ruins of the prehistoric Elden pueblo, Arizona, in 1928. The circumstances of the find suggest that the builders of the pueblo kept and venerated the meteorite as a sacred object after they had actually seen it fall. Winonaites are composed largely of fine-grained **pyroxenes**, with some magnesium-rich **olivine**, the iron-sulfide **troilite**, and nickel-iron metal. Recent research suggests that both the winonaites and the IAB group **iron meteorites** originated on the same parent body–a partially differentiated asteroid that was disrupted just as it began to form an iron core and a silicate-rich crust. This disrupting impact mixed silicates into molten nickel-iron forming the silicated IAB irons, and mixed olivine-rich residues of partial melts into unmelted silicates, forming the winonaites. A few winonaites have anomalous characteristics, however, that suggest they may have a different origin.

Wirtanen, Comet

A **Jupiter-family comet** discovered in 1948 at the Lick Observatory by Carl Wirtanen. Its orbit takes it almost to as close to the Sun as Earth, and almost as far away as Jupiter. Comet Wirtanen was to have been the target of **Rosetta**, but a delay in the probe's launch has forced a revision of the mission plans. Perihelion 1.06 AU, aphelion 5.13 AU, period 5.45 years.

Witch Head Nebula (IC 2118)

A very large, faint **reflection nebula** in **Eridanus** that shines primarily by light reflected from **Rigel**. Its blue color is caused not only by Rigel's color but because dust grains in the nebula reflect blue light more efficiently than red.

Apparent size:	160′ × 60′
Position:	R.A. 5h 4.4m, Dec −7° 11.5′

WIYN Telescope

A 3.5-m reflector that is the second largest optical telescope on Kitt Peak, Arizona, after the Mayall Telescope. The WIYN (pronounced "win") opened in 1994 and is operated jointly by the universities of Wisconsin, Indiana, and Yale and by the National Optical Astronomy Observatories. It is used mainly for wide-field spectroscopic observations.

Wolf, (Johann) Rudolf (1816–1893)

A Swiss astronomer who, following the discovery of the **sunspot cycle** by Heinrich **Schwabe**, amassed all available data on sunspot activity back as far as 1610 and calculated a period for the cycle of 11.1 years. He also codiscovered the link between the cycle and geomagnetic activity on Earth. In 1848 he devised a way of quantifying sunspot activity (see **relative sunspot number**). Wolf studied at the universities of Zürich, Vienna, and Berlin. He became professor of astronomy at the University of Bern in 1844 and director of the Bern Observatory in 1847. In 1855 he accepted a chair of astronomy at both the University of Zürich and the Federal Institute of Technology in Zürich.

Wolf, Max(imilian) Franz Joseph Cornelius (1863–1932)

A German astronomer who was a pioneer of astrophotography. He developed the dry plate in 1880 and the blink **comparator** in 1900, and, with such tools, discovered hundreds of variable stars and asteroids, and about 5,000 nebulae. Between 1919 and 1931, he searched for high **proper motion** stars, which tend to be relatively close by, and produced a catalog of 1,566 such objects. Wolf earned his Ph.D. at the University of Heidelberg, studied in Stockholm for two years, and then returned to spend the rest of his life at Heidelberg, where he founded and directed the Königstuhl Observatory and served as professor of astrophysics. He used wide-field photography to study the Milky Way and used statistical analyses of star counts to prove the existence of dark clouds of matter. He was among the first to show that spiral "nebulae" (now known to be galaxies) have absorption spectra typical of stars and thus differ from gaseous nebulae. Among his asteroid discoveries was 323 Brucia (named after the benefactress who paid for his telescope), the first asteroid to be found using photography (1891), and Achilles, the first found **Trojan**. Wolf also suggested the idea of the modern planetarium while advising on the new Deutsches Museum in Munich.

Wolf 359

The fifth nearest star (third nearest star system) to the Sun, after the **Alpha Centauri** trinary and **Barnard's Star**; it lies 7.78 light-years away in Leo. Discovered by Max **Wolf** in 1918, Wolf 359 is an extremely faint **red dwarf** that for 25 years was the least luminous star known. It is also a **flare star** with the variable star designation CN Leonis, and in 2001 became the first star, other than the Sun, to have its **corona** registered by near-optical ground-based detection when an **emission line** of highly ionized iron (Fe XIII at 3,388 Å) was reported. Spectral type M6Ve, visual magnitude 13.45, absolute magnitude 16.56, luminosity 1/50,000 L_{sun}. See **stars, nearest**.

Wolf 1055 (Gliese 752, Ross 652)

A binary system of two **red dwarfs** located 19.2 light-years away in **Aquila**, west of Altair (Alpha Aquilae) and north of Delta Aquilae. Both components, A and B, are **flare stars**; B is commonly referred to as Van Biesbröck's Star, after its discoverer, and was once thought to be low enough in mass to be a possible brown dwarf. In fact, it is one of the dimmest red dwarfs known with a surface temperature of only 2,700 K and a diameter little greater than that of Jupiter. In October 1994, while being observed with the Hubble Space Telescope's High Resolution Spectrograph, Van Biesbröck's Star produced a flare that temporarily heated its outer atmosphere to around 150,000 K. (See table, "The Component Stars of Wolf 1055.")

The Component Stars of Wolf 1055

	Spectral Type	Luminosity	Radius	Mass
Wolf 1055A	M3.5 Vne	0.02 L_{sun}	0.54 R_{sun}	0.48 M_{sun}
Wolf 1055B	M8 Ve	0.00011 L_{sun}	0.10 R_{sun}	0.09 M_{sun}

Wolf-Lundmark-Melotte Galaxy (WLM Galaxy)

A **dwarf irregular galaxy** in **Cetus** that is a member of the **Local Group**; though discovered in 1909 by Max **Wolf**, its nature as a galaxy was only established in 1926 by Knut Lundmark (1889–1958) and Philibert Melotte (1880–1961). WLM is quite isolated–its nearest neighbor, the dwarf galaxy IC 1613, is a full 1 million light-years away. Quite elongated, with a largest extension of about 10,000 light-years, WLM is about 10 times smaller than the Milky Way. Yet the discovery around it of many outlying red stars suggests that even such galactic minnows may have halos and that WLM may be similar to our own Galaxy in age.

Visual magnitude:	10.9
Absolute magnitude:	–14.7
Apparent size:	12′ × 4′
Distance:	3 million light-years
Position:	R.A. 0h 2.0m, Dec. –15° 28′

Wolf-Rayet star

A hot (25,000 to 50,000 K), massive (more than 25 M_{sun}), luminous star in an advanced stage of evolution, which is losing mass in the form a powerful **stellar wind**. Wolf-Rayets are believed to be **O stars** that have lost their hydrogen envelopes, leaving their helium cores exposed, often in a binary system, and that are doomed, within a few million years, to explode as Type Ib or Ic **supernovae**. There are two spectral subclasses of Wolf-Rayets: type WN, which have prominent **emission lines** of helium and nitrogen, and type WC in which carbon, oxygen, and helium lines dominate. They are named after the French astronomers Charles Wolf (1827–1918) and Georges Rayet (1839–1906) who studied the first example

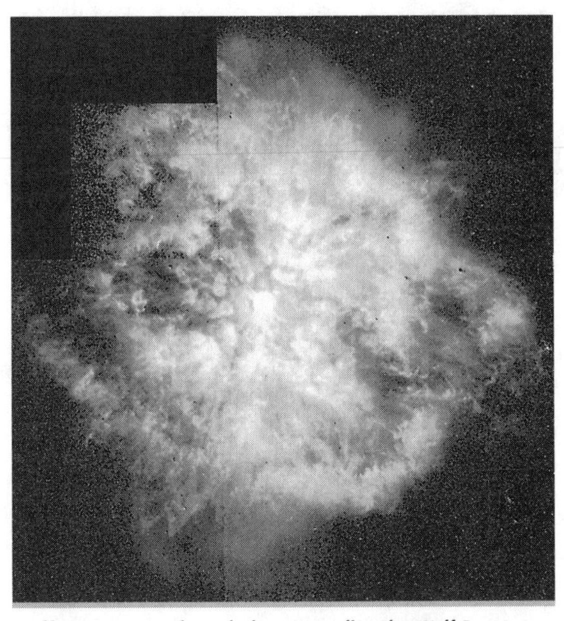

Wolf-Rayet star The nebula surrounding the Wolf-Rayet star WR124 in Sagittarius as seen by the Hubble Space Telescope. *NASA/STScI*

in 1867. A Wolf-Rayet phase is also present in some central stars of **planetary nebula**e. In these stars, which have lower masses and will evolve into white dwarfs, the outer envelope has been expelled in the red giant phase, exposing the hot core. Such stars show many of the characteristics of standard Wolf-Rayet stars and are referred to as *Wolf-Rayet-type stars.*

Wollaston, William Hyde (1766–1828)

An English chemist and physicist who, in 1801, discovered dark lines in the solar spectrum (the first observation of **spectral lines**), which were later investigated by Joseph **Fraunhofer**. After practicing as a physician, he confined himself to research work when he became partially blind in 1800. Having discovered a new element in 1802, he named it after the newly found asteroid Pallas and announced it by the curious expedient of offering it for sale anonymously. Wollaston also discovered rhodium, in 1804, the vibratory nature of muscular action, and a way of making platinum malleable. Among his many optical inventions were an apparatus for measuring the refractive power of solids, a modified sextant, a reflecting goniometer (for measuring the geometrical form of crystals), and the camera lucida, a double-image prism that subsequently proved indispensable in microscopy (see also **Wollaston prism**).

Wollaston prism

A device, invented by William **Wollaston**, that separates randomly polarized or unpolarized light into two orthogonal linearly polarized beams that exit the prism at an angle determined by the wavelength of the light and the length of the prism. Wollaston prisms are used in some kinds of **polarimeters**.

wormhole

A hypothetical tunnel through **space-time** that appears in solutions to the equations of the **general theory of relativity**.

Wright, Thomas (1711–1786)

An English amateur astronomer and instrument maker who, in *An Original Theory or New Hypothesis of the Universe* (1750), was the first to propose a reasonably modern view of the Milky Way in which the Sun is not centrally located. He suggested that the Milky Way is a slab of stars shaped like a mill wheel with the Sun lying between the two planes, though not necessarily at the hub. He further suggested that there might be other collections of stars (external galaxies) like the Milky Way, citing as evidence "the many cloudy spots [nebulae], just perceivable by us . . . in which . . . no one star or particular constituent can possibly be distinguished." His later description of the universe in terms of systems of spheres and rings of stars was taken up by Immanuel Kant.

WZ Sagittae star (WZ Sge star)

A variety of **SU Ursae Majoris star** (a type of dwarf nova) in which the interval between **super-outbursts** is unusually long, measured in decades, while normal outbursts are few and far between. WZ Sagittae itself has had super-outbursts at intervals of about 33, 32, and 23 years, and has never been seen to undergo a normal outburst. Following its first recorded brightening, in 1913, WZ Sge was classified as a classical nova and named Nova Sagittae 1913. When it erupted again, in 1946, it was recategorized as a recurrent nova. Further observations revealed an orbital period (of the white dwarf primary and its companion) of 81m 38s–much shorter that of any known recurrent nova–and a spectrum, lacking broad emission bands, that is more characteristic of a dwarf nova. Observations of the 1978 outburst also revealed **super-humps** in WZ Sge's light curve, which are the defining characteristics of SU UMa type dwarf novae; thus WZ Sge is now considered the prototype for a subset of the SU UMa class. Other WZ Sge stars include AL Com and EG Cnc, which have super-outburst intervals of around 20 years. With its roughly 30-year-long supercycle, WZ Sge ranks as the most inactive of the SU UMa type stars. The factor determining the different timescales appears to be mass-transfer rate. WZ Sge stars have a very low mass-transfer rate, perhaps only 10^{12} kg/s, so that it takes decades to build up enough material for a superoutburst. The puzzle of these stars, however, is why they show few or no normal outbursts during this interval. Even with a low mass-transfer rate, material should accumulate, drifting viscously into the inner disk, and trigger an outburst. One possible reason this does not happen is that the disk viscosity is very low. The material would then remain in the outer disc, where much more can be stored before an outburst is triggered. The problem with this idea, however, is to explain the extremely low viscosity level. A rival explanation involves the removal of the inner disk, to prevent outbursts starting there. This could occur through siphons or because of a magnetic field on the white dwarf.

X

XMM-Newton Observatory (X-ray Multi-Mirror)

European Space Agency orbiting X-ray observatory launched in December 1999. XMM-Newton is the most sensitive imaging X-ray observatory in the 250 eV to 12 keV range ever flown, exceeding the mirror area and energy range of **ROSAT**, ASCA (Advanced Satellite for Cosmology and Astrophysics), and even **Chandra X-ray Observatory**. It has three advanced X-ray telescopes, each containing 58 high-precision concentric mirrors nested to offer the largest possible collecting area. In addition, it carries five X-ray imaging cameras and spectrographs, and an optical monitoring telescope. The observatory moves in a highly elliptical orbit, traveling out to nearly one-third of the distance to the Moon and enabling long, uninterrupted observations of faint X-ray sources.

X-ray binary

An X-ray emitting binary star that consists of a **neutron star** or, rarely a **black hole**, and a normal stellar companion. The X rays come from matter, taken from the normal star, that has entered an **accretion disk** around the compact object and become heated to millions of degrees. X-ray binaries are classified according to the nature of the companion to the compact object. If its mass is smaller than or comparable to that of the Sun, as in the case of **Scorpius X-1**, it is known as a *low-mass X-ray binary* (LMXB). If the mass of the companion is greater than about 10 M_{sun}, as in the case of Cygnus X-1, it is called a *high mass X-ray binary* (HMXB). Systems in which the companion has an intermediate mass are very rare. Binaries containing a white dwarf and a low-mass companion aren't referred to as X-ray binaries even though they are X-ray sources. Instead, they are classified as **cataclysmic binaries**, because they show very large variations in their brightness, such as nova events, and are weaker X-ray sources than binaries containing neutron stars or black hole candidates. No X-ray source containing a white dwarf and a massive star is known.

X-ray burst

A sudden, powerful surge of X-ray emission from a certain **X-ray binary**. Such bursts start without warning, rise rapidly, then decline gradually, lasting in total only about 20 seconds. Like **nova**e, bursts involve mass transfer in close binary systems, but instead of the compact object being a **white dwarf**, as in the case of a nova, it is a **neutron star**. Material escaping from the ordinary companion star falls onto the neutron star. Most of the gas falling onto the neutron star is hydrogen, which becomes compressed against the hot surface of the star by the star's powerful surface gravity. In fact, temperatures and pressures in this accreting layer are so high that the arriving hydrogen is promptly converted into helium by the hydrogen-burning process. Constant hydrogen burning soon produces a layer of helium that covers the entire neutron star. Finally, when the helium layer is about 1 m thick, helium burning ignites explosively, and we observe a sudden burst of X rays. In other words, whereas explosive hydrogen burning on a white dwarf produces a nova, explosive helium burning on a neutron star produces a burster. In both cases, the burning is explosive because the fuel is so strongly compressed against the star's surface that it is degenerate, like the star itself. As with the

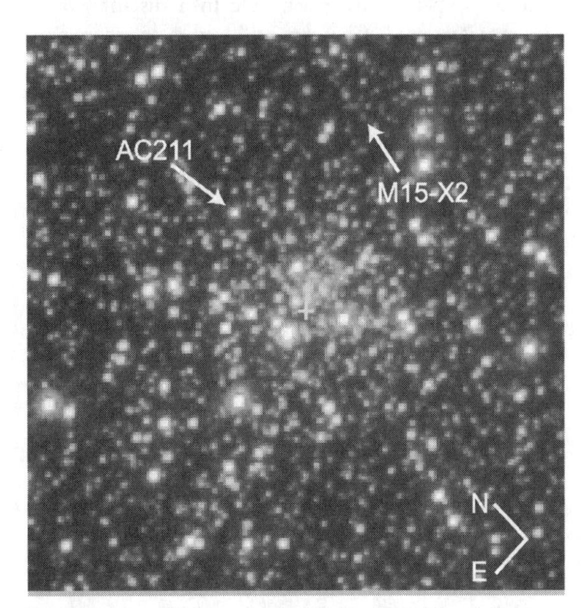

X-ray binary The innermost 1.6 light-years of the globular cluster M15, showing about 1,000 stars and the position of two X-ray binaries. One of these contains the neutron star 4U2127 orbiting around the normal star AC211, visible in this Hubble Space Telescope image; the other contains the neutron star M15-X2, likely associated with the second star indicated by the right arrow. *NASA/STScI*

helium flash inside red giants, ignition of a degenerate thermonuclear fuel always involves a sudden thermal runaway because the usual "safety valve" between temperature and pressure is not operating.

X-ray cluster

A cluster of galaxies with detectable X-ray emission from a hot diffuse medium pervading the cluster.

X-ray jet variable

An **X-ray binary** in which the matter falling onto the neutron star or black hole primary gives rise to **jets**. Observations of such systems in our Galaxy are shedding valuable light, not only on the binaries themselves but on vastly larger black-hole-powered jets in quasars and other remote active galaxies. In 2002, astronomers announced that they had tracked the life cycle of jets associated with a black hole, from initial explosion to fading, in the case of X-ray jet variable XTE J1550-564. A series of images from the Chandra X-ray Observatory revealed the two opposing jets traveled at near light speed for several years before slowing down and fading. In the space of just a few years, it was possible to watch developments that would have taken thousands of years around a supermassive black hole in a distant galaxy. Astronomers used Chandra and radio telescopes to observe the jets following an outburst, first detected in 1998 by the Rossi X-ray Timing Explorer. These jets, which require a continuous source of trillion-volt electrons to remain bright, were observed moving at about half the speed of light. Four years later, they were more than 3 light-years apart, slowing down and dimming.

X-ray nova

A short-lived X-ray source that appears suddenly in the sky and dramatically increases in strength over a period of a few days and then decreases, with an overall lifetime of a few months; it may have an optical counterpart. The outburst of an X-ray nova is caused when huge amounts of matter in the **accretion disk** of an **X-ray binary** suddenly fall on to the compact component, which is thought to be typically a **neutron star**. Less than 30 X-ray novae have been observed to date.

X-ray pulsar

An **X-ray binary** in which the compact object is a **neutron star** with a powerful magnetic field that gives rise to regular X-ray pulses. These pulses are thought to be caused by the magnetic field channeling the accreting gas onto the poles of the compact star, producing localized hotspots that move in and out of view as the star spins. **Hercules X-1** is an example. **Anomalous X-ray pulsars** (AXPs) make up an unusual subcategory.

X-ray transient

An X-ray source that rises rapidly and unpredictably from a normally quiescent state, gaining in intensity by a factor of 1,000 or more, typically over a period of a week, before gradually fading to quiescence. Unlike the more permanent, stable sources observed in the X-ray sky, transients may not flare up again for years. **X-ray novae** are a type of X-ray transient.

X rays

Electromagnetic radiation with wavelengths of about 0.01 to 10 nm (corresponding to energies of 0.1 to 100 keV), between those of ultraviolet light and gamma rays.

Y

Yarkovsky effect

The lopsided emission of radiation from a rotating object orbiting the Sun, due to uneven heating; named after the Russian engineer I. O. Yarkovsky, who first proposed the idea around 1900. Among other things, it has been used to explain why Earth is hit surprisingly often by meteorites. The uneven heating of a small rock in the asteroid belt by the Sun, which makes the day side warmer than the night side, creates a small force–enough to deflect the object into one of the resonance zones where the combined effect of Jupiter and Saturn's gravity pulls objects into the inner solar system. Combined with other, less subtle methods for getting fragments Earth-bound, including asteroid impacts, the newly proved method appears to account for the total yearly shower of space rocks.

year

The time taken for Earth to go once around the Sun. In astronomy, different kinds of year are distinguished by the reference point used to measure the period of revolution. A *sidereal year* is the time taken by Earth to make one complete circuit round the celestial sphere as seen from the Sun. (Or, equivalently, the time for the Sun to make one complete trip against the background stars as seen from the center of Earth.) A *tropical year* (or *solar year*) is the interval between successive vernal equinoxes. Because the equinoxes have an annual retrograde motion (due to precession), of 50.26″ relative to the stars, the tropical year is about 20 minutes shorter than the sidereal year. An *anomalistic year* is the interval between successive passages of Earth through perihelion or aphelion. An *eclipse year* is the time between successive returns of the Sun to the same node of the Moon's orbit. This period is keyed to the regular recurrence of both solar and lunar eclipses, which can only take place when the Sun and the Moon are close to the node. Nineteen eclipse years are 6585.78 days, which is almost exactly the same as the ancient **Saros** cycle of 6585.32 days–the period that separates eclipses in a given series. A *lunar year* is made of 12 lunations or synodic months (354.3672 days). A *civil year* has an exact number of days, which is determined by the **calendar** being used. (See table, "Different Types of Year.")

Yerkes Observatory

The University of Chicago's observatory at Williams Bay, Wisconsin, on the shores of Lake Geneva. It is home to the world's largest refractor, with an aperture of 40 in. (1.02 m), opened in 1897, which is still used for research including adaptive optics studies. Other instruments include a 41-inch reflector, opened in 1967, and a 24-inch Boller & Chivens reflector. The observatory was funded in 1892 by the businessman Charles Tyson Yerkes (1837–1905) and founded in 1897 by George **Hale**.

Yohkoh

A Japanese-led mission, launched in 1991, with collaboration from the United States and Britain, that is the first spacecraft to continuously observe the Sun in X rays over a whole **sunspot cycle**. Yohkoh is expected to stay in orbit until the next solar maximum, around 2010. In the coming years it will collaborate with RHESSI (Reuven-Ramatty High Energy Solar Spectroscopic Imager).

young

Descriptive of a planetary surface of which the visible features are of relatively recent origin; older features have been destroyed by erosion, lava flows, or other

Different Types of Year

Year	Reference Points	Length (days)
Sidereal	Background stars	365.25636
Tropical	Equinoxes	365.24219
Anomalistic	Apsides	365.25964
Eclipse	Moon's node	346.62003

processes. Young surfaces exhibit few **impact craters** and are typically varied and complex. In contrast, an "old" surface is one that has changed relatively little over geologic time. The surfaces of Earth, Io, and Europa, are young, whereas the surfaces of Mercury and Callisto are old.

young stellar object (YSO)

Any star that has evolved past the **protostar** stage (i.e., is shining by way of internal nuclear reactions) but has yet to arrive on the **main sequence**. YSOs come in a variety of forms depending on their age, mass, and environment, and include **Herbig Ae/Be stars**, **T Tauri stars**, and, in general, immature stars prone to irregular brightening, embedded in nebulosity, and associated with **bipolar outflows**.

YY Orionis star (YY Ori star)

An extremely young star, preceding even the T Tauri stage. YY Ori stars are very **young**, late-type, low-mass stars, still contracting gravitationally and accreting matter from the protostellar cloud. They show inverse **P Cygni profiles** corresponding to an infall of material at 300 to 400 km/s directly onto the star.

Z

Z Andromedae star (Z And star, ZAND)

A type of symbiotic star (which, in turn, is a type of **cataclysmic variable**) consisting of a close binary system in which a hot star ionizes part of an extended envelope of gas that has come from a cooler companion (of spectral class M, R, N, or S). The combined spectrum of the system shows the superposition of absorption and emission spectral features together with irregular variability (of up to 4 magnitudes in the visual region) that are characteristic of symbiotics.

The classification of symbiotics is confusing because of the great diversity of these stars and the many gaps in our knowledge about them. In the *General Catalogue of Variable Stars (GCVS)*, Z Andromedae stars (ZANDs) are the only recognized subcategory of symbiotics, so that Z And itself is officially regarded as not only the prototype ZAND but also the prototype symbiotic. However, objects such as RR Telescopii stars, R Aquarii, and **CH Cygni** have unique characteristics that set them apart within the overall symbiotic family.

The strange novalike spectral features and variability of Z Andromedae were discovered in 1901 by Williamina **Fleming** on Harvard College Observatory plates. When quiescent, which is most of the time, the only brightness changes shown are those of a small amplitude semiregular red variable with an M-type spectrum, a period of about 700 days, and an average magnitude of about 11. Every 10 to 20 years, however, Z And becomes very active, brightening by about 3 magnitudes. The large amplitude outbursts are then followed by smaller outbursts of decreasing amplitude, after which the star fades again to quiescence. The brighter state occurs either abruptly or is preceded by a smaller outburst. The brightest recorded magnitude was during the 1939 outburst when the star reached a mean visual magnitude of 7.9. During an outburst, the star gets bluer and the spectrum is dominated by the hot, compact star with a B-type shell spectrum. **P Cygni profiles** arise displaying blueshifted absorption lines, indicative of an expanding shell as seen in novae. As the shell-dominated spectrum slowly fades, the star gets redder, the P Cygni features disappear, the shell spectrum weakens and disappears, and the system returns to its slow and semiregular variations. Its lines almost disappear at minimum, and the titanium oxide bands from the red star become prominent. In the Z And system, the dwarf component appears to accrete mass from a wind emanating from the red giant component. Since it seems to be a detached system, **stellar wind** may be a key component during both quiescence and outburst. The accreted matter may then form a disk around the white dwarf component; however, the existence of a disk has not been confirmed. The outbursts are thought to derive from the blue star, but the red star appears to be variable as well. Many of the basic stellar parameters, such as stellar mass and inclination of the system, are unknown and prevent further knowledge and theoretical models of the system.

Z Camelopardalis star (Z Cam star)

One of three subcategories of **U Geminorum stars** (dwarf novae), the others being **SS Cygni stars** and **SU Ursae Majoris stars**. Z Cam stars are distinguished by, in addition to normal U Gem-type outbursts (which consist of a rise from quiescence of 2 to 6 magnitudes and 1- to 3-day durations), random *standstills*. A standstill usually starts at the end of an outburst and consists of a period of constant brightness, about one magnitude below maximum light that may last from a few days to 1,000 days. The average energy output in a standstill is larger than that during an outburst cycle. Standstills occur when the mass transfer rate from the secondary star into the accretion disk around the primary star is too large to produce normal outbursts. In the prototype system, Z Cam, a G-type (Sun-like) dwarf and a white dwarf or a blue subdwarf orbit around each other every 7h 21m; eruptions occur on average every 20 days. See also **UX Ursae Majoris star**.

Zaniah (Eta Virginis, η Vir)

A triple star system in **Virgo** that, in 2002, became the first object to be imaged by combining the light from multiple telescopes (the Navy Prototype Optical Interferometer at the **Lowell Observatory**), a major breakthrough in optical interferometry. The high-resolution image, which would have required a single instrument with a 50-m aperture,

Visual magnitude:	3.89
Absolute magnitude:	–0.53
Spectral type:	A2IV
Luminosity:	130 L_{sun}
Distance:	250 light-years

showed the three components of the system as separate entities for the first time. Zaniah consists of three **A stars**: a very close pair separated by just a few milliarcseconds (equivalent to less than half the Earth-Sun distance), and a third component about 0.06″ away.

Zeeman effect

The splitting of a spectral line into several polarized components when the source is in a strong magnetic field; it is named after the Dutch physicist Pieter Zeeman (1865–1943). Because the amount of Zeeman splitting and of polarization depend on the magnetic field strength, this effect provides a powerful tool for investigating cosmic magnetic fields. It was first used to map magnetic fields on the Sun, especially in and around sunspots where the local field strength can be as high as 0.4 tesla–thousands of times stronger than Earth's magnetic field. Across the Sun as a whole, and in the case of most other stars and cosmic objects, the field strengths are much lower and the spectral lines are broadened rather than actually split by the Zeeman effect. However, weak fields can still be studied by measuring the polarization of the *line wings,* the parts of a spectral line well away from the central peak. Zeeman splitting is observed in the case of some **magnetic cataclysmic variables,** which are associated with intense magnetic fields.

Zel'dovich, Yakov ("Ya") Borisovitch (1914–1987)

A Russian physicist and astrophysicist best known for his work on **shock waves** and gas dynamics. Having played a significant role in the development of Soviet nuclear and thermonuclear weapons, he turned in the 1950s to the theory of elementary particles, and in the 1960s to astrophysics and cosmology. He worked on the dynamics of neutron emission during the formation of black holes, the formation of galaxies and clusters, and the large scale structure of the universe. With Rashid Sunyaev (1943–) he proposed what is known as the **Sunyaev-Zel'dovich effect**, an important method for determining the **Hubble constant** from the effect of gas in galaxy clusters on the microwave background radiation. Zel'dovich was also a leader in attempts to relate particle physics to cosmology and to develop a quantum theory of gravity. From age 17 he worked at the Institute of Chemical Physics of the Academy of Sciences of the USSR in Leningrad (later in Moscow), where he essentially educated himself. In his later years he was also a professor at Moscow State University and head of the division of relativistic astrophysics at the Sternberg Astronomical Institute.

zenith

The point directly overhead; the opposite of *nadir.* The *zenith angle* is the angle between the overhead point for an observer and an object such as the Sun. The *solar zenith angle* is zero if the Sun is directly overhead, and is 90° when the Sun is on the horizon.

zenithal hourly rate (ZHR)

During a meteor shower, the number of meteors per hour an observer would see if his **limiting magnitude** (the faintest object visible) is 6.5 and the radiant is in his **zenith**. Since these are optimum conditions, the measured hourly rate is usually less than the ZHR.

zero-age main sequence (ZAMS)

The **main sequence** defined by a population of stars, such as a that of a large cluster, all the members of which have just evolved onto the main sequence.

Zeta Aurigae (ζ Aur)

A famous **eclipsing binary** in **Auriga** and one of the **Kids** (in Latin *Haedus*). It is known as Haedus I and by an astonishing coincidence, its constellation mate, **Epsilon Aurigae**, which is the second Kid (Haedus II), is also a well-known (and even more remarkable) eclipsing binary. Zeta consists of an orange supergiant in orbit around a blue main-sequence B star with a period of 972.2 days (2.66 years). Because the orientation of the orbit lies within 3° of our line of sight, every 2.66 years the smaller but brighter B star hides completely behind the larger K star (which is about as wide as Venus's orbit), and the combined visual light drops by 0.15 magnitude (about 15%). Averaging 4.2 AU apart, the two go around each other in an elliptical orbit that takes them from 5.9 AU to 2.5 AU. Both stars are about 80 million years old.

Visual magnitude:	3.69
Absolute magnitude:	–3.23
Spectral type:	K4Ib + B5V
Surface temperature:	3,950 K (K star), 15,300 K (B star)
Luminosity:	4,800 L_{sun} (K star), 1,000 L_{sun} (B star)
Radius:	148 R_{sun} (K star), 4.5 R_{sun} (B star)
Mass:	5.8 M_{sun} (K star), 4.8 M_{sun} (B star)
Distance:	787 light-years

Zeta Ophiuchi (ζ Oph)

The third brightest star in **Ophiuchus** and one of the brightest **O stars** on view in the night sky; it would be even more prominent if it were not partly obscured by clouds of gas and dust in its vicinity, one of which, an HII region known as S27, it illuminates. In the early 1970s, its spectrum was observed to vary in a way that suggests it has recently expelled a shell of matter. Zeta Oph is also a well-known **runaway star** having most likely been expelled from a binary system in the **Scorpius-Centaurus Association** when its partner exploded as a supernova. Evidently some supernova blasts are asymmetrical, causing the remnant core to hurtle away in one direction while any binary companion is thrown off the other way.

Visual magnitude:	2.54
Absolute magnitude:	–3.20
Spectral type:	O9.5V
Surface temperature:	32,500 K
Luminosity:	68,000 L_{sun}
Radius:	8 R_{sun}
Mass:	20 M_{sun}
Distance:	458 light-years

zodiac

A band around the sky about 18° wide, centered on the **ecliptic**, in which the Sun, Moon, and planets move. The band is divided into 12 **signs of the zodiac**, each 30° long, that were named by the ancient Greeks after the constellations that used to occupy these positions; "zodiac" means "circle of animals," and only Libra is inanimate. Over the past 2,000 years, precession has moved the constellations eastward by over 30° so that they no longer coincide with the old signs.

zodiacal cloud

A tenuous disk of **dust** orbiting within the inner part of the solar system out to about 5 AU (almost the distance of Jupiter). It is fed erratically by the disintegration of comets close to the Sun, and by dust from collisions in the asteroid belt. The cloud can be seen at certain times of the year as a second "Milky Way" along the ecliptic. Observable structures in this disk include dust bands associated with the major **Hirayama families** of asteroids and a circumsolar ring of particles in resonant lock with Earth. Observations by the Cosmic Background Explorer have also shown that the center of symmetry of the cloud is offset from the Sun and that the cloud is warped. The lifetimes of particles in the zodiacal cloud with diameters less than a few hundred microns are determined by the **Poynting-Robertson effect** rather than by mutual collisions and this gives time for significant orbital evolution of the dust. Orbital evolution allows for passage through secular resonances and for trapping into mean motion resonances and it is these interactions that give rise to the features of the disk.

zodiacal dust

Dust particles, 1 to 300 microns in diameter, that have come from comet tails and asteroid collisions, and occupy a flat cloud in the solar system, in the plane of the ecliptic, out to a radius of about 5 AU (just beyond Jupiter's orbit). A faint glow of light scattered off this dust is called the *zodiacal light;* it can sometimes be seen under

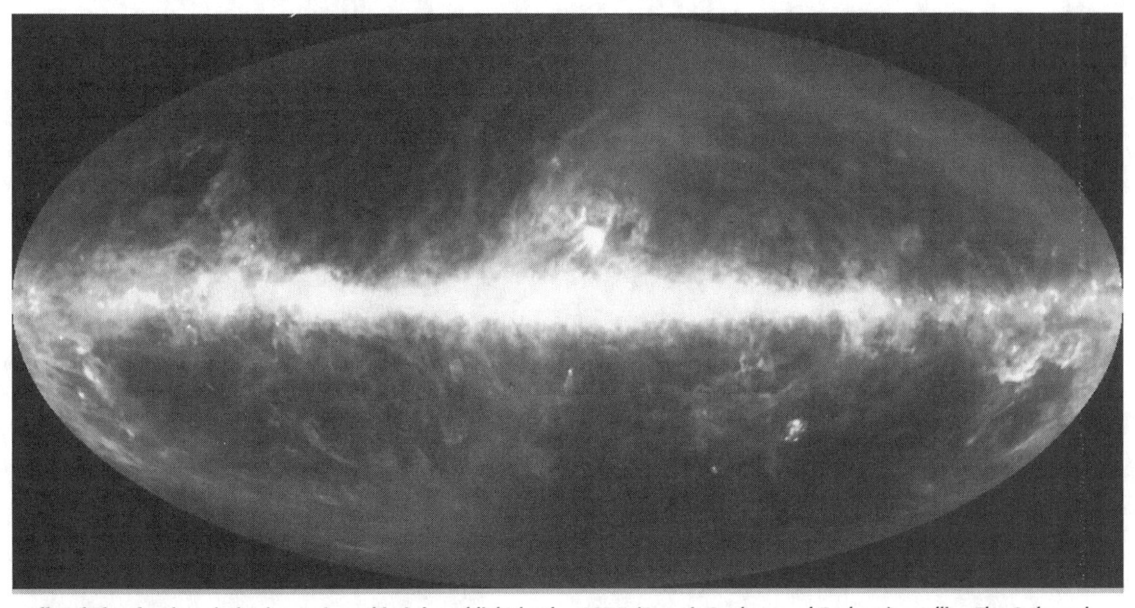

zodiacal cloud The whole sky as viewed in infrared light by the COBE (Cosmic Background Explorer) satellite. The S-shaped curve is due to infrared emission by the zodiacal cloud. *NASA*

very dark sky conditions, along the horizon, either just after dusk or before sunrise.

zone of avoidance

A band around the sky in which very few extragalactic objects can be seen at visible wavelengths because of the heavy absorption of light by dust in the plane of the Milky Way. The zone of avoidance is by no means regular, varying in width from about 38° in the direction of the center of our Galaxy (in Sagittarius) to about 12°. Additionally, there are a number of **galactic windows** in the zone that are relatively free of obscuring material and therefore offer a clear view of the outside universe.

Zwicky, Fritz (1898–1974)

A Bulgarian-born Swiss astronomer who worked in the United States and was well known for his work on supernovae, compact galaxies, and intergalactic matter. He made many significant breakthroughs, among them the discovery of the **dark matter** permeating the Coma Cluster of galaxies. He was also something of a character. On one occasion, while working at Palomar Observatory, he had the night assistant fire a bullet out the dome slit of the Hale Telescope in the direction the instrument was pointing to see if it improved the seeing (it didn't). In the January 19, 1934 edition of the *Los Angeles Times,* Zwicky was lampooned in an article accompanied by a cartoon entitled "Be Scientific with Ol' Doc Dabble." The article scoffed: "Cosmic rays are caused by exploding stars which burn with a fire equal to 100 million suns and then shrivel from ½ million mile diameters to little spheres 14 miles thick." In the polemical introduction to Zwicky's *Catalogue of Selected Compact Galaxies and of Post-Eruptive Galaxies* (see also **Zwicky Catalogue**), Zwicky quotes the cartoon and comments: "This, in all modesty, I claim to be one of the most concise triple predictions ever made in science." It correctly describes, he points out, the nature of origin of cosmic rays, supernovae, and the formation of neutron stars.

Zwicky Catalogue

Popular name for the *Catalogue of Galaxies and of Clusters of Galaxies* (also referred to simply as "The Red Book") published in six volumes in 1961–1968 by Fritz **Zwicky** and his colleagues. It contains 31,350 galaxies and 9,700 clusters recorded on plates of the ***Palomar Observatory Sky Survey*** and introduces a system of classifying clusters as compact, medium compact, or open that is still used. **Compact galaxies** identified in the Zwicky Catalogue are often known as *Zwicky galaxies.*

ZZ Ceti star (ZZ Cet star)

A **white dwarf** that is also a **pulsating variable** and changes in brightness by 0.001 to 0.2 magnitude over a period of 30 seconds to 25 minutes. **Nonradial pulsations** are involved with multiple periodicities, which result in only slight changes in the star's size but significant fluctuations in temperature. There are three subtypes: ZZA with DA (hydrogen absorption) spectra, ZZB with DB (helium absorption spectra), and ZZO, sometimes called *GW Virginis stars,* with extremely hot, DO spectra. Sometimes flares of about one magnitude are observed, but these may be due to UV Ceti–type companion stars.

Star Charts

The 22 star charts that follow provide complete coverage of the 88 constellations in the north and south celestial hemispheres. Included on each chart are the brightest stars and other prominent objects in the night sky, including star clusters, nebulae, and galaxies. Right ascension (in hours and minutes) runs along the horizontal axis, and declination (in degrees) along the vertical axis.

At the top of each chart are listed the constellations shown, beginning with the constellation that is most centrally placed, followed by other constellations that appear in full, followed by still other constellations (listed in parentheses) that appear partially. The star chart numbers correspond with those identified in the entries for constellations in the main text of the encyclopedia.

Star Chart 1

Andromeda—Cassiopeia—Lacerta—Pisces—Triangulum—(Aries—Pegasus)

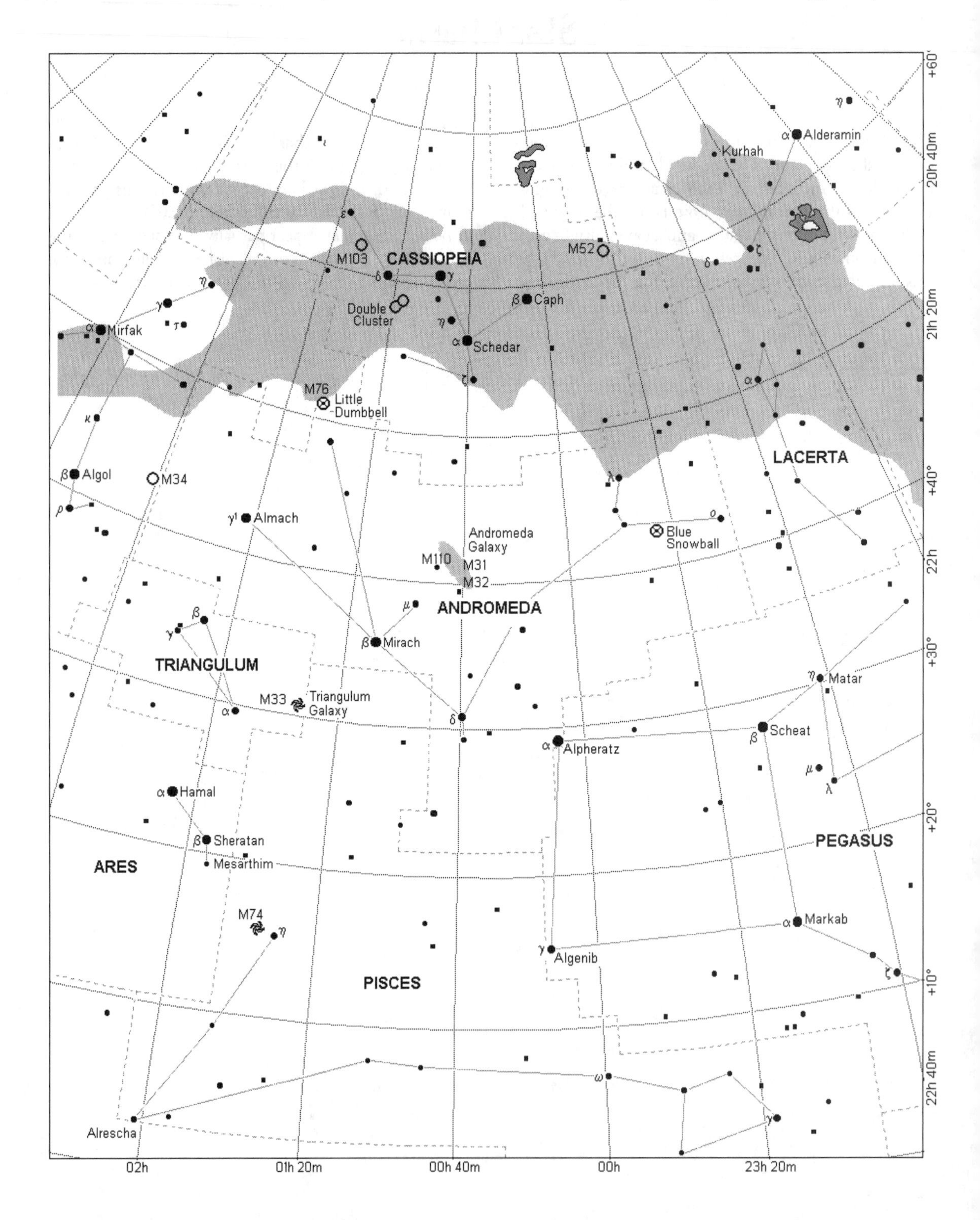

Star Chart 2

Auriga–Gemini–(Camelopardus–Lynx–Perseus–Taurus)

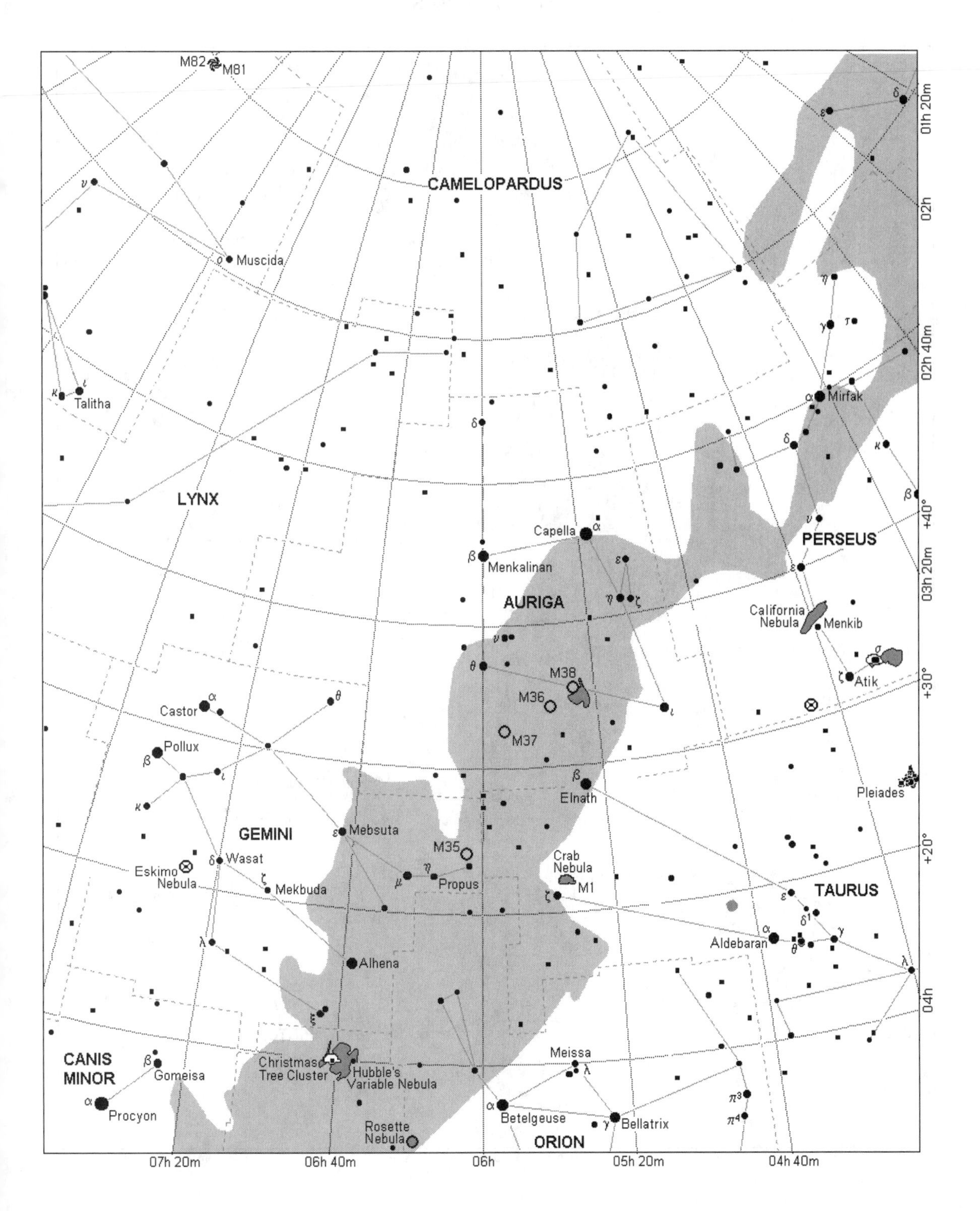

Star Chart 3

Canis Major—Caelum—Canis Minor—Columba—Lepus—Monoceros—Puppis—Pyxis

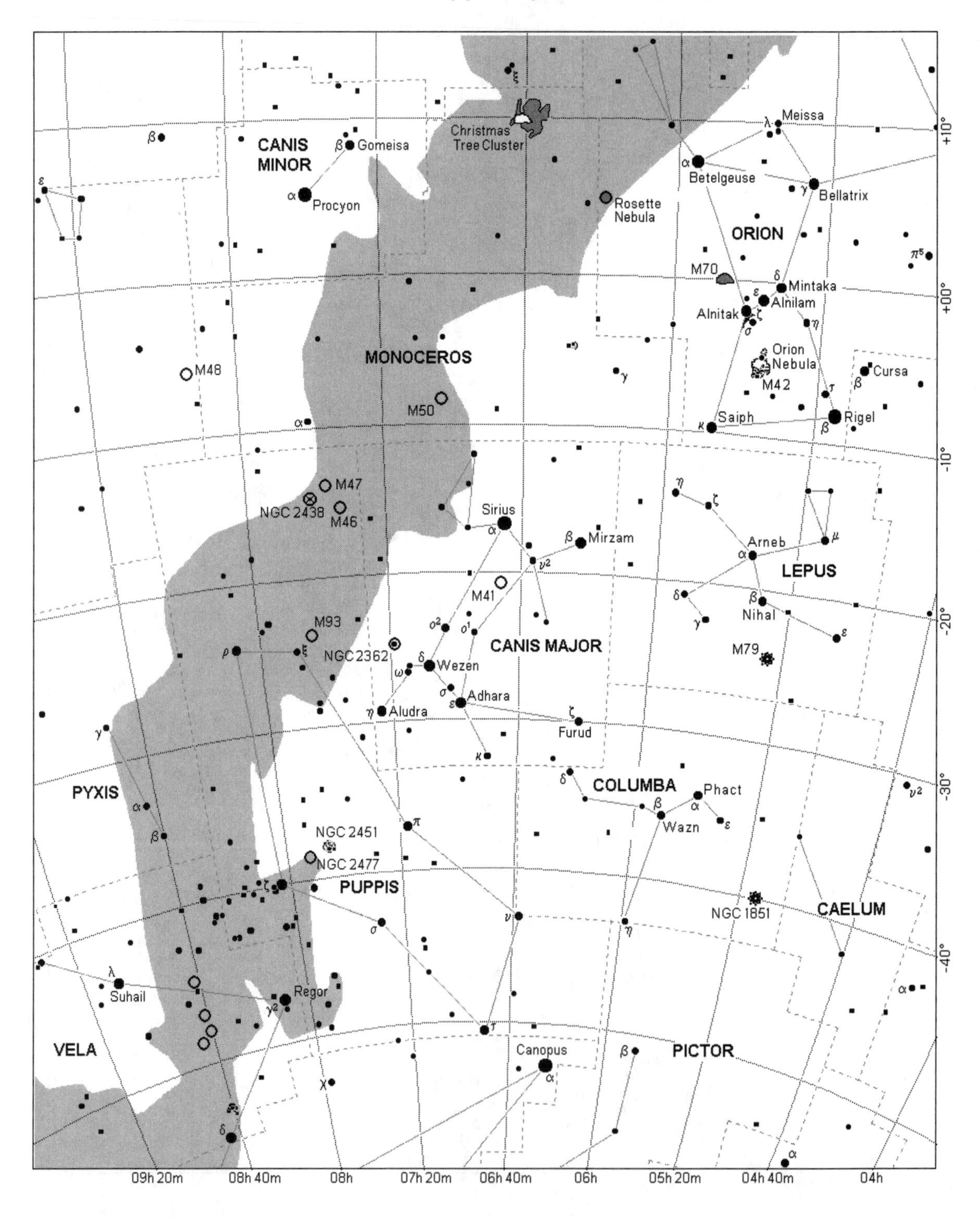

Star Chart 4

Carina—Antlia—Chameleon—Crux—Mensa—Musca—Vela—Volans—(Apus—Dorado—Octans—Puppis—Pyxis)

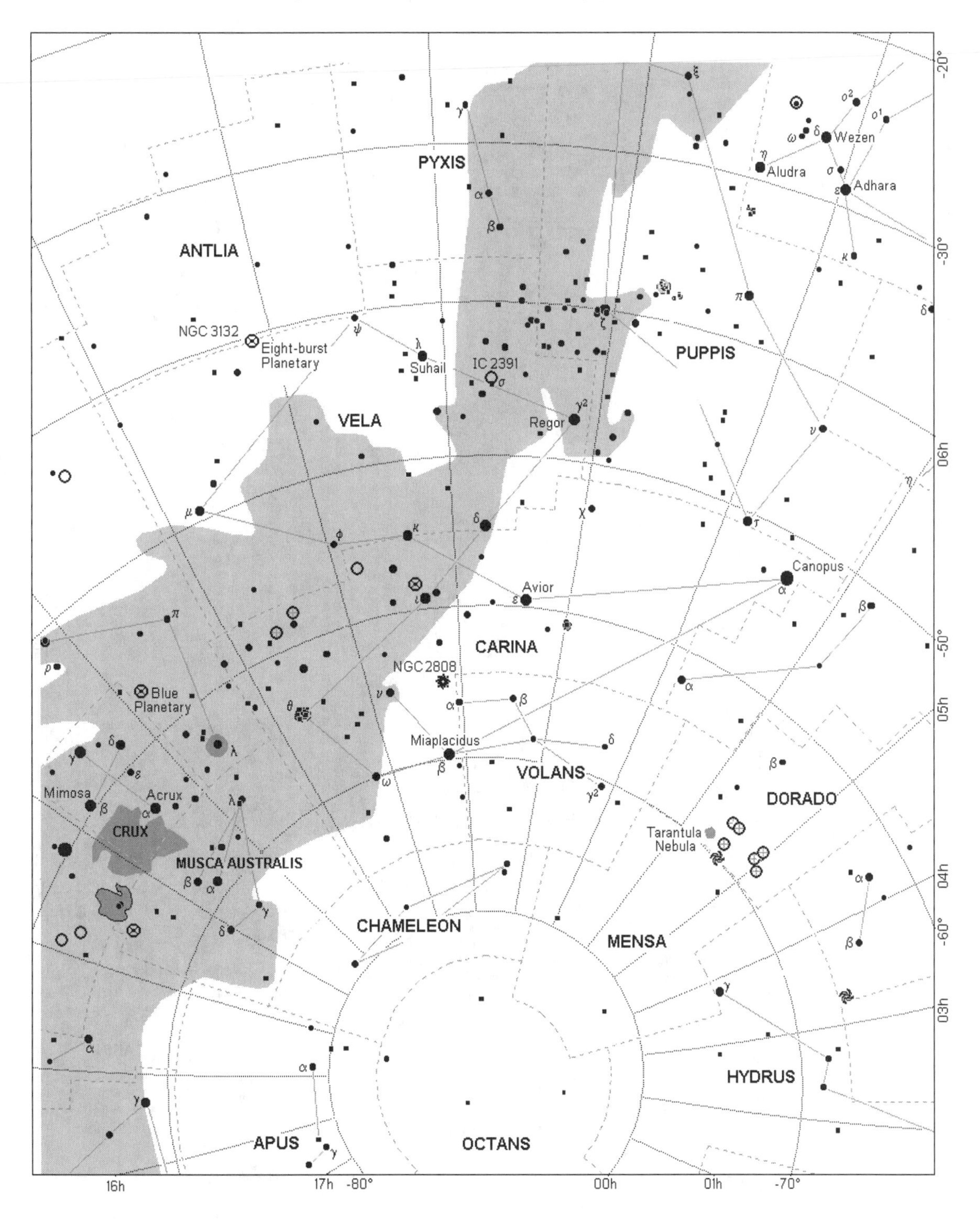

Star Chart 5

Centaurus—Circinus—Corvus—Crux—Musca—Triangulum Australe—(Apus—Carina—Crater—Lupus—Norma)

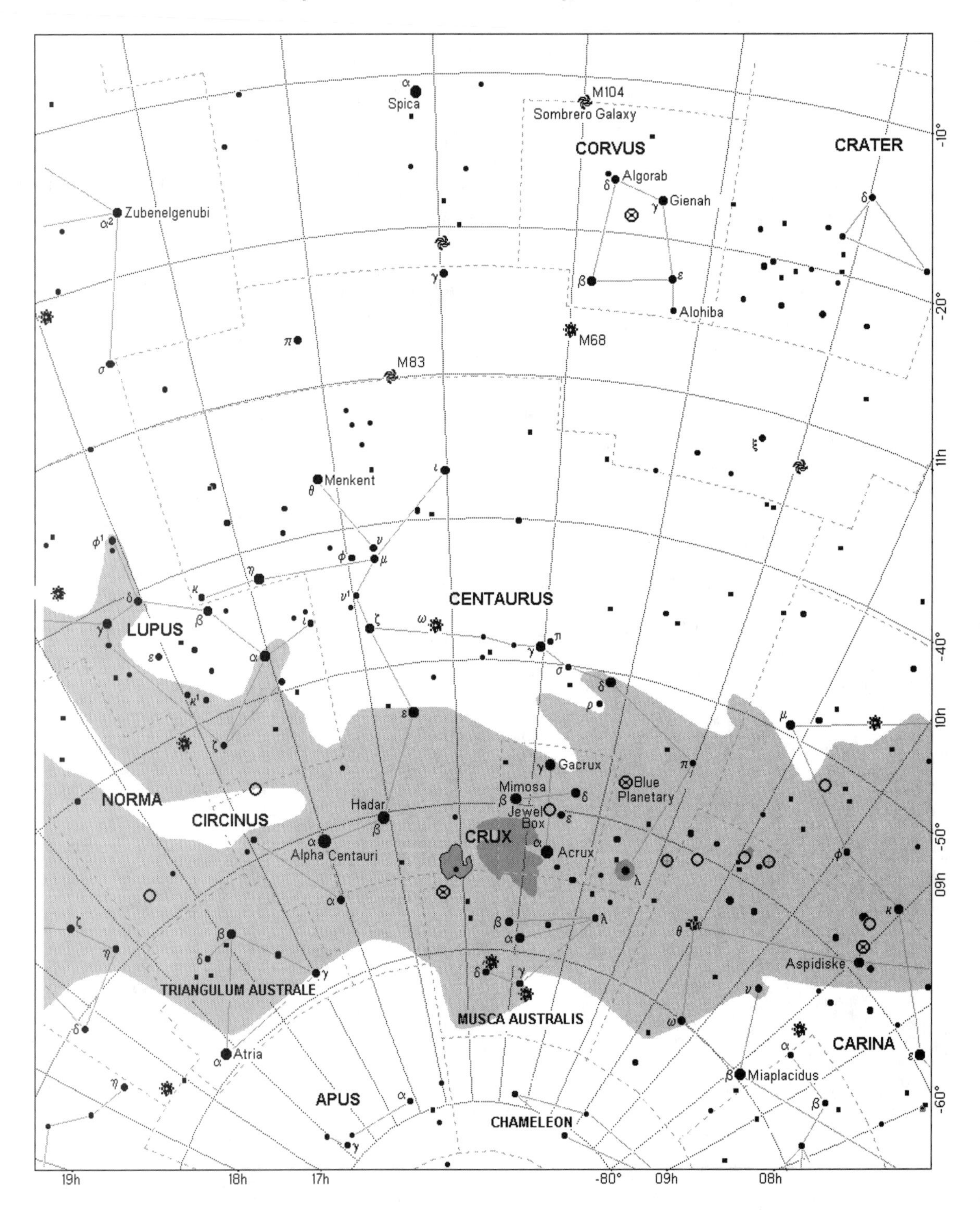

Star Chart 6

Cetus–Fornax–Sculptor–(Aries–Phoenix–Pisces)

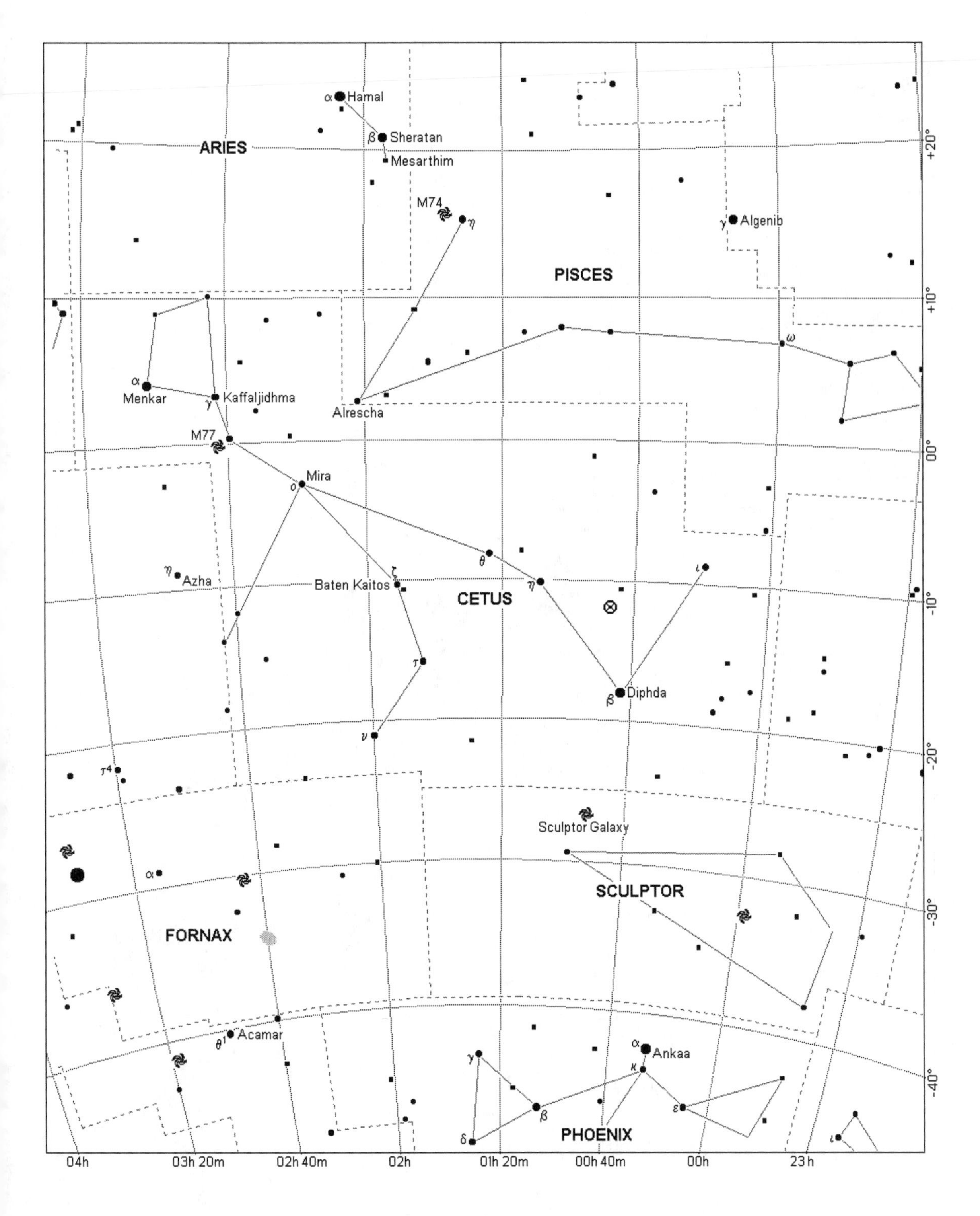

Star Chart 7

Cygnus—Lacerta—Lyra—Sagitta—Vulpecula—(Cepheus—Delphinus—Draco)

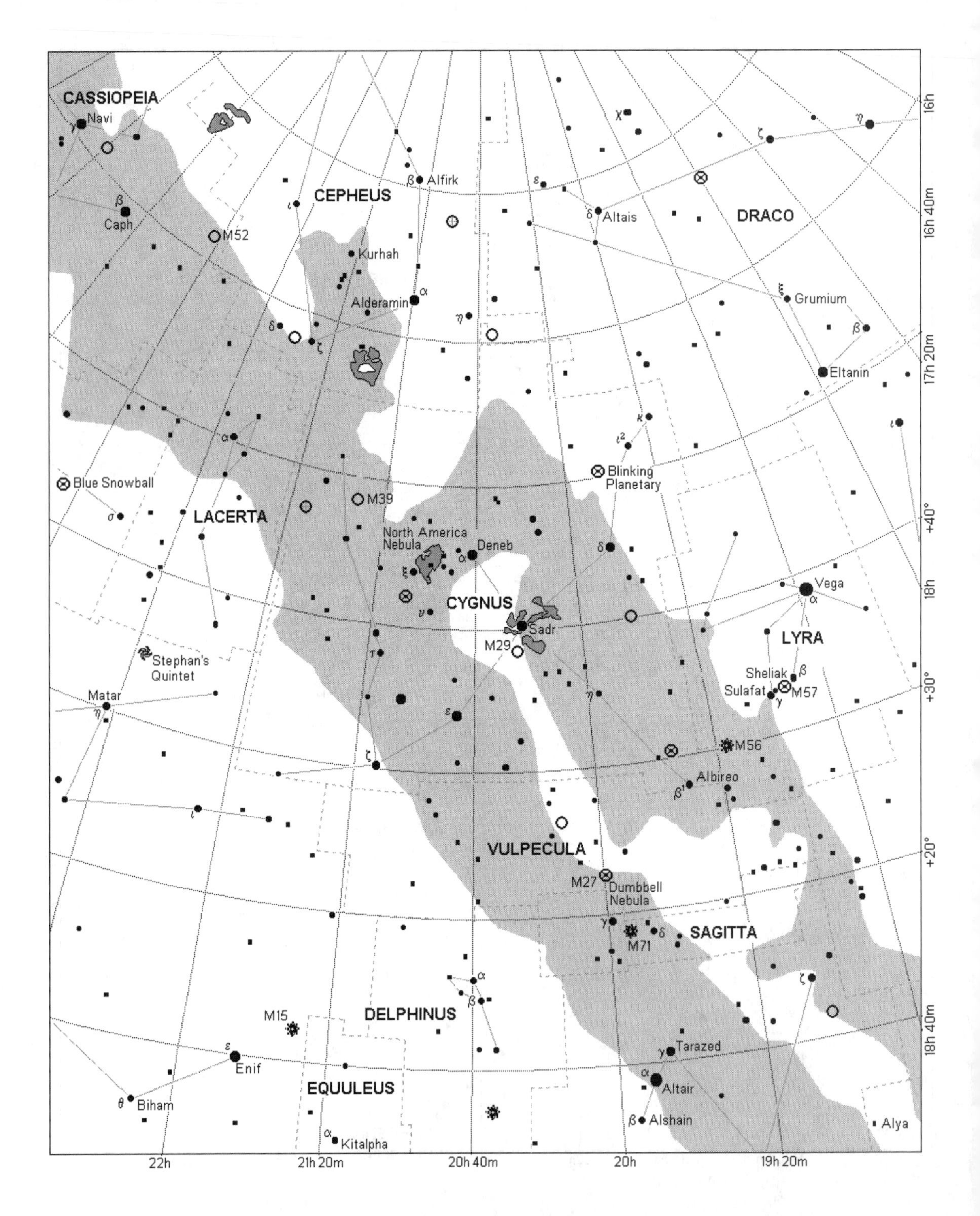

Star Chart 8

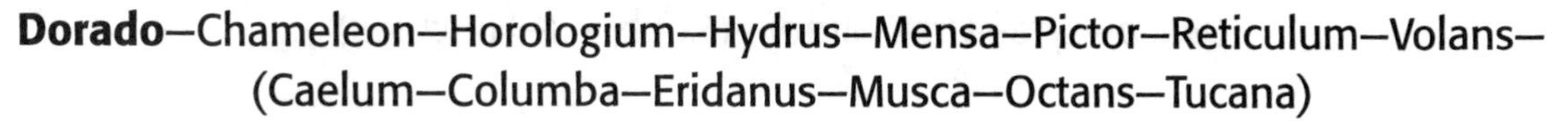

Dorado—Chameleon—Horologium—Hydrus—Mensa—Pictor—Reticulum—Volans—(Caelum—Columba—Eridanus—Musca—Octans—Tucana)

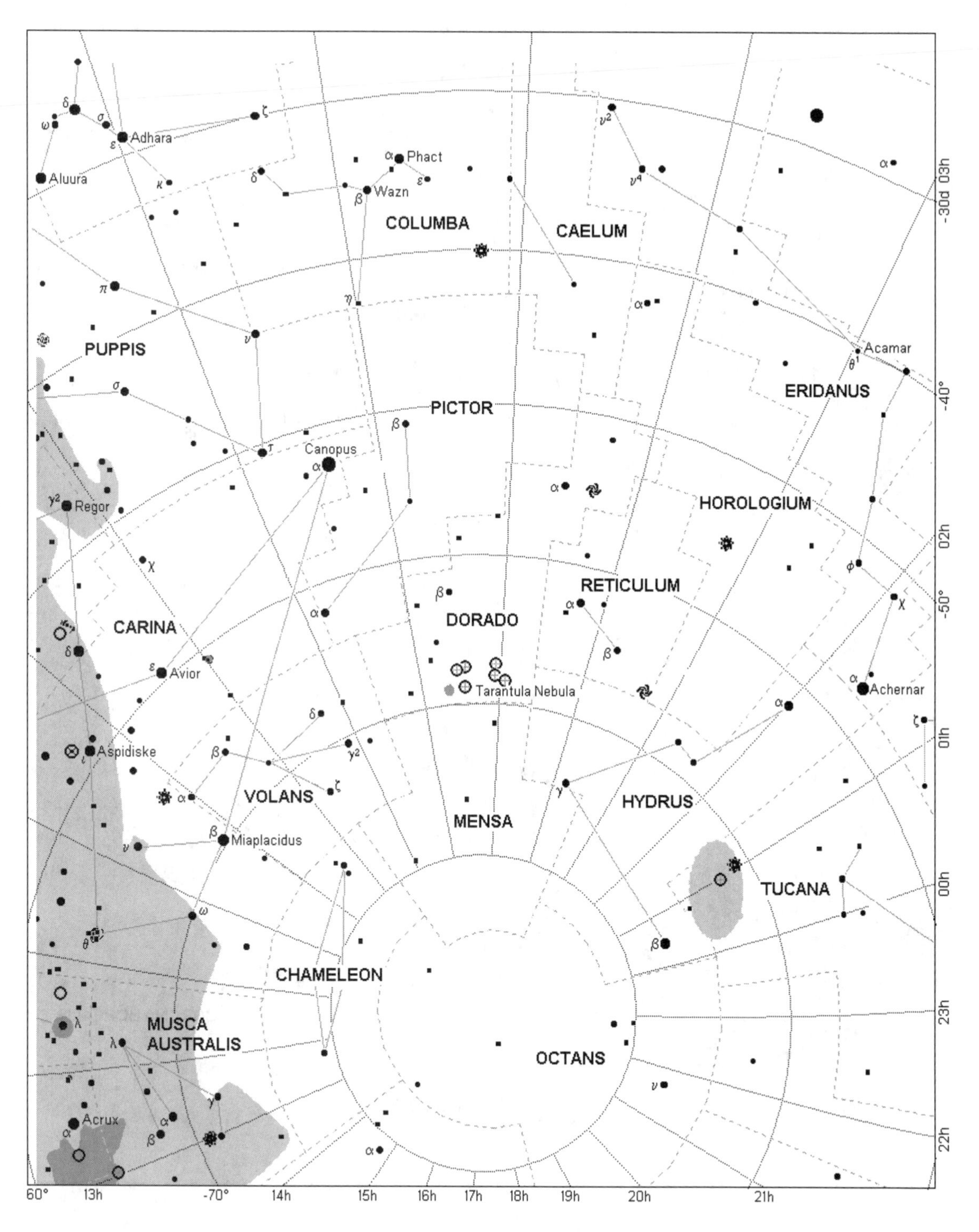

Star Chart 9

Draco—Lyra—Ursa Minor—(Cepheus—Hercules)

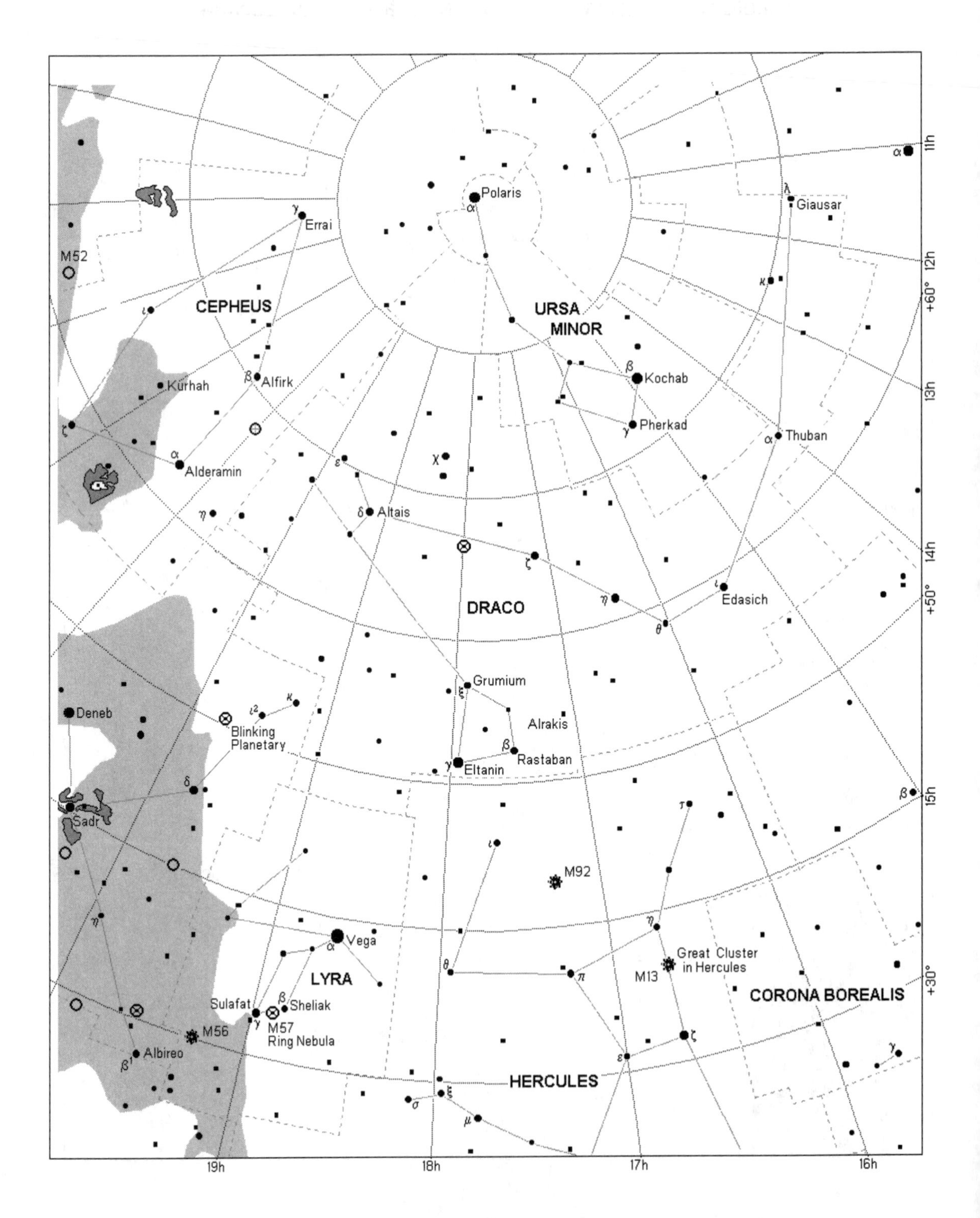

Star Chart 10

Eridanus—Caelum—Fornax—(Columba—Lepus)

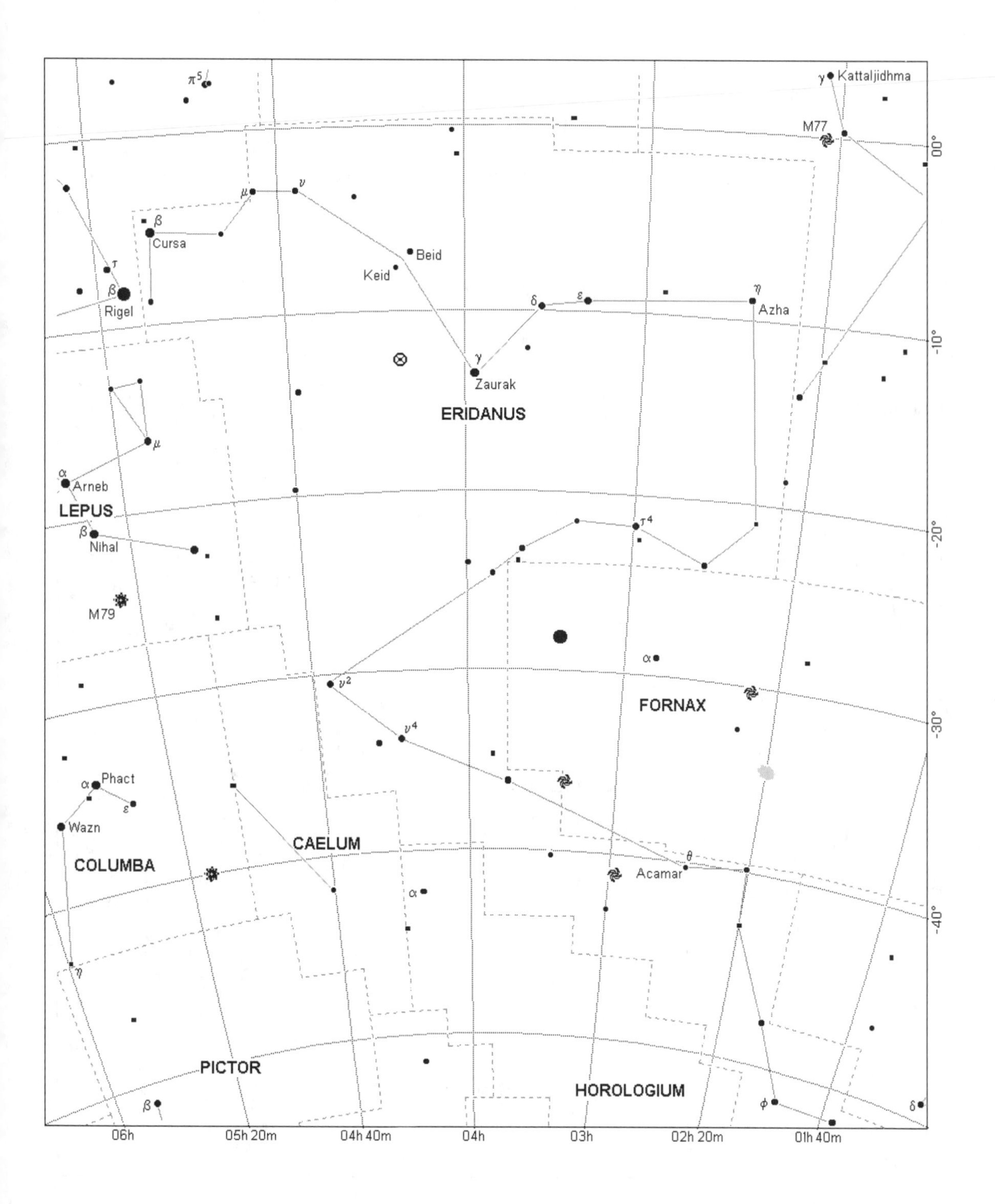

Star Chart 11

Hercules—Corona Borealis—Lyra—Serpens—(Draco—Ophiuchus)

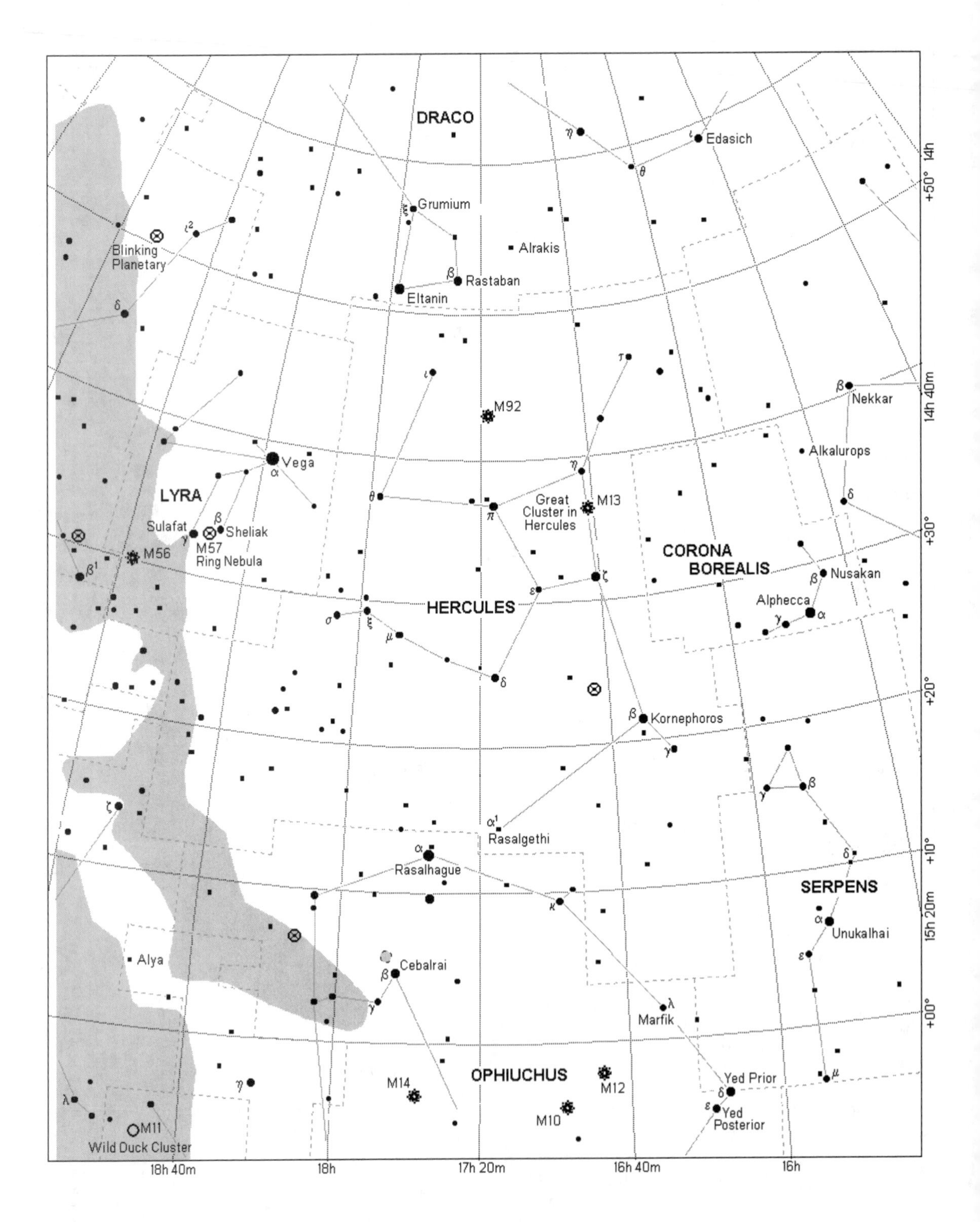

Star Chart 12

Leo—Leo Minor—Sextans—(Cancer—Crater—Hydra)

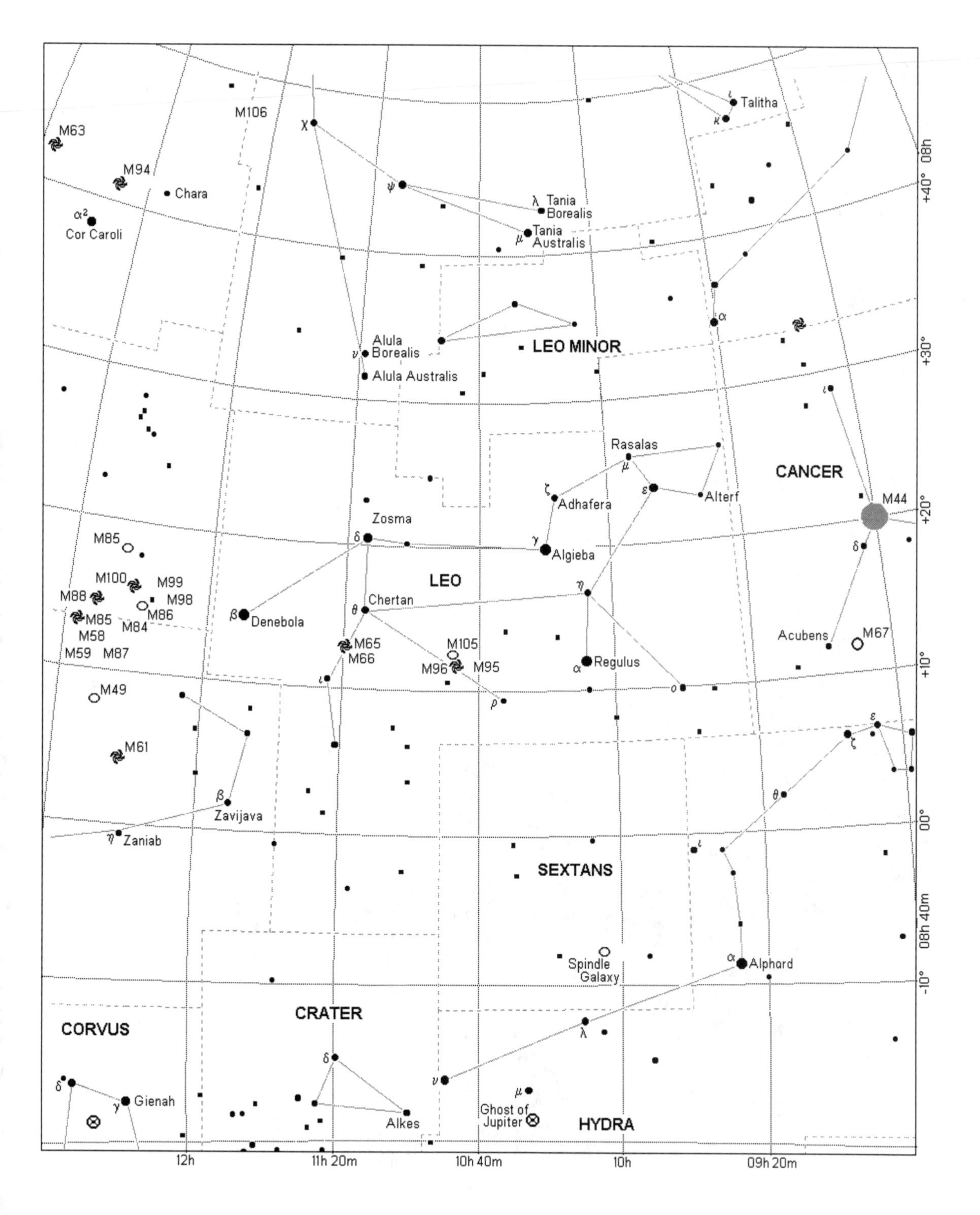

Star Chart 13

Ophiuchus–Scutum–Serpens–Serpens Cauda–(Sagittarius–Scorpius)

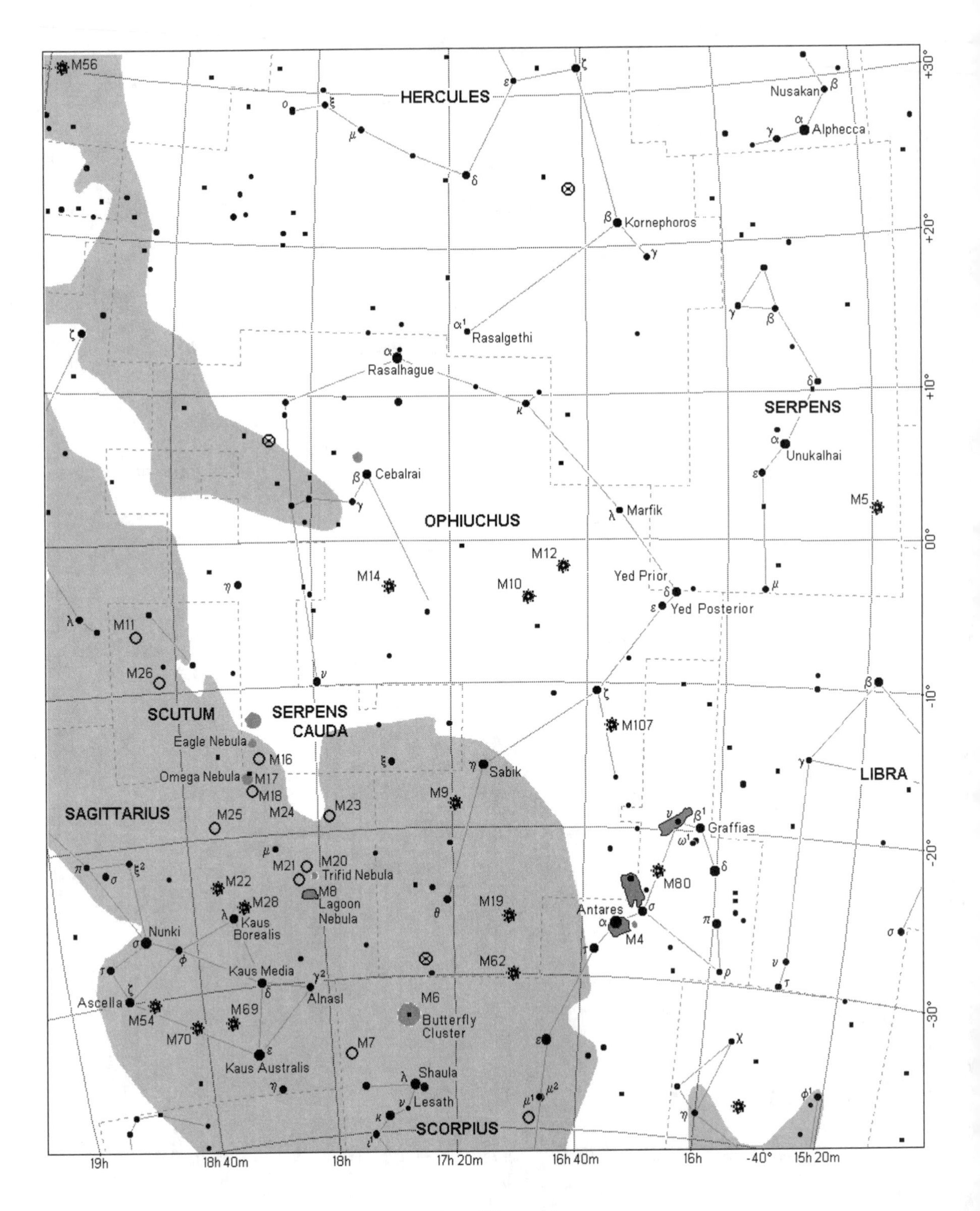

Star Chart 14

Sagittarius–Corona Australis–Indus–Microscopium–Serpens Cauda–Telescopium–Scutum–Ara–(Aquila–Capricornus)

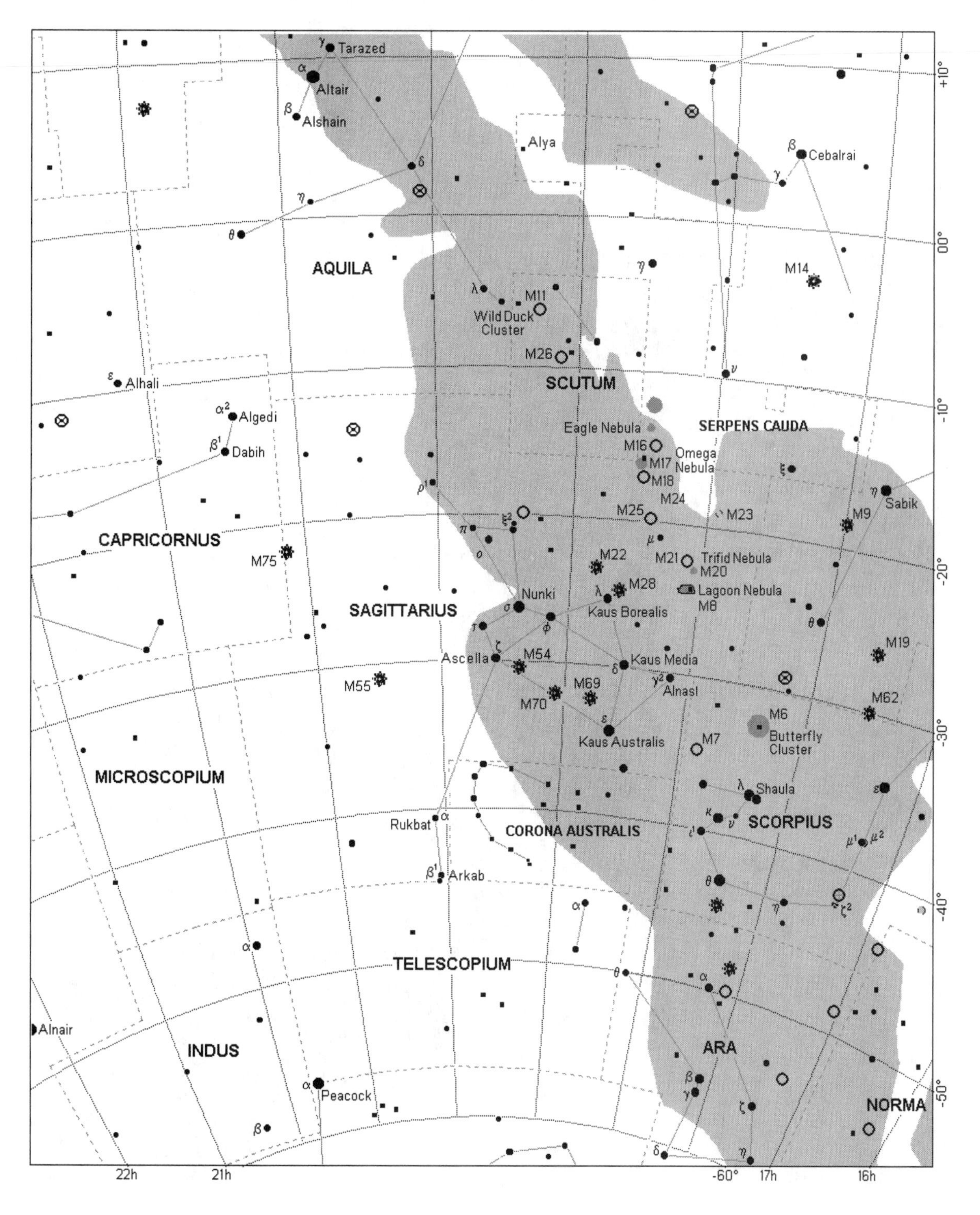

Star Chart 15

Scorpius—Ara—Circinus—Corona Australis—Lupus—Norma—Scutum—Triangulum Australe—(Libra—Pavo—Sagittarius—Telescopium)

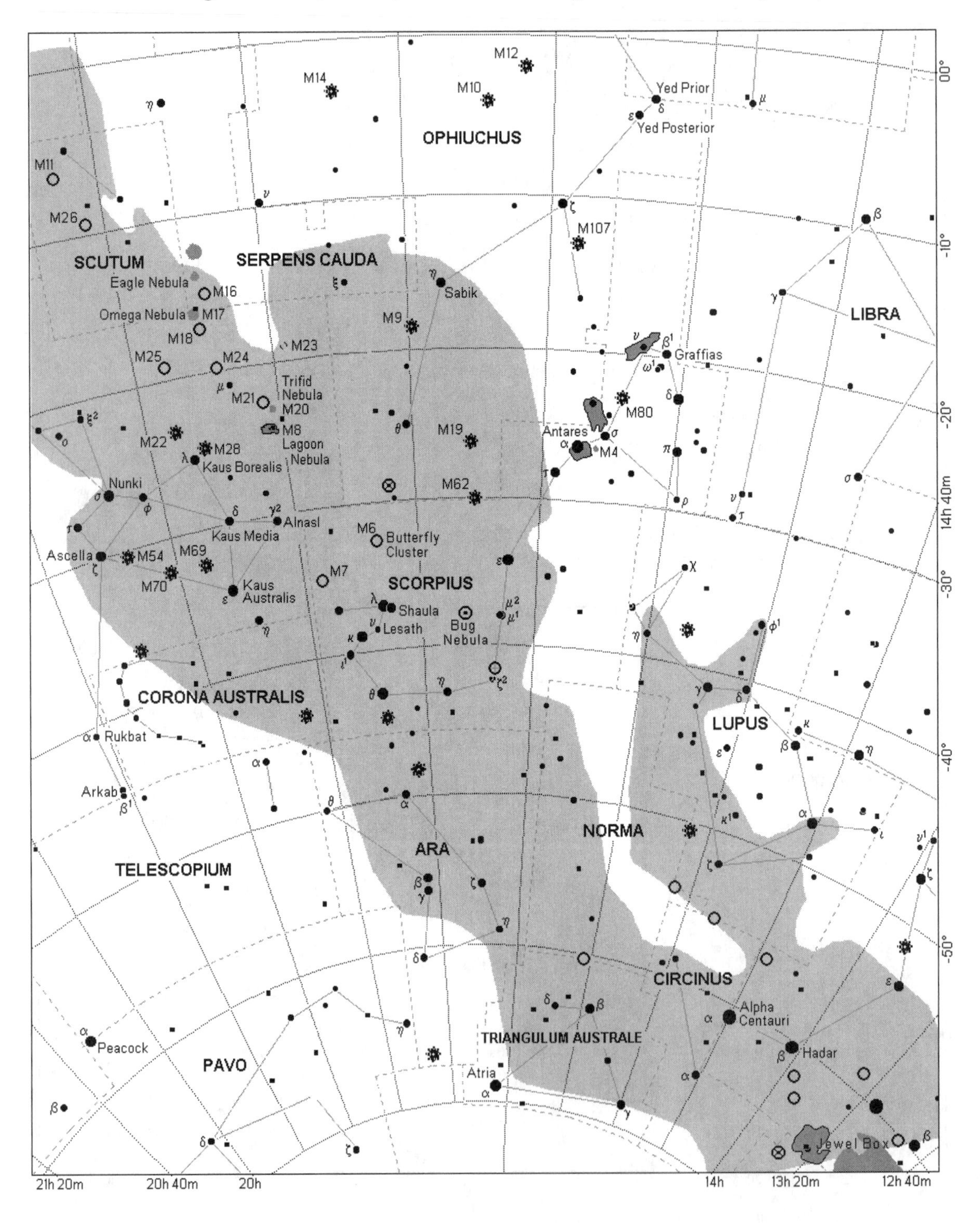

Star Chart 16

Taurus–Orion–(Auriga–Eridanus–Perseus)

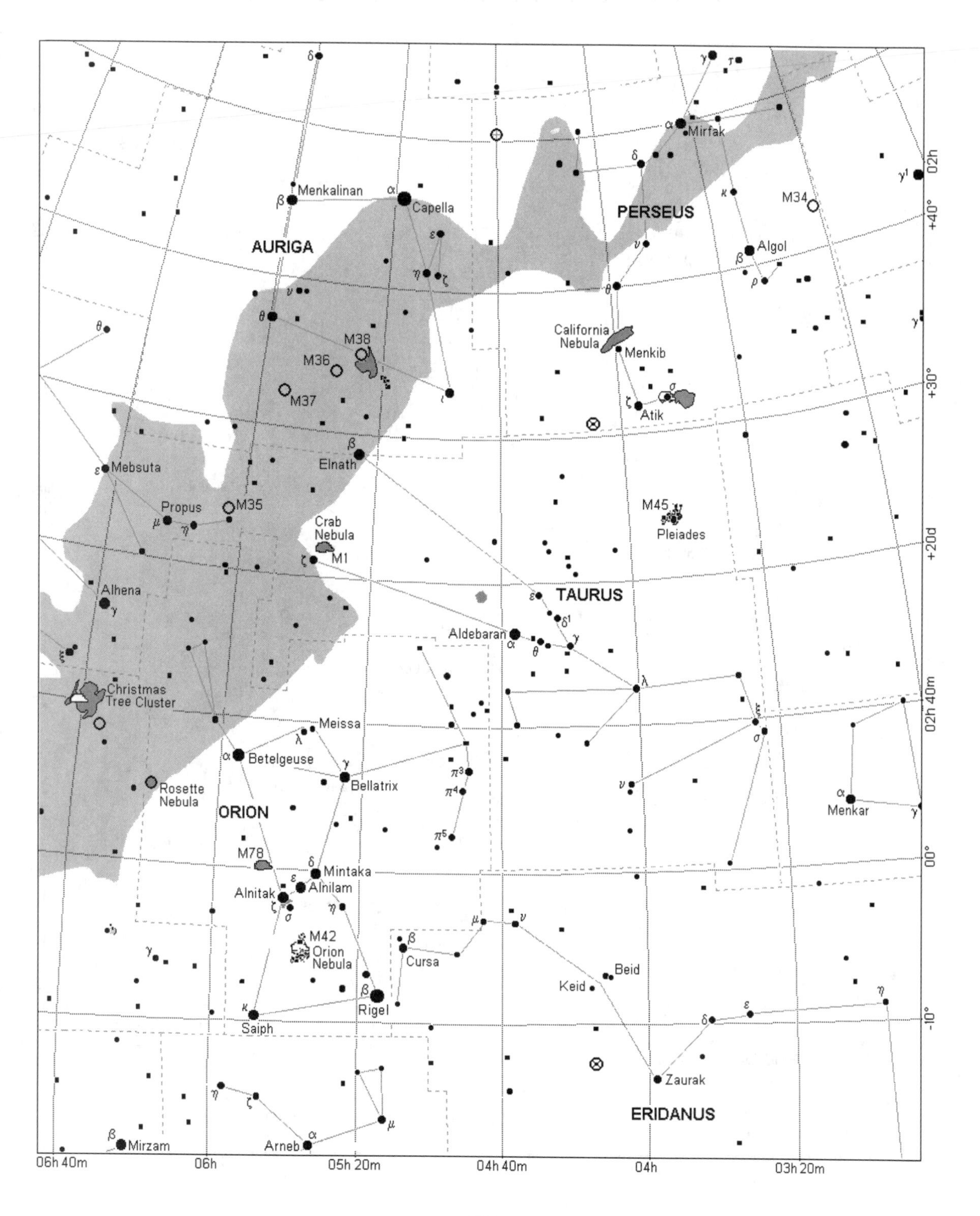

Star Chart 17

Ursa Major—Leo Minor—Ursa Minor—(Camelopardus–Canes Venatici–Coma Berenices)

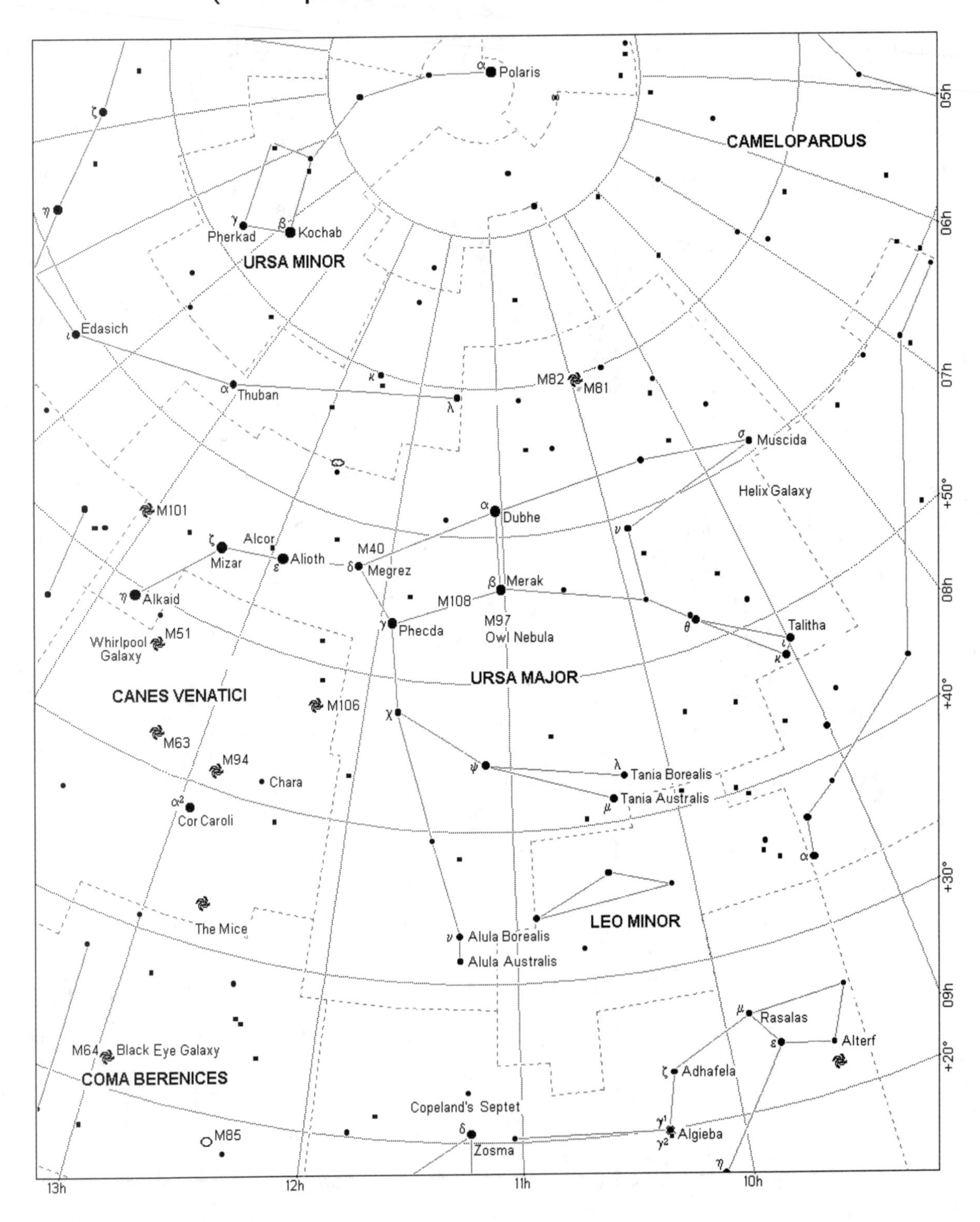

Star Chart 18

Virgo—Coma Berenices—Corvus—(Boötes—Crater—Hydra)

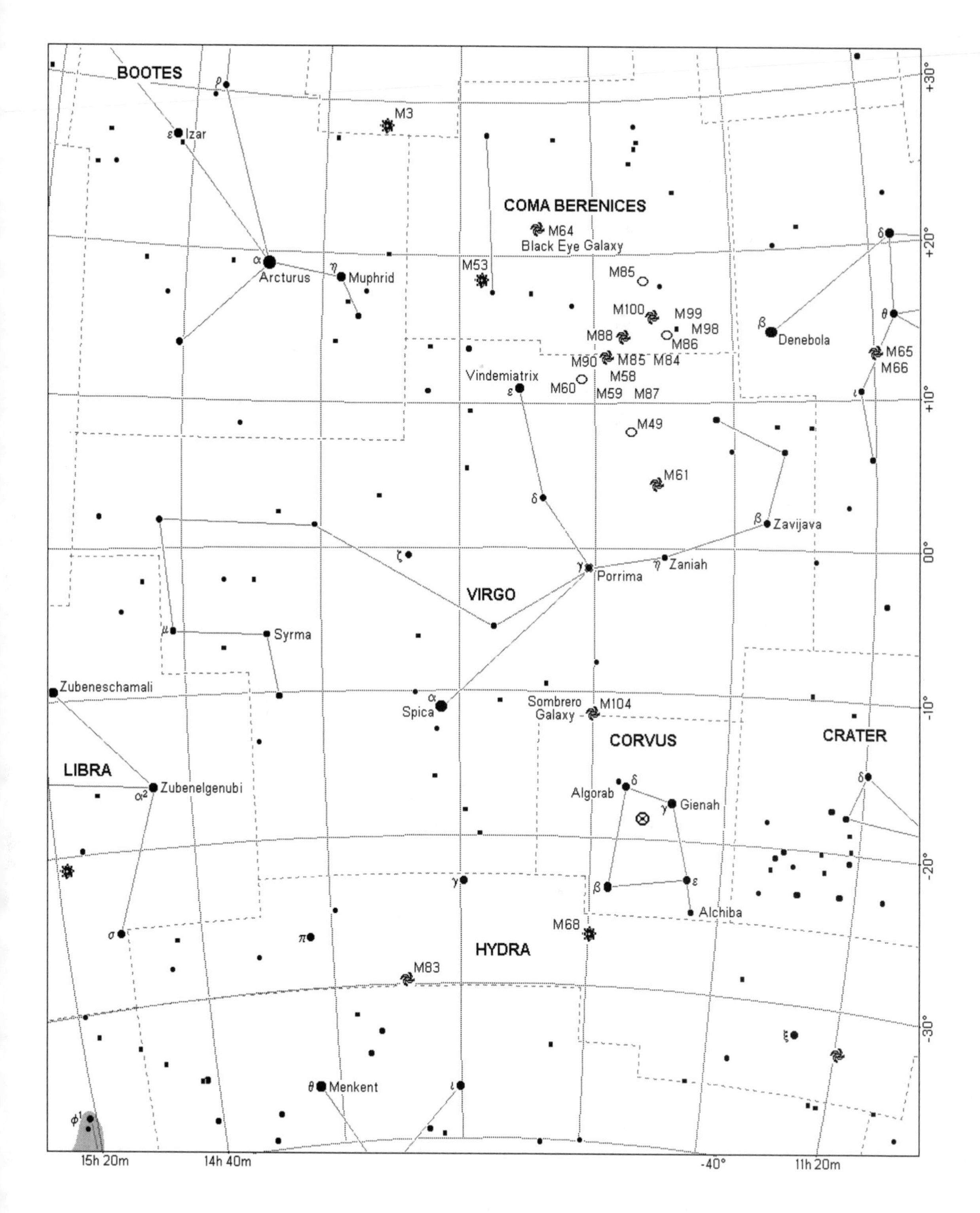

Category Index

Asteroids and other minor planets, individual

Asteroids and other minor planets, types and groups

Astrobiology

Astrochemistry

Astronomers and astrophysicists

Astronomy, branches

Astronomical and astrophysical quantities

Astrophysics

Atmospheric phenomena and structure

Black holes and neutron stars

Catalogs

Comets

Constellations and asterisms

Cosmology

Craters

Eclipses

Galaxies and groups of galaxies, individual

Galaxies, types and features

History of astronomy

See also *Astronomers*

Interstellar, interplanetary, and intergalactic matter

Meteors and meteorites

Moon

Nebulae

Observatories and instruments

Optics and optical phenomena

Orbital mechanics

Planets and planetary astronomy

Positional astronomy and time

Radio astronomy

Satellites

Societies

Spacecraft and satellites

Spectroscopy

Star clusters

Stars, individual

Stars, types

Stellar physics

Sun

Telescopes and other instruments, types and components